Probability and Mathematical Statistics (Continued)

PARZEN • Modern Probability Theory and Its Applications

PURI and SEN • Nonparametric Methods in General Linear Models

PURI and SEN • Nonparametric Methods in Multivariate Analysis

RANDLES and WOLFE • Introduction to the Theory of Nonparametric Statistics

RAO • Linear Statistical Inference and Its Applications, *Second Edition*

RAO • Real and Stochastic Analysis

RAO and SEDRANSK • W.G. Cochran's Impact on Statistics

ROHATGI • An Introduction to Probability Theory and Mathematical Statistics

ROHATGI • Statistical Inference

ROSS • Stochastic Processes

RUBINSTEIN • Simulation and The Monte Carlo Method

SCHEFFE • The Analysis of Variance

SEBER • Linear Regression Analysis

SEBER • Multivariate Observations

SEN • Sequential Nonparametrics: Invariance Principles and Statistical Inference

SERFLING • Approximation Theorems of Mathematical Statistics

SHORACK and WELLNER • Empirical Processes with Applications to Statistics

TJUR • Probability Based on Radon Measures

WILLIAMS • Diffusions, Markov Processes, and Martingales, Volume I: Foundations

ZACKS • Theory of Statistical Inference

Applied Probability and Statistics

ABRAHAM and LEDOLTER • Statistical Methods for Forecasting

AGRESTI • Analysis of Ordinal Categorical Data

AICKIN • Linear Statistical Analysis of Discrete Data

ANDERSON, AUQUIER, HAUCK, OAKES, VANDAELE, and WEISBERG • Statistical Methods for Comparative Studies

ARTHANARI and DODGE • Mathematical Programming in Statistics

BAILEY • The Elements of Stochastic Processes with Applications to the Natural Sciences

BAILEY • Mathematics, Statistics and Systems for Health

BARNETT • Interpreting Multivariate Data

BARNETT and LEWIS • Outliers in Statistical Data, *Second Edition*

BARTHOLOMEW • Stochastic Models for Social Processes, *Third Edition*

BARTHOLOMEW and FORBES • Statistical Techniques for Manpower Planning

BECK and ARNOLD • Parameter Estimation in Engineering and Science

BELSLEY, KUH, and WELSCH • Regression Diagnostics: Identifying Influential Data and Sources of Collinearity

BHAT • Elements of Applied Stochastic Processes, *Second Edition*

BLOOMFIELD • Fourier Analysis of Time Series: An Introduction

BOX • R. A. Fisher, The Life of a Scientist

BOX and DRAPER • Empirical Model-Building and Response Surfaces

BOX and DRAPER • Evolutionary Operation: A Statistical Method for Process Improvement

BOX, HUNTER, and HUNTER • Statistics for Experimenters: An Introduction to Design, Data Analysis, and Model Building

BROWN and HOLLANDER • Statistics: A Biomedical Introduction

BUNKE and BUNKE • Statistical Inference in Linear Models, Volume I

CHAMBERS • Computational Methods for Data Analysis

CHATTERJEE and PRICE • Regression Analysis by Example

CHOW • Econometric Analysis by Control Methods

CLARKE and DISNEY • Probability and Random Processes: A First Course with Applications, *Second Edition*

COCHRAN • Sampling Techniques, *Third Edition*

COCHRAN and COX • Experimental Designs, *Second Edition*
CONOVER • Practical Nonparametric Statistics, *Second Edition*
CONOVER and IMAN • Introduction to Modern Business Statistics
CORNELL • Experiments with Mixtures: Designs, Models and The Analysis of Mixture Data
COX • Planning of Experiments
DANIEL • Biostatistics: A Foundation for Analysis in the Health Sciences, *Third Edition*
DANIEL • Applications of Statistics to Industrial Experimentation
DANIEL and WOOD • Fitting Equations to Data: Computer Analysis of Multifactor Data, *Second Edition*
DAVID • Order Statistics, *Second Edition*
DAVISON • Multidimensional Scaling
DEGROOT, FIENBERG and KADANE • Statistics and the Law
DEMING • Sample Design in Business Research
DILLON and GOLDSTEIN • Multivariate Analysis: Methods and Applications
DODGE • Analysis of Experiments with Missing Data
DODGE and ROMIG • Sampling Inspection Tables, *Second Edition*
DOWDY and WEARDEN • Statistics for Research
DRAPER and SMITH • Applied Regression Analysis, *Second Edition*
DUNN • Basic Statistics: A Primer for the Biomedical Sciences, *Second Edition*
DUNN and CLARK • Applied Statistics: Analysis of Variance and Regression
ELANDT-JOHNSON and JOHNSON • Survival Models and Data Analysis
FLEISS • Statistical Methods for Rates and Proportions, *Second Edition*
FLEISS • The Design and Analysis of Clinical Experiments
FOX • Linear Statistical Models and Related Methods
FRANKEN, KÖNIG, ARNDT, and SCHMIDT • Queues and Point Processes
GALAMBOS • The Asymptotic Theory of Extreme Order Statistics
GIBBONS, OLKIN, and SOBEL • Selecting and Ordering Populations: A New Statistical Methodology
GNANADESIKAN • Methods for Statistical Data Analysis of Multivariate Observations
GOLDSTEIN and DILLON • Discrete Discriminant Analysis
GREENBERG and WEBSTER • Advanced Econometrics: A Bridge to the Literature
GROSS and CLARK • Survival Distributions: Reliability Applications in the Biomedical Sciences
GROSS and HARRIS • Fundamentals of Queueing Theory, *Second Edition*
GUPTA and PANCHAPAKESAN • Multiple Decision Procedures: Theory and Methodology of Selecting and Ranking Populations
GUTTMAN, WILKS, and HUNTER • Introductory Engineering Statistics, *Third Edition*
HAHN and SHAPIRO • Statistical Models in Engineering
HALD • Statistical Tables and Formulas
HALD • Statistical Theory with Engineering Applications
HAND • Discrimination and Classification
HILDEBRAND, LAING, and ROSENTHAL • Prediction Analysis of Cross Classifications
HOAGLIN, MOSTELLER and TUKEY • Exploring Data Tables, Trends and Shapes
HOAGLIN, MOSTELLER, and TUKEY • Understanding Robust and Exploratory Data Analysis
HOEL • Elementary Statistics, *Fourth Edition*
HOEL and JESSEN • Basic Statistics for Business and Economics, *Third Edition*

(*continued on back*)

Testing
Statistical Hypotheses

Testing
Statistical Hypotheses

Second Edition

E. L. LEHMANN
Professor of Statistics
University of California, Berkeley

JOHN WILEY & SONS
New York · Chichester · Brisbane · Toronto · Singapore

Library of Congress Cataloging-in-Publication Data:

Lehmann, E. L. (Erich Leo), 1917–
  Testing statistical hypotheses.

  Includes bibliographies and indexes.
  1. Statistical hypothesis testing.   I. Title.

QA277.L425 1986   519.5′6   85-29469
ISBN 0-471-84083-1

Printed in the United States of America

10 9 8 7 6 5 4 3 2

To Susanne

Preface

This new edition reflects the development of the field of hypothesis testing since the original book was published 27 years ago, but the basic structure has been retained. In particular, optimality considerations continue to provide the organizing principle. However, they are now tempered by a much stronger emphasis on the robustness properties of the resulting procedures. Other topics that receive greater attention than in the first edition are confidence intervals (which for technical reasons fit better here than in the companion volume on estimation, *TPE**), simultaneous inference procedures (which have become an important part of statistical methodology), and admissibility. A major criticism that has been leveled against the theory presented here relates to the choice of the reference set with respect to which performance is to be evaluated. A new chapter on conditional inference at the end of the book discusses some of the issues raised by this concern.

In order to accommodate the wealth of new results that have become available concerning the core material, it was necessary to impose some limitations. The most important omission is an adequate treatment of asymptotic optimality paralleling that given for estimation in *TPE*. Since the corresponding theory for testing is less satisfactory and would have required too much space, the earlier rather perfunctory treatment has been retained. Three sections of the first edition were devoted to sequential analysis. They are outdated and have been deleted, since it was not possible to do justice to the extensive and technically demanding expansion of this area. This is consistent with the decision not to include the theory of optimal experimental design. Together with sequential analysis and survey sampling, this topic should be treated in a separate book. Finally, although there is a section on Bayesian confidence intervals, Bayesian approaches to

Theory of Point Estimation [Lehmann (1983)].

vii

hypothesis testing are not discussed, since they play a less well-defined role here than do the corresponding techniques in estimation.

In addition to the major changes, many new comments and references have been included, numerous errors corrected, and some gaps filled. I am greatly indebted to Peter Bickel, John Pratt, and Fritz Scholz, who furnished me with lists of errors and improvements, and to Maryse Loranger and Carl Schaper who each read several chapters of the manuscript. For additional comments I should like to thank Jim Berger, Colin Blyth, Herbert Eisenberg, Jaap Fabius, Roger Farrell, Thomas Ferguson, Irving Glick, Jan Hemelrijk, Wassily Hoeffding, Kumar Jogdeo, the late Jack Kiefer, Olaf Krafft, William Kruskal, John Marden, John Rayner, Richard Savage, Robert Wijsman, and the many colleagues and students who made contributions of which I no longer have a record.

Another indebtedness I should like to acknowledge is to a number of books whose publication considerably eased the task of updating. Above all, there is the encyclopedic three-volume treatise by Kendall and Stuart, of which I consulted particularly the second volume, fourth edition (1979) innumerable times. The books by Ferguson (1967), Cox and Hinkley (1974), and Berger (1980) also were a great help. In the first edition, I provided references to tables and charts that were needed for the application of the tests whose theory was developed in the book. This has become less important in view of the four-volume work by Johnson and Kotz: *Distributions in Statistics* (1969–1972). Frequently I now simply refer to the appropriate chapter of this reference work.

There are two more books to which I must refer:

A complete set of solutions to the problems of the first edition was published as *Testing Statistical Hypotheses*: *Worked Solutions*. [Kallenberg et al. (1984)]. I am grateful to the group of Dutch authors for undertaking this labor and for furnishing me with a list of errors and corrections regarding both the statements of the problems and the hints to their solutions.

The other book is my *Theory of Point Estimation* [Lehmann (1983)], which combines with the present volume to provide a unified treatment of the classical theories of testing and estimation, both by confidence intervals and by point estimates. The two are independent of each other, but cross references indicate additional information on a given topic provided by the other book. Suggestions for ways in which the two books can be used to teach different courses are given in comments for instructors following this preface.

I owe very special thanks to two people. My wife, Juliet Shaffer, critically read the new sections and gave advice on many other points. Wei Yin Loh read an early version of the whole manuscript and checked many of the new problems. In addition, he joined me in the arduous task of reading the complete galley proofs. As a result, many errors and oversights were corrected.

The research required for this second edition was supported in part by the National Science Foundation, and I am grateful for the Foundation's continued support of my work. Finally, I should like to thank Linda Tiffany, who converted many illegible pages into beautifully typed ones.

REFERENCES

Berger, J. O.
  (1980). *Statistical Decision Theory*, Springer, New York.
Cox, D. R. and Hinkley, D. V.
  (1974). *Theoretical Statistics*, Chapman & Hall.
Ferguson, T. S.
  (1967). *Mathematical Statistics*, Academic, New York.
Johnson, N. L. and Kotz, S.
  (1969–1972). *Distributions in Statistics*, 4 vols., Wiley, New York.
Kallenberg, W. C. M., et al.
  (1984). *Testing Statistical Hypotheses: Worked Solutions*, Centrum voor Wiskunde en Informatica, Amsterdam.
Kendall, M. G. and Stuart, A.
  (1977, 1979). *The Advanced Theory of Statistics*, 4th ed., vols. 1, 2, Charles Griffin, London.
Kendall, M. G., Stuart, A., and Ord, J. K.
  (1983). *The Advanced Theory of Statistics*, 4th ed., vol. 3, Charles Griffin, London.
Lehmann, E. L.
  (1983). *Theory of Point Estimation*, Wiley, New York.

E. L. LEHMANN

Berkeley, California
February 1986

Preface to the First Edition

A mathematical theory of hypothesis testing in which tests are derived as solutions of clearly stated optimum problems was developed by Neyman and Pearson in the 1930s and since then has been considerably extended. The purpose of the present book is to give a systematic account of this theory and of the closely related theory of confidence sets, together with their principal applications. These include the standard one- and two-sample problems concerning normal, binomial, and Poisson distributions; some aspects of the analysis of variance and of regression analysis (linear hypothesis); certain multivariate and sequential problems. There is also an introduction to nonparametric tests, although here the theoretical approach has not yet been fully developed. One large area of methodology, the class of methods based on large-sample considerations, in particular χ^2 and likelihood-ratio tests, essentially has been omitted because the approach and the mathematical tools used are so different that an adequate treatment would require a separate volume. The theory of these tests is only briefly indicated at the end of Chapter 7.

At present the theory of hypothesis testing is undergoing important changes in at least two directions. One of these stems from the realization that the standard formulation constitutes a serious oversimplification of the problem. The theory is therefore being reexamined from the point of view of Wald's statistical decision functions. Although these investigations throw new light on the classical theory, they essentially confirm its findings. I have retained the Neyman–Pearson formulation in the main part of this book, but have included a discussion of the concepts of general decision theory in Chapter 1 to provide a basis for giving a broader justification of some of the results. It also serves as a background for the development of the theories of hypothesis testing and confidence sets.

Of much greater importance is the fact that many of the problems, which traditionally have been formulated in terms of hypothesis testing, are in reality multiple decision problems involving a choice between several decisions when the hypothesis is rejected. The development of suitable procedures for such problems is at present one of the most important tasks of statistics and is finding much attention in the current literature. However, since most of the work so far has been tentative, I have preferred to present the traditional tests even in cases in which the majority of the applications appear to call for a more elaborate procedure, adding only a warning regarding the limitations of this approach. Actually, it seems likely that the tests will remain useful because of their simplicity even when a more complete theory of multiple decision methods is available.

The natural mathematical framework for a systematic treatment of hypothesis testing is the theory of measure in abstract spaces. Since introductory courses in real variables or measure theory frequently present only Lebesgue measure, a brief orientation with regard to the abstract theory is given in Sections 1 and 2 of Chapter 2. Actually, much of the book can be read without knowledge of measure theory if the symbol $\int p(x)\,d\mu(x)$ is interpreted as meaning either $\int p(x)\,dx$ or $\Sigma p(x)$, and if the measure-theoretic aspects of certain proofs together with all occurrences of the letters a.e. (almost everywhere) are ignored. With respect to statistics, no specific requirements are made, all statistical concepts being developed from the beginning. On the other hand, since readers will usually have had previous experience with statistical methods, applications of each method are indicated in general terms, but concrete examples with data are not included. These are available in many of the standard textbooks.

The problems at the end of each chapter, many of them with outlines of solutions, provide exercises, further examples, and introductions to some additional topics. There is also given at the end of each chapter an annotated list of references regarding sources, both of ideas and of specific results. The notes are not intended to summarize the principal results of each paper cited but merely to indicate its significance for the chapter in question. In presenting these references I have not aimed for completeness but rather have tried to give a usable guide to the literature.

An outline of this book appeared in 1949 in the form of lecture notes taken by Colin Blyth during a summer course at the University of California. Since then, I have presented parts of the material in courses at Columbia, Princeton, and Stanford Universities and several times at the University of California. During these years I greatly benefited from comments of students, and I regret that I cannot here thank them individually. At different stages of the writing I received many helpful suggestions from W. Gautschi, A. Høyland, and L. J. Savage, and particularly from Mrs. C. Striebel, whose critical reading of the next to final version of the manuscript resulted in many improvements. Also, I should like to mention gratefully the benefit I derived from many long discussions with Charles Stein.

It is a pleasure to acknowledge the generous support of this work by the Office of Naval Research; without it the book would probably not have been written. Finally, I should like to thank Mrs. J. Rubalcava, who typed and retyped the various drafts of the manuscript with unfailing patience, accuracy, and speed.

E. L. LEHMANN

Berkeley, California
June 1959

Comments for Instructors

The two companion volumes, *Testing Statistical Hypotheses* (*TSH*) and *Theory of Point Estimation* (*TPE*), between them provide an introduction to classical statistics from a unified point of view. Different optimality criteria are considered, and methods for determining optimum procedures according to these criteria are developed. The application of the resulting theory to a variety of specific problems as an introduction to statistical methodology constitutes a second major theme.

On the other hand, the two books are essentially independent of each other. (As a result, there is some overlap in the preparatory chapters; also, each volume contains cross-references to related topics in the other.) They can therefore be taught in either order. However, *TPE* is somewhat more discursive and written at a slightly lower mathematical level, and for this reason may offer the better starting point.

The material of the two volumes combined somewhat exceeds what can be comfortably covered in a year's course meeting 3 hours a week, thus providing the instructor with some choice of topics to be emphasized. A one-semester course covering both estimation and testing can be obtained, for example, by deleting all large-sample considerations, all nonparametric material, the sections concerned with simultaneous estimation and testing, the minimax chapter of *TSH*, and some of the applications. Such a course might consist of the following sections: *TPE*: Chapter 2, Section 1 and a few examples from Sections 2, 3; Chapter 3, Sections 1-3; Chapter 4, Sections 1-4. *TSH*: Chapter 3, Sections 1-3, 5, 7 (without proof of Theorem 6); Chapter 4, Sections 1-7; Chapter 5, Sections 1-4, 6-8; Chapter 6, Sections 1-6, 11; Chapter 7, Sections 1-3, 5-8, 11, 12; together with material from the preparatory chapters (*TSH* Chapter 1, 2; *TPE* Chapter 1) as it is needed.

Contents

CHAPTER   PAGE

1 THE GENERAL DECISION PROBLEM 1

  1 Statistical inference and statistical decisions 1
  2 Specification of a decision problem 2
  3 Randomization; choice of experiment 6
  4 Optimum procedures 8
  5 Invariance and unbiasedness 10
  6 Bayes and minimax procedures 14
  7 Maximum likelihood 16
  8 Complete classes 17
  9 Sufficient statistics 18
  10 Problems.. 22
  11 References 28

2 THE PROBABILITY BACKGROUND 34

  1 Probability and measure 34
  2 Integration 37
  3 Statistics and subfields 41
  4 Conditional expectation and probability 43
  5 Conditional probability distributions 48
  6 Characterization of sufficiency 53
  7 Exponential families 57
  8 Problems.. 60
  9 References 66

3 UNIFORMLY MOST POWERFUL TESTS 68

  1 Stating the problem 68
  2 The Neyman–Pearson fundamental lemma 72
  3 Distributions with monotone likelihood ratio 78
  4 Comparison of experiments........................... 86
  5 Confidence bounds 89
  6 A generalization of the fundamental lemma 96

xvii

	7	Two-sided hypotheses	101
	8	Least favorable distributions	104
	9	Testing the mean and variance of a normal distribution	108
	10	Problems	111
	11	References	125

4 UNBIASEDNESS: THEORY AND FIRST APPLICATIONS ... 134

	1	Unbiasedness for hypothesis testing	134
	2	One-parameter exponential families	135
	3	Similarity and completeness	140
	4	UMP unbiased tests for multiparameter exponential families	145
	5	Comparing two Poisson or binomial populations	151
	6	Testing for independence in a 2×2 table	156
	7	Alternative models for 2×2 tables	159
	8	Some three-factor contingency tables	162
	9	The sign test	166
	10	Problems	170
	11	References	181

5 UNBIASEDNESS: APPLICATIONS TO NORMAL DISTRIBUTIONS; CONFIDENCE INTERVALS ... 188

	1	Statistics independent of a sufficient statistic	188
	2	Testing the parameters of a normal distribution	192
	3	Comparing the means and variances of two normal distributions	197
	4	Robustness	203
	5	Effect of dependence	209
	6	Confidence intervals and families of tests	213
	7	Unbiased confidence sets	216
	8	Regression	222
	9	Bayesian confidence sets	225
	10	Permutation tests	230
	11	Most powerful permutation tests	232
	12	Randomization as a basis for inference	237
	13	Permutation tests and randomization	240
	14	Randomization model and confidence intervals	245
	15	Testing for independence in a bivariate normal distribution	248
	16	Problems	253
	17	References	273

6 INVARIANCE ... 282

	1	Symmetry and invariance	282
	2	Maximal invariants	284
	3	Most powerful invariant tests	289
	4	Sample inspection by variables	293

5	Almost invariance	297
6	Unbiasedness and invariance	302
7	Admissibility	305
8	Rank tests	314
9	The two-sample problem	317
10	The hypothesis of symmetry	323
11	Equivariant confidence sets	326
12	Average smallest equivariant confidence sets	330
13	Confidence bands for a distribution function	334
14	Problems	337
15	References	357

7 LINEAR HYPOTHESES 365

1	A canonical form	365
2	Linear hypotheses and least squares	370
3	Tests of homogeneity	374
4	Multiple comparisons	380
5	Two-way layout: One observation per cell	388
6	Two-way layout: m observations per cell	392
7	Regression	396
8	Robustness against nonnormality	401
9	Scheffé's S-method: A special case	405
10	Scheffé's S-method for general linear models	411
11	Random-effects model: One-way classification	418
12	Nested classifications	422
13	Problems	427
14	References	444

8 MULTIVARIATE LINEAR HYPOTHESES 453

1	A canonical form	453
2	Reduction by invariance	456
3	The one- and two-sample problems	459
4	Multivariate analysis of variance (MANOVA)	462
5	Further applications	465
6	Simultaneous confidence intervals	471
7	χ^2-tests: Simple hypothesis and unrestricted alternatives	477
8	χ^2- and likelihood-ratio tests	480
9	Problems	488
10	References	498

9 THE MINIMAX PRINCIPLE 504

1	Tests with guaranteed power	504
2	Examples	508
3	Comparing two approximate hypotheses	512
4	Maximin tests and invariance	516

5   The Hunt–Stein theorem 519
6   Most stringent tests 525
7   Problems 527
8   References 535

10   CONDITIONAL INFERENCE 539
1   Mixtures of experiments 539
2   Ancillary statistics 542
3   Optimal conditional tests 549
4   Relevant subsets 553
5   Problems 559
6   References 564

APPENDIX 569
1   Equivalence relations; groups 569
2   Convergence of distributions 570
3   Dominated families of distributions 574
4   The weak compactness theorem 576
5   References 577

AUTHOR INDEX 579
SUBJECT INDEX 587

CHAPTER 1

The General Decision Problem

1. STATISTICAL INFERENCE AND STATISTICAL DECISIONS

The raw material of a statistical investigation is a set of observations; these are the values taken on by random variables X whose distribution P_θ is at least partly unknown. Of the parameter θ, which labels the distribution, it is assumed known only that it lies in a certain set Ω, the *parameter space*. *Statistical inference* is concerned with methods of using this observational material to obtain information concerning the distribution of X or the parameter θ with which it is labeled. To arrive at a more precise formulation of the problem we shall consider the purpose of the inference.

The need for statistical analysis stems from the fact that the distribution of X, and hence some aspect of the situation underlying the mathematical model, is not known. The consequence of such a lack of knowledge is uncertainty as to the best mode of behavior. To formalize this, suppose that a choice has to be made between a number of alternative actions. The observations, by providing information about the distribution from which they came, also provide guidance as to the best decision. The problem is to determine a rule which, for each set of values of the observations, specifies what decision should be taken. Mathematically such a rule is a function δ, which to each possible value x of the random variables assigns a decision $d = \delta(x)$, that is, a function whose domain is the set of values of X and whose range is the set of possible decisions.

In order to see how δ should be chosen, one must compare the consequences of using different rules. To this end suppose that the consequence of taking decision d when the distribution of X is P_θ is a *loss*, which can be expressed as a nonnegative real number $L(\theta, d)$. Then the long-term average loss that would result from the use of δ in a number of repetitions of the experiment is the expectation $E[L(\theta, \delta(X))]$ evaluated under the assumption that P_θ is the true distribution of X. This expectation, which depends on the decision rule δ and the distribution P_θ, is called the *risk function* of δ and will be denoted by $R(\theta, \delta)$. By basing the decision on the observations, the original problem of choosing a decision d with loss function $L(\theta, d)$ is thus replaced by that of choosing δ, where the loss is now $R(\theta, \delta)$.

The above discussion suggests that the aim of statistics is the selection of a decision function which minimizes the resulting risk. As will be seen later, this statement of aims is not sufficiently precise to be meaningful; its proper interpretation is in fact one of the basic problems of the theory.

2. SPECIFICATION OF A DECISION PROBLEM

The methods required for the solution of a specific statistical problem depend quite strongly on the three elements that define it: the class $\mathscr{P} = \{P_\theta, \theta \in \Omega\}$ to which the distribution of X is assumed to belong; the structure of the space D of possible decisions d; and the form of the loss function L. In order to obtain concrete results it is therefore necessary to make specific assumptions about these elements. On the other hand, if the theory is to be more than a collection of isolated results, the assumptions must be broad enough either to be of wide applicability or to define classes of problems for which a unified treatment is possible.

Consider first the specification of the class $\mathscr{P}$. Precise numerical assumptions concerning probabilities or probability distributions are usually not warranted. However, it is frequently possible to assume that certain events have equal probabilities and that certain others are statistically independent. Another type of assumption concerns the relative order of certain infinitesimal probabilities, for example the probability of occurrences in an interval of time or space as the length of the interval tends to zero. The following classes of distributions are derived on the basis of only such assumptions, and are therefore applicable in a great variety of situations.

The *binomial* distribution $b(p, n)$ with

$$(1) \quad P(X = x) = \binom{n}{x} p^x (1 - p)^{n-x}, \qquad x = 0, \dots, n, \quad 0 \le p \le 1.$$

This is the distribution of the total number of successes in n independent trials when the probability of success for each trial is p.

The *Poisson* distribution $P(\tau)$ with

$$(2) \qquad P(X = x) = \frac{\tau^x}{x!} e^{-\tau}, \qquad x = 0, 1, \dots, \quad 0 < \tau.$$

This is the distribution of the number of events occurring in a fixed interval of time or space if the probability of more than one occurrence in a very short interval is of smaller order of magnitude than that of a single occurrence, and if the numbers of events in nonoverlapping intervals are statistically independent. Under these assumptions, the process generating the events is called a *Poisson process*. Such processes are discussed, for example, in the books by Feller (1968), Karlin and Taylor (1975), and Ross (1980).

The *normal* distribution $N(\xi, \sigma^2)$ with probability density

$$(3) \quad p(x) = \frac{1}{\sqrt{2\pi}\,\sigma} \exp\left[-\frac{1}{2\sigma^2}(x - \xi)^2\right], \quad -\infty < x, \xi < \infty, \quad 0 < \sigma.$$

Under very general conditions, which are made precise by the central limit theorem, this is the approximate distribution of the sum of a large number of independent random variables when the relative contribution of each term to the sum is small.

We consider next the structure of the decision space D. The great variety of possibilities is indicated by the following examples.

Example 1. Let $X_1, \ldots, X_n$ be a *sample* from one of the distributions (1)–(3), that is, let the X's be distributed independently and identically according to one of these distributions. Let θ be p, τ, or the pair (ξ, σ) respectively, and let $\gamma = \gamma(\theta)$ be a real-valued function of θ.

(i)   If one wishes to decide whether or not γ exceeds some specified value γ_0, the choice lies between the two decisions $d_0 : \gamma > \gamma_0$ and $d_1 : \gamma \leq \gamma_0$. In specific applications these decisions might correspond to the acceptance or rejection of a lot of manufactured goods, of an experimental airplane as ready for flight testing, of a new treatment as an improvement over a standard one, and so on. The loss function of course depends on the application to be made. Typically, the loss is 0 if the correct decision is chosen, while for an incorrect decision the losses $L(\gamma, d_0)$ and $L(\gamma, d_1)$ are increasing functions of $|\gamma - \gamma_0|$.

(ii)   At the other end of the scale is the much more detailed problem of obtaining a numerical estimate of γ. Here a decision d of the statistician is a real number, the estimate of γ, and the losses might be $L(\gamma, d) = v(\gamma)w(|d - \gamma|)$, where w is a strictly increasing function of the error $|d - \gamma|$.

(iii)   An intermediate case is the choice between the three alternatives $d_0 : \gamma < \gamma_0$, $d_1 : \gamma > \gamma_1$, $d_2 : \gamma_0 \leq \gamma \leq \gamma_1$, for example accepting a new treatment, rejecting it, or recommending it for further study.

The distinction illustrated by this example is the basis for one of the principal classifications of statistical methods. Two-decision problems such as (i) are usually formulated in terms of *testing a hypothesis* which is to be accepted or rejected (see Chapter 3). It is the theory of this class of problems with which we shall be mainly concerned here. The other principal branch of statistics is the theory of *point estimation* dealing with problems such as (ii). This is the subject of *TPE*. The intermediate problem (iii) is a special case of a *multiple decision procedure*. Some problems of this kind are treated in Ferguson (1967, Chapter 6); a discussion of some others is given in Chapter 7, Section 4.

Example 2. Suppose that the data consist of samples X_{ij}, $j = 1, \ldots, n_i$, from normal populations $N(\xi_i, \sigma^2)$, $i = 1, \ldots, s$.

(i) Consider first the case $s = 2$ and the question of whether or not there is a material difference between the two populations. This has the same structure as problem (iii) of the previous example. Here the choice lies between the three decisions $d_0 : |\xi_2 - \xi_1| \leq \Delta, d_1 : \xi_2 > \xi_1 + \Delta, d_2 : \xi_2 < \xi_1 - \Delta$, where Δ is pre-assigned. An analogous problem, involving $k + 1$ possible decisions, occurs in the general case of k populations. In this case one must choose between the decision that the k distributions do not differ materially, $d_0 : \max|\xi_j - \xi_i| \leq \Delta$, and the decisions $d_k : \max|\xi_j - \xi_i| > \Delta$ and ξ_k is the largest of the means.

(ii) A related problem is that of ranking the distributions in increasing order of their mean ξ.

(iii) Alternatively, a standard ξ_0 may be given and the problem is to decide which, if any, of the population means exceed the standard.

Example 3. Consider two distributions—to be specific, two Poisson distributions $P(\tau_1)$, $P(\tau_2)$—and suppose that τ_1 is known to be less than τ_2 but that otherwise the τ's are unknown. Let $Z_1, \ldots, Z_n$ be independently distributed, each according to either $P(\tau_1)$ or $P(\tau_2)$. Then each Z is to be classified as to which of the two distributions it comes from. Here the loss might be the number of Z's that are incorrectly classified, multiplied by a suitable function of τ_1 and τ_2. An example of the complexity that such problems can attain and the conceptual as well as mathematical difficulties that they may involve is provided by the efforts of anthropologists to classify the human population into a number of homogeneous races by studying the frequencies of the various blood groups and of other genetic characters.

All the problems considered so far could be termed *action problems*. It was assumed in all of them that if θ were known a unique correct decision would be available, that is, given any θ, there exists a unique d for which $L(\theta, d) = 0$. However, not all statistical problems are so clear-cut. Frequently it is a question of providing a convenient summary of the data or indicating what information is available concerning the unknown parameter or distribution. This information will be used for guidance in various considerations but will not provide the sole basis for any specific decisions. In such cases the emphasis is on the inference rather than on the decision aspect of the problem. Although formally it can still be considered a decision problem if the inferential statement itself is interpreted as the decision to be taken, the distinction is of conceptual and practical significance despite the fact that frequently it is ignored.* An important class of such problems, estimation by interval, is illustrated by the following example. (For the more usual formulation in terms of confidence intervals, see Chapter 3, Section 5, and Chapter 5, Sections 4 and 5.)

Example 4. Let $X = (X_1, \ldots, X_n)$ be a sample from $N(\xi, \sigma^2)$ and let a decision consist in selecting an interval $[\underline{L}, \overline{L}]$ and stating that it contains ξ. Suppose that decision procedures are restricted to intervals $[\underline{L}(X), \overline{L}(X)]$ whose expected length for all ξ and σ does not exceed $k\sigma$ where k is some preassigned constant. An appropriate loss function would be 0 if the decision is correct and would otherwise depend on the relative position of the interval to the true value of ξ. In this case there are many correct decisions corresponding to a given distribution $N(\xi, \sigma^2)$.

It remains to discuss the choice of loss function,[†] and of the three elements defining the problem this is perhaps the most difficult to specify. Even in the simplest case, where all losses eventually reduce to financial ones, it can hardly be expected that one will be able to evaluate all the short- and long-term consequences of an action. Frequently it is possible to simplify the formulation by taking into account only certain aspects of the loss function. As an illustration consider Example 1(i) and let $L(\theta, d_0) = a$ for $\gamma(\theta) \leq \gamma_0$ and $L(\theta, d_1) = b$ for $\gamma(\theta) > \gamma_0$. The risk function becomes

$$(4) \qquad R(\theta, \delta) = \begin{cases} aP_\theta\{\delta(X) = d_0\} & \text{if} \quad \gamma \leq \gamma_0, \\ bP_\theta\{\delta(X) = d_1\} & \text{if} \quad \gamma > \gamma_0, \end{cases}$$

and is seen to involve only the two probabilities of error, with weights which can be adjusted according to the relative importance of these errors. Similarly, in Example 3 one may wish to restrict attention to the number of misclassifications.

Unfortunately, such a natural simplification is not always available, and in the absence of specific knowledge it becomes necessary to select the loss function in some conventional way, with mathematical simplicity usually an important consideration. In point estimation problems such as that considered in Example 1(ii), if one is interested in estimating a real-valued function $\gamma = \gamma(\theta)$ it is customary to take the square of the error, or somewhat more generally to put

$$(5) \qquad L(\theta, d) = v(\theta)(d - \gamma)^2.$$

---

*For a more detailed discussion of this distinction see, for example, Cox (1958), Blyth (1970), and Barnett (1982).

†Some aspects of the choice of model and loss function are discussed in Lehmann (1984, 1985).

Besides being particularly simple mathematically, this can be considered as an approximation to the true loss function L provided that for each fixed θ, $L(\theta, d)$ is twice differentiable in d, that $L(\theta, \gamma(\theta)) = 0$ for all θ, and that the error is not large.

It is frequently found that, within one problem, quite different types of losses may occur, which are difficult to measure on a common scale. Consider once more Example 1(i) and suppose that γ_0 is the value of γ when a standard treatment is applied to a situation in medicine, agriculture, or industry. The problem is that of comparing some new process with unknown γ to the standard one. Turning down the new method when it is actually superior, or adopting it when it is not, clearly entails quite different consequences. In such cases it is sometimes convenient to treat the various loss components, say $L_1, L_2, \ldots, L_r$, separately. Suppose in particular that $r = 2$ and that L_1 represents the more serious possibility. One can then assign a bound to this risk component, that is, impose the condition

$$(6) \qquad EL_1(\theta, \delta(X)) \leq \alpha,$$

and subject to this condition minimize the other component of the risk. Example 4 provides an illustration of this procedure. The length of the interval $[\underline{L}, \overline{L}]$ (measured in σ-units) is one component of the loss function, the other being the loss that results if the interval does not cover the true ξ.

3. RANDOMIZATION; CHOICE OF EXPERIMENT

The description of the general decision problem given so far is still too narrow in certain respects. It has been assumed that for each possible value of the random variables a definite decision must be chosen. Instead, it is convenient to permit the selection of one out of a number of decisions according to stated probabilities, or more generally the selection of a decision according to a probability distribution defined over the decision space; which distribution depends of course on what x is observed. One way to describe such a randomized procedure is in terms of a nonrandomized procedure depending on X and a random variable Y whose values lie in the decision space and whose conditional distribution given x is independent of θ.

Although it may run counter to one's intuition that such extra randomization should have any value, there is no harm in permitting this greater freedom of choice. If the intuitive misgivings are correct, it will turn out that the optimum procedures always are of the simple nonrandomized kind. Actually, the introduction of randomized procedures leads to an important mathematical simplification by enlarging the class of risk functions so that it becomes convex. In addition, there are problems in which some features of the risk function such as its maximum can be improved by using a randomized procedure.

Another assumption that tacitly has been made so far is that a definite experiment has already been decided upon so that it is known what observations will be taken. However, the statistical considerations involved in designing an experiment are no less important than those concerning its analysis. One question in particular that must be decided before an investigation is undertaken is how many observations should be taken so that the risk resulting from wrong decisions will not be excessive. Frequently it turns out that the required sample size depends on the unknown distribution and therefore cannot be determined in advance as a fixed number. Instead it is then specified as a function of the observations and the decision whether or not to continue experimentation is made *sequentially* at each stage of the experiment on the basis of the observations taken up to that point.

Example 5. On the basis of a sample $X_1, \ldots, X_n$ from a normal distribution $N(\xi, \sigma^2)$ one wishes to estimate ξ. Here the risk function of an estimate, for example its expected squared error, depends on σ. For large σ the sample contains only little information in the sense that two distributions $N(\xi_1, \sigma^2)$ and $N(\xi_2, \sigma^2)$ with fixed difference $\xi_2 - \xi_1$ become indistinguishable as $\sigma \to \infty$, with the result that the risk tends to infinity. Conversely, the risk approaches zero as $\sigma \to 0$, since then effectively the mean becomes known. Thus the number of observations needed to control the risk at a given level is unknown. However, as soon as some observations have been taken, it is possible to estimate σ^2 and hence to determine the additional number of observations required.

Example 6. In a sequence of trials with constant probability p of success, one wishes to decide whether $p \leq \frac{1}{2}$ or $p > \frac{1}{2}$. It will usually be possible to reach a decision at an early stage if p is close to 0 or 1 so that practically all observations are of one kind, while a larger sample will be needed for intermediate values of p. This difference may be partially balanced by the fact that for intermediate values a loss resulting from a wrong decision is presumably less serious than for the more extreme values.

Example 7. The possibility of determining the sample size sequentially is important not only because the distributions P_θ can be more or less informative but also because the same is true of the observations themselves. Consider, for example, observations from the uniform distribution over the interval $(\theta - \frac{1}{2}, \theta + \frac{1}{2})$ and the problem of estimating θ. Here there is no difference in the amount of information provided by the different distributions P_θ. However, a sample $X_1, X_2, \ldots, X_n$ can practically pinpoint θ if $\max|X_j - X_i|$ is sufficiently close to 1, or it can give essentially no more information than a single observation if $\max|X_j - X_i|$ is close to 0. Again the required sample size should be determined sequentially.

Except in the simplest situations, the determination of the appropriate sample size is only one aspect of the design problem. In general, one must decide not only how many but also what kind of observations to take. In clinical trials, for example, when a new treatment is being compared with a standard procedure, a protocol is required which specifies to which of the two treatments each of the successive incoming patients is to be assigned. Formally, such questions can be subsumed under the general decision problem described at the beginning of the chapter, by interpreting X as the set of all available variables, by introducing the decisions whether or not to stop experimentation at the various stages, by specifying in case of continuance which type of variable to observe next, and by including the cost of observation in the loss function.

The determination of optimum sequential stopping rules and experimental designs is outside the scope of this book. Introductions to these subjects are provided, for example, by Chernoff (1972), Ghosh (1970), and Govindarajulu (1981).

4. OPTIMUM PROCEDURES

At the end of Section 1 the aim of statistical theory was stated to be the determination of a decision function δ which minimizes the risk function

$$(7) \qquad R(\theta, \delta) = E_\theta[L(\theta, \delta(X))].$$

Unfortunately, in general the minimizing δ depends on θ, which is unknown. Consider, for example, some particular decision d_0, and the decision procedure $\delta(x) \equiv d_0$ according to which decision d_0 is taken regardless of the outcome of the experiment. Suppose that d_0 is the correct decision for some θ_0, so that $L(\theta_0, d_0) = 0$. Then δ minimizes the risk at θ_0 since $R(\theta_0, \delta) = 0$, but presumably at the cost of a high risk for other values of θ.

In the absence of a decision function that minimizes the risk for all θ, the mathematical problem is still not defined, since it is not clear what is meant by a best procedure. Although it does not seem possible to give a definition of optimality that will be appropriate in all situations, the following two methods of approach frequently are satisfactory.

The nonexistence of an optimum decision rule is a consequence of the possibility that a procedure devotes too much of its attention to a single parameter value at the cost of neglecting the various other values that might arise. This suggests the restriction to decision procedures which possess a certain degree of impartiality, and the possibility that within such a restricted class there may exist a procedure with uniformly smallest risk. Two conditions of this kind, invariance and unbiasedness, will be discussed in the next section.

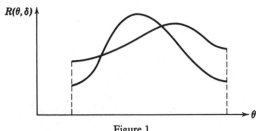

Figure 1

Instead of restricting the class of procedures, one can approach the problem somewhat differently. Consider the risk functions corresponding to two different decision rules δ_1 and δ_2. If $R(\theta, \delta_1) < R(\theta, \delta_2)$ for all θ, then δ_1 is clearly preferable to δ_2, since its use will lead to a smaller risk no matter what the true value of θ is. However, the situation is not clear when the two risk functions intersect as in Figure 1. What is needed is a principle which in such cases establishes a preference of one of the two risk functions over the other, that is, which introduces an ordering into the set of all risk functions. A procedure will then be optimum if its risk function is best according to this ordering. Some criteria that have been suggested for ordering risk functions will be discussed in Section 6.

A weakness of the theory of optimum procedures sketched above is its dependence on an extraneous restricting or ordering principle, and on knowledge concerning the loss function and the distributions of the observable random variables which in applications is frequently unavailable or unreliable. These difficulties, which may raise doubt concerning the value of an optimum theory resting on such shaky foundations, are in principle no different from those arising in any application of mathematics to reality. Mathematical formulations always involve simplification and approximation, so that solutions obtained through their use cannot be relied upon without additional checking. In the present case a check consists in an overall evaluation of the performance of the procedure that the theory produces, and an investigation of its sensitivity to departure from the assumptions under which it was derived.

The optimum theory discussed in this book should therefore not be understood to be prescriptive. The fact that a procedure δ is optimal according to some optimality criterion does not necessarily mean that it is the right procedure to use, or even a satisfactory procedure. It does show how well one can do in this particular direction and how much is lost when other aspects have to be taken into account.

The aspect of the formulation that typically has the greatest influence on the solution of the optimality problem is the family $\mathscr{P}$ to which the distribution of the observations is assumed to belong. The investigation of the *robustness* of a proposed procedure to departures from the specified model is an indispensable feature of a suitable statistical procedure, and although optimality (exact or asymptotic) may provide a good starting point, modifications are often necessary before an acceptable solution is found. It is possible to extend the decision-theoretic framework to include robustness as well as optimality. Suppose robustness is desired against some class $\mathscr{P}'$ of distributions which is larger (possibly much larger) than the given $\mathscr{P}$. Then one may assign a bound M to the risk to be tolerated over $\mathscr{P}'$. Within the class of procedures satisfying this restriction, one can then optimize the risk over $\mathscr{P}$ as before. Such an approach has been proposed and applied to a number of specific problems by Bickel (1984).

Another possible extension concerns the actual choice of the family $\mathscr{P}$, the model used to represent the actual physical situation. The problem of choosing a model which provides an adequate description of the situation without being unnecessarily complex can be treated within the decision-theoretic formulation of Section 1 by adding to the loss function a component representing the complexity of the proposed model. For a discussion of such an approach to *model selection*, see Stone (1981).

5. INVARIANCE AND UNBIASEDNESS*

A natural definition of impartiality suggests itself in situations which are symmetric with respect to the various parameter values of interest: *The procedure is then required to act symmetrically with respect to these values.*

Example 8. Suppose two treatments are to be compared and that each is applied n times. The resulting observations $X_{11}, \ldots, X_{1n}$ and $X_{21}, \ldots, X_{2n}$ are samples from $N(\xi_1, \sigma^2)$ and $N(\xi_2, \sigma^2)$ respectively. The three available decisions are $d_0 : |\xi_2 - \xi_1| \leq \Delta$, $d_1 : \xi_2 > \xi_1 + \Delta$, $d_2 : \xi_2 < \xi_1 - \Delta$, and the loss is w_{ij} if decision d_j is taken when d_i would have been correct. If the treatments are to be compared solely in terms of the ξ's and no outside considerations are involved, the losses are symmetric with respect to the two treatments so that $w_{01} = w_{02}$, $w_{10} = w_{20}$, $w_{12} = w_{21}$. Suppose now that the labeling of the two treatments as 1 and 2 is reversed, and correspondingly also the labeling of the X's, the ξ's, and the decisions d_1 and d_2. This changes the meaning of the symbols, but the formal decision problem, because of its symmetry, remains unaltered. It is then natural to require the corresponding symmetry from the procedure δ and ask that $\delta(x_{11}, \ldots, x_{1n}, x_{21}, \ldots, x_{2n}) = d_0$, d_1, or d_2 as $\delta(x_{21}, \ldots, x_{2n}, x_{11}, \ldots, x_{1n}) = d_0$, d_2, or d_1 respectively. If this condition were not satisfied, the decision as to which population

---

*The concepts discussed here for general decision theory will be developed in more specialized form in later chapters. The present section may therefore be omitted at first reading.

has the greater mean would depend on the presumably quite accidental and irrelevant labeling of the samples. Similar remarks apply to a number of further symmetries that are present in this problem.

Example 9. Consider a sample $X_1, \ldots, X_n$ from a distribution with density $\sigma^{-1} f[(x - \xi)/\sigma]$ and the problem of estimating the location parameter ξ, say the mean of the X's, when the loss is $(d - \xi)^2/\sigma^2$, the square of the error expressed in σ-units. Suppose that the observations are originally expressed in feet, and let $X_i' = aX_i$ with $a = 12$ be the corresponding observations in inches. In the transformed problem the density is $\sigma'^{-1} f[(x' - \xi')/\sigma']$ with $\xi' = a\xi$, $\sigma' = a\sigma$. Since $(d' - \xi')^2/\sigma'^2 = (d - \xi)^2/\sigma^2$, the problem is formally unchanged. The same estimation procedure that is used for the original observations is therefore appropriate after the transformation and leads to $\delta(aX_1, \ldots, aX_n)$ as an estimate of $\xi' = a\xi$, the parameter ξ expressed in inches. On reconverting the estimate into feet one finds that if the result is to be independent of the scale of measurements, δ must satisfy the condition of scale invariance

$$\frac{\delta(aX_1, \ldots, aX_n)}{a} = \delta(X_1, \ldots, X_n).$$

The general mathematical expression of symmetry is invariance under a suitable group of transformations. A group G of transformations g of the sample space is said to leave a statistical decision problem invariant if it satisfies the following conditions:

(i) It leaves invariant the family of distributions $\mathscr{P} = \{ P_\theta, \theta \in \Omega \}$, that is, for any possible distribution P_θ of X the distribution of gX, say $P_{\theta'}$, is also in $\mathscr{P}$. The resulting mapping $\theta' = \bar{g}\theta$ of Ω is assumed to be onto[†] Ω and $1:1$.

(ii) To each $g \in G$, there corresponds a transformation $g^* = h(g)$ of the decision space D onto itself such that h is a homomorphism, that is, satisfies the relation $h(g_1 g_2) = h(g_1) h(g_2)$, and the loss function L is unchanged under the transformation, so that

$$L(\bar{g}\theta, g^*d) = L(\theta, d).$$

Under these assumptions the transformed problem, in terms of $X' = gX$, $\theta' = \bar{g}\theta$, and $d' = g^*d$, is formally identical with the original problem in terms of X, θ, and d. Given a decision procedure δ for the latter, this is therefore still appropriate after the transformation. Interpreting the transformation as a change of coordinate system and hence of the names of the elements, one would, on observing x', select the decision which in the new

---

[†] The term *onto* is used to indicate that $\bar{g}\Omega$ is not only contained in but actually equals Ω; that is, given any θ' in Ω, there exists θ in Ω such that $\bar{g}\theta = \theta'$.

system has the name $\delta(x')$, so that its old name is $g^{*-1}\delta(x')$. If the decision taken is to be independent of the particular coordinate system adopted, this should coincide with the original decision $\delta(x)$, that is, the procedure must satisfy the *invariance* condition

$$(8) \qquad \delta(gx) = g^*\delta(x) \qquad \text{for all} \quad x \in X, \quad g \in G.$$

Example 10. The model described in Example 8 is invariant also under the transformations $X'_{ij} = X_{ij} + c$, $\xi'_i = \xi_i + c$. Since the decisions d_0, d_1, and d_2 concern only the differences $\xi_2 - \xi_1$, they should remain unchanged under these transformations, so that one would expect to have $g^*d_i = d_i$ for $i = 0, 1, 2$. It is in fact easily seen that the loss function does satisfy $L(\bar{g}\theta, d) = L(\theta, d)$, and hence that $g^*d = d$. A decision procedure therefore remains invariant in the present case if it satisfies $\delta(gx) = \delta(x)$ for all $g \in G$, $x \in X$.

It is helpful to make a terminological distinction between situations like that of Example 10 in which $g^*d = d$ for all d, and those like Examples 8 and 9 where invariance considerations require $\delta(gx)$ to vary with g. In the former case the decision procedure remains unchanged under the transformations $X' = gX$ and is thus truly invariant; in the latter, the procedure varies with g and may then more appropriately be called *equivariant* rather than invariant.[†] Typically, hypothesis testing leads to procedures that are invariant in this sense; estimation problems (whether by point or interval estimation), to equivariant ones. Invariant tests and equivariant confidence sets will be discussed in Chapter 6. For a brief discussion of equivariant point estimation, see Bondessen (1983); a fuller treatment is given in *TPE*, Chapter 3.

Invariance considerations are applicable only when a problem exhibits certain symmetries. An alternative impartiality restriction which is applicable to other types of problems is the following condition of unbiasedness. Suppose the problem is such that for each θ there exists a unique correct decision and that each decision is correct for some θ. Assume further that $L(\theta_1, d) = L(\theta_2, d)$ for all d whenever the same decision is correct for both θ_1 and θ_2. Then the loss $L(\theta, d')$ depends only on the actual decision taken, say d', and the correct decision d. The loss can thus be denoted by $L(d, d')$ and this function measures how far apart d and d' are. Under these assumptions a decision function δ is said to be unbiased with respect to the loss function L, or L-unbiased, if for all θ and d'

$$E_\theta L(d', \delta(X)) \geq E_\theta L(d, \delta(X))$$

where the subscript θ indicates the distribution with respect to which the

[†] This distinction is not adopted by all authors.

expectation is taken and where d is the decision that is correct for θ. Thus δ is unbiased if on the average $\delta(X)$ comes closer to the correct decision than to any wrong one. Extending this definition, δ is said to be *L-unbiased* for an arbitrary decision problem if for all θ and θ'

$$(9) \qquad E_\theta L(\theta', \delta(X)) \geq E_\theta L(\theta, \delta(X)).$$

Example 11. Suppose that in the problem of estimating a real-valued parameter θ by confidence intervals, as in Example 4, the loss is 0 or 1 as the interval $[\underline{L}, \overline{L}]$ does or does not cover the true θ. Then the set of intervals $[\underline{L}(X), \overline{L}(X)]$ is unbiased if the probability of covering the true value is greater than or equal to the probability of covering any false value.

Example 12. In a two-decision problem such as that of Example 1(i), let ω_0 and ω_1 be the sets of θ-values for which d_0 and d_1 are the correct decisions. Assume that the loss is 0 when the correct decision is taken, and otherwise is given by $L(\theta, d_0) = a$ for $\theta \in \omega_1$, and $L(\theta, d_1) = b$ for $\theta \in \omega_0$. Then

$$E_\theta L(\theta', \delta(X)) = \begin{cases} aP_\theta\{\delta(X) = d_0\} & \text{if } \theta' \in \omega_1, \\ bP_\theta\{\delta(X) = d_1\} & \text{if } \theta' \in \omega_0, \end{cases}$$

so that (9) reduces to

$$aP_\theta\{\delta(X) = d_0\} \geq bP_\theta\{\delta(X) = d_1\} \qquad \text{for } \theta \in \omega_0,$$

with the reverse inequality holding for $\theta \in \omega_1$. Since $P_\theta\{\delta(X) = d_0\} + P_\theta\{\delta(X) = d_1\} = 1$, the unbiasedness condition (9) becomes

$$(10) \qquad P_\theta\{\delta(X) = d_1\} \leq \frac{a}{a+b} \qquad \text{for } \theta \in \omega_0,$$

$$P_\theta\{\delta(X) = d_1\} \geq \frac{a}{a+b} \qquad \text{for } \theta \in \omega_1.$$

Example 13. In the problem of estimating a real-valued function $\gamma(\theta)$ with the square of the error as loss, the condition of unbiasedness becomes

$$E_\theta[\delta(X) - \gamma(\theta')]^2 \geq E_\theta[\delta(X) - \gamma(\theta)]^2 \qquad \text{for all } \theta, \theta'.$$

On adding and subtracting $h(\theta) = E_\theta\delta(X)$ inside the brackets on both sides, this reduces to

$$[h(\theta) - \gamma(\theta')]^2 \geq [h(\theta) - \gamma(\theta)]^2 \qquad \text{for all } \theta, \theta'.$$

If $h(\theta)$ is one of the possible values of the function γ, this condition holds if and only if

$$(11) \qquad E_\theta\delta(X) = \gamma(\theta).$$

In the theory of point estimation, (11) is customarily taken as the definition of unbiasedness. Except under rather pathological conditions, it is both a necessary and sufficient condition for δ to satisfy (9). (See Problem 2.)

6. BAYES AND MINIMAX PROCEDURES

We now turn to a discussion of some preference orderings of decision procedures and their risk functions. One such ordering is obtained by assuming that in repeated experiments the parameter itself is a random variable Θ, the distribution of which is known. If for the sake of simplicity one supposes that this distribution has a probability density $\rho(\theta)$, the overall average loss resulting from the use of a decision procedure δ is

$$(12) \qquad r(\rho, \delta) = \int E_\theta L(\theta, \delta(X)) \rho(\theta) \, d\theta = \int R(\theta, \delta) \rho(\theta) \, d\theta$$

and the smaller $r(\rho, \delta)$, the better is δ. An optimum procedure is one that minimizes $r(\rho, \delta)$ and is called a *Bayes solution* of the given decision problem corresponding to the a priori density ρ. The resulting minimum of $r(\rho, \delta)$ is called the *Bayes risk* of δ.

Unfortunately, in order to apply this principle it is necessary to assume not only that θ is a random variable but also that its distribution is known. This assumption is usually not warranted in applications. Alternatively, the right-hand side of (12) can be considered as a weighted average of the risks; for $\rho(\theta) \equiv 1$ in particular, it is then the area under the risk curve. With this interpretation the choice of a weight function ρ expresses the importance the experimenter attaches to the various values of θ. A systematic Bayes theory has been developed which interprets ρ as describing the state of mind of the investigator towards θ. For an account of this approach see, for example, Berger (1985).

If no prior information regarding θ is available, one might consider the maximum of the risk function its most important feature. Of two risk functions the one with the smaller maximum is then preferable, and the optimum procedures are those with the *minimax* property of minimizing the maximum risk. Since this maximum represents the worst (average) loss that can result from the use of a given procedure, a minimax solution is one that gives the greatest possible protection against large losses. That such a principle may sometimes be quite unreasonable is indicated in Figure 2, where under most circumstances one would prefer δ_1 to δ_2 although its risk function has the larger maximum.

Perhaps the most common situation is one intermediate to the two just described. On the one hand, past experience with the same or similar kind

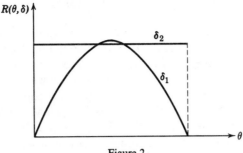

Figure 2

of experiment is available and provides an indication of what values of θ to expect; on the other, this information is neither sufficiently precise nor sufficiently reliable to warrant the assumptions that the Bayes approach requires. In such circumstances it seems desirable to make use of the available information without trusting it to such an extent that catastrophically high risks might result if it is inaccurate or misleading. To achieve this one can place a bound on the risk and restrict consideration to decision procedures δ for which

$$(13) \qquad\qquad R(\theta, \delta) \le C \qquad \text{for all } \theta.$$

[Here the constant C will have to be larger than the maximum risk C_0 of the minimax procedure, since otherwise there will exist no procedures satisfying (13).] Having thus assured that the risk can under no circumstances get out of hand, the experimenter can now safely exploit his knowledge of the situation, which may be based on theoretical considerations as well as on past experience; he can follow his hunches and guess at a distribution ρ for θ. This leads to the selection of a procedure δ (a *restricted Bayes solution*), which minimizes the average risk (12) for this a priori distribution subject to (13). The more certain one is of ρ, the larger one will select C, thereby running a greater risk in case of a poor guess but improving the risk if the guess is good.

Instead of specifying an ordering directly, one can postulate conditions that the ordering should satisfy. Various systems of such conditions have been investigated and have generally led to the conclusion that the only orderings satisfying these systems are those which order the procedures according to their Bayes risk with respect to some prior distribution of θ. For details, see for example Blackwell and Girshick (1954), Ferguson (1967), Savage (1972), and Berger (1985).

7. MAXIMUM LIKELIHOOD

Another approach, which is based on considerations somewhat different from those of the preceding sections, is the method of maximum likelihood. It has led to reasonable procedures in a great variety of problems, and is still playing a dominant role in the development of new tests and estimates. Suppose for a moment that X can take on only a countable set of values $x_1, x_2, \ldots$, with $P_\theta(x) = P_\theta\{X = x\}$, and that one wishes to determine the correct value of θ, that is, the value that produced the observed x. This suggests considering for each possible θ how probable the observed x would be if θ were the true value. The higher this probability, the more one is attracted to the explanation that the θ in question produced x, and the more likely the value of θ appears. Therefore, the expression $P_\theta(x)$ considered for fixed x as a function of θ has been called the *likelihood* of θ. To indicate the change in point of view, let it be denoted by $L_x(\theta)$. Suppose now that one is concerned with an action problem involving a countable number of decisions, and that it is formulated in terms of a gain function (instead of the usual loss function), which is 0 if the decision taken is incorrect and is $a(\theta) > 0$ if the decision taken is correct and θ is the true value. Then it seems natural to weight the likelihood $L_x(\theta)$ by the amount that can be gained if θ is true, to determine the value of θ that maximizes $a(\theta)L_x(\theta)$ and to select the decision that would be correct if this were the true value of θ. Essentially the same remarks apply in the case in which $P_\theta(x)$ is a probability density rather than a discrete probability.

In problems of point estimation, one usually assumes that $a(\theta)$ is independent of θ. This leads to estimating θ by the value that maximizes the likelihood $L_x(\theta)$, the *maximum-likelihood estimate* of θ. Another case of interest is the class of two-decision problems illustrated by Example 1(i). Let ω_0 and ω_1 denote the sets of θ-values for which d_0 and d_1 are the correct decisions, and assume that $a(\theta) = a_0$ or a_1 as θ belongs to ω_0 or ω_1 respectively. Then decision d_0 or d_1 is taken as $a_1 \sup_{\theta \in \omega_1} L_x(\theta) <$ or $> a_0 \sup_{\theta \in \omega_0} L_x(\theta)$, that is, as

$$(14) \qquad \frac{\sup\limits_{\theta \in \omega_0} L_x(\theta)}{\sup\limits_{\theta \in \omega_1} L_x(\theta)} > \quad \text{or} \quad < \frac{a_1}{a_0}.$$

This is known as a *likelihood-ratio procedure.*[*]

---

[*]This definition differs slightly from the usual one where in the denominator on the left-hand side of (14) the supremum is taken over the set $\omega_0 \cup \omega_1$. The two definitions agree whenever the left-hand side of (14) is ≤ 1, and the procedures therefore agree if $a_1 < a_0$.

Although the maximum-likelihood principle is not based on any clearly defined optimum considerations, it has been very successful in leading to satisfactory procedures in many specific problems. For wide classes of problems, maximum-likelihood procedures have also been shown to possess various asymptotic optimum properties as the sample size tends to infinity. [An asymptotic theory of likelihood-ratio tests has been developed by Wald (1943) and Le Cam (1953, 1979); an overview with additional references is given by Cox and Hinkley (1974). The corresponding theory of maximum-likelihood estimators is treated in Chapter 6 of *TPE*.] On the other hand, there exist examples for which the maximum-likelihood procedure is worse than useless; where it is, in fact, so bad that one can do better without making any use of the observations (see Chapter 6, Problem 18).

8. COMPLETE CLASSES

None of the approaches described so far is reliable in the sense that the resulting procedure is necessarily satisfactory. There are problems in which a decision procedure δ_0 exists with uniformly minimum risk among all unbiased or invariant procedures, but where there exists a procedure δ_1 not possessing this particular impartiality property and preferable to δ_0. (Cf. Problems 14 and 16.) As was seen earlier, minimax procedures can also be quite undesirable, while the success of Bayes and restricted Bayes solutions depends on a priori information which is usually not very reliable if it is available at all. In fact, it seems that in the absence of reliable a priori information no principle leading to a unique solution can be entirely satisfactory.

This suggests the possibility, at least as a first step, of not insisting on a unique solution but asking only how far a decision problem can be reduced without loss of relevant information. It has already been seen that a decision procedure δ can sometimes be eliminated from consideration because there exists a procedure δ' *dominating* it in the sense that

$$R(\theta, \delta') \leq R(\theta, \delta) \quad \text{for all } \theta$$
(15)
$$R(\theta, \delta') < R(\theta, \delta) \quad \text{for some } \theta.$$

In this case δ is said to be *inadmissible*; δ is called *admissible* if no such dominating δ' exists. A class $\mathscr{C}$ of decision procedures is said to be *complete* if for any δ not in $\mathscr{C}$ there exists δ' in $\mathscr{C}$ dominating it. A complete class is *minimal* if it does not contain a complete subclass. If a minimal complete class exists, as is typically the case, it consists exactly of the totality of admissible procedures.

It is convenient to define also the following variant of the complete class notion. A class $\mathscr{C}$ is said to be *essentially complete* if for any procedure δ there exists δ' in $\mathscr{C}$ such that $R(\theta, \delta') \leq R(\theta, \delta)$ for all θ. Clearly, any complete class is also essentially complete. In fact, the two definitions differ only in their treatment of equivalent decision rules, that is, decision rules with identical risk function. If δ belongs to the minimal complete class $\mathscr{C}$, any equivalent decision rule must also belong to $\mathscr{C}$. On the other hand, a minimal essentially complete class need contain only one member from such a set of equivalent procedures.

In a certain sense a minimal essentially complete class provides the maximum possible reduction of a decision problem. On the one hand, there is no reason to consider any of the procedures that have been weeded out. For each of them, there is included one in $\mathscr{C}$ that is as good or better. On the other hand, it is not possible to reduce the class further. Given any two procedures in $\mathscr{C}$, each of them is better in places than the other, so that without additional information it is not known which of the two is preferable.

The primary concern in statistics has been with the explicit determination of procedures, or classes of procedures, for various specific decision problems. Those studied most extensively have been estimation problems, and problems involving a choice between only two decisions (hypothesis testing), the theory of which constitutes the subject of the present volume. However, certain conclusions are possible without such specialization. In particular, two results concerning the structure of complete classes and minimax procedures have been proved to hold under very general assumptions:*

(i)   The totality of Bayes solutions and limits of Bayes solutions constitute a complete class.

(ii)   Minimax procedures are Bayes solutions with respect to a *least favorable* a priori distribution, that is, an a priori distribution that maximizes the associated Bayes risk, and the minimax risk equals this maximum Bayes risk. Somewhat more generally, if there exists no least favorable a priori distribution but only a sequence for which the Bayes risk tends to the maximum, the minimax procedures are limits of the associated sequence of Bayes solutions.

9. SUFFICIENT STATISTICS

A minimal complete class was seen in the preceding section to provide the maximum possible reduction of a decision problem without loss of informa-

---

*Precise statements and proofs of these results are given in the book by Wald (1950). See also Ferguson (1967) and Berger (1985).

tion. Frequently it is possible to obtain a less extensive reduction of the data, which applies simultaneously to all problems relating to a given class $\mathscr{P} = \{P_\theta, \ \theta \in \Omega\}$ of distributions of the given random variable X. It consists essentially in discarding that part of the data which contains no information regarding the unknown distribution P_θ, and which is therefore of no value for any decision problem concerning θ.

Example 14. Trials are performed with constant unknown probability p of success. If X_i is 1 or 0 as the ith trial is a success or failure, the sample $(X_1, \ldots, X_n)$ shows how many successes there were and in which trials they occurred. The second of these pieces of information contains no evidence as to the value of p. Once the total number of successes ΣX_i is known to be equal to t, each of the $\binom{n}{t}$ possible positions of these successes is equally likely regardless of p. It follows that knowing ΣX_i but neither the individual X_i nor p, one can, from a table of random numbers, construct a set of random variables $X'_1, \ldots, X'_n$ whose joint distribution is the same as that of $X_1, \ldots, X_n$. Therefore, the information contained in the X_i is the same as that contained in ΣX_i and a table of random numbers.

Example 15. If $X_1, \ldots, X_n$ are independently normally distributed with zero mean and variance σ^2, the conditional distribution of the sample point over each of the spheres, $\Sigma X_i^2 = $ constant, is uniform irrespective of σ^2. One can therefore construct an equivalent sample $X'_1, \ldots, X'_n$ from a knowledge of ΣX_i^2 and a mechanism that can produce a point randomly distributed over a sphere.

More generally, a statistic T is said to be *sufficient* for the family $\mathscr{P} = \{P_\theta, \ \theta \in \Omega\}$ (or sufficient for θ, if it is clear from the context what set Ω is being considered) if the conditional distribution of X given $T = t$ is independent of θ. As in the two examples it then follows under mild assumptions* that it is not necessary to utilize the original observations X. If one is permitted to observe only T instead of X, this does not restrict the class of available decision procedures. For any value t of T let X_t be a random variable possessing the conditional distribution of X given t. Such a variable can, at least theoretically, be constructed by means of a suitable random mechanism. If one then observes T to be t and X_t to be x', the random variable X' defined through this two-stage process has the same distribution as X. Thus, given any procedure based on X, it is possible to construct an equivalent one based on X' which can be viewed as a randomized procedure based solely on T. Hence if randomization is permitted (and we shall assume throughout that this is the case), there is no loss of generality in restricting consideration to a sufficient statistic.

It is inconvenient to have to compute the conditional distribution of X given t in order to determine whether or not T is sufficient. A simple check is provided by the following *factorization criterion*.

---

*These are connected with difficulties concerning the behavior of conditional probabilities. For a discussion of these difficulties see Chapter 2, Sections 3–5.

Consider first the case that X is discrete, and let $P_\theta(x) = P_\theta\{X = x\}$. Then a necessary and sufficient condition for T to be sufficient for θ is that there exists a factorization

(16) $$P_\theta(x) = g_\theta[T(x)] h(x),$$

where the first factor may depend on θ but depends on x only through $T(x)$, while the second factor is independent of θ.

Suppose that (16) holds, and let $T(x) = t$. Then $P_\theta\{T = t\} = \Sigma P_\theta(x')$ summed over all points x' with $T(x') = t$, and the conditional probability

$$P_\theta\{X = x | T = t\} = \frac{P_\theta(x)}{P_\theta\{T = t\}} = \frac{h(x)}{\Sigma h(x')}$$

is independent of θ. Conversely, if this conditional distribution does not depend on θ and is equal to, say $k(x, t)$, then $P_\theta(x) = P_\theta\{T = t\} k(x, t)$, so that (16) holds.

Example 16. Let $X_1, \ldots, X_n$ be independently and identically distributed according to the Poisson distribution (2). Then

$$P_\tau(x_1, \ldots, x_n) = \frac{\tau^{\Sigma x_i} e^{-n\tau}}{\prod_{j=1}^{n} x_j!},$$

and it follows that ΣX_i is a sufficient statistic for τ.

In the case that the distribution of X is continuous and has probability density $p_\theta^X(x)$, let X and T be vector-valued, $X = (X_1, \ldots, X_n)$ and $T = (T_1, \ldots, T_r)$ say. Suppose that there exist functions $Y = (Y_1, \ldots, Y_{n-r})$ on the sample space such that the transformation

(17) $$(x_1, \ldots, x_n) \leftrightarrow (T_1(x), \ldots, T_r(x), Y_1(x), \ldots, Y_{n-r}(x))$$

is $1 : 1$ on a suitable domain, and that the joint density of T and Y exists and is related to that of X by the usual formula

(18) $$p_\theta^X(x) = p_\theta^{T, Y}(T(x), Y(x)) \cdot |J|,$$

where J is the Jacobian of $(T_1, \ldots, T_r, Y_1, \ldots, Y_{n-r})$ with respect to $(x_1, \ldots, x_n)$. Thus in Example 15, $T = \sqrt{\Sigma X_i^2}$, $Y_1, \ldots, Y_{n-1}$ can be taken to be the polar coordinates of the sample point. From the joint density $p_\theta^{T, Y}(t, y)$ of T and Y, the conditional density of Y given $T = t$ is obtained as

(19) $$p_\theta^{Y|t}(y) = \frac{p_\theta^{T, Y}(t, y)}{\int p_\theta^{T, Y}(t, y') \, dy'}$$

provided the denominator is different from zero. Regularity conditions for the validity of (18) are given by Tukey (1958).

Since in the conditional distribution given t only the Y's vary, T is sufficient for θ if the conditional distribution of Y given t is independent of θ. Suppose that T satisfies (19). Then analogously to the discrete case, a necessary and sufficient condition for T to be sufficient is a factorization of the density of the form

$$(20) \qquad p_\theta^X(x) = g_\theta[T(x)] h(x).$$

(See Problem 19.) The following two examples illustrate the application of the criterion in this case. In both examples the existence of functions Y satisfying (17)–(19) will be assumed but not proved. As will be shown later (Chapter 2, Section 6), this assumption is actually not needed for the validity of the factorization criterion.

Example 17. Let $X_1, \ldots, X_n$ be independently distributed with normal probability density

$$p_{\xi, \sigma}(x) = (2\pi\sigma^2)^{-n/2} \exp\left(-\frac{1}{2\sigma^2} \sum x_i^2 + \frac{\xi}{\sigma^2} \sum x_i - \frac{n}{2\sigma^2} \xi^2 \right).$$

Then the factorization criterion shows $(\sum X_i, \sum X_i^2)$ to be sufficient for (ξ, σ).

Example 18. Let $X_1, \ldots, X_n$ be independently distributed according to the uniform distribution $U(0, \theta)$ over the interval $(0, \theta)$. Then $p_\theta(x) = \theta^{-n} u(\max x_i, \theta)$, where $u(a, b)$ is 1 or 0 as $a \le b$ or $a > b$, and hence max X_i is sufficient for θ.

An alternative criterion of *Bayes sufficiency*, due to Kolmogorov (1942), provides a direct connection between this concept and some of the basic notions of decision theory. As in the theory of Bayes solutions, consider the unknown parameter θ as a random variable Θ with an a priori distribution, and assume for simplicity that it has a density $\rho(\theta)$. Then if T is sufficient, the conditional distribution of Θ given $X = x$ depends only on $T(x)$. Conversely, if $\rho(\theta) \neq 0$ for all θ and if the conditional distribution of Θ given x depends only on $T(x)$, then T is sufficient for θ.

In fact, under the assumptions made, the joint density of X and Θ is $p_\theta(x)\rho(\theta)$. If T is sufficient, it follows from (20) that the conditional density of Θ given x depends only on $T(x)$. Suppose, on the other hand, that for some a priori distribution for which $\rho(\theta) \neq 0$ for all θ the conditional distribution of Θ given x depends only on $T(x)$. Then

$$\frac{p_\theta(x)\rho(\theta)}{\int p_{\theta'}(x)\rho(\theta')\, d\theta'} = f_\theta[T(x)]$$

and by solving for $p_\theta(x)$ it is seen that T is sufficient.

Any Bayes solution depends only on the conditional distribution of Θ given x (see Problem 8) and hence on $T(x)$. Since typically Bayes solutions together with their limits form an essentially complete class, it follows that this is also true of the decision procedures based on T. The same conclusion had already been reached more directly at the beginning of the section.

For a discussion of the relation of these different aspects of sufficiency in more general circumstances and references to the literature see Le Cam (1964) and Roy and Ramamoorthi (1979). An example of a statistic which is Bayes sufficient in the Kolmogorov sense but not according to the definition given at the beginning of this section is provided by Blackwell and Ramamoorthi (1982).

By restricting attention to a sufficient statistic, one obtains a reduction of the data, and it is then desirable to carry this reduction as far as possible. To illustrate the different possibilities, consider once more the binomial Example 14. If m is any integer less than n and $T_1 = \sum_{i=1}^{m} X_i$, $T_2 = \sum_{i=m+1}^{n} X_i$, then (T_1, T_2) constitutes a sufficient statistic, since the conditional distribution of $X_1, \ldots, X_n$ given $T_1 = t_1$, $T_2 = t_2$ is independent of p. For the same reason, the full sample $(X_1, \ldots, X_n)$ itself is also a sufficient statistic. However, $T = \sum_{i=1}^{n} X_i$ provides a more thorough reduction than either of these and than various others that can be constructed. A sufficient statistic T is said to be *minimal sufficient* if the data cannot be reduced beyond T without losing sufficiency. For the binomial example in particular, $\sum_{i=1}^{n} X_i$ can be shown to be minimal (Problem 17). This illustrates the fact that in specific examples the sufficient statistic determined by inspection through the factorization criterion usually turns out to be minimal. Explicit procedures for constructing minimal sufficient statistics are discussed in Section 1.5 of *TPE*.

10. PROBLEMS

Section 2

1. The following distributions arise on the basis of assumptions similar to those leading to (1)–(3).

    (i) Independent trials with constant probability p of success are carried out until a preassigned number m of successes has been obtained. If the number of trials required is $X + m$, then X has the *negative binomial* distribution $Nb(p, m)$:

    $$P\{X = x\} = \binom{m + x - 1}{x} p^m (1 - p)^x, \qquad x = 0, 1, 2 \ldots .$$

    (ii) In a sequence of random events, the number of events occurring in any time interval of length τ has the Poisson distribution $P(\lambda\tau)$, and the numbers of events in nonoverlapping time intervals are independent. Then the "waiting time" T, which elapses from the starting point, say $t = 0$, until the first event occurs, has the *exponential* probability density

$$p(t) = \lambda e^{-\lambda t}, \qquad t \geq 0.$$

Let T_i, $i \geq 2$, be the time elapsing from the occurrence of the $(i - 1)$st event to that of the ith event. Then it is also true, although more difficult to prove, that $T_1, T_2, \ldots$ are identically and independently distributed. A proof is given, for example, in Karlin and Taylor (1975).

(iii)   A point X is selected "at random" in the interval (a, b), that is, the probability of X falling in any subinterval of (a, b) depends only on the length of the subinterval, not on its position. Then X has the *uniform* distribution $U(a, b)$ with probability density

$$p(x) = 1/(b - a), \qquad a < x < b.$$

[(ii): If $t > 0$, then $T > t$ if and only if no event occurs in the time interval $(0, t)$.]

Section 5

2. *Unbiasedness in point estimation.* Suppose that γ is a continuous real-valued function defined over Ω which is not constant in any open subset of Ω, and that the expectation $h(\theta) = E_\theta \delta(X)$ is a continuous function of θ for every estimate $\delta(X)$ of $\gamma(\theta)$. Then (11) is a necessary and sufficient condition for $\delta(X)$ to be unbiased when the loss function is the square of the error.
[Unbiasedness implies that $\gamma^2(\theta') - \gamma^2(\theta) \geq 2h(\theta)[\gamma(\theta') - \gamma(\theta)]$ for all θ, θ'. If θ is neither a relative minimum or maximum of γ, it follows that there exist points θ' arbitrarily close to θ both such that $\gamma(\theta) + \gamma(\theta') \geq$ and $\leq 2h(\theta)$, and hence that $\gamma(\theta) = h(\theta)$. That this equality also holds for an extremum of γ follows by continuity, since γ is not constant in any open set.]

3. *Median unbiasedness.*

(i)   A real number m is a median for the random variable Y if $P\{Y \geq m\} \geq \frac{1}{2}$, $P\{Y \leq m\} \geq \frac{1}{2}$. Then all real a_1, a_2 such that $m \leq a_1 \leq a_2$ or $m \geq a_1 \geq a_2$ satisfy $E|Y - a_1| \leq E|Y - a_2|$.

(ii)   For any estimate $\delta(X)$ of $\gamma(\theta)$, let $m^-(\theta)$ and $m^+(\theta)$ denote the infimum and supremum of the medians of $\delta(X)$, and suppose that they are continuous functions of θ. Let $\gamma(\theta)$ be continuous and not constant in any open subset of Ω. Then the estimate $\delta(X)$ of $\gamma(\theta)$ is unbiased with respect to the loss function $L(\theta, d) = |\gamma(\theta) - d|$ if and only if $\gamma(\theta)$ is a median of $\delta(X)$ for each θ. An estimate with this property is said to be *median-unbiased*.

4. *Nonexistence of unbiased procedures.* Let $X_1, \ldots, X_n$ be independently distributed with density $(1/a)f((x - \xi)/a)$, and let $\theta = (\xi, a)$. Then no estima-

tor of ξ exists which is unbiased with respect to the loss function $(d - \xi)^k/a^k$.
Note. For more general results concerning the nonexistence of unbiased procedures see Rojo (1983).

5. Let $\mathscr{C}$ be any class of procedures that is closed under the transformations of a group G in the sense that $\delta \in \mathscr{C}$ implies $g^*\delta g^{-1} \in \mathscr{C}$ for all $g \in G$. If there exists a unique procedure δ_0 that uniformly minimizes the risk within the class $\mathscr{C}$, then δ_0 is invariant.[†] If δ_0 is unique only up to sets of measure zero, then it is *almost invariant*, that is, for each g it satisfies the equation $\delta(gx) = g^*\delta(x)$ except on a set N_g of measure 0.

6. *Relation of unbiasedness and invariance.*

   (i) If δ_0 is the unique (up to sets of measure 0) unbiased procedure with uniformly minimum risk, it is almost invariant.

   (ii) If $\overline{G}$ is transitive and G^* commutative, and if among all invariant (almost invariant) procedures there exists a procedure δ_0 with uniformly minimum risk, then it is unbiased.

   (iii) That conclusion (ii) need not hold without the assumptions concerning G^* and $\overline{G}$ is shown by the problem of estimating the mean ξ of a normal distribution $N(\xi, \sigma^2)$ with loss function $(\xi - d)^2/\sigma^2$. This remains invariant under the groups $G_1 : gx = x + b$, $-\infty < b < \infty$ and $G_2 : gx = ax + b, 0 < a < \infty, -\infty < b < \infty$. The best invariant estimate relative to both groups is X, but there does not exist an estimate which is unbiased with respect to the given loss function.

[(i): This follows from the preceding problem and the fact that when δ is unbiased so is $g^*\delta g^{-1}$.
(ii): It is the defining property of transitivity that given θ, θ' there exists $\overline{g}$ such that $\theta' = \overline{g}\theta$. Hence for any θ, θ'

$$E_\theta L(\theta', \delta_0(X)) = E_\theta L(\overline{g}\theta, \delta_0(X)) = E_\theta L(\theta, g^{*-1}\delta_0(X)).$$

Since G^* is commutative, $g^{*-1}\delta_0$ is invariant, so that

$$R(\theta, g^{*-1}\delta_0) \geq R(\theta, \delta_0) = E_\theta L(\theta, \delta_0(X)).]$$

Section 6

7. *Unbiasedness in interval estimation.* Confidence intervals $I = (\underline{L}, \overline{L})$ are unbiased for estimating θ with loss function $L(\theta, I) = (\theta - \underline{L})^2 + (\overline{L} - \theta)^2$ provided $E[\frac{1}{2}(\underline{L} + \overline{L})] = \theta$ for all θ, that is, provided the midpoint of I is an unbiased estimate of θ in the sense of (11).

[†] Here and in Problems 6, 7, 11, 15, and 16 the term "invariant" is used in the general sense (8) of "invariant or equivariant".

8. *Structure of Bayes solutions.*

   (i)   Let Θ be an unobservable random quantity with probability density $\rho(\theta)$, and let the probability density of X be $p_\theta(x)$ when $\Theta = \theta$. Then δ is a Bayes solution of a given decision problem if for each x the decision $\delta(x)$ is chosen so as to minimize $\int L(\theta, \delta(x))\pi(\theta|x)\,d\theta$, where $\pi(\theta|x) = \rho(\theta)p_\theta(x)/\int\rho(\theta')p_{\theta'}(x)\,d\theta'$ is the conditional (a posteriori) probability density of Θ given x.

   (ii)   Let the problem be a two-decision problem with the losses as given in Example 12. Then the Bayes solution consists in choosing decision d_0 if

$$aP\{\Theta \in \omega_1|x\} < bP\{\Theta \in \omega_0|x\}$$

and decision d_1 if the reverse inequality holds. The choice of decision is immaterial in case of equality.

   (iii)   In the case of point estimation of a real-valued function $g(\theta)$ with loss function $L(\theta, d) = (g(\theta) - d)^2$, the Bayes solution becomes $\delta(x) = E[g(\Theta)|x]$. When instead the loss function is $L(\theta, d) = |g(\theta) - d|$, the Bayes estimate $\delta(x)$ is any median of the conditional distribution of $g(\Theta)$ given x.

[(i): The Bayes risk $r(\rho, \delta)$ can be written as $\int[\int L(\theta, \delta(x))\pi(\theta|x)\,d\theta] \times p(x)\,dx$, where $p(x) = \int\rho(\theta')p_{\theta'}(x)\,d\theta'$.
(ii): The conditional expectation $\int L(\theta, d_0)\pi(\theta|x)\,d\theta$ reduces to $aP\{\Theta \in \omega_1|x\}$, and similarly for d_1.]

9.   (i)   As an example in which randomization reduces the maximum risk, suppose that a coin is known to be either standard (HT) or to have heads on both sides (HH). The nature of the coin is to be decided on the basis of a single toss, the loss being 1 for an incorrect decision and 0 for a correct one. Let the decision be HT when T is observed, whereas in the contrary case the decision is made at random, with probability ρ for HT and $1 - \rho$ for HH. Then the maximum risk is minimized for $\rho = \frac{1}{3}$.

   (ii)   A genetic setting in which such a problem might arise is that of a couple, of which the husband is either dominant homozygous (AA) or heterozygous (Aa) with respect to a certain characteristic, and the wife is homozygous recessive (aa). Their child is heterozygous, and it is of importance to determine to which genetic type the husband belongs. However, in such cases an a priori probability is usually available for the two possibilities. One is then dealing with a Bayes problem, and randomization is no longer required. In fact, if the a priori probability is p that the husband is dominant, then the Bayes procedure classifies him as such if $p > \frac{1}{3}$ and takes the contrary decision if $p < \frac{1}{3}$.

10. *Unbiasedness and minimax.* Let $\Omega = \Omega_0 \cup \Omega_1$ where Ω_0, Ω_1 are mutually exclusive, and consider a two-decision problem with loss function $L(\theta, d_i) = a_i$ for $\theta \in \Omega_j$ ($j \neq i$) and $L(\theta, d_i) = 0$ for $\theta \in \Omega_i$ ($i = 0, 1$).

(i) Any minimax procedure is unbiased.

(ii) The converse of (i) holds provided $P_\theta(A)$ is a continuous function of θ for all A, and if the sets Ω_0 and Ω_1 have at least one common boundary point.

[(i): The condition of unbiasedness in this case is equivalent to $\sup R_\delta(\theta) \leq a_0 a_1 / (a_0 + a_1)$. That this is satisfied by any minimax procedure is seen by comparison with the procedure $\delta(x) = d_0$ or $= d_1$ with probabilities $a_1 / (a_0 + a_1)$ and $a_0 / (a_0 + a_1)$ respectively.

(ii): If θ_0 is a common boundary point, continuity of the risk function implies that any unbiased procedure satisfies $R_\delta(\theta_0) = a_0 a_1 / (a_0 + a_1)$ and hence $\sup R_\delta(\theta) = a_0 a_1 / (a_0 + a_1)$.]

11. *Invariance and minimax.* Let a problem remain invariant relative to the groups G, $\bar{G}$, and G^* over the spaces $\mathcal{X}$, Ω, and D respectively. Then a randomized procedure Y_x is defined to be invariant if for all x and g the conditional distribution of Y_x given x is the same as that of $g^{*-1} Y_{gx}$.

(i) Consider a decision procedure which remains invariant under a finite group $G = \{g_1, \ldots, g_N\}$. If a minimax procedure exists, then there exists one that is invariant.

(ii) This conclusion does not necessarily hold for infinite groups, as is shown by the following example. Let the parameter space Ω consist of all elements θ of the free group with two generators, that is, the totality of formal products $\pi_1 \ldots \pi_n$ ($n = 0, 1, 2, \ldots$) where each π_i is one of the elements a, a^{-1}, b, b^{-1} and in which all products aa^{-1}, $a^{-1}a$, bb^{-1}, and $b^{-1}b$ have been canceled. The empty product ($n = 0$) is denoted by e. The sample point X is obtained by multiplying θ on the right by one of the four elements a, a^{-1}, b, b^{-1} with probability $\frac{1}{4}$ each, and canceling if necessary, that is, if the random factor equals π_n^{-1}. The problem of estimating θ with $L(\theta, d)$ equal to 0 if $d = \theta$ and equal to 1 otherwise remains invariant under multiplication of X, θ, and d on the left by an arbitrary sequence $\pi_{-m} \ldots \pi_{-2} \pi_{-1}$ ($m = 0, 1, \ldots$). The invariant procedure that minimizes the maximum risk has risk function $R(\theta, \delta) \equiv \frac{3}{4}$. However, there exists a noninvariant procedure with maximum risk $\frac{1}{4}$.

[(i): If Y_x is a (possibly randomized) minimax procedure, an invariant minimax procedure Y_x' is defined by $P(Y_x' = d) = \sum_{i=1}^N P(Y_{g_i x} = g_i^* d)/N$.

(ii): The better procedure consists in estimating θ to be $\pi_1 \ldots \pi_{k-1}$ when $\pi_1 \ldots \pi_k$ is observed ($k \geq 1$), and estimating θ to be a, a^{-1}, b, b^{-1} with probability $\frac{1}{4}$ each in case the identity is observed. The estimate will be correct unless the last element of X was canceled, and hence will be correct with probability $\geq \frac{3}{4}$.]

Section 7

12. (i) Let X have probability density $p_\theta(x)$ with θ one of the values $\theta_1, \ldots, \theta_n$, and consider the problem of determining the correct value of θ, so that the choice lies between the n decisions $d_1 = \theta_1, \ldots, d_n = \theta_n$ with gain $a(\theta_i)$ if $d_i = \theta_i$ and 0 otherwise. Then the Bayes solution (which maximizes the average gain) when θ is a random variable taking on each of the n values with probability $1/n$ coincides with the maximum-likelihood procedure.

    (ii) Let X have probability density $p_\theta(x)$ with $0 \le \theta \le 1$. Then the maximum-likelihood estimate is the mode (maximum value) of the a posteriori density of Θ given x when Θ is uniformly distributed over $(0, 1)$.

13. (i) Let $X_1, \ldots, X_n$ be a sample from $N(\xi, \sigma^2)$, and consider the problem of deciding between $\omega_0 : \xi < 0$ and $\omega_1 : \xi \ge 0$. If $\bar{x} = \Sigma x_i/n$ and $C = (a_1/a_0)^{2/n}$, the likelihood-ratio procedure takes decision d_0 or d_1 as

$$\frac{\sqrt{n}\,\bar{x}}{\sqrt{\Sigma(x_i - \bar{x})^2}} < k \quad \text{or} \quad > k,$$

where $k = -\sqrt{C-1}$ if $C > 1$ and $k = \sqrt{(1-C)/C}$ if $C < 1$.

    (ii) For the problem of deciding between $\omega_0 : \sigma < \sigma_0$ and $\omega_1 : \sigma \ge \sigma_0$, the likelihood ratio procedure takes decision d_0 or d_1 as

$$\frac{\Sigma(x_i - \bar{x})^2}{n\sigma_0^2} < \quad \text{or} \quad > k,$$

where k is the smaller root of the equation $Cx = e^{x-1}$ if $C > 1$, and the larger root of $x = Ce^{x-1}$ if $C < 1$, where C is defined as in (i).

Section 8

14. *Admissibility of unbiased procedures.*

    (i) Under the assumptions of Problem 10, if among the unbiased procedures there exists one with uniformly minimum risk, it is admissible.

    (ii) That in general an unbiased procedure with uniformly minimum risk need not be admissible is seen by the following example. Let X have a Poisson distribution truncated at 0, so that $P_\theta\{X = x\} = \theta^x e^{-\theta}/[x!(1 - e^{-\theta})]$ for $x = 1, 2, \ldots$. For estimating $\gamma(\theta) = e^{-\theta}$ with loss function $L(\theta, d) = (d - e^{-\theta})^2$, there exists a unique unbiased estimate, and it is not admissible.

[(ii): The unique unbiased estimate $\delta_0(x) = (-1)^{x+1}$ is dominated by $\delta_1(x) = 0$ or 1 as x is even or odd.]

15. *Admissibility of invariant procedures.* If a decision problem remains invariant under a finite group, and if there exists a procedure δ_0 that uniformly minimizes the risk among all invariant procedures, then δ_0 is admissible.
[This follows from the identity $R(\theta, \delta) = R(\bar{g}\theta, g^*\delta g^{-1})$ and the hint given in Problem 11(i).]

16. (i) Let X take on the values $\theta - 1$ and $\theta + 1$ with probability $\frac{1}{2}$ each. The problem of estimating θ with loss function $L(\theta, d) = \min(|\theta - d|, 1)$ remains invariant under the transformation $gX = X + c$, $\bar{g}\theta = \theta + c$, $g^*d = d + c$. Among invariant estimates, those taking on the values $X - 1$ and $X + 1$ with probabilities p and q (independent of X) uniformly minimize the risk.

(ii) That the conclusion of Problem 15 need not hold when G is infinite follows by comparing the best invariant estimates of (i) with the estimate $\delta_1(x)$ which is $X + 1$ when $X < 0$ and $X - 1$ when $X \geq 0$.

Section 9

17. In n independent trials with constant probability p of success, let $X_i = 1$ or 0 as the ith trial is a success or not. Then $\sum_{i=1}^n X_i$ is minimal sufficient.
[Let $T = \sum X_i$ and suppose that $U = f(T)$ is sufficient and that $f(k_1) = \cdots = f(k_r) = u$. Then $P\{T = t | U = u\}$ depends on p.]

18. (i) Let $X_1, \ldots, X_n$ be a sample from the uniform distribution $U(0, \theta)$, $0 < \theta < \infty$, and let $T = \max(X_1, \ldots, X_n)$. Show that T is sufficient, once by using the definition of sufficiency and once by using the factorization criterion and assuming the existence of statistics Y_i satisfying (17)–(19).

(ii) Let $X_1, \ldots, X_n$ be a sample from the exponential distribution $E(a, b)$ with density $(1/b)e^{-(x-a)/b}$ when $x \geq a$ ($-\infty < a < \infty$, $0 < b$). Use the factorization criterion to prove that $(\min(X_1, \ldots, X_n), \sum_{i=1}^n X_i)$ is sufficient for a, b, assuming the existence of statistics Y_i satisfying (17)–(19).

19. A statistic T satisfying (17)–(19) is sufficient if and only if it satisfies (20).

11. REFERENCES

Some of the basic concepts of statistical theory were initiated during the first quarter of the 19th century by Laplace in his fundamental Théorie Analytique des Probabilités (1812), and by Gauss in his papers on the method of least squares. Loss and risk functions are mentioned in their discussions of the problem of point estimation, for which Gauss also introduced the condition of unbiasedness.

A period of intensive development of statistical methods began toward the end of the century with the work of Karl Pearson. In particular, two areas were explored in the researches of R. A. Fisher, J. Neyman, and many others: estimation and the testing of hypotheses. The work of Fisher can be found in his books (1925, 1935, 1956) and in the five volumes of his collected papers (1971–1973). An interesting review of Fisher's contributions is provided by Savage (1976), and his life and work are recounted in the biography by his daughter Joan Fisher Box (1978). Many of Neyman's principal ideas are summarized in his Lectures and Conferences (1983b). Collections of his early papers and of his joint papers with E. S. Pearson have been published [Neyman (1967) and Neyman and Pearson (1967)], and Constance Reid (1982) has written his biography: from life. An influential synthesis of the work of this period by Cramér appeared in 1946. More recent surveys of the modern theories of estimation and testing are contained, for example, in the books by Bickel and Doksum (1977), Cox and Hinkley (1974), Kendall and Stuart (1979), and Schmetterer (1974).

A formal unification of the theories of estimation and hypothesis testing, which also contains the possibility of many other specializations, was achieved by Wald in his general theory of decision procedures. An account of this theory, which is closely related to von Neumann's theory of games, is found in Wald's book (1950) and in those of Blackwell and Girshick (1954), Ferguson (1967), and Berger (1985).

Barnett, V.
  (1982). *Comparative Statistical Inference.* 2nd ed. Wiley, New York.

Berger, J. O.
  (1985). *Statistical Decision Theory and Bayesian Analysis.* 2nd ed. Springer, New York.

Bickel, P. J.
  (1984). "Parametric robustness: small biases can be worthwhile." *Ann. Statist.* 12, 864–879.

Bickel, P. J. and Doksum, K. A.
  (1977). *Mathematical Statistics.* Holden-Day, San Francisco.

Blackwell, D. and Girshick, M. A.
  (1954). *Theory of Games and Statistical Decisions.* Wiley, New York.

Blackwell, D. and Ramamoorthi, R. V.
  (1982). "A Bayes but not classically sufficient statistic." *Ann. Statist.* 10, 1025–1026.

Blyth, C. R.
  (1970). "On the inference and decision models of statistics" (with discussion). *Ann. Statist.* 41, 1034–1058.

Bondessen, L.
  (1983). "Equivariant estimators." in *Encyclopedia of Statistical Sciences*, Vol. 2. Wiley, New York.

Box, J. F.
  (1978). *R. A. Fisher: The Life of a Scientist*, Wiley, New York.

Brown, G.
  (1947). "On small sample estimation." *Ann. Math. Statist.* 18, 582–585.
  [Definition of median unbiasedness.]

Chernoff, H.

(1972). *Sequential Analysis and Optimal Design*, SIAM, Philadelphia.

Cox, D. R.

(1958). "Some problems connected with statistical inference." *Ann. Math. Statist.* **29**, 357–372.

Cox, D. R. and Hinkley, D. V.

(1974). *Theoretical Statistics*, Chapman and Hall, London.

Cramér, H.

(1946). *Mathematical Methods of Statistics*, Princeton Univ. Press, Princeton, N.J.

Edgeworth, F. Y.

(1908–09). "On the probable errors of frequency constants." *J. Roy. Statist. Soc.* **71**, 381–397, 499–512, 651–678; **72**, 81–90.

[Edgeworth's work on maximum-likelihood estimation and its relation to the results of Fisher in the same area is reviewed by Pratt (1976). Stigler (1978) provides a systematic account of Edgeworth's many other important contributions to statistics.]

Feller, W.

(1968). *An Introduction to Probability Theory and its Applications*, 3rd ed., Vol. 1, Wiley, New York.

Ferguson, T. S.

(1967). *Mathematical Statistics*, Academic, New York.

Fisher, R. A.

(1920). "A mathematical examination of the methods of determining the accuracy of an observation by the mean error and by the mean square error." *Monthly Notices Roy. Astron. Soc.* **80**, 758–770.

(1922). "On the mathematical foundations of theoretical statistics." *Phil. Trans. Roy. Soc. Ser. A* **222**, 309–368.

(1925). "Theory of statistical estimation." *Proc. Cambridge Phil. Soc.* **22**, 700–725.

[These papers develop a theory of point estimation (based on the maximum likelihood principle) and the concept of sufficiency. The factorization theorem is given in a form which is formally weaker but essentially equivalent to (20).]

(1925). *Statistical Methods for Research Workers*, 1st ed. (14th ed., 1970), Oliver and Boyd, Edinburgh.

(1935). *The Design of Experiments*, 1st ed. (8th ed., 1966), Oliver and Boyd, Edinburgh.

(1956). *Statistical Methods and Scientific Inference*, Oliver and Boyd, Edinburgh (3rd ed., Hafner, New York, 1973).

(1971–1973). *Collected Papers* (J. H. Bennett, ed.), Univ. of Adelaide.

Ghosh, B. K. (1970).

Sequential Tests of Statistical Hypotheses, Addison-Wesley, Reading, Mass.

Govindarajulu, Z. (1981).

The Sequential Statistical Analysis, American Sciences Press, Columbus, Ohio.

Hodges, J. L., Jr., and Lehmann, E. L.

(1952). "The use of previous experience in reaching statistical decisions." *Ann. Math. Statist.* **23**, 396–407.

[Theory of restricted Bayes solutions.]

Hotelling, H.

(1936). "Relations between two sets of variates." *Biometrika* **28**, 321–377.

[One of the early papers making explicit use of invariance considerations.]

Hunt, G. and Stein, C.
(1946). Most stringent tests of statistical hypotheses.
[In this paper, which unfortunately was never published, a general theory of invariance is developed for hypothesis testing.]

Karlin, S. and Taylor, H. M.
(1975). *A First Course in Stochastic Processes*, Academic, New York.

Kendall, M. and Stuart, A.
(1979). *The Advanced Theory of Statistics, Vol. 2, 4th ed.*, Macmillan, New York.

Kiefer, J.
(1957). "Invariance, minimax sequential estimation, and continuous time processes." *Ann. Math. Statist.* **28**, 573–601.
(1966). "Multivariate optimality results." In *Multivariate Analysis* (Krishnaiah, ed.), Academic, New York.

Kolmogorov, A.
(1942). "Sur l'estimation statistique des paramètres de la loi de Gauss." *Bull. Acad. Sci. URSS Ser. Math.* **6**, 3–32. (Russian—French summary.)
[Definition of sufficiency in terms of distributions for the parameters.]

Kudo, H.
(1955). "On minimax invariant estimates of the transformation parameter." *Nat. Sci. Rept. Ochanomizu Univ., Tokyo* **6**, 31–73.

Laplace, P. S.
(1812). *Théorie Analytique des Probabilités*, Paris.

Le Cam, L.
(1953). "On some asymptotic properties of maximum likelihood estimates and related Bayes estimates." In *Univ. Calif. Publs. Statistics*, Vol. 1, pp. 277–329, Univ. of California Press, Berkeley and Los Angeles.
[Rigorous and very general treatment of the large-sample theory of maximum-likelihood estimates, with a survey of the large previous literature on the subject.]
(1964). "Sufficiency and approximate sufficiency." *Ann. Math. Statist.* **35**, 1419–1455.
(1979). "On a theorem of J. Hájek." In *Contributions to Statistics: J. Hájek Memorial Volume* (Jureckova, ed.), Academia, Prague.

Lehmann, E. L.
(1947). "On families of admissible tests." *Ann. Math. Statist.* **18**, 97–104.
[Introduction of the complete class concept in connection with a special class of testing problems.]
(1950). "Some principles of the theory of hypothesis testing." *Ann. Math. Statist.* **21**, 1–26.
(1951). "A general concept of unbiasedness." *Ann. Math. Statist.* **22**, 587–597.
[Definition (8); Problems 2, 3, 4, 6, 7, and 14.]
(1984). "Specification problems in the Neyman-Pearson-Wald theory." In *Statistics: An Appraisal* (David and David, eds.), Iowa State Univ. Press, Ames.
(1985). "The Neyman–Pearson theory after 50 years." In *Proc. Neyman–Kiefer Conference* (LeCam and Olshen, eds.), Wadsworth, Belmont, Cal.

Neyman, J.
(1935). "Sur un teorema concernente le cosidette statistiche sufficienti." *Giorn. Ist. Ital. Att.* **6**, 320–334.
[Obtains the factorization theorem in the form (20).]
(1938a). "L'estimation statistique traitée comme un problème classique de probabilité." *Actualités Sci. et Ind.* **739**, 25–57.

[Puts forth the point of view that statistics is primarily concerned with how to behave under uncertainty rather than with determining the values of unknown parameters, with inductive behavior rather than with inductive inference.]

(1938b). *Lectures and Conferences on Mathematical Statistics and Probability*, 1st ed. (2nd ed., 1952), Graduate School, U.S. Dept. of Agriculture, Washington.

(1967). *A Selection of Early Statistical Papers of J. Neyman*, Univ. of California Press, Berkeley.

Neyman, J. and Pearson, E. S.

(1928). "On the use and interpretation of certain test criteria for purposes of statistical inference." *Biometrika* **20A**, 175–240, 263–295.

[Proposes the likelihood-ratio principle for obtaining reasonable tests, and applies it to a number of specific problems.]

(1933). "On the testing of statistical hypotheses in relation to probability a priori." *Proc. Cambridge Phil. Soc.* **29**, 492–510.

[In connection with the problem of hypothesis testing, suggests assigning weights for the various possible wrong decisions and the use of the minimax principle.]

(1967). *Joint Statistical Papers of J. Neyman and E. S. Pearson*, Univ. of California Press, Berkeley.

Pearson, E. S.

(1929). "Some notes on sampling tests with two variables." *Biometrika* **21**, 337–360.

Peisakoff, M.

(1951). *Transformation of Parameters*, unpublished thesis. Princeton Univ.

[Extends the Hunt-Stein theory of invariance to more general classes of decision problems; see Problem 11(ii). The theory is generalized further in Kiefer (1957, 1966) and Kudo (1955).]

Pitman, E. J. G.

(1938). "Location and scale parameters." *Biometrika* **30**, 391–421.

(1939). "Tests of hypotheses concerning location and scale parameters." *Biometrika* **31**, 200–215.

[In these papers the restriction to invariant procedures is introduced for estimation and testing problems involving location and scale parameters.]

Pratt, J. W.

(1976). "F. Y. Edgeworth and R. A. Fisher on the efficiency of maximum likelihood estimation." *Ann. Statist.* **4**, 501–514.

Reid, C.

(1982). *Neyman from Life*, Springer, New York.

Rojo, J.

(1983). *On Lehmann's General Concept of Unbiasedness and Some of Its Applications*, Ph.D. Thesis, Univ. of California, Berkeley.

Ross, S.

(1980). *Introduction to Probability Models*, 2nd ed., Academic, New York.

Roy, K. K. and Ramamoorthi, R. V.

(1979). "Relationship between Bayes, classical and decision theoretic sufficiency." Sankhyā **41**, 48–58.

Savage, L. J.

(1972). *The Foundations of Statistics*, 2nd ed., Dover, New York.

(1976). "On rereading R. A. Fisher" (with discussion), *Ann. Statist.* **4**, 441–500.

Schmetterer, L.

(1974). *Introduction to Mathematical Statistics*, 2nd ed., Springer, Berlin.

Silvey, S. D.
(1980). *Optimal Design*: *An Introduction to the Theory of Parameter Estimation*, Chapman and Hall, London.

Stigler, S. M.
(1978). "Francis Ysidro Edgeworth, Statistician" (with discussion). *J. Roy. Statist. Soc. (A)* **141**, 287–322.

Stone, C. J.
(1981). "Admissible selection of an accurate and parsimonious normal linear regression model." *Ann. Statist.* **9**, 475–485.

Tukey, J. W.
(1958). "A smooth invertibility theorem." *Ann. Math. Statist.* **29**, 581–584.

Wald, A.
(1939). "Contributions to the theory of statistical estimation and testing hypotheses." *Ann. Math. Statist.* **10**, 299–326.
[A general formulation of statistical problems containing estimation and testing problems as special cases. Discussion of Bayes and minimax procedures.]

Wald, A.
(1943). "Tests of statistical hypotheses concerning several parameters when the number of observations is large." *Trans. Am. Math. Soc.* **54**, 462–482.
(1947). "An essentially complete class of admissible decision functions." *Ann. Math. Statist.* **18**, 549–555.
[Defines and characterizes complete classes of decision procedures for general decision problems. The ideas of this and the preceding paper were developed further in a series of papers culminating in Wald's book (1950).]
(1950). *Statistical Decision Functions*, Wiley, New York.
(1958). *Selected Papers in Statistics and Probability by Abraham Wald*. Stanford Univ. Press.

CHAPTER 2

The Probability
Background

1. PROBABILITY AND MEASURE

The mathematical framework for statistical decision theory is provided by the theory of probability, which in turn has its foundations in the theory of measure and integration. The present and following sections serve to define some of the basic concepts of these theories, to establish some notation, and to state without proof some of the principal results. In the remainder of the chapter, certain special topics are treated in more detail.

Probability theory is concerned with situations which may result in different outcomes. The totality of these possible outcomes is represented abstractly by the totality of points in a space $\mathscr{X}$. Since the events to be studied are aggregates of such outcomes, they are represented by subsets of $\mathscr{X}$. The union of two sets C_1, C_2 will be denoted by $C_1 \cup C_2$, their intersection by $C_1 \cap C_2$, the complement of C by $\tilde{C} = \mathscr{X} - C$, and the empty set by 0. The probability $P(C)$ of an event C is a real number between 0 and 1; in particular

$$(1) \qquad P(0) = 0 \quad \text{and} \quad P(\mathscr{X}) = 1.$$

Probabilities have the property of *countable additivity*,

$$(2) \qquad P\left(\bigcup C_i\right) = \sum P(C_i) \quad \text{if} \quad C_i \cap C_j = 0 \quad \text{for all} \quad i \neq j.$$

Unfortunately it turns out that the set functions with which we shall be concerned usually cannot be defined in a reasonable manner for all subsets of $\mathscr{X}$ if they are to satisfy (2). It is, for example, not possible to give a reasonable definition of "area" for all subsets of a unit square in the plane.

34

The sets for which the probability function P will be defined are said to be "measurable". The domain of definition of P should include with any set C its complement $\tilde{C}$, and with any countable number of events their union. By (1), it should also include $\mathscr{X}$. A class of sets that contains $\mathscr{X}$ and is closed under complementation and countable unions is a σ-field. Such a class is automatically also closed under countable intersections.

The starting point of any probabilistic considerations is therefore a space $\mathscr{X}$, representing the possible outcomes, and a σ-field $\mathscr{C}$ of subsets of $\mathscr{X}$, representing the events whose probability is to be defined. Such a couple $(\mathscr{X}, \mathscr{C})$ is called a *measurable space*, and the elements of $\mathscr{C}$ constitute the *measurable sets*. A countably additive nonnegative (not necessarily finite) set function μ defined over $\mathscr{C}$ and such that $\mu(0) = 0$ is called a *measure*. If it assigns the value 1 to $\mathscr{X}$, it is a *probability measure*. More generally, μ is *finite* if $\mu(\mathscr{X}) < \infty$ and σ-*finite* if there exist $C_1, C_2, \ldots$ in $\mathscr{C}$ (which may always be taken to be mutually exclusive) such that $\bigcup C_i = \mathscr{X}$ and $\mu(C_i) < \infty$ for $i = 1, 2, \ldots$. Important special cases are provided by the following examples.

Example 1. Lebesgue measure. Let $\mathscr{X}$ be the n-dimensional Euclidean space E_n, and $\mathscr{C}$ the smallest σ-field containing all rectangles*

$$R = \left\{ (z_1, \ldots, z_n) : a_i < z_i \leq b_i, i = 1, \ldots, n \right\}.$$

The elements of $\mathscr{C}$ are called the *Borel sets* of E_n. Over $\mathscr{C}$ a unique measure μ can be defined, which to any rectangle R assigns as its measure the volume of R,

$$\mu(R) = \prod_{i=1}^{n} (b_i - a_i).$$

The measure μ can be *completed* by adjoining to $\mathscr{C}$ all subsets of sets of measure zero. The domain of μ is thereby enlarged to a σ-field $\mathscr{C}'$, the class of *Lebesgue-measurable* sets. The term *Lebesgue measure* is used for μ both when it is defined over the Borel sets and when it is defined over the Lebesgue-measurable sets.

This example can be generalized to any nonnegative set function ν, which is defined and countably additive over the class of rectangles R. There exists then, as before, a unique measure μ over $(\mathscr{X}, \mathscr{C})$ that agrees with ν for all R. This measure can again be completed; however, the resulting σ-field depends on μ and need not agree with the σ-field $\mathscr{C}'$ obtained above.

Example 2. Counting measure. Suppose that $\mathscr{X}$ is countable, and let $\mathscr{C}$ be the class of all subsets of $\mathscr{X}$. For any set C, define $\mu(C)$ as the number of elements of C

---

*If $\pi(z)$ is a statement concerning certain objects z, then $\{z : \pi(z)\}$ denotes the set of all those z for which $\pi(z)$ is true.

if that number is finite, and otherwise as $+\infty$. This measure is sometimes called *counting measure*.

In applications, the probabilities over $(\mathscr{Z}, \mathscr{C})$ refer to random experiments or observations, the possible outcomes of which are the points $z \in \mathscr{Z}$. When recording the results of an experiment, one is usually interested only in certain of its aspects, typically some counts or measurements. These may be represented by a function T taking values in some space $\mathscr{T}$.

Such a function generates in $\mathscr{T}$ the σ-field $\mathscr{B}'$ of sets B whose inverse image

$$C = T^{-1}(B) = \{ z : z \in \mathscr{Z}, T(z) \in B \}$$

is in $\mathscr{C}$, and for any given probability measure P over $(\mathscr{Z}, \mathscr{C})$ a probability measure Q over $(\mathscr{T}, \mathscr{B}')$ defined by

(3) $$Q(B) = P(T^{-1}(B)).$$

Frequently, there is given a σ-field $\mathscr{B}$ of sets in $\mathscr{T}$ such that the probability of B should be defined if and only if $B \in \mathscr{B}$. This requires that $T^{-1}(B) \in \mathscr{C}$ for all $B \in \mathscr{B}$, and the function (or transformation) T from $(\mathscr{Z}, \mathscr{C})$ into* $(\mathscr{T}, \mathscr{B})$ is then said to be $\mathscr{C}$-measurable. Another implication is the sometimes convenient restriction of probability statements to the sets $B \in \mathscr{B}$ even though there may exist sets $B \notin \mathscr{B}$ for which $T^{-1}(B) \in \mathscr{C}$ and whose probability therefore could be defined.

Of particular interest is the case of a single measurement in which the function T is real-valued. Let us denote it by X, and let $\mathscr{A}$ be the class of Borel sets on the real line $\mathscr{X}$. Such a measurable real-valued X is called a *random variable*, and the probability measure it generates over $(\mathscr{X}, \mathscr{A})$ will be denoted by P^X and called the probability distribution of X. The value this measure assigns to a set $A \in \mathscr{A}$ will be denoted interchangeably by $P^X(A)$ and $P(X \in A)$. Since the intervals $\{ x : x \leq a \}$ are in $\mathscr{A}$, the probabilities $F(a) = P(X \leq a)$ are defined for all a. The function F, the *cumulative distribution function* (cdf) of X, is nondecreasing and continuous on the right, and $F(-\infty) = 0$, $F(+\infty) = 1$. Conversely, if F is any function with these properties, a measure can be defined over the intervals by $P\{ a < X \leq b \} = F(b) - F(a)$. It follows from Example 1 that this measure uniquely determines a probability distribution over the Borel sets. Thus the probability distribution P^X and the cumulative distribution function F uniquely determine each other. These remarks extend to probability

---

*The term *into* indicates that the range of T is in $\mathscr{T}$; if $T(\mathscr{Z}) = \mathscr{T}$, the transformation is said to be from $\mathscr{Z}$ onto $\mathscr{T}$.

distributions over an n-dimensional Euclidean space, where the cumulative distribution function is defined by

$$F(a_1, \ldots, a_n) = P\{ X_1 \leq a_1, \ldots, X_n \leq a_n \}.$$

In concrete problems, the space $(\mathcal{Z}, \mathcal{C})$, corresponding to the totality of possible outcomes, is usually not specified and remains in the background. The real starting point is the set X of observations (typically vector-valued) that are being recorded and which constitute the *data*, and the associated measurable space $(\mathcal{X}, \mathcal{A})$, the *sample space*. Random variables or vectors that are measurable transformations T from $(\mathcal{X}, \mathcal{A})$ into some $(\mathcal{T}, \mathcal{B})$ are called *statistics*. The distribution of T is then given by (3) applied to all $B \in \mathcal{B}$. With this definition, a statistic is specified by the function T and the σ-field $\mathcal{B}$. We shall, however, adopt the convention that when a function T takes on its values in a Euclidean space, unless otherwise stated the σ-field $\mathcal{B}$ of measurable sets will be taken to be the class of Borel sets. It then becomes unnecessary to mention it explicitly or to indicate it in the notation.

The distinction between statistics and random variables as defined here is slight. The term statistic is used to indicate that the quantity is a function of more basic observations; all statistics in a given problem are functions defined over the same sample space $(\mathcal{X}, \mathcal{A})$. On the other hand, any real-valued statistic T is a random variable, since it has a distribution over $(\mathcal{T}, \mathcal{B})$, and it will be referred to as a random variable when its origin is irrelevant. Which term is used therefore depends on the point of view and to some extent is arbitrary.

2. INTEGRATION

According to the convention of the preceding section, a real-valued function f defined over $(\mathcal{X}, \mathcal{A})$ is measurable if $f^{-1}(B) \in \mathcal{A}$ for every Borel set B on the real line. Such a function f is said to be *simple* if it takes on only a finite number of values. Let μ be a measure defined over $(\mathcal{X}, \mathcal{A})$, and let f be a simple function taking on the distinct values $a_1, \ldots, a_m$ on the sets $A_1, \ldots, A_m$, which are in $\mathcal{A}$, since f is measurable. If $\mu(A_i) < \infty$ when $a_i \neq 0$, the integral of f with respect to μ is defined by

$$(4) \qquad \int f \, d\mu = \sum a_i \mu(A_i).$$

Given any nonnegative measurable function f, there exists a nondecreasing sequence of simple functions f_n converging to f. Then the integral of f

is defined as

(5)
$$\int f\, d\mu = \lim_{n\to\infty} \int f_n\, d\mu,$$

which can be shown to be independent of the particular sequence of f_n's chosen. For any measurable function f its positive and negative parts

(6) $f^+(x) = \max[f(x), 0]$ and $f^-(x) = \max[-f(x), 0]$

are also measurable, and

$$f(x) = f^+(x) - f^-(x).$$

If the integrals of f^+ and f^- are both finite, then f is said to be *integrable*, and its integral is defined as

$$\int f\, d\mu = \int f^+\, d\mu - \int f^-\, d\mu.$$

If of the two integrals one is finite and one infinite, then the integral of f is defined to be the appropriate infinite value; if both are infinite, the integral is not defined.

   Example 3.   Let $\mathcal{X}$ be the closed interval $[a, b]$, $\mathcal{A}$ be the class of Borel sets or of Lebesgue measurable sets in $\mathcal{X}$, and μ be Lebesgue measure. Then the integral of f with respect to μ is written as $\int_a^b f(x)\, dx$, and is called the Lebesgue integral of f. This integral generalizes the Riemann integral in that it exists and agrees with the Riemann integral of f whenever the latter exists.

   Example 4.   Let $\mathcal{X}$ be countable and consist of the points $x_1, x_2, \ldots$; let $\mathcal{A}$ be the class of all subsets of $\mathcal{X}$, and let μ assign measure b_i to the point x_i. Then f is integrable provided $\sum f(x_i) b_i$ converges absolutely, and $\int f\, d\mu$ is given by this sum.

   Let P^X be the probability distribution of a random variable X, and let T be a real-valued statistic. If the function $T(x)$ is integrable, its *expectation* is defined by

(7) $$E(T) = \int T(x)\, dP^X(x).$$

It will be seen from Lemma 2 in Section 3 below that the integration can be carried out alternatively in t-space with respect to the distribution of T defined by (3), so that also

(8) $$E(T) = \int t\, dP^T(t).$$

The definition (5) of the integral permits the basic convergence theorems:

Theorem 1. *Let f_n be a sequence of measurable functions, and let $f_n(x) \to f(x)$ for all x. Then*

$$\int f_n \, d\mu \to \int f \, d\mu$$

if either one of the following conditions holds:

(i)   Lebesgue monotone-convergence theorem: *the f_n's are nonnegative and the sequence is nondecreasing*;

or

(ii)   Lebesgue dominated-convergence theorem: *there exists an integrable function g such that $|f_n(x)| \le g(x)$ for all n and x.*

For any set $A \in \mathscr{A}$, let I_A be its *indicator function* defined by

$$(9) \qquad\qquad I_A(x) = 1 \text{ or } 0 \qquad as \quad x \in A \text{ or } x \in \tilde{A},$$

and let

$$(10) \qquad\qquad \int_A f \, d\mu = \int f I_A \, d\mu.$$

If μ is a measure and f a nonnegative measurable function over $(\mathscr{X}, \mathscr{A})$, then

$$(11) \qquad\qquad \nu(A) = \int_A f \, d\mu$$

defines a new measure over $(\mathscr{X}, \mathscr{A})$. The fact that (11) holds for all $A \in \mathscr{A}$ is expressed by writing

$$(12) \qquad\qquad d\nu = f \, d\mu \quad \text{or} \quad f = \frac{d\nu}{d\mu}.$$

Let μ and ν be two given σ-finite measures over $(\mathscr{X}, \mathscr{A})$. If there exists a function f satisfying (12), it is determined through this relation up to sets of measure zero, since

$$\int_A f \, d\mu = \int_A g \, d\mu \qquad \text{for all} \quad A \in \mathscr{A}$$

implies that $f = g$ a.e. μ.* Such an f is called the *Radon-Nikodym derivative* of ν with respect to μ, and in the particular case that ν is a probability measure, the *probability density* of ν with respect to μ.

The question of existence of a function f satisfying (12) for given measures μ and ν is answered in terms of the following definition. A measure ν is *absolutely continuous* with respect to μ if

$$\mu(A) = 0 \quad \text{implies} \quad \nu(A) = 0.$$

Theorem 2. (*Radon-Nikodym.*) *If μ and ν are σ-finite measures over $(\mathscr{X}, \mathscr{A})$, then there exists a measurable function f satisfying (12) if and only if ν is absolutely continuous with respect to μ.*

The *direct* (or *Cartesian*) *product* $A \times B$ of two sets A and B is the set of all pairs (x, y) with $x \in A$, $y \in B$. Let $(\mathscr{X}, \mathscr{A})$ and $(\mathscr{Y}, \mathscr{B})$ be two measurable spaces, and let $\mathscr{A} \times \mathscr{B}$ be the smallest σ-field containing all sets $A \times B$ with $A \in \mathscr{A}$ and $B \in \mathscr{B}$. If μ and ν are two σ-finite measures over $(\mathscr{X}, \mathscr{A})$ and $(\mathscr{Y}, \mathscr{B})$ respectively, then there exists a unique measure $\lambda = \mu \times \nu$ over $(\mathscr{X} \times \mathscr{Y}, \mathscr{A} \times \mathscr{B})$, the *product* of μ and ν, such that for any $A \in \mathscr{A}$, $B \in \mathscr{B}$,

(13) $$\lambda(A \times B) = \mu(A)\nu(B).$$

Example 5. Let $\mathscr{X}, \mathscr{Y}$ be Euclidean spaces of m and n dimensions, and let $\mathscr{A}, \mathscr{B}$ be the σ-fields of Borel sets in these spaces. Then $\mathscr{X} \times \mathscr{Y}$ is an $(m + n)$-dimensional Euclidean space, and $\mathscr{A} \times \mathscr{B}$ the class of its Borel sets.

Example 6. Let $Z = (X, Y)$ be a random variable defined over $(\mathscr{X} \times \mathscr{Y}, \mathscr{A} \times \mathscr{B})$, and suppose that the random variables X and Y have distributions P^X, P^Y over $(\mathscr{X}, \mathscr{A})$ and $(\mathscr{Y}, \mathscr{B})$. Then X and Y are said to be *independent* if the probability distribution P^Z of Z is the product $P^X \times P^Y$.

In terms of these concepts the reduction of a double integral to a repeated one is given by the following theorem.

Theorem 3. (*Fubini.*) *Let μ and ν be σ-finite measures over $(\mathscr{X}, \mathscr{A})$ and $(\mathscr{Y}, \mathscr{B})$ respectively, and let $\lambda = \mu \times \nu$. If $f(x, y)$ is integrable with respect to λ, then*

(i) *for almost all (ν) fixed y, the function $f(x, y)$ is integrable with respect to μ,*

(ii) *the function $\int f(x, y) \, d\mu(x)$ is integrable with respect to ν, and*

(14) $$\int f(x, y) \, d\lambda(x, y) = \int \left[\int f(x, y) \, d\mu(x) \right] d\nu(y).$$

---

*A statement that holds for all points x except possibly on a set of μ-measure zero is said to hold a.e. μ; or to hold $(\mathscr{A}, \mu)$ if it is desirable to indicate the σ-field over which μ is defined.

3. STATISTICS AND SUBFIELDS

According to the definition of Section 1, a statistic is a measurable transformation T from the sample space $(\mathcal{X}, \mathcal{A})$ into a measurable space $(\mathcal{T}, \mathcal{B})$. Such a transformation induces in the original sample space the subfield*

$$(15) \qquad \mathcal{A}_0 = T^{-1}(\mathcal{B}) = \{ T^{-1}(B) : B \in \mathcal{B} \}.$$

Since the set $T^{-1}[T(A)]$ contains A but is not necessarily equal to A, the σ-field $\mathcal{A}_0$ need not coincide with $\mathcal{A}$ and hence can be a proper subfield of $\mathcal{A}$. On the other hand, suppose for a moment that $\mathcal{T} = T(\mathcal{X})$, that is, that the transformation T is onto rather than into $\mathcal{T}$. Then

$$(16) \qquad T[T^{-1}(B)] = B \qquad \text{for all} \quad B \in \mathcal{B},$$

so that the relationship $A_0 = T^{-1}(B)$ establishes a $1:1$ correspondence between the sets of $\mathcal{A}_0$ and $\mathcal{B}$, which is an isomorphism—that is, which preserves the set operations of intersection, union, and complementation. For most purposes it is therefore immaterial whether one works in the space $(\mathcal{X}, \mathcal{A}_0)$ or in $(\mathcal{T}, \mathcal{B})$. These generate two equivalent classes of events, and therefore of measurable functions, possible decision procedures, etc. If the transformation T is only into $\mathcal{T}$, the above $1:1$ correspondence applies to the class $\mathcal{B}'$ of subsets of $\mathcal{T}' = T(\mathcal{X})$ which belong to $\mathcal{B}$, rather than to $\mathcal{B}$ itself. However, any set $B \in \mathcal{B}$ is equivalent to $B' = B \cap \mathcal{T}'$ in the sense that any measure over $(\mathcal{X}, \mathcal{A})$ assigns the same measure to B' as to B. Considered as classes of events, $\mathcal{A}_0$ and $\mathcal{B}$ therefore continue continue to be equivalent, with the only difference that $\mathcal{B}$ contains several (equivalent) representations of the same event.

As an example, let $\mathcal{X}$ be the real line and $\mathcal{A}$ the class of Borel sets, and let $T(x) = x^2$. Let $\mathcal{T}$ be either the positive real axis or the whole real axis, and let $\mathcal{B}$ be the class of Borel subsets of $\mathcal{T}$. Then $\mathcal{A}_0$ is the class of Borel sets that are symmetric with respect to the origin. When considering, for example, real-valued measurable functions, one would, when working in $\mathcal{T}$-space, restrict attention to measurable functions of x^2. Instead, one could remain in the original space, where the restriction would be to the class of even measurable functions of x. The equivalence is clear. Which representation is more convenient depends on the situation.

That the correspondence between the sets $A_0 = T^{-1}(B) \in \mathcal{A}_0$ and $B \in \mathcal{B}$ establishes an analogous correspondence between measurable functions defined over $(\mathcal{X}, \mathcal{A}_0)$ and $(\mathcal{T}, \mathcal{B})$ is shown by the following lemma.

---

*We shall use this term in place of the more cumbersome "sub-σ-field".

Lemma 1.   *Let the statistic T from $(\mathscr{X}, \mathscr{A})$ into $(\mathscr{T}, \mathscr{B})$ induce the subfield $\mathscr{A}_0$. Then a real-valued $\mathscr{A}$-measurable function f is $\mathscr{A}_0$-measurable if and only if there exists a $\mathscr{B}$-measurable function g such that*

$$f(x) = g[T(x)]$$

for all x.

Proof.   Suppose first that such a function g exists. Then the set

$$\{x : f(x) < r\} = T^{-1}(\{t : g(t) < r\})$$

is in $\mathscr{A}_0$, and f is $\mathscr{A}_0$-measurable. Conversely, if f is $\mathscr{A}_0$-measurable, then the sets

$$A_{in} = \left\{ x : \frac{i}{2^n} < f(x) \le \frac{i+1}{2^n} \right\}, \qquad i = 0, \pm 1, \pm 2, \ldots,$$

are (for fixed n) disjoint sets in $\mathscr{A}_0$ whose union is $\mathscr{X}$, and there exist $B_{in} \in \mathscr{B}$ such that $A_{in} = T^{-1}(B_{in})$. Let

$$B_{in}^* = B_{in} \cap \overline{\bigcup_{j \neq i} B_{jn}}.$$

Since A_{in} and A_{jn} are mutually exclusive for $i \neq j$, the set $T^{-1}(B_{in} \cap B_{jn})$ is empty and so is the set $T^{-1}(B_{in} \cap \widetilde{B_{in}^*})$. Hence, for fixed n, the sets B_{in}^* are disjoint, and still satisfy $A_{in} = T^{-1}(B_{in}^*)$. Defining

$$f_n(x) = \frac{i}{2^n} \quad \text{if } x \in A_{in}, \qquad i = 0, \pm 1, \pm 2, \ldots,$$

one can write

$$f_n(x) = g_n[T(x)],$$

where

$$g_n(t) = \begin{cases} \dfrac{i}{2^n} & \text{for } t \in B_{in}^*, \quad i = 0, \pm 1, \pm 2, \ldots, \\ 0 & \text{otherwise.} \end{cases}$$

Since the functions g_n are $\mathscr{B}$-measurable, the set B on which $g_n(t)$ converges to a finite limit is in $\mathscr{B}$. Let $R = T(\mathscr{X})$ be the range of T. Then for

$t \in R$,

$$\lim g_n[T(x)] = \lim f_n(x) = f(x)$$

for all $x \in \mathcal{X}$, so that R is contained in B. Therefore, the function g defined by $g(t) = \lim g_n(t)$ for $t \in B$ and $g(t) = 0$ otherwise possesses the required properties.

The relationship between integrals of the functions f and g above is given by the following lemma.

Lemma 2. *Let T be a measurable transformation from $(\mathcal{X}, \mathcal{A})$ into $(\mathcal{T}, \mathcal{B})$, μ a σ-finite measure over $(\mathcal{X}, \mathcal{A})$, and g a real-valued measurable function of t. If μ^* is the measure defined over $(\mathcal{T}, \mathcal{B})$ by*

$$(17) \qquad \mu^*(B) = \mu[T^{-1}(B)] \qquad \text{for all} \quad B \in \mathcal{B},$$

then for any $B \in \mathcal{B}$,

$$(18) \qquad \int_{T^{-1}(B)} g[T(x)] \, d\mu(x) = \int_B g(t) \, d\mu^*(t)$$

in the sense that if either integral exists, so does the other and the two are equal.

Proof. Without loss of generality let B be the whole space $\mathcal{T}$. If g is the indicator of a set $B_0 \in \mathcal{B}$, the lemma holds, since the left- and right-hand sides of (18) reduce respectively to $\mu[T^{-1}(B_0)]$ and $\mu^*(B_0)$, which are equal by the definition of μ^*. It follows that (18) holds successively for all simple functions, for all nonnegative measurable functions, and hence finally for all integrable functions.

4. CONDITIONAL EXPECTATION AND PROBABILITY

If two statistics induce the same subfield $\mathcal{A}_0$, they are equivalent in the sense of leading to equivalent classes of measurable events. This equivalence is particularly relevant to considerations of conditional probability. Thus if X is normally distributed with zero mean, the information carried by the statistics $|X|$, X^2, e^{-X^2}, and so on, is the same. Given that $|X| = t$, $X^2 = t^2$, $e^{-X^2} = e^{-t^2}$, it follows that X is $\pm t$, and any reasonable definition of conditional probability will assign probability $\frac{1}{2}$ to each of these values. The general definition of conditional probability to be given below will in fact involve essentially only $\mathcal{A}_0$ and not the range space $\mathcal{T}$ of T. However, when referred to $\mathcal{A}_0$ alone the concept loses much of its intuitive meaning, and the gap between the elementary definition and that of the general case becomes unnecessarily wide. For these reasons it is frequently more convenient to work with a particular representation of a statistic, involving a definite range space $(\mathcal{T}, \mathcal{B})$.

Let P be a probability measure over $(\mathcal{X}, \mathcal{A})$, T a statistic with range space $(\mathcal{T}, \mathcal{B})$, and $\mathcal{A}_0$ the subfield it induces. Consider a nonnegative function f which is integrable $(\mathcal{A}, P)$, that is, $\mathcal{A}$-measurable and P-integrable. Then $\int_A f \, dP$ is defined for all $A \in \mathcal{A}$ and therefore for all $A_0 \in \mathcal{A}_0$. It follows from the Radon–Nikodym theorem (Theorem 2) that there exists a function f_0 which is integrable $(\mathcal{A}_0, P)$ and such that

$$(19) \qquad \int_{A_0} f \, dP = \int_{A_0} f_0 \, dP \qquad \text{for all} \quad A_0 \in \mathcal{A}_0,$$

and that f_0 is unique $(\mathcal{A}_0, P)$. By Lemma 1, f_0 depends on x only through $T(x)$. In the example of a normally distributed variable X with zero mean, and $T = X^2$, the function f_0 is determined by (19) holding for all sets A_0 that are symmetric with respect to the origin, so that $f_0(x) = \frac{1}{2}[f(x) + f(-x)]$.

The function f_0 defined through (19) is determined by two properties:

(i) Its average value over any set A_0 with respect to P is the same as that of f;

(ii) It depends on x only through $T(x)$ and hence is constant on the sets D_x over which T is constant.

Intuitively, what one attempts to do in order to construct such a function is to define $f_0(x)$ as the conditional P-average of f over the set D_x. One would thereby replace the single averaging process of integrating f represented by the left-hand side with a two-stage averaging process such as an iterated integral. Such a construction can actually be carried out when X is a discrete variable and in the regular case considered in Chapter 1, Section 9; $f_0(x)$ is then just the conditional expectation of $f(X)$ given $T(x)$. In general, it is not clear how to define this conditional expectation directly. Since it should, however, possess properties (i) and (ii), and since these through (19) determine f_0 uniquely $(\mathcal{A}_0, P)$, we shall take $f_0(x)$ of (19) as the general definition of the *conditional expectation* $E[f(X)|T(x)]$. Equivalently, if $f_0(x) = g[T(x)]$ one can write

$$E[f(X)|t] = E[f(X)|T = t] = g(t),$$

so that $E[f(X)|t]$ is a $\mathcal{B}$-measurable function defined up to equivalence

$(\mathscr{B}, P^T)$. In the relationship of integrals given in Lemma 2, if $\mu = P^X$ then $\mu^* = P^T$, and it is seen that the function g can be defined directly in terms of f through

$$(20) \qquad \int_{T^{-1}(B)} f(x)\, dP^X(x) = \int_B g(t)\, dP^T(t) \qquad \text{for all} \quad B \in \mathscr{B},$$

which is equivalent to (19).

So far, f has been assumed to be nonnegative. In the general case, the conditional expectation of f is defined as

$$E[f(x)|t] = E[f^+(X)|t] - E[f^-(X)|t].$$

Example 7. Order statistics. Let $X_1, \ldots, X_n$ be identically and independently distributed random variables with a continuous distribution function, and let

$$T(x_1, \ldots, x_n) = (x_{(1)}, \ldots, x_{(n)})$$

where $x_{(1)} \le \cdots \le x_{(n)}$ denote the ordered x's. Without loss of generality one can restrict attention to the points with $x_{(1)} < \cdots < x_{(n)}$, since the probability of two coordinates being equal is 0. Then $\mathscr{X}$ is the set of all n-tuples with distinct coordinates, $\mathscr{T}$ the set of all ordered n-tuples, and $\mathscr{A}$ and $\mathscr{B}$ are the classes of Borel subsets of $\mathscr{X}$ and $\mathscr{T}$. Under T^{-1} the set consisting of the single point $a = (a_1, \ldots, a_n)$ is transformed into the set consisting of the $n!$ points $(a_{i_1}, \ldots, a_{i_n})$ that are obtained from a by permuting the coordinates in all possible ways. It follows that $\mathscr{A}_0$ is the class of all sets that are symmetric in the sense that if A_0 contains a point $x = (x_1, \ldots, x_n)$, then it also contains all points $(x_{i_1}, \ldots, x_{i_n})$.

For any integrable function f, let

$$f_0(x) = \frac{1}{n!} \sum f(x_{i_1}, \ldots, x_{i_n}),$$

where the summation extends over the $n!$ permutations of $(x_1, \ldots, x_n)$. Then f_0 is $\mathscr{A}_0$-measurable, since it is symmetric in its n arguments. Also

$$\int_{A_0} f(x_1, \ldots, x_n)\, dP(x_1) \ldots dP(x_n) = \int_{A_0} f(x_{i_1}, \ldots, x_{i_n})\, dP(x_1) \ldots dP(x_n),$$

so that f_0 satisfies (19). It follows that $f_0(x)$ is the conditional expectation of $f(X)$ given $T(x)$.

The conditional expectation of $f(X)$ given the above statistic $T(x)$ can also be found without assuming the X's to be identically and independently distributed. Suppose that X has a density $h(x)$ with respect to a measure μ (such as Lebesgue measure), which is symmetric in the variables $x_1, \ldots, x_n$ in the sense that for any $A \in \mathscr{A}$ it assigns to the set $\{x : (x_{i_1}, \ldots, x_{i_n}) \in A\}$ the same measure for all permutations $(i_1, \ldots, i_n)$. Let

$$f_0(x_1, \ldots, x_n) = \frac{\sum f(x_{i_1}, \ldots, x_{i_n}) h(x_{i_1}, \ldots, x_{i_n})}{\sum h(x_{i_1}, \ldots, x_{i_n})};$$

here and in the sums below the summation extends over the $n!$ permutations of $(x_1, \ldots, x_n)$. The function f_0 is symmetric in its n arguments and hence $\mathscr{A}_0$-measurable. For any symmetric set A_0, the integral

$$\int_{A_0} f_0(x_1, \ldots, x_n) h(x_{j_1}, \ldots, x_{j_n}) \, d\mu(x_1, \ldots, x_n)$$

has the same value for each permutation $(x_{j_1}, \ldots, x_{j_n})$, and therefore

$$\int_{A_0} f_0(x_1, \ldots, x_n) h(x_1, \ldots, x_n) \, d\mu(x_1, \ldots, x_n)$$

$$= \int_{A_0} f_0(x_1, \ldots, x_n) \frac{1}{n!} \sum h(x_{i_1}, \ldots, x_{i_n}) \, d\mu(x_1, \ldots, x_n)$$

$$= \int_{A_0} f(x_1, \ldots, x_n) h(x_1, \ldots, x_n) \, d\mu(x_1, \ldots, x_n).$$

It follows that $f_0(x) = E[f(X)|T(x)]$.

Equivalent to the statistic $T(x) = (x_{(1)}, \ldots, x_{(n)})$, the set of *order statistics*, is $U(x) = (\Sigma x_i, \Sigma x_i^2, \ldots, \Sigma x_i^n)$. This is an immediate consequence of the fact, to be shown below, that if $T(x^0) = t^0$ and $U(x^0) = u^0$, then

$$T^{-1}(\{t^0\}) = U^{-1}(\{u^0\}) = S$$

where $\{t^0\}$ and $\{u^0\}$ denote the sets consisting of the single point t^0 and u^0 respectively, and where S consists of the totality of points $x = (x_1, \ldots, x_n)$ obtained by permuting the coordinates of $x^0 = (x_1^0, \ldots, x_n^0)$ in all possible ways.

That $T^{-1}(\{t^0\}) = S$ is obvious. To see the corresponding fact for U^{-1}, let

$$V(x) = \left(\sum_i x_i, \sum_{i<j} x_i x_j, \sum_{i<j<k} x_i x_j x_k, \ldots, x_1 x_2 \cdots x_n \right),$$

so that the components of $V(x)$ are the elementary symmetric functions $v_1 = \Sigma x_i, \ldots, v_n = x_1 \ldots x_n$ of the n arguments $x_1, \ldots, x_n$. Then

$$(x - x_1) \ldots (x - x_n) = x^n - v_1 x^{n-1} + v_2 x^{n-2} - \cdots + (-1)^n v_n.$$

Hence $V(x^0) = v^0 = (v_1^0, \ldots, v_n^0)$ implies that $V^{-1}(\{v^0\}) = S$. That then also

$U^{-1}(\{u^0\}) = S$ follows from the $1:1$ correspondence between u and v established by the relations (known as Newton's identities),*

$$u_k - v_1 u_{k-1} + v_2 u_{k-2} - \cdots + (-1)^{k-1} v_{k-1} u_1 + (-1)^k k v_k = 0, \qquad 1 \le k \le n.$$

It is easily verified from the above definition that conditional expectation possesses most of the usual properties of expectation. It follows of course from the nonuniqueness of the definition that these properties can hold only $(\mathscr{B}, P^T)$. We state this formally in the following lemma.

Lemma 3. *If T is a statistic and the functions $f, g, \ldots$ are integrable $(\mathscr{A}, P)$, then a.e. $(\mathscr{B}, P^T)$*

(i) $E[af(X) + bg(X)|t] = aE[f(X)|t] + bE[g(X)|t]$;

(ii) $E[h(T)f(X)|t] = h(t)E[f(X)|t]$;

(iii) $a \le f(x) \le b(\mathscr{A}, P)$ *implies* $a \le E[f(X)|t] \le b$;

(iv) $|f_n| \le g, f_n(x) \to f(x)(\mathscr{A}, P)$ *implies* $E[f_n(X)|t] \to E[f(X)|t]$.

A further useful result is obtained by specializing (20) to the case that B is the whole space $\mathscr{T}$. One then has

Lemma 4. *If $E|f(X)| < \infty$, and if $g(t) = E[f(X)|t]$, then*

$$(21) \qquad\qquad Ef(X) = Eg(T),$$

that is, the expectation can be obtained as the expected value of the conditional expectation.

Since $P\{X \in A\} = E[I_A(X)]$, where I_A denotes the indicator of the set A, it is natural to define the *conditional probability* of A given $T = t$ by

$$(22) \qquad\qquad P(A|t) = E\big[I_A(X)|t\big].$$

In view of (20) the defining equation for $P(A|t)$ can therefore be written as

$$(23) \quad P^X(A \cap T^{-1}(B)) = \int_{A \cap T^{-1}(B)} dP^X(x)$$

$$= \int_B P(A|t)\, dP^T(t) \qquad \text{for all} \quad B \in \mathscr{B}.$$

It is an immediate consequence of Lemma 3 that subject to the appropriate

---

*For a proof of these relations see for example Turnbull (1952), *Theory of Equations*, 5th ed., Oliver and Boyd, Edinburgh, Section 32.

null-set* qualifications, $P(A|t)$ possesses the usual properties of probabilities, as summarized in the following lemma.

Lemma 5. *If T is a statistic with range space $(\mathcal{T}, \mathcal{B})$, and $A, B, A_1, A_2, \ldots$ are sets belonging to $\mathcal{A}$, then a.e. $(\mathcal{B}, P^T)$*

(i) $0 \leq P(A|t) \leq 1$;

(ii)   *if the sets $A_1, A_2, \ldots$ are mutually exclusive,*

$$P\left(\bigcup A_i \middle| t\right) = \sum P(A_i|t);$$

(iii) $A \subset B$ *implies* $P(A|t) \leq P(B|t)$.

According to the definition (22), the conditional probability $P(A|t)$ must be considered for fixed A as a $\mathcal{B}$-measurable function of t. This is in contrast to the elementary definition in which one takes t as fixed and considers $P(A|t)$ for varying A as a set function over $\mathcal{A}$. Lemma 5 suggests the possibility that the interpretation of $P(A|t)$ for fixed t as a probability distribution over $\mathcal{A}$ may be valid also in the general case. However, the equality $P(A_1 \cup A_2|t) = P(A_1|t) + P(A_2|t)$, for example, can break down on a null set that may vary with A_1 and A_2, and the union of all these null sets need no longer have measure zero.

For an important class of cases, this difficulty can be overcome through the nonuniqueness of the functions $P(A|t)$, which for each fixed A are determined only up to sets of measure zero in t. Since all determinations of these functions are equivalent, it is enough to find a specific determination for each A so that for each fixed t these determinations jointly constitute a probability distribution over $\mathcal{A}$. This possibility is illustrated by Example 7, in which the conditional probability distribution given $T(x) = t$ can be taken to assign probability $1/n!$ to each of the $n!$ points satisfying $T(x) = t$. Sufficient conditions for the existence of such conditional distributions will be given in the next section. For counterexamples see Blackwell and Dubins (1975).

5. CONDITIONAL PROBABILITY DISTRIBUTIONS[†]

We shall now investigate the existence of conditional probability distributions under the assumption, satisfied in most statistical applications, that $\mathcal{X}$ is a Borel set in a Euclidean space. We shall then say for short that $\mathcal{X}$ is

*This term is used as an alternative to the more cumbersome "set of measure zero."

[†] This section may be omitted at first reading. Its principal application is in the proof of Lemma 8(ii) in Section 7, which in turn is used only in the proof of Theorem 3 of Chapter 4.

Euclidean and assume that, unless otherwise stated, $\mathscr{A}$ is the class of Borel subsets of $\mathscr{X}$.

Theorem 4. *If $\mathscr{X}$ is Euclidean, there exist determinations of the functions $P(A|t)$ such that for each t, $P(A|t)$ is a probability measure over $\mathscr{A}$.*

Proof. By setting equal to 0 the probability of any Borel set in the complement of $\mathscr{X}$, one can extend the given probability measure to the class of all Borel sets and can therefore assume without loss of generality that $\mathscr{X}$ is the full Euclidean space. For simplicity we shall give the proof only in the one-dimensional case. For each real x put $F(x, t) = P((-\infty, x]|t)$ for some version of this conditional probability function, and let $r_1, r_2, \ldots$ denote the set of all rational numbers in some order. Then $r_i < r_j$ implies that $F(r_i, t) \le F(r_j, t)$ for all t except those in a null set N_{ij}, and hence that $F(x, t)$ is nondecreasing in x over the rationals for all t outside of the null set $N' = \cup N_{ij}$. Similarly, it follows from Lemma 3(iv) that for all t not in a null set N'', as n tends to infinity $\lim F(r_i + 1/n, t) = F(r_i, t)$ for $i = 1, 2, \ldots$, $\lim F(n, t) = 1$, and $\lim F(-n, t) = 0$. Therefore, for all t outside of the null set $N' \cup N''$, $F(x, t)$ considered as a function of x is properly normalized, monotone, and continuous on the right over the rationals. For t not in $N' \cup N''$ let $F^*(x, t)$ be the unique function that is continuous on the right in x and agrees with $F(x, t)$ for all rational x. Then $F^*(x, t)$ is a cumulative distribution function and therefore determines a probability measure $P^*(A|t)$ over $\mathscr{A}$. We shall now show that $P^*(A|t)$ is a conditional probability of A given t, by showing that for each fixed A it is a $\mathscr{B}$-measurable function of t satisfying (23). This will be accomplished by proving that for each fixed $A \in \mathscr{A}$

$$P^*(A|t) = P(A|t) \qquad (\mathscr{B}, P^T).$$

By definition of P^* this is true whenever A is one of the sets $(-\infty, x]$ with x rational. It holds next when A is an interval $(a, b] = (-\infty, b] - (-\infty, a]$ with a, b rational, since P^* is a measure and P satisfies Lemma 5(ii). Therefore, the desired equation holds for the field $\mathscr{F}$ of all sets A which are finite unions of intervals $(a_i, b_i]$ with rational end points. Finally, the class of sets for which the equation holds is a monotone class (see Problem 1) and hence contains the smallest σ-field containing $\mathscr{F}$, which is $\mathscr{A}$. The measure $P^*(A|t)$ over $\mathscr{A}$ was defined above for all t not in $N' \cup N''$. However, since neither the measurability of a function nor the values of its integrals is affected by its values on a null set, one can take arbitrary probability measures over $\mathscr{A}$ for t in $N' \cup N''$ and thereby complete the determination.

If X is a vector-valued random variable with probability distribution P^X and T is a statistic defined over $(\mathscr{X}, \mathscr{A})$, let $P^{X|t}$ denote any version of the family of conditional distributions $P(A|t)$ over $\mathcal{A}$ guaranteed by Theorem 4. The connection with conditional expectation is given by the following theorem.

Theorem 5. *If X is a vector-valued random variable and $E|f(X)| < \infty$, then*

$$(24) \qquad E[f(X)|t] = \int f(x)\, dP^{X|t}(x) \qquad (\mathcal{B}, P^T).$$

Proof. Equation (24) holds if f is the indicator of any set $A \in \mathcal{A}$. It then follows from Lemma 3 that it also holds for any simple function and hence for any integrable function.

The determination of the conditional expectation $E[f(X)|t]$ given by the right-hand side of (24) possesses for each t the usual properties of an expectation, (i), (iii), and (iv) of Lemma 3, which previously could be asserted only up to sets of measure zero depending on the functions $f, g, \ldots$ involved. Under the assumptions of Theorem 4 a similar strengthening is possible with respect to (ii) of Lemma 3, which can be shown to hold except possibly on a null set N not depending on the function h. It will be sufficient for the present purpose to prove this under the additional assumption that the range space of the statistic T is also Euclidean. For a proof without this restriction see for example Billingsley (1979).

Theorem 6. *If T is a statistic with Euclidean domain and range spaces $(\mathcal{X}, \mathcal{A})$ and $(\mathcal{T}, \mathcal{B})$, there exists a determination $P^{X|t}$ of the conditional probability distribution and a null set N such that the conditional expectation computed by*

$$E[f(X)|t] = \int f(x)\, dP^{X|t}(x)$$

satisfies for all $t \notin N$

$$(25) \qquad E[h(T)f(X)|t] = h(t)E[f(X)|t].$$

Proof. For the sake of simplicity and without essential loss of generality suppose that T is real-valued. Let $P^{X|t}(A)$ be a probability distribution over $\mathcal{A}$ for each t, the existence of which is guaranteed by Theorem 4. For $B \in \mathcal{B}$, the indicator function $I_B(t)$ is $\mathcal{B}$-measurable and

$$\int_{B'} I_B(t)\, dP^T(t) = P^T(B' \cap B) = P^X(T^{-1}B' \cap T^{-1}B)$$

$$\text{for all} \quad B' \in \mathcal{B}.$$

Thus by (20)

$$I_B(t) = P^{X|t}(T^{-1}B) \qquad \text{a.e. } P^T.$$

Let B_n, $n = 1, 2, \ldots$, be the intervals of $\mathcal{T}$ with rational end points. Then there exists a P-null set $N = \bigcup N_n$ such that for $t \notin N$

$$I_{B_n}(t) = P^{X|t}(T^{-1}B_n)$$

for all n. For fixed $t \notin N$, the two set functions $P^{X|t}(T^{-1}B)$ and $I_B(t)$ are probability distributions over $\mathcal{B}$, the latter assigning probability 1 or 0 to a set as it does or does not contain the point t. Since these distributions agree over the rational intervals B_n, they agree for all $B \in \mathcal{B}$. In particular, for $t \notin N$, the set consisting of the single point t is in $\mathcal{B}$, and if

$$A^{(t)} = \{ x : T(x) = t \},$$

it follows that for all $t \notin N$

(26) $$P^{X|t}(A^{(t)}) = 1.$$

Thus

$$\int h[T(x)] f(x) \, dP^{X|t}(x) = \int_{A^{(t)}} h[T(x)] f(x) \, dP^{X|t}(x)$$

$$= h(t) \int f(x) \, dP^{X|t}(x)$$

for $t \notin N$, as was to be proved.

It is a consequence of Theorem 6 that for all $t \notin N$, $E[h(T)|t] = h(t)$ and hence in particular $P(T \in B|t) = 1$ or 0 as $t \in B$ or $t \notin B$.

The conditional distributions $P^{X|t}$ still differ from those of the elementary case considered in Chapter 1, Section 9, in being defined over $(\mathcal{X}, \mathcal{A})$ rather than over the set $A^{(t)}$ and the σ-field $\mathcal{A}^{(t)}$ of its Borel subsets. However, (26) implies that for $t \notin N$

$$P^{X|t}(A) = P^{X|t}(A \cap A^{(t)}).$$

The calculations of conditional probabilities and expectations are therefore unchanged if for $t \notin N$, $P^{X|t}$ is replaced by the distribution $\bar{P}^{X|t}$, which is defined over $(A^{(t)}, \mathcal{A}^{(t)})$ and which assigns to any subset of $A^{(t)}$ the same probability as $P^{X|t}$.

Theorem 6 establishes for all $t \notin N$ the existence of conditional probability distributions $\overline{P}^{X|t}$, which are defined over $(A^{(t)}, \mathscr{A}^{(t)})$ and which by Lemma 4 satisfy

$$(27) \qquad E[f(X)] = \int_{\mathscr{T}-N}\left[\int_{A^{(t)}} f(x)\, dP^{X|t}(x)\right] dP^{T}(t)$$

for all integrable functions f. Conversely, consider any family of distributions satisfying (27), and the experiment of observing first T, and then, if $T = t$, a random quantity with distribution $\overline{P}^{X|t}$. The result of this two-stage procedure is a point distributed over $(\mathscr{X}, \mathscr{A})$ with the same distribution as the original X. Thus $\overline{P}^{X|t}$ satisfies this "functional" definition of conditional probability.

If $(\mathscr{X}, \mathscr{A})$ is a product space $(\mathscr{T} \times \mathscr{Y}, \mathscr{B} \times \mathscr{C})$, then $A^{(t)}$ is the product of $\mathscr{Y}$ with the set consisting of the single point t. For $t \notin N$, the conditional distribution $\overline{P}^{X|t}$ then induces a distribution over $(\mathscr{Y}, \mathscr{C})$, which in analogy with the elementary case will be denoted by $P^{Y|t}$. In this case the definition can be extended to all of $\mathscr{T}$ by letting $P^{Y|t}$ assign probability 1 to a common specified point y_0, for all $t \in N$. With this definition, (27) becomes

$$(28) \qquad Ef(T, Y) = \int_{\mathscr{T}}\left[\int_{\mathscr{Y}} f(t, y)\, dP^{Y|t}(y)\right] dP^{T}(t).$$

As an application, we shall prove the following lemma, which will be used in Section 7.

Lemma 6.   *Let $(\mathscr{T}, \mathscr{B})$ and $(\mathscr{Y}, \mathscr{C})$ be Euclidean spaces, and let $P_0^{T, Y}$ be a distribution over the product space $(\mathscr{X}, \mathscr{A}) = (\mathscr{T} \times \mathscr{Y}, \mathscr{B} \times \mathscr{C})$. Suppose that another distribution P_1 over $(\mathscr{X}, \mathscr{A})$ is such that*

$$dP_1(t, y) = a(y)b(t)\, dP_0(t, y),$$

with $a(y) > 0$ for all y. Then under P_1 the marginal distribution of T and a version of the conditional distribution of Y given t are given by

$$dP_1^{T}(t) = b(t)\left[\int a(y)\, dP_0^{Y|t}(y)\right] dP_0^{T}(t)$$

and

$$dP_1^{Y|t}(y) = \frac{a(y)\, dP_0^{Y|t}(y)}{\int_{\mathscr{Y}} a(y')\, dP_0^{Y|t}(y')}.$$

Proof. The first statement of the lemma follows from the equation

$$P_1\{T \in B\} = E_1[I_B(T)] = E_0[I_B(T)a(Y)b(T)]$$

$$= \int_B b(t)\left[\int_{\mathcal{Y}} a(y)\, dP_0^{Y|t}(y)\right] dP_0^T(t).$$

To check the second statement, one need only show that for any integrable f the expectation $E_1 f(Y, T)$ satisfies (28), which is immediate. The denominator of $dP_1^{Y|t}$ is positive, since $a(y) > 0$ for all y.

6. CHARACTERIZATION OF SUFFICIENCY

We can now generalize the definition of sufficiency given in Chapter 1, Section 9. If $\mathcal{P} = \{P_\theta,\ \theta \in \Omega\}$ is any family of distributions defined over a common sample space $(\mathcal{X}, \mathcal{A})$, a statistic T is *sufficient* for $\mathcal{P}$ (or for θ) if for each A in $\mathcal{A}$ there exists a determination of the conditional probability function $P_\theta(A|t)$ that is independent of θ. As an example suppose that $X_1, \ldots, X_n$ are identically and independently distributed with continuous distribution function F_θ, $\theta \in \Omega$. Then it follows from Example 7 that the set of order statistics $T(X) = (X_{(1)}, \ldots, X_{(n)})$ is sufficient for θ.

Theorem 7. *If $\mathcal{X}$ is Euclidean, and if the statistic T is sufficient for $\mathcal{P}$, then there exist determinations of the conditional probability distributions $P_\theta(A|t)$ which are independent of θ and such that for each fixed t, $P(A|t)$ is a probability measure over $\mathcal{A}$.*

Proof. This is seen from the proof of Theorem 4. By the definition of sufficiency one can, for each rational number r, take the functions $F(r, t)$ to be independent of θ, and the resulting conditional distributions will then also not depend on θ.

In Chapter 1 the definition of sufficiency was justified by showing that in a certain sense a sufficient statistic contains all the available information. In view of Theorem 7 the same justification applies quite generally when the sample space is Euclidean. With the help of a random mechanism one can then construct from a sufficient statistic T a random vector X' having the same distribution as the original sample vector X. Another generalization of the earlier result, not involving the restriction to a Euclidean sample space, is given in Problem 12.

The factorization criterion of sufficiency, derived in Chapter 1, can be extended to any *dominated* family of distributions, that is, any family $\mathcal{P} = \{P_\theta,\ \theta \in \Omega\}$ possessing probability densities p_θ with respect to some

σ-finite measure μ over $(\mathcal{X}, \mathcal{A})$. The proof of this statement is based on the existence of a probability distribution $\lambda = \Sigma c_i P_{\theta_i}$ (Theorem 2 of the Appendix), which is *equivalent* to $\mathcal{P}$ in the sense that for any $A \in \mathcal{A}$

(29)　　　$\lambda(A) = 0$　if and only if　$P_\theta(A) = 0$　for all $\theta \in \Omega$.

Theorem 8. *Let $\mathcal{P} = \{ P_\theta, \theta \in \Omega \}$ be a dominated family of probability distributions over $(\mathcal{X}, \mathcal{A})$, and let $\lambda = \Sigma c_i P_{\theta_i}$ satisfy (29). Then a statistic T with range space $(\mathcal{T}, \mathcal{B})$ is sufficient for $\mathcal{P}$ if and only if there exist nonnegative $\mathcal{B}$-measurable functions $g_\theta(t)$ such that*

(30)　　　　　　　$dP_\theta(x) = g_\theta[T(x)]\, d\lambda(x)$

for all $\theta \in \Omega$.

Proof. Let $\mathcal{A}_0$ be the subfield induced by T, and suppose that T is sufficient for θ. Then for all $\theta \in \Omega$, $A_0 \in \mathcal{A}_0$, and $A \in \mathcal{A}$

$$\int_{A_0} P(A|T(x))\, dP_\theta(x) = P_\theta(A \cap A_0);$$

and since $\lambda = \Sigma c_i P_{\theta_i}$,

$$\int_{A_0} P(A|T(x))\, d\lambda(x) = \lambda(A \cap A_0),$$

so that $P(A|T(x))$ serves as conditional probability function also for λ. Let $g_\theta(T(x))$ be the Radon–Nikodym derivative $dP_\theta(x)/d\lambda(x)$ for $(\mathcal{A}_0, \lambda)$. To prove (30) it is necessary to show that $g_\theta(T(x))$ is also the derivative of P_θ for $(\mathcal{A}, \lambda)$. If A_0 is put equal to $\mathcal{X}$ in the first displayed equation, this follows from the relation

$$P_\theta(A) = \int P(A|T(x))\, dP_\theta(x) = \int E_\lambda[I_A(x)|T(x)]\, dP_\theta(x)$$

$$= \int E_\lambda[I_A(x)|T(x)] g_\theta(T(x))\, d\lambda(x)$$

$$= \int E_\lambda[g_\theta(T(x))I_A(x)|T(x)]\, d\lambda(x)$$

$$= \int g_\theta(T(x))I_A(x)\, d\lambda(x) = \int_A g_\theta(T(x))\, d\lambda(x).$$

Here the second equality uses the fact, established at the beginning of the proof, that $P(A|T(x))$ is also the conditional probability for λ; the third equality holds because the function being integrated is $\mathscr{A}_0$-measurable and because $dP_\theta = g_\theta \, d\lambda$ for $(\mathscr{A}_0, \lambda)$; the fourth is an application of Lemma 3(ii); and the fifth employs the defining property of conditional expectation.

Suppose conversely that (30) holds. We shall then prove that the conditional probability function $P_\lambda(A|t)$ serves as a conditional probability function for all $P \in \mathscr{P}$. Let $g_\theta(T(x)) = dP_\theta(x)/d\lambda(x)$ on $\mathscr{A}$ and for fixed A and θ define a measure ν over $\mathscr{A}$ by the equation $d\nu = I_A \, dP_\theta$. Then over $\mathscr{A}_0$, $d\nu(x)/dP_\theta(x) = E_\theta[I_A(X)|T(x)]$, and therefore

$$\frac{d\nu(x)}{d\lambda(x)} = P_\theta[A|T(x)] g_\theta(T(x)) \qquad \text{over } \mathscr{A}_0.$$

On the other hand, $d\nu(x)/d\lambda(x) = I_A(x)g_\theta(T(x))$ over $\mathscr{A}$, and hence

$$\frac{d\nu(x)}{d\lambda(x)} = E_\lambda\big[I_A(X)g_\theta(T(X))|T(x)\big]$$

$$= P_\lambda[A|T(x)] g_\theta(T(x)) \qquad \text{over } \mathscr{A}_0.$$

It follows that $P_\lambda(A|T(x))g_\theta(T(x)) = P_\theta(A|T(x))g_\theta(T(x))$ $(\mathscr{A}_0, \lambda)$ and hence $(\mathscr{A}_0, P_\theta)$. Since $g_\theta(T(x)) \neq 0$ $(\mathscr{A}_0, P_\theta)$, this shows that $P_\theta(A|T(x))$ $= P_\lambda(A|T(x))$ $(\mathscr{A}_0, P_\theta)$, and hence that $P_\lambda(A|T(x))$ is a determination of $P_\theta(A|T(x))$.

Instead of the above formulation, which explicitly involves the distribution λ, it is sometimes more convenient to state the result with respect to a given dominating measure μ.

Corollary 1. (*Factorization theorem.*)   If the distributions P_θ of $\mathscr{P}$ have probability densities $p_\theta = dP_\theta/d\mu$ with respect to a σ-finite measure μ, then T is sufficient for $\mathscr{P}$ if and only if there exist nonnegative $\mathscr{B}$-measurable functions g_θ on T and a nonnegative $\mathscr{A}$-measurable function h on $\mathscr{X}$ such that

$$(31) \qquad p_\theta(x) = g_\theta[T(x)] h(x) \qquad (\mathscr{A}, \mu).$$

Proof. Let $\lambda = \Sigma c_i P_{\theta_i}$ satisfy (29). Then if T is sufficient, (31) follows from (30) with $h = d\lambda/d\mu$. Conversely, if (31) holds,

$$d\lambda(x) = \Sigma c_i g_{\theta_i}[T(x)] h(x) \, d\mu(x) = k[T(x)] h(x) \, d\mu(x)$$

and therefore $dP_\theta(x) = g_\theta^*(T(x)) \, d\lambda(x)$, where $g_\theta^*(t) = g_\theta(t)/k(t)$ when $k(t) > 0$ and may be defined arbitrarily when $k(t) = 0$.

For extensions of the factorization theorem to undominated families, see Ghosh, Morimoto, and Yamada (1981) and the literature cited there.

7. EXPONENTIAL FAMILIES

An important family of distributions which admits a reduction by means of sufficient statistics is the *exponential family*, defined by probability densities of the form

$$(32) \qquad p_\theta(x) = C(\theta)\exp\left[\sum_{j=1}^{k} Q_j(\theta)T_j(x)\right]h(x)$$

with respect to a σ-finite measure μ over a Euclidean sample space $(\mathscr{X}, \mathscr{A})$. Particular cases are the distributions of a sample $X = (X_1, \ldots, X_n)$ from a binomial, Poisson, or normal distribution. In the binomial case, for example, the density (with respect to counting measure) is

$$\binom{n}{x}p^x(1-p)^{n-x} = (1-p)^n\exp\left[x\log\left(\frac{p}{1-p}\right)\right]\binom{n}{x}.$$

Example 8. If $Y_1, \ldots, Y_n$ are independently distributed, each with density (with respect to Lebesgue measure)

$$(33) \qquad p_\sigma(y) = \frac{y^{[(f/2)-1]}\exp[-y/(2\sigma^2)]}{(2\sigma^2)^{f/2}\Gamma(f/2)}, \qquad y > 0,$$

then the joint distribution of the Y's constitutes an exponential family. For $\sigma = 1$, (33) is the density of the χ^2-distribution with f degrees of freedom; in particular, for f an integer this is the density of $\sum_{j=1}^{f}X_j^2$, where the X's are a sample from the normal distribution $N(0,1)$.

Example 9. Consider n independent trials, each of them resulting in one of the s outcomes $E_1, \ldots, E_s$ with probabilities $p_1, \ldots, p_s$ respectively. If X_{ij} is 1 when the outcome of the ith trial is E_j and 0 otherwise, the joint distribution of the X's is

$$P\{X_{11} = x_{11}, \ldots, X_{ns} = x_{ns}\} = p_1^{\sum x_{i1}}p_2^{\sum x_{i2}} \ldots p_s^{\sum x_{is}},$$

where all $x_{ij} = 0$ or 1 and $\sum_j x_{ij} = 1$. This forms an exponential family with $T_j(x) = \sum_{i=1}^{n}x_{ij}$ $(j = 1, \ldots, s - 1)$. The joint distribution of the T's is the multinomial distribution $M(n; p_1, \ldots, p_s)$ given by

$$(34) \qquad P\{T_1 = t_1, \ldots, T_{s-1} = t_{s-1}\}$$

$$= \frac{n!}{t_1! \ldots t_{s-1}!(n - t_1 - \cdots - t_{s-1})!}$$

$$\times p_1^{t_1} \ldots p_{s-1}^{t_{s-1}}(1 - p_1 - \cdots - p_{s-1})^{n-t_1-\cdots-t_{s-1}}.$$

If $X_1, \ldots, X_n$ is a sample from a distribution with density (32), the joint distribution of the X's constitutes an exponential family with the sufficient statistics $\sum_{i=1}^{n} T_j(X_i)$, $j = 1, \ldots, k$. Thus there exists a k-dimensional sufficient statistic for $(X_1, \ldots, X_n)$ regardless of the sample size. Suppose conversely that $X_1, \ldots, X_n$ is a sample from a distribution with some density $p_\theta(x)$ and that the set over which this density is positive is independent of θ. Then under regularity assumptions which make the concept of dimensionality meaningful, if there exists a k-dimensional sufficient statistic with $k < n$, the densities $p_\theta(x)$ constitute an exponential family. For a proof and discussion of regularity conditions see, for example, Barankin and Maitra (1963), Brown (1964), Barndorff-Nielsen and Pedersen (1968), and Hipp (1974).

Employing a more natural parametrization and absorbing the factor $h(x)$ into μ, we shall write an exponential family in the form $dP_\theta(x) = p_\theta(x) \, d\mu(x)$ with

$$(35) \qquad p_\theta(x) = C(\theta) \exp\left[\sum_{j=1}^{k} \theta_j T_j(x) \right].$$

For suitable choice of the constant $C(\theta)$, the right-hand side of (35) is a probability density provided its integral is finite. The set Ω of parameter points $\theta = (\theta_1, \ldots, \theta_k)$ for which this is the case is the *natural parameter space* of the exponential family (35).

Optimum tests of certain hypotheses concerning any θ_j are obtained in Chapter 4. We shall now consider some properties of exponential families required for this purpose.

Lemma 7. *The natural parameter space of an exponential family is convex.*

Proof. Let $(\theta_1, \ldots, \theta_k)$ and $(\theta_1', \ldots, \theta_k')$ be two parameter points for which the integral of (35) is finite. Then by Hölder's inequality,

$$\int \exp\left[\sum \left[\alpha\theta_j + (1 - \alpha)\theta_j' \right] T_j(x) \right] d\mu(x)$$

$$\leq \left[\int \exp\left[\sum \theta_j T_j(x) \right] d\mu(x) \right]^{\alpha} \left[\int \exp\left[\sum \theta_j' T_j(x) \right] d\mu(x) \right]^{1-\alpha} < \infty$$

for any $0 < \alpha < 1$.

If the convex set Ω lies in a linear space of dimension $< k$, then (35) can be rewritten in a form involving fewer than k components of T. We shall therefore, without loss of generality, assume Ω to be k-dimensional.

It follows from the factorization theorem that $T(x) = (T_1(x), \ldots, T_k(x))$ is sufficient for $\mathscr{P} = \{P_\theta, \theta \in \Omega\}$.

Lemma 8. *Let X be distributed according to the exponential family*

$$dP_{\theta,\vartheta}^X(x) = C(\theta, \vartheta)\exp\left[\sum_{i=1}^{r} \theta_i U_i(x) + \sum_{j=1}^{s} \vartheta_j T_j(x)\right] d\mu(x).$$

Then there exist measures λ_θ and ν_t over s- and r-dimensional Euclidean space respectively such that

(i) *the distribution of $T = (T_1, \ldots, T_s)$ is an exponential family of the form*

$$(36) \qquad dP_{\theta,\vartheta}^T(t) = C(\theta, \vartheta)\exp\left(\sum_{j=1}^{s} \vartheta_j t_j\right) d\lambda_\theta(t),$$

(ii) *the conditional distribution of $U = (U_1, \ldots, U_r)$ given $T = t$ is an exponential family of the form*

$$(37) \qquad dP_\theta^{U|t}(u) = C_t(\theta)\exp\left(\sum_{i=1}^{r} \theta_i u_i\right) d\nu_t(u),$$

and hence in particular is independent of ϑ.

Proof. Let (θ^0, ϑ^0) be a point of the natural parameter space, and let $\mu^* = P_{\theta^0,\vartheta^0}^X$. Then

$$dP_{\theta,\vartheta}^X(x) = \frac{C(\theta, \vartheta)}{C(\theta^0, \vartheta^0)}$$

$$\times \exp\left[\sum_{i=1}^{r} (\theta_i - \theta_i^0)U_i(x) + \sum_{j=1}^{s} (\vartheta_j - \vartheta_j^0)T_j(x)\right] d\mu^*(x),$$

and the result follows from Lemma 6, with

$$d\lambda_\theta(t) = \exp\left(-\sum \vartheta_i^0 t_i\right)\left[\int \exp\left[\sum_{i=1}^{r} (\theta_i - \theta_i^0)u_i\right] dP_{\theta^0,\vartheta^0}^{U|t}(u)\right] dP_{\theta^0,\vartheta^0}^T(t)$$

and

$$d\nu_t(u) = \exp\left(-\sum \theta_i^0 u_i\right) dP_{\theta^0,\vartheta^0}^{U|t}(u).$$

Theorem 9. *Let ϕ be any function on $(\mathscr{X}, \mathscr{A})$ for which the integral*

$$(38) \qquad \int \phi(x) \exp\left[\sum_{j=1}^{k} \theta_j T_j(x)\right] d\mu(x)$$

considered as a function of the complex variables $\theta_j = \xi_j + i\eta_j$ ($j = 1, \ldots, k$) exists for all $(\xi_1, \ldots, \xi_k) \in \Omega$ and is finite. Then

(i) *the integral is an analytic function of each of the θ's in the region R of parameter points for which $(\xi_1, \ldots, \xi_k)$ is an interior point of the natural parameter space Ω;*

(ii) *the derivatives of all orders with respect to the θ's of the integral (38) can be computed under the integral sign.*

Proof. Let $(\xi_1^0, \ldots, \xi_k^0)$ be any fixed point in the interior of Ω, and consider one of the variables in question, say θ_1. Breaking up the factor

$$\phi(x) \exp\left[\left(\xi_2^0 + i\eta_2^0\right) T_2(x) + \cdots + \left(\xi_k^0 + i\eta_k^0\right) T_k(x)\right]$$

into its real and complex part and each of these into its positive and negative part, and absorbing this factor in each of the four terms thus obtained into the measure μ, one sees that as a function of θ_1 the integral (38) can be written as

$$\int \exp[\theta_1 T_1(x)] \, d\mu_1(x) - \int \exp[\theta_1 T_1(x)] \, d\mu_2(x)$$

$$+ i \int \exp[\theta_1 T_1(x)] \, d\mu_3(x) - i \int \exp[\theta_1 T_1(x)] \, d\mu_4(x).$$

It is therefore sufficient to prove the result for integrals of the form

$$\psi(\theta_1) = \int \exp[\theta_1 T_1(x)] \, d\mu(x).$$

Since $(\xi_1^0, \ldots, \xi_k^0)$ is in the interior of Ω, there exists $\delta > 0$ such that $\psi(\theta_1)$ exists and is finite for all θ_1 with $|\xi_1 - \xi_1^0| \leq \delta$. Consider the difference quotient

$$\frac{\psi(\theta_1) - \psi(\theta_1^0)}{\theta_1 - \theta_1^0} = \int \frac{\exp[\theta_1 T_1(x)] - \exp[\theta_1^0 T_1(x)]}{\theta_1 - \theta_1^0} \, d\mu(x).$$

The integrand can be written as

$$\exp\left[\theta_1^0 T_1(x)\right] \left[\frac{\exp\left[(\theta_1 - \theta_1^0)T_1(x)\right] - 1}{\theta_1 - \theta_1^0}\right].$$

Applying to the second factor the inequality

$$\left|\frac{\exp(az) - 1}{z}\right| \leq \frac{\exp(\delta|a|)}{\delta} \quad \text{for} \quad |z| \leq \delta,$$

the integrand is seen to be bounded above in absolute value by

$$\frac{1}{\delta}\left|\exp\left(\theta_1^0 T_1 + \delta|T_1|\right)\right| \leq \frac{1}{\delta}\left|\exp\left[(\theta_1^0 + \delta)T_1\right] + \exp\left[(\theta_1^0 - \delta)T_1\right]\right|$$

for $|\theta_1 - \theta_1^0| \leq \delta$. Since the right-hand side is integrable, it follows from the Lebesgue dominated-convergence theorem [Theorem 1(ii)] that for any sequence of points $\theta_1^{(n)}$ tending to θ_1^0, the difference quotient of ψ tends to

$$\int T_1(x)\exp\left[\theta_1^0 T_1(x)\right] d\mu(x).$$

This completes the proof of (i), and proves (ii) for the first derivative. The proof for the higher derivatives is by induction and is completely analogous.

8. PROBLEMS

Section 1

1. *Monotone class.* A class $\mathcal{F}$ of subsets of a space is a *field* if it contains the whole space and is closed under complementation and under finite unions; a class $\mathcal{M}$ is *monotone* if the union and intersection of every increasing and decreasing sequence of sets of $\mathcal{M}$ is again in $\mathcal{M}$. The smallest monotone class $\mathcal{M}_0$ containing a given field $\mathcal{F}$ coincides with the smallest σ-field $\mathcal{A}$ containing $\mathcal{F}$.

   [One proves first that $\mathcal{M}_0$ is a field. To show, for example, that $A \cap B \in \mathcal{M}_0$ when A and B are in $\mathcal{M}_0$, consider, for a fixed set $A \in \mathcal{F}$, the class $\mathcal{M}_A$ of all B in $\mathcal{M}_0$ for which $A \cap B \in \mathcal{M}_0$. Then $\mathcal{M}_A$ is a monotone class containing $\mathcal{F}$, and hence $\mathcal{M}_A = \mathcal{M}_0$. Thus $A \cap B \in \mathcal{M}_A$ for all B. The argument can now be repeated with a fixed set $B \in \mathcal{M}_0$ and the class $\mathcal{M}_B$ of sets A in $\mathcal{M}_0$ for which $A \cap B \in \mathcal{M}_0$. Since $\mathcal{M}_0$ is a field and monotone, it is a σ-field containing $\mathcal{F}$ and hence contains $\mathcal{A}$. But any σ-field is a monotone class so that also $\mathcal{M}_0$ is contained in $\mathcal{A}$.]

Section 2

2. *Radon–Nikodym derivatives.*

   (i) If λ and μ are σ-finite measures over $(\mathscr{X}, \mathscr{A})$ and μ is absolutely continuous with respect to λ, then

$$\int f d\mu = \int f \frac{d\mu}{d\lambda} d\lambda$$

   for any μ-integrable function f.

   (ii) If λ, μ, and ν are σ-finite measures over $(\mathscr{X}, \mathscr{A})$ such that ν is absolutely continuous with respect to μ and μ with respect to λ, then

$$\frac{d\nu}{d\lambda} = \frac{d\nu}{d\mu} \frac{d\mu}{d\lambda} \qquad \text{a.e. } \lambda.$$

   (iii) If μ and ν are σ-finite measures, which are *equivalent* in the sense that each is absolutely continuous with respect to the other, then

$$\frac{d\nu}{d\mu} = \left(\frac{d\mu}{d\nu}\right)^{-1} \qquad \text{a.e. } \mu, \nu.$$

   (iv) If μ_k, $k = 1, 2, \ldots$, and μ are finite measures over $(\mathscr{X}, \mathscr{A})$ such that $\sum_{k=1}^{\infty} \mu_k(A) = \mu(A)$ for all $A \in \mathscr{A}$, and if the μ_k are absolutely continuous with respect to a σ-finite measure λ, then μ is absolutely continuous with respect to λ, and

$$\frac{d \sum_{k=1}^{n} \mu_k}{d\lambda} = \sum_{k=1}^{n} \frac{d\mu_k}{d\lambda}, \quad \lim_{n \to \infty} \frac{d \sum_{k=1}^{n} \mu_k}{d\lambda} = \frac{d\mu}{d\lambda} \qquad \text{a.e. } \lambda.$$

   [(i): The equation in question holds when f is the indicator of a set, hence when f is simple, and therefore for all integrable f.
   (ii): Apply (i) with $f = d\nu/d\mu$.]

3. If $f(x) > 0$ for all $x \in S$ and μ is σ-finite, then $\int_S f d\mu = 0$ implies $\mu(S) = 0$. [Let S_n be the subset of S on which $f(x) \geq 1/n$. Then $\mu(S) \leq \Sigma\mu(S_n)$ and $\mu(S_n) \leq n\int_{S_n} f d\mu \leq n\int_S f d\mu = 0$.]

Section 3

4. Let $(\mathscr{X}, \mathscr{A})$ be a measurable space, and $\mathscr{A}_0$ a σ-field contained in $\mathscr{A}$. Suppose that for any function T, the σ-field $\mathscr{B}$ is taken as the totality of sets B such that $T^{-1}(B) \in \mathscr{A}$. Then it is not necessarily true that there exists a function T such that $T^{-1}(\mathscr{B}) = \mathscr{A}_0$.
   [An example is furnished by any $\mathscr{A}_0$ such that for all x the set consisting of the single point x is in $\mathscr{A}_0$.]

Section 4

5.   (i)   Let $\mathscr{P}$ be any family of distributions $X = (X_1, \ldots, X_n)$ such that

$$P\{(X_i, X_{i+1}, \ldots, X_n, X_1, \ldots, X_{i-1}) \in A\} = P\{(X_1, \ldots, X_n) \in A\}$$

for all Borel sets A and all $i = 1, \ldots, n$. For any sample point $(x_1, \ldots, x_n)$ define $(y_1, \ldots, y_n) = (x_i, x_{i+1}, \ldots, x_n, x_1, \ldots, x_{i-1})$, where $x_i = x_{(1)} = \min(x_1, \ldots, x_n)$. Then the conditional expectation of $f(X)$ given $Y = y$ is

$$f_0(y_1, \ldots, y_n) = \frac{1}{n}\big[\, f(y_1, \ldots, y_n) + f(y_2, \ldots, y_n, y_1)$$

$$+ \cdots + \big(f(y_n, y_1, \ldots, y_{n-1})\big)\big].$$

(ii)   Let $G = \{g_1, \ldots, g_r\}$ be any group of permutations of the coordinates $x_1, \ldots, x_n$ of a point x in n-space, and denote by gx the point obtained by applying g to the coordinates of x. Let $\mathscr{P}$ be any family of distributions P of $X = (X_1, \ldots, X_n)$ such that

$$(39) \qquad P\{gX \in A\} = P\{X \in A\} \qquad \text{for all} \quad g \in G.$$

For any point x let $t = T(x)$ be any rule that selects a unique point from the r points $g_k x$, $k = 1, \ldots, r$ (for example the smallest first coordinate if this defines it uniquely, otherwise also the smallest second coordinate, etc.). Then

$$E[f(X)|t] = \frac{1}{r}\sum_{k=1}^{r} f(g_k t).$$

(iii)   Suppose that in (ii) the distributions P do not satisfy the invariance condition (39) but are given by

$$dP(x) = h(x)\, d\mu(x),$$

where μ is invariant in the sense that $\mu\{x: gx \in A\} = \mu(A)$. Then

$$E[f(X)|t] = \frac{\displaystyle\sum_{k=1}^{r} f(g_k t)\, h(g_k t)}{\displaystyle\sum_{k=1}^{r} h(g_k t)}.$$

Section 5

6.   Prove Theorem 4 for the case of an n-dimensional sample space.
[The condition that the cumulative distribution function is nondecreasing is replaced by $P\{x_1 < X_1 \le x_1', \ldots, x_n < X_n \le x_n'\} \ge 0$; the condition that it is

continuous on the right can be stated as $\lim_{m \to \infty} F(x_1 + 1/m, \ldots, x_n + 1/m) = F(x_1, \ldots, x_n).]$

7. Let $\mathcal{X} = \mathcal{Y} \times \mathcal{T}$, and suppose that P_0, P_1 are two probability distributions given by

$$dP_0(y, t) = f(y)g(t) \, d\mu(y) \, d\nu(t),$$

$$dP_1(y, t) = h(y, t) \, d\mu(y) \, d\nu(t),$$

where $h(y, t)/f(y)g(t) < \infty$. Then under P_1 the probability density of Y with respect to μ is

$$p_1^Y(y) = f(y)E_0\left[\frac{h(y, T)}{f(y)g(T)}\bigg| Y = y\right].$$

[We have

$$p_1^Y(y) = \int_{\mathcal{T}} h(y, t) \, d\nu(t) = f(y)\int_{\mathcal{T}} \frac{h(y, t)}{f(y)g(t)} g(t) \, d\nu(t).]$$

Section 6

8. *Symmetric distributions.*

(i) Let $\mathscr{P}$ be any family of distributions of $X = (X_1, \ldots, X_n)$ which are symmetric in the sense that

$$P\{(X_{i_1}, \ldots, X_{i_n}) \in A\} = P\{(X_1, \ldots, X_n) \in A\}$$

for all Borel sets A and all permutations $(i_1, \ldots, i_n)$ of $(1, \ldots, n)$. Then the statistic T of Example 7 is sufficient for $\mathscr{P}$, and the formula given in the first part of the example for the conditional expectation $E[f(X)|T(x)]$ is valid.

(ii) The statistic Y of Problem 5 is sufficient.

(iii) Let $X_1, \ldots, X_n$ be identically and independently distributed according to a continuous distribution $P \in \mathscr{P}$, and suppose that the distributions of $\mathscr{P}$ are symmetric with respect to the origin. Let $V_i = |X_i|$ and $W_i = V_{(i)}$. Then $(W_1, \ldots, W_n)$ is sufficient for $\mathscr{P}$.

9. *Sufficiency of likelihood ratios.* Let P_0, P_1 be two distributions with densities p_0, p_1. Then $T(x) = p_1(x)/p_0(x)$ is sufficient for $\mathscr{P} = \{P_0, P_1\}$.
[This follows from the factorization criterion by writing $p_1 = T \cdot p_0$, $p_0 = 1 \cdot p_0$.]

10. *Pairwise sufficiency.* A statistic T is pairwise sufficient for $\mathscr{P}$ if it is sufficient for every pair of distributions in $\mathscr{P}$.

(i) If $\mathscr{P}$ is countable and T is pairwise sufficient for $\mathscr{P}$, then T is sufficient for $\mathscr{P}$.

(ii) If $\mathscr{P}$ is a dominated family and T is pairwise sufficient for $\mathscr{P}$, then T is sufficient for $\mathscr{P}$.

[(i): Let $\mathscr{P} = \{ P_0, P_1, \ldots \}$, and let $\mathscr{A}_0$ be the sufficient subfield induced by T. Let $\lambda = \Sigma c_i P_i$ ($c_i > 0$) be equivalent to $\mathscr{P}$. For each $j = 1, 2, \ldots$ the probability measure λ_j that is proportional to $(c_0/n) P_0 + c_j P_j$ is equivalent to $\{ P_0, P_j \}$. Thus by pairwise sufficiency, the derivative $f_j = dP_0/[(c_0/n) dP_0 + c_j dP_j)]$ is $\mathscr{A}_0$-measurable. Let $S_j = \{ x : f_j(x) = 0 \}$ and $S = \bigcup_{j=1}^n S_j$. Then $S \in \mathscr{A}_0$, $P_0(S) = 0$, and on $\mathscr{X} - S$ the derivative $dP_0/d\Sigma_{j=1}^n c_j P_j$ equals $(\Sigma_{j=1}^n 1/f_j)^{-1}$ which is $\mathscr{A}_0$-measurable. It then follows from Problem 2 that

$$\frac{dP_0}{d\lambda} = \frac{dP_0}{d\sum\limits_{j=0}^n c_j P_j} \frac{d\sum\limits_{j=0}^n c_j P_j}{d\lambda}$$

is also $\mathscr{A}_0$-measurable.

(ii): Let $\lambda = \Sigma_{j=1}^\infty c_j P_{\theta_j}$ be equivalent to $\mathscr{P}$. Then pairwise sufficiency of T implies for any θ_0 that $dP_{\theta_0}/(dP_{\theta_0} + d\lambda)$ and hence $dP_{\theta_0}/d\lambda$ is a measurable function of T.]

11. If a statistic T is sufficient for $\mathscr{P}$, then for every function f which is $(\mathscr{A}, P_\theta)$-integrable for all $\theta \in \Omega$ there exists a determination of the conditional expectation function $E_\theta[f(X)|t]$ that is independent of θ.

[If $\mathscr{X}$ is Euclidean, this follows from Theorems 5 and 7. In general, if f is nonnegative there exists a nondecreasing sequence of simple nonnegative functions f_n tending to f. Since the conditional expectation of a simple function can be taken to be independent of θ by Lemma 3(i), the desired result follows from Lemma 3(iv).]

12. For a decision problem with a finite number of decisions, the class of procedures depending on a sufficient statistic T only is essentially complete.

[For Euclidean sample spaces this follows from Theorem 4 without any restriction on the decision space. For the present case, let a decision procedure be given by $\delta(x) = (\delta^{(1)}(x), \ldots, \delta^{(m)}(x))$ where $\delta^{(i)}(x)$ is the probability with which decision d_i is taken when x is observed. If T is sufficient and $\eta^{(i)}(t) = E[\delta^{(i)}(X)|t]$, the procedures δ and η have identical risk functions.]

[More general versions of this result are discussed, for example, by Elfving (1952), Bahadur (1955), Burkholder (1961), LeCam (1964), and Roy and Ramamoorthi (1979).]

Section 7

13. Let X_i $(i = 1, \ldots, s)$ be independently distributed with Poisson distribution $P(\lambda_i)$, and let $T_0 = \Sigma X_j$, $T_i = X_i$, $\lambda = \Sigma \lambda_j$. Then T_0 has the Poisson distribution $P(\lambda)$, and the conditional distribution of $T_1, \ldots, T_{s-1}$ given $T_0 = t_0$ is the multinomial distribution (34) with $n = t_0$ and $p_i = \lambda_i / \lambda$. [Direct computation.]

14. *Life testing.* Let $X_1, \ldots, X_n$ be independently distributed with exponential density $(2\theta)^{-1} e^{-x/2\theta}$ for $x \geq 0$, and let the ordered X's be denoted by $Y_1 \leq Y_2 \leq \cdots \leq Y_n$. It is assumed that Y_1 becomes available first, then Y_2, and so on, and that observation is continued until Y_r has been observed. This might arise, for example, in life testing where each X measures the length of life of, say, an electron tube, and n tubes are being tested simultaneously. Another application is to the disintegration of radioactive material, where n is the number of atoms, and observation is continued until r α-particles have been emitted.

(i) The joint distribution of $Y_1, \ldots, Y_r$ is an exponential family with density

$$\frac{1}{(2\theta)^r} \frac{n!}{(n-r)!} \exp\left[-\frac{\sum_{i=1}^{r} y_i + (n-r) y_r}{2\theta} \right], \qquad 0 \leq y_1 \leq \cdots \leq y_r.$$

(ii) The distribution of $[\Sigma_{i=1}^{r} Y_i + (n-r) Y_r] / \theta$ is χ^2 with $2r$ degrees of freedom.

(iii) Let $Y_1, Y_2, \ldots$ denote the time required until the first, second, ... event occurs in a Poisson process with parameter $1/2\theta'$ (see Chapter 1, Problem 1). Then $Z_1 = Y_1/\theta'$, $Z_2 = (Y_2 - Y_1)/\theta'$, $Z_3 = (Y_3 - Y_2)/\theta', \ldots$ are independently distributed as χ^2 with 2 degrees of freedom, and the joint density of $Y_1, \ldots, Y_r$ is an exponential family with density

$$\frac{1}{(2\theta')^r} \exp\left(-\frac{y_r}{2\theta'} \right), \qquad 0 \leq y_1 \leq \cdots \leq y_r.$$

The distribution of Y_r/θ' is again χ^2 with $2r$ degrees of freedom.

(iv) The same model arises in the application to life testing if the number n of tubes is held constant by replacing each burned-out tube with a new one, and if Y_1 denotes the time at which the first tube burns out, Y_2 the time at which the second tube burns out, and so on, measured from some fixed time.

[(ii): The random variables $Z_i = (n - i + 1)(Y_i - Y_{i-1})/\theta$ $(i = 1, \ldots, r)$ are independently distributed as χ^2 with 2 degrees of freedom, and $[\Sigma_{i=1}^{r} Y_i + (n - r) Y_r]/\theta = \Sigma_{i=1}^{r} Z_i.$]

15. For any θ which is an interior point of the natural parameter space, the expectations and covariances of the statistics T_j in the exponential family (35) are given by

$$E[T_j(X)] = -\frac{\partial \log C(\theta)}{\partial \theta_j} \qquad (j = 1, \ldots, k),$$

$$E[T_i(X)T_j(X)] - [ET_i(X)ET_j(X)] = -\frac{\partial^2 \log C(\theta)}{\partial \theta_i \, \partial \theta_j} \qquad (i, j = 1, \ldots, k).$$

16. Let Ω be the natural parameter space of the exponential family (35), and for any fixed $t_{r+1}, \ldots, t_k$ $(r < k)$ let $\Omega'_{\theta_1, \ldots, \theta_r}$ be the natural parameter space of the family of conditional distributions given $T_{r+1} = t_{r+1}, \ldots, T_k = t_k$.

(i) Then $\Omega'_{\theta_1, \ldots, \theta_r}$ contains the projection $\Omega_{\theta_1, \ldots, \theta_r}$ of Ω onto $\theta_1, \ldots, \theta_r$.
(ii) An example in which $\Omega_{\theta_1, \ldots, \theta_r}$ is a proper subset of $\Omega'_{\theta_1, \ldots, \theta_r}$ is the family of densities

$$p_{\theta_1\theta_2}(x, y) = C(\theta_1, \theta_2)\exp(\theta_1 x + \theta_2 y - xy), \qquad x, y > 0.$$

9. REFERENCES

The theory of measure and integration in abstract spaces and its application to probability theory, including in particular conditional probability and expectation, is treated in a number of books, among them Loève (1977–78) and Billingsley (1979). The material on sufficient statistics and exponential families is complemented by the corresponding sections in *TPE*. A much fuller treatment of exponential families is provided by Barndorff-Nielsen (1978), who also discusses various generalizations of sufficiency.

Bahadur, R. R.
(1954). "Sufficiency and statistical decision functions." *Ann. Math. Statist.* **25**, 423–462.
[A detailed abstract treatment of sufficient statistics, including the factorization theorem, the structure theorem for minimal sufficient statistics, and a discussion of sufficiency for the case of sequential experiments.]
(1955). "A characterization of sufficiency." *Ann. Math. Statist.* **26**, 286–293.

Bahadur, R. R. and Lehmann, E. L.
(1955). "Two comments on 'sufficiency and statistical decision functions'." *Ann. Math. Statist.* **26**, 139–142.
[Problem 4.]

Barankin, E. W. and Maitra, A. P.
(1963). "Generalizations of the Fisher–Darmois–Koopman–Pitman theorem on sufficient statistics." *Sankhyā (A)* **25**, 217–244.

Barndorff-Nielsen, O.
(1978). *Information and Exponential Families in Statistical Theory*, Wiley, New York.

Barndorff-Nielsen, O. and Pedersen, K.
(1968). "Sufficient data reduction and exponential families." *Math. Scand.* **2**, 197–202.

Billingsley, P.
(1979). *Probability and Measure*, Wiley, New York.

Blackwell, D. and Dubins, L. E.
(1975). "On existence and non-existence of proper, regular conditional distributions," *Ann. Probab.* **3**, 741–752.

Blackwell, D. and Ryll-Nardzewski, C.
(1963). "Non-existence of everywhere proper conditional distributions." *Ann. Math. Statist.* **34**, 223–225.

Brown, L.
(1964). "Sufficient statistics in the case of independent random variables." *Ann. Math. Statist.* **35**, 1456–1474.

Burkholder, D. L.
(1961). "Sufficiency in the undominated case." *Ann. Math. Statist.* **32**, 1191–1200.

Elfving, G.
(1952). "Sufficiency and completeness." *Ann. Acad. Sci. Fennicae* (*A*), No. 135.

Epstein, B. and Sobel, M.
(1954). "Some theorems relevant to life testing from an exponential distribution." *Ann. Math. Statist.* **25**, 373–381.
[Problem 14.]

Ghosh, J. K., Morimoto, H., and Yamada, S.
(1981). "Neyman factorization and minimality of pairwise sufficient subfields." *Ann. Statist.* **9**, 514–530.

Halmos, P. R. and Savage, L. J.
(1949). "Application of the Radon–Nikodym theorem to the theory of sufficient statistics." *Ann. Math. Statist.* **20**, 225–241.
[First abstract treatment of sufficient statistics; the factorization theorem. Problem 10.]

Hipp, C.
(1974). "Sufficient statistics and exponential families." *Ann. Statist.* **2**, 1283–1292.

Johansen, S.
(1979). *Introduction to the Theory of Regular Exponential Families*, Lecture Notes, No. 3, Inst. of Math. Statist., Univ. of Copenhagen.
[Provides a good introduction to exponential families and their statistical applications.]

LeCam, L.
(1964). "Sufficiency and approximate sufficiency." *Ann. Math. Statist.* **35**, 1419–1455.

Loève, M.
(1977–78). *Probability Theory*, 4th ed. (2 vols.), Springer, Berlin.

Roy, K. K. and Ramamoorthi, R. V.
(1979). "Relationship between Bayes, classical and decision theoretic sufficiency." *Sankhyā* (*A*) **41**, 48–58.

CHAPTER 3

Uniformly Most Powerful Tests

1. STATING THE PROBLEM

We now begin the study of the statistical problem that forms the principal subject of this book,* the problem of hypothesis testing. As the term suggests, one wishes to decide whether or not some hypothesis that has been formulated is correct. The choice here lies between only two decisions: accepting or rejecting the hypothesis. A decision procedure for such a problem is called a *test* of the hypothesis in question.

The decision is to be based on the value of a certain random variable X, the distribution P_θ of which is known to belong to a class $\mathscr{P} = \{P_\theta, \theta \in \Omega\}$. We shall assume that if θ were known, one would also know whether or not the hypothesis is true. The distributions of $\mathscr{P}$ can then be classified into those for which the hypothesis is true and those for which it is false. The resulting two mutually exclusive classes are denoted by H and K, and the corresponding subsets of Ω by Ω_H and Ω_K respectively, so that $H \cup K = \mathscr{P}$ and $\Omega_H \cup \Omega_K = \Omega$. Mathematically, the hypothesis is equivalent to the statement that P_θ is an element of H. It is therefore convenient to identify the hypothesis with this statement and to use the letter H also to denote the hypothesis. Analogously we call the distributions in K the alternatives to H, so that K is the *class of alternatives*.

Let the decisions of accepting or rejecting H be denoted by d_0 and d_1 respectively. A nonrandomized test procedure assigns to each possible value x of X one of these two decisions and thereby divides the sample space into two complementary regions S_0 and S_1. If X falls into S_0 the hypothesis is accepted; otherwise it is rejected. The set S_0 is called the region of acceptance, and the set S_1 the region of rejection or *critical* region.

*The related subject of confidence intervals is treated in Chapter 3, Section 5; Chapter 5, Sections 6, 7; Chapter 6, Sections 11–13; Chapter 7, Section 8; Chapter 8, Section 6; and Chapter 10, Section 4.

68

When performing a test one may arrive at the correct decision, or one may commit one of two errors: rejecting the hypothesis when it is true (error of the first kind) or accepting it when it is false (error of the second kind). The consequences of these are often quite different. For example, if one tests for the presence of some disease, incorrectly deciding on the necessity of treatment may cause the patient discomfort and financial loss. On the other hand, failure to diagnose the presence of the ailment may lead to the patient's death.

It is desirable to carry out the test in a manner which keeps the probabilities of the two types of error to a minimum. Unfortunately, when the number of observations is given, both probabilities cannot be controlled simultaneously. It is customary therefore to assign a bound to the probability of incorrectly rejecting H when it is true, and to attempt to minimize the other probability subject to this condition. Thus one selects a number α between 0 and 1, called the *level of significance*, and imposes the condition that

$$(1) \qquad P_\theta\{\delta(X) = d_1\} = P_\theta\{X \in S_1\} \leq \alpha \qquad \text{for all} \quad \theta \in \Omega_H.$$

Subject to this condition, it is desired to minimize $P_\theta\{\delta(X) = d_0\}$ for θ in Ω_K or, equivalently, to maximize

$$(2) \qquad P_\theta\{\delta(X) = d_1\} = P_\theta\{X \in S_1\} \qquad \text{for all} \quad \theta \in \Omega_K.$$

Although usually (2) implies that

$$(3) \qquad \sup_{\Omega_H} P_\theta\{X \in S_1\} = \alpha,$$

it is convenient to introduce a term for the left-hand side of (3): it is called the *size* of the test or critical region S_1. The condition (1) therefore restricts consideration to tests whose size does not exceed the given level of significance. The probability of rejection (2) evaluated for a given θ in Ω_K is called the *power* of the test against the alternative θ. Considered as a function of θ for all $\theta \in \Omega$, the probability (2) is called the *power function* of the test and is denoted by $\beta(\theta)$.

The choice of a level of significance α will usually be somewhat arbitrary, since in most situations there is no precise limit to the probability of an error of the first kind that can be tolerated. Standard values, such as .01 or .05, were originally chosen to effect a reduction in the tables needed for carrying out various tests. By habit, and because of the convenience of standardization in providing a common frame of reference, these values gradually became entrenched as the conventional levels to use. This is unfortunate, since the choice of significance level should also take into consideration the power that the test will achieve against the alternatives of interest. There is little point in carrying out an experiment which has only a small chance of detecting the effect being sought when it exists. Surveys by Cohen (1962) and Freiman et al. (1978) suggest that this is in fact the case for many studies. Ideally, the sample size should then be increased to permit adequate values for both significance level and power. If that is not feasible, one may wish to use higher values of α than the customary ones. The opposite possibility, that one would like to decrease α, arises when the latter is so close to 1 that α can be lowered appreciably without a significant loss of power (cf. Problem 50). Rules for choosing α in relation to the attainable power are discussed by Lehmann (1958), Arrow (1960), and Sanathanan (1974), and from a Bayesian point of view by Savage (1962, pp. 64–66). See also Rosenthal and Rubin (1985).

Another consideration that may enter into the specification of a significance level is the attitude toward the hypothesis before the experiment is performed. If one firmly believes the hypothesis to be true, extremely convincing evidence will be required before one is willing to give up this belief, and the significance level will accordingly be set very low. (A low significance level results in the hypothesis being rejected only for a set of values of the observations whose total probability under the hypothesis is small, so that such values would be most unlikely to occur if H were true.)

In applications, there is usually available a nested family of rejection regions, corresponding to different significance levels. It is then good practice to determine not only whether the hypothesis is accepted or rejected at the given significance level, but also to determine the smallest significance level $\hat{\alpha} = \hat{\alpha}(x)$, the *significance probability* or *p-value*,* at which the hypothesis would be rejected for the given observation. This number gives an idea of how strongly the data contradict the hypothesis, and enables others to reach a verdict based on the significance level of their choice (cf. Problem 9 and Chapter 4, Problem 2). For various questions of interpretation and some extensions of the concept, see Dempster and Schatzoff (1965), Stone (1969), Gibbons and Pratt (1975), Cox (1977), Pratt and Gibbons (1981, Chapter 1) and Thompson (1985). The large-sample behavior of p-values is discussed in Lambert and Hall (1982), and their sensitivity to changes in the model in Lambert (1982). A graphical procedure for assessing the p-values of simultaneous tests of several hypotheses is proposed by Schweder and Spjøtvoll (1982).

---

*For a related concept, which compares the "acceptability" of two or more parameter values, see Spjøtvoll (1983).

Significance probabilities, with the additional information they provide, are typically more appropriate than fixed levels in scientific problems, whereas a fixed predetermined α is unavoidable when acceptance or rejection of H implies an imminent concrete decision. A review of some of the issues arising in this context, with references to the literature, is given in Kruskal (1978).

A decision making aspect is often imposed on problems of scientific inference by the tendency of journals to publish papers only if the reported results are significant at a conventional level such as 5%. The unfortunate consequences of such a policy have been explored, among others, by Sterling (1959) and Greenwald (1975).

Let us next consider the structure of a randomized test. For any value x such a test chooses between the two decisions, rejection or acceptance, with certain probabilities that depend on x and will be denoted by $\phi(x)$ and $1 - \phi(x)$ respectively. If the value of X is x, a random experiment is performed with two possible outcomes R and $\bar{R}$, the probabilities of which are $\phi(x)$ and $1 - \phi(x)$. If in this experiment R occurs, the hypothesis is rejected, otherwise it is accepted. A randomized test is therefore completely characterized by a function ϕ, the *critical function*, with $0 \leq \phi(x) \leq 1$ for all x. If ϕ takes on only the values 1 and 0, one is back in the case of a nonrandomized test. The set of points x for which $\phi(x) = 1$ is then just the region of rejection, so that in a nonrandomized test ϕ is simply the indicator function of the critical region.

If the distribution of X is P_θ, and the critical function ϕ is used, the probability of rejection is

$$E_\theta \phi(X) = \int \phi(x)\, dP_\theta(x),$$

the conditional probability $\phi(x)$ of rejection given x, integrated with respect to the probability distribution of X. The problem is to select ϕ so as to maximize the power

(4) $\beta_\phi(\theta) = E_\theta \phi(X)$ for all $\theta \in \Omega_K$

subject to the condition

(5) $E_\theta \phi(X) \leq \alpha$ for all $\theta \in \Omega_H$.

The same difficulty now arises that presented itself in the general discussion of Chapter 1. Typically, the test that maximizes the power against a particular alternative in K depends on this alternative, so that some additional principle has to be introduced to define what is meant by an optimum test. There is one important exception: if K contains only one distribution, that is, if one is concerned with a single alternative, the problem is completely specified by (4) and (5). It then reduces to the mathematical problem of maximizing an integral subject to certain side conditions. The theory of this problem, and its statistical applications, constitutes the principal subject of the present chapter. In special cases it may of course turn out that the same test maximizes the power for all alternatives in K even when there is more than one. Examples of such *uniformly most powerful* (UMP) tests will be given in Sections 3 and 7.

In the above formulation the problem can be considered as a special case of the general decision problem with two types of losses. Corresponding to the two kinds of error, one can introduce the two component loss functions,

$$L_1(\theta, d_1) = 1 \ or \ 0 \quad \text{as} \quad \theta \in \Omega_H \text{ or } \theta \in \Omega_K,$$

$$L_1(\theta, d_0) = 0 \quad \text{for all } \theta$$

and

$$L_2(\theta, d_0) = 0 \ or \ 1 \quad \text{as} \quad \theta \in \Omega_H \text{ or } \theta \in \Omega_K,$$

$$L_2(\theta, d_1) = 0 \quad \text{for all } \theta.$$

With this definition the minimization of $EL_2(\theta, \delta(X))$ subject to the restriction $EL_1(\theta, \delta(X)) \leq \alpha$ is exactly equivalent to the problem of hypothesis testing as given above.

The formal loss functions L_1 and L_2 clearly do not represent in general the true losses. The loss resulting from an incorrect acceptance of the hypothesis, for example, will not be the same for all alternatives. The more the alternative differs from the hypothesis, the more serious are the consequences of such an error. As was discussed earlier, we have purposely forgone the more detailed approach implied by this criticism. Rather than working with a loss function which in practice one does not know, it seems preferable to base the theory on the simpler and intuitively appealing notion of error. It will be seen later that at least some of the results can be justified also in the more elaborate formulation.

2. THE NEYMAN–PEARSON FUNDAMENTAL LEMMA

A class of distributions is called *simple* if it contains only a single distribution, and otherwise is said to be *composite*. The problem of hypothesis testing is completely specified by (4) and (5) if K is simple. Its solution is easiest and can be given explicitly when the same is true of H. Let the distributions under a simple hypothesis H and alternative K be P_0 and P_1, and suppose for a moment that these distributions are discrete with $P_i\{X = x\} = P_i(x)$ for $i = 0, 1$. If at first one restricts attention to nonrandomized tests, the optimum test is defined as the critical region S satisfying

(6)
$$\sum_{x \in S} P_0(x) \le \alpha$$

and

$$\sum_{x \in S} P_1(x) = \text{maximum}.$$

It is easy to see which points should be included in S. To each point are attached two values, its probability under P_0 and under P_1. The selected points are to have a total value not exceeding α on the one scale, and as large as possible on the other. This is a situation that occurs in many contexts. A buyer with a limited budget who wants to get "the most for his money" will rate the items according to their *value per dollar*. In order to travel a given distance in the shortest possible time, one must choose the speediest mode of transportation, that is, the one that yields the largest number of *miles per hour*. Analogously in the present problem the most valuable points x are those with the highest value of

$$r(x) = \frac{P_1(x)}{P_0(x)}.$$

The points are therefore rated according to the value of this ratio and selected for S in this order, as many as one can afford under restriction (6). Formally this means that S is the set of all points x for which $r(x) > c$, where c is determined by the condition

$$P_0\{X \in S\} = \sum_{x \,:\, r(x) > c} P_0(x) = \alpha.$$

Here a difficulty is seen to arise. It may happen that when a certain point is included, the value α has not yet been reached but that it would be exceeded if the next point were also included. The exact value α can then either not be achieved at all, or it can be attained only by breaking the preference order established by $r(x)$. The resulting optimization problem has no explicit solution. (Algorithms for obtaining the maximizing set S are given by the theory of linear programming.) The difficulty can be avoided, however, by a modification which does not require violation of the r-order and which does lead to a simple explicit solution, namely by permitting randomization.* This makes it possible to split the next point, including only a portion of it, and thereby to obtain the exact value α without breaking the order of preference that has been established for inclusion of the various sample points. These considerations are formalized in the following theorem, the *fundamental lemma of Neyman and Pearson*.

Theorem 1. *Let P_0 and P_1 be probability distributions possessing densities p_0 and p_1 respectively with respect to a measure μ.[†]*

(i) Existence. *For testing $H: p_0$ against the alternative $K: p_1$ there exists a test ϕ and a constant k such that*

$$(7) \qquad\qquad E_0\phi(X) = \alpha$$

and

$$(8) \qquad\qquad \phi(x) = \begin{cases} 1 & when \quad p_1(x) > kp_0(x), \\ 0 & when \quad p_1(x) < kp_0(x). \end{cases}$$

(ii) Sufficient condition for a most powerful test. *If a test satisfies (7) and (8) for some k, then it is most powerful for testing p_0 against p_1 at level α.*

(iii) Necessary condition for a most powerful test. *If ϕ is most powerful at level α for testing p_0 against p_1, then for some k it satisfies (8) a.e. μ. It also satisfies (7) unless there exists a test of size $< \alpha$ and with power 1.*

Proof. For $\alpha = 0$ and $\alpha = 1$ the theorem is easily seen to be true provided the value $k = +\infty$ is admitted in (8) and $0 \cdot \infty$ is interpreted as 0. Throughout the proof we shall therefore assume $0 < \alpha < 1$.

(i): Let $\alpha(c) = P_0\{ p_1(X) > cp_0(X)\}$. Since the probability is computed under P_0, the inequality need be considered only for the set where $p_0(x) > 0$, so that $\alpha(c)$ is the probability that the random variable $p_1(X)/p_0(X)$ exceeds c. Thus $1 - \alpha(c)$ is a cumulative distribution function, and $\alpha(c)$ is nonincreasing and continuous on the right, $\alpha(c - 0) - \alpha(c) = P_0\{ p_1(X)/p_0(X) = c\}$, $\alpha(-\infty) = 1$, and $\alpha(\infty) = 0$. Given any $0 < \alpha < 1$, let c_0 be such that $\alpha(c_0) \le \alpha \le \alpha(c_0 - 0)$, and consider the test ϕ defined

---

*In practice, typically neither the breaking of the r-order nor randomization is considered acceptable. The common solution, instead, is to adopt a value of α that can be attained exactly and therefore does not present this problem.

[†] There is no loss of generality in this assumption, since one can take $\mu = P_0 + P_1$.

by

$$\phi(x) = \begin{cases} 1 & \text{when} \quad p_1(x) > c_0 p_0(x), \\ \dfrac{\alpha - \alpha(c_0)}{\alpha(c_0 - 0) - \alpha(c_0)} & \text{when} \quad p_1(x) = c_0 p_0(x), \\ 0 & \text{when} \quad p_1(x) < c_0 p_0(x). \end{cases}$$

Here the middle expression is meaningful unless $\alpha(c_0) = \alpha(c_0 - 0)$; since then $P_0\{p_1(X) = c_0 p_0(X)\} = 0$, ϕ is defined a.e. The size of ϕ is

$$E_0\phi(X) = P_0\left\{\frac{p_1(X)}{p_0(X)} > c_0\right\} + \frac{\alpha - \alpha(c_0)}{\alpha(c_0 - 0) - \alpha(c_0)} P_0\left\{\frac{p_1(X)}{p_0(X)} = c_0\right\} = \alpha,$$

so that c_0 can be taken as the k of the theorem.

It is of interest to note that c_0 is essentially unique. The only exception is the case that an interval of c's exists for which $\alpha(c) = \alpha$. If (c', c'') is such an interval, and

$$C = \left\{x : p_0(x) > 0 \text{ and } c' < \frac{p_1(x)}{p_0(x)} < c''\right\},$$

then $P_0(C) = \alpha(c') - \alpha(c'' - 0) = 0$. By Problem 3 of Chapter 2, this implies $\mu(C) = 0$ and hence $P_1(C) = 0$. Thus the sets corresponding to two different values of c differ only in a set of points which has probability 0 under both distributions, that is, points that could be excluded from the sample space.

(ii): Suppose that ϕ is a test satisfying (7) and (8) and that ϕ^* is any other test with $E_0\phi^*(X) \le \alpha$. Denote by S^+ and S^- the sets in the sample space where $\phi(x) - \phi^*(x) > 0$ and < 0 respectively. If x is in S^+, $\phi(x)$ must be > 0 and $p_1(x) \ge kp_0(x)$. In the same way $p_1(x) \le kp_0(x)$ for all x in S^-, and hence

$$\int(\phi - \phi^*)(p_1 - kp_0)\, d\mu = \int_{S^+ \cup S^-}(\phi - \phi^*)(p_1 - kp_0)\, d\mu \ge 0.$$

The difference in power between ϕ and ϕ^* therefore satisfies

$$\int(\phi - \phi^*)p_1\, d\mu \ge k\int(\phi - \phi^*)p_0\, d\mu \ge 0,$$

as was to be proved.

(iii): Let ϕ^* be most powerful at level α for testing p_0 against p_1, and let ϕ satisfy (7) and (8). Let S be the intersection of the set $S^+ \cup S^-$, on which ϕ and ϕ^* differ, with the set $\{x : p_1(x) \neq kp_0(x)\}$, and suppose that $\mu(S) > 0$. Since $(\phi - \phi^*)(p_1 - kp_0)$ is positive on S, it follows from Problem 3 of Chapter 2 that

$$\int_{S^+ \cup S^-} (\phi - \phi^*)(p_1 - kp_0)\, d\mu = \int_S (\phi - \phi^*)(p_1 - kp_0)\, d\mu > 0$$

and hence that ϕ is more powerful against p_1 than ϕ^*. This is a contradiction, and therefore $\mu(S) = 0$, as was to be proved.

If ϕ^* were of size $< \alpha$ and power < 1, it would be possible to include in the rejection region additional points or portions of points and thereby to increase the power until either the power is 1 or the size is α. Thus either $E_0\phi^*(X) = \alpha$ or $E_1\phi^*(X) = 1$.

The proof of part (iii) shows that the most powerful test is uniquely determined by (7) and (8) except on the set on which $p_1(x) = kp_0(x)$. On this set, ϕ can be defined arbitrarily provided the resulting test has size α. Actually, we have shown that it is always possible to define ϕ to be constant over this boundary set. In the trivial case that there exists a test of power 1, the constant k of (8) is 0, and one will accept H for all points for which $p_1(x) = kp_0(x)$ even though the test may then have size $< \alpha$.

It follows from these remarks that the most powerful test is determined uniquely (up to sets of measure zero) by (7) and (8) whenever the set on which $p_1(x) = kp_0(x)$ has μ-measure zero. This unique test is then clearly nonrandomized. More generally, it is seen that randomization is not required except possibly on the boundary set, where it may be necessary to randomize in order to get the size equal to α. When there exists a test of power 1, (7) and (8) will determine a most powerful test, but it may not be unique in that there may exist a test also most powerful and satisfying (7) and (8) for some $\alpha' < \alpha$.

Corollary 1. *Let β denote the power of the most powerful level-α test $(0 < \alpha < 1)$ for testing P_0 against P_1. Then $\alpha < \beta$ unless $P_0 = P_1$.*

Proof. Since the level-α test given by $\phi(x) \equiv \alpha$ has power α, it is seen that $\alpha \leq \beta$. If $\alpha = \beta < 1$, the test $\phi(x) \equiv \alpha$ is most powerful and by Theorem 1(iii) must satisfy (8). Then $p_0(x) = p_1(x)$ a.e. μ, and hence $P_0 = P_1$.

An alternative method for proving the results of this section is based on the following geometric representation of the problem of testing a simple hypothesis against a simple alternative. Let N be the set of all points (α, β)

for which there exists a test ϕ such that

$$\alpha = E_0\phi(X), \qquad \beta = E_1\phi(X).$$

This set is convex, contains the points $(0,0)$ and $(1,1)$, and is symmetric with respect to the point $(\frac{1}{2}, \frac{1}{2})$ in the sense that with any point (α, β) it also contains the point $(1 - \alpha, 1 - \beta)$. In addition, the set N is closed. [This follows from the weak compactness theorem for critical functions, Theorem 3 of the Appendix; the argument is the same as that in the proof of Theorem 5(i).]

For each value $0 < \alpha_0 < 1$, the level-α_0 tests are represented by the points whose abscissa is $\leq \alpha_0$. The most powerful of these tests (whose existence follows from the fact that N is closed) corresponds to the point on the upper boundary of N with abscissa α_0. This is the only point corresponding to a most powerful level-α_0 test unless there exists a point $(\alpha, 1)$ in N with $\alpha < \alpha_0$ (Figure 1b).

As an example of this geometric approach, consider the following alternative proof of Corollary 1. Suppose that for some $0 < \alpha_0 < 1$ the power of the most powerful level-α_0 test is α_0. Then it follows from the convexity of N that $(\alpha, \beta) \in N$ implies $\beta \leq \alpha$, and hence from the symmetry of N that N consists exactly of the line segment connecting the points $(0,0)$ and $(1,1)$. This means that $\int \phi p_0 \, d\mu = \int \phi p_1 \, d\mu$ for all ϕ and hence that $p_0 = p_1$ (a.e. μ), as was to be proved. A proof of Theorem 1 along these lines is given in a more general setting in the proof of Theorem 5.

The Neyman–Pearson lemma has been generalized in many directions. An extension to the case of several side conditions is given in Section 6, and this result is further generalized in Section 8. A sequential version, due to

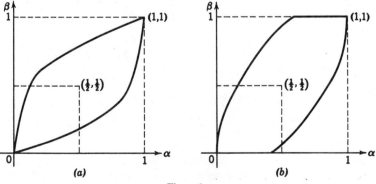

Figure 1

Wald and Wolfowitz (1948, 1950), plays a fundamental role in sequential analysis [see, for example, Ghosh (1970)]. Extensions to stochastic processes are discussed by Grenander (1950) and Dvoretzky, Kiefer, and Wolfowitz (1953), and a version for abstract spaces by Grenander (1981, Section 3.1). A modification due to Huber, in which the distributions are known only approximately, is presented in Section 3 of Chapter 9.

An extension to a selection problem, proposed by Birnbaum and Chapman (1950), is sketched in Problem 23. Generalizations to a variety of decision problems with a finite number of actions can be found, for example, in Hoel and Peterson (1949), Karlin and Rubin (1956), Karlin and Truax (1960), Lehmann (1961), Hall and Kudo (1968) and Spjøtvoll (1972).

3. DISTRIBUTIONS WITH MONOTONE LIKELIHOOD RATIO

The case that both the hypothesis and the class of alternatives are simple is mainly of theoretical interest, since problems arising in applications typically involve a parametric family of distributions depending on one or more parameters. In the simplest situation of this kind the distributions depend on a single real-valued parameter θ, and the hypothesis is one-sided, say $H : \theta \leq \theta_0$. In general, the most powerful test of H against an alternative $\theta_1 > \theta_0$ depends on θ_1 and is then not UMP. However, a UMP test does exist if an additional assumption is satisfied. The real-parameter family of densities $p_\theta(x)$ is said to have *monotone likelihood ratio** if there exists a real-valued function $T(x)$ such that for any $\theta < \theta'$ the distributions P_θ and $P_{\theta'}$ are distinct, and the ratio $p_{\theta'}(x)/p_\theta(x)$ is a nondecreasing function of $T(x)$.

Theorem 2. *Let θ be a real parameter, and let the random variable X have probability density $p_\theta(x)$ with monotone likelihood ratio in $T(x)$.*

(i) *For testing $H : \theta \leq \theta_0$ against $K : \theta > \theta_0$, there exists a UMP test, which is given by*

(9)
$$\phi(x) = \begin{cases} 1 & when \quad T(x) > C, \\ \gamma & when \quad T(x) = C, \\ 0 & when \quad T(x) < C, \end{cases}$$

---

*This definition is in terms of specific versions of the densities p_θ. If instead the definition is to be given in terms of the distributions P_θ, various null-set considerations enter which are discussed in Pfanzagl (1967).

where C and γ are determined by

(10)
$$E_{\theta_0}\phi(X) = \alpha.$$

(ii)   *The power function*

$$\beta(\theta) = E_\theta\phi(X)$$

of this test is strictly increasing for all points θ for which $0 < \beta(\theta) < 1$.

(iii)   *For all θ', the test determined by (9) and (10) is UMP for testing* $H': \theta \leq \theta'$ *against* $K': \theta > \theta'$ *at level* $\alpha' = \beta(\theta')$.

(iv)   *For any* $\theta < \theta_0$ *the test minimizes* $\beta(\theta)$ (*the probability of an error of the first kind*) *among all tests satisfying (10)*.

Proof.   (i) and (ii): Consider first the hypothesis $H_0: \theta = \theta_0$ and some simple alternative $\theta_1 > \theta_0$. The most desirable points for rejection are those for which $r(x) = p_{\theta_1}(x)/p_{\theta_0}(x) = g[T(x)]$ is sufficiently large. If $T(x) < T(x')$, then $r(x) \leq r(x')$ and x' is at least as desirable as x. Thus the test which rejects for large values of $T(x)$ is most powerful. As in the proof of Theorem 1(i), it is seen that there exist C and γ such that (9) and (10) hold. By Theorem 1(ii), the resulting test is also most powerful for testing $P_{\theta'}$ against $P_{\theta''}$ at level $\alpha' = \beta(\theta')$ provided $\theta' < \theta''$. Part (ii) of the present theorem now follows from Corollary 1. Since $\beta(\theta)$ is therefore nondecreasing, the test satisfies

(11)
$$E_\theta\phi(X) \leq \alpha \quad \text{for} \quad \theta \leq \theta_0.$$

The class of tests satisfying (11) is contained in the class satisfying $E_{\theta_0}\phi(X) \leq \alpha$. Since the given test maximizes $\beta(\theta_1)$ within this wider class, it also maximizes $\beta(\theta_1)$ subject to (11); since it is independent of the particular alternative $\theta_1 > \theta_0$ chosen, it is UMP against K.

(iii) is proved by an analogous argument.

(iv) follows from the fact that the test which minimizes the power for testing a simple hypothesis against a simple alternative is obtained by applying the fundamental lemma (Theorem 1) with all inequalities reversed.

By interchanging inequalities throughout, one obtains in an obvious manner the solution of the dual problem, $H: \theta \geq \theta_0$, $K: \theta < \theta_0$.

The proof of (i) and (ii) exhibits the basic property of families with monotone likelihood ratio: every pair of parameter values $\theta_0 < \theta_1$ establishes essentially the same preference order of the sample points (in the sense of the preceding section). A few examples of such families, and hence of UMP one-sided tests, will be given below. However, the main applications of Theorem 2 will come later, when such families appear as the set of conditional distributions given a sufficient statistic (Chapters 4 and 5) and as distributions of a maximal invariant (Chapters 6, 7, and 8).

Example 1. Hypergeometric. From a lot containing N items of a manufactured product, a sample of size n is selected at random, and each item in the sample is inspected. If the total number of defective items in the lot is D, the number X of defectives found in the sample has the *hypergeometric* distribution

$$P\{X = x\} = P_D(x) = \frac{\binom{D}{x}\binom{N - D}{n - x}}{\binom{N}{n}}, \quad \max(0, n + D - N) \le x \le \min(n, D).$$

Interpreting $P_D(x)$ as a density with respect to the measure μ that assigns to any set on the real line as measure the number of integers $0, 1, 2, \ldots$ that it contains, and noting that for values of x within its range

$$\frac{P_{D+1}(x)}{P_D(x)} = \begin{cases} \dfrac{D + 1}{N - D} \dfrac{N - D - n + x}{D + 1 - x} & \text{if} \quad n + D + 1 - N \le x \le D, \\ 0 \text{ or } \infty & \text{if} \quad x = n + D - N \text{ or } D + 1, \end{cases}$$

it is seen that the distributions satisfy the assumption of monotone likelihood ratios with $T(x) = x$. Therefore there exists a UMP test for testing the hypothesis $H : D \le D_0$ against $K : D > D_0$, which rejects H when X is too large, and an analogous test for testing $H' : D \ge D_0$.

An important class of families of distributions that satisfy the assumptions of Theorem 2 are the *one-parameter exponential families*.

Corollary 2. *Let θ be a real parameter, and let X have probability density (with respect to some measure μ)*

$$(12) \qquad\qquad p_\theta(x) = C(\theta)e^{Q(\theta)T(x)}h(x),$$

where Q is strictly monotone. Then there exists a UMP test ϕ for testing $H : \theta \le \theta_0$ against $K : \theta > \theta_0$. If Q is increasing,

$$\phi(x) = 1, \gamma, 0 \qquad as \quad T(x) > , = , < C,$$

where C and γ are determined by $E_{\theta_0}\phi(X) = \alpha$. If Q is decreasing, the inequalities are reversed.

A converse of Corollary 2 is given by Pfanzagl (1968), who shows under weak regularity conditions that the existence of UMP tests against one-sided alternatives for all sample sizes and one value of α implies an exponential family.

As in Example 1, we shall denote the right-hand side of (12) by $P_\theta(x)$ instead of $p_\theta(x)$ when it is a probability, that is, when X is discrete and μ is counting measure.

Example 2. Binomial The binomial distributions $b(p, n)$ with

$$P_p(x) = \binom{n}{x} p^x (1 - p)^{n - x}$$

satisfy (12) with $T(x) = x$, $\theta = p$, $Q(p) = \log[p/(1 - p)]$. The problem of testing $H : p \geq p_0$ arises, for instance, in the situation of Example 1 if one supposes that the production process is in statistical control, so that the various items constitute independent trials with constant probability p of being defective. The number of defectives X in a sample of size n is then a sufficient statistic for the distribution of the variables X_i ($i = 1, \ldots, n$), where X_i is 1 or 0 as the ith item drawn is defective or not, and X is distributed as $b(p, n)$. There exists therefore a UMP test of H, which rejects H when X is too small.

An alternative sampling plan which is sometimes used in binomial situations is *inverse binomial sampling*. Here the experiment is continued until a specified number m of successes—for example, cures effected by some new medical treatment—have been obtained. If Y_i denotes the number of trials after the $(i - 1)$st success up to but not including the ith success, the probability that $Y_i = y$ is pq^y for $y = 0, 1, \ldots$, so that the joint distribution of $Y_1, \ldots, Y_m$ is

$$P_p(y_1, \ldots, y_m) = p^m q^{\Sigma y_i}, \qquad y_k = 0, 1, \ldots, \quad k = 1, \ldots, m.$$

This is an exponential family with $T(y) = \Sigma y_i$ and $Q(p) = \log(1 - p)$. Since $Q(p)$ is a decreasing function of p, the UMP test of $H : p \leq p_0$ rejects H when T is too small. This is what one would expect, since the realization of m successes in only a few more than m trials indicates a high value of p. The test statistic T, which is the number of trials required in excess of m to get m successes, has the negative binomial distribution [Chapter 1, Problem 1(i)]

$$P(t) = \binom{m + t - 1}{m - 1} p^m q^t, \qquad t = 0, 1, \ldots .$$

Example 3. Poisson. If $X_1, \ldots, X_n$ are independent Poisson variables with $E(X_i) = \lambda$, their joint distribution is

$$P_\lambda(x_1, \ldots, x_n) = \frac{\lambda^{x_1 + \cdots + x_n}}{x_1! \cdots x_n!} e^{-n\lambda}.$$

This constitutes an exponential family with $T(x) = \Sigma x_i$, and $Q(\lambda) = \log \lambda$. One-sided hypotheses concerning λ might arise if λ is a bacterial density and the X's are a number of bacterial counts, or if the X's denote the number of α-particles produced in equal time intervals by a radioactive substance, etc. The UMP test of the hypothesis $\lambda \leq \lambda_0$ rejects when ΣX_i is too large. Here the test statistic ΣX_i has itself a Poisson distribution with parameter $n\lambda$.

Instead of observing the radioactive material for given time periods or counting the number of bacteria in given areas of a slide, one can adopt an inverse sampling method. The experiment is then continued, or the area over which the bacteria are counted is enlarged, until a count of m has been obtained. The observations consist of the times $T_1, \ldots, T_m$ that it takes for the first occurrence, from the first to the second, and so on. If one is dealing with a Poisson process and the number of occurrences in a time or space interval τ has the distribution

$$P(x) = \frac{(\lambda \tau)^x}{x!} e^{-\lambda \tau}, \qquad x = 0, 1, \ldots,$$

then the observed times are independently distributed, each with the exponential probability density $\lambda e^{-\lambda t}$ for $t \geq 0$ [Problem 1(ii) of Chapter 1]. The joint densities

$$p_\lambda(t_1, \ldots, t_m) = \lambda^m \exp\left(-\lambda \sum_{i=1}^m t_i\right), \qquad t_1, \ldots, t_m \geq 0,$$

form an exponential family with $T(t_1, \ldots, t_m) = \Sigma t_i$ and $Q(\lambda) = -\lambda$. The UMP test of $H : \lambda \leq \lambda_0$ rejects when $T = \Sigma T_i$ is too small. Since $2\lambda T_i$ has density $\frac{1}{2} e^{-u/2}$ for $u \geq 0$, which is the density of a χ^2-distribution with 2 degrees of freedom, $2\lambda T$ has a χ^2-distribution with $2m$ degrees of freedom. The boundary of the rejection region can therefore be determined from a table of χ^2.

The formulation of the problem of hypothesis testing given at the beginning of the chapter takes account of the losses resulting from wrong decisions only in terms of the two types of error. To obtain a more detailed description of the problem of testing $H : \theta \leq \theta_0$ against the alternatives $\theta > \theta_0$, one can consider it as a decision problem with the decisions d_0 and d_1 of accepting and rejecting H and a loss function $L(\theta, d_i) = L_i(\theta)$. Typically, $L_0(\theta)$ will be 0 for $\theta \leq \theta_0$ and strictly increasing for $\theta \geq \theta_0$, and $L_1(\theta)$ will be strictly decreasing for $\theta \leq \theta_0$ and equal to 0 for $\theta \geq \theta_0$. The difference then satisfies

(13) $$L_1(\theta) - L_0(\theta) \gtrless 0 \qquad \text{as} \quad \theta \lessgtr \theta_0.$$

The following theorem is a special case of complete class results of Karlin and Rubin (1956) and Brown, Cohen, and Strawderman (1976).

Theorem 3.

(i) *Under the assumptions of Theorem 2, the family of tests given by* (9) *and* (10) *with* $0 \leq \alpha \leq 1$ *is essentially complete provided the loss function satisfies* (13).

(ii) *This family is also minimal essentially complete if the set of points x for which $p_\theta(x) > 0$ is independent of θ.*

Proof.   (i): The risk function of any test ϕ is

$$R(\theta, \phi) = \int p_\theta(x)\{\phi(x)L_1(\theta) + [1 - \phi(x)]L_0(\theta)\}\, d\mu(x)$$

$$= \int p_\theta(x)\{L_0(\theta) + [L_1(\theta) - L_0(\theta)]\phi(x)\}\, d\mu(x),$$

and hence the difference of two risk functions is

$$R(\theta, \phi') - R(\theta, \phi) = [L_1(\theta) - L_0(\theta)]\int(\phi' - \phi)p_\theta\, d\mu.$$

This is ≤ 0 for all θ if

$$\beta_{\phi'}(\theta) - \beta_\phi(\theta) = \int(\phi' - \phi)p_\theta\, d\mu \gtreqless 0 \qquad \text{for} \quad \theta \gtreqless \theta_0.$$

Given any test ϕ, let $E_{\theta_0}\phi(X) = \alpha$. It follows from Theorem 2(i) that there exists a UMP level-α test ϕ' for testing $\theta = \theta_0$ against $\theta > \theta_0$, which satisfies (9) and (10). By Theorem 2(iv), ϕ' also minimizes the power for $\theta < \theta_0$. Thus the two risk functions satisfy $R(\theta, \phi') \leq R(\theta, \phi)$ for all θ, as was to be proved.

(ii): Let ϕ_α and $\phi_{\alpha'}$ be of sizes $\alpha < \alpha'$ and UMP for testing θ_0 against $\theta > \theta_0$. Then $\beta_{\phi_\alpha}(\theta) < \beta_{\phi_{\alpha'}}(\theta)$ for all $\theta > \theta_0$ unless $\beta_{\phi_\alpha}(\theta) = 1$. By considering the problem of testing $\theta = \theta_0$ against $\theta < \theta_0$ it is seen analogously that this inequality also holds for all $\theta < \theta_0$ unless $\beta_{\phi_{\alpha'}}(\theta) = 0$. Since the exceptional possibilities are excluded by the assumptions, it follows that $R(\theta, \phi') \lesseqgtr R(\theta, \phi)$ as $\theta \gtreqless \theta_0$. Hence each of the two risk functions is better than the other for some values of θ.

The class of tests previously derived as UMP at the various significance levels α is now seen to constitute an essentially complete class for a much more general decision problem, in which the loss function is only required to satisfy certain broad qualitative conditions. From this point of view, the formulation involving the specification of a level of significance can be considered as a simple way of selecting a particular procedure from an essentially complete family.

The property of monotone likelihood ratio defines a very strong ordering of a family of distributions. For later use, we consider also the following somewhat weaker definition. A family of cumulative distribution functions

F_θ on the real line is said to be *stochastically increasing* (and the same term is applied to random variables possessing these distributions) if the distributions are distinct and if $\theta < \theta'$ implies $F_\theta(x) \geq F_{\theta'}(x)$ for all x. If then X and X' have distributions F_θ and $F_{\theta'}'$ respectively, it follows that $P\{X > x\} \leq P\{X' > x\}$ for all x, so that X' tends to have larger values than X. In this case the variable X' is said to be *stochastically larger* than X. This relationship is made more intuitive by the following characterization of the stochastic ordering of two distributions.

Lemma 1. *Let F_0 and F_1 be two cumulative distribution functions on the real line. Then $F_1(x) \leq F_0(x)$ for all x if and only if there exist two nondecreasing functions f_0 and f_1, and a random variable V, such that* (a) *$f_0(v) \leq f_1(v)$ for all v, and* (b) *the distributions of $f_0(V)$ and $f_1(V)$ are F_0 and F_1 respectively.*

Proof. Suppose first that the required f_0, f_1, and V exist. Then

$$F_1(x) = P\{f_1(V) \leq x\} \leq P\{f_0(V) \leq x\} = F_0(x)$$

for all x. Conversely, suppose that $F_1(x) \leq F_0(x)$ for all x, and let $f_i(y) = \inf\{x : F_i(x - 0) \leq y \leq F_i(x)\}$, $i = 0, 1$. These functions are nondecreasing and for $f_i = f$, $F_i = F$ satisfy

$$f[F(x)] \leq x \text{ and } F[f(y)] \geq y \quad \text{for all } x \text{ and } y.$$

It follows that $y \leq F(x_0)$ implies $f(y) \leq f[F(x_0)] \leq x_0$ and that conversely $f(y) \leq x_0$ implies $F[f(y)] \leq F(x_0)$ and hence $y \leq F(x_0)$, so that the two inequalities $f(y) \leq x_0$ and $y \leq F(x_0)$ are equivalent. Let V be uniformly distributed on $(0, 1)$. Then $P\{f_i(V) \leq x\} = P\{V \leq F_i(x)\} = F_i(x)$. Since $F_1(x) \leq F_0(x)$ for all x implies $f_0(y) \leq f_1(y)$ for all y, this completes the proof.

One of the simplest examples of a stochastically ordered family is a location parameter family, that is, a family satisfying

$$F_\theta(x) = F(x - \theta).$$

To see that this is stochastically increasing, let X be a random variable with distribution $F(x)$. Then $\theta < \theta'$ implies

$$F(x - \theta) = P\{X \leq x - \theta\} \geq P\{X \leq x - \theta'\} = F(x - \theta'),$$

as was to be shown.

Another example is furnished by families with monotone likelihood ratio. This is seen from the following lemma, which establishes some basic properties of these families.

Lemma 2. *Let $p_\theta(x)$ be a family of densities on the real line with monotone likelihood ratio in x.*

(i) *If ψ is a nondecreasing function of x, then $E_\theta\psi(X)$ is a nondecreasing function of θ; if $X_1, \ldots, X_n$ are independently distributed with density p_θ and ψ' is a function of $x_1, \ldots, x_n$ which is nondecreasing in each of its arguments, then $E_\theta\psi'(X_1, \ldots, X_n)$ is a nondecreasing function of θ.*

(ii) *For any $\theta < \theta'$, the cumulative distribution functions of X under θ and θ' satisfy*

$$F_{\theta'}(x) \leq F_\theta(x) \qquad \text{for all } x.$$

(iii) *Let ψ be a function with a single change of sign. More specifically, suppose there exists a value x_0 such that $\psi(x) \leq 0$ for $x < x_0$ and $\psi(x) \geq 0$ for $x \geq x_0$. Then there exists θ_0 such that $E_\theta\psi(X) \leq 0$ for $\theta < \theta_0$ and $E_\theta\psi(X) \geq 0$ for $\theta > \theta_0$, unless $E_\theta\psi(X)$ is either positive for all θ or negative for all θ.*

(iv) *Suppose that $p_\theta(x)$ is positive for all θ and all x, that $p_{\theta'}(x)/p_\theta(x)$ is strictly increasing in x for $\theta < \theta'$, and that $\psi(x)$ is as in (iii) and is $\neq 0$ with positive probability. If $E_{\theta_0}\psi(X) = 0$, then $E_\theta\psi(X) < 0$ for $\theta < \theta_0$ and > 0 for $\theta > \theta_0$.*

Proof. (i): Let $\theta < \theta'$, and let A and B be the sets for which $p_{\theta'}(x) < p_\theta(x)$ and $p_{\theta'}(x) > p_\theta(x)$ respectively. If $a = \sup_A \psi(x)$ and $b = \inf_B \psi(x)$, then $b - a \geq 0$ and

$$\int \psi(p_{\theta'} - p_\theta)\, d\mu \geq a\int_A (p_{\theta'} - p_\theta)\, d\mu + b\int_B (p_{\theta'} - p_\theta)\, d\mu$$

$$= (b - a)\int_B (p_{\theta'} - p_\theta)\, d\mu \geq 0,$$

which proves the first assertion. The result for general n follows by induction.

(ii): This follows from (i) by letting $\psi(x) = 1$ for $x > x_0$ and $\psi(x) = 0$ otherwise.

(iii): We shall show first that for any $\theta' < \theta''$, $E_{\theta'}\psi(X) > 0$ implies $E_{\theta''}\psi(X) \geq 0$. If $p_{\theta''}(x_0)/p_{\theta'}(x_0) = \infty$, then $p_{\theta'}(x) = 0$ for $x \geq x_0$ and hence $E_{\theta'}\psi(X) \leq 0$. Suppose therefore that $p_{\theta''}(x_0)/p_{\theta'}(x_0) = c < \infty$.

Then $\psi(x) \geq 0$ on the set $S = \{x : p_{\theta'}(x) = 0 \text{ and } p_{\theta''}(x) > 0\}$, and

$$E_{\theta''}\psi(X) \geq \int_{\tilde{S}} \psi \frac{p_{\theta''}}{p_{\theta'}} p_{\theta'} \, d\mu$$

$$\geq \int_{-\infty}^{x_0^-} c\psi p_{\theta'} \, d\mu + \int_{x_0}^{\infty} c\psi p_{\theta'} \, d\mu = cE_{\theta'}\psi(X) \geq 0.$$

The result now follows by letting $\theta_0 = \inf\{\theta : E_\theta \psi(X) > 0\}$.

(iv): The proof is analogous to that of (iii).

Part (ii) of the lemma shows that any family of distributions with monotone likelihood ratio in x is stochastically increasing. That the converse does not hold is shown for example by the Cauchy densities

$$\frac{1}{\pi} \frac{1}{1 + (x - \theta)^2}.$$

The family is stochastically increasing, since θ is a location parameter; however, the likelihood ratio is not monotone. Conditions under which a location parameter family possesses monotone likelihood ratio are given in Chapter 9, Example 1.

Lemma 2 is a special case of a theorem of Karlin (1957, 1968) relating the number of sign changes of $E_\theta \psi(X)$ to those of $\psi(x)$ when the densities $p_\theta(x)$ are *totally positive* (defined in Problem 27). The application of totally positive—or equivalently, variation diminishing—distributions to statistics is discussed by Brown, Johnstone, and MacGibbon (1981); see also Problem 30.

4. COMPARISON OF EXPERIMENTS*

Suppose that different experiments are available for testing a simple hypothesis H against a simple alternative K. One experiment results in a random variable X, which has probability densities f and g under H and K respectively; the other leads to the observation of X' with densities f' and g'. Let $\beta(\alpha)$ and $\beta'(\alpha)$ denote the power of the most powerful level-α test based on X and X'. In general, the relationship between $\beta(\alpha)$ and $\beta'(\alpha)$ will depend on α. However, if $\beta'(\alpha) \leq \beta(\alpha)$ for all α, then X or the experiment (f, g) is said to be *more informative* than X'. As an example, suppose that the family of densities $p_\theta(x)$ is the exponential family (12) and

---

*This section constitutes a digression and may be omitted.

that $f = f' = p_{\theta_0}$, $g = p_{\theta_2}$, $g' = p_{\theta_1}$, where $\theta_0 < \theta_1 < \theta_2$. Then (f, g) is more informative than (f', g') by Theorem 2.

A simple sufficient condition* for X to be more informative than X' is the existence of a function $h(x, u)$ and a random quantity U, independent of X and having a known distribution, such that the density of $Y = h(X, U)$ is f' or g' as that of X is f or g. This follows, as in the theory of sufficient statistics, from the fact that one can then construct from X (with the help of U) a variable Y which is equivalent to X'. One can also argue more specifically that if $\phi(x')$ is the most powerful level-α test for testing f' against g' and if $\psi(x) = E\phi[h(x, U)]$, then $E\psi(X) = E\phi(X')$ under both H and K. The test $\psi(x)$ is therefore a level-α test with power $\beta'(\alpha)$, and hence $\beta(\alpha) \geq \beta'(\alpha)$.

When such a transformation h exists, the experiment (f, g) is said to be *sufficient* for (f', g'). If then $X_1, \ldots, X_n$ and $X_1', \ldots, X_n'$ are samples from X and X' respectively, the first of these samples is more informative than the second one. It is also more informative than $(Z_1, \ldots, Z_n)$ where each Z_i is either X_i or X_i' with certain probabilities.

Example 4. 2 × 2 Table. Two characteristics A and B, which each member of a population may or may not possess, are to be tested for independence. The probabilities $p = P(A)$ and $\pi = P(B)$, that is, the proportions of individuals possessing properties A and B, are assumed to be known. This might be the case, for example, if the characteristics have previously been studied separately but not in conjunction. The probabilities of the four possible combinations AB, $A\tilde{B}$, $\tilde{A}B$, and $\tilde{A}\tilde{B}$ under the hypothesis of independence and under the alternative that $P(AB)$ has a specified value ρ are

	Under H:		Under K:	
	B	$\tilde{B}$	B	$\tilde{B}$
A	$p\pi$	$p(1 - \pi)$	ρ	$p - \rho$
$\tilde{A}$	$(1 - p)\pi$	$(1 - p)(1 - \pi)$	$\pi - \rho$	$1 - p - \pi + \rho$

The experimental material is to consist of a sample of size s. This can be selected, for example, at random from those members of the population possessing property A. One then observes for each member of the sample whether or not it possesses property B, and hence is dealing with a sample from a binomial distribution with probabilities

$$H: P(B|A) = \pi \quad \text{and} \quad K: P(B|A) = \frac{\rho}{p}.$$

Alternatively, one can draw the sample from one of the other categories B, $\tilde{B}$, or $\tilde{A}$,

---

*For a proof that this condition is also necessary see Blackwell (1951b).

obtaining in each case a sample from a binomial distribution with probabilities given by the following table:

Population Sampled	Probability	H	K
A	$P(B\|A)$	π	ρ/p
B	$P(A\|B)$	p	ρ/π
$\tilde{B}$	$P(A\|\tilde{B})$	p	$(p - \rho)/(1 - \pi)$
$\tilde{A}$	$P(B\|\tilde{A})$	π	$(\pi - \rho)/(1 - p)$

Without loss of generality let the categories A, $\tilde{A}$, B, and $\tilde{B}$ be labeled so that $p \leq \pi \leq \frac{1}{2}$. We shall now show that of the four experiments, which consist in observing an individual from one of the four categories, the first one (sampling from A) is most informative and in fact is sufficient for each of the others.

To compare A with B, let X and X' be 1 or 0, and let the probabilities of their being equal to 1 be given by the first and the second row of the table respectively. Let U be uniformly distributed on $(0,1)$ and independent of X, and let $Y = h(X, U) = 1$ when $X = 1$ and $U \leq p/\pi$, and $Y = 0$ otherwise. Then $P\{Y = 1\}$ is p under H and ρ/π under K, so that Y has the same distribution as X'. This proves that X is sufficient for X', and hence is the more informative of the two. For the comparison of A with $\tilde{B}$ define Y to be 1 when $X = 0$ and $U \leq p/(1 - \pi)$, and to be 0 otherwise. Then the probability that $Y = 1$ coincides with the third row of the table. Finally, the probability that $Y = 1$ is given by the last row of the table if one defines Y to be equal to 1 when $X = 1$ and $U \leq (\pi - p)/(1 - p)$ and when $X = 0$ and $U > (1 - \pi - p)/(1 - p)$.

It follows from the general remarks preceding the example that if the experimental material is to consist of s individuals, these should be drawn from category A, that is, the rarest of the four categories, in preference to any of the others. This is preferable also to drawing the s from the population at large, since the latter procedure is equivalent to drawing each of them from either A or $\tilde{A}$ with probabilities p and $1 - p$ respectively.

The comparison between these various experiments is independent not only of α but also of ρ. Furthermore, if a sample is taken from A, there exists by Corollary 2 a UMP test of H against the one-sided alternatives of positive dependence, $P(B\|A) > \pi$ and hence $\rho > p\pi$, according to which the probabilities of AB and $\tilde{A}\tilde{B}$ are larger, and those of $A\tilde{B}$ and $\tilde{A}B$ smaller, than under the assumption of independence. This test therefore provides the best power that can be obtained for the hypothesis of independence on the basis of a sample of size s.

Example 5. In a Poisson process the number of events occurring in a time interval of length v has the Poisson distribution $P(\lambda v)$. The problem of testing λ_0 against λ_1 for these distributions arises also for spatial distributions of particles where one is concerned with the number of particles in a region of volume v. To see that the experiment is the more informative the longer the interval v, let $v < w$ and denote by X and Y the number of occurrences in the intervals $(t, t + v)$ and $(t + v, t + w)$. Then X and Y are independent Poisson variables and $Z = X + Y$ is a sufficient statistic for λ. Thus any test based on X can be duplicated by one based on Z, and Z is more informative than X. That it is in fact strictly more informative in an obvious sense is seen from the fact that the unique most powerful test for testing λ_0 against λ_1 depends on $X + Y$ and therefore cannot be duplicated from X alone.

Sometimes it is not possible to count the number of occurrences but only to determine whether or not at least one event has taken place. In the dilution method in bacteriology, for example, a bacterial culture is diluted in a certain volume of water, from which a number of samples of fixed size are taken and tested for the presence or absence of bacteria. In general, one observes then for each of n intervals whether an event occurred. The result is a binomial variable with probability of success (at least one occurrence)

$$p = 1 - e^{-\lambda v}.$$

Since a very large or small interval leads to nearly certain success or failure, one might suspect that for testing λ_0 against λ_1 intermediate values of v would be more informative than extreme ones. However, it turns out that the experiments $(\lambda_0 v, \lambda_1 v)$ and $(\lambda_0 w, \lambda_1 w)$ are not comparable for any values of v and w. (See Problem 19.) For a discussion of how to select v in this and similar situations see Hodges (1949).

The definition of an experiment $\mathscr{E}$ being more informative than an experiment $\mathscr{E}'$ can be extended in a natural way to probability models containing more than two distributions by requiring that for any decision problem a risk function that is obtainable on the basis of $\mathscr{E}'$ can be matched or improved upon by one based on $\mathscr{E}$. Unfortunately, interesting pairs of experiments permitting such a strong ordering are rare. (For an example, see Problems 11 and 12 of Chapter 7). LeCam (1964) initiated a more generally applicable method of comparison by defining a measure of the extent to which one experiment is more informative than another. A survey of some of the principal concepts and results of this theory is given by Torgersen (1976).

5. CONFIDENCE BOUNDS

The theory of UMP one-sided tests can be applied to the problem of obtaining a lower or upper bound for a real-valued parameter θ. The problem of setting a lower bound arises, for example, when θ is the breaking strength of a new alloy; that of setting an upper bound, when θ is the toxicity of a drug or the probability of an undesirable event. The discussion of lower and upper bounds is completely parallel, and it is therefore enough to consider the case of a lower bound, say $\underline{\theta}$.

Since $\underline{\theta} = \underline{\theta}(X)$ will be a function of the observations, it cannot be required to fall below θ with certainty, but only with specified high probability. One selects a number $1 - \alpha$, the *confidence level*, and restricts attention to bounds $\underline{\theta}$ satisfying

(14) $$P_\theta\{\underline{\theta}(X) \leq \theta\} \geq 1 - \alpha \quad \text{for all } \theta.$$

The function $\underline{\theta}$ is called a lower *confidence bound* for θ at confidence level $1 - \alpha$; the infimum of the left-hand side of (14), which in practice will be equal to $1 - \alpha$, is called the *confidence coefficient* of $\underline{\theta}$.

Subject to (14), $\underline{\theta}$ should underestimate θ by as little as possible. One can ask, for example, that the probability of $\underline{\theta}$ falling below any $\theta' < \theta$ should be a minimum. A function $\underline{\theta}$ for which

(15) $$P_{\theta}\{\underline{\theta}(X) \leq \theta'\} = \text{minimum}$$

for all $\theta' < \theta$ subject to (14) is a uniformly *most accurate* lower confidence bound for θ at confidence level $1 - \alpha$.

Let $L(\theta, \underline{\theta})$ be a measure of the loss resulting from underestimating θ, so that for each fixed θ the function $L(\theta, \underline{\theta})$ is defined and nonnegative for $\underline{\theta} < \theta$, and is nonincreasing in its second argument. One would then wish to minimize

(16) $$E_{\theta}L(\theta, \underline{\theta})$$

subject to (14). It can be shown that a uniformly most accurate lower confidence bound $\underline{\theta}$ minimizes (16) subject to (14) for every such loss function L. (See Problem 21.)

The derivation of uniformly most accurate confidence bounds is facilitated by introducing the following more general concept, which will be considered in more detail in Chapter 5. A family of subsets $S(x)$ of the parameter space Ω is said to constitute a family of *confidence sets* at confidence level $1 - \alpha$ if

(17) $$P_{\theta}\{\theta \in S(X)\} \geq 1 - \alpha \qquad \text{for all} \quad \theta \in \Omega,$$

that is, if the random set $S(X)$ covers the true parameter point with probability $\geq 1 - \alpha$. A lower confidence bound corresponds to the special case that $S(x)$ is a one-sided interval

$$S(x) = \{\theta : \underline{\theta}(x) \leq \theta < \infty\}.$$

Theorem 4.

(i)   *For each $\theta_0 \in \Omega$ let $A(\theta_0)$ be the acceptance region of a level-α test for testing $H(\theta_0) : \theta = \theta_0$, and for each sample point x let $S(x)$ denote the set of parameter values*

$$S(x) = \{\theta : x \in A(\theta), \theta \in \Omega\}.$$

Then $S(x)$ is a family of confidence sets for θ at confidence level $1 - \alpha$.

(ii) *If for all θ_0, $A(\theta_0)$ is UMP for testing $H(\theta_0)$ at level α against the alternatives $K(\theta_0)$, then for each θ_0 in Ω, $S(X)$ minimizes the probability*

$$P_\theta\{\theta_0 \in S(X)\} \qquad \text{for all} \quad \theta \in K(\theta_0)$$

among all level-$(1 - \alpha)$ families of confidence sets for θ.

Proof. (i): By definition of $S(x)$,

(18) $\theta \in S(x)$ if and only if $x \in A(\theta)$,

and hence

$$P_\theta\{\theta \in S(X)\} = P_\theta\{X \in A(\theta)\} \geq 1 - \alpha.$$

(ii): If $S^*(x)$ is any other family of confidence sets at level $1 - \alpha$, and if $A^*(\theta) = \{x : \theta \in S^*(x)\}$, then

$$P_\theta\{X \in A^*(\theta)\} = P_\theta\{\theta \in S^*(X)\} \geq 1 - \alpha,$$

so that $A^*(\theta_0)$ is the acceptance region of a level-α test of $H(\theta_0)$. It follows from the assumed property of $A(\theta_0)$ that for any $\theta \in K(\theta_0)$

$$P_\theta\{X \in A^*(\theta_0)\} \geq P_\theta\{X \in A(\theta_0)\}$$

and hence that

$$P_\theta\{\theta_0 \in S^*(X)\} \geq P_\theta\{\theta_0 \in S(X)\},$$

as was to be proved.

The equivalence (18) shows the structure of the confidence sets $S(x)$ as the totality of parameter values θ for which the hypothesis $H(\theta)$ is accepted when x is observed. A confidence set can therefore be viewed as a combined statement regarding the tests of the various hypotheses $H(\theta)$, which exhibits the values for which the hypothesis is accepted [$\theta \in S(x)$] and those for which it is rejected [$\theta \in \overline{S(x)}$].

Corollary 3. *Let the family of densities $p_\theta(x)$, $\theta \in \Omega$, have monotone likelihood ratio in $T(x)$, and suppose that the cumulative distribution function $F_\theta(t)$ of $T = T(X)$ is a continuous function in each of the variables t and θ when the other is fixed.*

(i) *There exists a uniformly most accurate confidence bound $\underline{\theta}$ for θ at each confidence level $1 - \alpha$.*

(ii) *If x denotes the observed values of X and $t = T(x)$, and if the equation*

(19) $$F_\theta(t) = 1 - \alpha$$

has a solution $\theta = \hat{\theta}$ in Ω, then this solution is unique and $\underline{\theta}(x) = \hat{\theta}$.

Proof. (i): There exists for each θ_0 a constant $C(\theta_0)$ such that

$$P_{\theta_0}\{T > C(\theta_0)\} = \alpha,$$

and by Theorem 2, $T > C(\theta_0)$ is a UMP level-α rejection region for testing $\theta = \theta_0$ against $\theta > \theta_0$. By Corollary 1, the power of this test against any alternative $\theta_1 > \theta_0$ exceeds α, and hence $C(\theta_0) < C(\theta_1)$ so that the function C is strictly increasing; it is also continuous. Let $A(\theta_0)$ denote the acceptance region $T \leq C(\theta_0)$, and let $S(x)$ be defined by (18). It follows from the monotonicity of the function C that $S(x)$ consists of those values $\theta \in \Omega$ which satisfy $\underline{\theta} \leq \theta$, where

$$\underline{\theta} = \inf\{\theta : T(x) \leq C(\theta)\}.$$

By Theorem 4, the sets $\{\theta : \underline{\theta}(x) \leq \theta\}$, restricted to possible values of the parameter, thus constitute a family of confidence sets at level $1 - \alpha$, which minimize $P_\theta\{\underline{\theta} \leq \theta'\}$ for all $\theta \in K(\theta')$, that is, for all $\theta > \theta'$. This shows $\underline{\theta}$ to be a uniformly most accurate confidence bound for θ.

(ii): It follows from Corollary 1 that $F_\theta(t)$ is a strictly decreasing function of θ at any point t for which $0 < F_\theta(t) < 1$, and hence that (19) can have at most one solution. Suppose now that t is the observed value of T and that the equation $F_\theta(t) = 1 - \alpha$ has the solution $\hat{\theta} \in \Omega$. Then $F_{\hat{\theta}}(t) = 1 - \alpha$, and by definition of the function C, $C(\hat{\theta}) = t$. The inequality $t \leq C(\theta)$ is then equivalent to $C(\hat{\theta}) \leq C(\theta)$ and hence to $\hat{\theta} \leq \theta$. It follows that $\underline{\theta} = \hat{\theta}$, as was to be proved.

Under the same assumptions, the corresponding upper confidence bound with confidence coefficient $1 - \alpha$ is the solution $\bar{\theta}$ of the equation $P_\theta\{T \geq t\} = 1 - \alpha$ or equivalently of $F_\theta(t) = \alpha$.

Example 6. Exponential waiting times. To determine an upper bound for the degree of radioactivity λ of a radioactive substance, the substance is observed until a count of m has been obtained on a Geiger counter. Under the assumptions of Example 3, the joint probability density of the times $T_i (i = 1, \ldots, m)$ elapsing between the $(i - 1)$st count and the ith one is

$$p(t_1, \ldots, t_m) = \lambda^m e^{-\lambda \Sigma t_i}, \qquad t_1, \ldots, t_m \geq 0.$$

If $T = \Sigma T_i$ denotes the total time of observation, then $2\lambda T$ has a χ^2-distribution with $2m$ degrees of freedom, and, as was shown in Example 3, the acceptance region of the most powerful test of $H(\lambda_0): \lambda = \lambda_0$ against $\lambda < \lambda_0$ is $2\lambda_0 T \leq C$, where C is determined by the equation

$$\int_0^C \chi_{2m}^2 = 1 - \alpha.$$

The set $S(t_1, \ldots, t_m)$ defined by (18) is then the set of values λ such that $\lambda \leq C/2T$, and it follows from Theorem 4 that $\bar{\lambda} = C/2T$ is a uniformly most accurate upper confidence bound for λ. This result can also be obtained through Corollary 3.

If the variables X or T are discrete, Corollary 3 cannot be applied directly, since the distribution functions $F_\theta(t)$ are not continuous, and for most values θ_0 the optimum tests of $H: \theta = \theta_0$ are randomized. However, any randomized test based on X has the following representation as a nonrandomized test depending on X and an independent variable U distributed uniformly over $(0, 1)$. Given a critical function ϕ, consider the rejection region

$$R = \{(x, u): u \leq \phi(x)\}.$$

Then

$$P\{(X, U) \in R\} = P\{U \leq \phi(X)\} = E\phi(X),$$

whatever the distribution of X, so that R has the same power function as ϕ and the two tests are equivalent. The pair of variables (X, U) has a particularly simple representation when X is integer-valued. In this case the statistic

$$T = X + U$$

is equivalent to the pair (X, U), since with probability 1

$$X = [T], \qquad U = T - [T],$$

where $[T]$ denotes the largest integer $\leq T$. The distribution of T is continuous, and confidence bounds can be based on this statistic.

Example 7. Binomial. An upper bound is required for a binomial probability p—for example, the probability that a batch of polio vaccine manufactured according to a certain procedure contains any live virus. Let $X_1, \ldots, X_n$ denote the outcomes of n trials, X_i being 1 or 0 with probabilities p and q respectively, and let $X = \sum X_i$. Then $T = X + U$ has probability density

$$\binom{n}{[t]} p^{[t]} q^{n-[t]}, \qquad 0 \leq t < n + 1.$$

This satisfies the conditions of Corollary 3, and the upper confidence bound $\bar{p}$ is therefore the solution, if it exists, of the equation

$$P_p\{T < t\} = \alpha,$$

where t is the observed value of T. A solution does exist for all values $\alpha \leq t \leq n + \alpha$. For $n + \alpha < t$, the hypothesis $H(p_0): p = p_0$ is accepted against the alternatives $p < p_0$ for all values of p_0 and hence $\bar{p} = 1$. For $t < \alpha$, $H(p_0)$ is rejected for all values of p_0 and the confidence set $S(t)$ is therefore empty. Consider instead the sets $S^*(t)$ which are equal to $S(t)$ for $t \geq \alpha$ and which for $t < \alpha$ consist of the single point $p = 0$. They are also confidence sets at level $1 - \alpha$, since for all p,

$$P_p\{p \in S^*(T)\} \geq P_p\{p \in S(T)\} = 1 - \alpha.$$

On the other hand, $P_p\{p' \in S^*(T)\} = P_p\{p' \in S(T)\}$ for all $p' > 0$ and hence

$$P_p\{p' \in S^*(T)\} = P_p\{p' \in S(T)\} \qquad \text{for all} \quad p' > p.$$

Thus the family of sets $S^*(t)$ minimizes the probability of covering p' for all $p' > p$ at confidence level $1 - \alpha$. The associated confidence bound $\bar{p}^*(t) = \bar{p}(t)$ for $t \geq \alpha$ and $\bar{p}^*(t) = 0$ for $t < \alpha$ is therefore a uniformly most accurate upper confidence bound for p at level $1 - \alpha$.

In practice, so as to avoid randomization and obtain a bound not dependent on the extraneous variable U, one usually replaces T by $X + 1 = [T] + 1$. Since $\bar{p}^*(t)$ is a nondecreasing function of t, the resulting upper confidence bound $\bar{p}^*([t] + 1)$ is then somewhat larger than necessary; as a compensation it also gives a correspondingly higher probability of not falling below the true p.

References to tables for the confidence bounds and a careful discussion of various approximations can be found in Hall (1982) and Blyth (1984).

Let $\underline{\theta}$ and $\bar{\theta}$ be lower and upper bounds for θ with confidence coefficients $1 - \alpha_1$ and $1 - \alpha_2$, and suppose that $\underline{\theta}(x) < \bar{\theta}(x)$ for all x. This will be the case under the assumptions of Corollary 3 if $\alpha_1 + \alpha_2 < 1$. The intervals $(\underline{\theta}, \bar{\theta})$ are then *confidence intervals* for θ with confidence coefficient $1 - \alpha_1 - \alpha_2$; that is, they contain the true parameter value with probability $1 - \alpha_1 - \alpha_2$, since

$$P_\theta\{\underline{\theta} \leq \theta \leq \bar{\theta}\} = 1 - \alpha_1 - \alpha_2 \qquad \text{for all } \theta.$$

If $\underline{\theta}$ and $\bar{\theta}$ are uniformly most accurate, they minimize $E_\theta L_1(\theta, \underline{\theta})$ and $E_\theta L_2(\theta, \bar{\theta})$ at their respective levels for any function L_1 that is nonincreasing in $\underline{\theta}$ for $\underline{\theta} < \theta$ and 0 for $\underline{\theta} \geq \theta$ and any L_2 that is nondecreasing in $\bar{\theta}$ for $\bar{\theta} > \theta$ and 0 for $\bar{\theta} \leq \theta$. Letting

$$L(\theta; \underline{\theta}, \bar{\theta}) = L_1(\theta, \underline{\theta}) + L_2(\theta, \bar{\theta}),$$

the intervals $(\underline{\theta}, \bar{\theta})$ therefore minimize $E_\theta L(\theta; \underline{\theta}, \bar{\theta})$ subject to

$$P_\theta\{\underline{\theta} > \theta\} \leq \alpha_1, \qquad P_\theta\{\bar{\theta} < \theta\} \leq \alpha_2.$$

An example of such a loss function is

$$L(\theta; \underline{\theta}, \bar{\theta}) = \begin{cases} \bar{\theta} - \underline{\theta} & \text{if } \underline{\theta} \leq \theta \leq \bar{\theta}, \\ \bar{\theta} - \theta & \text{if } \theta < \underline{\theta}, \\ \theta - \underline{\theta} & \text{if } \bar{\theta} < \theta, \end{cases}$$

which provides a natural measure of the accuracy of the intervals. Other possible measures are the actual length $\bar{\theta} - \underline{\theta}$ of the intervals, or, for example, $a(\theta - \underline{\theta})^2 + b(\bar{\theta} - \theta)^2$, which gives an indication of the distance of the two end points from the true value.*

An important limiting case corresponds to the levels $\alpha_1 = \alpha_2 = \frac{1}{2}$. Under the assumptions of Corollary 3 and if the region of positive density is independent of θ so that tests of power 1 are impossible when $\alpha < 1$, the upper and lower confidence bounds $\bar{\theta}$ and $\underline{\theta}$ coincide in this case. The common bound satisfies

$$P_\theta\{\underline{\theta} \leq \theta\} = P_\theta\{\underline{\theta} \geq \theta\} = \frac{1}{2},$$

and the estimate $\underline{\theta}$ of θ is therefore as likely to underestimate as to overestimate the true value. An estimate with this property is said to be *median unbiased*. (For the relation of this to other concepts of unbiasedness, see Chapter 1, Problem 3.) It follows from the above result for arbitrary α_1 and α_2 that among all median unbiased estimates, $\underline{\theta}$ minimizes $EL(\theta, \underline{\theta})$ for any *monotone* loss function, that is, any loss function which for fixed θ has a minimum of 0 at $\underline{\theta} = \theta$ and is nondecreasing as $\underline{\theta}$ moves away from θ in either direction. By taking in particular $L(\theta, \underline{\theta}) = 0$ when $|\theta - \underline{\theta}| \leq \Delta$ and $= 1$ otherwise, it is seen that among all median unbiased estimates, $\underline{\theta}$ minimizes the probability of differing from θ by more than any given amount; more generally it maximizes the probability

$$P_\theta\{-\Delta_1 \leq \theta - \underline{\theta} \leq \Delta_2\}$$

for any $\Delta_1, \Delta_2 \geq 0$.

A more detailed assessment of the position of θ than that provided by confidence bounds or intervals corresponding to a fixed level $\gamma = 1 - \alpha$ is obtained by stating confidence bounds for a number of levels, for example

*Proposed by Wolfowitz (1950).

upper confidence bounds corresponding to values such as $\gamma = .05, .1, .25, .5,$ .75, .9, .95. These constitute a set of *standard confidence bounds,** from which different specific intervals or bounds can be obtained in the obvious manner.

6. A GENERALIZATION OF THE FUNDAMENTAL LEMMA

The following is a useful extension of Theorem 1 to the case of more than one side condition.

Theorem 5. *Let* $f_1, \ldots, f_{m+1}$ *be real-valued functions defined on a Euclidean space* $\mathfrak{X}$ *and integrable* μ, *and suppose that for given constants* $c_1, \ldots, c_m$ *there exists a critical function* ϕ *satisfying*

$$(20) \qquad \int \phi f_i \, d\mu = c_i, \qquad i = 1, \ldots, m.$$

Let $\mathscr{C}$ *be the class of critical functions* ϕ *for which* (20) *holds.*

(i)   *Among all members of* $\mathscr{C}$ *there exists one that maximizes*

$$\int \phi f_{m+1} \, d\mu.$$

(ii)   *A sufficient condition for a member of* $\mathscr{C}$ *to maximize*

$$\int \phi f_{m+1} \, d\mu$$

is the existence of constants $k_1, \ldots, k_m$ *such that*

$$(21) \qquad \begin{aligned} \phi(x) &= 1 \quad \text{when} \quad f_{m+1}(x) > \sum_{i=1}^{m} k_i f_i(x), \\[2mm] \phi(x) &= 0 \quad \text{when} \quad f_{m+1}(x) < \sum_{i=1}^{m} k_i f_i(x). \end{aligned}$$

(iii)   *If a member of* $\mathscr{C}$ *satisfies* (21) *with* $k_1, \ldots, k_m \geq 0$, *then it maximizes*

$$\int \phi f_{m+1} \, d\mu$$

*Suggested by Tukey (1949).

among all critical functions satisfying

$$(22) \qquad \int \phi f_i \, d\mu \leq c_i, \qquad i = 1, \ldots, m.$$

(iv) *The set M of points in m-dimensional space whose coordinates are*

$$\left(\int \phi f_1 \, d\mu, \ldots, \int \phi f_m \, d\mu \right)$$

for some critical function ϕ is convex and closed. If $(c_1, \ldots, c_m)$ is an inner point* of M, then there exist constants $k_1, \ldots, k_m$ and a test ϕ satisfying (20) and (21), and a necessary condition for a member of $\mathscr{C}$ to maximize

$$\int \phi f_{m+1} \, d\mu$$

is that (21) holds a.e. μ.

Here the term "inner point of M" in statement (iv) can be interpreted as meaning a point interior to M relative to m-space or relative to the smallest linear space (of dimension $\leq m$) containing M. The theorem is correct with both interpretations but is stronger with respect to the latter, for which it will be proved.

We also note that exactly analogous results hold for the minimization of $\int \phi f_{m+1} \, d\mu$.

Proof. (i): Let $\{\phi_n\}$ be a sequence of functions in $\mathscr{C}$ such that $\int \phi_n f_{m+1} \, d\mu$ tends to $\sup_\phi \int \phi f_{m+1} \, d\mu$. By the weak compactness theorem for critical functions (Theorem 3 of the Appendix), there exists a subsequence $\{\phi_{n_i}\}$ and a critical function ϕ such that

$$\int \phi_{n_i} f_k \, d\mu \to \int \phi f_k \, d\mu \qquad \text{for} \quad k = 1, \ldots, m + 1.$$

It follows that ϕ is in $\mathscr{C}$ and maximizes the integral with respect to $f_{m+1} \, d\mu$ within $\mathscr{C}$.

(ii) and (iii) are proved exactly as was part (ii) of Theorem 1.

(iv): That M is closed follows again from the weak compactness theorem, and its convexity is a consequence of the fact that if ϕ_1 and ϕ_2 are critical functions, so is $\alpha \phi_1 + (1 - \alpha)\phi_2$ for any $0 \leq \alpha \leq 1$. If N (see Figure 2) is

*A discussion of the problem when this assumption is not satisfied is given by Dantzig and Wald (1951).

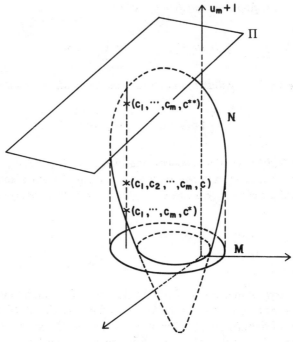

Figure 2

the totality of points in $(m + 1)$-dimensional space with coordinates

$$\left(\int \phi f_1 \, d\mu, \ldots, \int \phi f_{m+1} \, d\mu \right),$$

where ϕ ranges over the class of all critical functions, then N is convex and closed by the same argument. Denote the coordinates of a general point in M and N by $(u_1, \ldots, u_m)$ and $(u_1, \ldots, u_{m+1})$ respectively. The points of N, the first m coordinates of which are $c_1, \ldots, c_m$, form a closed interval $[c^*, c^{**}]$.

Assume first that $c^* < c^{**}$. Since $(c_1, \ldots, c_m, c^{**})$ is a boundary point of N, there exists a hyperplane Π through it such that every point of N lies below or on Π. Let the equation of Π be

$$\sum_{i=1}^{m+1} k_i u_i = \sum_{i=1}^{m} k_i c_i + k_{m+1} c^{**}.$$

Since $(c_1, \ldots, c_m)$ is an inner point of M, the coefficient $k_{m+1} \neq 0$. To see this, let $c^* < c < c^{**}$, so that $(c_1, \ldots, c_m, c)$ is an inner point of N. Then there exists a sphere with this point as center lying entirely in N and hence below Π. It follows that the point $(c_1, \ldots, c_m, c)$ does not lie on Π and hence that $k_{m+1} \neq 0$. We may therefore take $k_{m+1} = -1$ and see that for any point of N

$$u_{m+1} - \sum_{i=1}^{m} k_i u_i \leq c_{m+1}^{**} - \sum_{i=1}^{m} k_i c_i.$$

That is, all critical functions ϕ satisfy

$$\int \phi \left(f_{m+1} - \sum_{i=1}^{m} k_i f_i \right) d\mu \leq \int \phi^{**} \left(f_{m+1} - \sum_{i=1}^{m} k_i f_i \right) d\mu,$$

where ϕ^{**} is the test giving rise to the point $(c_1, \ldots, c_m, c^{**})$. Thus ϕ^{**} is the critical function that maximizes the left-hand side of this inequality. Since the integral in question is maximized by putting ϕ equal to 1 when the integrand is positive and equal to 0 when it is negative, ϕ^{**} satisfies (21) a.e. μ.

If $c^* = c^{**}$, let $(c_1', \ldots, c_m')$ be any point of M other than $(c_1, \ldots, c_m)$. We shall show now that there exists exactly one real number c' such that $(c_1', \ldots, c_m', c')$ is in N. Suppose to the contrary that $(c_1' \ldots, c_m', \underline{c}')$ and $(c_1', \ldots, c_m', \bar{c}')$ are both in N, and consider any point $(c_1'', \ldots, c_m'', c'')$ of N such that $(c_1, \ldots, c_m)$ is an interior point of the line segment joining $(c_1', \ldots, c_m')$ and $(c_1'', \ldots, c_m'')$. Such a point exists since $(c_1, \ldots, c_m)$ is an inner point of M. Then the convex set spanned by the three points $(c_1', \ldots, c_m', \underline{c}')$, $(c_1', \ldots, c_m', \bar{c}')$, and $(c_1'', \ldots, c_m'', c'')$ is contained in N and contains points $(c_1, \ldots, c_m, \underline{c})$ and $(c_1, \ldots, c_m, \bar{c})$ with $\underline{c} < \bar{c}$, which is a contradiction. Since N is convex, contains the origin, and has at most one point on any vertical line $u_1 = c_1', \ldots, u_m = c_m'$, it is contained in a hyperplane, which passes through the origin and is not parallel to the u_{m+1}-axis. It follows that

$$\int \phi f_{m+1} \, d\mu = \sum_{i=1}^{m} k_i \int \phi f_i \, d\mu$$

for all ϕ. This arises of course only in the trivial case that

$$f_{m+1} = \sum_{i=1}^{m} k_i f_i \qquad \text{a.e. } \mu,$$

and (21) is satisfied vacuously.

Corollary 4. *Let* $p_1, \ldots, p_m, p_{m+1}$ *be probability densities with respect to a measure* μ, *and let* $0 < \alpha < 1$. *Then there exists a test* ϕ *such that* $E_i\phi(X) = \alpha$ $(i = 1, \ldots, m)$ *and* $E_{m+1}\phi(X) > \alpha$, *unless* $p_{m+1} = \sum_{i=1}^{m} k_i p_i$, *a.e.* μ.

Proof. The proof will be by induction over m. For $m = 1$ the result reduces to Corollary 1. Assume now that it has been proved for any set of m distributions, and consider the case of $m + 1$ densities $p_1, \ldots, p_{m+1}$. If $p_1, \ldots, p_m$ are linearly dependent, the number of p_i can be reduced and the result follows from the induction hypothesis. Assume therefore that $p_1, \ldots, p_m$ are linearly independent. Then for each $j = 1, \ldots, m$ there exist by the induction hypothesis tests ϕ_j and ϕ'_j such that $E_i\phi_j(X) = E_i\phi'_j(X) = \alpha$ for all $i = 1, \ldots, j - 1, j + 1, \ldots, m$ and $E_j\phi_j(X) < \alpha < E_j\phi'_j(X)$. It follows that the point of m-space for which all m coordinates are equal to α is an inner point of M, so that Theorem 5(iv) is applicable. The test $\phi(x) \equiv \alpha$ is such that $E_i\phi(X) = \alpha$ for $i = 1, \ldots, m$. If among all tests satisfying the side conditions this one is most powerful, it has to satisfy (21). Since $0 < \alpha < 1$, this implies

$$p_{m+1} = \sum_{i=1}^{m} k_i p_i \qquad \text{a.e. } \mu,$$

as was to be proved.

The most useful parts of Theorems 1 and 5 are the parts (ii), which give sufficient conditions for a critical function to maximize an integral subject to certain side conditions. These results can be derived very easily as follows by the method of undetermined multipliers.

Lemma 3. *Let* $F_1, \ldots, F_{m+1}$ *be real-valued functions defined over a space* U, *and consider the problem of maximizing* $F_{m+1}(u)$ *subject to* $F_i(u) = c_i$ $(i = 1, \ldots, m)$. *A sufficient condition for a point* u^0 *satisfying the side conditions to be a solution of the given problem is that among all points of* U *it maximizes*

$$F_{m+1}(u) - \sum_{i=1}^{m} k_i F_i(u)$$

for some $k_1, \ldots, k_m$.

When applying the lemma one usually carries out the maximization for arbitrary k's, and then determines the constants so as to satisfy the side conditions.

Proof. If u is any point satisfying the side conditions, then

$$F_{m+1}(u) - \sum_{i=1}^{m} k_i F_i(u) \le F_{m+1}(u^0) - \sum_{i=1}^{m} k_i F_i(u^0),$$

and hence $F_{m+1}(u) \le F_{m+1}(u^0)$.

As an application consider the problem treated in Theorem 5. Let U be the space of critical functions ϕ, and let $F_i(\phi) = \int \phi f_i \, d\mu$. Then a sufficient condition for ϕ to maximize $F_{m+1}(\phi)$, subject to $F_i(\phi) = c_i$, is that it maximizes $F_{m+1}(\phi) - \Sigma k_i F_i(\phi) = \int (f_{m+1} - \Sigma k_i f_i) \phi \, d\mu$. This is achieved by setting $\phi(x) = 1$ or 0 as $f_{m+1}(x) >$ or $< \Sigma k_i f_i(x)$.

7. TWO-SIDED HYPOTHESES

UMP tests exist not only for one-sided but also for certain two-sided hypotheses of the form

$$(23) \qquad H : \theta \le \theta_1 \text{ or } \theta \ge \theta_2 \qquad (\theta_1 < \theta_2).$$

Such testing problems occur when one wishes to determine whether given specifications have been met concerning the proportion of an ingredient in a drug or some other compound, or whether a measuring instrument, for example a scale, is properly balanced. One then sets up the hypothesis that θ does not lie within the required limits, so that an error of the first kind consists in declaring θ to be satisfactory when in fact it is not. In practice, the decision to accept H will typically be accompanied by a statement of whether θ is believed to be $\le \theta_1$ or $\ge \theta_2$. The implications of H are, however, frequently sufficiently important so that acceptance will in any case be followed by a more detailed investigation. If a manufacturer tests each precision instrument before releasing it and the test indicates an instrument to be out of balance, further work will be done to get it properly adjusted. If in a scientific investigation the inequalities $\theta \le \theta_1$ and $\theta \ge \theta_2$ contradict some assumptions that have been formulated, a more complex theory may be needed and further experimentation will be required. In such situations there may be only two basic choices, to act as if $\theta_1 < \theta < \theta_2$ or to carry out some further investigation, and the formulation of the problem as that of testing the hypothesis H may be appropriate. In the present section the existence of a UMP test of H will be proved for exponential families.

Theorem 6.

(i) *For testing the hypothesis* $H : \theta \le \theta_1$ *or* $\theta \ge \theta_2$ $(\theta_1 < \theta_2)$ *against the alternatives* $K : \theta_1 < \theta < \theta_2$ *in the one-parameter exponential family* (12) *there exists a UMP test given by*

$$\text{(24)} \qquad \phi(x) = \begin{cases} 1 & \text{when} \quad C_1 < T(x) < C_2 \quad (C_1 < C_2), \\ \gamma_i & \text{when} \quad T(x) = C_i, \quad i = 1, 2, \\ 0 & \text{when} \quad T(x) < C_1 \text{ or } > C_2. \end{cases}$$

where the C's and γ's are determined by

$$\text{(25)} \qquad E_{\theta_1}\phi(X) = E_{\theta_2}\phi(X) = \alpha.$$

(ii)   *This test minimizes $E_\theta\phi(X)$ subject to (25) for all $\theta < \theta_1$ and $> \theta_2$.*

(iii)   *For $0 < \alpha < 1$ the power function of this test has a maximum at a point θ_0 between θ_1 and θ_2 and decreases strictly as θ tends away from θ_0 in either direction, unless there exist two values t_1, t_2 such that $P_\theta\{T(X) = t_1\} + P_\theta\{T(X) = t_2\} = 1$ for all θ.*

Proof. (i): One can restrict attention to the sufficient statistic $T = T(X)$, the distribution of which by Lemma 8 of Chapter 2 is

$$dP_\theta(t) = C(\theta) e^{Q(\theta)t} d\nu(t),$$

where $Q(\theta)$ is assumed to be strictly increasing. Let $\theta_1 < \theta' < \theta_2$, and consider first the problem of maximizing $E_{\theta'}\psi(T)$ subject to (25) with $\phi(x) = \psi[T(x)]$. If M denotes the set of all points $(E_{\theta_1}\psi(T), E_{\theta_2}\psi(T))$ as ψ ranges over the totality of critical functions, then the point (α, α) is an inner point of M. This follows from the fact that by Corollary 1 the set M contains points (α, u_1) and (α, u_2) with $u_1 < \alpha < u_2$ and that it contains all points (u, u) with $0 < u < 1$. Hence by part (iv) of Theorem 5 there exist constants k_1, k_2 and a test $\psi_0(t)$ such that $\phi_0(x) = \psi_0[T(x)]$ satisfies (25) and that $\psi_0(t) = 1$ when

$$k_1 C(\theta_1) e^{Q(\theta_1)t} + k_2 C(\theta_2) e^{Q(\theta_2)t} < C(\theta') e^{Q(\theta')t}$$

and therefore when

$$a_1 e^{b_1 t} + a_2 e^{b_2 t} < 1 \qquad (b_1 < 0 < b_2),$$

and $\psi_0(t) = 0$ when the left-hand side is > 1. Here the a's cannot both be ≤ 0, since then the test would always reject. If one of the a's is ≤ 0 and the other one is > 0, then the left-hand side is strictly monotone, and the test is of the one-sided type considered in Corollary 2, which has a strictly monotone power function and hence cannot satisfy (25). Since therefore both a's are positive, the test satisfies (24). It follows from Lemma 4 below that the C's and γ's are uniquely determined by (24) and (25), and hence from Theorem 5(iii) that the test is UMP subject to the weaker restriction $E_{\theta_i}\psi(T) \leq \alpha$ $(i = 1, 2)$. To complete the proof that this test is UMP for testing H, it is necessary to show that it satisfies $E_\theta\psi(T) \leq \alpha$ for $\theta \leq \theta_1$ and $\theta \geq \theta_2$. This follows from (ii) by comparison with the test $\psi(t) \equiv \alpha$.

(ii): Let $\theta' < \theta_1$, and apply Theorem 5(iv) to minimize $E_{\theta'}\phi(X)$ subject to (25). Dividing through by $e^{Q(\theta_1)t}$, the desired test is seen to have a rejection region of the form

$$a_1 e^{b_1 t} + a_2 e^{b_2 t} < 1 \qquad (b_1 < 0 < b_2).$$

Thus it coincides with the test $\psi_0(t)$ obtained in (i). By Theorem 5(iv) the first and third conditions of (24) are also necessary, and the optimum test is therefore unique provided $P\{T = C_i\} = 0$.

(iii): Without loss of generality let $Q(\theta) = \theta$. It follows from (i) and the continuity of $\beta(\theta) = E_\theta\phi(X)$ that either $\beta(\theta)$ satisfies (iii) or there exist three points $\theta' < \theta'' < \theta'''$ such that $\beta(\theta'') \leq \beta(\theta') = \beta(\theta''') = c$, say. Then $0 < c < 1$, since $\beta(\theta') = 0$ (or 1) implies $\phi(t) = 0$ (or 1) a.e. ν and this is excluded by (25). As is seen by the proof of (i), the test maximizes $E_{\theta''}\phi(X)$ subject to $E_{\theta'}\phi(X) = E_{\theta'''}\phi(X) = c$ for all $\theta' < \theta'' < \theta'''$. However, unless T takes on at most two values with probability 1 or all θ, $p_{\theta'}, p_{\theta''}, p_{\theta'''}$ are linearly independent, which by Corollary 4 implies $\beta(\theta'') > c$.

In order to determine the C's and γ's, one will in practice start with some trial values C_1^*, γ_1^*, find C_2^*, γ_2^* such that $\beta^*(\theta_1) = \alpha$, and compute $\beta^*(\theta_2)$, which will usually be either too large or too small. For the selection of the next trial values it is then helpful to note that if $\beta^*(\theta_2) < \alpha$, the correct acceptance region is to the right of the one chosen, that is, it satisfies either $C_1 > C_1^*$ or $C_1 = C_1^*$ and $\gamma_1 < \gamma_1^*$, and that the converse holds if $\beta^*(\theta_2) > \alpha$. This is a consequence of the following lemma.

Lemma 4. *Let $p_\theta(x)$ satisfy the assumptions of Lemma 2(iv).*

(i) *If ϕ and ϕ^* are two tests satisfying (24) and $E_{\theta_1}\phi(T) = E_{\theta_1}\phi^*(T)$, and if ϕ^* is to the right of ϕ, then $\beta(\theta) < $ or $> \beta^*(\theta)$ as $\theta > \theta_1$ or $< \theta_1$.*

(ii) *If ϕ and ϕ^* satisfy (24) and (25), then $\phi = \phi^*$ with probability one.*

Proof. (i): The result follows from Lemma 2(iv) with $\psi = \phi^* - \phi$.

(ii): Since $E_{\theta_1}\phi(T) = E_{\theta_1}\phi^*(T)$, ϕ^* lies either to the left or the right of ϕ, and application of (i) completes the proof.

Although a UMP test exists for testing that $\theta \leq \theta_1$ or $\geq \theta_2$ in an exponential family, the same is not true for the dual hypothesis H: $\theta_1 \leq \theta \leq \theta_2$ or for testing $\theta = \theta_0$ (Problem 31). There do, however, exist UMP unbiased tests of these hypotheses, as will be shown in Chapter 4.

8. LEAST FAVORABLE DISTRIBUTIONS

It is a consequence of Theorem 1 that there always exists a most powerful test for testing a simple hypothesis against a simple alternative. More generally, consider the case of a Euclidean sample space; probability densities f_θ, $\theta \in \omega$, and g with respect to a measure μ; and the problem of testing $H: f_\theta$, $\theta \in \omega$, against the simple alternative $K: g$. The existence of a most powerful level-α test then follows from the weak compactness theorem for critical functions (Theorem 3 of the Appendix) as in Theorem 5(i).

Theorem 1 also provides an explicit construction for the most powerful test in the case of a simple hypothesis. We shall now extend this theorem to composite hypotheses in the direction of Theorem 5 by the method of undetermined multipliers. However, in the process of extension the result becomes much less explicit. Essentially it leaves open the determination of the multipliers, which now take the form of an arbitrary distribution. In specific problems this usually still involves considerable difficulty.

From another point of view the method of attack, as throughout the theory of hypothesis testing, is to reduce the composite hypothesis to a simple one. This is achieved by considering weighted averages of the distributions of H. The composite hypothesis H is replaced by the simple hypothesis H_Λ that the probability density of X is given by

$$h_\Lambda(x) = \int_\omega f_\theta(x) \, d\Lambda(\theta),$$

where Λ is a probability distribution over ω. The problem of finding a suitable Λ is frequently made easier by the following consideration. Since H provides no information concerning θ and since H_Λ is to be equivalent to H for the purpose of testing against g, knowledge of the distribution Λ should provide as little help for this task as possible. To make this precise suppose that θ is known to have a distribution Λ. Then the maximum power β_Λ that can be attained against g is that of the most powerful test ϕ_Λ for testing H_Λ against g. The distribution Λ is said to be *least favorable* (at level α) if for all Λ' the inequality $\beta_\Lambda \leq \beta_{\Lambda'}$ holds.

Theorem 7. *Let a σ-field be defined over ω such that the densities $f_\theta(x)$ are jointly measurable in θ and x. Suppose that over this σ-field there exists a* *probability distribution* Λ *such that the most powerful level-α test* ϕ_Λ *for testing* H_Λ *against g is of size* $\leq \alpha$ *also with respect to the original hypothesis* H.

(i)   *The test* ϕ_Λ *is most powerful for testing H against g.*

(ii)   *If* ϕ_Λ *is the unique most powerful level-α test for testing* H_Λ *against g, it is also the unique most powerful test of H against g.*

(iii)   *The distribution* Λ *is least favorable.*

Proof.   We note first that h_Λ is again a density with respect to μ, since by Fubini's theorem (Theorem 3 of Chapter 2)

$$\int h_\Lambda(x)\, d\mu(x) = \int_\omega d\Lambda(\theta) \int f_\theta(x)\, d\mu(x) = \int_\omega d\Lambda(\theta) = 1.$$

Suppose that ϕ_Λ is a level-α test for testing H, and let ϕ^* be any other level-α test. Then since $E_\theta \phi^*(X) \leq \alpha$ for all $\theta \in \omega$, we have

$$\int \phi^*(x) h_\Lambda(x)\, d\mu(x) = \int_\omega E_\theta \phi^*(X)\, d\Lambda(\theta) \leq \alpha.$$

Therefore ϕ^* is a level-α test also for testing H_Λ and its power cannot exceed that of ϕ_Λ. This proves (i) and (ii). If Λ' is any distribution, it follows further that ϕ_Λ is a level-α test also for testing $H_{\Lambda'}$, and hence that its power against g cannot exceed that of the most powerful test, which by definition is $\beta_{\Lambda'}$.

The conditions of this theorem can be given a somewhat different form by noting that ϕ_Λ can satisfy $\int_\omega E_\theta \phi_\Lambda(X)\, d\Lambda(\theta) = \alpha$ and $E_\theta \phi_\Lambda(X) \leq \alpha$ for all $\theta \in \omega$ only if the set of θ's with $E_\theta \phi_\Lambda(X) = \alpha$ has Λ-measure one.

Corollary 5.   *Suppose that* Λ *is a probability distribution over* ω *and that* ω' *is a subset of* ω *with* $\Lambda(\omega') = 1$. *Let* ϕ_Λ *be a test such that*

$$(26) \qquad \phi_\Lambda(x) = \begin{cases} 1 & \text{if } g(x) > k \int f_\theta(x)\, d\Lambda(\theta), \\[2mm] 0 & \text{if } g(x) < k \int f_\theta(x)\, d\Lambda(\theta). \end{cases}$$

Then ϕ_Λ *is a most powerful level-α test for testing H against g provided*

$$(27) \qquad E_{\theta'} \phi_\Lambda(X) = \sup_{\theta \in \omega} E_\theta \phi_\Lambda(X) = \alpha \qquad \text{for } \theta' \in \omega'.$$

Theorems 2 and 6 constitute two simple applications of Theorem 7. The set ω' over which the least favorable distribution Λ is concentrated consists of the single point θ_0 in the first of these examples and of the two points θ_1 and θ_2 in the second. This is what one might expect, since in both cases these are the distributions of H that appear to be "closest" to K. Another example in which the least favorable distribution is concentrated at a single point is the following.

Example 8. Sign test. The quality of items produced by a manufacturing process is measured by a characteristic X such as the tensile strength of a piece of material, or the length of life or brightness of a light bulb. For an item to be satisfactory X must exceed a given constant u, and one wishes to test the hypothesis $H : p \geq p_0$, where

$$p = P\{ X \leq u \}$$

is the probability of an item being defective. Let $X_1, \ldots, X_n$ be the measurements of n sample items, so that the X's are independently distributed with common distribution about which no knowledge is assumed. Any distribution on the real line can be characterized by the probability p together with the conditional probability distributions P_- and P_+ of X given $X \leq u$ and $X > u$ respectively. If the distributions P_- and P_+ have probability densities p_- and p_+, for example with respect to $\mu = P_- + P_+$, then the joint density of $X_1, \ldots, X_n$ at a sample point $x_1, \ldots, x_n$ satisfying

$$x_{i_1}, \ldots, x_{i_m} \leq u < x_{j_1}, \ldots, x_{j_{n-m}}$$

is

$$p^m (1 - p)^{n-m} p_-(x_{i_1}) \cdots p_-(x_{i_m}) p_+(x_{j_1}) \cdots p_+(x_{j_{n-m}}).$$

Consider now a fixed alternative to H, say (p_1, P_-, P_+), with $p_1 < p_0$. One would then expect the least favorable distribution Λ over H to assign probability 1 to the distribution (p_0, P_-, P_+) since this appears to be closest to the selected alternative. With this choice of Λ, the test (26) becomes

$$\phi_\Lambda(x) = 1 \text{ or } 0 \quad \text{as} \quad \left(\frac{p_1}{p_0}\right)^m \left(\frac{q_1}{q_0}\right)^{n-m} > \text{ or } < C,$$

and hence as $m <$ or $> C$. The test therefore rejects when the number M of defectives is sufficiently small, or more pecisely, when $M < C$ and with probability γ when $M = C$, where

(28) $$P\{ M < C \} + \gamma P\{ M = C \} = \alpha \quad \text{for} \quad p = p_0.$$

The distribution of M is the binomial distribution $b(p, n)$, and does not depend on P_+ and P_-. As a consequence, the power function of the test depends only on p

and is a decreasing function of p, so that under H it takes on its maximum for $p = p_0$. This proves Λ to be least favorable and ϕ_Λ to be most powerful. Since the test is independent of the particular alternative chosen, it is UMP.

Expressed in terms of the variables $Z_i = X_i - u$, the test statistic M is the number of variables ≤ 0, and the test is the so-called *sign test* (cf. Chapter 4, Section 9). It is an example of a *nonparametric* test, since it is derived without assuming a given functional form for the distribution of the X's such as the normal, uniform, or Poisson, in which only certain parameters are unknown.

The above argument applies, with only the obvious modifications, to the case that an item is satisfactory if X lies within certain limits: $u < X < v$. This occurs, for example, if X is the length of a metal part or the proportion of an ingredient in a chemical compound, for which certain tolerances have been specified. More generally the argument applies also to the situation in which X is vector-valued. Suppose that an item is satisfactory only when X lies in a certain set S, for example, if all the dimensions of a metal part or the proportions of several ingredients lie within specified limits. The probability of a defective is then

$$p = P\{ X \in \tilde{S} \},$$

and P_- and P_+ denote the conditional distributions of X given $X \in S$ and $X \in \tilde{S}$ respectively. As before, there exists a UMP test of $H: p \geq p_0$, and it rejects H when the number M of defectives is sufficiently small, with the boundary of the test being determined by (28).

A distribution Λ satisfying the conditions of Theorem 7 exists in most of the usual statistical problems, and in particular under the following assumptions. Let the sample space be Euclidean, let ω be a closed Borel set in s-dimensional Euclidean space, and suppose that $f_\theta(x)$ is a continuous function of θ for almost all x. Then given any g there exists a distribution Λ satisfying the conditions of Theorem 7 provided

$$\lim_{n \to \infty} \int_S f_{\theta_n}(x) \, d\mu(x) = 0$$

for every bounded set S in the sample space and for every sequence of vectors θ_n whose distance from the origin tends to infinity.

From this it follows, as did Corollaries 1 and 4 from Theorems 1 and 5, that if the above conditions hold and if $0 < \alpha < 1$, there exists a test of power $\beta > \alpha$ for testing $H: f_\theta$, $\theta \in \omega$, against g unless $g = \int f_\theta \, d\Lambda(\theta)$ for some Λ. An example of the latter possibility is obtained by letting f_θ and g be the normal densities $N(\theta, \sigma_0^2)$ and $N(0, \sigma_1^2)$ respectively with $\sigma_0^2 < \sigma_1^2$. (See the following section.)

The above and related results concerning the existence and structure of least favorable distributions are given in Lehmann (1952) (with the requirement that ω be closed mistakenly omitted), in Reinhardt (1961), and in Krafft and Witting (1967), where the relation to linear programming is explored.

9. TESTING THE MEAN AND VARIANCE OF A NORMAL DISTRIBUTION

Because of their wide applicability, the problems of testing the mean ξ and variance σ^2 of a normal distribution are of particular importance. Here and in similar problems later, the parameter not being tested is assumed to be unknown, but will not be shown explicitly in a statement of the hypothesis. We shall write, for example, $\sigma \leq \sigma_0$ instead of the more complete statement $\sigma \leq \sigma_0$, $-\infty < \xi < \infty$. The standard (likelihood-ratio) tests of the two hypotheses $\sigma \leq \sigma_0$ and $\xi \leq \xi_0$ are given by the rejection regions

$$(29) \qquad \sum (x_i - \bar{x})^2 \geq C$$

and

$$(30) \qquad \frac{\sqrt{n}\,(\bar{x} - \xi_0)}{\sqrt{\dfrac{1}{n-1}\sum(x_i - \bar{x})^2}} \geq C.$$

The corresponding tests for the hypotheses $\sigma \geq \sigma_0$ and $\xi \geq \xi_0$ are obtained from the rejection regions (29) and (30) by reversing the inequalities. As will be shown in later chapters, these four tests are UMP both within the class of unbiased and within the class of invariant tests (but see Chapter 5, Section 4 for problems arising when the assumption of normality does not hold exactly). However, at the usual significance levels only the first of them is actually UMP.

Let $X_1, \ldots, X_n$ be a sample from $N(\xi, \sigma^2)$, and consider first the hypotheses $H_1: \sigma \geq \sigma_0$ and $H_2: \sigma \leq \sigma_0$, and a simple alternative $K: \xi = \xi_1$, $\sigma = \sigma_1$. It seems reasonable to suppose that the least favorable distribution Λ in the (ξ, σ)-plane is concentrated on the line $\sigma = \sigma_0$. Since $Y = \sum X_i / n = \bar{X}$ and $U = \sum (X_i - \bar{X})^2$ are sufficient statistics for the parameters (ξ, σ), attention can be restricted to these variables. Their joint density under H_Λ is

$$C_0 u^{(n-3)/2} \exp\left(-\frac{u}{2\sigma_0^2}\right) \int \exp\left[-\frac{n}{2\sigma_0^2}(y - \xi)^2\right] d\Lambda(\xi),$$

while under K it is

$$C_1 u^{(n-3)/2} \exp\left(-\frac{u}{2\sigma_1^2}\right) \exp\left[-\frac{n}{2\sigma_1^2}(y - \xi_1)^2\right].$$

The choice of Λ is seen to affect only the distribution of Y. A least favorable Λ should therefore have the property that the density of Y under H_Λ,

$$\int \frac{\sqrt{n}}{\sqrt{2\pi\sigma_0^2}} \exp\left[-\frac{n}{2\sigma_0^2}(y - \xi)^2 \right] d\Lambda(\xi),$$

comes as close as possible to the alternative density,

$$\frac{\sqrt{n}}{\sqrt{2\pi\sigma_1^2}} \exp\left[-\frac{n}{2\sigma_1^2}(y - \xi_1)^2 \right].$$

At this point one must distinguish between H_1 and H_2. In the first case $\sigma_1 < \sigma_0$. By suitable choice of Λ the mean of Y can be made equal to ξ_1, but the variance will if anything be increased over its initial value σ_0^2. This suggests that the least favorable distribution assigns probability 1 to the point $\xi = \xi_1$, since in this way the distribution of Y is normal both under H and K with the same mean in both cases and the smallest possible difference between the variances. The situation is somewhat different for H_2, for which $\sigma_0 < \sigma_1$. If the least favorable distribution Λ has a density, say Λ', the density of Y under H_Λ becomes

$$\int_{-\infty}^{\infty} \frac{\sqrt{n}}{\sqrt{2\pi\sigma_0}} \exp\left[-\frac{n}{2\sigma_0^2}(y - \xi)^2 \right] \Lambda'(\xi)\, d\xi.$$

This is the probability density of the sum of two independent random variables, one distributed as $N(0, \sigma_0^2/n)$ and the other with density $\Lambda'(\xi)$. If Λ is taken to be $N(\xi_1, (\sigma_1^2 - \sigma_0^2)/n)$, the distribution of Y under H_Λ becomes $N(\xi_1, \sigma_1^2/n)$, the same as under K.

We now apply Corollary 5 with the distributions Λ suggested above. For H_1 it is more convenient to work with the original variables than with Y and U. Substitution in (26) gives $\phi(x) = 1$ when

$$\frac{\left(2\pi\sigma_1^2\right)^{-n/2} \exp\left[-\frac{1}{2\sigma_1^2} \sum (x_i - \xi_1)^2 \right]}{\left(2\pi\sigma_0^2\right)^{-n/2} \exp\left[-\frac{1}{2\sigma_0^2} \sum (x_i - \xi_1)^2 \right]} > C,$$

that is, when

$$(31) \qquad \sum (x_i - \xi_1)^2 \le C.$$

To justify the choice of Λ, one must show that

$$P\left\{ \sum (X_i - \xi_1)^2 \le C \,\middle|\, \xi, \sigma \right\}$$

takes on its maximum over the half plane $\sigma \ge \sigma_0$ at the point $\xi = \xi_1$, $\sigma = \sigma_0$. For any fixed σ, the above is the probability of the sample point falling in a sphere of fixed radius, computed under the assumption that the X's are independently distributed as $N(\xi, \sigma^2)$. This probability is maximized when the center of the sphere coincides with that of the distribution, that is, when $\xi = \xi_1$. (This follows for example from Problem 25 of Chapter 7.) The probability then becomes

$$P\left\{ \sum \left(\frac{X_i - \xi_1}{\sigma} \right)^2 \le \frac{C}{\sigma^2} \,\middle|\, \xi_1, \sigma \right\} = P\left\{ \sum V_i^2 \le \frac{C}{\sigma^2} \right\},$$

where $V_1, \ldots, V_n$ are independently distributed as $N(0, 1)$. This is a decreasing function of σ and therefore takes on its maximum when $\sigma = \sigma_0$.

In the case of H_2, application of Corollary 5 to the sufficient statistics (Y, U) gives $\phi(y, u) = 1$ when

$$\frac{C_1 u^{(n-3)/2} \exp\left(-\dfrac{u}{2\sigma_1^2} \right) \exp\left[-\dfrac{n}{2\sigma_1^2} (y - \xi_1)^2 \right]}{C_0 u^{(n-3)/2} \exp\left(-\dfrac{u}{2\sigma_0^2} \right) \int \exp\left[-\dfrac{n}{2\sigma_0^2} (y - \xi)^2 \right] \Lambda'(\xi)\, d\xi}$$

$$= C' \exp\left[-\frac{u}{2}\left(\frac{1}{\sigma_1^2} - \frac{1}{\sigma_0^2} \right) \right] \ge C,$$

that is, when

$$(32) \qquad u = \sum (x_i - \bar{x})^2 \ge C.$$

Since the distribution of $\sum (X_i - \bar{X})^2 / \sigma^2$ does not depend on ξ or σ, the probability $P\{ \sum (X_i - \bar{X})^2 \ge C \mid \xi, \sigma \}$ is independent of ξ and increases with σ, so that the conditions of Corollary 5 are satisfied. The test (32), being independent of ξ_1 and σ_1, is UMP for testing $\sigma \le \sigma_0$ against $\sigma > \sigma_0$. It is also seen to coincide with the likelihood-ratio test (29). On the other hand, the most powerful test (31) for testing $\sigma \ge \sigma_0$ against $\sigma < \sigma_0$ does depend on the value ξ_1 of ξ under the alternative.

It has been tacitly assumed so far that $n > 1$. If $n = 1$, the argument applies without change with respect to H_1, leading to (31) with $n = 1$. However, in the discussion of H_2 the statistic U now drops out, and Y coincides with the single observation X. Using the same Λ as before, one sees that X has the same distribution under H_Λ as under K, and the test ϕ_Λ therefore becomes $\phi_\Lambda(x) \equiv \alpha$. This satisfies the conditions of Corollary 5 and is therefore the most powerful test for the given problem. It follows that a single observation is of no value for testing the hypothesis H_2, as seems intuitively obvious, but that it could be used to test H_1 if the class of alternatives were sufficiently restricted.

The corresponding derivation for the hypothesis $\xi \le \xi_0$ is less straight-forward. It turns out* that Student's test given by (30) is most powerful if the level of significance α is $\ge \frac{1}{2}$, regardless of the alternative $\xi_1 > \xi_0, \sigma_1$. This test is therefore UMP for $\alpha \ge \frac{1}{2}$. On the other hand, when $\alpha < \frac{1}{2}$ the most powerful test of H rejects when $\Sigma(x_i - a)^2 \le b$, where the constants a and b depend on the alternative (ξ_1, σ_1) and on α. Thus for the significance levels that are of interest, a UMP test of H does not exist. No new problem arises for the hypothesis $\xi \ge \xi_0$, since this reduces to the case just considered through the transformation $Y_i = \xi_0 - (X_i - \xi_0)$.

10. PROBLEMS

Section 2

1. Let $X_1, \ldots, X_n$ be a sample from the normal distribution $N(\xi, \sigma^2)$.

   (i) If $\sigma = \sigma_0$ (known), there exists a UMP test for testing $H: \xi \le \xi_0$ against $\xi > \xi_0$, which rejects when $\Sigma(X_i - \xi_0)$ is too large.

   (ii) If $\xi = \xi_0$ (known), there exists a UMP test for testing $H: \sigma \le \sigma_0$ against $K: \sigma > \sigma_0$, which rejects when $\Sigma(X_i - \xi_0)^2$ is too large.

2. *UMP test for* $U(0, \theta)$. Let $X = (X_1, \ldots, X_n)$ be a sample from the uniform distribution on $(0, \theta)$.

   (i) For testing $H: \theta \le \theta_0$ against $K: \theta > \theta_0$ any test is UMP at level α for which $E_{\theta_0}\phi(X) = \alpha$, $E_\theta\phi(X) \le \alpha$ for $\theta \le \theta_0$, and $\phi(x) = 1$ when $\max(x_1, \ldots, x_n) > \theta_0$.

   (ii) For testing $H: \theta = \theta_0$ against $K: \theta \ne \theta_0$ a unique UMP test exists, and is given by $\phi(x) = 1$ when $\max(x_1, \ldots, x_n) > \theta_0$ or $\max(x_1, \ldots, x_n) \le \theta_0 \sqrt[n]{\alpha}$, and $\phi(x) = 0$ otherwise.

*See Lehmann and Stein (1948).

[(i): For each $\theta > \theta_0$ determine the ordering established by $r(x) = p_\theta(x)/p_{\theta_0}(x)$ and use the fact that many points are equivalent under this ordering.
(ii): Determine the UMP tests for testing $\theta = \theta_0$ against $\theta < \theta_0$ and combine this result with that of part (i).]

3. *UMP test for exponential densities.* Let $X_1, \ldots, X_n$ be a sample from the exponential distribution $E(a, b)$ of Chapter 1, Problem 18, and let $X_{(1)} = \min(X_1, \ldots, X_n)$.

   (i)   Determine the UMP test for testing $H: a = a_0$ against $K: a \neq a_0$ when b is assumed known.

   (ii) The power of any MP level-α test of $H: a = a_0$ against $K: a = a_1 < a_0$ is given by

$$\beta^*(a_1) = 1 - (1 - \alpha) e^{-n(a_0 - a_1)/b}.$$

   (iii) For the problem of part (i), when b is unknown, the power of any level α test which rejects when

$$\frac{X_{(1)} - a_0}{\Sigma [X_i - X_{(1)}]} \leq C_1 \text{ or } \geq C_2$$

against any alternative (a_1, b) with $a_1 < a_0$ is equal to $\beta^*(a_1)$ of part (ii) (independent of the particular choice of C_1 and C_2).

   (iv) The test of part (iii) is a UMP level-α test of $H: a = a_0$ against $K: a \neq a_0$ (b unknown).

   (v)   Determine the UMP test for testing $H: a = a_0$, $b = b_0$ against the alternatives $a < a_0$, $b < b_0$.

   (vi) Explain the (very unusual) existence in this case of a UMP test in the presence of a nuisance parameter [part (iv)] and for a hypothesis specifying two parameters [part (v)].

[(i): the variables $Y_i = e^{-X_i/b}$ are a sample from the uniform distribution on $(0, e^{-a/b})$.]

Note. For more general versions of parts (ii)–(iv) see Takeuchi (1969) and Kabe and Laurent (1981).

4. The following example shows that the power of a test can sometimes be increased by selecting a random rather than a fixed sample size even when the randomization does not depend on the observations. Let $X_1, \ldots, X_n$ be independently distributed as $N(\theta, 1)$, and consider the problem of testing $H: \theta = 0$ against $K: \theta = \theta_1 > 0$.

   (i)   The power of the most powerful test as a function of the sample size n is not necessarily concave.

(ii)   In particular for $\alpha = .005$, $\theta_1 = \frac{1}{2}$, better power is obtained by taking 2 or 16 observations with probability $\frac{1}{2}$ each than by taking a fixed sample of 9 observations.

(iii) The power can be increased further if the test is permitted to have different significance levels α_1 and α_2 for the two sample sizes and it is required only that the expected significance level be equal to $\alpha = .005$. Examples are: (a) with probability $\frac{1}{2}$ take $n_1 = 2$ observations and perform the test of significance at level $\alpha_1 = .001$, or take $n_2 = 16$ observations and perform the test at level $\alpha_2 = .009$; (b) with probability $\frac{1}{2}$ take $n_1 = 0$ or $n_2 = 18$ observations and let the respective significance levels be $\alpha_1 = 0$, $\alpha_2 = .01$.

Note.   This and related examples were discussed by Kruskal in a seminar held at Columbia University in 1954. A more detailed investigation of the phenomenon has been undertaken by Cohen (1958).

5. If the sample space $\mathcal{X}$ is Euclidean and P_0, P_1 have densities with respect to Lebesgue measure, there exists a nonrandomized most powerful test for testing P_0 against P_1 at every significance level α.*
    [This is a consequence of Theorem 1 and the following lemma.† Let $f \geq 0$ and $\int_A f(x)\, dx = a$. Given any $0 \leq b \leq a$, there exists a subset B of A such that $\int_B f(x)\, dx = b$.]

6. *Fully informative statistics.* A statistic T is *fully informative* if for every decision problem the decision procedures based only on T form an essentially complete class. If $\mathcal{P}$ is dominated and T is fully informative, then T is sufficient.
    [Consider any pair of distributions P_0, $P_1 \in \mathcal{P}$ with densities p_0, p_1, and let $g_i = p_i/(p_0 + p_1)$. Suppose that T is fully informative, and let $\mathcal{A}_0$ be the subfield induced by T. Then $\mathcal{A}_0$ contains the subfield induced by (g_0, g_1) since it contains every rejection region which is unique most powerful for testing P_0 against P_1 (or P_1 against P_0) at some level α. Therefore, T is sufficient for every pair of distributions (P_0, P_1), and hence by Problem 10 of Chapter 2 it is sufficient for $\mathcal{P}$.]

Section 3

7. Let X be the number of successes in n independent trials with probability p of success, and let $\phi(x)$ be the UMP test (9) for testing $p \leq p_0$ against $p > p_0$ at level of significance α.

    (i)   For $n = 6$, $p_0 = .25$ and the levels $\alpha = .05, .1, .2$ determine C and γ, and find the power of the test against $p_1 = .3, .4, .5, .6, .7$.

---

*For more general results concerning the possibility of dispensing with randomized procedures, see Dvoretzky, Wald, and Wolfowitz (1951).

†For a proof of this lemma see Halmos (1974, p. 174.) The lemma is a special case of a theorem of Lyapounov (see Blackwell (1951a).)

(ii) If $p_0 = .2$ and $\alpha = .05$, and it is desired to have power $\beta \geq .9$ against $p_1 = .4$, determine the necessary sample size (a) by using tables of the binomial distribution, (b) by using the normal approximation.*

(iii) Use the normal approximation to determine the sample size required when $\alpha = .05$, $\beta = .9$, $p_0 = .01$, $p_1 = .02$.

8. (i) A necessary and sufficient condition for densities $p_\theta(x)$ to have monotone likelihood ratio in x, if the mixed second derivative $\partial^2 \log p_\theta(x)/\partial\theta\,\partial x$ exists, is that this derivative is ≥ 0 for all θ and x.

(ii) An equivalent condition is that

$$p_\theta(x)\frac{\partial^2 p_\theta(x)}{\partial\theta\,\partial x} \geq \frac{\partial p_\theta(x)}{\partial\theta}\frac{\partial p_\theta(x)}{\partial x} \qquad \text{for all } \theta \text{ and } x.$$

9. Let the probability density p_θ of X have monotone likelihood ratio in $T(x)$, and consider the problem of testing $H: \theta \leq \theta_0$ against $\theta > \theta_0$. If the distribution of T is continuous, the p-value $\hat{\alpha}$ of the UMP test is given by $\hat{\alpha} = P_{\theta_0}\{T \geq t\}$, where t is the observed value of T. This holds also without the assumption of continuity if for randomized tests $\hat{\alpha}$ is defined as the smallest significance level at which the hypothesis is rejected with probability 1.

10. Let $X_1, \ldots, X_n$ be independently distributed with density $(2\theta)^{-1}e^{-x/2\theta}$, $x \geq 0$, and let $Y_1 \leq \cdots \leq Y_n$ be the ordered X's. Assume that Y_1 becomes available first, then Y_2, and so on, and that observation is continued until Y_r has been observed. On the basis of $Y_1, \ldots, Y_r$ it is desired to test $H: \theta \geq \theta_0 = 1000$ at level $\alpha = .05$ against $\theta < \theta_0$.

(i) Determine the rejection region when $r = 4$, and find the power of the test against $\theta_1 = 500$.

(ii) Find the value of r required to get power $\beta \geq .95$ against this alternative.

[In Problem 14, Chapter 2, the distribution of $[\Sigma_{i=1}^r Y_i + (n-r)Y_r]/\theta$ was found to be χ^2 with $2r$ degrees of freedom.]

11. When a Poisson process with rate λ is observed for a time interval of length τ, the number X of events occurring has the Poisson distribution $P(\lambda\tau)$. Under an alternative scheme, the process is observed until r events have occurred, and the time T of observation is then a random variable such that $2\lambda T$ has a χ^2-distribution with $2r$ degrees of freedom. For testing $H: \lambda \leq \lambda_0$ at level α one can, under either design, obtain a specified power β against an alternative λ_1 by choosing τ and r sufficiently large.

(i) The ratio of the time of observation required for this purpose under the first design to the expected time required under the second is $\lambda\tau/r$.

(ii) Determine for which values of λ each of the two designs is preferable when $\lambda_0 = 1$, $\lambda_1 = 2$, $\alpha = .05$, $\beta = .9$.

*Tables and approximations are discussed, for example, in Chapter 3 of Johnson and Kotz (1969).

12. Let $X = (X_1, \ldots, X_n)$ be a sample from the uniform distribution $U(\theta, \theta + 1)$.

   (i) For testing $H: \theta \leq \theta_0$ against $K: \theta > \theta_0$ at level α there exists a UMP test which rejects when $\min(X_1, \ldots, X_n) > \theta_0 + C(\alpha)$ or $\max(X_1, \ldots, X_n) > \theta_0 + 1$ for suitable $C(\alpha)$.

   (ii) The family $U(\theta, \theta + 1)$ does not have monotone likelihood ratio. [Additional results for this family are given in Birnbaum (1954) and Pratt (1958).]

   [(ii) By Theorem 2, monotone likelihood ratio implies that the family of UMP tests of $H: \theta \leq \theta_0$ against $K: \theta > \theta_0$ generated as α varies from 0 to 1 is independent of θ_0].

13. Let X be a single observation from the Cauchy density given at the end of Section 3.

   (i) Show that no UMP test exists for testing $\theta = 0$ against $\theta > 0$.

   (ii) Determine the totality of different shapes the MP level-α rejection region for testing $\theta = \theta_0$ against $\theta = \theta_1$ can take on for varying α and $\theta_1 - \theta_0$.

14. *Extension of Lemma 2.* Let P_0 and P_1 be two distributions with densities p_0, p_1 such that $p_1(x)/p_0(x)$ is a nondecreasing function of a real-valued statistic $T(x)$.

   (i)   If T has probability density p_i' when the original distribution is P_i, then $p_1'(t)/p_0'(t)$ is nondecreasing in t.

   (ii) $E_0\psi(T) \leq E_1\psi(T)$ for any nondecreasing function ψ.

   (iii) If $p_1(x)/p_0(x)$ is a strictly increasing function of $t = T(x)$, so is $p_1'(t)/p_0'(t)$, and $E_0\psi(T) < E_1\psi(T)$ unless $\psi[T(x)]$ is constant a.e. $(P_0 + P_1)$ or $E_0\psi(T) = E_1\psi(T) = \pm\infty$.

   (iv) For any distinct distributions with densities p_0, p_1,

$$-\infty \leq E_0\log\left[\frac{p_1(X)}{p_0(X)}\right] < E_1\log\left[\frac{p_1(X)}{p_0(X)}\right] \leq \infty.$$

[(i): Without loss of generality suppose that $p_1(x)/p_0(x) = T(x)$. Then for any integrable ϕ,

$$\int \phi(t)p_1'(t)\, d\nu(t) = \int \phi[T(x)]T(x)p_0(x)\, d\mu(x) = \int \phi(t)tp_0'(t)\, d\nu(t),$$

and hence $p_1'(t)/p_0'(t) = t$ a.e.

(iv): The possibility $E_0\log[p_1(X)/p_0(X)] = \infty$ is excluded, since by the convexity of the function log,

$$E_0\log\left[\frac{p_1(X)}{p_0(X)}\right] \leq \log E_0\left[\frac{p_1(X)}{p_0(X)}\right] = 0.$$

Similarly for E_1. The strict inequality now follows from (iii) with $T(x) = p_1(x)/p_0(x)$.]

15. If F_0, F_1 are two cumulative distribution functions on the real line, then $F_1(x) \leq F_0(x)$ for all x if and only if $E_0\psi(X) \leq E_1\psi(X)$ for any nondecreasing function ψ.

Section 4

16. If the experiment (f, g) is more informative than (f', g'), then (g, f) is more informative than (g', f').

17. *Conditions for comparability.*

   (i) Let X and X' be two random variables taking on the values 1 and 0, and suppose that $P\{X = 1\} = p_0$, $P\{X' = 1\} = p_0'$ or that $P\{X = 1\} = p_1$, $P\{X' = 1\} = p_1'$. Without loss of generality let $p_0 < p_0'$, $p_0 < p_1$, $p_0' < p_1'$. (This can be achieved by exchanging X with X' and by exchanging the values 0 and 1 of one or both of the variables.) Then X is more informative than X' if and only if $(1 - p_1)(1 - p_0') \leq (1 - p_0)(1 - p_1')$.

   (ii) Let U_0, U_1 be independently uniformly distributed over $(0,1)$, and let $Y = 1$ if $X = 1$ and $U_1 \leq \gamma_1$ and if $X = 0$ and $U_0 \leq \gamma_0$ and $Y = 0$ otherwise. Under the assumptions of (i) there exist $0 \leq \gamma_0, \gamma_1 \leq 1$ such that $P\{Y = 1\} = p_i'$ when $P\{X = 1\} = p_i$ $(i = 0, 1)$ provided $(1 - p_1)(1 - p_0') \leq (1 - p_0)(1 - p_1')$. This inequality, which is therefore sufficient for a sample $X_1, \ldots, X_n$ from X to be more informative than a sample $X_1', \ldots, X_n'$ from X', is also necessary. Similarly, the condition $p_0'p_1 \leq p_0p_1'$ is necessary and sufficient for a sample from X' to be more informative than one from X.

   [(i): The power $\beta(\alpha)$ of the most powerful level-α test of p_0 against p_1 based on X is $\alpha p_1/p_0$ if $\alpha \leq p_0$, and $p_1 + q_1 q_0^{-1}(\alpha - p_0)$ if $p_0 \leq \alpha$. One obtains the desired result by comparing the graphs of $\beta(\alpha)$ and $\beta'(\alpha)$.
   (ii): The last part of (ii) follows from a comparison of the power $\beta_n(\alpha)$ and $\beta_n'(\alpha)$ of the most powerful level α tests based on ΣX_i and $\Sigma X_i'$ for α close to 1. The dual condition is obtained from Problem 16.]

18. For the 2×2 table described in Example 4, and under the assumption $p \leq \pi \leq \frac{1}{2}$ made there, a sample from $\tilde{B}$ is more informative than one from $\tilde{A}$. On the other hand, samples from B and $\tilde{B}$ are not comparable.
   [A necessary and sufficient condition for comparability is given in the preceding problem.]

19. In the experiment discussed in Example 5, n binomial trials with probability of success $p = 1 - e^{-\lambda v}$ are performed for the purpose of testing $\lambda = \lambda_0$ against $\lambda = \lambda_1$. Experiments corresponding to two different values of v are not comparable.

Section 5

20. (i) For $n = 5$, 10 and $1 - \alpha = .95$, graph the upper confidence limits $\bar{p}$ and $\bar{p}^*$ of Example 7 as functions of $t = x + u$.

    (ii) For the same values of n and $\alpha_1 = \alpha_2 = .05$, graph the lower and upper confidence limits $\underline{p}$ and $\bar{p}$.

21. *Confidence bounds with minimum risk.* Let $L(\theta, \underline{\theta})$ be nonnegative and nonincreasing in its second argument for $\underline{\theta} < \theta$, and equal to 0 for $\underline{\theta} \geq \theta$. If $\underline{\theta}$ and $\underline{\theta}^*$ are two lower confidence bounds for θ such that

$$P_\theta \{ \underline{\theta} \leq \theta' \} \leq P_\theta \{ \underline{\theta}^* \leq \theta' \} \qquad \text{for all} \quad \theta' \leq \theta,$$

then

$$E_\theta L(\theta, \underline{\theta}) \leq E_\theta L(\theta, \underline{\theta}^*).$$

[Define two cumulative distribution functions F and F^* by $F(u) = P_\theta \{ \underline{\theta} \leq u \} / P_\theta \{ \underline{\theta}^* \leq \theta \}$, $F^*(u) = P_\theta \{ \underline{\theta}^* \leq u \} / P_\theta \{ \underline{\theta}^* \leq \theta \}$ for $u < \theta$, and $F(u) = F^*(u) = 1$ for $u \geq \theta$. Then $F(u) \leq F^*(u)$ for all u, and it follows from Problem 15 that

$$E_\theta [L(\theta, \underline{\theta})] = P_\theta \{ \underline{\theta}^* \leq \theta \} \int L(\theta, u) \, dF(u)$$

$$\leq P_\theta \{ \underline{\theta}^* \leq \theta \} \int L(\theta, u) \, dF^*(u) = E_\theta [L(\theta, \underline{\theta}^*)].]$$

Section 6

22. If $\beta(\theta)$ denotes the power function of the UMP test of Corollary 2, and if the function Q of (12) is differentiable, then $\beta'(\theta) > 0$ for all θ for which $Q'(\theta) > 0$.

    [To show that $\beta'(\theta_0) > 0$, consider the problem of maximizing, subject to $E_{\theta_0} \phi(X) = \alpha$, the derivative $\beta'(\theta_0)$ or equivalently the quantity $E_{\theta_0} [T(X) \phi(X)]$.]

23. *Optimum selection procedures.* On each member of a population n measurements $(X_1, \ldots, X_n) = X$ are taken, for example the scores of n aptitude tests which are administered to judge the qualifications of candidates for a certain training program. A future measurement Y such as the score in a final test at the end of the program is of interest but unavailable. The joint distribution of X and Y is assumed known.

    (i) One wishes to select a given proportion α of the candidates in such a way as to maximize the expectation of Y for the selected group. This is achieved by selecting the candidates for which $E(Y|x) \geq C$, where C is determined by the condition that the probability of a member being selected is α. When $E(Y|x) = C$, it may be necessary to randomize in order to get the exact value α.

(ii) If instead the problem is to maximize the probability with which in the selected population Y is greater than or equal to some preassigned score y_0, one selects the candidates for which the conditional probability $P\{Y \geq y_0|x\}$ is sufficiently large.

[(i): Let $\phi(x)$ denote the probability with which a candidate with measurements x is to be selected. Then the problem is that of maximizing

$$\int \left[\int yp^{Y|x}(y)\phi(x)\, dy \right] p^X(x)\, dx$$

subject to

$$\int \phi(x) p^X(x)\, dx = \alpha.]$$

24. The following example shows that Corollary 4 does not extend to a countably infinite family of distributions. Let p_n be the uniform probability density on $[0, 1 + 1/n]$, and p_0 the uniform density on $(0, 1)$.

(i)   Then p_0 is linearly independent of $(p_1, p_2, \ldots)$, that is, there do not exist constants $c_1, c_2, \ldots$ such that $p_0 = \Sigma c_n p_n$.

(ii) There does not exist a test ϕ such that $\int \phi p_n = \alpha$ for $n = 1, 2, \ldots$ but $\int \phi p_0 > \alpha$.

25. Let $F_1, \ldots, F_{m+1}$ be real-valued functions defined over a space U. A sufficient condition for u_0 to maximize F_{m+1} subject to $F_i(u) \leq c_i$ $(i = 1, \ldots, m)$ is that it satisfies these side conditions, that it maximizes $F_{m+1}(u) - \Sigma k_i F_i(u)$ for some constants $k_i \geq 0$, and that $F_i(u_0) = c_i$ for those values i for which $k_i > 0$.

Section 7

26. For a random variable X with binomial distribution $b(p, n)$, determine the constants C_i, γ_i $(i = 1, 2)$ in the UMP test (24) for testing $H : p \leq .2$ or $\leq .7$ when $\alpha = .1$ and $n = 15$. Find the power of the test against the alternative $p = .4$.

27. *Totally positive families.* A family of distributions with probability densities $p_\theta(x)$, θ and x real-valued and varying over Ω and $\mathcal{X}$ respectively, is said to be totally positive of order r (TP$_r$) if for all $x_1 < \cdots < x_n$ and $\theta_1 < \cdots < \theta_n$

$$(33) \quad \Delta_n = \begin{vmatrix} p_{\theta_1}(x_1) & \cdots & p_{\theta_1}(x_n) \\ p_{\theta_n}(x_1) & \cdots & p_{\theta_n}(x_n) \end{vmatrix} \geq 0 \qquad \text{for all} \quad n = 1, 2, \ldots, r.$$

It is said to be strictly totally positive of order r (STP$_r$) if strict inequality holds in (33). The family is said to be (strictly) totally positive of order infinity if (33) holds for all $n = 1, 2, \ldots$. These definitions apply not only to probability densities but to any real-valued functions $p_\theta(x)$ of two real variables.

(i)   For $r = 1$, (33) states that $p_\theta(x) \geq 0$; for $r = 2$, that $p_\theta(x)$ has monotone likelihood ratio in x.

(ii) If $a(\theta) > 0$, $b(x) > 0$, and $p_\theta(x)$ is STP$_r$, then so is $a(\theta)b(x)p_\theta(x)$.

(iii) If a and b are real-valued functions mapping Ω and $\mathscr{X}$ onto Ω' and $\mathscr{X}'$ and are strictly monotone in the same direction, and if $p_\theta(x)$ is (S)TP$_r$, then $p_{\theta'}(x')$ with $\theta' = a^{-1}(\theta)$ and $x' = b^{-1}(x)$ is (S)TP$_r$ over $(\Omega', \mathscr{X}')$.

28. *Exponential families.* The exponential family (12) with $T(x) = x$ and $Q(\theta) = \theta$ is STP$_\infty$, with Ω the natural parameter space and $\mathscr{X} = (-\infty, \infty)$.
     [That the determinant $|e^{\theta_i x_j}|$, $i, j = 1, \ldots, n$, is positive can be proved by induction. Divide the ith column by $e^{\theta_1 x_i}$, $i = 1, \ldots, n$; subtract in the resulting determinant the $(n-1)$st column from the nth, the $(n-2)$nd from the $(n-1)$st, $\ldots$, the 1st from the 2nd; and expand the determinant obtained in this way by the first row. Then Δ_n is seen to have the same sign as

$$\Delta'_n = \left| e^{\eta_i x_j} - e^{\eta_i x_{j-1}} \right|, \qquad i, j = 2, \ldots, n,$$

where $\eta_i = \theta_i - \theta_1$. If this determinant is expanded by the first column one obtains a sum of the form

$$a_2\left(e^{\eta_2 x_2} - e^{\eta_2 x_1}\right) + \cdots + a_n\left(e^{\eta_n x_2} - e^{\eta_n x_1}\right) = h(x_2) - h(x_1)$$

$$= (x_2 - x_1)h'(y_2),$$

where $x_1 < y_2 < x_2$. Rewriting $h'(y_2)$ as a determinant of which all columns but the first coincide with those of Δ'_n and proceeding in the same manner with the other columns, one reduces the determinant to $|e^{\eta_i y_j}|$, $i, j = 2, \ldots, n$, which is positive by the induction hypothesis.]

29. STP$_3$. Let θ and x be real-valued, and suppose that the probability densities $p_\theta(x)$ are such that $p_{\theta'}(x)/p_\theta(x)$ is strictly increasing in x for $\theta < \theta'$. Then the following two conditions are equivalent: (a) For $\theta_1 < \theta_2 < \theta_3$ and $k_1, k_2, k_3 > 0$, let

$$g(x) = k_1 p_{\theta_1}(x) - k_2 p_{\theta_2}(x) + k_3 p_{\theta_3}(x).$$

If $g(x_1) = g(x_3) = 0$, then the function g is positive outside the interval (x_1, x_3) and negative inside. (b) The determinant Δ_3 given by (33) is positive for all $\theta_1 < \theta_2 < \theta_3$, $x_1 < x_2 < x_3$. [It follows from (a) that the equation $g(x) = 0$ has at most two solutions.]

[That (b) implies (a) can be seen for $x_1 < x_2 < x_3$ by considering the determinant

$$\begin{vmatrix} g(x_1) & g(x_2) & g(x_3) \\ p_{\theta_2}(x_1) & p_{\theta_2}(x_2) & p_{\theta_2}(x_3) \\ p_{\theta_3}(x_1) & p_{\theta_3}(x_2) & p_{\theta_3}(x_3) \end{vmatrix}.$$

Suppose conversely that (a) holds. Monotonicity of the likelihood ratios implies that the rank of Δ_3 is at least two, so that there exist constants k_1, k_2, k_3 such that $g(x_1) = g(x_3) = 0$. That the k's are positive follows again from the monotonicity of the likelihood ratios.]

30. *Extension of Theorem 6.* The conclusions of Theorem 6 remain valid if the density of a sufficient statistic T (which without loss of generality will be taken to be X), say $p_\theta(x)$, is STP$_3$ and is continuous in x for each θ.
[The two properties of exponential families that are used in the proof of Theorem 6 are continuity in x and (a) of the preceding problem.]

31. For testing the hypothesis $H' : \theta_1 \le \theta \le \theta_2$ $(\theta_1 \le \theta_2)$ against the alternatives $\theta < \theta_1$ or $\theta > \theta_2$, or the hypothesis $\theta = \theta_0$ against the alternatives $\theta \ne \theta_0$, in an exponential family or more generally in a family of distributions satisfying the assumptions of Problem 30, a UMP test does not exist.
[This follows from a consideration of the UMP tests for the one-sided hypotheses $H_1 : \theta \ge \theta_1$ and $H_2 : \theta \le \theta_2$.]

Section 8

32. Let the variables X_i $(i = 1, \dots, s)$ be independently distributed with Poisson distribution $P(\lambda_i)$. For testing the hypothesis $H : \Sigma \lambda_j \le a$ (for example, that the combined radioactivity of a number of pieces of radioactive material does not exceed a), there exists a UMP test, which rejects when $\Sigma X_j > C$.
[If the joint distribution of the X's is factored into the marginal distribution of ΣX_j (Poisson with mean $\Sigma \lambda_j$) times the conditional distribution of the variables $Y_i = X_i / \Sigma X_j$ given ΣX_j (multinomial with probabilities $p_i = \lambda_i / \Sigma \lambda_j$), the argument is analogous to that given in Example 8.]

33. *Confidence bounds for a median.* Let $X_1, \dots, X_n$ be a sample from a continuous cumulative distribution function F. Let ξ be the unique median of F if it exists, or more generally let $\xi = \inf\{\xi' : F(\xi') = \frac{1}{2}\}$.

(i) If the ordered X's are $X_{(1)} < \cdots < X_{(n)}$, a uniformly most accurate lower confidence bound for ξ is $\underline{\xi} = X_{(k)}$ with probability ρ, $\underline{\xi} = X_{(k+1)}$ with probability $1 - \rho$, where k and ρ are determined by

$$\rho \sum_{j=k}^{n} \binom{n}{j} \frac{1}{2^n} + (1 - \rho) \sum_{j=k+1}^{n} \binom{n}{j} \frac{1}{2^n} = 1 - \alpha.$$

    (ii)   This bound has confidence coefficient $1 - \alpha$ for any median of F.

    (iii) Determine most accurate lower confidence bounds for the $100p$-percentile ξ of F defined by $\xi = \inf\{\xi' : F(\xi') = p\}$.

[For fixed ξ_0 the problem of testing $H : \xi = \xi_0$ against $K : \xi > \xi_0$ is equivalent to testing $H' : p = \frac{1}{2}$ against $K' : p < \frac{1}{2}$.]

34.   *A counterexample.* Typically, as α varies the most powerful level-α tests for testing a hypothesis H against a simple alternative are nested in the sense that the associated rejection regions, say R_α, satisfy $R_\alpha \subset R_{\alpha'}$ for any $\alpha < \alpha'$. This relation always holds when H is simple, but the following example shows that it need not be satisfied for composite H. Let X take on the values $1, 2, 3, 4$ with probabilities under distributions P_0, P_1, Q:

	1	2	3	4
P_0	$\frac{2}{13}$	$\frac{4}{13}$	$\frac{3}{13}$	$\frac{4}{13}$
P_1	$\frac{4}{13}$	$\frac{2}{13}$	$\frac{1}{13}$	$\frac{6}{13}$
Q	$\frac{4}{13}$	$\frac{3}{13}$	$\frac{2}{13}$	$\frac{4}{13}$

Then the most powerful test for testing the hypothesis that the distribution of X is P_0 or P_1 against the alternative that it is Q rejects at level $\alpha = \frac{5}{13}$ when $X = 1$ or 3, and at level $\alpha = \frac{6}{13}$ when $X = 1$ or 2.

35.   Let X and Y be the number of successes in two sets of n binomial trials with probabilities p_1 and p_2 of success.

    (i)   The most powerful test of the hypothesis $H : p_2 \leq p_1$ against an alternative (p_1', p_2') with $p_1' < p_2'$ and $p_1' + p_2' = 1$ at level $\alpha < \frac{1}{2}$ rejects when $Y - X > C$ and with probability γ when $Y - X = C$.

    (ii) This test is not UMP against the alternatives $p_1 < p_2$.

[(i): Take the distribution Λ assigning probability 1 to the point $p_1 = p_2 = \frac{1}{2}$ as an a priori distribution over H. The most powerful test against (p_1', p_2') is then the one proposed above. To see that Λ is least favorable, consider the probability of rejection $\beta(p_1, p_2)$ for $p_1 = p_2 = p$. By symmetry this is given by

$$2\beta(p, p) = P\{|Y - X| > C\} + \gamma P\{|Y - X| = C\}.$$

Let X_i be 1 or 0 as the ith trial in the first series is a success or failure, and let Y_i be defined analogously with respect to the second series. Then $Y - X = \sum_{i=1}^{n}(Y_i - X_i)$, and the fact that $2\beta(p, p)$ attains its maximum for $p = \frac{1}{2}$ can be proved by induction over n.

(ii): Since $\beta(p, p) < \alpha$ for $p \neq \frac{1}{2}$, the power $\beta(p_1, p_2)$ is $< \alpha$ for alternatives $p_1 < p_2$ sufficiently close to the line $p_1 = p_2$. That the test is not UMP now follows from a comparison with $\phi(x, y) \equiv \alpha$.]

36. *Sufficient statistics with nuisance parameters.*

   (i) A statistic T is said to be *partially sufficient* for θ in the presence of a nuisance parameter η if the parameter space is the direct product of the set of possible θ- and η-values, and if the following two conditions hold: (a) the conditional distribution given $T = t$ depends only on η; (b) the marginal distribution of T depends only on θ. If these conditions are satisfied, there exists a UMP test for testing the composite hypothesis $H: \theta = \theta_0$ against the composite class of alternatives $\theta = \theta_1$, which depends only on T.

   (ii) Part (i) provides an alternative proof that the test of Example 8 is UMP.

   [Let $\psi_0(t)$ be the most powerful level α test for testing θ_0 against θ_1 that depends only on t, let $\phi(x)$ be any level-α test, and let $\psi(t) = E_{\eta_1}[\phi(X)|t]$. Since $E_{\theta_i}\psi(T) = E_{\theta_i, \eta_1}\phi(X)$, it follows that ψ is a level-α test of H and its power, and therefore the power of ϕ, does not exceed the power of ψ_0.]

   Note. For further discussion of this and related concepts of partial sufficiency see Dawid (1975), Sprott (1975), Basu (1978), and Barndorff-Nielsen (1978).

Section 9

37. Let $X_1, \ldots, X_m$ and $Y_1, \ldots, Y_n$ be independent samples from $N(\xi, 1)$ and $N(\eta, 1)$, and consider the hypothesis $H: \eta \leq \xi$ against $K: \eta > \xi$. There exists a UMP test, and it rejects the hypothesis when $\overline{Y} - \overline{X}$ is too large. [If $\xi_1 < \eta_1$ is a particular alternative, the distribution assigning probability 1 to the point $\eta = \xi = (m\xi_1 + n\eta_1)/(m + n)$ is least favorable.]

38. Let $X_1, \ldots, X_m; Y_1, \ldots, Y_n$ be independently, normally distributed with means ξ and η, and variances σ^2 and τ^2 respectively, and consider the hypothesis $H: \tau \leq \sigma$ against $K: \sigma < \tau$.

   (i) If ξ and η are known, there exists a UMP test given by the rejection region $\Sigma(Y_j - \eta)^2/\Sigma(X_i - \xi)^2 \geq C$.

   (ii) No UMP test exists when ξ and η are unknown.

Additional Problems

39. Let P_0, P_1, P_2 be the probability distributions assigning to the integers $1, \ldots, 6$ the following probabilities:

	1	2	3	4	5	6
P_0	.03	.02	.02	.01	0	.92
P_1	.06	.05	.08	.02	.01	.78
P_2	.09	.05	.12	0	.02	.72

Determine whether there exists a level-α test of $H: P = P_0$ which is UMP against the alternatives P_1 and P_2 when (i) $\alpha = .01$; (ii) $\alpha = .05$; (iii) $\alpha = .07$.

40. Let the distribution of X be given by

x	0	1	2	3
$P_\theta(X = x)$	θ	2θ	$.9 - 2\theta$	$.1 - \theta$

where $0 < \theta < .1$. For testing $H: \theta = .05$ against $\theta > .05$ at level $\alpha = .05$, determine which of the following tests (if any) is UMP:

(i) $\phi(0) = 1$, $\phi(1) = \phi(2) = \phi(3) = 0$;

(ii) $\phi(1) = .5$, $\phi(0) = \phi(2) = \phi(3) = 0$;

(iii) $\phi(3) = 1$, $\phi(0) = \phi(1) = \phi(2) = 0$.

41. Let $X_1, \ldots, X_n$ be independently distributed, each uniformly over the integers $1, 2, \ldots, \theta$. Determine whether there exists a UMP test for testing $H: \theta = \theta_0$ at level $1/\theta_0^n$ against the alternatives (i) $\theta > \theta_0$; (ii) $\theta < \theta_0$; (iii) $\theta \neq \theta_0$.

42. Let X_i be independently distributed as $N(i\Delta, 1)$, $i = 1, \ldots, n$. Show that there exists a UMP test of $H: \Delta \leq 0$ against $K: \Delta > 0$, and determine it as explicitly as possible.

Note. The following problems (and some of the Additional Problems in later chapters) refer to the gamma, Pareto, Weibull, and inverse Gaussian distributions. For more information about these distributions, see Chapter 17, 19, 20, and 25 respectively of Johnson and Kotz (1970).

43. Let $X_1, \ldots, X_n$ be a sample from the *gamma distribution* $\Gamma(g, b)$ with density

$$\frac{1}{\Gamma(g)b^g} x^{g-1} e^{-x/b}, \quad 0 < x, \quad 0 < b, g.$$

Show that there exist a UMP test for testing

(i) $H: b \leq b_0$ against $b > b_0$ when g is known;

(ii) $H: g \leq g_0$ against $g > g_0$ when b is known.

In each case give the form of the rejection region.

44. A random variable X has the *Pareto distribution* $P(c, \tau)$ if its density is $c\tau^c / x^{c+1}$, $0 < \tau < x$, $0 < c$.

(i) Show that this defines a probability density.

(ii) If X has distribution $P(c, \tau)$, then $Y = \log X$ has exponential distribution $E(\xi, b)$ with $\xi = \log \tau$, $b = 1/c$.

(iii) If $X_1, \ldots, X_n$ is a sample from $P(c, \tau)$, use (ii) and Problem 3 to obtain UMP tests of (a) $H: \tau = \tau_0$ against $\tau \neq \tau_0$ when b is known; (b) $H: c = c_0$, $\tau = \tau_0$ against $c > c_0$, $\tau < \tau_0$.

45. A random variable X has the *Weibull distribution* $W(b, c)$ if its density is

$$\frac{c}{b}\left(\frac{x}{b}\right)^{c-1} e^{-(x/b)^c}, \qquad x > 0, \quad b, c > 0.$$

(i) Show that this defines a probability density.

(ii) If $X_1, \ldots, X_n$ is a sample from $W(b, c)$, with the shape parameter c known, show that there exists a UMP test of $H: b \leq b_0$ against $b > b_0$ and give its form.

46. Consider a single observation X from $W(1, c)$.

(i) The family of distributions does not have monotone likelihood ratio in x.

(ii) The most powerful test of $H: c = 1$ against $c = 2$ rejects when $X < k_1$ and when $X > k_2$. Show how to determine k_1 and k_2.

(iii) Generalize (ii) to arbitrary alternatives $c_1 > 1$, and show that a UMP test of $H: c = 1$ against $c > 1$ does not exist.

(iv) For any $c_1 > 1$, the power function of the MP test of $H: c = 1$ against $c = c_1$ is an increasing function of c.

47. Let $X_1, \ldots, X_n$ be a sample from the *inverse Gaussian* distribution $I(\mu, \tau)$ with density

$$\sqrt{\frac{\tau}{2\pi x^3}} \exp\left(-\frac{\tau}{2x\mu^2}(x - \mu)^2\right), \qquad x > 0, \quad \tau, \mu > 0.$$

Show that there exists a UMP test for testing

(i) $H: \mu \leq \mu_0$ against $\mu > \mu_0$ when τ is known;

(ii) $H: \tau \leq \tau_0$ against $\tau > \tau_0$ when μ is known.

In each case give the form of the rejection region.

(iii) The distribution of $V = \tau(X_i - \mu)^2/X_i\mu^2$ is χ_1^2, and hence that of $\tau\Sigma[(X_i - \mu)^2/X_i\mu^2]$ is χ_n^2.

[Let $Y = \min(X_i, \mu^2/X_i)$, $Z = \tau(Y - \mu)^2/\mu^2 Y$. Then $Z = V$ and Z is χ_1^2 [Shuster (1968)].]

Note. The UMP test for (ii) is discussed in Chhikara and Folks (1976).

48. Let X be distributed according to P_θ, $\theta \in \Omega$, and let T be sufficient for θ. If $\varphi(X)$ is any test of a hypothesis concerning θ, then $\psi(T)$ given by $\psi(t) = E[\varphi(X)|t]$ is a test depending on T only, an its power function is identical with that of $\varphi(X)$.

49. In the notation of Section 2, consider the problem of testing $H_0 : P = P_0$ against $H_1 : P = P_1$, and suppose that known probabilities $\pi_0 = \pi$ and $\pi_1 = 1 - \pi$ can be assigned to H_0 and H_1 prior to the experiment.

    (i) The overall probability of an error resulting from the use of a test φ is

$$\pi E_0 \varphi(X) + (1 - \pi) E_1[1 - \varphi(X)].$$

    (ii) The *Bayes test* minimizing this probability is given by (8) with $k = \pi_0/\pi_1$.

    (iii) The conditional probability of H_i given $X = x$, the *posterior probability* of H_i is

$$\frac{\pi_i p_i(x)}{\pi_0 p_0(x) + \pi_1 p_1(x)},$$

and the Bayes test therefore decides in favor of the hypothesis with the larger posterior probability.

50. (i) For testing $H_0 : \theta = 0$ against $H_1 : \theta = \theta_1$ when X is $N(\theta, 1)$, given any $0 < \alpha < 1$ and any $0 < \pi < 1$ (in the notation of the preceding problem), there exists θ_1 and x such that (a) H_0 is rejected when $X = x$ but (b) $P(H_0|x)$ is arbitrarily close to 1.

    (ii) The paradox of part (i) is due to the fact that α is held constant while the power against θ_1 is permitted to get arbitrarily close to 1. The paradox disappears if α is determined so that the probabilities of type I and type II error are equal [but see Berger and Sellke (1984)].

[For a discussion of such paradoxes, see Lindley (1957), Bartlett (1957) and Schafer (1982).]

51. Let $X_1, \ldots, X_n$ be i.i.d. with density p_0 or p_1, so that the MP level-α test of $H : p_0$ rejects when $\prod_{i=1}^{n} r(X_i) \geq C_n$, where $r(X_i) = p_1(X_i)/p_0(X_i)$, or equivalently when

(34)
$$\frac{1}{\sqrt{n}} \left\{ \sum \log r(X_i) - E_0[\log r(X_i)] \right\} \geq k_n.$$

    (i) It follows from the central limit theorem (Chapter 5, Theorem 3) that under H the left side of (34) tends in law to $N(0, \sigma^2)$ with $\sigma^2 = \text{Var}_0[\log r(X_i)]$ provided $\sigma^2 < \infty$.

    (ii) From (i) it follows that $k_n \to \sigma u_\alpha$ where $\Phi(u_\alpha) = 1 - \alpha$.

    (iii) The power of the test (34) agaisnt p_1 tends to 1 as $n \to \infty$.

[(iii): Problem 14(iv).]

52. Let $X_1, \ldots, X_n$ be independent $N(\theta, \gamma)$, $0 < \gamma < 1$ known, and $Y_1, \ldots, Y_n$ independent $N(\theta, 1)$. Then X is more informative than Y according to the definition at the end of Section 4.

[If V_i is $N(0, 1 - \gamma)$, then $X_i + V_i$ has the same distribution as Y_i.]

Note. If σ is unknown, it is not true that a sample from $N(\theta, \gamma\sigma^2)$, $0 < \gamma < 1$, is more informative than one from $N(\theta, \sigma^2)$; see Hansen ad Torgersen (1974).

53. Let f, g be two probability densities with respect to μ. For testing the hypothesis $H: \theta \leq \theta_0$ or $\theta \geq \theta_1$ $(0 < \theta_0 < \theta_1 < 1)$ against the alternatives $\theta_0 < \theta < \theta_1$ in the family $\mathscr{P} = \{\theta f(x) + (1 - \theta)g(x), 0 \leq \theta \leq 1\}$, the test $\varphi(x) \equiv \alpha$ is UMP at level α.

11. REFERENCES

Hypothesis testing developed gradually, with early instances frequently being rather vague statements of the significance or nonsignificance of a set of observations. Isolated applications are found in the 18th century [Arbuthnot (1710), Daniel Bernoulli (1734), and Laplace (1773), for example] and centuries earlier in the Royal Mint's Trial of the Pyx [discussed by Stigler (1977)]. They became more frequent in the 19th century in the writings of such authors as Gavarret (1840), Lexis (1875, 1877), and Edgeworth (1885). Systematic use of hypothesis testing began with the work of Karl Pearson, particularly his χ^2 paper of 1900.

The first authors to recognize that the rational choice of a test must involve consideration not only of the hypothesis but also of the alternatives against which it is being tested were Neyman and E. S. Pearson (1928). They introduced the distinction between errors of the first and second kind, and thereby motivated their proposal of the likelihood-ratio criterion as a general method of test construction. These considerations were carried to their logical conclusion by Neyman and Pearson in their paper of 1933, in which they developed the theory of UMP tests. Accounts of their collaboration can be found in Pearson's recollections (1966), and in the biography of Neyman by Reid (1982).

The earliest example of confidence intervals appears to occur in the work of Laplace (1812), who points out how an (approximate) probability statement concerning the difference between an observed frequency and a binomial probability p can be inverted to obtain an associated interval for p. Other examples can be found in the work of Gauss (1816), Fourier (1826), and Lexis (1875). However, in all these cases, although the statements made are formally correct, the authors appear to consider the parameter as the variable which with the stated probability falls in the fixed confidence interval. The proper interpretation seems to have been pointed out for the first time by E. B. Wilson (1927). About the same time two examples of exact confidence statements were given by Working and Hotelling (1929) and Hotelling (1931).

A general method for obtaining exact confidence bounds for a real-valued parameter in a continuous distribution was proposed by Fisher (1930), who however later disavowed this interpretation of his work. For a discussion of Fisher's controversial concept of fiducial probability, see Chapter 5, Section 9. At about the same time,* a completely general theory of confidence statements was developed by Neyman and shown by him to be intimately related to the theory of hypothesis testing. A detailed account of this work, which underlies the treatment given here, was published by Neyman in his papers of 1937 and 1938.

Arbuthnot, J.

(1710). "An argument for Divine Providence, taken from the constant regularity observ'd in the births of both sexes." *Phil. Trans.* **27**, 186–190.

Arrow, K.

(1960). "Decision theory and the choice of a level of significance for the *t*-test." In *Contributions to Probability and Statistics* (Olkin et al., eds.) Stanford U.P., Stanford, Calif.

Barndorff-Nielsen, O.

(1978). *Information and Exponential Families*, Wiley, New York.

Barnett, V.

(1982). *Comparative Statistical Inference*, 2nd ed., Wiley, New York.

Bartlett, M. S.

(1957). "A comment on D. V. Lindley's statistical paradox." *Biometrika* **44**, 533–534.

Basu, D.

(1978). "On partial sufficiency: A review." *J. Statist. Planning and Inference* **2**, 1–13.

Berger, J. and Sellke, T.

(1984). "Testing a point null-hypothesis: The irreconcilability of significance levels and evidence." Tech. Report #84-27, Purdue University.

Bernoulli, D.

(1734). "Quelle est la cause physique de l'inclinaison des planètes . . . ." *Recueil des Pièces qui ont Remporté le Prix de l'Académie Royale des Sciences* **3**, 95–122.

Birnbaum, A.

(1954). "Admissible test for the mean of a rectangular distribution." *Ann. Math. Statist.* **25**, 157–161.

Birnbaum, Z. W. and Chapman, D. G.

(1950). "On optimum selections from multinormal populations." *Ann. Math. Statist.* **21**, 433–447.
[Problem 23.]

Blackwell, D.

(1951a). "On a theorem of Lyapunov." *Ann. Math. Statist.* **22**, 112–114.

(1951b). "Comparison of experiments." In *Proc. Second Berkeley Symposium on Mathematical Statistics and Probability*, Univ. of California Press, Berkeley, 93–102.

(1953). "Equivalent comparisons of experiments." *Ann. Math. Statist.* **24**, 265–272.
[Theory, Example 4, and Problems of Section 4.]

*Cf. Neyman (1941).

Blyth, C.

(1984). "Approximate binomial confidence limits." Queen's Math. Preprint. #1984-6 Queen's Univ. Kingston, Ontario.

Brown, L. D., Cohen, A., and Strawderman, W. E.

(1976). "A complete class theorem for strict monotone likelihood ratio with applications." *Ann. Statist.* **4**, 712–722.

Brown, L. D., Johnstone, I. M. and MacGibbon, K. B.

(1981). "Variation diminishing transformations: A direct approach to total positivity and its statistical applications." *J. Amer. Statist. Assoc.* **76**, 824–832.

Buehler, R.

(1980). "Fiducial inference." in *R. A. Fisher: An Appreciation* (Fienberg and Hinkley, eds.), Lecture Notes in Statistics, Vol. 1, Springer, New York.

(1983). "Fiducial inference." In *Encyclopedia of Statistical Sciences*, Vol. 3, Wiley, New York.

Chernoff, H. and Scheffé, H.

(1952). "A generalization of the Neyman–Pearson fundamental lemma." *Ann. Math. Statist.* **23**, 213–225.

Chhikara, R. S. and Folks, J. L.

(1976). "Optimum test procedures for the mean of first passage time distribution in Brownian motion with positive drift." *Technometrics* **18**, 189–193.

Cohen, J.

(1962). "The statistical power of abnormal–social psychological research: A review." *J. Abnormal and Soc. Psychology* **65**, 145–153.

Cohen, L.

(1958). "On mixed single sample experiments." *Ann. Math. Statist.* **29**, 947–971.

Cox, D.

(1977). "The role of significance tests." *Scand. J. Statist.* **4**, 49–62.

Dantzig, G. B. and Wald, A.

(1951). "On the fundamental lemma of Neyman and Pearson." *Ann. Math. Statist.* **22**, 87–93.

[Gives necessary conditions, including those of Theorem 5, for a critical function which maximizes an integral subject to a number of integral side conditions, to satisfy (21).]

Dawid, A. P.

(1975). "On the concepts of sufficiency and ancillarity in the presence of nuisance parameters." *J. Roy. Statist. Soc. (B)* **37**, 248–258.

Dawid, A. P. and Stone, M.

(1982). "The functional-model basis of fiducial inference (with discussion)." *Ann. Statist.* **10**, 1054–1073.

Dempster, A. P. and Schatzoff, M.

(1965). "Expected significance level as sensitivity index for test statistics." *J. Amer. Statist. Assoc.* **60**, 420–436.

Dvoretzky, A., Kiefer, J., and Wolfowitz, J.

(1953). "Sequential decision problems for processes with continuous time parameter. Testing hypotheses." *Ann. Math. Statist.* **24**, 254–264.

Dvoretzky, A., Wald, A., and Wolfowitz, J.

(1951). "Elimination of randomization in certain statistical decision procedures and zero-sum two-person games." *Ann. Math. Statist.* **22**, 1–21.

Edgeworth, F. Y.

(1885). *Methods of Statistics*, Jubilee volume of the Statist. Soc., E. Stanford, London.

Edwards, A. W. F.

(1983). "Fiducial distributions." In *Encyclopedia of Statistical Sciences*, Vol. 3, Wiley, New York.

Epstein, B. and Sobel, M.

(1953). "Life testing." *J. Amer. Statist. Assoc.* **48**, 486–502.

[Problem 10]

Fisher, R. A.

(1930). "Inverse probability." *Proc. Cambridge Phil. Soc.* **26**, 528–535.

Fourier, J. B. J.

(1826). *Recherches Statistiques sur la Ville de Paris et le Département de la Seine*, Vol. 3.

Fraser, D. A. S.

(1953). "Non-parametric theory: Scale and location parameters." *Canad. J. Math.* **6**, 46–68.

[Example 8.]

(1956). "Sufficient statistics with nuisance parameters." *Ann. Math. Statist.* **27**, 838–842.

[Problem 36.]

Freiman, J. A., Chalmers, T. C., Smith, H., and Kuebler, R. R.

(1978). "The importance of beta, the type II error and sample size in the design and interpretation of the randomized control trial." *New England J. Med.* **299**, 690–694.

Gauss, C. F.

(1816). "Bestimmung der Genauigkeit der Beobachtungen." *Z. Astron. und Verw. Wiss* **1**. (Reprinted in Gauss' collected works, Vol. 4, pp. 109–119.)

Gavarret, J.

(1840). *Principes Généraux de Statistique Médicale*, Paris.

Ghosh, B. K.

(1970). *Sequential Tests of Statistical Hypotheses*, Addison-Wesley, Reading, Mass.

Gibbons, J. D. and Pratt, J. W.

(1975). "P-values: Interpretation and methodology." *Amer. Statist.* **29**, 20–24.

Godambe, V. P.

(1980). "On sufficiency and ancillarity in the presence of a nuisance parameter." *Biometrika* **67**, 155–162.

Greenwald, A. G.

(1975). "Consequences of prejudice against the null hypothesis." *Psych. Bull.* **82**, 1–20.

Grenander, U.

(1950). "Stochastic processes and statistical inference." *Ark. Mat.* **1**, 195–277.

[Application of the fundamental lemma to problems in stochastic processes.]

(1981). *Abstract Inference*, Wiley, New York.

Hall, I. J. and Kudo, A.

(1968). "On slippage tests—(I) A generalization of Neyman–Pearson's lemma." *Ann. Math. Statist.* **39**, 2029–2037.

Hall, P.

(1982). "Improving the normal approximation when constructing one-sided confidence intervals for binomial or Poisson parameters." *Biometrika* **69**, 647–652.

Halmos, P.

(1974). *Measure Theory*, Springer, New York.

Hansen, O. H. and Torgersen, E. N.
(1974). "Comparison of linear normal experiments." *Ann. Statist.* **2**, 367–373.

Hodges, J. L., Jr.
(1949). "The choice of inspection stringency in acceptance sampling by attributes." *Univ. Calif. Publ. Statist.* **1**, 1–14.

Hoel, P. G. and Peterson, R. P.
(1949). "A solution to the problem of optimum classification." *Ann. Math. Statist.* **20**, 433–438.

Hotelling, H.
(1931). "The generalization of Student's ratio." *Ann. Math. Statist.* **2**, 360–378.

Jogdeo, K. and Bohrer, R.
(1973). "Some simple examples and counterexamples about the existence of optimum tests." *J. Amer. Statist. Assoc.* **68**, 679–682.
[Problems 13 and 53.]

Johnson, N. L. and Kotz, S.
(1969). *Distributions in Statistics: Discrete Distributions*, Houghton Mifflin, New York.
(1970). *Distributions in Statistics: Continuous Univariate Distributions*, Vol. 1, Houghton Mifflin, Boston.

Kabe, D. G. and Laurent, A. G.
(1981). "On some nuisance parameter free uniformly most powerful tests." *Biom. J.* **23**, 245–250.

Karlin, S.
(1955). "Decision theory for Pólya type distributions, Case of two actions. I." In *Proc. Third Berkeley Symposium on Mathematical Statistics and Probability*, Vol. 1, Univ. of Calif. Press, Berkeley, 115–129.
(1957). "Pólya type distributions. II." *Ann. Math. Statist.* **28**, 281–308.
[Properties of TP distributions, including Problems 27–30.]
(1968). *Total Positivity*, Vol. I, Stanford U.P. Stanford, Calif.

Karlin, S. and Rubin, H.
(1956). "The theory of decision procedures for distributions with monotone likelihood ratio." *Ann. Math. Statist.* **27**, 272–299.
[General theory of families with monotone likelihood ratio, including Theorem 3. For further developments of this theory, see Brown, Cohen, and Strawderman (1976).]

Karlin, S. and Truax, D. R.
(1960). "Slippage problems." *Ann. Math. Statist.* **31**, 296–323.

Krafft, O. and Witting, H.
(1967). "Optimale tests und ungünstigste Verteilungen." *Z. Wahrsch.* **7**, 289–302.

Kruskal, W. H.
(1978). "Significance, Tests of." In *International Encyclopedia of Statistics*, Free Press and Macmillan, New York and London.

Lambert, D.
(1982). "Qualitative robustness of tests." *J. Amer. Statist. Assoc.* **77**, 352–357.

Lambert. D. and Hall, W. J.
(1982). "Asymptotic lognormality of P-values." *Ann. Statist.* **10**, 44–64.

Laplace, P. S.
(1773). "Mémoire sur l'inclinaison moyenne des orbites des comètes." *Mem. Acad. Roy. Sci. Paris* **7** (1776), 503–524.
(1812). *Théorie Analytique des Probabilités*, Paris. (The 3rd edition of 1820 is reprinted as Vol. 7 of Laplace's collected works.)

Le Cam, L.
(1964). "Sufficiency and approximate sufficiency." *Ann. Math. Statist.* **35**, 1419–1455.

Lehmann, E. L.
(1952). "On the existence of least favorable distributions." *Ann. Math. Statist.* **23**, 408–416.
(1955). "Ordered families of distributions." *Ann. Math. Statist.* **26**, 399–419.
[Lemmas 1, 2, and 4.]
(1958). "Significance level and power." *Ann. Math. Statist.* **29**, 1167–1176.
(1961). "Some model I problems of selection." *Ann. Math. Statist.* **32**, 990–1012.

Lehmann, E. L. and Stein, C.
(1948). "Most powerful tests of composite hypotheses." *Ann. Math. Statist.* **19**, 495–516.
[Theorem 7 and applications.]

Lexis, W.
(1875). *Einleitung in die Theorie der Bevölkerungsstatistik*, Strassburg.
(1877). *Zur Theorie der Massenerscheinungen in der Menschlichen Gesellschaft*, Freiburg.

Lindley, D. V.
(1957). "A statistical paradox." *Biometrika* **44**, 187–192.

Lyapounov, A. M.
(1940). Sur les fonctions–vecteurs complètement additives, *Izv. Akad. Nauk SSSR Ser. Mat.* **4**, 465–478.

Neyman, J.
(1937). "Outline of a theory of statistical estimation based on the classical theory of probability." *Phil. Trans. Roy. Soc. Ser. A.* **236**, 333–380.
[Develops the theory of optimum confidence sets so that it reduces to the determination of optimum tests of associated classes of hypotheses.]
(1938). "L'estimation statistique traitée comme un problème classique de probabilité." *Actualités Sci. et Ind.* **739**, 25–57.
(1941). "Fiducial argument and the theory of confidence intervals." *Biometrika* **32**, 128–150.
(1952). *Lectures and Conferences on Mathematical Statistics*, 2nd ed., Washington Graduate School, U.S. Dept. of Agriculture, 43–66.
[An account of various approaches to the problem of hypothesis testing.]

Neyman, J. and Pearson, E. S.
(1928). "On the use and interpretation of certain test criteria." *Biometrika* **20A**, 175–240, 263–294.
(1933). "On the problem of the most efficient tests of statistical hypotheses." *Phil. Trans. Roy. Soc. Ser. A.* **231**, 289–337.
[The basic paper on the theory of hypothesis testing. Formulates the problem in terms of the two kinds of error, and develops a body of theory including the fundamental lemma. Applications including Problem 2.]
(1936a). "Contributions to the theory of testing statistical hypotheses. I. Unbiased critical regions of type A and type A_1." *Statist. Res. Mem.* **1**, 1–37.
[Generalization of the fundamental lemma to more than one side condition.]
(1936b). "Sufficient statistics and uniformly most powerful tests of statistical hypotheses." *Statist. Res. Mem.* **1**, 113–137.
[Problem 3(ii).]

Paulson, E.
(1952). "An optimum solution to the k-sample slippage problem for the normal distribution." *Ann. Math. Statist.* **23**, 610–616.

Pearson, E. S.
(1966). "The Neyman–Pearson story: 1926–1934." In *Research Papers in Statistics: Festschrift for J. Neyman* (F. N. David, ed.), Wiley, New York.

Pearson, K.

(1900). "On the criterion that a given system of deviations from the probable in the case of a correlated system of variables is such that it can be reasonably supposed to have arisen from random sampling." *Phil. Mag.* **5:50**, 157–172.

Pedersen, J. G.

(1976.) "Fiducial inference." *Internat. Statist. Rev.* **46**, 147–170.

Pfanzagl, J.

(1967). "A technical lemma for monotone likelihood ratio families." *Ann. Math. Statist.* **38**, 611–613.

(1968). "A characterization of the one parameter exponential family by existence of uniformly most powerful tests." *Sankhyā (A)* **30**, 147–156.

Pratt, J. W.

(1958). "Admissible one-sided tests for the mean of a rectangular distribution." *Ann. Math. Statist.* **29**, 1268–1271.

Pratt, J. W. and Gibbons, J. D.

(1981). *Concepts of Nonparametric Theory*, Springer, New York.

Reid, C.

(1982). *Neyman from Life*, Springer, New York.

Reinhardt, H. E.

(1961). "The use of least favorable distributions in testing composite hypotheses." *Ann. Math. Statist.* **32**, 1034–1041.

Rosenthal, R. and Rubin, D. B.

(1985). "Statistical Analysis: Summarizing evidence versus establishing facts." *Psych. Bull.* **97**, 527–529.

Sanathanan, L.

(1974). "Critical power function and decision making." *J. Amer. Statist. Assoc.* **69**, 398–402.

Savage, L. J.

(1962). *The Foundations of Statistical Inference*, Methuen, London.

(1976). "On rereading R. A. Fisher" (with discussion). *Ann. Statist.* **4**, 441–500.

Schafer, G.

(1982). "Lindley's paradox" (with discussion). *J. Amer. Statist. Assoc.* **77**, 325–351.

Schweder, T. and Spjøtvoll, E.

(1982). "Plots of P-values to evaluate many tests simultaneously." *Biometrika* **69**, 493–502.

Shuster, J.

(1968). "On the inverse Gaussian distribution function." *J. Amer. Statist. Assoc.* **63**, 1514–1516.

Spjøtvoll, E.

(1972). "On the optimality of some multiple comparison procedures." *Ann. Math. Statist.* **43**, 398–411.

(1983). "Preference functions." In *A Festschrift for Erich L. Lehmann* (P. J. Bickel, K. Doksum, and J. L. Hodges Jr., eds.), Wadsworth, Belmont, Calif.

Sprott, D. A.

(1975). "Marginal and conditional sufficiency." *Biometrika* **62**, 599–605.

Stein, C. M.

(1951). "A property of some tests of composite hypotheses." *Ann. Math. Statist.* **22**, 475–476. [Problem 34.]

Sterling, T. D.

(1959). "Publication decisions and their possible effects on inferences drawn from tests of significance—or vice versa." *J. Amer. Statist. Assoc.* **54**, 30–34.

Stigler, S. M.

(1977). "Eight centuries of sampling inspection: The trial of the Pyx." *J. Amer. Statist. Assoc.* **72**, 493–500.

Stone, M.

(1969). "The role of significance testing: Some data with a message." *Biometrika* **56**, 485–493.

(1983). "Fiducial probability." In *Encyclopedia of Statistical Sciences*, Vol. 3, Wiley, New York.

Takeuchi, K.

(1969). "A note on the test for the location parameter of an exponential distribution." *Ann. Math. Statist.* **40**, 1838–1839.

Thompson, W. A., Jr.

(1985). "Optimal significance procedures for simple hypotheses." *Biometrika* **72**, 230–232.

Thompson, W. R.

(1936). "On confidence ranges for the median and other expectation distributions for populations of unknown distribution form." *Ann. Math. Statist.* **7**, 122–128.
[Problem 33.]

Torgersen, E. N.

(1976). "Comparison of statistical experiments." *Scand. J. Statist.* **3**, 186–208.

Tukey, J. W.

(1949). "Standard confidence points." Unpublished Report 16, Statist. Res. Group, Princeton Univ. (To be published in Tukey's *Collected Works*, Wadsworth, Belmont, Calif.)

(1957). "Some examples with fiducial relevance." *Ann. Math. Statist.* **28**, 687–695.

Wald, A. and Wolfowitz, J.

(1948). "Optimum character of the sequential probability ratio test." *Ann. Math. Statist.* **19**, 326–339.

(1950). "Bayes solutions of sequential decision problems." *Ann. Math. Statist.* **21**, 82–89.

Wilkinson, G. N.

(1977). "On resolving the controversy in statistical inference" (with discussion). *J. Roy. Statist. Soc. B* **39**, 119–171.

Wilson, E. B.

(1927). "Probable inference, the law of succession, and statistical inference." *J. Amer. Statist. Assoc.* **22**, 209–212.

Wolfowitz, J.

(1950). "Minimax estimates of the mean of a normal distribution with known variance." *Ann. Math. Statist.* **21**, 218–230.

Working, H. and Hotelling, H.

(1929). "Applications of the theory of error to the interpretation of trends." *J. Amer. Statist. Assoc., Suppl.* **24**, 73–85.

CHAPTER 4

Unbiasedness: Theory and First Applications

1. UNBIASEDNESS FOR HYPOTHESIS TESTING

A simple condition that one may wish to impose on tests of the hypothesis $H: \theta \in \Omega_H$ against the composite class of alternatives $K: \theta \in \Omega_K$ is that for no alternative in K should the probability of rejection be less than the size of the test. Unless this condition is satisfied, there will exist alternatives under which acceptance of the hypothesis is more likely than in some cases in which the hypothesis is true. A test ϕ for which the above condition holds, that is, for which the power function $\beta_\phi(\theta) = E_\theta \phi(X)$ satisfies

(1)
$$\beta_\phi(\theta) \leq \alpha \quad \text{if} \quad \theta \in \Omega_H,$$

$$\beta_\phi(\theta) \geq \alpha \quad \text{if} \quad \theta \in \Omega_K,$$

is said to be *unbiased*. For an appropriate loss function this was seen in Chapter 1 to be a particular case of the general definition of unbiasedness given there. Whenever a UMP test exists, it is unbiased, since its power cannot fall below that of the test $\phi(x) \equiv \alpha$.

For a large class of problems for which a UMP test does not exist, there does exist a UMP unbiased test. This is the case in particular for certain hypotheses of the form $\theta \leq \theta_0$ or $\theta = \theta_0$, where the distribution of the random observables depends on other parameters besides θ.

When $\beta_\phi(\theta)$ is a continuous function of θ, unbiasedness implies

(2)
$$\beta_\phi(\theta) = \alpha \quad \text{for all} \quad \theta \text{ in } \omega,$$

where ω is the common boundary of Ω_H and Ω_K, that is, the set of points θ that are points or limit points of both Ω_H and Ω_K. Tests satisfying this condition are said to be *similar on the boundary* (of H and K). Since it is more convenient to work with (2) than with (1), the following lemma plays an important role in the determination of UMP unbiased tests.

Lemma 1. *If the distributions P_θ are such that the power function of every test is continuous, and if ϕ_0 is UMP among all tests satisfying (2) and is a level-α test of H, then ϕ_0 is UMP unbiased.*

Proof. The class of tests satisfying (2) contains the class of unbiased tests, and hence ϕ_0 is uniformly at least as powerful as any unbiased test. On the other hand, ϕ_0 is unbiased, since it is uniformly at least as powerful as $\phi(x) \equiv \alpha$.

2. ONE-PARAMETER EXPONENTIAL FAMILIES

Let θ be a real parameter, and $X = (X_1, \ldots, X_n)$ a random vector with probability density (with respect to some measure μ)

$$p_\theta(x) = C(\theta)e^{\theta T(x)}h(x).$$

It was seen in Chapter 3 that a UMP test exists when the hypothesis H and the class K of alternatives are given by (i) $H: \theta \le \theta_0$, $K: \theta > \theta_0$ (Corollary 2) and (ii) $H: \theta \le \theta_1$ or $\theta \ge \theta_2$ $(\theta_1 < \theta_2)$, $K: \theta_1 < \theta < \theta_2$ (Theorem 6), but not for (iii) $H: \theta_1 \le \theta \le \theta_2$, $K: \theta < \theta_1$ or $\theta > \theta_2$. We shall now show that in case (iii) there does exist a UMP unbiased test given by

$$(3) \qquad \phi(x) = \begin{cases} 1 & \text{when} \quad T(x) < C_1 \text{ or } > C_2, \\ \gamma_i & \text{when} \quad T(x) = C_i, \quad i = 1, 2, \\ 0 & \text{when} \quad C_1 < T(x) < C_2, \end{cases}$$

where the C's and γ's are determined by

$$(4) \qquad E_{\theta_1}\phi(X) = E_{\theta_2}\phi(X) = \alpha.$$

The power function $E_\theta\phi(X)$ is continuous by Theorem 9 of Chapter 2, so that Lemma 1 is applicable. The set ω consists of the two points θ_1 and θ_2, and we therefore consider first the problem of maximizing $E_{\theta'}\phi(X)$ for some θ' outside the interval $[\theta_1, \theta_2]$, subject to (4). If this problem is restated in terms of $1 - \phi(x)$, it follows from part (ii) of Theorem 6, Chapter 3, that its solution is given by (3) and (4). This test is therefore UMP among those satisfying (4), and hence UMP unbiased by Lemma 1. It further follows from part (iii) of the theorem that the power function of the test has a minimum at a point between θ_1 and θ_2, and is strictly increasing as θ tends away from this minimum in either direction.

A closely related problem is that of testing (iv) $H: \theta = \theta_0$ against the alternatives $\theta \neq \theta_0$. For this there also exists a UMP unbiased test given by (3), but the constants are now determined by

(5)
$$E_{\theta_0}[\phi(X)] = \alpha$$

and

(6)
$$E_{\theta_0}[T(X)\phi(X)] = E_{\theta_0}[T(X)]\alpha.$$

To see this, let θ' be any particular alternative, and restrict attention to the sufficient statistic T, the distribution of which by Chapter 2, Lemma 8, is of the form

$$dP_\theta(t) = C(\theta) e^{\theta t} d\nu(t).$$

Unbiasedness of a test $\psi(t)$ implies (5) with $\phi(x) = \psi[T(x)]$; also that the power function $\beta(\theta) = E_\theta[\psi(T)]$ must have a minimum at $\theta = \theta_0$. By Theorem 9 of Chapter 2 the function $\beta(\theta)$ is differentiable, and the derivative can be computed by differentiating $E_\theta\psi(T)$ under the expectation sign, so that for all tests $\psi(t)$

$$\beta'(\theta) = E_\theta[T\psi(T)] + \frac{C'(\theta)}{C(\theta)}E_\theta[\psi(T)].$$

For $\psi(t) \equiv \alpha$, this equation becomes

$$0 = E_\theta(T) + \frac{C'(\theta)}{C(\theta)}.$$

Substituting this in the expression for $\beta'(\theta)$ gives

$$\beta'(\theta) = E_\theta[T\psi(T)] - E_\theta(T)E_\theta[\psi(T)],$$

and hence unbiasedness implies (6) in addition to (5).

Let M be the set of points $(E_{\theta_0}[\psi(T)], E_{\theta_0}[T\psi(T)])$ as ψ ranges over the totality of critical functions. Then M is convex and contains all points $(u, uE_{\theta_0}(T))$ with $0 < u < 1$. It also contains points (α, u_2) with $u_2 > \alpha E_{\theta_0}(T)$. This follows from the fact that there exist tests with $E_{\theta_0}[\psi(T)] = \alpha$ and $\beta'(\theta_0) > 0$ (see Problem 22 of Chapter 3). Since similarly M contains points (α, u_1) with $u_1 < \alpha E_{\theta_0}(T)$, the point $(\alpha, \alpha E_{\theta_0}(T))$ is an inner point of M. Therefore, by Theorem 5(iv) of Chapter 3 there exist constants k_1, k_2 and a test $\psi(t)$ satisfying (5) and (6) with $\phi(x) = \psi[T(x)]$, such that $\psi(t) = 1$ when

$$C(\theta_0)(k_1 + k_2 t)\, e^{\theta_0 t} < C(\theta')\, e^{\theta' t}$$

and therefore when

$$a_1 + a_2 t < e^{bt}.$$

This region is either one-sided or the outside of an interval. By Theorem 2 of Chapter 3 a one-sided test has a strictly monotone power function and therefore cannot satisfy (6). Thus $\psi(t)$ is 1 when $t < C_1$ or $> C_2$, and the most powerful test subject to (5) and (6) is given by (3). This test is unbiased, as is seen by comparing it with $\phi(x) \equiv \alpha$. It is then also UMP unbiased, since the class of tests satisfying (5) and (6) includes the class of unbiased tests.

A simplification of this test is possible if for $\theta = \theta_0$ the distribution of T is symmetric about some point a, that is, if $P_{\theta_0}\{T < a - u\} = P_{\theta_0}\{T > a + u\}$ for all real u. Any test which is symmetric about a and satisfies (5) must also satisfy (6), since $E_{\theta_0}[T\psi(T)] = E_{\theta_0}[(T - a)\psi(T)] + aE_{\theta_0}\psi(T) = a\alpha = E_{\theta_0}(T)\alpha$. The C's and γ's are therefore determined by

$$P_{\theta_0}\{T < C_1\} + \gamma_1 P_{\theta_0}\{T = C_1\} = \frac{\alpha}{2},$$

$$C_2 = 2a - C_1, \qquad \gamma_2 = \gamma_1.$$

The above tests of the hypotheses $\theta_1 \le \theta \le \theta_2$ and $\theta = \theta_0$ are *strictly unbiased* in the sense that the power is $> \alpha$ for all alternatives θ. For the first of these tests, given by (3) and (4), strict unbiasedness is an immediate consequence of Theorem 6(iii) of Chapter 3. This states in fact that the power of the test has a minimum at a point θ_0 between θ_1 and θ_2 and increases strictly as θ tends away from θ_0 in either direction. The second of the tests, determined by (3), (5), and (6), has a continuous power function with a minimum of α at $\theta = \theta_0$. Thus there exist $\theta_1 < \theta_0 < \theta_2$ such that $\beta(\theta_1) = \beta(\theta_2) = c$ where $\alpha \le c < 1$. The test therefore coincides with the UMP unbiased level-c test of the hypothesis $\theta_1 \le \theta \le \theta_2$, and the power increases strictly as θ moves away from θ_0 in either direction. This proves the desired result.

Example 1. **Binomial.** Let X be the number of successes in n binomial trials with probability p of success. A theory to be tested assigns to p the value p_0, so that one wishes to test the hypothesis $H: p = p_0$. When rejecting H one will usually wish to state also whether p appears to be less or greater than p_0. If, however, the conclusion that $p \neq p_0$ in any case requires further investigation, the preliminary decision is essentially between the two possibilities that the data do or do not contradict the hypothesis $p = p_0$. The formulation of the problem as one of hypothesis testing may then be appropriate.

The UMP unbiased test of H is given by (3) with $T(X) = X$. The condition (5) becomes

$$\sum_{x=C_1+1}^{C_2-1} \binom{n}{x} p_0^x q_0^{n-x} + \sum_{i=1}^{2} (1 - \gamma_i)\binom{n}{C_i} p_0^{C_i} q_0^{n-C_i} = 1 - \alpha,$$

and the left-hand side of this can be obtained from tables of the individual probabilities and cumulative distribution function of X. The condition (6), with the help of the identity

$$x\binom{n}{x} p_0^x q_0^{n-x} = np_0 \binom{n-1}{x-1} p_0^{x-1} q_0^{(n-1)-(x-1)}$$

reduces to

$$\sum_{x=C_1+1}^{C_2-1} \binom{n-1}{x-1} p_0^{x-1} q_0^{(n-1)-(x-1)}$$

$$+ \sum_{i=1}^{2} (1 - \gamma_i)\binom{n-1}{C_i-1} p_0^{C_i-1} q_0^{(n-1)-(C_i-1)} = 1 - \alpha,$$

the left-hand side of which can be computed from the binomial tables.

As n increases, the distribution of $(X - np_0)/\sqrt{np_0 q_0}$ tends to the normal distribution $N(0,1)$. For sample sizes which are not too small, and values of p_0 which are not too close to 0 or 1, the distribution of X is therefore approximately symmetric. In this case, the much simpler "equal tails" test, for which the C's and γ's are determined by

$$\sum_{x=0}^{C_1-1} \binom{n}{x} p_0^x q_0^{n-x} + \gamma_1 \binom{n}{C_1} p_0^{C_1} q_0^{n-C_1}$$

$$= \gamma_2 \binom{n}{C_2} p_0^{C_2} q_0^{n-C_2} + \sum_{x=C_2+1}^{n} \binom{n}{x} p_0^x q_0^{n-x} = \frac{\alpha}{2},$$

is approximately unbiased, and constitutes a reasonable approximation to the unbiased test. Of course, when n is sufficiently large, the constants can be determined directly from the normal distribution.

Example 2. Normal variance. Let $X = (X_1, \ldots, X_n)$ be a sample from a normal distribution with mean 0 and variance σ^2, so that the density of the X's is

$$\left(\frac{1}{\sqrt{2\pi}\sigma}\right)^n \exp\left(-\frac{1}{2\sigma^2}\sum x_i^2\right).$$

Then $T(x) = \sum x_i^2$ is sufficient for σ^2, and has probability density $(1/\sigma^2)f_n(y/\sigma^2)$, where

$$f_n(y) = \frac{1}{2^{n/2}\Gamma(n/2)} y^{(n/2)-1} e^{-(y/2)}, \qquad y > 0,$$

is the density of a χ^2-distribution with n degrees of freedom. For varying σ, these distributions form an exponential family, which arises also in problems of life testing (see Problem 14 of Chapter 2), and concerning normally distributed variables with unknown mean and variance (Section 3 of Chapter 5). The acceptance region of the UMP unbiased test of the hypothesis $H : \sigma = \sigma_0$ is

$$C_1 \leq \sum \frac{x_i^2}{\sigma_0^2} \leq C_2$$

with

$$\int_{C_1}^{C_2} f_n(y) \, dy = 1 - \alpha$$

and

$$\int_{C_1}^{C_2} y f_n(y) \, dy = \frac{(1-\alpha) E_{\sigma_0}\left(\sum X_i^2\right)}{\sigma_0^2} = n(1-\alpha).$$

For the determination of the constants from tables of the χ^2-distribution, it is convenient to use the identity

$$y f_n(y) = n f_{n+2}(y),$$

to rewrite the second condition as

$$\int_{C_1}^{C_2} f_{n+2}(y) \, dy = 1 - \alpha.$$

Alternatively, one can integrate $\int_{C_1}^{C_2} y f_n(y) \, dy$ by parts to reduce the second condition to

$$C_1^{n/2} e^{-C_1/2} = C_2^{n/2} e^{-C_2/2}.$$

[For tables giving C_1 and C_2 see Pachares (1961).] Actually, unless n is very small or σ_0 very close to 0 or ∞, the equal-tails test given by

$$\int_0^{C_1} f_n(y)\,dy = \int_{C_2}^{\infty} f_n(y)\,dy = \frac{\alpha}{2}$$

is a good approximation to the unbiased test. This follows from the fact that T, suitably normalized, tends to be normally and hence symmetrically distributed for large n.

UMP unbiased tests of the hypotheses (iii) $H : \theta_1 \le \theta \le \theta_2$ and (iv) $H : \theta = \theta_0$ against two-sided alternatives exist not only when the family $p_\theta(x)$ is exponential but also more generally when it is strictly totally positive (STP_∞). A proof of (iv) in this case is given in Brown, Johnstone, and MacGibbon (1981); the proof of (iii) follows from Chapter 3, Problem 30.

3. SIMILARITY AND COMPLETENESS

In many important testing problems, the hypothesis concerns a single real-valued parameter, but the distribution of the observable random variables depends in addition on certain nuisance parameters. For a large class of such problems a UMP unbiased test exists and can be found through the method indicated by Lemma 1. This requires the characterization of the tests ϕ, which satisfy

$$E_\theta \phi(X) = \alpha$$

for all distributions of X belonging to a given family $\mathscr{P}^X = \{P_\theta, \theta \in \omega\}$. Such tests are called *similar* with respect to $\mathscr{P}^X$ or ω, since if ϕ is nonrandomized with critical region S, the latter is "similar to the sample space" $\mathscr{X}$ in that both the probability $P_\theta\{X \in S\}$ and $P_\theta\{X \in \mathscr{X}\}$ are independent of $\theta \in \omega$.

Let T be a sufficient statistic for $\mathscr{P}^X$, and let $\mathscr{P}^T$ denote the family $\{P_\theta^T, \theta \in \omega\}$ of distributions of T as θ ranges over ω. Then any test satisfying

(7) $E[\phi(X)|t] = \alpha$ a.e. $\mathscr{P}^{T*}$

is similar with respect to $\mathscr{P}^X$, since then

$$E_\theta[\phi(X)] = E_\theta\{E[\phi(X)|T]\} = \alpha \qquad \text{for all} \quad \theta \in \omega.$$

---

*A statement is said to hold a.e. $\mathscr{P}$ if it holds except on a set N with $P(N) = 0$ for all $P \in \mathscr{P}$.

A test satisfying (7) is said to have *Neyman structure* with respect to T. It is characterized by the fact that the conditional probability of rejection is α on each of the surfaces $T = t$. Since the distribution on each such surface is independent of θ for $\theta \in \omega$, the condition (7) essentially reduces the problem to that of testing a simple hypothesis for each value of t. It is frequently easy to obtain a most powerful test among those having Neyman structure, by solving the optimum problem on each surface separately. The resulting test is then most powerful among all similar tests provided every similar test has Neyman structure. A condition for this to be the case can be given in terms of the following definition.

A family $\mathscr{P}$ of probability distributions P is *complete* if

$$(8) \qquad\qquad E_P[f(X)] = 0 \qquad \text{for all} \quad P \in \mathscr{P}$$

implies

$$(9) \qquad\qquad\qquad f(x) = 0 \qquad \text{a.e.} \ \mathscr{P}.$$

In applications, $\mathscr{P}$ will be the family of distributions of a sufficient statistic.

Example 3. Consider n independent trials with probability p of success, and let X_i be 1 or 0 as the ith trial is a success or failure. Then $T = X_1 + \cdots + X_n$ is a sufficient statistic for p, and the family of its possible distributions is $\mathscr{P} = \{b(p, n),$ $0 \le p \le 1\}$. For this family (8) implies that

$$\sum_{t=0}^{n} f(t)\binom{n}{t}\rho^t = 0 \qquad \text{for all} \ \ 0 < \rho < \infty,$$

where $\rho = p/(1 - p)$. The left-hand side is a polynomial in ρ, all the coefficients of which must be zero. Hence $f(t) = 0$ for $t = 0, \ldots, n$ and the binomial family of distributions of T is complete.

Example 4. Let $X_1, \ldots, X_n$ be a sample from the uniform distribution $U(0, \theta)$, $0 < \theta < \infty$. Then $T = \max(X_1, \ldots, X_n)$ is a sufficient statistic for θ, and (8) becomes

$$\int f(t)\, dP_\theta^T(t) = n\theta^{-n}\int_0^\theta f(t) \cdot t^{n-1}\, dt = 0 \qquad \text{for all} \ \ \theta.$$

Let $f(t) = f^+(t) - f^-(t)$ where f^+ and f^- denote the positive and negative parts of f respectively. Then

$$\nu^+(A) = \int_A f^+(t) t^{n-1}\, dt \quad \text{and} \quad \nu^-(A) = \int_A f^-(t) t^{n-1}\, dt$$

are two measures over the Borel sets on $(0, \infty)$, which agree for all intervals and hence for all A. This implies $f^+(t) = f^-(t)$ except possibly on a set of Lebesgue measure zero, and hence $f(t) = 0$ a.e. $\mathscr{P}^T$.

Example 5. Let $X_1, \ldots, X_m; Y_1, \ldots, Y_n$ be independently normally distributed as $N(\xi, \sigma^2)$ and $N(\xi, \tau^2)$ respectively. Then the joint density of the variables is

$$C(\xi, \sigma, \tau)\exp\left(-\frac{1}{2\sigma^2}\sum x_i^2 + \frac{\xi}{\sigma^2}\sum x_i - \frac{1}{2\tau^2}\sum y_j^2 + \frac{\xi}{\tau^2}\sum y_j\right).$$

The statistic

$$T = \left(\sum X_i, \sum X_i^2, \sum Y_j, \sum Y_j^2\right)$$

is sufficient; it is, however, not complete, since $E(\sum Y_j/n - \sum X_i/m)$ is identically zero. If the Y's are instead distributed with a mean $E(Y) = \eta$ which varies independently of ξ, the set of possible values of the parameters $\theta_1 = -1/2\sigma^2$, $\theta_2 = \xi/\sigma^2$, $\theta_3 = -1/2\tau^2$, $\theta_4 = \eta/\tau^2$ contains a four-dimensional rectangle, and it follows from Theorem 1 below that $\mathscr{P}^T$ is complete.

Completeness of a large class of families of distributions including that of Example 3 is covered by the following theorem.

Theorem 1. *Let X be a random vector with probability distribution*

$$dP_\theta(x) = C(\theta)\exp\left[\sum_{j=1}^k \theta_j T_j(x)\right] d\mu(x),$$

and let $\mathscr{P}^T$ be the family of distributions of $T = (T_1(X), \ldots, T_k(X))$ as θ ranges over the set ω. Then $\mathscr{P}^T$ is complete provided ω contains a k-dimensional rectangle.

Proof. By making a translation of the parameter space one can assume without loss of generality that ω contains the rectangle

$$I = \left\{(\theta_1, \ldots, \theta_k) : -a \leq \theta_j \leq a, \ j = 1, \ldots, k\right\}.$$

Let $f(t) = f^+(t) - f^-(t)$ be such that

$$E_\theta f(T) = 0 \qquad \text{for all} \quad \theta \in \omega.$$

Then for all $\theta \in I$, if ν denotes the measure induced in T-space by the measure μ,

$$\int e^{\sum \theta_j t_j} f^+(t) \, d\nu(t) = \int e^{\sum \theta_j t_j} f^-(t) \, d\nu(t)$$

and hence in particular

$$\int f^+(t)\, d\nu(t) = \int f^-(t)\, d\nu(t).$$

Dividing f by a constant, one can take the common value of these two integrals to be 1, so that

$$dP^+(t) = f^+(t)\, d\nu(t) \quad \text{and} \quad dP^-(t) = f^-(t)\, d\nu(t)$$

are probability measures, and

$$\int e^{\Sigma \theta_j t_j}\, dP^+(t) = \int e^{\Sigma \theta_j t_j}\, dP^-(t)$$

for all θ in I. Changing the point of view, consider these integrals now as functions of the complex variables $\theta_j = \xi_j + i\eta_j$, $j = 1, \ldots, k$. For any fixed $\theta_1, \ldots, \theta_{j-1}, \theta_{j+1}, \ldots, \theta_k$, with real parts strictly between $-a$ and $+a$, they are by Theorem 9 of Chapter 2 analytic functions of θ_j in the strip $R_j: -a < \xi_j < a$, $-\infty < \eta_j < \infty$ of the complex plane. For $\theta_2, \ldots, \theta_k$ fixed, real, and between $-a$ and a, equality of the integrals holds on the line segment $\{(\xi_1, \eta_1): -a < \xi_1 < a,\ \eta_1 = 0\}$ and can therefore be extended to the strip R_1, in which the integrals are analytic. By induction the equality can be extended to the complex region $\{(\theta_1, \ldots, \theta_k): (\xi_j, \eta_j) \in R_j$ for $j = 1, \ldots, k\}$. It follows in particular that for all real $(\eta_1, \ldots, \eta_k)$

$$\int e^{i\Sigma \eta_j t_j}\, dP^+(t) = \int e^{i\Sigma \eta_j t_j}\, dP^-(t).$$

These integrals are the characteristic functions of the distributions P^+ and P^- respectively, and by the uniqueness theorem for characteristic functions,[*] the two distributions P^+ and P^- coincide. From the definition of these distributions it then follows that $f^+(t) = f^-(t)$, a.e. ν, and hence that $f(t) = 0$ a.e. $\mathcal{P}^T$, as was to be proved.

Example 6. Nonparametric completeness. Let $X_1, \ldots, X_N$ be independently and identically distributed with cumulative distribution function $F \in \mathcal{F}$, where $\mathcal{F}$ is the family of all absolutely continuous distributions. Then the set of order statistics $T(X) = (X_{(1)}, \ldots, X_{(N)})$ was shown to be sufficient for $\mathcal{F}$ in Chapter 2, Section 6. We shall now prove it to be complete. Since, by Example 7 of Chapter 2, $T'(X) = (\Sigma X_i, \Sigma X_i^2, \ldots, \Sigma X_i^N)$ is equivalent to $T(X)$ in the sense that both induce the same subfield of the sample space, $T'(X)$ is also sufficient and is complete if

---

[*] See for example Section 26 of Billingsley (1979).

and only if $T(X)$ is complete. To prove the completeness of $T'(X)$ and thereby that of $T(X)$, consider the family of densities

$$f(x) = C(\theta_1, \ldots, \theta_N)\exp\left(-x^{2N} + \theta_1 x + \cdots + \theta_N x^N\right),$$

where C is a normalizing constant. These densities are defined for all values of the θ's since the integral of the exponential is finite, and their distributions belong to $\mathcal{F}$. The density of a sample of size N is

$$C^N\exp\left(-\sum x_j^{2N} + \theta_1\sum x_j + \cdots + \theta_N\sum x_j^N\right)$$

and these densities constitute an exponential family $\mathcal{F}_0$. By Theorem 1, $T'(X)$ is complete for $\mathcal{F}_0$, and hence also for $\mathcal{F}$, as was to be proved.

The same method of proof establishes also the following more general result. Let X_{ij}, $j = 1, \ldots, N_i$, $i = 1, \ldots, c$, be independently distributed with absolutely continuous distributions F_i, and let $X_i^{(1)} < \cdots < X_i^{(N_i)}$ denote the N_i observations $X_{i1}, \ldots, X_{iN_i}$ arranged in increasing order. Then the set of order statistics

$$\left(X_1^{(1)}, \ldots, X_1^{(N_1)}, \ldots, X_c^{(1)}, \ldots, X_c^{(N_c)}\right)$$

is sufficient and complete for the family of distributions obtained by letting $F_1, \ldots, F_c$ range over all distributions of $\mathcal{F}$. Here completeness is proved by considering the subfamily $\mathcal{F}_0$ of $\mathcal{F}$ in which the distributions F_i have densities of the form

$$f_i(x) = C_i\left(\theta_{i1}, \ldots, \theta_{iN_i}\right)\exp\left(-x^{2N_i} + \theta_{i1}x + \cdots + \theta_{iN_i}x^{N_i}\right).$$

The result remains true if $\mathcal{F}$ is replaced by the family $\mathcal{F}_1$ of continuous distributions. For a proof see Problem 12 or Bell, Blackwell, and Breiman (1960).

For the present purpose the slightly weaker property of bounded completeness is appropriate, a family $\mathcal{P}$ of probability distributions being *boundedly complete* if for all bounded functions f, (8) implies (9). If $\mathcal{P}$ is complete it is a fortiori boundedly complete.

Theorem 2.   *Let X be a random variable with distribution $P \in \mathcal{P}$, and let T be a sufficient statistic for $\mathcal{P}$. Then a necessary and sufficient condition for all similar tests to have Neyman structure with respect to T is that the family $\mathcal{P}^T$ of distributions of T is boundedly complete.*

Proof.   Suppose first that $\mathcal{P}^T$ is boundedly complete, and let $\phi(X)$ be similar with respect to $\mathcal{P}$. Then

$$E[\phi(X) - \alpha] = 0 \qquad \text{for all} \quad P \in \mathcal{P}$$

and hence, if $\psi(t)$ denotes the conditional expectation of $\phi(X) - \alpha$ given t,

$$E\psi(T) = 0 \qquad \text{for all} \quad P^T \in \mathcal{P}^T.$$

Since $\psi(t)$ can be taken to be bounded by Lemma 3 of Chapter 2, it follows from the bounded completeness of $\mathscr{P}^T$ that $\psi(t) = 0$ and hence $E[\phi(X)|t] = \alpha$ a.e. $\mathscr{P}^T$, as was to be proved.

Conversely suppose that $\mathscr{P}^T$ is not boundedly complete. Then there exists a function f such that $|f(t)| \leq M$ for some M, that $Ef(T) = 0$ for all $P^T \in \mathscr{P}^T$, and $f(T) \neq 0$ with positive probability for some $P^T \in \mathscr{P}^T$. Let $\phi(t) = cf(t) + \alpha$, where $c = \min(\alpha, 1 - \alpha)/M$. Then ϕ is a critical function, since $0 \leq \phi(t) \leq 1$, and it is a similar test, since $E\phi(T) = \alpha$ for all $P^T \in \mathscr{P}^T$. But ϕ does not have Neyman structure, since $\phi(T) \neq \alpha$ with positive probability for at least some distribution in $\mathscr{P}^T$.

4. UMP UNBIASED TESTS FOR MULTIPARAMETER EXPONENTIAL FAMILIES

An important class of hypotheses concerns a real-valued parameter in an exponential family, with the remaining parameters occurring as unspecified nuisance parameters. In many of these cases, UMP unbiased tests exist and can be constructed by means of the theory of the preceding section.

Let X be distributed according to

(10)

$$dP^X_{\theta, \vartheta}(x) = C(\theta, \vartheta)\exp\left[\theta U(x) + \sum_{i=1}^{k} \vartheta_i T_i(x)\right] d\mu(x), \qquad (\theta, \vartheta) \in \Omega,$$

and let $\vartheta = (\vartheta_1, \ldots, \vartheta_k)$ and $T = (T_1, \ldots, T_k)$. We shall consider the problems* of testing the following hypotheses H_j against the alternatives K_j, $j = 1, \ldots, 4$:

$$
\begin{array}{ll}
H_1 : \theta \leq \theta_0 & K_1 : \theta > \theta_0 \\
H_2 : \theta \leq \theta_1 \text{ or } \theta \geq \theta_2 & K_2 : \theta_1 < \theta < \theta_2 \\
H_3 : \theta_1 \leq \theta \leq \theta_2 & K_3 : \theta < \theta_1 \text{ or } \theta > \theta_2 \\
H_4 : \theta = \theta_0 & K_4 : \theta \neq \theta_0.
\end{array}
$$

We shall assume that the parameter space Ω is convex, and that it has dimension $k + 1$, that is, that it is not contained in a linear space of dimension $< k + 1$. This is the case in particular when Ω is the natural parameter space of the exponential family. We shall also assume that there are points in Ω with θ both $<$ and $> \theta_0$, θ_1, and θ_2 respectively.

---

*Such problems are also treated in Johansen (1979), which in addition discusses large-sample tests of hypotheses specifying more than one parameter.

Attention can be restricted to the sufficient statistics (U, T) which have the joint distribution

$$(11) \quad dP_{\theta, \vartheta}^{U, T}(u, t) = C(\theta, \vartheta) \exp\left(\theta u + \sum_{i=1}^{k} \vartheta_i t_i \right) d\nu(u, t), \quad (\theta, \vartheta) \in \Omega.$$

When $T = t$ is given, U is the only remaining variable and by Lemma 8 of Chapter 2 the conditional distribution of U given t constitutes an exponential family

$$dP_\theta^{U|t}(u) = C_t(\theta) e^{\theta u} d\nu_t(u).$$

In this conditional situation there exists by Corollary 2 of Chapter 3 a UMP test for testing H_1 with critical function ϕ_1 satisfying

$$(12) \quad \phi(u, t) = \begin{cases} 1 & \text{when} \quad u > C_0(t), \\ \gamma_0(t) & \text{when} \quad u = C_0(t), \\ 0 & \text{when} \quad u < C_0(t), \end{cases}$$

where the functions C_0 and γ_0 are determined by

$$(13) \quad E_{\theta_0}[\phi_1(U, T)|t] = \alpha \quad \text{for all } t.$$

For testing H_2 in the conditional family there exists by Theorem 6 of Chapter 3 a UMP test with critical function

$$(14) \quad \phi(u, t) = \begin{cases} 1 & \text{when} \quad C_1(t) < u < C_2(t), \\ \gamma_i(t) & \text{when} \quad u = C_i(t), \quad i = 1, 2, \\ 0 & \text{when} \quad u < C_1(t) \text{ or } > C_2(t), \end{cases}$$

where the C's and γ's are determined by

$$(15) \quad E_{\theta_1}[\phi_2(U, T)|t] = E_{\theta_2}[\phi_2(U, T)|t] = \alpha.$$

Consider next the test ϕ_3 satisfying

$$(16) \quad \phi(u, t) = \begin{cases} 1 & \text{when} \quad u < C_1(t) \text{ or } > C_2(t), \\ \gamma_i(t) & \text{when} \quad u = C_i(t), \quad i = 1, 2, \\ 0 & \text{when} \quad C_1(t) < u < C_2(t), \end{cases}$$

with the C's and γ's determined by

(17) $$E_{\theta_1}[\phi_3(U, T)|t] = E_{\theta_2}[\phi_3(U, T)|t] = \alpha.$$

When $T = t$ is given, this is (by Section 2 of the present chapter) UMP unbiased for testing H_3 and UMP among all tests satisfying (17).

Finally, let ϕ_4 be a critical function satisfying (16) with the C's and γ's determined by

(18) $$E_{\theta_0}[\phi_4(U, T)|t] = \alpha$$

and

(19) $$E_{\theta_0}[U\phi_4(U, T)|t] = \alpha E_{\theta_0}[U|t].$$

Then given $T = t$, it follows again from the results of Section 2 that ϕ_4 is UMP unbiased for testing H_4 and UMP among all tests satisfying (18) and (19).

So far, the critical functions ϕ_j have been considered as conditional tests given $T = t$. Reinterpreting them now as tests depending on U and T for the hypotheses concerning the distribution of X (or the joint distribution of U and T) as originally stated, we have the following main theorem.*

Theorem 3. *Define the critical functions ϕ_1 by (12) and (13); ϕ_2 by (14) and (15); ϕ_3 by (16) and (17); ϕ_4 by (16), (18), and (19). These constitute UMP unbiased level-α tests for testing the hypotheses $H_1, \ldots, H_4$ respectively when the joint distribution of U and T is given by (11).*

Proof. The statistic T is sufficient for ϑ if θ has any fixed value, and hence T is sufficient for each

$$\omega_j = \{(\theta, \vartheta) : (\theta, \vartheta) \in \Omega, \theta = \theta_j\}, \qquad j = 0, 1, 2.$$

By Lemma 8 of Chapter 2, the associated family of distributions of T is given by

$$dP_{\theta_j, \vartheta}^T(t) = C(\theta_j, \vartheta)\exp\left(\sum_{i=1}^{k} \vartheta_i t_i\right) d\nu_{\theta_j}(t), \quad (\theta_j, \vartheta) \in \omega_j, \qquad j = 0, 1, 2.$$

Since by assumption Ω is convex and of dimension $k + 1$ and contains

---

*A somewhat different asymptotic optimality property of these tests is established by Michel (1979).

points on both sides of $\theta = \theta_j$, it follows that ω_j is convex and of dimension k. Thus ω_j contains a k-dimensional rectangle; by Theorem 1 the family

$$\mathscr{P}_j^T = \left\{ P_{\theta_j, \vartheta}^T : (\theta, \vartheta) \in \omega_j \right\}$$

is complete; and similarity of a test ϕ on ω_j implies

$$E_{\theta_j}[\phi(U, T)|t] = \alpha.$$

(1)   Consider first H_1. By Theorem 9 of Chapter 2 the power function of all tests is continuous for an exponential family. It is therefore enough to prove ϕ_1 to be UMP among all tests that are similar on ω_0 (Lemma 1), and hence among those satisfying (13). On the other hand, the overall power of a test ϕ against an alternative (θ, ϑ) is

$$(20) \qquad E_{\theta, \vartheta}[\phi(U, T)] = \int \left[\int \phi(u, t) \, dP_\theta^{U|t}(u) \right] dP_{\theta, \vartheta}^T(t).$$

One therefore maximizes the overall power by maximizing the power of the conditional test, given by the expression in brackets, separately for each t. Since ϕ_1 has the property of maximizing the conditional power against any $\theta > \theta_0$ subject to (13), this establishes the desired result.

(2)   The proof for H_2 and H_3 is completely analogous. By Lemma 1, it is enough to prove ϕ_2 and ϕ_3 to be UMP among all tests that are similar on both ω_1 and ω_2, and hence among all tests satisfying (15). For each t, ϕ_2 and ϕ_3 maximize the conditional power for their respective problems subject to this condition and therefore also the unconditional power.

(3)   Unbiasedness of a test of H_4 implies similarity on ω_0 and

$$\frac{\partial}{\partial \theta} \left[E_{\theta, \vartheta} \phi(U, T) \right] = 0 \qquad \text{on } \omega_0.$$

The differentiation on the left-hand side of this equation can be carried out under the expectation sign, and by the computation which earlier led to (6), the equation is seen to be equivalent to

$$E_{\theta, \vartheta}[U\phi(U, T) - \alpha U] = 0 \qquad \text{on } \omega_0.$$

Therefore, since $\mathscr{P}_0^T$ is complete, unbiasedness implies (18) and (19). As in the preceding cases, the test, which in addition satisfies (16), is UMP among all tests satisfying these two conditions. That it is UMP unbiased now follows, as in the proof of Lemma 1, by comparison with the test $\phi(u, t) \equiv \alpha$.

(4)   The functions $\phi_1, \ldots, \phi_4$ were obtained above for each fixed t as a function of u. To complete the proof it is necessary to show that they are jointly measurable in u and t, so that the expectation (20) exists. We shall prove this here for the case of ϕ_1; the proof for the other cases is sketched in Problems 14 and 15. To establish the measurability of ϕ_1, one needs to show that the functions $C_0(t)$ and $\gamma_0(t)$ defined by (12) and (13) are t-measurable. Omitting the subscript 0, and denoting the conditional distribution function of U given $T = t$ and for $\theta = \theta_0$ by

$$F_t(u) = P_{\theta_0}\{U \le u \,|\, t\},$$

one can rewrite (13) as

$$F_t(C) - \gamma[F_t(C) - F_t(C - 0)] = 1 - \alpha.$$

Here $C = C(t)$ is such that $F_t(C - 0) \le 1 - \alpha \le F_t(C)$, and hence

$$C(t) = F_t^{-1}(1 - \alpha)$$

where $F_t^{-1}(y) = \inf\{u : F_t(u) \ge y\}$. It follows that $C(t)$ and $\gamma(t)$ will both be measurable provided $F_t(u)$ and $F_t(u - 0)$ are jointly measurable in u and t and $F_t^{-1}(1 - \alpha)$ is measurable in t.

For each fixed u the function $F_t(u)$ is a measurable function of t, and for each fixed t it is a cumulative distribution function and therefore in particular nondecreasing and continuous on the right. From the second property it follows that $F_t(u) \ge c$ if and only if for each n there exists a rational number r such that $u \le r < u + 1/n$ and $F_t(r) \ge c$. Therefore, if the rationals are denoted by $r_1, r_2, \ldots$,

$$\{(u, t): F_t(u) \ge c\} = \bigcap_n \bigcup_i \left\{(u, t): 0 \le r_i - u < \frac{1}{n}, F_t(r_i) \ge c\right\}.$$

This shows that $F_t(u)$ is jointly measurable in u and t. The proof for $F_t(u - 0)$ is completely analogous. Since $F_t^{-1}(y) \le u$ if and only if $F_t(u) \ge y$, $F_t^{-1}(y)$ is t-measurable for any fixed y and this completes the proof.

The test ϕ_1 of the above theorem is also UMP unbiased if Ω is replaced by the set $\Omega' = \Omega \cap \{(\theta, \vartheta): \theta \ge \theta_0\}$, and hence for testing $H': \theta = \theta_0$ against $\theta > \theta_0$. The assumption that Ω should contain points with $\theta < \theta_0$ was in fact used only to prove that the boundary set ω_0 contains a k-dimensional rectangle, and this remains valid if Ω is replaced by Ω'.

The remainder of this chapter as well as the next chapter will be concerned mainly with applications of the preceding theorem to various statistical problems. While this provides the most expeditious proof that the tests in all these cases are UMP unbiased, there is available also a variation of the approach, which is more elementary. The proof of Theorem 3 is quite elementary except for the following points: (i) the fact that the conditional distributions of U given $T = t$ constitute an exponential family, (ii) that the family of distributions of T is complete, (iii) that the derivative of $E_{\theta, \vartheta}\phi(U, T)$ exists and can be computed by differentiating under the expectation sign, (iv) that the functions $\phi_1, \ldots, \phi_4$ are measurable. Instead of verifying (i) through (iv) in general, as was done in the above proof, it is possible in applications of the theorem to check these conditions directly for each specific problem, which in some cases is quite easy.

Through a transformation of parameters, Theorem 3 can be extended to cover hypotheses concerning parameters of the form

$$\theta^* = a_0\theta + \sum_{i=1}^{k} a_i\vartheta_i, \qquad a_0 \neq 0.$$

This transformation is formally given by the following lemma, the proof of which is immediate.

Lemma 2. *The exponential family of distributions* (10) *can also be written as*

$$dP^X_{\theta, \vartheta}(x) = K(\theta^*, \vartheta)\exp\left[\theta^*U^*(x) + \sum\vartheta_i T_i^*(x)\right] d\mu(x)$$

where

$$U^* = \frac{U}{a_0}, \qquad T_i^* = T_i - \frac{a_i}{a_0}U.$$

Application of Theorem 3 to the form of the distributions given in the lemma leads to UMP unbiased tests of the hypothesis $H_1^* : \theta^* \leq \theta_0$ and the analogously defined hypotheses H_2^*, H_3^*, H_4^*.

When testing one of the hypotheses H_j one is frequently interested in the power $\beta(\theta', \vartheta)$ of ϕ_j against some alternative θ'. As is indicated by the notation and is seen from (20), this power will usually depend on the unknown nuisance parameters ϑ. On the other hand, the power of the conditional test given $T = t$,

$$\beta(\theta'|t) = E_{\theta'}[\phi(U, T)|t],$$

is independent of ϑ and therefore has a known value.

The quantity $\beta(\theta'|t)$ can be interpreted in two ways: (i) It is the probability of rejecting H when $T = t$. Once T has been observed to have the value t, it may be felt, at least in certain problems, that this is a more appropriate expression of the power in the given situation than $\beta(\theta', \vartheta)$, which is obtained by averaging $\beta(\theta'|t)$ with respect to other values of t not relevant to the situation at hand. This argument leads to difficulties, since in many cases the conditioning could be carried even further and it is not clear where the process should stop. (ii) A more clear-cut interpretation is obtained by considering $\beta(\theta'|t)$ as an estimate of $\beta(\theta', \vartheta)$. Since

$$E_{\theta', \vartheta}[\beta(\theta'|T)] = \beta(\theta', \vartheta),$$

this estimate is unbiased in the sense of Chapter 1, equation (11). It follows further from the theory of unbiased estimation and the completeness of the exponential family that among all unbiased estimates of $\beta(\theta', \vartheta)$ the present one has the smallest variance. (See TPE, Chapter 2.)

Regardless of the interpretation, $\beta(\theta'|t)$ has the disadvantage compared with an unconditional power that it becomes available only after the observations have been taken. It therefore cannot be used to plan the experiment and in particular to determine the sample size, if this must be done prior to the experiment. On the other hand, a simple sequential procedure guaranteeing a specified power β against the alternatives $\theta = \theta'$ is obtained by continuing taking observations until the conditional power $\beta(\theta'|t)$ is $\geq \beta$.

The general question of whether to interpret measures of performance such as the power of a test or coverage probability of a family of confidence statements conditionally, and if so, conditionally on what aspects of the data, will be considered in Chapter 10.

5. COMPARING TWO POISSON OR BINOMIAL POPULATIONS

A problem arising in many different contexts is the comparison of two treatments or of one treatment with a control situation in which no treatment is applied. If the observations consist of the number of successes in a sequence of trials for each treatment, for example the number of cures of a certain disease, the problem becomes that of testing the equality of two binomial probabilities. If the basic distributions are Poisson, for example in a comparison of the radioactivity of two substances, one will be testing the equality of two Poisson distributions.

When testing whether a treatment has a beneficial effect by comparing it with the control situation of no treatment, the problem is of the one-sided type. If ξ_2 and ξ_1 denote the parameter values when the treatment is or is not applied, the class of alternatives is $K : \xi_2 > \xi_1$. The hypothesis is $\xi_2 = \xi_1$ if it is known a priori that there is either no effect or a beneficial one; it is $\xi_2 \leq \xi_1$ if the possibility is admitted that the treatment may actually be harmful. Since the test is the same for the two hypotheses, the second somewhat safer hypothesis would seem preferable in most cases.

A one-sided formulation is sometimes appropriate also when a new treatment or process is being compared with a standard one, where the new treatment is of interest only if it presents an improvement. On the other hand, if the two treatments are on an equal footing, the hypothesis $\xi_2 = \xi_1$ of equality of two treatments is tested against the two-sided alternatives $\xi_2 \neq \xi_1$. The formulation of this problem as one of hypothesis testing is usually quite artificial, since in case of rejection of the hypothesis one will obviously wish to know which of the treatments is better.* Such two-sided tests do, however, have important applications to the problem of obtaining confidence limits for the extent by which one treatment is better than the other. They also arise when the parameter ξ does not measure a treatment effect but refers to an auxiliary variable which one hopes can be ignored. For example, ξ_1 and ξ_2 may refer to the effect of two different hospitals in a medical investigation in which one would like to combine the patients into a single study group. (In this connection, see also Chapter 7, Section 3.)

To apply Theorem 3 to this comparison problem it is necessary to express the distributions in an exponential form with $\theta = f(\xi_1, \xi_2)$, for example $\theta = \xi_2 - \xi_1$ or ξ_2/ξ_1, such that the hypotheses of interest become equivalent to those of Theorem 3. In the present section the problem will be considered for Poisson and binomial distributions; the case of normal distributions will be taken up in Chapter 5.

We consider first the Poisson problem in which X and Y are independently distributed according to $P(\lambda)$ and $P(\mu)$, so that their joint distribution can be written as

$$P\{ X = x, Y = y \} = \frac{e^{-(\lambda + \mu)}}{x! y!} \exp\left[y \log \frac{\mu}{\lambda} + (x + y) \log \lambda \right].$$

By Theorem 3 there exist UMP unbiased tests of the four hypotheses $H_1, \ldots, H_4$ concerning the parameter $\theta = \log(\mu/\lambda)$ or equivalently concerning the ratio $\rho = \mu/\lambda$. This includes in particular the hypotheses $\mu \leq \lambda$ (or $\mu = \lambda$) against the alternatives $\mu > \lambda$, and $\mu = \lambda$ against $\mu \neq \lambda$. Comparing the distribution of (X, Y) with (10), one has $U = Y$ and $T = X + Y$, and by Theorem 3 the tests are performed conditionally on the integer points of the

---

*For a discussion of the comparison of two treatments as a three-decision problem, see Bahadur (1952) and Lehmann (1957).

line segment $X + Y = t$ in the positive quadrant of the (x, y) plane. The conditional distribution of Y given $X + Y = t$ is (Problem 13 of Chapter 2)

$$P\{Y = y \mid X + Y = t\} = \binom{t}{y} \left(\frac{\mu}{\lambda + \mu}\right)^y \left(\frac{\lambda}{\lambda + \mu}\right)^{t-y}, \qquad y = 0, 1, \ldots, t,$$

the binomial distribution corresponding to t trials and probability $p = \mu/(\lambda + \mu)$ of success. The original hypotheses therefore reduce to the corresponding ones about the parameter p of a binomial distribution. The hypothesis $H: \mu \leq a\lambda$, for example, becomes $H: p \leq a/(a + 1)$, which is rejected when Y is too large. The cutoff point depends of course, in addition to a, also on t. It can be determined from tables of the binomial, and for large t approximately from tables of the normal distribution.

In many applications the ratio $\rho = \mu/\lambda$ is a reasonable measure of the extent to which the two Poisson populations differ, since the parameters λ and μ measure the rates (in time or space) at which two Poisson processes produce the events in question. One might therefore hope that the power of the above tests depends only on this ratio, but this is not the case. On the contrary, for each fixed value of ρ corresponding to an alternative to the hypothesis being tested, the power $\beta(\lambda, \mu) = \beta(\lambda, \rho\lambda)$ is an increasing function of λ, which tends to 1 as $\lambda \to \infty$ and to α as $\lambda \to 0$. To see this consider the power $\beta(\rho|t)$ of the conditional test given t. This is an increasing function of t, since it is the power of the optimum test based on t binomial trials. The conditioning variable T has a Poisson distribution with parameter $\lambda(1 + \rho)$, and its distribution for varying λ forms an exponential family. It follows (Lemma 2 of Chapter 3) that the overall power $E[\beta(\rho|T)]$ is an increasing function of λ. As $\lambda \to 0$ or ∞, T tends in probability to 0 or ∞, and the power against a fixed alternative ρ tends to α or 1.

The above test is also applicable to samples $X_1, \ldots, X_m$ and $Y_1, \ldots, Y_n$ from two Poisson distributions. The statistics $X = \sum_{i=1}^{m} X_i$ and $Y = \sum_{j=1}^{n} Y_j$ are then sufficient for λ and μ, and have Poisson distributions with parameters $m\lambda$ and $n\mu$ respectively. In planning an experiment one might wish to determine $m = n$ so large that the test of, say, $H: \rho \leq \rho_0$ has power against a specified alternative ρ_1 greater than or equal to some preassigned β. However, it follows from the discussion of the power function for $n = 1$, which applies equally to any other n, that this cannot be achieved for any fixed n, no matter how large. This is seen more directly by noting that as $\lambda \to 0$, for both $\rho = \rho_0$ and $\rho = \rho_1$ the probability of the event $X = Y = 0$ tends to 1. Therefore, the power of any level-α test against $\rho = \rho_1$ and for varying λ cannot be bounded away from α. This difficulty can be overcome only by permitting observations to be taken sequentially. One can for example determine t_0 so large that the test of the hypothesis $p \leq \rho_0/(1 + \rho_0)$ on the basis of t_0 binomial trials has power $\geq \beta$ against the alternative $p_1 = \rho_1/(1 + \rho_1)$. By observing (X_1, Y_1), (X_2, Y_2),... and continuing until $\Sigma(X_i + Y_i) \geq t_0$, one obtains a test with power $\geq \beta$ against all alternatives with $\rho \geq \rho_1$.*

The corresponding comparison of two binomial probabilities is quite similar. Let X and Y be independent binomial variables with joint distribution

$$P\{X = x, Y = y\} = \binom{m}{x} p_1^x q_1^{m-x} \binom{n}{y} p_2^y q_2^{n-y}$$

$$= \binom{m}{x}\binom{n}{y} q_1^m q_2^n \exp\left[y\left(\log\frac{p_2}{q_2} - \log\frac{p_1}{q_1}\right)\right.$$

$$\left. + (x + y)\log\frac{p_1}{q_1}\right].$$

The four hypotheses $H_1, \ldots, H_4$ can then be tested concerning the parameter

$$\theta = \log\left(\frac{p_2}{q_2} \Big/ \frac{p_1}{q_1}\right),$$

or equivalently concerning the *odds ratio* (also called *cross-product ratio*)

$$\rho = \frac{p_2}{q_2} \Big/ \frac{p_1}{q_1}.$$

This includes in particular the problems of testing $H_1' : p_2 \leq p_1$ against $p_2 > p_1$ and $H_4' : p_2 = p_1$ against $p_2 \neq p_1$. As in the Poisson case, $U = Y$ and $T = X + Y$, and the test is carried out in terms of the conditional distribution of Y on the line segment $X + Y = t$. This distribution is given by

(21) $P\{Y = y \mid X + Y = t\} = C_t(\rho)\binom{m}{t - y}\binom{n}{y}\rho^y,$ $y = 0, 1, \ldots, t,$

*A discussion of this and alternative procedures for achieving the same aim is given by Birnbaum (1954).

where

$$C_t(\rho) = \frac{1}{\sum\limits_{y'=0}^{t} \binom{m}{t-y'}\binom{n}{y'}\rho^{y'}}.$$

In the particular case of the hypotheses H_1' and H_4', the boundary value θ_0 of (13), (18), and (19) is 0, and the corresponding value of ρ is $\rho_0 = 1$. The conditional distribution then reduces to

$$P\{Y = y \mid X + Y = t\} = \frac{\binom{m}{t-y}\binom{n}{y}}{\binom{m+n}{t}},$$

which is the hypergeometric distribution.

Tables of critical values by Finney (1948) are reprinted in *Biometrika Tables for Statisticians*, Vol. 1, Table 38 and are extended in Finney, Latscha, Bennett, Hsu, and Horst (1963, 1966). Somewhat different ranges are covered in Armsen (1955), and related charts are provided by Bross and Kasten (1957). Extensive tables of the hypergeometric distributions have been computed by Lieberman and Owen (1961). Various approximations are discussed in Johnson and Kotz (1969, Section 6.5) and by Ling and Pratt (1984); see also Cressie (1978).

The UMP unbiased test of $p_1 = p_2$, which is based on the (conditional) hypergeometric distribution, requires randomization to obtain an exact conditional level α for each t of the sufficient statistic T. Since in practice randomization is usually unacceptable, the one-sided test is frequently performed by rejecting when $Y \geq C(T)$, where $C(t)$ is the smallest integer for which $P\{Y \geq C(T) \mid T = t\} \leq \alpha$. This conservative test is called *Fisher's exact test* [after the treatment given in Fisher (1934)], since the probabilities are calculated from the exact hypergeometric rather than an approximate normal distribution. The resulting conditional levels (and hence the unconditional level) are often considerably smaller than α, and this results in a substantial loss of power. An approximate test whose overall level tends to be closer to α is obtained by using the normal approximation to the hypergeometric distribution *without* continuity correction. [For a comparison of this test with some competitors, see e.g. Garside and Mack (1976).] A nonrandomized test that provides a conservative overall level, but that is less conservative than the "exact" test, is described by Boschloo (1970) and by McDonald, Davis, and Milliken (1977). Convenient entries into the extensive literature on these and related aspects of 2×2 tables can be found in Conover (1974), Kempthorne (1979), and Cox and Plackett (1980); see also Haber (1980), Barnard (1982), Overall and Starbuck (1983), and Yates (1984). For extensions to $r \times c$ tables, see Mehta and Patel (1983) and the literature cited there.

6. TESTING FOR INDEPENDENCE IN A 2 × 2 TABLE

The problem of deciding whether two characteristics A and B are independent in a population was discussed in Section 4 of Chapter 3 (Example 4), under the assumption that the marginal probabilities $p(A)$ and $p(B)$ are known. The most informative sample of size s was found to be one selected entirely from that one of the four categories A, $\tilde{A}$, B, or $\tilde{B}$, say A, which is rarest in the population. The problem then reduces to testing the hypothesis $H: p = p(B)$ in a binomial distribution $b(p, s)$.

In the more usual situation that $p(A)$ and $p(B)$ are not known, a sample from one of the categories such as A does not provide a basis for distinguishing between the hypothesis and the alternatives. This follows from the fact that the number in the sample possessing characteristic B then constitutes a binomial variable with probability $p(B|A)$, which is completely unknown both when the hypothesis is true and when it is false. The hypothesis can, however, be tested if samples are taken both from categories A and $\tilde{A}$ or both from B and $\tilde{B}$. In the latter case, for example, if the sample sizes are m and n, the numbers of cases possessing characteristic A in the two samples constitute independent variables with binomial distributions $b(p_1, m)$ and $b(p_2, n)$ respectively, where $p_1 = P(A|B)$ and $p_2 = P(A|\tilde{B})$. The hypothesis of independence of the two characteristics, $p(A|B) = p(A)$, is then equivalent to the hypothesis $p_1 = p_2$, and the problem reduces to that treated in the preceding section.

Instead of selecting samples from two of the categories, it is frequently more convenient to take the sample at random from the population as a whole. The results of such a sample can be summarized in the following 2×2 contingency table, the entries of which give the numbers in the various categories:

	A	$\tilde{A}$	
B	X	X'	M
$\tilde{B}$	Y	Y'	N
	T	T'	s

The joint distribution of the variables X, X', Y, and Y' is multinomial, and is given by

$$P\{ X = x, \ X' = x', \ Y = y, \ Y' = y'\}$$

$$= \frac{s!}{x!x'!y!y'!} p_{AB}^x p_{\bar{A}B}^{x'} p_{A\bar{B}}^y p_{\bar{A}\bar{B}}^{y'}$$

$$= \frac{s!}{x!x'!y!y'!} p_{\bar{A}\bar{B}}^s \exp\left(x \log \frac{p_{AB}}{p_{\bar{A}\bar{B}}} + x' \log \frac{p_{\bar{A}B}}{p_{\bar{A}\bar{B}}} + y \log \frac{p_{A\bar{B}}}{p_{\bar{A}\bar{B}}} \right).$$

Lemma 2 and Theorem 3 are therefore applicable to any parameter of the form

$$\theta^* = a_0 \log \frac{p_{AB}}{p_{\bar{A}\bar{B}}} + a_1 \log \frac{p_{\bar{A}B}}{p_{\bar{A}\bar{B}}} + a_2 \log \frac{p_{A\bar{B}}}{p_{\bar{A}\bar{B}}}.$$

Putting $a_1 = a_2 = 1$, $a_0 = -1$, $\Delta = e^{\theta^*} = (p_{\bar{A}B} p_{A\bar{B}})/(p_{AB} p_{\bar{A}\bar{B}})$, and denoting the probabilities of A and B in the population by $p_A = p_{AB} + p_{A\bar{B}}$, $p_B = p_{AB} + p_{\bar{A}B}$, one finds

$$p_{AB} = p_A p_B + \frac{1 - \Delta}{\Delta} p_{\bar{A}B} p_{A\bar{B}},$$

$$p_{\bar{A}B} = p_{\bar{A}} p_B - \frac{1 - \Delta}{\Delta} p_{\bar{A}B} p_{A\bar{B}},$$

$$p_{A\bar{B}} = p_A p_{\bar{B}} - \frac{1 - \Delta}{\Delta} p_{\bar{A}B} p_{A\bar{B}},$$

$$p_{\bar{A}\bar{B}} = p_{\bar{A}} p_{\bar{B}} + \frac{1 - \Delta}{\Delta} p_{\bar{A}B} p_{A\bar{B}}.$$

Independence of A and B is therefore equivalent to $\Delta = 1$, and $\Delta < 1$ and $\Delta > 1$ correspond to positive and negative dependence respectively.[†]

The test of the hypothesis of independence, or any of the four hypotheses concerning Δ, is carried out in terms of the conditional distribution of X given $X + X' = m$, $X + Y = t$. Instead of computing this distribution

---

[†] Δ is equivalent to Yule's measure of association, which is $Q = (1 - \Delta)/(1 + \Delta)$. For a discussion of this and related measures see Goodman and Kruskal (1954, 1959), Edwards (1963), and Haberman (1982).

directly, consider first the conditional distribution subject only to the condition $X + X' = m$, and hence $Y + Y' = s - m = n$. This is seen to be

$$P\{ X = x, Y = y \mid X + X' = m \}$$

$$= \binom{m}{x}\binom{n}{y}\left(\frac{p_{AB}}{p_B}\right)^x \left(\frac{p_{\tilde{A}B}}{p_B}\right)^{m-x} \left(\frac{p_{A\tilde{B}}}{p_{\tilde{B}}}\right)^y \left(\frac{p_{\tilde{A}\tilde{B}}}{p_{\tilde{B}}}\right)^{n-y},$$

which is the distribution of two independent binomial variables, the number of successes in m and n trials with probability $p_1 = p_{AB}/p_B$ and $p_2 = p_{A\tilde{B}}/p_{\tilde{B}}$. Actually, this is clear without computation, since we are now dealing with samples of fixed size m and n from the subpopulations B and $\tilde{B}$, and the probability of A in these subpopulations is p_1 and p_2. If now the additional restriction $X + Y = t$ is imposed, the conditional distribution of X subject to the two conditions $X + X' = m$ and $X + Y = t$ is the same as that of X given $X + Y = t$ in the case of two independent binomials considered in the previous section. It is therefore given by

$$P\{ X = x \mid X + X' = m, X + Y = t \} = C_t(\rho)\binom{m}{x}\binom{n}{t-x}\rho^{t-x},$$

$$x = 0, \ldots, t,$$

that is, by (21) expressed in terms of x instead of y. (Here the choice of X as testing variable is quite arbitrary; we could equally well again have chosen Y.) For the parameter ρ one finds

$$\rho = \frac{p_2}{q_2} \Big/ \frac{p_1}{q_1} = \frac{p_{\tilde{A}B} p_{A\tilde{B}}}{p_{AB} p_{\tilde{A}\tilde{B}}} = \Delta.$$

From these considerations it follows that the conditional test given $X + X' = m$, $X + Y = t$, for testing any of the hypotheses concerning Δ is identical with the conditional test given $X + Y = t$ of the same hypothesis concerning $\rho = \Delta$ in the preceding section, in which $X + X' = m$ was given a priori. In particular, the conditional test for testing the hypothesis of independence $\Delta = 1$, Fisher's exact test, is the same as that of testing the equality of two binomial p's and is therefore given in terms of the hypergeometric distribution.

    At the beginning of the section it was pointed out that the hypothesis of independence can be tested on the basis of samples obtained in a number of different ways. Either samples of fixed size can be taken from A and $\tilde{A}$ or from B and $\tilde{B}$, or the sample can be selected at random from the population at large. Which of these designs is most efficient depends on the cost of sampling from the various categories and from the population at large, and also on the cost of performing the necessary classification of a selected individual with respect to the characteristics in question. Suppose, however, for a moment that these considerations are neglected and that the designs are compared solely in terms of the power that the resulting tests achieve against a common alternative. Then the following results* can be shown to hold asymptotically as the total sample size s tends to infinity:

(i) If samples of size m and n $(m + n = s)$ are taken from B and $\tilde{B}$ or from A and $\tilde{A}$, the best choice of m and n is $m = n = s/2$.

(ii) It is better to select samples of equal size $s/2$ from B and $\tilde{B}$ than from A and $\tilde{A}$ provided $|p_B - \frac{1}{2}| > |p_A - \frac{1}{2}|$.

(iii) Selecting the sample at random from the population at large is worse than taking equal samples either from A and $\tilde{A}$ or from B and $\tilde{B}$.

These statements, which we shall not prove here, can be established by using the normal approximation for the distribution of the binomial variables X and Y when m and n are fixed, and by noting that under random sampling from the population at large, M/s and N/s tend in probability to p_B and $p_{\tilde{B}}$ respectively.

7. ALTERNATIVE MODELS FOR 2×2 TABLES

Conditioning of the multinomial model for the 2×2 table on the row (or column) totals was seen in the last section to lead to the two-binomial model of Section 5. Similarly, the multinomial model itself can be obtained as a conditional model in some situations in which not only the marginal totals M, N, T, and T' are random but the total sample size s is also a random variable. Suppose that the occurrence of events (e.g. patients presenting themselves for treatment) is observed over a given period of time, and that the events belonging to each of the categories AB, $\tilde{A}B$, $A\tilde{B}$, $\tilde{A}\tilde{B}$ are governed by independent Poisson processes, so that by (2) of Chapter 1 the numbers X, X', Y, Y' are independent Poisson variables with expectations λ_{AB}, $\lambda_{\tilde{A}B}$, $\lambda_{A\tilde{B}}$, $\lambda_{\tilde{A}\tilde{B}}$, and hence s is a Poisson variable with expectation $\lambda = \lambda_{AB} + \lambda_{\tilde{A}B} + \lambda_{A\tilde{B}} + \lambda_{\tilde{A}\tilde{B}}$.

It may then be of interest to compare the ratio $\lambda_{AB}/\lambda_{\tilde{A}B}$ with $\lambda_{A\tilde{B}}/\lambda_{\tilde{A}\tilde{B}}$ and in particular to test the hypothesis $H: \lambda_{AB}/\lambda_{\tilde{A}B} \leq \lambda_{A\tilde{B}}/\lambda_{\tilde{A}\tilde{B}}$. The joint distribution of X, X', Y, Y' constitutes a four-parameter exponential family,

---

*These results were conjectured by Berkson and proved by Neyman in a course on χ^2.

which can be written as

$$P(X = x, X' = x', Y = y, Y' = y')$$

$$= \frac{1}{x!x'!y!y'!} \exp\left\{ x \log\left(\frac{\lambda_{AB}\lambda_{\tilde{A}\tilde{B}}}{\lambda_{A\tilde{B}}\lambda_{\tilde{A}B}} \right) + (x' + x)\log \lambda_{\tilde{A}B} \right.$$

$$\left. + (y + x)\log \lambda_{A\tilde{B}} + (y' - x)\log \lambda_{\tilde{A}\tilde{B}} \right\}.$$

Thus, UMP unbiased tests exist of the usual one- and two-sided hypotheses concerning the parameter $\theta = \lambda_{AB}\lambda_{\tilde{A}\tilde{B}}/\lambda_{\tilde{A}B}\lambda_{A\tilde{B}}$. These are carried out in terms of the conditional distribution of X given

$$X' + X = m, \qquad Y + X = t, \qquad X + X' + Y + Y' = s,$$

where the last condition follows from the fact that given the first two it is equivalent to $Y' - X = s - t - m$. By Problem 13 of Chapter 2, the conditional distribution of X, X', Y given $X + X' + Y + Y' = s$ is the multinomial distribution of Section 6 with

$$p_{AB} = \frac{\lambda_{AB}}{\lambda}, \quad p_{\tilde{A}B} = \frac{\lambda_{\tilde{A}B}}{\lambda}, \quad p_{A\tilde{B}} = \frac{\lambda_{A\tilde{B}}}{\lambda}, \quad p_{\tilde{A}\tilde{B}} = \frac{\lambda_{\tilde{A}\tilde{B}}}{\lambda}.$$

The tests therefore reduce to those derived in Section 6.

The three models discussed so far involve different sampling schemes. However, frequently the subjects for study are not obtained by any sampling but are the only ones readily available to the experimenter. To create a probabilistic basis for a test in such situations, suppose that B and $\tilde{B}$ are two treatments, either of which can be assigned to each subject, and that A and $\tilde{A}$ denote success or failure (e.g. survival, relief of pain, etc.). The hypothesis of no difference in the effectiveness of the two treatments (i.e. independence of A and B) can then be tested by assigning the subjects to the treatments, say m to B and n to $\tilde{B}$, at random, i.e. in such a way that all possible $\binom{s}{m}$ assignments are equally likely. It is now this random assignment which takes the place of the sampling process in creating a probability model, thus making it possible to calculate significance.

Under the hypothesis H of no treatment difference, the success or failure of a subject is independent of the treatment to which it is assigned. If the numbers of subjects in categories A and $\tilde{A}$ are t and t' respectively $(t + t' = s)$, the values of t and t' are therefore fixed, so that we are now dealing with a 2×2 table in which all four margins t, t', m, n are fixed.

Then any one of the four cell counts X, X', Y, Y' determines the other three. Under H, the distribution of Y is the hypergeometric distribution derived as the conditional null distribution of Y given $X + Y = t$ at the end of Section 5. The hypothesis is rejected in favor of the alternative that treatment $\tilde{B}$ enhances success if Y is sufficiently large. Although this is the natural test under the given circumstances, no optimum property can be claimed for it, since no clear alternative model to H has been formulated.*

Consider finally the situation in which the subjects are again given rather than sampled, but B and $\tilde{B}$ are attributes (for example, male or female, smoker or nonsmoker) which cannot be assigned to the subjects at will. Then there exists no stochastic basis for answering the question whether observed differences in the rates X/M and Y/N correspond to differences between B and $\tilde{B}$, or whether they are accidental. An approach to the testing of such hypotheses in a nonstochastic setting has been proposed by Freedman and Lane (1982).

The various models for the 2×2 table discussed in Sections 6 and 7 may be characterized by indicating which elements are random and which fixed:

(i)   All margins and s random (Poisson).

(ii) All margins are random, s fixed (multinomial sampling).

(iii) One set of margins random, the other (and then a fortiori s) fixed (binomial sampling).

(iv) All margins fixed. Sampling replaced by random assignment of subjects to treatments.

(v)   All aspects fixed; no element of randomness.

In the first three cases there exist UMP unbiased one- and two-sided tests of the hypothesis of independence of A and B. These tests are carried out by conditioning on the values of all elements in (i)–(iii) that are random, so that in the conditional model all margins are fixed. The remaining randomness in the table can be described by any one of the four cell entries; once it is known, the others are determined by the margins. The distribution of such an entry under H has the hypergeometric distribution given at the end of Section 5.

The models (i)–(iii) have a common feature. The subjects under observation have been obtained by sampling from a population, and the inference corresponding to acceptance or rejection of H refers to that population. This is not true in cases (iv) and (v).

---

*The one-sided test is of course UMP against the class of alternatives defined by the right side of (21), but no reasonable assumptions have been proposed that would lead to this class. For suggestions of a different kind of alternative see Gokhale and Johnson (1978).

In (iv) the subjects are given, and a probabilistic basis is created by assigning them at random, m to B and n to $\tilde{B}$. Under the hypothesis H of no treatment difference, the four margins are fixed without any conditioning, and the four cell entries are again determined by any one of them, which under H has the same hypergeometric distribution as before. The present situation differs from the earlier three in that the inference cannot be extended beyond the subjects at hand.*

The situation (v) is outside the scope of this book, since it contains no basis for the type of probability calculations considered here. Problems of this kind are however of great importance, since they arise in many observational (as opposed to experimental) studies. For a related discussion, see Finch (1979).

8. SOME THREE-FACTOR CONTINGENCY TABLES

When an association between A and B exists in a 2×2 table, it does not follow that one of the factors has a causal influence on the other. Instead, the explanation may, for example, lie in the fact that both factors are causally affected by a third factor C. If C has K possible outcomes $C_1, \ldots, C_K$, one may then be faced with the apparently paradoxical situation that A and B are independent under each of the conditions C_k $(k = 1, \ldots, K)$ but exhibit positive (or negative) association when the tables are aggregated over C, that is, when the K separate 2×2 tables are combined into a single one showing the total counts of the four categories. [An interesting example is discussed by Bickel et al. (1977); see also Lindley and Novick (1981).] In order to determine whether the association of A and B in the aggregated table is indeed "spurious", one would test the hypothesis, (which arises also in other contexts) that A and B are conditionally independent given C_k for all $k = 1, \ldots, K$, against the alternative that there is an association for at least some k.

Let X_k, X_k', Y_k, Y_k' denote the counts in the $4K$ cells of the $2 \times 2 \times K$ table which extends the 2×2 table of Section 6 to the present case.

Again, several sampling schemes are possible. Consider first a random sample of size s from the population at large. The joint distribution of the $4K$ cellcounts then is multinomial with probabilities p_{ABC_k}, $p_{\bar{A}BC_k}, p_{A\bar{B}C_k}, p_{\bar{A}\bar{B}C_k}$ for the outcomes indicated by the subscripts. If Δ_k

---

*For a more detailed treatment of the distinction between population models [such as (i)–(iii)] and randomization models [such as (iv)], see Lehmann (1975).

denotes the AB odds ratio for C_k defined by

$$\Delta_k = \frac{p_{A\tilde{B}C_k} p_{\tilde{A}BC_k}}{p_{ABC_k} p_{\tilde{A}\tilde{B}C_k}} = \frac{p_{A\tilde{B}|C_k} p_{\tilde{A}B|C_k}}{p_{AB|C_k} p_{\tilde{A}\tilde{B}|C_k}},$$

where $p_{AB|C_k}, \ldots$ denotes the conditional probability of the indicated event given C_k, then the hypothesis to be tested is $\Delta_k = 1$ for all k.

A second scheme takes samples of size s_k from C_k and classifies the subjects as AB, $\tilde{A}B$, $A\tilde{B}$, or $\tilde{A}\tilde{B}$. This is the case of K independent 2×2 tables, in which one is dealing with K quadrinomial distributions of the kind considered in the preceding sections. Since the kth of these distributions is also that of the same four outcomes in the first model conditionally given C_k, we shall denote the probabilities of these outcomes in the present model again by $p_{AB|C_k}, \ldots$.

To motivate the next sampling scheme, suppose that A and $\tilde{A}$ represent success or failure of a medical treatment, $\tilde{B}$ and B that the treatment is applied or the subject is used as a control, and C_k the kth hospital taking part in this study. If samples of size n_k and m_k are obtained and are assigned to treatment and control respectively, we are dealing with K pairs of binomial distributions. Letting Y_k and X_k denote the number of successes obtained by the treatment subjects and controls in the kth hospital, the joint distribution of these variables by Section 5 is

$$\left[\prod \binom{m_k}{x_k} \binom{n_k}{y_k} q_{1k}^{m_k} q_{2k}^{n_k} \right] \exp\left(\sum y_k \log \Delta_k + \sum (x_k + y_k) \log \frac{p_{1k}}{q_{1k}} \right),$$

where p_{1k} and q_{1k}, (p_{2k} and q_{2k}) denote the probabilities of success and failure under B (under $\tilde{B}$).

The above three sampling schemes lead to $2 \times 2 \times K$ tables in which respectively none, one, or two of the margins are fixed. Alternatively, in some situations a model may be appropriate in which the $4K$ variables X_k, X_k', Y_k, Y_k' are independent Poisson with expectations $\lambda_{ABC_k}, \ldots$. In this case, the total sample size s is also random.

For a test of the hypothesis of conditional independence of A and B given C_k for all k (i.e. that $\Delta_1 = \cdots = \Delta_k = 1$), see Problem 43 of Chapter 8. Here we shall consider the problem under the simplifying assumption that the Δ_k have a common value Δ, so that the hypothesis reduces to $H: \Delta = 1$. Applying Theorem 3 to the third model (K pairs of binomials) and assuming the alternatives to be $\Delta > 1$, we see that a UMP unbiased test exists and rejects H when $\sum Y_k > C(X_1 + Y_1, \ldots, X_K + Y_K)$, where C is determined so that the conditional probability of rejection, given that $X_k + Y_k = t_k$, is α for all $k = 1, \ldots, K$. It follows from Section 5 that the conditional joint distribution of the Y_k under H is

$$P_H[Y_1 = y_1, \ldots, Y_K = y_K \mid X_k + Y_k = t_k, k = 1, \ldots, K]$$

$$= \prod \frac{\binom{m_k}{t_k - y_k}\binom{n_k}{y_k}}{\binom{m_k + n_k}{t_k}}$$

The conditional distribution of ΣY_k can now be obtained by adding the probabilities over all $(y_1, \ldots, y_K)$ whose sum has a given value. Unless the numbers are very small, this is impractical and approximations must be used [see Cox (1966) and Gart (1970)].

The assumption $H': \Delta_1 = \cdots = \Delta_K = \Delta$ has a simple interpretation when the successes and failures of the binomial trials are obtained by dichotomizing underlying unobservable continuous response variables. In a single such trial, suppose the underlying variable is Z and that success occurs when $Z > 0$ and failure when $Z \leq 0$. If Z is distributed as $F(Z - \zeta)$ with location parameter ζ, we have $p = 1 - F(-\zeta)$ and $q = F(-\zeta)$. Of particular interest is the logistic distribution, for which $F(x) = 1/(1 + e^{-x})$. In this case $p = e^{\zeta}/(1 + e^{\zeta})$, $q = 1/(1 + e^{\zeta})$, and hence $\log(p/q) = \zeta$. Applying this fact to the success probabilities

$$p_{1k} = 1 - F(-\zeta_{1k}), \qquad p_{2k} = 1 - F(-\zeta_{2k}),$$

we find that

$$\theta_k = \log \Delta_k = \log\left(\frac{p_{2k}}{q_{2k}} \middle/ \frac{p_{1k}}{q_{1k}}\right) = \zeta_{2k} - \zeta_{1k},$$

so that $\zeta_{2k} = \zeta_{1k} + \theta_k$. In this model, H' thus reduces to the assumption that $\zeta_{2k} = \zeta_{1k} + \theta$, that is, that the treatment shifts the distribution of the underlying response by a constant amount θ.

If it is assumed that F is normal rather than logistic, $F(x) = \Phi(x)$ say, then $\zeta = \Phi^{-1}(p)$, and constancy of $\zeta_{2k} - \zeta_{1k}$ requires the much more cumbersome condition $\Phi^{-1}(p_{2k}) - \Phi^{-1}(p_{1k}) = $ constant. However, the functions $\log(p/q)$ and $\Phi^{-1}(p)$ agree quite well in the range $.1 \leq p \leq .9$ [see Cox (1970, p. 28)], and the assumption of constant Δ_k in the logistic response model is therefore close to the corresponding assumption for an underlying normal response.* [The so-called loglinear models, which for contingency tables correspond to the linear models to be considered in Chapter 7 but with a logistic rather than a normal response variable, provide the most widely used approach to contingency tables. See, for example, the books by Cox (1970), Haberman (1974), Bishop, Fienberg, and Holland (1975), Fienberg (1980), Plackett (1981), and Agresti (1984).]

The UMP unbiased test, derived above for the case that the B- and C-margins are fixed, applies equally when any two margins, any one margin, or no margins are fixed, with the understanding that in all cases the test is carried out conditionally, given the values of all random margins.

The test is also used (but no longer UMP unbiased) for testing $H: \Delta_1 = \cdots = \Delta_K = 1$ when the Δ's are not assumed to be equal but when the $\Delta_k - 1$ can be assumed to have the same sign, so that the departure from independence is in the same direction for all the 2×2 tables. A one- or two-sided version is appropriate as the alternatives do or do not specify the direction. For a discussion of this test, the Cochran–Mantel–Haenszel test, and some of its extensions see the reviews by Landis, Heyman, and Koch (1978), Darroch (1981), and Somes and O'Brien (1985).

Consider now the case $K = 2$, with m_k and n_k fixed, and the problem of testing $H': \Delta_2 = \Delta_1$ rather than assuming it. The joint distribution of the X's and Y's given earlier can then be written as

$$\left[\prod_{k=1}^{2} \binom{m_k}{x_k}\binom{n_k}{y_k} q_{1k}^{m_k} q_{2k}^{n_k} \right]$$

$$\times \exp\left(y_2 \log\frac{\Delta_2}{\Delta_1} + (y_1 + y_2)\log \Delta_1 + \sum (x_i + y_i)\log\frac{p_{1i}}{q_{1i}} \right),$$

and H' is rejected in favor of $\Delta_2 > \Delta_1$ if $Y_2 > C$, where C depends on $Y_1 + Y_2$, $X_1 + Y_1$ and $X_2 + Y_2$, and is determined so that the conditional probability of rejection given $Y_1 + Y_2 = w$, $X_1 + Y_1 = t_1$, $X_2 + Y_2 = t_2$ is α. The conditional null distribution of Y_1 and Y_2, given $X_k + Y_k = t_k$ $(k = 1, 2)$, by (21) with Δ in place of ρ is

$$C_{t_1}(\Delta)C_{t_2}(\Delta)\binom{m_1}{t_1 - y_1}\binom{n_1}{y_1}\binom{m_2}{t_2 - y_2}\binom{n_2}{y_2}\Delta^{y_1 + y_2},$$

and hence the conditional distribution of Y_2, given in addition that $Y_1 + Y_2$

---

*The problem of discriminating between a logistic and normal response model is discussed by Chambers and Cox (1967).

$= w$, is of the form

$$k(t_1, t_2, w)\binom{m_1}{y + t_1 - w}\binom{n_1}{w - y}\binom{m_2}{t_2 - y}\binom{n_2}{y}.$$

Some approximations to the critical value of this test are discussed by Birch (1964); see also Venable and Bhapkar (1978). [Optimum large-sample tests of some other hypotheses in $2 \times 2 \times 2$ tables are obtained by Cohen, Gatsonis, and Marden (1983).]

9. THE SIGN TEST

To test consumer preferences between two products, a sample of n subjects are asked to state their preferences. Each subject is recorded as plus or minus as it favors product B or A. The total number Y of plus signs is then a binomial variable with distribution $b(p, n)$. Consider the problem of testing the hypothesis $p = \frac{1}{2}$ of no difference against the alternatives $p \neq \frac{1}{2}$. (As in previous such problems, we disregard here that in case of rejection it will be necessary to decide which of the two products is preferred.) The appropriate test is the two-sided *sign test*, which rejects when $|Y - \frac{1}{2}n|$ is too large. This is UMP unbiased (Section 2).

Sometimes the subjects are also given the possibility of declaring themselves as undecided. If p_-, p_+, and p_0 denote the probabilities of preference for product A, product B, and of no preference respectively, the numbers X, Y, and Z of decisions in favor of these three possibilities are distributed according to the multinomial distribution

$$(22) \qquad \frac{n!}{x!y!z!}p_-^x p_+^y p_0^z \qquad (x + y + z = n),$$

and the hypothesis to be tested is $H: p_+ = p_-$. The distribution (22) can also be written as

$$(23) \qquad \frac{n!}{x!y!z!}\left(\frac{p_+}{1 - p_0 - p_+}\right)^y\left(\frac{p_0}{1 - p_0 - p_+}\right)^z(1 - p_0 - p_+)^n,$$

and is then seen to constitute an exponential family with $U = Y$, $T = Z$, $\theta = \log[p_+/(1 - p_0 - p_+)]$, $\vartheta = \log[p_0/(1 - p_0 - p_+)]$. Rewriting the hypothesis H as $p_+ = 1 - p_0 - p_+$, it is seen to be equivalent to $\theta = 0$. There exists therefore a UMP unbiased test of H, which is obtained by considering z as fixed and determining the best unbiased conditional test of H given

$Z = z$. Since the conditional distribution of Y given z is a binomial distribution $b(p, n - z)$ with $p = p_+/(p_+ + p_-)$, the problem reduces to that of testing the hypothesis $p = \frac{1}{2}$ in a binomial distribution with $n - z$ trials, for which the rejection region is $|Y - \frac{1}{2}(n - z)| > C(z)$. The UMP unbiased test is therefore obtained by disregarding the number of cases in which no preference is expressed (the number of *ties*), and applying the sign test to the remaining data.

The power of the test depends strongly on p_0, which governs the distribution of Z. For large p_0, the number $n - z$ of trials in the conditional binomial distribution can be expected to be small, and the test will thus have little power. This may be an advantage in the present case, since a sufficiently high value of p_0, regardless of the value of p_+/p_-, implies that the population as a whole is largely indifferent with respect to the products.

The above conditional sign test applies to any situation in which the observations are the result of n independent trials, each of which is either a success $(+)$, a failure $(-)$, or a tie. As an alternative treatment of ties, it is sometimes proposed to assign each tie at random (with probability $\frac{1}{2}$ each) to either plus or minus. The total number Y' of plus signs after the ties have been broken is then a binomial variable with distribution $b(\pi, n)$, where $\pi = p_+ + \frac{1}{2}p_0$. The hypothesis H becomes $\pi = \frac{1}{2}$, and is rejected when $|Y' - \frac{1}{2}n| > C$, where the probability of rejection is α when $\pi = \frac{1}{2}$. This test can be viewed also as a randomized test based on X, Y, and Z, and it is unbiased for testing H in its original form, since p_+ is $=$ or $\neq p_-$ as π is $=$ or $\neq \frac{1}{2}$. Since the test involves randomization other than on the boundaries of the rejection region, it is less powerful than the UMP unbiased test for this situation, so that the random breaking of ties results in a loss of power.

This remark might be thought to throw some light on the question of whether in the determination of consumer preferences it is better to permit the subject to remain undecided or to force an expression of preference. However, here the assumption of a completely random assignment in case of a tie does not apply. Even when the subject is not conscious of a definite preference, there will usually be a slight inclination toward one of the two possibilities, which in a majority of the cases will be brought out by a forced decision. This will be balanced in part by the fact that such forced decisions are more variable than those reached voluntarily. Which of these two factors dominates depends on the strength of the preference.

Frequently, the question of preference arises between a standard product and a possible modification or a new product. If each subject is required to express a definite preference, the hypothesis of interest is usually the one-sided hypothesis $p_+ \leq p_-$, where $+$ denotes a preference for the modification. However, if an expression of indifference is permitted, the hypothesis to be tested is not $p_+ \le p_-$ but rather $p_+ \le p_0 + p_-$, since typically the modification is of interest only if it is actually preferred. As was shown in Chapter 3, Example 8, the one-sided sign test which rejects when the number of plus signs is too large is UMP for this problem.

In some investigations, the subject is asked not only to express a preference but to give a more detailed evaluation, such as a score on some numerical scale. Depending on the situation, the hypothesis can then take on one of two forms. One may be interested in the hypothesis that there is no difference in the consumer's reaction to the two products. Formally, this states that the distribution of the scores $X_1, \ldots, X_n$ expressing the degree of preference of the n subjects for the modified product is symmetric about the origin. This problem, for which a UMP unbiased test does not exist without further assumptions, will be considered in Chapter 6, Section 10.

Alternatively, the hypothesis of interest may continue to be $H: p_+ = p_-$. Since $p_- = P\{X < 0\}$ and $p_+ = P\{X > 0\}$, this now becomes

$$H: P\{X > 0\} = P\{X < 0\}.$$

Here symmetry of X is no longer assumed even when $P\{X < 0\} = P\{X > 0\}$. If no assumptions are made concerning the distribution of X beyond the fact that the set of its possible values is given, the sign test based on the number of X's that are positive and negative continues to be UMP unbiased.

To see this, note that any distribution of X can be specified by the probabilities

$$p_- = P\{X < 0\}, \qquad p_+ = P\{X > 0\}, \qquad p_0 = P\{X = 0\},$$

and the conditional distributions F_- and F_+ of X given $X < 0$ and $X > 0$ respectively. Consider any fixed distributions F'_-, F'_+, and denote by $\mathscr{F}_0$ the family of all distributions with $F_- = F'_-$, $F_+ = F'_+$ and arbitrary p_-, p_+, p_0. Any test that is unbiased for testing H in the original family of distributions $\mathscr{F}$ in which F_- and F_+ are unknown is also unbiased for testing H in the smaller family $\mathscr{F}_0$. We shall show below that there exists a UMP unbiased test ϕ_0 of H in $\mathscr{F}_0$. It turns out that ϕ_0 is also unbiased for testing H in $\mathscr{F}$ and is independent of F'_-, F'_+. Let ϕ be any other unbiased test of H in $\mathscr{F}$, and consider any fixed alternative, which without loss of generality can be assumed to be in $\mathscr{F}_0$. Since ϕ is unbiased for $\mathscr{F}$, it is unbiased for testing $p_+ = p_-$ in $\mathscr{F}_0$; the power of ϕ_0 against the particular alternative is therefore at least as good as that of ϕ. Hence ϕ_0 is UMP unbiased.

To determine the UMP unbiased test of H in $\mathscr{F}_0$, let the densities of F'_- and F'_+ with respect to some measure μ be f'_- and f'_+. The joint density of the X's at a point $(x_1, \ldots, x_n)$ with

$$x_{i_1}, \ldots, x_{i_r} < 0 = x_{j_1} = \cdots = x_{j_s} < x_{k_1}, \ldots, x_{k_m}$$

is

$$p_-^r p_0^s p_+^m f_-'(x_{i_1}) \ldots f_-'(x_{i_r}) f_+'(x_{k_1}) \ldots f_+'(x_{k_m}).$$

The set of statistics (r, s, m) is sufficient for (p_-, p_0, p_+), and its distribution is given by (22) with $x = r$, $y = m$, $z = s$. The sign test is therefore seen to be UMP unbiased as before.

A different application of the sign test arises in the context of a 2×2 table for matched pairs. In Section 5, success probabilities for two treatments were compared on the basis of two independent random samples. Unless the population of subjects from which these samples are drawn is fairly homogeneous, a more powerful test can often be obtained by using a sample of matched pairs (for example, twins or the same subject given the treatments at different times). For each pair there are then four possible outcomes: $(0,0)$, $(0,1)$, $(1,0)$, and $(1,1)$, where 1 and 0 stand for success and failure, and the first and second number in each pair of responses refer to the subject receiving treatment 1 or 2 respectively.

The results of such a study are sometimes displayed in a 2×2 table,

2nd \ 1st	0	1
0	X	X'
1	Y	Y'

which despite the formal similarity differs from that considered in Section 6. If a sample of s pairs is drawn, the joint distribution of X, Y, X', Y' as before is multinomial, with probabilities $p_{00}, p_{01}, p_{10}, p_{11}$. The success probabilities of the two treatments are $\pi_1 = p_{10} + p_{11}$ for the first and $\pi_2 = p_{01} + p_{11}$ for the second treatment, and the hypothesis to be tested is $H: \pi_1 = \pi_2$ or equivalently $p_{10} = p_{01}$, rather than $p_{10}p_{01} = p_{00}p_{11}$ as it was earlier.

In exponential form, the joint distribution can be written as

$$(24) \qquad \frac{s! p_{11}^s}{x! x'! y! y'!} \exp\left(y \log \frac{p_{01}}{p_{10}} + (x' + y) \log \frac{p_{10}}{p_{11}} + x \log \frac{p_{00}}{p_{11}} \right).$$

There exists a UMP unbiased test, *McNemar's test*, which rejects H in favor of the alternatives $p_{10} < p_{01}$ when $Y > C(X' + Y, X)$, where the conditional probability of rejection given $X' + Y = d$ and $X = x$ is α for all d and x. Under this condition, the numbers of pairs $(0, 0)$ and $(1, 1)$ are fixed, and the only remaining variables are Y and $X' = d - Y$ which specify the division of the d cases with mixed response between the outcomes $(0, 1)$ and $(1, 0)$. Conditionally, one is dealing with d binomial trials with success probability $p = p_{01}/(p_{01} + p_{10})$, H becomes $p = \frac{1}{2}$, and the UMP unbiased test reduces to the sign test. [The issue of conditional versus unconditional power for this test is discussed by Frisén (1980).]

The situation is completely analogous to that of the sign test in the presence of undecided opinions, with the only difference that there are now two types of ties, $(0, 0)$ and $(1, 1)$, both of which are disregarded in performing the test.

10. PROBLEMS

Section 1

1. *Admissibility.* Any UMP unbiased test ϕ_0 is admissible in the sense that there cannot exist another test ϕ_1 which is at least as powerful as ϕ_0 against all alternatives and more powerful against some.
   [If ϕ is unbiased and ϕ' is uniformly at least as powerful as ϕ, then ϕ' is also unbiased.]

2. *p-values.* Consider a family of tests of $H: \theta = \theta_0$ (or $\theta \leq \theta_0$), with level-α rejection regions S_α such that (a) $P_{\theta_0}\{X \in S_\alpha\} = \alpha$ for all $0 < \alpha < 1$, and (b) $S_{\alpha_0} = \bigcap_{\alpha > \alpha_0} S_\alpha$ for all $0 < \alpha_0 < 1$, which in particular implies $S_\alpha \subset S_{\alpha'}$ for $\alpha < \alpha'$.

   (i)   Then the p-value $\hat{\alpha}$ is given by $\hat{\alpha} = \hat{\alpha}(x) = \inf\{\alpha : x \in S_\alpha\}$.

   (ii) When $\theta = \theta_0$, the distribution of $\hat{\alpha}$ is the uniform distribution over $(0, 1)$.

   (iii) If the tests S_α are unbiased, the distribution of $\hat{\alpha}$ under any alternative θ satisfies

   $$P_\theta\{\hat{\alpha} \leq \alpha\} \geq P_{\theta_0}\{\hat{\alpha} \leq \alpha\} = \alpha,$$

   so that it is shifted toward the origin.

   If p-values are available from a number of independent experiments, they can be combined by (ii) and (iii) to provide an overall test* of the hypothesis.
   [$\hat{\alpha} \leq \alpha$ if and only if $x \in S_\alpha$, and hence $P_\theta\{\hat{\alpha} \leq \alpha\} = P_\theta\{X \in S_\alpha\} = \beta_\alpha(\theta)$, which is α for $\theta = \theta_0$ and $\geq \alpha$ if θ is an alternative to H.]

---

*For discussions of such tests see for example Koziol and Perlman (1978), Berk and Cohen (1979), Mudholkar and George (1979), Scholz (1982), and the related work of Marden (1982). Associated confidence intervals are proposed by Littell and Louv (1981).

Section 2

3. Let X have the binomial distribution $b(p, n)$, and consider the hypothesis $H : p = p_0$ at level of significance α. Determine the boundary values of the UMP unbiased test for $n = 10$ with $\alpha = .1$, $p_0 = .2$ and with $\alpha = .05$, $p_0 = .4$, and in each case graph the power functions of both the unbiased and the equal-tails test.

4. Let X have the Poisson distribution $P(\tau)$, and consider the hypothesis $H : \tau = \tau_0$. Then condition (6) reduces to

$$\sum_{x = C_1 + 1}^{C_2 - 1} \frac{\tau_0^{x-1}}{(x-1)!} e^{-\tau_0} + \sum_{i=1}^{2} (1 - \gamma_i) \frac{\tau_0^{C_i - 1}}{(C_i - 1)!} e^{-\tau_0} = 1 - \alpha,$$

provided $C_1 > 1$.

5. Let T_n/θ have a χ^2-distribution with n degrees of freedom. For testing $H : \theta = 1$ at level of significance $\alpha = .05$, find n so large that the power of the UMP unbiased test is $\geq .9$ against both $\theta \geq 2$ and $\theta \leq \frac{1}{2}$. How large does n have to be if the test is not required to be unbiased?

6. Let X and Y be independently distributed according to one-parameter exponential families, so that their joint distribution is given by

$$dP_{\theta_1, \theta_2}(x, y) = C(\theta_1) e^{\theta_1 T(x)} d\mu(x) K(\theta_2) e^{\theta_2 U(y)} d\nu(y).$$

Suppose that with probability 1 the statistics T and U each take on at least three values and that (a, b) is an interior point of the natural parameter space. Then a UMP unbiased test does not exist for testing $H : \theta_1 = a$, $\theta_2 = b$ against the alternatives $\theta_1 \neq a$ or $\theta_2 \neq b$.*
[The most powerful unbiased tests against the alternatives $\theta_1 \neq a$, $\theta_2 = b$ and $\theta_1 = a$, $\theta_2 \neq b$ have acceptance regions $C_1 < T(x) < C_2$ and $K_1 < U(y) < K_2$ respectively. These tests are also unbiased against the wider class of alternatives $K : \theta_1 \neq a$ or $\theta_2 \neq b$ or both.]

7. Let (X, Y) be distributed according to the exponential family

$$dP_{\theta_1, \theta_2}(x, y) = C(\theta_1, \theta_2) e^{\theta_1 x + \theta_2 y} d\mu(x, y).$$

The only unbiased test for testing $H : \theta_1 \leq a$, $\theta_2 \leq b$ against $K : \theta_1 > a$ or $\theta_2 > b$ or both is $\phi(x, y) \equiv \alpha$.

---

*For counterexamples when the conditions of the problem are not satisfied, see Kallenberg (1984).

[Take $a = b = 0$, and let $\beta(\theta_1, \theta_2)$ be the power function of any level-α test. Unbiasedness implies $\beta(0, \theta_2) = \alpha$ for $\theta_2 < 0$ and hence for all θ_2, since $\beta(0, \theta_2)$ is an analytic function of θ_2. For fixed $\theta_2 > 0$, $\beta(\theta_1, \theta_2)$ considered as a function of θ_1 therefore has a minimum at $\theta_1 = 0$, so that $\partial\beta(\theta_1, \theta_2)/\partial\theta_1$ vanishes at $\theta_1 = 0$ for all positive θ_2, and hence for all θ_2. By considering alternatively positive and negative values of θ_2 and using the fact that the partial derivatives of all orders of $\beta(\theta_1, \theta_2)$ with respect to θ_1 are analytic, one finds that for each fixed θ_2 these derivatives all vanish at $\theta_1 = 0$ and hence that the function β must be a constant. Because of the completeness of (X, Y), $\beta(\theta_1, \theta_2) \equiv \alpha$ implies $\phi(x, y) \equiv \alpha$.]

8. For testing the hypothesis $H : \theta = \theta_0$ (θ_0 an interior point of Ω) in the one-parameter exponential family of Section 2, let $\mathscr{C}$ be the totality of tests satisfying (3) and (5) for some $-\infty \le C_1 \le C_2 \le \infty$ and $0 \le \gamma_1, \gamma_2 \le 1$.

    (i) $\mathscr{C}$ is complete in the sense that given any level-α test ϕ_0 of H there exists $\phi \in \mathscr{C}$ such that ϕ is uniformly at least as powerful as ϕ_0.

    (ii) If $\phi_1, \phi_2 \in \mathscr{C}$, then neither of the two tests is uniformly more powerful than the other.

    (iii) Let the problem be considered as a two-decision problem, with decisions d_0 and d_1 corresponding to acceptance and rejection of H, and with loss function $L(\theta, d_i) = L_i(\theta)$, $i = 0, 1$. Then $\mathscr{C}$ is minimal essentially complete provided $L_1(\theta) < L_0(\theta)$ for all $\theta \ne \theta_0$.

    (iv) Extend the result of part (iii) to the hypothesis $H' : \theta_1 \le \theta \le \theta_2$.

    [(i): Let the derivative of the power function of ϕ_0 at θ_0 be $\beta'_{\phi_0}(\theta_0) = \rho$. Then there exists $\phi \in \mathscr{C}$ such that $\beta'_\phi(\theta_0) = \rho$ and ϕ is UMP among all tests satisfying this condition.
    (ii): See Chapter 3, end of Section 7.
    (iii): See Chapter 3, proof of Theorem 3.]

Section 3

9. Let $X_1, \ldots, X_n$ be a sample from (i) the normal distribution $N(a\sigma, \sigma^2)$, with a fixed and $0 < \sigma < \infty$; (ii) the uniform distribution $U(\theta - \frac{1}{2}, \theta + \frac{1}{2})$, $-\infty < \theta < \infty$; (iii) the uniform distribution $U(\theta_1, \theta_2)$, $-\infty < \theta_1 < \theta_2 < \infty$. For these three families of distributions the following statistics are sufficient: (i), $T = (\Sigma X_i, \Sigma X_i^2)$; (ii) and (iii), $T = (\min(X_1, \ldots, X_n), \max(X_1, \ldots, X_n))$. The family of distributions of T is complete for case (iii), but for (i) and (ii) it is not complete or even boundedly complete.
    [(i): The distribution of $\Sigma X_i / \sqrt{\Sigma X_i^2}$ does not depend on σ.]

10. Let $X_1, \ldots, X_m$ and $Y_1, \ldots, Y_n$ be samples from $N(\xi, \sigma^2)$ and $N(\xi, \tau^2)$. Then $T = (\Sigma X_i, \Sigma Y_j, \Sigma X_i^2, \Sigma Y_j^2)$, which in Example 5 was seen not to be complete, is also not boundedly complete.
    [Let $f(t)$ be 1 or -1 as $\bar{y} - \bar{x}$ is positive or not.]

11. *Counterexample.* Let X be a random variable taking on the values $-1, 0, 1, 2, \ldots$ with probabilities

$$P_\theta\{X = -1\} = \theta; \qquad P_\theta\{X = x\} = (1 - \theta)^2 \theta^x, \quad x = 0, 1, \ldots .$$

Then $\mathscr{P} = \{P_\theta, 0 < \theta < 1\}$ is boundedly complete but not complete.

12. The completeness of the order statistics in Example 6 remains true if the family $\mathscr{F}$ is replaced by the family $\mathscr{F}_1$ of all continuous distributions.
[To show that for any integrable symmetric function ϕ, $\int \phi(x_1, \ldots, x_n)\, dF(x_1) \ldots dF(x_n) = 0$ for all continuous F implies $\phi = 0$ a.e., replace F by $\alpha_1 F_1 + \cdots + \alpha_n F_n$, where $0 < \alpha_i < 1$, $\Sigma \alpha_i = 1$. By considering the left side of the resulting identity as a polynomial in the α's one sees that $\int \phi(x_1, \ldots, x_n)\, dF_1(x_1) \ldots dF_n(x_n) = 0$ for all continuous F_i. This last equation remains valid if the F_i are replaced by $I_{a_i}(x)F(x)$, where $I_{a_i}(x) = 1$ if $x \le a_i$ and $= 0$ otherwise. This implies that $\phi = 0$ except on a set which has measure 0 under $F \times \cdots \times F$ for all continuous F.]

13. Determine whether T is complete for each of the following situations:

    (i) $X_1, \ldots, X_n$ are independently distributed according to the uniform distribution over the integers $1, 2, \ldots, \theta$ and $T = \max(X_1, \ldots, X_n)$.

    (ii) X takes on the values $1, 2, 3, 4$ with probabilities $pq, p^2q, pq^2, 1 - 2pq$ respectively, and $T = X$.

Section 4

14. *Measurability of tests of Theorem 3.*   The function ϕ_3 defined by (16) and (17) is jointly measurable in u and t.
[With $C_1 = v$ and $C_2 = w$, the determining equations for v, w, γ_1, γ_2 are

$$(25) \qquad F_t(v-) + [1 - F_t(w)] + \gamma_1[F_t(v) - F_t(v-)]$$

$$+ \gamma_2[F_t(w) - F_t(w-)] = \alpha$$

and

$$(26) \qquad G_t(v-) + [1 - G_t(w)] + \gamma_1[G_t(v) - G_t(v-)]$$

$$+ \gamma_2[G_t(w) - G_t(w-)] = \alpha,$$

where

$$(27) \quad F_t(u) = \int_{-\infty}^{u} C_t(\theta_1) e^{\theta_1 y}\, d\nu_t(y), \qquad G_t(u) = \int_{-\infty}^{u} C_t(\theta_2) e^{\theta_2 y}\, d\nu_t(y)$$

denote the conditional cumulative distribution function of U given t when $\theta = \theta_1$ and $\theta = \theta_2$ respectively.

(1)   For each $0 \leq y \leq \alpha$ let $v(y, t) = F_t^{-1}(y)$ and $w(y, t) = F_t^{-1}(1 - \alpha + y)$, where the inverse function is defined as in the proof of Theorem 3. Define $\gamma_1(y, t)$ and $\gamma_2(y, t)$ so that for $v = v(y, t)$ and $w = w(y, t)$,

$$F_t(v -) + \gamma_1[F_t(v) - F_t(v -)] = y,$$

$$1 - F_t(w) + \gamma_2[F_t(w) - F_t(w -)] = \alpha - y.$$

(2)   Let $H(y, t)$ denote the left-hand side of (26), with $v = v(y, t)$, etc. Then $H(0, t) > \alpha$ and $H(\alpha, t) < \alpha$. This follows by Theorem 2 of Chapter 3 from the fact that $v(0, t) = -\infty$ and $w(\alpha, t) = \infty$ (which shows the conditional tests corresponding to $y = 0$ and $y = \alpha$ to be one-sided), and that the left-hand side of (26) for any y is the power of this conditional test.

(3)   For fixed t, the functions

$$H_1(y, t) = G_t(v -) + \gamma_1[G_t(v) - G_t(v -)]$$

and

$$H_2(y, t) = 1 - G_t(w) + \gamma_2[G_t(w) - G_t(w -)]$$

are continuous functions of y. This is a consequence of the fact, which follows from (27), that a.e. $\mathscr{P}^T$ the discontinuities and flat stretches of F_t and G_t coincide.

(4)   The function $H(y, t)$ is jointly measurable in y and t. This follows from the continuity of H by an argument similar to the proof of measurability of $F_t(u)$ in the text. Define

$$y(t) = \inf\{ y : H(y, t) < \alpha \},$$

and let $v(t) = v[y(t), t]$, etc. Then (25) and (26) are satisfied for all t. The measurability of $v(t)$, $w(t)$, $\gamma_1(t)$, and $\gamma_2(t)$ defined in this manner will follow from measurability in t of $y(t)$ and $F_t^{-1}[y(t)]$. This is a consequence of the relations, which hold for all real c,

$$\{ t : y(t) < c \} = \bigcup_{r < c} \{ t : H(r, t) < \alpha \},$$

where r indicates a rational, and

$$\{ t : F_t^{-1}[y(t)] \leq c \} = \{ t : y(t) - F_t(c) \leq 0 \}.]$$

15.   *Continuation.*   The function ϕ_4 defined by (16), (18), and (19) is jointly measurable in u and t.

[The proof, which otherwise is essentially like that outlined in the preceding problem, requires the measurability in z and t of the integral

$$g(z, t) = \int_{-\infty}^{z-} u \, dF_t(u).$$

This integral is absolutely convergent for all t, since F_t is a distribution belonging to an exponential family. For any $z < \infty$, $g(z, t) = \lim g_n(z, t)$, where

$$g_n(z, t) = \sum_{j=1}^{\infty} \left(z - \frac{j}{2^n} \right) \left[F_t \left(z - \frac{j-1}{2^n} - 0 \right) - F_t \left(z - \frac{j}{2^n} - 0 \right) \right],$$

and the measurability of g follows from that of the functions g_n. The inequalities corresponding to those obtained in step (2) of the preceding problem result from the property of the conditional one-sided tests established in Problem 22 of Chapter 3.]

16. The UMP unbiased tests of the hypotheses $H_1, \ldots, H_4$ of Theorem 3 are unique if attention is restricted to tests depending on U and the T's.

Section 5

17. Let X and Y be independently distributed with Poisson distributions $P(\lambda)$ and $P(\mu)$. Find the power of the UMP unbiased test of $H : \mu \le \lambda$, against the alternatives $\lambda = .1$, $\mu = .2$; $\lambda = 1$, $\mu = 2$; $\lambda = 10$, $\mu = 20$; $\lambda = .1$, $\mu = .4$; at level of significance $\alpha = .1$.
[Since $T = X + Y$ has the Poisson distribution $P(\lambda + \mu)$, the power is

$$\beta = \sum_{t=0}^{\infty} \beta(t) \frac{(\lambda + \mu)^t}{t!} e^{-(\lambda + \mu)},$$

where $\beta(t)$ is the power of the conditional test given t against the alternative in question.]

18. *Sequential comparison of two binomials.* Consider two sequences of binomial trials with probabilities of success p_1 and p_2 respectively, and let $\rho = (p_2/q_2) \div (p_1/q_1)$.

   (i) If $\alpha < \beta$, no test with fixed numbers of trials m and n for testing $H : \rho = \rho_0$ can have power $\ge \beta$ against all alternatives with $\rho = \rho_1$.

   (ii) The following is a simple sequential sampling scheme leading to the desired result. Let the trials be performed in pairs of one of each kind, and restrict attention to those pairs in which one of the trials is a success and the other a failure. If experimentation is continued until N such pairs have been observed, the number of pairs in which the successful trial belonged to the first series has the binomial distribution $b(\pi, N)$ with $\pi = p_1 q_2 / (p_1 q_2 + p_2 q_1) = 1/(1 + \rho)$. A test of arbitrarily high power against ρ_1 is therefore obtained by taking N large enough.

(iii)   If $p_1/p_2 = \lambda$, use inverse binomial sampling to devise a test of $H : \lambda = \lambda_0$ against $K : \lambda > \lambda_0$.

19.   *Positive dependence.* Two random variables (X, Y) with c.d.f. $F(x, y)$ are said to be *positively quadrant dependent* if $F(x, y) \geq F(x, \infty) F(\infty, y)$ for all x, y.* For the case that (X, Y) takes on the four pairs of values $(0, 0)$, $(0, 1)$, $(1, 0)$, $(1, 1)$ with probabilities $p_{00}, p_{01}, p_{10}, p_{11}$, (X, Y) are positively quadrant dependent if and only if the odds ratio $\Delta = p_{01} p_{10} / p_{00} p_{11} \leq 1$.

20.   *Runs.* Consider a sequence of N dependent trials, and let X_i be 1 or 0 as the ith trial is a success or failure. Suppose that the sequence has the *Markov property*[†]

$$P\{X_i = 1 | x_1, \ldots, x_{i-1}\} = P\{X_i = 1 | x_{i-1}\}$$

and the property of *stationarity* according to which $P\{X_i = 1\}$ and $P\{X_i = 1 | x_{i-1}\}$ are independent of i. The distribution of the X's is then specified by the probabilities

$$p_1 = P\{X_i = 1 | X_{i-1} = 1\} \quad \text{and} \quad p_0 = P\{X_i = 1 | X_{i-1} = 0\}$$

and by the initial probabilities

$$\pi_1 = P\{X_1 = 1\} \quad \text{and} \quad \pi_0 = 1 - \pi_1 = P\{X_1 = 0\}.$$

(i)   Stationarity implies that

$$\pi_1 = \frac{p_0}{p_0 + q_1}, \qquad \pi_0 = \frac{q_1}{p_0 + q_1}.$$

(ii)   A set of successive outcomes $x_i, x_{i+1}, \ldots, x_{i+j}$ is said to form a *run* of zeros if $x_i = x_{i+1} = \cdots = x_{i+j} = 0$, and $x_{i-1} = 1$ or $i = 1$, and $x_{i+j+1} = 1$ or $i + j = N$. A run of ones is defined analogously. The probability of any particular sequence of outcomes $(x_1, \ldots, x_N)$ is

$$\frac{1}{p_0 + q_1} p_0^v p_1^{n-v} q_1^u q_0^{m-u},$$

---

*For a systematic discussion of this and other concepts of dependence, see Tong (1980, Chapter 5).

[†]Statistical inference in these and more general Markov chains is discussed, for example, in Anderson and Goodman (1957), Goodman (1958), Billingsley (1961), Denny and Wright (1978), and Denny and Yakowitz (1978).

where m and n denote the numbers of zeros and ones, and u and v the numbers of runs of zeros and ones in the sequence.

21. *Continuation.* For testing the hypothesis of independence of the X's, $H: p_0 = p_1$, against the alternatives $K: p_0 < p_1$, consider the *run test*, which rejects H when the total number of runs $R = U + V$ is less than a constant $C(m)$ depending on the number m of zeros in the sequence. When $R = C(m)$, the hypothesis is rejected with probability $\gamma(m)$, where C and γ are determined by

$$P_H\{R < C(m)|m\} + \gamma(m)P_H\{R = C(m)|m\} = \alpha.$$

(i) Against any alternative of K the most powerful similar test (which is at least as powerful as the most powerful unbiased test) coincides with the run test in that it rejects H when $R < C(m)$. Only the supplementary rule for bringing the conditional probability of rejection (given m) up to α depends on the specific alternative under consideration.

(ii) The run test is unbiased against the alternatives K.

(iii) The conditional distribution of R given m, when H is true, is*

$$P\{R = 2r\} = \frac{2\binom{m-1}{r-1}\binom{n-1}{r-1}}{\binom{m+n}{m}},$$

$$P\{R = 2r+1\} = \frac{\binom{m-1}{r-1}\binom{n-1}{r} + \binom{m-1}{r}\binom{n-1}{r-1}}{\binom{m+n}{m}}.$$

[(i): Unbiasedness implies that the conditional probability of rejection given m is α for all m. The most powerful conditional level-α test rejects H for those sample sequences for which $\Delta(u, v) = (p_0/p_1)^v(q_1/q_0)^u$ is too large. Since $p_0 < p_1$ and $q_1 < q_0$ and since $|v - u|$ can only take on the values 0 and 1, it follows that

$$\Delta(1,1) > \Delta(1,2), \qquad \Delta(2,1) > \Delta(2,2) > \Delta(2,3), \qquad \Delta(3,2) > \cdots.$$

Thus only the relation between $\Delta(i, i+1)$ and $\Delta(i+1, i)$ depends on the specific alternative, and this establishes the desired result.

(ii): That the above conditional test is unbiased for each m is seen by writing its power as

$$\beta(p_0, p_1|m) = (1 - \gamma)P\{R < C(m)|m\} + \gamma P\{R \le C(m)|m\},$$

---

*This distribution is tabled by Swed and Eisenhart (1943) and can be obtained from the hypergeometric distribution [Guenther (1978)]. For further discussion of the run test, see Wolfowitz (1943).

since by (i) the rejection regions $R < C(m)$ and $R < C(m) + 1$ are both UMP at their respective conditional levels.

(iii): When H is true, the conditional probability given m of any set of m zeros and n ones is $1/\binom{m+n}{m}$. The number of ways of dividing n ones into r groups is $\binom{n-1}{r-1}$, and that of dividing m zeros into $r + 1$ groups is $\binom{m-1}{r}$. The conditional probability of getting $r + 1$ runs of zeros and r runs of ones is therefore

$$\frac{\binom{m-1}{r}\binom{n-1}{r-1}}{\binom{m+n}{m}}.$$

To complete the proof, note that the total number of runs is $2r + 1$ if and only if there are either $r + 1$ runs of zeros and r runs of ones or r runs of zeros and $r + 1$ runs of ones.]

22. (i) Based on the conditional distribution of $X_2, \ldots, X_n$ given $X_1 = x_1$ in the model of Problem 20, there exists a UMP unbiased test of $H: p_0 = p_1$ against $p_1 > p_0$ for every α.

(ii) For the same testing problem, without conditioning on X_1 there exists a UMP unbiased test if the initial probability π_1 is assumed to be completely unknown instead of being given by the value stated in (i) of Problem 20.

[The conditional distribution of $X_2, \ldots, X_n$ given x_1 is of the form

$$C(x_1; p_0, p_1, q_0, q_1) p_1^{y_1} p_0^{y_0} q_1^{z_1} q_0^{z_0} h(y_1, y_2, z_1, z_2),$$

where y_1 is the number of times a 1 follows a 1, y_0 the number of times a 1 follows a 0, and so on, in the sequence $x_1, X_2, \ldots, X_n$. [See Billingsley (1961, p. 14).]

23. *Rank-sum test.* Let $Y_1, \ldots, Y_N$ be independently distributed according to the binomial distributions $b(p_i, n_i)$, $i = 1, \ldots, N$, where

$$p_i = \frac{1}{1 + e^{-(\alpha + \beta x_i)}}.$$

This is the model frequently assumed in bioassay, where x_i denotes the dose, or some function of the dose such as its logarithm, of a drug given to n_i experimental subjects, and where Y_i is the number among these subjects which respond to the drug at level x_i. Here the x_i are known, and α and β are unknown parameters.

(i) The joint distribution of the Y's constitutes an exponential family, and UMP unbiased tests exist for the four hypotheses of Theorem 3, concerning both α and β.

(ii) Suppose in particular that $x_i = \Delta i$, where Δ is known, and that $n_i = 1$ for all i. Let n be the number of successes in the N trials, and let these successes occur in the s_1st, s_2nd,..., s_nth trial, where $s_1 < s_2 < \cdots < s_n$. Then the UMP unbiased test for testing $H : \beta = 0$ against the alternatives $\beta > 0$ is carried out conditionally, given n, and rejects when the *rank sum* $\sum_{i=1}^{n} s_i$ is too large.

(iii) Let $Y_1,\ldots, Y_M$ and $Z_1,\ldots, Z_N$ be two independent sets of experiments of the type described at the beginning of the problem, corresponding, say, to two different drugs. If Y_i is distributed as $b(p_i, m_i)$ and Z_j as $b(\pi_j, n_j)$, with

$$p_i = \frac{1}{1 + e^{-(\alpha + \beta u_i)}}, \qquad \pi_j = \frac{1}{1 + e^{-(\gamma + \delta v_j)}},$$

then UMP unbiased tests exist for the four hypotheses concerning $\gamma - \alpha$ and $\delta - \beta$.

Section 8

24. In a $2 \times 2 \times 2$ table with $m_1 = 3$, $n_1 = 4$; $m_2 = 4$, $n_2 = 4$; and $t_1 = 3$, $t_1' = 4$, $t_2 = t_2' = 4$, determine the probabilities that $P(Y_1 + Y_2 \le k | X_i + Y_i = t_i$, $i = 1, 2)$ for $k = 0, 1, 2, 3$.

25. In a $2 \times 2 \times K$ table with $\Delta_k = \Delta$, the test derived in the text as UMP unbiased for the case that the B and C margins are fixed has the same property when any two, one, or no margins are fixed.

26. Let X_{ijkl} $(i, j, k = 0, 1, l = 1,\ldots, L)$ denote the entries in a $2 \times 2 \times 2 \times L$ table with factors A, B, C, and D, and let

$$\Gamma_l = \frac{p_{A\bar{B}CD_l} p_{\bar{A}BCD_l} p_{A B\bar{C}D_l} p_{\bar{A}B\bar{C}D_l}}{p_{ABCD_l} p_{\bar{A}\bar{B}CD_l} p_{AB\bar{C}D_l} p_{\bar{A}\bar{B}\bar{C}D_l}}.$$

Then

(i) under the assumption $\Gamma_l = \Gamma$ there exists a UMP unbiased test of the hypothesis $\Gamma \le \Gamma_0$ for any fixed Γ_0;

(ii) When $l = 2$, there exists a UMP unbiased test of the hypothesis $\Gamma_1 = \Gamma_2$

—in both cases regardless of whether 0, 1, 2 or 3 of the sets of margins are fixed.

Section 9

27. In the 2×2 table for matched pairs, show by formal computation that the conditional distribution of Y given $X' + Y = d$ and $X = x$ is binomial with the indicated p.

28. Consider the comparison of two success probabilities in (a) the two-binomial situation of Section 5 with $m = n$, and (b) the matched-pairs situation of Section 9. Suppose the matching is completely at random, that is, a random sample of $2n$ subjects, obtained from a population of size N $(2n \leq N)$, is divided at random into n pairs, and the two treatments B and $\tilde{B}$ are assigned at random within each pair.

   (i) The UMP unbiased test for design (a) (Fisher's exact test) is always more powerful than the UMP unbiased test for design (b) (McNemar's test).

   (ii) Let X_i (respectively Y_i) be 1 or 0 as the 1st (respectively 2nd) member of the ith pair is a success or failure. Then the correlation coefficient of X_i and Y_i can be positive or negative and tends to zero as $N \to \infty$.

   [(ii): Assume that the kth member of the population has probability of success $p_A^{(k)}$ under treatment A and $p_{\tilde{A}}^{(k)}$ under $\tilde{A}$.]

29. In the 2×2 table for matched pairs, in the notation of Section 9, the correlation between the responses of the two members of a pair is

$$\rho = \frac{p_{11} - \pi_1 \pi_2}{\sqrt{\pi_1(1 - \pi_1)\pi_2(1 - \pi_2)}}.$$

   For any given values of $\pi_1 < \pi_2$, the power of the one-sided McNemar test of $H: \pi_1 = \pi_2$ is an increasing function of ρ.
   [The conditional power of the test given $X + Y = d$, $X = x$ is an increasing function $p = p_{01}/(p_{01} + p_{10})$.]
   Note. The correlation ρ increases with the effectiveness of the matching, and McNemar's test under (b) of Problem 28 soon becomes more powerful than Fisher's test under (a). For detailed numerical comparisons see Wacholder and Weinberg (1982) and the references given there.

Additional Problems

30. Let X, Y be independent binomial $b(p, m)$ and $b(p^2, n)$ respectively. Determine whether (X, Y) is complete when

   (i) $m = n = 1$,
   (ii) $m = 2, n = 1$.

31. Let $X_1, \ldots, X_n$ be a sample from the uniform distribution over the integers $1, \ldots, \theta$, and let a be a positive integer.

   (i) The sufficient statistic $X_{(n)}$ is complete when the parameter space is $\Omega = \{\theta : \theta \leq a\}$.
   (ii) Show that $X_{(n)}$ is not complete when $\Omega = \{\theta : \theta \geq a\}$, $a \geq 2$, and find a complete sufficient statistic in this case.

32. *Negative binomial.* Let X, Y be independently distributed according to negative binomial distributions $Nb(p_1, m)$ and $Nb(p_2, n)$ respectively, and let $q_i = 1 - p_i$.

    (i) There exists a UMP unbiased test for testing $H: \theta = q_2/q_1 \leq \theta_0$ and hence in particular $H': p_1 \leq p_2$.
    (ii) Determine the conditional distribution required for testing H' when $m = n = 1$.

33. Let X_i $(i = 1, 2)$ be independently distributed according to distributions from the exponential families (12) of Chapter 3 with C, Q, T, and h replaced by C_i, Q_i, T_i, and h_i. Then there exists a UMP unbiased test of

    (i) $H: Q_2(\theta_2) - Q_1(\theta_1) \leq c$ and hence in particular of $Q_2(\theta_2) \leq Q_1(\theta_1)$;
    (ii) $H: Q_2(\theta_2) + Q_1(\theta_1) \leq c$.

34. Let X, Y, Z be independent Poisson variables with means λ, μ, ν. Then there exists a UMP unbiased test of $H: \lambda\mu \leq \nu^2$.

35. *Random sample size.* Let N be a random variable with a *power-series* distribution

$$P(N = n) = \frac{a(n)\lambda^n}{C(\lambda)}, \quad n = 0, 1, \ldots \quad (\lambda > 0, \text{unknown}).$$

    When $N = n$, a sample $X_1, \ldots, X_n$ from the exponential family (12) of Chapter 3 is observed. On the basis of $(N, X_1, \ldots, X_N)$ there exists a UMP unbiased test of $H: Q(\theta) \leq c$.

36. The UMP unbiased test of $H: \Delta = 1$ derived in Section 8 for the case that the B- and C-margins are fixed (where the conditioning now extends to all random margins) is also UMP unbiased when

    (i) only one of the margins is fixed;
    (ii) the entries in the $4K$ cells are independent Poisson variables with means $\lambda_{ABC}, \ldots$, and Δ is replaced by the corresponding cross-ratio of the λ's.

11. REFERENCES

Agresti, A.
  (1984). *Analysis of ordinal categorical data.* Wiley, 1984.
Anderson, T. W. and Goodman, L. A.
  (1957). "Statistical inference about Markov chains." *Ann. Math. Statist.* **28**, 89–110.
Armsen, P.
  (1955). "Tables for significance tests of 2×2 contingency tables." *Biometrika* **42**, 494–511.
Bahadur, R. R.
  (1952). "A property of the t-statistic." *Sankhyā* **12**, 79–88.

Barnard, G. A.
(1982). "Conditionality versus similarity in the analysis of 2 × 2 tables." In *Statistics and Probability*: *Essays in Honor of C. R. Rao* (Kallianpur et al., eds.), North Holland, Amsterdam.

Bartlett, M. S.
(1937). "Properties of sufficiency and statistical tests." *Proc. Roy. Soc. London, Ser. A* **160**, 268–282.
[Points out that *exact* (that is, similar) tests can be obtained by combining the conditional tests given the different values of a sufficient statistic. Applications.]

Bell, C. B., Blackwell, D., and Breiman, L.
(1960). "On the completeness of order statistics." *Ann. Math. Statist.* **31**, 794–797.

Berk, R. H. and Cohen, A.
(1979). "Asymptotically optimal methods of combining tests." *J. Amer. Statist. Assoc.* **74**, 812–814.

Bickel, P. J., Hammel, E. A., and O'Connell, W.
(1977). "Sex bias in graduate admissions: data from Berkeley" (with discussion). In *Statistics and Public Policy* (Fairly and Mosteller, eds.), Addison-Wesley, Reading, Mass.

Billingsley, P.
(1961). "Statistical methods in Markov chains." *Ann. Math. Statist.* **32**, 12–40.
(1979). *Probability and Measure*, Wiley, New York.

Birch, M. W.
(1964). "The detection of partial association, I: The 2 × 2 case." *J. Roy. Statist. Soc. (B)* **26**, 313–324.

Birnbaum, A.
(1954). "Statistical methods for Poisson processes and exponential populations." *J. Amer. Statist. Assoc.* **49**, 254–266.

Bishop, Y. M. M., Fienberg, S. E., and Holland, P. W.
(1975). *Discrete Multivariate Analysis: Theory and Practice*, M.I.T. Press, Cambridge, Mass.

Boschloo, R. D.
(1970). "Raised conditional level of significance for the 2 × 2 table when testing the equality of two probabilities." *Statist. Neerl.* **24**, 1–35.

Bross, I. D. J. and Kasten, E. L.
(1957). "Rapid analysis of 2 × 2 tables." *J. Amer. Statist. Assoc.* **52**, 18–28.

Brown, L. D., Johnstone, I. M., and MacGibbon, K. B.
(1981). "Variation diminishing transformations: A direct approach to total positivity and its statistical applications." *J. Amer. Statist. Assoc.* **76**, 824–832.

Chambers, E. A. and Cox, D. R.
(1967). "Discrimination between alternative binary response models." *Biometrika* **54**, 573–578.

Chapman, J. W.
(1976). "A comparison of X^2, $-2 \log R$, and multinomial probability criteria for significance tests when expected frequencies are small." *J. Amer. Statist. Assoc.* **71**, 854–863.

Chen, H. J.
(1984). "Sample size determinations when two binomial proportions are very small." *Comm. Statist.—Theor. Meth.* **13**, 2707–2712.

Cohen, A., Gatsonis, C., and Marden, J.
(1983). "Hypothesis tests and optimality properties in discrete multivariate analysis." In *Studies in Econometrics, Time Series, and Multivariate Statistics* (Karlin et al., eds.), Academic.

Conover, W. J.

(1974). "Some reasons for not using the Yates continuity correction on 2×2 contingency tables" (with discussion). *J. Amer. Statist. Assoc.* **69**, 374–382.

Cox, D. R.

(1966). "A simple example of a comparison involving quantal data." *Biometrika* **53**, 215–220.

(1970). *The Analysis of Binary Data*, Methuen, London.

[An introduction to the problems treated in Sections 6 and 7 and some of their extensions.]

Cox, M. A. and Plackett, R. L.

(1980). "Small samples in contingency tables." *Biometrika* **67**, 1–13.

Cressie, N.

(1978). "Testing for the equality of two binomial proportions." *Ann. Inst. Statist. Math.* **30**, 421–427.

Darroch, J. N.

(1981). "The Mantel–Haenszel test and tests of marginal symmetry: Fixed effects and mixed models for a categorical response." *Int. Statist. Rev.* **49**, 285–307.

David, F. N.

(1947). "A power function for tests of randomness in a sequence." *Biometrika* **34**, 335–339.

[Discusses the run test in connection with the model of Problem 20.]

Denny, J. L. and Wright, A. L.

(1978). "On tests for Markov dependence." *Z. Wahrsch.* **43**, 331–338.

Denny, J. L. and Yakowitz, S. J.

(1978). "Admissible run-contingency type test for independence." *J. Amer. Statist. Assoc.* **73**, 177–181.

Eberhardt, K. R. and Fligner, M. A.

(1977). "A comparison of two tests for equality of two proportions." *Amer. Statistician* **31**, 151–155.

Edwards, A. W. F.

(1963). "The measure of association in a 2×2 table." *J. Roy. Statist. Soc.* (A) **126**, 109–114.

Feller, W.

(1936). "Note on regions similar to the sample space." *Statist. Res. Mem.* **2**, 117–125.

[Obtains a result which implies the completeness of order statistics.]

Fienberg, S. E.

(1980). *The Analysis of Cross-Classified Categorical Data*, 2nd ed., MIT Press, Cambridge, Mass.

Finch, P. D.

(1979). "Description and analogy in the practice of statistics" (with discussion). *Biometrika* **66**, 195–208.

Finney, D. J.

(1948). "The Fisher–Yates test of significance in 2×2 contingency tables." *Biometrika* **35**, 145–156.

Finney, D. J., Latscha, R., Bennett, B., Hsu, P., and Horst, C.

(1963, 1966). *Tables for Testing Significance in a 2×2 Contingency Table*, Cambridge U.P.

Fisher, R. A.

(1934). *Statistical Methods for Research Workers*, 5th and subsequent eds., Oliver and Boyd, Edinburgh, Section 21.02.

[Proposes the conditional tests for the hypothesis of independence in a 2×2 table.]

Fraser, D. A. S.
(1953). "Completeness of order statistics." *Canad. J. Math.* **6**, 42–45.
[Problem 12.]

Freedman, D. A. and Lane, D.
(1982). "Significance testing in a nonstochastic setting." In *Festschrift for Erich L. Lehmann* (Bickel, Doksum, and Hodges, eds.), Wadsworth, Belmont, Calif.

Frisén, M.
(1980). "Consequences of the use of conditional inference in the analysis of a correlated contingency table." *Biometrika* **67**, 23–30.

Garside, G. R. and Mack, C.
(1976). "Actual type 1 error probabilities for various tests in the homogeneity case of the 2 × 2 contingency table." *Amer. Statistician* **30**, 18–21.

Gart, J. J.
(1970). "Point and interval estimation of the common odds ratio in the combination of 2 × 2 tables with fixed marginals." *Biometrika* **57**, 471–475.

Ghosh, M. N.
(1948). "On the problem of similar regions." *Sankhyā* **8**, 329–338.
[Theorem 1.]

Girschick, M. A., Mosteller, F., and Savage, L. J.
(1946). "Unbiased estimates for certain binomial sampling problems with applications." *Ann. Math. Statist.* **17**, 13–23.
[Problem 11.]

Gokhale, D. V. and Johnson, N. S.
(1978). "A class of alternatives to independence in contingency tables." *J. Amer. Statist. Assoc.* **73**, 800–804.

Goodman, L. A.
(1958). "Simplified runs tests and likelihood ratio tests for Markoff chains." *Biometrika* **45**, 181–197.
(1964). "Simple methods for analyzing three-factor interaction in contingency tables." *J. Amer. Statist. Assoc.* **59**, 319–352.

Goodman, L. A. and Kruskal, W.
(1954, 1959). "Measures of association for cross classification." *J. Amer. Statist. Assoc.* **49**, 732–764; **54**, 123–163.

Guenther, W. C.
(1978). "Some remarks on the runs tests and the use of the hypergeometric distribution." *Amer. Statistician* **32**, 71–73.

Haber, M.
(1980). "A comparison of some continuity corrections for the chi-squared test on 2 × 2 tables." *J. Amer. Statist. Assoc.* **75**, 510–515.

Haberman, S. J.
(1974). *The Analysis of Frequency Data.* Univ. of Chicago Press.
(1982). "Association, Measures of." In *Encycl. Statist. Sci.*, Vol. 1, Wiley, New York, 130–136.

Haldane, J. B. S. and Smith, C. A. B.
(1948). "A simple exact test for birth-order effect." *Ann. Eugenics* **14**, 117–124.
[Proposes the rank-sum test in a setting similar to that of Problem 23.]

Hoel, P. G.
(1945). "Testing the homogeneity of Poisson frequencies." *Ann. Math. Statist.* **16**, 362–368.
[First example of Section 5.]
(1948). "On the uniqueness of similar regions." *Ann. Math. Statist.* **19**, 66–71.
[Theorem 1 under regularity assumptions.]

Johansen, S.
(1979). *Introduction to the Theory of Regular Exponential Families*, Univ. of Copenhagen.

Johnson, N. L. and Kotz, S.
(1969). *Discrete Distributions*, Houghton Mifflin, Boston.

Kallenberg, W. C. M. et al.
(1984). *Testing Statistical Hypotheses: Worked Solutions*, Mathematische Centrum, Amsterdam.

Kempthorne, O.
(1979). "In dispraise of the exact test: Reactions." *J. Statist. Planning and Inf.* **3**, 199–213.

Koziol, J. A. and Perlman, M. D.
(1978). "Combining independent chi-squared tests." *J. Amer. Statist. Assoc.* **73**, 753–763.

Kruskal, W. H.
(1957). "Historical notes on the Wilcoxon unpaired two-sample test." *J. Amer. Statist. Assoc.* **52**, 356–360.

Landis, J. R., Heyman, E. R., and Koch, G. G.
(1978). "Average partial association in three-way contingency tables: A review and discussion of alternative tests." *Int. Statist. Rev.* **46**, 237–254.

Lehmann, E. L.
(1947). "On families of admissible tests." *Ann. Math. Statist.* **18**, 97–104.
[Problem 8.]
(1950). "Some principles of the theory of testing hypotheses." *Ann. Math. Statist.* **21**, 1–26.
[Lemma 1.]
(1952). "Testing multiparameter hypotheses." *Ann. Math. Statist.* **23**, 541–552.
[Problem 7.]
(1957). "A theory of some multiple decision procedures." *Ann. Math. Statist.* **28**, 1–25, 547–572.
(1975). *Nonparametrics: Statistical Methods Based on Ranks*, Holden-Day, San Francisco.

Lehmann, E. L. and Scheffé, H.
(1950, 1955). "Completeness, similar regions, and unbiased estimation." *Sankhyā* **10**, 305–340; **15**, 219–236.
[Introduces the concept of completeness. Theorem 3 and applications.]

Lieberman, G. J. and Owen, D. B.
(1961). *Tables of the Hypergeometric Probability Distribution*, Stanford U.P.

Lindley, D. V. and Novick, M. R.
(1981). The role of exchangeability in inference." *Ann. Statist.* **9**, 45–58.

Ling, R. F. and Pratt, J. W.
(1984). "The accuracy of Peizer approximations to the hypergeometric distribution, with comparisons to some other approximations." *J. Amer. Statist. Assoc.* **79**, 49–60.

Littell, R. C. and Louv, W. C.
(1981). "Confidence regions based on methods of combining test statistics." *J. Amer. Statist. Assoc.* **76**, 125–130.

McDonald, L. L., Davis, B. M., and Milliken, G. A.
(1977). "A nonrandomized unconditional test for comparing two proportions in 2×2 contingency tables." *Technometrics* **19**, 145–158.

Marden, J. I.
(1982). "Combining independent noncentral chi-squared or *F*-tests." *Ann. Statist.* **10**, 266–270.

Mehta, C. R. and Patel, N. R.
(1983). "A network algorithm for performing Fisher's exact tests in $r \times c$ contingency tables." *J. Amer. Statist. Assoc.* **78**, 427–434.

Michel, R.
(1979). "On the asymptotic efficiency of conditional tests for exponential families." *Ann. Statist.* **7**, 1256–1263.

Mudholkar, G. S. and George, E. O.
(1979). "The logit statistic for combining probabilities—an overview." In *Optimizing Methods in Statistics* (Rustagi, ed.), Academic, New York.

Nandi, H. K.
(1951). "On type B_1 and type B regions." *Sankhyā* **11**, 13–22.
[One of the cases of Theorem 3, under regularity assumptions.]

Neyman, J.
(1935). "Sur la vérification des hypothèses statistiques composées." *Bull. Soc. Math. France* **63**, 246–266.
[Theory of tests of composite hypotheses that are locally unbiased and locally most powerful.]
(1941). "On a statistical problem arising in routine analyses and in sampling inspection of mass distributions." *Ann. Math. Statist.* **12**, 46–76.

Neyman, J. and Pearson, E. S.
(1933). "On the problem of the most efficient tests of statistical hypotheses." *Phil. Trans. Roy. Soc., Ser. A* **231**, 289–337.
[Introduces the concept of similarity and develops a method for determining the totality of similar regions.]
(1936, 1938). "Contributions to the theory of testing statistical hypotheses." *Statist. Res. Mem.* **1**, 1–37; **2**, 25–57.
[Defines unbiasedness and determines both locally and UMP unbiased tests of certain classes of simple hypotheses.]

Overall, J. E. and Starbuck, R. R.
(1983). "*F*-test alternatives to Fisher's exact test and to the chi-square test of homogeneity in 2×2 tables." *J. Educ. Statist.* **8**, 59–73.

Pachares, J.
(1961). "Tables for unbiased tests on the variance of a normal population." *Ann. Math. Statist.* **32**, 84–87.

Plackett, R. L.
(1981). *The Analysis of Categorical Data*, 2nd ed., MacMillan, New York.

Przyborowski, J. and Wilenski, H.
(1939). "Homogeneity of results in testing samples from Poisson series." *Biometrika* **31**, 313–323.
[Derives the UMP similar test for the equality of two Poisson parameters.]

Putter, J.
(1955). "The treatment of ties in some nonparametric tests." *Ann. Math. Statist.* **26**, 368–386.
[Discusses the treatment of ties in the sign test.]

Scheffé, H.

(1943). "On a measure problem arising in the theory of non-parametric tests." *Ann. Math. Statist.* **14**, 227–233.

[Proves the completeness of order statistics.]

Scholz, F. W.

(1982). "Combining independent *P*-values." In *A Festschrift for Erich L. Lehmann* (Bickel, Doksum, and Hodges, eds.), Wadsworth, Belmont, Calif.

Somes, G. W. and O'Brien, K. F.

(1985). "Mantel-Haenszel statistic." In *Encycl. Statist. Sci.*, Vol. 5, Wiley, New York.

Sverdrup, E.

(1953). "Similarity, unbiasedness, minimaxibility and admissibility of statistical test procedures." *Skand. Aktuar. Tidskrift* **36**, 64–86.

[Theorem 1 and results of the type of Theorem 3. Applications including the 2×2 table.]

Swed, F. S. and Eisenhart, C.

(1943). "Tables for testing randomness of grouping in a sequence of alternatives." *Ann. Math. Statist.* **14**, 66–87.

Tocher, K. D.

(1950). "Extension of Neyman–Pearson theory of tests to discontinuous variates." *Biometrika* **37**, 130–144.

[Proves the optimum property of Fisher's exact test.]

Tong, Y. L.

(1980). *Probability Inequalities in Multivariate Distributions*, Academic, New York.

Venable, T. C. and Bhapkar, V. P.

(1978). "Gart's test of interaction in a $2 \times 2 \times 2$ contingency table for small samples." *Biometrika* **65**, 669–672.

Wacholder, S. and Weinberg, C. R.

(1982). "Paired versus two-sample design for a clinical trial of treatments with dichotomous outcome: Power considerations." *Biometrics* **38**, 801–812.

Walsh, J. E.

(1949). "Some significance tests for the median which are valid under very general conditions." *Ann. Math. Statist.* **20**, 64–81.

[Contains a result related to Problem 12.]

Wolfowitz, J.

(1943). "On the theory of runs with some applications to quality control." *Ann. Math. Statist.* **14**, 280–288.

Yates, F.

(1984). "Tests of significance for 2×2 contingency tables" (with discussion). *J. Roy. Statist. Soc. (A)* **147**, 426–463.

CHAPTER 5

Unbiasedness: Applications
to Normal Distributions;
Confidence Intervals

1. STATISTICS INDEPENDENT OF A SUFFICIENT STATISTIC

A general expression for the UMP unbiased tests of the hypotheses $H_1 : \theta \leq \theta_0$ and $H_4 : \theta = \theta_0$ in the exponential family

$$(1) \qquad dP_{\theta,\vartheta}(x) = C(\theta, \vartheta)\exp\left[\theta U(x) + \sum \vartheta_i T_i(x)\right] d\mu(x)$$

was given in Theorem 3 of the preceding chapter. However, this turns out to be inconvenient in the applications to normal and certain other families of continuous distributions, with which we shall be concerned in the present chapter. In these applications, the tests can be given a more convenient form, in which they no longer appear as conditional tests in terms of U given t, but are expressed unconditionally in terms of a single test statistic. The following are three general methods of achieving this.

(i) In many of the problems to be considered below, the UMP unbiased test ϕ_0 is also UMP invariant, as will be shown in Chapter 6. From Theorem 6 of Chapter 6 it is then possible to conclude that ϕ_0 is UMP unbiased. This approach, in which the latter property must be taken on faith during the discussion of the test in the present chapter, is the most economical of the three, and has the additional advantage that it derives the test instead of verifying a guessed solution as is the case with methods (ii) and (iii).

(ii) The conditional descriptions (12), (14), and (16) of Chapter 4 can be replaced by equivalent unconditional ones, and it is then enough to find an unbiased test which has the indicated structure. This approach is discussed in Pratt (1962).

(iii)   Finally, it is often possible to show the equivalence of the test given by Theorem 3 of Chapter 4 to a test suspected to be optimal, by means of Theorem 2 below. This is the course we shall follow here; the alternative derivation (i) will be discussed in Chapter 6.

The reduction by method (iii) depends on the existence of a statistic $V = h(U, T)$, which is independent of T when $\theta = \theta_0$, and which for each fixed t is monotone in U for H_1 and linear in U for H_4. The critical function ϕ_1 for testing H_1 then satisfies

(2)
$$\phi(v) = \begin{cases} 1 & \text{when} \quad v > C_0, \\ \gamma_0 & \text{when} \quad v = C_0, \\ 0 & \text{when} \quad v < C_0, \end{cases}$$

where C_0 and γ_0 are no longer dependent on t, and are determined by

(3)
$$E_{\theta_0}\phi_1(V) = \alpha.$$

Similarly the test ϕ_4 of H_4 reduces to

(4)
$$\phi(v) = \begin{cases} 1 & \text{when} \quad v < C_1 \text{ or } v > C_2, \\ \gamma_i & \text{when} \quad v = C_i, \quad i = 1, 2, \\ 0 & \text{when} \quad C_1 < v < C_2, \end{cases}$$

where the C's and γ's are determined by

(5)
$$E_{\theta_0}[\phi_4(V)] = \alpha$$

and

(6)
$$E_{\theta_0}[V\phi_4(V)] = \alpha E_{\theta_0}(V).$$

The corresponding reduction for the hypotheses $H_2 : \theta \le \theta_1$ or $\theta \ge \theta_2$ and $H_3 : \theta_1 \le \theta \le \theta_2$ requires that V be monotone in U for each fixed t, and be independent of T when $\theta = \theta_1$ and $\theta = \theta_2$. The test ϕ_3 is then given by (4) with the C's and γ's determined by

(7)
$$E_{\theta_1}\phi_3(V) = E_{\theta_2}\phi_3(V) = \alpha.$$

The test for H_2 as before has the critical function

$$\phi_2(v; \alpha) = 1 - \phi_3(v; 1 - \alpha).$$

This is summarized in the following theorem.

Theorem 1. *Suppose that the distribution of X is given by* (1) *and that* $V = h(U, T)$ *is independent of T when $\theta = \theta_0$. Then ϕ_1 is UMP unbiased for testing H_1 provided the function h is increasing in u for each t, and ϕ_4 is UMP unbiased for H_4 provided*

$$h(u, t) = a(t)u + b(t) \quad \text{with} \quad a(t) > 0.$$

The tests ϕ_2 and ϕ_3 are UMP unbiased for H_2 and H_3 if V is independent of T when $\theta = \theta_1$ and θ_2, and if h is increasing in u for each t.

Proof. The test of H_1 defined by (12) and (13) of Chapter 4 is equivalent to that given by (2), with the constants determined by

$$P_{\theta_0}\{V > C_0(t)|t\} + \gamma_0(t)P_{\theta_0}\{V = C_0(t)|t\} = \alpha.$$

By assumption, V is independent of T when $\theta = \theta_0$, and C_0 and γ_0 therefore do not depend on t. This completes the proof for H_1, and that for H_2 and H_3 is quite analogous.

The test of H_4 given in Section 4 of Chapter 4 is equivalent to that defined by (4) with the constants C_i and γ_i determined by $E_{\theta_0}[\phi_4(V, t)|t] = \alpha$ and

$$E_{\theta_0}\left[\phi_4(V, t)\frac{V - b(t)}{a(t)}\bigg|t\right] = \alpha E_{\theta_0}\left[\frac{V - b(t)}{a(t)}\bigg|t\right],$$

which reduces to

$$E_{\theta_0}[V\phi_4(V, t)|t] = \alpha E_{\theta_0}[V|t].$$

Since V is independent of T for $\theta = \theta_0$, so are the C's and γ's as was to be proved.

To prove the required independence of V and T in applications of Theorem 1 to special cases, the standard methods of distribution theory are available: transformation of variables, characteristic functions, and the geometric method. Frequently, an alternative approach, which is particularly useful also in determining a suitable statistic V, is provided by the following theorem.

Theorem 2. (*Basu*).   *Let the family of possible distributions of X be* $\mathscr{P} = \{P_\vartheta, \vartheta \in \omega\}$, *let T be sufficient for* $\mathscr{P}$, *and suppose that the family* $\mathscr{P}^T$ *of distributions of T is boundedly complete. If V is any statistic whose distribution does not depend on* ϑ, *then V is independent of T.*

Proof.   For any critical function ϕ, the expectation $E_\vartheta\phi(V)$ is by assumption independent of ϑ. It therefore follows from Theorem 2 of Chapter 4 that $E[\phi(V)|t]$ is constant (a.e. $\mathscr{P}^T$) for every critical function ϕ, and hence that V is independent of T.

For converse aspects of this theorem see Basu (1958), Koehn and Thomas (1975), Bahadur (1979), and Lehmann (1980).

Corollary 1.   *Let* $\mathscr{P}$ *be the exponential family obtained from* (1) *by letting* θ *have some fixed value. Then a statistic V is independent of T for all* ϑ *provided the distribution of V does not depend on* ϑ.

Proof.   It follows from Theorem 1 of Chapter 4 that $\mathscr{P}^T$ is complete and hence boundedly complete, and the preceding theorem is therefore applicable.

Example 1.   Let $X_1, \ldots, X_n$ be independently, normally distributed with mean ξ and variance σ^2. Suppose first that σ^2 is fixed at σ_0^2. Then the assumptions of Corollary 1 hold with $T = \bar{X} = \Sigma X_i/n$ and ϑ proportional to ξ. Let f be any function satisfying

$$f(x_1 + c, \ldots, x_n + c) = f(x_1, \ldots, x_n) \qquad \text{for all real } c.$$

If

$$V = f(X_1, \ldots, X_n),$$

then also $V = f(X_1 - \xi, \ldots, X_n - \xi)$. Since the variables $X_i - \xi$ are distributed as $N(0, \sigma_0^2)$, which does not involve ξ, the distribution of V does not depend on ξ. It follows from Corollary 1 that any such statistic V, and therefore in particular $V = \Sigma(X_i - \bar{X})^2$, is independent of $\bar{X}$. This is true for all σ.

Suppose, on the other hand, that ξ is fixed at ξ_0. Then Corollary 1 applies with $T = \Sigma(X_i - \xi_0)^2$ and $\vartheta = -1/2\sigma^2$. Let f be any function such that

$$f(cx_1, \ldots, cx_n) = f(x_1, \ldots, x_n) \qquad \text{for all } c > 0,$$

and let

$$V = f(X_1 - \xi_0, \ldots, X_n - \xi_0).$$

Then V is unchanged if each $X_i - \xi_0$ is replaced by $(X_i - \xi_0)/\sigma$, and since these variables are normally distributed with zero mean and unit variance, the distribution of V does not depend on σ. It follows that all such statistics V, and hence for example

$$\frac{\bar{X} - \xi_0}{\sqrt{\Sigma(X_i - \bar{X})^2}} \quad \text{and} \quad \frac{\bar{X} - \xi_0}{\sqrt{\Sigma(X_i - \xi_0)^2}},$$

are independent of $\Sigma(X_i - \xi_0)^2$. This, however, does not hold for all ξ, but only when $\xi = \xi_0$.

Example 2. Let U_1/σ_1^2 and U_2/σ_2^2 be independently distributed according to χ^2-distributions with f_1 and f_2 degrees of freedom respectively, and suppose that $\sigma_2^2/\sigma_1^2 = a$. The joint density of the U's is then

$$Cu_1^{(f_1/2)-1}u_2^{(f_2/2)-1}\exp\left[-\frac{1}{2\sigma_2^2}(au_1 + u_2)\right]$$

so that Corollary 1 is applicable with $T = aU_1 + U_2$ and $\vartheta = -1/2\sigma_2^2$. Since the distribution of

$$V = \frac{U_2}{U_1} = a\frac{U_2/\sigma_2^2}{U_1/\sigma_1^2}$$

does not depend on σ_2, V is independent of $aU_1 + U_2$. For the particular case that $\sigma_2 = \sigma_1$, this proves the independence of U_2/U_1 and $U_1 + U_2$.

Example 3. Let $(X_1, \ldots, X_n)$ and $(Y_1, \ldots, Y_n)$ be samples from normal distributions $N(\xi, \sigma^2)$ and $N(\eta, \tau^2)$ respectively. Then $T = (\bar{X}, \Sigma X_i^2, \bar{Y}, \Sigma Y_i^2)$ is sufficient for $(\xi, \sigma^2, \eta, \tau^2)$ and the family of distributions of T is complete. Since

$$V = \frac{\Sigma(X_i - \bar{X})(Y_i - \bar{Y})}{\sqrt{\Sigma(X_i - \bar{X})^2\Sigma(Y_i - \bar{Y})^2}}$$

is unchanged when X_i and Y_i are replaced by $(X_i - \xi)/\sigma$ and $(Y_i - \eta)/\tau$, the distribution of V does not depend on any of the parameters, and Theorem 2 shows V to be independent of T.

2. TESTING THE PARAMETERS OF A NORMAL DISTRIBUTION

The four hypotheses $\sigma \leq \sigma_0$, $\sigma \geq \sigma_0$, $\xi \leq \xi_0$, $\xi \geq \xi_0$ concerning the variance σ^2 and mean ξ of a normal distribution were discussed in Chapter 3, Section 9, and it was pointed out there that at the usual significance levels there exists a UMP test only for the first one. We shall now show that the standard (likelihood-ratio) tests are UMP unbiased for the above four hypotheses as well as for some of the corresponding two-sided problems.

For varying ξ and σ, the densities

$$(8) \qquad (2\pi\sigma^2)^{-n/2}\exp\left(-\frac{n\xi^2}{2\sigma^2}\right)\exp\left(-\frac{1}{2\sigma^2}\sum x_i^2 + \frac{\xi}{\sigma^2}\sum x_i\right)$$

of a sample $X_1,\ldots,X_n$ from $N(\xi,\sigma^2)$ constitute a two-parameter exponential family, which coincides with (1) for

$$\theta = -\frac{1}{2\sigma^2}, \quad \vartheta = \frac{n\xi}{\sigma^2}, \quad U(x) = \sum x_i^2, \quad T(x) = \bar{x} = \frac{\sum x_i}{n}.$$

By Theorem 3 of Chapter 4 there exists therefore a UMP unbiased test of the hypothesis $\theta \geq \theta_0$, which for $\theta_0 = -1/2\sigma_0^2$ is equivalent to $H: \sigma \geq \sigma_0$. The rejection region of this test can be obtained from (12) of Chapter 4, with the inequalities reversed because the hypothesis is now $\theta \geq \theta_0$. In the present case this becomes

$$\sum x_i^2 \leq C_0(\bar{x})$$

where

$$P_{\sigma_0}\left\{\sum X_i^2 \leq C_0(\bar{x})|\bar{x}\right\} = \alpha.$$

If this is written as

$$\sum x_i^2 - n\bar{x}^2 \leq C_0'(\bar{x}),$$

it follows from the independence of $\sum X_i^2 - n\bar{X}^2 = \sum(X_i - \bar{X})^2$ and $\bar{X}$ (Example 1) that $C_0'(\bar{x})$ does not depend on $\bar{x}$. The test therefore rejects when $\sum(x_i - \bar{x})^2 \leq C_0'$, or equivalently when

$$(9) \qquad \frac{\sum(x_i - \bar{x})^2}{\sigma_0^2} \leq C_0,$$

with C_0 determined by $P_{\sigma_0}\{\sum(X_i - \bar{X})^2/\sigma_0^2 \leq C_0\} = \alpha$. Since $\sum(X_i - \bar{X})^2/\sigma_0^2$ has a χ^2-distribution with $n-1$ degrees of freedom, the determining condition for C_0 is

$$(10) \qquad \int_0^{C_0} \chi_{n-1}^2(y)\, dy = \alpha$$

where χ_{n-1}^2 denotes the density of a χ^2 variable with $n-1$ degrees of freedom.

The same result can be obtained through Theorem 1. A statistic $V = h(U, T)$ of the kind required by the theorem—that is, independent of $\bar{X}$ for $\sigma = \sigma_0$ and all ξ—is

$$V = \sum (X_i - \bar{X})^2 = U - nT^2.$$

This is in fact independent of $\bar{X}$ for all ξ and σ^2. Since $h(u, t)$ is an increasing function of u for each t, it follows that the UMP unbiased test has a rejection region of the form $V \leq C_0'$.

This derivation also shows that the UMP unbiased rejection region for $H : \sigma \leq \sigma_1$ or $\sigma \geq \sigma_2$ is

$$(11) \qquad C_1 < \sum (x_i - \bar{x})^2 < C_2$$

where the C's are given by

$$(12) \qquad \int_{C_1/\sigma_1^2}^{C_2/\sigma_1^2} \chi_{n-1}^2(y)\, dy = \int_{C_1/\sigma_2^2}^{C_2/\sigma_2^2} \chi_{n-1}^2(y)\, dy = \alpha.$$

Since $h(u, t)$ is linear in u, it is further seen that the UMP unbiased test of $H : \sigma = \sigma_0$ has the acceptance region

$$(13) \qquad C_1' < \frac{\sum (x_i - \bar{x})^2}{\sigma_0^2} < C_2'$$

with the constants determined by

$$(14) \qquad \int_{C_1'}^{C_2'} \chi_{n-1}^2(y)\, dy = \frac{1}{n-1} \int_{C_1'}^{C_2'} y\chi_{n-1}^2(y)\, dy = 1 - \alpha.$$

This is just the test obtained in Example 2 of Chapter 4 with $\sum (x_i - \bar{x})^2$ in place of $\sum x_i^2$ and $n - 1$ degrees of freedom instead of n, as could have been foreseen. Theorem 1 shows for this and the other hypotheses considered that the UMP unbiased test depends only on V. Since the distributions of V do not depend on ξ, and constitute an exponential family in σ, the problems are thereby reduced to the corresponding ones for a one-parameter exponential family, which were solved previously.

The power of the above tests can be obtained explicitly in terms of the χ^2-distribution. In the case of the one-sided test (9) for example, it is given by

$$\beta(\sigma) = P_\sigma \left\{ \frac{\sum (X_i - \bar{X})^2}{\sigma^2} \leq \frac{C_0\sigma_0^2}{\sigma^2} \right\} = \int_0^{C_0\sigma_0^2/\sigma^2} \chi_{n-1}^2(y)\, dy.$$

The same method can be applied to the problems of testing the hypotheses $\xi \le \xi_0$ against $\xi > \xi_0$ and $\xi = \xi_0$ against $\xi \ne \xi_0$. As is seen by transforming to the variables $X_i - \xi_0$, there is no loss of generality in assuming that $\xi_0 = 0$. It is convenient here to make the identification of (8) with (1) through the correspondence

$$\theta = \frac{n\xi}{\sigma^2}, \quad \vartheta = -\frac{1}{2\sigma^2}, \quad U(x) = \bar{x}, \quad T(x) = \sum x_i^2.$$

Theorem 3 of Chapter 4 then shows that UMP unbiased tests exist for the hypotheses $\theta \le 0$ and $\theta = 0$, which are equivalent to $\xi \le 0$ and $\xi = 0$. Since

$$V = \frac{\bar{X}}{\sqrt{\sum(X_i - \bar{X})^2}} = \frac{U}{\sqrt{T - nU^2}}$$

is independent of $T = \sum X_i^2$ when $\xi = 0$ (Example 1), it follows from Theorem 1 that the UMP unbiased rejection region for $H : \xi \le 0$ is $V \ge C_0'$ or equivalently

(15) $$t(x) \ge C_0,$$

where

(16) $$t(x) = \frac{\sqrt{n}\,\bar{x}}{\sqrt{\dfrac{1}{n-1}\sum(x_i - \bar{x})^2}}.$$

In order to apply the theorem to $H' : \xi = 0$, let $W = \bar{X}/\sqrt{\sum X_i^2}$. This is also independent of $\sum X_i^2$ when $\xi = 0$, and in addition is linear in $U = \bar{X}$. The distribution of W is symmetric about 0 when $\xi = 0$, and conditions (4), (5), (6) with W in place of V are therefore satisfied for the rejection region $|w| \ge C'$ with $P_{\xi=0}\{|W| \ge C'\} = \alpha$. Since

$$t(x) = \frac{\sqrt{(n-1)n}\,W(x)}{\sqrt{1 - nW^2(x)}},$$

the absolute value of $t(x)$ is an increasing function of $|W(x)|$, and the rejection region is equivalent to

(17) $$|t(x)| \ge C.$$

From (16) it is seen that $t(X)$ is the ratio of the two independent random variables $\sqrt{n}\,\bar{X}/\sigma$ and $\sqrt{\Sigma(X_i - \bar{X})^2/(n-1)\sigma^2}$. The denominator is distributed as the square root of a χ^2-variable with $n-1$ degrees of freedom, divided by $n-1$; the distribution of the numerator, when $\xi = 0$, is the normal distribution $N(0, 1)$. The distribution of such a ratio is *Student's* t-distribution with $n-1$ degrees of freedom, which has probability density

$$(18) \qquad t_{n-1}(y) = \frac{1}{\sqrt{\pi(n-1)}} \frac{\Gamma(\tfrac{1}{2}n)}{\Gamma[\tfrac{1}{2}(n-1)]} \frac{1}{\left(1 + \dfrac{y^2}{n-1}\right)^{\frac{1}{2}n}}.$$

The distribution is symmetric about 0, and the constants C_0 and C of the one- and two-sided tests are determined by

$$(19) \qquad \int_{C_0}^{\infty} t_{n-1}(y)\, dy = \alpha \quad \text{and} \quad \int_{C}^{\infty} t_{n-1}(y)\, dy = \frac{\alpha}{2}.$$

For $\xi \neq 0$, the distribution of $t(X)$ is the so-called *noncentral t-distribution*, which is derived in Problem 3. Some properties of the power function of the one- and two-sided t-test are given in Problems 1, 2, and 4. We note here that the distribution of $t(X)$, and therefore the power of the above tests, depends only on the noncentrality parameter $\delta = \sqrt{n}\,\xi/\sigma$. This is seen from the expression of the probability density given in Problem 3, but can also be shown by the following direct argument. Suppose that $\xi'/\sigma' = \xi/\sigma \neq 0$, and denote the common value of ξ'/ξ and σ'/σ by c, which is then also different from zero. If $X_i' = cX_i$ and the X_i are distributed as $N(\xi, \sigma^2)$, the variables X_i' have distribution $N(\xi', \sigma'^2)$. Also $t(X) = t(X')$, and hence $t(X')$ has the same distribution as $t(X)$, as was to be proved. [Tables of the power of the t-test are discussed, for example, in Chapter 31, Section 7 of Johnson and Kotz (1970, Vol. 2).]

If ξ_1 denotes any alternative value to $\xi = 0$, the power $\beta(\xi, \sigma) = f(\delta)$ depends on σ. As $\sigma \to \infty$, $\delta \to 0$, and

$$\beta(\xi_1, \sigma) \to f(0) = \beta(0, \sigma) = \alpha,$$

since f is continuous by Theorem 9 of Chapter 2. Therefore, regardless of the sample size, the probability of detecting the hypothesis to be false when $\xi \geq \xi_1 > 0$ cannot be made $\geq \beta > \alpha$ for all σ. This is not surprising, since the distributions $N(0, \sigma^2)$ and $N(\xi_1, \sigma^2)$ become practically indistinguishable when σ is sufficiently large. To obtain a procedure with guaranteed power for $\xi \geq \xi_1$, the sample size must be made to depend on σ. This can be achieved by a sequential procedure, with the stopping rule depending on an estimate of σ, but not with a procedure of fixed sample size. (See Problems 26 and 28).

The tests of the more general hypotheses $\xi \leq \xi_0$ and $\xi = \xi_0$ are reduced to those above by transforming to the variables $X_i - \xi_0$. The rejection regions for these hypotheses are given as before by (15), (17), and (19), but now with

$$t(x) = \frac{\sqrt{n}(\bar{x} - \xi_0)}{\sqrt{\dfrac{1}{n-1} \sum (x_i - \bar{x})^2}}.$$

It is seen from the representation of (8) as an exponential family with $\theta = n\xi/\sigma^2$ that there exists a UMP unbiased test of the hypothesis $a \leq \xi/\sigma^2 \leq b$, but the method does not apply to the more interesting hypothesis $a \leq \xi \leq b$;* nor is it applicable to the corresponding hypothesis for the mean expressed in σ-units: $a \leq \xi/\sigma \leq b$, which will be discussed in Chapter 6.

When testing the mean ξ of a normal distribution, one may from extensive past experience believe σ to be essentially known. If in fact σ is known to be equal to σ_0, it follows from Problem 1 of Chapter 3 that there exists a UMP test ϕ_0 of $H : \xi \leq \xi_0$ against $K : \xi > \xi_0$, which rejects when $(\bar{X} - \xi_0)/\sigma_0$ is sufficiently large, and this test is then uniformly more powerful than the t-test (15). On the other hand, if the assumption $\sigma = \sigma_0$ is in error, the size of ϕ_0 will differ from α and may greatly exceed it. Whether to take such a risk depends on one's confidence in the assumption and the gain resulting from the use of ϕ_0 when σ is equal to σ_0. A measure of this gain is the *deficiency d* of the t-test with respect to ϕ_0, the number of additional observations required by the t-test to match the power of ϕ_0 when $\sigma = \sigma_0$. Except for very small n, d is essentially independent of sample size and for typical values of α is of the order of 1 to 3 additional observations. [For details see Hodges and Lehmann (1970). Other approaches to such comparisons are reviewed, for example, in Rothenberg (1984).]

3. COMPARING THE MEANS AND VARIANCES OF TWO NORMAL DISTRIBUTIONS

The problem of comparing the parameters of two normal distributions arises in the comparison of two treatments, products, etc., under conditions similar to those discussed in Chapter 4 at the beginning of Section 5. We consider first the comparison of two variances σ^2 and τ^2, which occurs for example when one is concerned with the variability of analyses made by two

---

*This problem is discussed in Section 3 of Hodges and Lehmann (1954).

different laboratories or by two different methods, and specifically the hypotheses $H: \tau^2/\sigma^2 \le \Delta_0$ and $H': \tau^2/\sigma^2 = \Delta_0$.

Let $X = (X_1, \ldots, X_m)$ and $Y = (Y_1, \ldots, Y_n)$ be samples from the normal distributions $N(\xi, \sigma^2)$ and $N(\eta, \tau^2)$ with joint density

$$C(\xi, \eta, \sigma, \tau)\exp\left(-\frac{1}{2\sigma^2}\sum x_i^2 - \frac{1}{2\tau^2}\sum y_j^2 + \frac{m\xi}{\sigma^2}\bar{x} + \frac{n\eta}{\tau^2}\bar{y}\right).$$

This is an exponential family with the four parameters

$$\theta = -\frac{1}{2\tau^2}, \qquad \vartheta_1 = -\frac{1}{2\sigma^2}, \qquad \vartheta_2 = \frac{n\eta}{\tau^2}, \qquad \vartheta_3 = \frac{m\xi}{\sigma^2}$$

and the sufficient statistics

$$U = \sum Y_j^2, \qquad T_1 = \sum X_i^2, \qquad T_2 = \bar{Y}, \qquad T_3 = \bar{X}.$$

It can be expressed equivalently (see Lemma 2 of Chapter 4), in terms of the parameters

$$\theta^* = -\frac{1}{2\tau^2} + \frac{1}{2\Delta_0\sigma^2}, \qquad \vartheta_i^* = \vartheta_i \quad (i = 1, 2, 3)$$

and the statistics

$$U^* = \sum Y_j^2, \qquad T_1^* = \sum X_i^2 + \frac{1}{\Delta_0}\sum Y_j^2, \qquad T_2^* = \bar{Y}, \qquad T_3^* = \bar{X}.$$

The hypotheses $\theta^* \le 0$ and $\theta^* = 0$, which are equivalent to H and H' respectively, therefore possess UMP unbiased tests by Theorem 3 of Chapter 4.

When $\tau^2 = \Delta_0\sigma^2$, the distribution of the statistic

$$V = \frac{\sum(Y_j - \bar{Y})^2/\Delta_0}{\sum(X_i - \bar{X})^2} = \frac{\sum(Y_j - \bar{Y})^2/\tau^2}{\sum(X_i - \bar{X})^2/\sigma^2}$$

does not depend on σ, ξ, or η, and it follows from Corollary 1 that V is independent of (T_1^*, T_2^*, T_3^*). The UMP unbiased test of H is therefore given by (2) and (3), so that the rejection region can be written as

$$(20) \qquad \frac{\sum(Y_j - \bar{Y})^2/\Delta_0(n - 1)}{\sum(X_i - \bar{X})^2/(m - 1)} \geq C_0.$$

When $\tau^2 = \Delta_0\sigma^2$, the statistic on the left-hand side of (20) is the ratio of the two independent χ^2 variables $\sum(Y_j - \bar{Y})^2/\tau^2$ and $\sum(X_i - \bar{X})^2/\sigma^2$, each divided by the number of its degrees of freedom. The distribution of such a ratio is the *F-distribution* with $n - 1$ and $m - 1$ degrees of freedom, which has the density

$$(21) \quad F_{n-1,\,m-1}(y) = \frac{\Gamma[\frac{1}{2}(m + n - 2)]}{\Gamma[\frac{1}{2}(m - 1)]\Gamma[\frac{1}{2}(n - 1)]} \left(\frac{n - 1}{m - 1}\right)^{\frac{1}{2}(n-1)}$$

$$\times \frac{y^{\frac{1}{2}(n-1)-1}}{\left(1 + \dfrac{n - 1}{m - 1}y\right)^{\frac{1}{2}(m+n-2)}}.$$

The constant C_0 of (20) is then determined by

$$(22) \qquad \int_{C_0}^{\infty} F_{n-1,\,m-1}(y)\,dy = \alpha.$$

In order to apply Theorem 1 to H' let

$$W = \frac{\sum(Y_j - \bar{Y})^2/\Delta_0}{\sum(X_i - \bar{X})^2 + (1/\Delta_0)\sum(Y_j - \bar{Y})^2}.$$

This is also independent of $T^* = (T_1^*, T_2^*, T_3^*)$ when $\tau^2 = \Delta_0\sigma^2$, and is linear in U^*. The UMP unbiased acceptance region of H' is therefore

$$(23) \qquad\qquad\qquad C_1 \leq W \leq C_2$$

with the constants determined by (5) and (6) where V is replaced by W. On dividing numerator and denominator of W by σ^2 it is seen that for $\tau^2 = \Delta_0\sigma^2$, the statistic W is a ratio of the form $W_1/(W_1 + W_2)$, where W_1 and W_2 are independent χ^2 variables with $n - 1$ and $m - 1$ degrees of freedom respectively. Equivalently, $W = Y/(1 + Y)$, where $Y = W_1/W_2$ and where $(m - 1)Y/(n - 1)$ has the distribution $F_{n-1,\,m-1}$. The distribution of W is the *beta-distribution** with density

(24)

$$B_{\frac{1}{2}(n-1),\frac{1}{2}(m-1)}(w) = \frac{\Gamma[\frac{1}{2}(m+n-2)]}{\Gamma[\frac{1}{2}(m-1)]\Gamma[\frac{1}{2}(n-1)]} w^{\frac{1}{2}(n-3)}(1-w)^{\frac{1}{2}(m-3)},$$

$$0 < w < 1.$$

The conditions (5) and (6), by means of the relations

$$E(W) = \frac{n-1}{m+n-2}$$

and

$$wB_{\frac{1}{2}(n-1),\frac{1}{2}(m-1)}(w) = \frac{n-1}{m+n-2} B_{\frac{1}{2}(n+1),\frac{1}{2}(m-1)}(w),$$

become

(25) $\displaystyle \int_{C_1}^{C_2} B_{\frac{1}{2}(n-1),\frac{1}{2}(m-1)}(w)\,dw = \int_{C_1}^{C_2} B_{\frac{1}{2}(n+1),\frac{1}{2}(m-1)}(w)\,dw = 1 - \alpha.$

The definition of V shows that its distribution depends only on the ratio τ^2/σ^2, and so does the distribution of W. The power of the tests (20) and (23) is therefore also a function only of the variable $\Delta = \tau^2/\sigma^2$; it can be expressed explicitly in terms of the F-distribution, for example in the first case by

$$\beta(\Delta) = P\left\{ \frac{\sum(Y_j - \bar{Y})^2/\tau^2(n-1)}{\sum(X_i - \bar{X})^2/\sigma^2(m-1)} \geq \frac{C_0\Delta_0}{\Delta} \right\}$$

$$= \int_{C_0\Delta_0/\Delta}^{\infty} F_{n-1,\,m-1}(y)\,dy.$$

The hypothesis of equality of the means ξ, η of two normal distributions with unknown variances σ^2 and τ^2, the so-called *Behrens–Fisher problem*, is

---

*The relationship $W = Y/(1 + Y)$ shows the F- and beta-distributions to be equivalent. Tables of these distributions are discussed in Chapters 24 and 26 of Johnson and Kotz (1970, Vol. 2). Critical values of F are tabled by Mardia and Zamroch (1978), who also provide algorithms for the associated computations.

not accessible by the present method. (See Example 5 of Chapter 4; for a discussion of this problem see the next section and Chapter 6, Section 6.) We shall therefore consider only the simpler case in which the two variances are assumed to be equal. The joint density of the X's and Y's is then

$$(26) \quad C(\xi, \eta, \sigma)\exp\left[-\frac{1}{2\sigma^2}\left(\sum x_i^2 + \sum y_j^2\right) + \frac{\xi}{\sigma^2}\sum x_i + \frac{\eta}{\sigma^2}\sum y_j\right],$$

which is an exponential family with parameters

$$\theta = \frac{\eta}{\sigma^2}, \qquad \vartheta_1 = \frac{\xi}{\sigma^2}, \qquad \vartheta_2 = -\frac{1}{2\sigma^2}$$

and the sufficient statistics

$$U = \sum Y_j, \qquad T_1 = \sum X_i, \qquad T_2 = \sum X_i^2 + \sum Y_j^2.$$

For testing the hypotheses

$$H : \eta - \xi \leq 0 \quad \text{and} \quad H' : \eta - \xi = 0$$

it is more convenient to represent the densities as an exponential family with the parameters

$$\theta^* = \frac{\eta - \xi}{\left(\dfrac{1}{m} + \dfrac{1}{n}\right)\sigma^2}, \qquad \vartheta_1^* = \frac{m\xi + n\eta}{(m + n)\sigma^2}, \qquad \vartheta_2^* = \vartheta_2$$

and the sufficient statistics

$$U^* = \overline{Y} - \overline{X}, \qquad T_1^* = m\overline{X} + n\overline{Y}, \qquad T_2^* = \sum X_i^2 + \sum Y_j^2.$$

That this is possible is seen from the identity

$$m\xi\overline{x} + n\eta\overline{y} = \frac{(\overline{y} - \overline{x})(\eta - \xi)}{\dfrac{1}{m} + \dfrac{1}{n}} + \frac{(m\overline{x} + n\overline{y})(m\xi + n\eta)}{m + n}.$$

It follows from Theorem 3 of Chapter 4 that UMP unbiased tests exist for the hypotheses $\theta^* \leq 0$ and $\theta^* = 0$, and hence for H and H'.

When $\eta = \xi$, the distribution of

$$V = \frac{\bar{Y} - \bar{X}}{\sqrt{\sum(X_i - \bar{X})^2 + \sum(Y_j - \bar{Y})^2}}$$

$$= \frac{U^*}{\sqrt{T_2^* - \dfrac{1}{m+n}T_1^{*2} - \dfrac{mn}{m+n}U^{*2}}}$$

does not depend on the common mean ξ or on σ, as is seen by replacing X_i with $(X_i - \xi)/\sigma$ and Y_j with $(Y_j - \xi)/\sigma$ in the expression for V, and V is independent of (T_1^*, T_2^*). The rejection region of the UMP unbiased test of H can therefore be written as $V \geq C_0'$ or

$$(27) \qquad\qquad t(X, Y) \geq C_0,$$

where

$$(28) \quad t(X, Y) = \frac{(\bar{Y} - \bar{X})\Big/\sqrt{\dfrac{1}{m} + \dfrac{1}{n}}}{\sqrt{\Big[\sum(X_i - \bar{X})^2 + \sum(Y_j - \bar{Y})^2\Big]\Big/(m + n - 2)}}.$$

The statistic $t(X, Y)$ is the ratio of the two independent variables

$$\frac{\bar{Y} - \bar{X}}{\sqrt{\left(\dfrac{1}{m} + \dfrac{1}{n}\right)\sigma^2}} \quad \text{and} \quad \sqrt{\frac{\sum(X_i - \bar{X})^2 + \sum(Y_j - \bar{Y})^2}{(m + n - 2)\sigma^2}}.$$

The numerator is normally distributed with mean $(\eta - \xi)/\sqrt{m^{-1} + n^{-1}}\,\sigma$ and unit variance; the denominator, as the square root of a χ^2 variable with $m + n - 2$ degrees of freedom, divided by $m + n - 2$. Hence $t(X, Y)$ has a noncentral t-distribution with $m + n - 2$ degrees of freedom and noncentrality parameter

$$\delta = \frac{\eta - \xi}{\sqrt{\dfrac{1}{m} + \dfrac{1}{n}}\,\sigma}.$$

When in particular $\eta - \xi = 0$, the distribution of $t(X, Y)$ is Student's t-distribution, and the constant C_0 is determined by

(29) $$\int_{C_0}^{\infty} t_{m+n-2}(y)\, dy = \alpha.$$

As before, the assumptions required by Theorem 1 for H' are not satisfied by V itself but by a function of V,

$$W = \frac{\overline{Y} - \overline{X}}{\sqrt{\sum X_i^2 + \sum Y_j^2 - \dfrac{\left(\sum X_i + \sum Y_j\right)^2}{m + n}}},$$

which is related to V through

$$V = \frac{W}{\sqrt{1 - \dfrac{mn}{m + n}W^2}}.$$

Since W is a function of V, it is also independent of (T_1^*, T_2^*) when $\eta = \xi$; in addition it is a linear function of U^* with coefficients dependent only on T^*. The distribution of W being symmetric about 0 when $\eta = \xi$, it follows, as in the derivation of the corresponding rejection region (17) for the one-sample problem, that the UMP unbiased test of H' rejects when $|W|$ is too large, or equivalently when

(30) $$|t(X, Y)| > C.$$

The constant C is determined by

$$\int_C^{\infty} t_{m+n-2}(y)\, dy = \frac{\alpha}{2}.$$

The power of the tests (27) and (30) depends only on $(\eta - \xi)/\sigma$ and is given in terms of the noncentral t-distribution. Its properties are analogous to those of the one-sample t-test (Problems 1, 2, and 4).

4. ROBUSTNESS

Optimality theory postulates a statistical model and then attempts to determine a best procedure for that model. Since model assumptions tend to be unreliable, it is necessary to go a step further and ask how sensitive the procedure and its optimality are to the assumptions. In the normal models of the preceding section, three assumptions are made: Independence, identity of distribution, and normality. In the two-sample t-test, there is the additional assumption of equality of variance. We shall consider the effects of nonnormality and inequality of variance in the present section, and that of dependence in the next.

The natural first question to ask about the robustness of a test concerns the behavior of the significance level. If an assumption is violated, is the significance level still approximately valid? Such questions are typically answered by combining two methods of attack: The actual significance level under some alternative distributions is either calculated exactly or, more usually, estimated by simulation. In addition, asymptotic results are obtained which provide approximations to the true significance level for a wide variety of models.

We here restrict ourselves to a brief sketch of the latter approach. For this purpose we require the following basic results from probability theory. [For a more detailed discussion, see for example Cramér (1946); *TPE*, Chapter 5; and Serfling (1980).] The first is the simplest form of the central limit theorem.

Theorem 3. (*Central limit theorem.*) *Let* $X_1, \ldots, X_n$ *be independently identically distributed with mean* $E(X_i) = \xi$ *and* $\mathrm{Var}(X_i) = \sigma^2 < \infty$. *Then for all real* t

$$P\left\{ \frac{\sqrt{n}\,(\overline{X} - \xi)}{\sigma} \leq t \right\} \to \Phi(t),$$

where Φ *denotes the cumulative distribution function of the standard normal distribution* $N(0, 1)$.

When the cumulative distribution functions of a sequence of random variables T_n tend to a continuous limiting cumulative distribution function G as above, we shall say that T_n *converges to* G *in law*. If T_n and T_n' are independent and converge to $N(a, b^2)$ and $N(a', b'^2)$ respectively, then $T_n \pm T_n'$ converges to $N(a \pm a', b^2 + b'^2)$.

If T_n converges in law to $N(0, 1)$, then $bT_n + a$ ($b \neq 0$) converges in law to $N(a, b^2)$. The following result concerns the corresponding limit behavior when a and b are replaced by random variables which tend to a and b in probability.

Theorem 4. *If* T_n *converges in law to some distribution* G *and if* A_n, B_n *are random variables converging in probability to* a *and* $b \neq 0$ *respectively, then* $B_n T_n + A_n$ *has the same limit distribution as* $bT_n + a$.

Corollary 2. *If T_n tends in law to G (continuous) and if $c_n \to G$, then*

$$P\{T_n \le c_n\} \to G(c).$$

The last of the auxiliary results concerns the asymptotic behavior of functions of asymptotically normal variables.

Theorem 5. *If T_n is a sequence of random variables for which $\sqrt{n}\,(T_n - \theta)$ tends in law to $N(0, \tau^2)$, then for any function f for which $f'(\theta)$ exists and is $\neq 0$,*

$$\sqrt{n}\,[f(T_n) - f(\theta)]$$

tends in law to $N(0, \tau^2[f'(\theta)]^2)$.

Consider now the one-sample problem of Section 2, so that $X_1, \ldots, X_n$ are independently distributed as $N(\xi, \sigma^2)$. Tests of $H : \xi = \xi_0$ are based on the test statistic

$$t(X) = \frac{\sqrt{n}\,(\overline{X} - \xi_0)}{S} = \frac{\sqrt{n}\,(\overline{X} - \xi_0)}{\sigma} \bigg/ \frac{S}{\sigma},$$

where $S^2 = \Sigma(X_i - \overline{X})^2/(n - 1)$. When $\xi = \xi_0$ and the X's are normal, $t(X)$ has the t-distribution with $n - 1$ degrees of freedom. Suppose, however, that the normality assumption fails and the X's instead are distributed according to some other distribution F with mean ξ_0 and finite variance. Then by Theorem 3, $\sqrt{n}\,(\overline{X} - \xi_0)/\sigma$ has the limit distribution $N(0,1)$; furthermore S/σ tends to 1 in probability (see, for example, *TPE*, Chapter 5). By Theorem 4, $t(X)$ therefore has the limit distribution $N(0,1)$ regardless of F. This shows in particular that the t-distribution tends to $N(0,1)$ as $n \to \infty$.

To be specific, consider the one-sided t-test which rejects when $t(X) \ge C_n$, where $P\{t(X) \ge C_n\} = \alpha$ when F is normal. It follows from Corollary 2 and the asymptotic normality of the t-distribution that

$$C_n \to u_\alpha = \Phi^{-1}(1 - \alpha).$$

(If this were not the case, a subsequence of the C_n would converge to a different limit, and this would lead to a contradiction.)

Let $\alpha_n(F)$ be the true probability of the rejection region $t \ge C_n$ when the distribution of the X's is F. Then $\alpha_n(F) = P_F\{t \ge C_n\}$ has the same limit as $P_\Phi\{t \ge u_\alpha\}$, which is α. For sufficiently large n, the actual size $\alpha_n(F)$ will therefore be close to the nominal level α; how close depends on F and

n. For entries to the literature dealing with this dependence, see Cressie (1980), Tan (1982), and Benjamini (1983).

To study the corresponding test of variance, suppose first that the mean ξ is 0. When F is normal, the UMP test of $H : \sigma = \sigma_0$ against $\sigma > \sigma_0$ rejects when $\Sigma X_i^2 / \sigma_0^2$ is too large, where the null distribution of $\Sigma X_i^2 / \sigma_0^2$ is χ_n^2. By Theorem 3, $\sqrt{n}(\Sigma X_i^2 - n\sigma_0^2)/n$ tends in law to $N(0, 2\sigma_0^4)$ as $n \to \infty$, since $\mathrm{Var}(X_i^2) = 2\sigma_0^4$. If the rejection region is written as

$$\frac{\Sigma X_i^2 - n\sigma_0^2}{\sqrt{2n}\,\sigma_0^2} \geq C_n,$$

it follows that $C_n \to u_\alpha$.

Suppose now instead that the X's are distributed according to a distribution F with $E(X_i) = 0$, $E(X_i^2) = \mathrm{Var}\, X_i = \sigma^2$, and $\mathrm{Var}\, X_i^2 = \gamma^2$. Then $\Sigma(X_i^2 - n\sigma_0^2)/\sqrt{n}$ tends in law to $N(0, \gamma^2)$ when $\sigma = \sigma_0$, and the size $\alpha_n(F)$ of the test tends to

$$\lim P\left\{ \frac{\Sigma X_i^2 - n\sigma_0^2}{\sqrt{2n}\,\sigma_0^2} \geq u_\alpha \right\} = 1 - \Phi\left(\frac{u_\alpha \sqrt{2}\,\sigma_0^2}{\gamma} \right).$$

Depending on γ, which can take on any positive value, the sequence $\alpha_n(F)$ can thus tend to any limit $< \frac{1}{2}$. Even asymptotically and under rather small departures from normality (if they lead to big changes in γ), the size of the χ^2-test is thus completely uncontrolled.

For sufficiently large n, the difficulty is easy to overcome. Let $Y_i = X_i^2$, $E(Y_i) = \eta = \sigma^2$. The test statistic then reduces to $\sqrt{n}(\bar{Y} - \eta_0)$. To obtain an asymptotically valid test, it is only necessary to divide by a suitable estimator of $\sqrt{\mathrm{Var}\, Y_i}$ such as $\sqrt{\Sigma(Y_i - \bar{Y})^2/n}$. (However, since $Y_i^2 = X_i^4$, small changes in the tail of X_i may have large effects on Y_i^2, and n may have to be rather large for the asymptotic result to give a good approximation.)

When ξ is unknown, the normal theory test for σ^2 is based on $\Sigma(X_i - \bar{X})^2$, and the sequence

$$\frac{1}{\sqrt{n}}\left[\Sigma(X_i - \bar{X})^2 - n\sigma_0^2 \right] = \frac{1}{\sqrt{n}}\left(\Sigma X_i^2 - n\sigma_0^2 \right) - \frac{1}{\sqrt{n}} n\bar{X}^2$$

again has the limit distribution $N(0, \gamma^2)$. To see this, note that the distribution of $\Sigma(X_i - \bar{X})^2$ is independent of ξ, and put $\xi = 0$. Since $\sqrt{n}\,\bar{X}$ has a (normal) limit distribution, $n\bar{X}^2$ is bounded in probability,* and $n\bar{X}^2/\sqrt{n}$ tends to zero in probability. The result now follows from that for $\xi = 0$ and Theorem 4.

The above results carry over to the corresponding two-sample problems. For the t-test, an extension of the one-sample argument shows that as $m, n \to \infty$, $(\bar{Y} - \bar{X})/\sqrt{1/m + 1/n}\,\sigma$ tends in law to $N(0, 1)$ while $[\Sigma(X_i - \bar{X})^2 + \Sigma(Y_j - \bar{Y})^2]/(m + n - 2)\sigma^2$ tends in probability to 1 for samples $X_1, \ldots, X_m; Y_1, \ldots, Y_n$ from any common distribution F with finite variance. Thus, the actual size $\alpha_{m,n}(F)$ tends to α for any such F.

On the other hand, the F-test for variances, just like the one-sample χ^2-test, is extremely sensitive to the assumption of normality. To see this, express the rejection region in terms of $\log S_Y^2 - \log S_X^2$, where $S_X^2 = \Sigma(X_i - \bar{X})^2/(m - 1)$ and $S_Y^2 = \Sigma(Y_j - \bar{Y})^2/(n - 1)$, and suppose that as m and $n \to \infty$, $m/(m + n)$ remains fixed at ρ. By the result for the one-sample problem and Theorem 5 with $f(u) = \log u$, it is seen that $\sqrt{m}\,[\log S_X^2 - \log \sigma^2]$ and $\sqrt{n}\,[\log S_Y^2 - \log \sigma^2]$ both tend in law to $N(0, \gamma^2/\sigma^4)$ when the X's and Y's are distributed as F, and hence that $\sqrt{m + n}\,[\log S_Y^2 - \log S_X^2]$ tends in law to the normal distribution with mean 0 and variance

$$\frac{\gamma^2}{\sigma^4}\left(\frac{1}{\rho} + \frac{1}{1 - \rho}\right) = \frac{\gamma^2}{\rho(1 - \rho)\sigma^4}.$$

In the particular case that F is normal, $\gamma^2 = 2\sigma^4$ and the variance of the limit distribution is $2/\rho(1 - \rho)$. For other distributions γ^2/σ^4 can take on any positive value and, as in the one-sample case, $\alpha_n(F)$ can tend to any limit $< \frac{1}{2}$. [For an entry into the extensive literature on more robust alternatives, see for example Conover, Johnson, and Johnson (1981) and Tiku and Balakrishnan (1984).]

Having found that the size of the one- and two-sample t-tests is relatively insensitive to nonnormality (at least for large samples), let us turn to the corresponding question concerning the power of these tests. By similar asymptotic calculations, it can be shown that the same conclusion holds: Power values of the t-tests obtained under normality are asymptotically valid also for all other distributions with finite variance. This is a useful result if it has been decided to employ a t-test and one wishes to know what power it will have against a given alternative ξ/σ or $(\eta - \xi)/\sigma$, or what sample sizes are required to obtain a given power.

It is interesting to note that there exists a modification of the t-test, whose size is independent of F not only asymptotically but exactly, and

*See, for example, *TPE*, Chapter 5, Problem 1.24.

whose asymptotic power is equal to that of the t-test. This *permutation version* of the t-test will be discussed in Sections 10–14. It may seem that such a test has all the properties one could hope for. However, this overlooks the basic question of whether the t-test itself, which is optimal under normality, will retain a high standing with respect to its competitors under other distributions. The t-tests are in fact not robust in this sense. Tests which are preferable when a broad spectrum of distributions F is considered possible will be discussed in Chapter 6, Section 9. A permutation test with this property has been proposed by Lambert (1985).

The above distinction between robustness of the performance of a given test and robustness of its relative efficiency with respect to alternative tests has been pointed out by Tukey and McLaughlin (1963) and Box and Tiao (1964), who have described these concepts as robustness of validity or criterion robustness, and as robustness of efficiency or inference robustness, respectively.

As a last problem, consider the level of the two-sample t-test when the variances $\mathrm{Var}(X_i) = \sigma^2$ and $\mathrm{Var}(Y_j) = \tau^2$ are in fact not equal. As before, one finds that $(\overline{Y} - \overline{X})/\sqrt{\sigma^2/m + \tau^2/n}$ tends in law to $N(0, 1)$ as $m, n \to \infty$, while $S_X^2 = \Sigma(X_i - \overline{X})^2/(m - 1)$ and $S_Y^2 = \Sigma(Y_j - \overline{Y})^2/(n - 1)$ respectively tend to σ^2 and τ^2 in probability. If m and n tend to ∞ through a sequence with fixed proportion $m/(m + n) = \rho$, the squared denominator of t,

$$D^2 = \frac{m - 1}{m + n - 2} S_X^2 + \frac{n - 1}{m + n - 2} S_Y^2,$$

tends in probability to $\rho\sigma^2 + (1 - \rho)\tau^2$, and the limit of

$$t = \frac{1}{\sqrt{\dfrac{1}{m} + \dfrac{1}{n}}} \left(\frac{\overline{Y} - \overline{X}}{\sqrt{\dfrac{\sigma^2}{m} + \dfrac{\tau^2}{n}}} \cdot \frac{\sqrt{\dfrac{\sigma^2}{m} + \dfrac{\tau^2}{n}}}{D} \right)$$

is normal with mean zero and variance

(31)
$$\frac{(1 - \rho)\sigma^2 + \rho\tau^2}{\rho\sigma^2 + (1 - \rho)\tau^2}.$$

When $m = n$, so that $\rho = \frac{1}{2}$, the t-test thus has approximately the right level even if σ and τ are far apart. The accuracy of this approximation for different values of $m = n$ and τ/σ is discussed by Ramsey (1980) and Posten, Yeh, and Owen (1982). However, when $\rho \neq \frac{1}{2}$, the actual size of the test can differ greatly from the nominal level α even for large m and n. An approximate test of the hypothesis $H: \eta = \xi$ when σ, τ are not assumed equal (the Behrens–Fisher problem), which asymptotically is free of this difficulty, can be obtained through Studentization*, i.e., by replacing D^2 with $(1/m)S_X^2 + (1/n)S_Y^2$ and referring the resulting statistic to the standard normal distribution. This approximation is very crude, and not reliable unless m and n are fairly large. A refinement, the *Welch approximate t-test*, refers the resulting statistic not to the standard normal but to the t-distribution with a random number of degrees of freedom f given by

$$\frac{1}{f} = \left(\frac{R}{1 + R} \right)^2 \frac{1}{m - 1} + \frac{1}{(1 + R)^2} \cdot \frac{1}{n - 1},$$

where

$$R = \frac{(1/m)S_X^2}{(1/n)S_Y^2}.†$$

When the X's and Y's are normal, the actual level of this test has been shown to be quite close to the nominal level for sample sizes as small as $m = 4$, $n = 8$ and $m = n = 6$ [see Wang (1971)]. A further refinement will be mentioned in Chapter 6, Section 6.

The robustness of the level of Welch's test against nonnormality is studied by Yuen (1974), who shows that for heavy-tailed distributions the actual level tends to be considerably smaller than the nominal level (which leads to an undesirable loss of power), and who proposes an alternative. Some additional results are discussed in Scheffé (1970) and in Tiku and Singh (1981). The robustness of some quite different competitors of the t-test is investigated in Pratt (1964).

5. EFFECT OF DEPENDENCE

The one-sample t-test arises when a sequence of measurements $X_1, \ldots, X_n$ is taken of a quantity ξ, and the X's are assumed to be independently distributed as $N(\xi, \sigma^2)$. The effect of nonnormality on the level of the test was discussed in the preceding section. Independence may seem like a more innocuous assumption. However, it has been found that observations occur-

*Studentization is defined in a more general context at the end of Chapter 7, Section 3.
†For a variant, see Fenstad (1983).

ring close in time or space are often positively correlated [Student (1927), Hotelling (1961), Cochran (1968)]. The present section will therefore be concerned with the effect of this type of dependence.

Lemma 1. Let $X_1, \ldots, X_n$ be jointly normally distributed with common marginal distribution $N(0, \sigma^2)$ and with correlation coefficients $\rho_{ij} = \mathrm{corr}(X_i, X_j)$. As $n \to \infty$, suppose that

(a)
$$\mathrm{Var}\,\bar{X} = \frac{\sigma^2}{n^2} \sum_{i=1}^{n} \sum_{j=1}^{n} \rho_{ij} \to 0,$$

(b)
$$\mathrm{Var}\left(\frac{1}{n} \sum X_i^2 \right) \to 0$$

and

(c)
$$\frac{1}{n} \sum\sum_{i \neq j} \rho_{ij} \to \gamma.$$

Then

(i)   the distribution of the t-statistic (16) tends to the normal distribution $N(0, 1 + \gamma)$;

(ii)   if $\gamma \neq 0$, the level of the t-test is not robust even asymptotically as $n \to \infty$. Specifically, if $\gamma > 0$, the asymptotic level of the t-test carried out at nominal level α is

$$1 - \Phi\left(\frac{u_\alpha}{\sqrt{1 + \gamma}} \right) > 1 - \Phi(u_\alpha) = \alpha.$$

Proof.   (i): Since the X_i are jointly normal, the numerator $\sqrt{n}\,\bar{X}$ of t is also normal, with mean zero and variance

$$\mathrm{Var}(\sqrt{n}\,\bar{X}) = \sigma^2\left[1 + \frac{1}{n} \sum\sum_{i \neq j} \rho_{ij} \right],$$

and hence tends in law to $N(0, \sigma^2(1 + \gamma))$. The denominator of t is the square root of

$$D^2 = \frac{1}{n-1} \sum X_i^2 - \frac{n}{n-1} \bar{X}^2.$$

It follows from the Chebyshev inequality (Problem 18) that $\Sigma X_i^2/(n-1)$ tends in probability to $E(X_i^2) = \sigma^2$ and $[n/(n-1)]\overline{X}^2$ to zero, so that $D \to \sigma$ in probability. By Theorem 4, the distribution of t therefore tends to $N(0, 1 + \gamma)$.

The implications (ii) are obvious.

Under the assumptions of Lemma 1, the joint distribution of the X's is determined by σ^2 and the correlation coefficients ρ_{ij}, with the asymptotic level of the t-test depending only on γ. The following examples illustrating different correlation structures show that even under rather weak dependence of the observations, the assumptions of Lemma 1 are satisfied with $\gamma \neq 0$, and hence that the level of the t-test is quite sensitive to the assumption of independence.

MODEL A. (CLUSTER SAMPLING). Suppose the observations occur in s groups (or clusters) of size m, and that any two observations within a group have a common correlation coefficient ρ, while those in different groups are independent. (This may be the case, for instance, when the observations within a group are those taken on the same day or by the same observer, or involve some other common factor.) Then (Problem 20)

$$\text{Var } \overline{X} = \frac{\sigma^2}{ms}[1 + (m-1)\rho],$$

which tends to zero as $s \to \infty$; and analogously assumption (b) is seen to hold. Since $\gamma = (m-1)\rho$, the level of the t-test is not asymptotically robust as $s \to \infty$. In particular, the test overstates the significance of the results when $\rho > 0$.

To provide a specific structure leading to this model, denote the observations in the ith group by X_{ij} ($j = 1, \ldots, m$), and suppose that $X_{ij} = A_i + U_{ij}$, where A_i is a factor common to the observations in the ith group. If the A's and U's (none of which are observable) are all independent with normal distributions $N(\xi, \sigma_A^2)$ and $N(0, \sigma_0^2)$ respectively, then the joint distribution of the X's is that prescribed by Model A with $\sigma^2 = \sigma_A^2 + \sigma_0^2$ and $\rho = \sigma_A^2/\sigma^2$.

MODEL B. (MOVING-AVERAGE PROCESS). When the dependence of nearby observations is not due to grouping as in Model A, it is often reasonable to assume that ρ_{ij} depends only on $|j - i|$ and is nonincreasing in $|j - i|$. Let $\rho_{i, i+k}$ then be denoted by ρ_k, and suppose that the correlation between X_i and X_{i+k} is negligible for $k > m$ (m an integer $< n$), so that one can put $\rho_k = 0$ for $k > m$. Then the conditions for Lemma 1 are satisfied (Problem 22) with

$$\gamma = 2 \sum_{k=1}^{m} \rho_k.$$

In particular, if $\rho_1, \ldots, \rho_m$ are all positive, the t-test is again too liberal.

A specific structure leading to Model B is given by the moving-average process

$$X_i = \xi + \sum_{j=0}^{m} \beta_j U_{i+j},$$

where the U's are independent $N(0, \sigma_0^2)$. The variance σ^2 of the X's is then $\sigma^2 = \sigma_0^2 \sum_{j=0}^{m} \beta_j^2$ and

$$\rho_k = \begin{cases} \dfrac{\displaystyle\sum_{i=0}^{m-k} \beta_i \beta_{i+k}}{\displaystyle\sum_{j=0}^{m} \beta_j^2} & \text{for } k \le m, \\[2em] 0 & \text{for } k > m. \end{cases}$$

MODEL C. (FIRST-ORDER AUTOREGRESSIVE PROCESS). A simple model for dependence in which the $|\rho_k|$ are decreasing in k but $\ne 0$ for all k is the *first-order autoregressive process* defined by

$$X_{i+1} = \xi + \beta(X_i - \xi) + U_{i+1}, \qquad |\beta| < 1, \quad i = 1, \ldots, n,$$

with the U_i independent $N(0, \sigma_0^2)$. If X_1 is $N(\xi, \tau^2)$, the marginal distribution of X_i for $i > 1$ is normal with mean ξ and variance $\sigma_i^2 = \beta^2 \sigma_{i-1}^2 + \sigma_0^2$. The variance of X_i will thus be independent of i provided $\tau^2 = \sigma_0^2/(1 - \beta^2)$. For the sake of simplicity, we shall assume this to be the case, and take ξ to be zero. From

$$X_{i+k} = \beta^k X_i + \beta^{k-1} U_{i+1} + \beta^{k-2} U_{i+2} + \cdots + \beta U_{i+k-1} + U_{i+k}$$

it then follows that $\rho_k = \beta^k$, so that the correlation between X_i and X_j decreases exponentially with increasing $|j - i|$. The assumptions of Lemma 1 are again satisfied, and $\gamma = 2\beta/(1 - \beta)$. Thus, in this case too, the level of the t-test is not asymptotically robust. [Some values of the actual asymptotic level when the nominal level is .05 or .01 are given by Gastwirth and Rubin (1971).]

It is seen that in general the effect of dependence on the level of the t-test is more serious than that of nonnormality. Unfortunately, it is not possible to robustify the test against general dependence through Studentization, as can be done for unequal variances in the two-sample case. This would require consistent estimation of γ and hence of the ρ_{ij}, which is unavailable, since the number of unknown parameters far exceeds the number of observations.

The difficulty can be overcome if enough information is available to reduce the general model to one, such as A–C,* depending only on a finite number of parameters which can then be estimated consistently. Some specific procedures of this type are discussed by Albers (1978), [and for an associated sign test by Falk and Kohne (1984)]. Such robust procedures will in fact often also be insensitive to the assumption of normality, as can be shown by appealing to an appropriate central limit theorem for dependent variables [see e.g. Billingsley (1979)]. The validity of these procedures is of course limited to the particular model assumed, including the value of a parameter such as m in Models A and B.

The results of the present section easily extend to the case of the two-sample t-test, when each of the two series of observations shows dependence of the kind considered here.

6. CONFIDENCE INTERVALS AND FAMILIES OF TESTS

Confidence bounds for a parameter θ corresponding to a confidence level $1 - \alpha$ were defined in Chapter 3, Section 5, for the case that the distribution of the random variable X depends only on θ. When nuisance parameters ϑ are present the defining condition for a lower confidence bound $\underline{\theta}$ becomes

$$(32) \qquad P_{\theta, \vartheta}\{\underline{\theta}(X) \le \theta\} \ge 1 - \alpha \qquad \text{for all } \theta, \vartheta.$$

Similarly, confidence intervals for θ at confidence level $1 - \alpha$ are defined as a set of random intervals with end points $\underline{\theta}(X), \bar{\theta}(X)$ such that

$$(33) \qquad P_{\theta, \vartheta}\{\underline{\theta}(X) \le \theta \le \bar{\theta}(X)\} \ge 1 - \alpha \qquad \text{for all } \theta, \vartheta.$$

The infinum over (θ, ϑ) of the left-hand side of (32) and (33) is the *confidence coefficient* associated with these statements.

As was already indicated in Chapter 3, confidence statements permit a dual interpretation. Directly, they provide bounds for the unknown parame-

---

*Models of a sequence of dependent observations with various covariance structures are discussed in books on time series such as Anderson (1971) and Box and Jenkins (1970).

ter θ and thereby a solution to the problem of estimating θ. The statement $\underline{\theta} \leq \theta \leq \bar{\theta}$ is not as precise as a point estimate, but it has the advantage that the probability of it being correct can be guaranteed to be at least $1 - \alpha$. Similarly, a lower confidence bound can be thought of as an estimate $\underline{\theta}$ which overestimates the true parameter value with probability $\leq \alpha$. In particular for $\alpha = \frac{1}{2}$, if $\underline{\theta}$ satisfies

$$P_{\theta, \vartheta}\{\underline{\theta} \leq \theta\} = P_{\theta, \vartheta}\{\underline{\theta} \geq \theta\} = \tfrac{1}{2},$$

the estimate is as likely to underestimate as to overestimate and is then said to be *median unbiased*. (See Chapter 1, Problem 3, for the relation of this property to a more general concept of unbiasedness.) For an exponential family given by (10) of Chapter 4 there exists an estimator of θ which among all median unbiased estimators uniformly minimizes the risk for any loss function $L(\theta, d)$ that is monotone in the sense of the last paragraph of Chapter 3, Section 5. A full treatment of this result including some probabilistic and measure-theoretic complications, is given by Pfanzagl (1979).

Alternatively, as was shown in Chapter 3, confidence statements can be viewed as equivalent to a family of tests. The following is essentially a review of the discussion of this relationship in Chapter 3, made slightly more specific by restricting attention to the two-sided case. For each θ_0 let $A(\theta_0)$ denote the acceptance region of a level-α test (assumed for the moment to be nonrandomized) of the hypothesis $H(\theta_0): \theta = \theta_0$. If

$$S(x) = \{\theta : x \in A(\theta)\}$$

then

(34) $\theta \in S(x)$ if and only if $x \in A(\theta)$,

and hence

(35) $P_{\theta, \vartheta}\{\theta \in S(X)\} \geq 1 - \alpha$ for all θ, ϑ.

Thus any family of level-α acceptance regions, through the correspondence (34), leads to a family of confidence sets at confidence level $1 - \alpha$.

Conversely, given any class of confidence sets $S(x)$ satisfying (35), let

(36) $A(\theta) = \{x : \theta \in S(x)\}.$

Then the sets $A(\theta_0)$ are level-α acceptance regions for testing the hypotheses $H(\theta_0): \theta = \theta_0$, and the confidence sets $S(x)$ show for each θ_0 whether for the particular x observed the hypothesis $\theta = \theta_0$ is accepted or rejected at level α.

Exactly the same arguments apply if the sets $A(\theta_0)$ are acceptance regions for the hypotheses $\theta \leq \theta_0$. As will be seen below, one- and two-sided tests typically, although not always, lead to one-sided confidence bounds and to confidence intervals respectively.

Example 4. Normal mean. Confidence intervals for the mean ξ of a normal distribution with unknown variance can be obtained from the acceptance regions $A(\xi_0)$ of the hypothesis $H : \xi = \xi_0$. These are given by

$$\frac{\left|\sqrt{n}\,(\bar{x} - \xi_0)\right|}{\sqrt{\sum (x_i - \bar{x})^2 / (n - 1)}} \leq C,$$

where C is determined from the t-distribution so that the probability of this inequality is $1 - \alpha$ when $\xi = \xi_0$. [See (17) and (19) of Section 2.] The set $S(x)$ is then the set of ξ's satisfying this inequality with $\xi = \xi_0$, that is, the interval

$$(37) \quad \bar{x} - \frac{C}{\sqrt{n}} \sqrt{\frac{1}{n - 1} \sum (x_i - \bar{x})^2} \leq \xi \leq \bar{x} + \frac{C}{\sqrt{n}} \sqrt{\frac{1}{n - 1} \sum (x_i - \bar{x})^2}.$$

The class of these intervals therefore constitutes confidence intervals for ξ with confidence coefficient $1 - \alpha$.

The length of the intervals (37) is proportional to $\sqrt{\sum (x_i - \bar{x})^2}$, and their expected length to σ. For large σ, the intervals will therefore provide little information concerning the unknown ξ. This is a consequence of the fact, which led to similar difficulties for the corresponding testing problem, that two normal distributions $N(\xi_0, \sigma^2)$ and $N(\xi_1, \sigma^2)$ with fixed difference of means become indistinguishable as σ tends to infinity. In order to obtain confidence intervals for ξ whose length does not tend to infinity with σ, it is necessary to determine the number of observations sequentially so that it can be adjusted to σ. A sequential procedure leading to confidence intervals of prescribed length is given in Problems 26 and 27.

However, even such a sequential procedure does not really dispose of the difficulty, but only shifts the lack of control from the length of the interval to the number of observations. As $\sigma \to \infty$, the number of observations required to obtain confidence intervals of bounded length also tends to infinity. Actually, in practice one will frequently have an idea of the order of magnitude of σ. With a sample either of fixed size or obtained sequentially, it is then necessary to establish a balance between the desired confidence $1 - \alpha$, the accuracy given by the length l of the interval, and the number of observations n one is willing to expend. In such an arrangement two of the three quantities $1 - \alpha$, l, and n will be fixed, while the third is a random variable whose distribution depends on σ, so that it will be less well controlled than the others. If $1 - \alpha$ is taken as fixed, the choice between a sequential scheme and one of fixed sample size thus depends essentially on whether it is more important to control l or n.

To obtain lower confidence limits for ξ, consider the acceptance regions

$$\frac{\sqrt{n}\,(\bar{x} - \xi_0)}{\sqrt{\sum(x_i - \bar{x})^2/(n-1)}} \le C_0$$

for testing $\xi \le \xi_0$ against $\xi > \xi_0$. The sets $S(x)$ are then the one-sided intervals

$$\bar{x} - \frac{C_0}{\sqrt{n}}\sqrt{\frac{1}{n-1}\sum(x_i - \bar{x})^2} \le \xi,$$

the left-hand sides of which therefore constitute the desired lower bounds ξ. If $\alpha = \frac{1}{2}$, the constant C_0 is 0; the resulting confidence bound $\xi = \bar{X}$ is a median unbiased estimate of ξ, and among all such estimates it uniformly maximizes

$$P\left\{-\Delta_1 \le \xi - \underline{\xi} \le \Delta_2\right\} \qquad \text{for all} \quad \Delta_1, \Delta_2 \ge 0.$$

(For a proof see Chapter 3, Section 5.)

7. UNBIASED CONFIDENCE SETS

Confidence sets can be viewed as a family of tests of the hypotheses $\theta \in H(\theta')$ against alternatives $\theta \in K(\theta')$ for varying θ'. A confidence level of $1 - \alpha$ then simply expresses the fact that all the tests are to be at level α, and the condition therefore becomes

$$(38) \qquad P_{\theta,\vartheta}\{\theta' \in S(X)\} \ge 1 - \alpha \qquad \text{for all } \theta \in H(\theta') \text{ and all } \vartheta.$$

In the case that $H(\theta')$ is the hypothesis $\theta = \theta'$ and $S(X)$ is the interval $[\underline{\theta}(X), \bar{\theta}(X)]$, this agrees with (33). In the one-sided case in which $H(\theta')$ is the hypothesis $\theta \le \theta'$ and $S(X) = \{\theta : \underline{\theta}(X) \le \theta\}$, the condition reduces to $P_{\theta,\vartheta}\{\underline{\theta}(X) \le \theta'\} \ge 1 - \alpha$ for all $\theta' \ge \theta$, and this is seen to be equivalent to (32). With this interpretation of confidence sets, the probabilities

$$(39) \qquad\qquad P_{\theta,\vartheta}\{\theta' \in S(X)\}, \qquad \theta \in K(\theta'),$$

are the probabilities of false acceptance of $H(\theta')$ (error of the second kind). The smaller these probabilities are, the more desirable are the tests.

From the point of view of estimation, on the other hand, (39) is the probability of covering the wrong value θ'. With a controlled probability of covering the true value, the confidence sets will be more informative the less likely they are to cover false values of the parameter. In this sense the probabilities (39) provide a measure of the accuracy of the confidence sets. A justification of (39) in terms of loss functions was given for the one-sided case in Chapter 3, Section 5.

In the presence of nuisance parameters, UMP tests usually do not exist, and this implies the nonexistence of confidence sets that are uniformly most accurate in the sense of minimizing (39) for all θ' such that $\theta \in K(\theta')$ and for all ϑ. This suggests restricting attention to confidence sets which in a suitable sense are unbiased. In analogy with the corresponding definition for tests, a family of confidence sets at confidence level $1 - \alpha$ is said to be *unbiased* if

(40) $P_{\theta,\vartheta}\{\theta' \in S(X)\} \le 1 - \alpha$

$$\text{for all } \theta' \text{ such that } \theta \in K(\theta') \quad \text{and} \quad \text{for all } \vartheta \text{ and } \theta,$$

so that the probability of covering these false values does not exceed the confidence level.

In the two- and one-sided cases mentioned above, the condition (40) reduces to

$$P_{\theta,\vartheta}\{\underline{\theta} \le \theta' \le \bar{\theta}\} \le 1 - \alpha \qquad \text{for all } \theta' \ne \theta \text{ and all } \vartheta$$

and

$$P_{\theta,\vartheta}\{\underline{\theta} \le \theta'\} \le 1 - \alpha \qquad \text{for all } \theta' < \theta \text{ and all } \vartheta.$$

With this definition of unbiasedness, unbiased families of tests lead to unbiased confidence sets and conversely. A family of confidence sets is uniformly most accurate unbiased at confidence level $1 - \alpha$ if it minimizes the probabilities

$$P_{\theta,\vartheta}\{\theta' \in S(X)\} \text{ for all } \theta' \text{ such that } \theta \in K(\theta') \text{ and for all } \vartheta \text{ and } \theta,$$

subject to (38) and (40). The confidence sets obtained on the basis of the UMP unbiased tests of the present and preceding chapter are therefore uniformly most accurate unbiased. This applies in particular to the confidence intervals obtained in the preceding sections. Some further examples are the following.

Example 5. Normal variance. If $X_1, \ldots, X_n$ is a sample from $N(\xi, \sigma^2)$, the UMP unbiased test of the hypothesis $\sigma = \sigma_0$ is given by the acceptance region (13)

$$C_1' \le \frac{\sum(x_i - \bar{x})^2}{\sigma_0^2} \le C_2',$$

where C_1' and C_2' are determined by (14). The most accurate unbiased confidence intervals for σ^2 are therefore

$$\frac{1}{C_2'}\sum(x_i - \bar{x})^2 \le \sigma^2 \le \frac{1}{C_1'}\sum(x_i - \bar{x})^2.$$

[Tables of C_1' and C_2' are provided by Tate and Klett (1959).] Similarly, from (9) and (10) the most accurate unbiased upper confidence limits for σ^2 are

$$\sigma^2 \le \frac{1}{C_0}\sum(x_i - \bar{x})^2,$$

where

$$\int_{C_0}^{\infty} \chi_{n-1}^2(y)\,dy = 1 - \alpha.$$

The corresponding lower confidence limits are uniformly most accurate (without the restriction of unbiasedness) by Chapter 3, Section 9.

Example 6.   Difference of means.   Confidence intervals for the difference $\Delta = \eta - \xi$ of the means of two normal distributions with common variance are obtained from tests of the hypothesis $\eta - \xi = \Delta_0$. If $X_1, \ldots, X_m$ and $Y_1, \ldots, Y_n$ are distributed as $N(\xi, \sigma^2)$ and $N(\eta, \sigma^2)$ respectively, and if $Y_j' = Y_j - \Delta_0$, $\eta' = \eta - \Delta_0$, the hypothesis can be expressed in terms of the variables X_i and Y_j' as $\eta' - \xi = 0$. From (28) and (30) the UMP unbiased acceptance region is then seen to be

$$\frac{|(\bar{y} - \bar{x} - \Delta_0)| / \sqrt{\dfrac{1}{m} + \dfrac{1}{n}}}{\sqrt{\left[\sum(x_i - \bar{x})^2 + \sum(y_j - \bar{y})^2\right] / (m + n - 2)}} \le C,$$

where C is determined by the equation following (30). The most accurate unbiased confidence intervals for $\eta - \xi$ are therefore

(41) $(\bar{y} - \bar{x}) - CS \le \eta - \xi \le (\bar{y} - \bar{x}) + CS$

where

$$S^2 = \left(\frac{1}{m} + \frac{1}{n}\right)\frac{\sum(x_i - \bar{x})^2 + \sum(y_j - \bar{y})^2}{m + n - 2}.$$

The one-sided intervals are obtained analogously.

Example 7.   Ratio of variances.   If $X_1, \ldots, X_m$ and $Y_1, \ldots, Y_n$ are samples from $N(\xi, \sigma^2)$ and $N(\eta, \tau^2)$, most accurate unbiased confidence intervals for $\Delta = \tau^2/\sigma^2$

are derived from the acceptance region (23) as

$$(42) \qquad \frac{1 - C_2}{C_2} \frac{\sum (y_j - \bar{y})^2}{\sum (x_i - \bar{x})^2} \leq \frac{\tau^2}{\sigma^2} \leq \frac{1 - C_1}{C_1} \frac{\sum (y_j - \bar{y})^2}{\sum (x_i - \bar{x})^2},$$

where C_1 and C_2 are determined from (25).* In the particular case that $m = n$, the intervals take on the simpler form

$$(43) \qquad \frac{1}{k} \frac{\sum (y_j - \bar{y})^2}{\sum (x_i - \bar{x})^2} \leq \frac{\tau^2}{\sigma^2} \leq k \frac{\sum (y_j - \bar{y})^2}{\sum (x_i - \bar{x})^2},$$

where k is determined from the F-distribution. Most accurate unbiased lower confidence limits for the variance ratio are

$$(44) \qquad \underline{\Delta} = \frac{1}{C_0} \frac{\sum (y_j - \bar{y})^2 / (n - 1)}{\sum (x_i - \bar{x})^2 / (m - 1)} \leq \frac{\tau^2}{\sigma^2}$$

with C_0 given by (22). If in (22) α is taken to be $\frac{1}{2}$, this lower confidence limit $\underline{\Delta}$ becomes a median unbiased estimate of τ^2/σ^2. Among all such estimates it uniformly minimizes

$$P\left\{ -\Delta_1 \leq \frac{\tau^2}{\sigma^2} - \underline{\Delta} \leq \Delta_2 \right\} \qquad \text{for all} \quad \Delta_1, \Delta_2 \geq 0.$$

(For a proof see Chapter 3, Section 5).

So far it has been assumed that the tests from which the confidence sets are obtained are nonrandomized. The modifications that are necessary when this assumption is not satisfied were discussed in Chapter 3. The randomized tests can then be interpreted as being nonrandomized in the space of X and an auxiliary variable V which is uniformly distributed on the unit interval. If in particular X is integer-valued as in the binomial or Poisson case, the tests can be represented in terms of the continuous variable $X + V$. In this way, most accurate unbiased confidence intervals can be obtained, for example, for a binomial probability p from the UMP unbiased tests of $H: p = p_0$ (Example 1 of Chapter 4). It is not clear a priori that the resulting confidence sets for p will necessarily by intervals. This is, however, a consequence of the following Lemma.

---

*A comparison of these limits with those obtained from the equal-tails test is given by Scheffé (1942); some values of C_1 and C_2 are provided by Ramachandran (1958).

Lemma 2. *Let X be a real-valued random variable with probability density $p_\theta(x)$ which has monotone likelihood ratio in x. Suppose that UMP unbiased tests of the hypotheses $H(\theta_0): \theta = \theta_0$ exist and are given by the acceptance regions*

$$C_1(\theta_0) \le x \le C_2(\theta_0),$$

and that they are strictly unbiased. Then the functions $C_i(\theta)$ are strictly increasing in θ, and the most accurate unbiased confidence intervals for θ are

$$C_2^{-1}(x) \le \theta \le C_1^{-1}(x).$$

Proof. Let $\theta_0 < \theta_1$, and let $\beta_0(\theta)$ and $\beta_1(\theta)$ denote the power functions of the above tests ϕ_0 and ϕ_1 for testing $\theta = \theta_0$ and $\theta = \theta_1$. It follows from the strict unbiasedness of the tests that

$$E_{\theta_0}[\phi_1(X) - \phi_0(X)] = \beta_1(\theta_0) - \alpha > 0 > \alpha - \beta_0(\theta_1)$$

$$= E_{\theta_1}[\phi_1(X) - \phi_0(X)].$$

Thus neither of the two intervals $[C_1(\theta_i), C_2(\theta_i)]$ ($i = 0, 1$) contains the other, and it is seen from Lemma 2(iii) of Chapter 3 that $C_i(\theta_0) < C_i(\theta_1)$ for $i = 1, 2$. The functions C_i therefore have inverses, and the inequalities defining the acceptance region for $H(\theta)$ are equivalent to $C_2^{-1}(x) \le \theta \le C_1^{-1}(x)$, as was to be proved.

The situation is indicated in Figure 1. From the boundaries $x = C_1(\theta)$ and $x = C_2(\theta)$ of the acceptance regions $A(\theta)$ one obtains for each fixed value of x the confidence set $S(x)$ as the interval of θ's for which $C_1(\theta) \le x \le C_2(\theta)$.

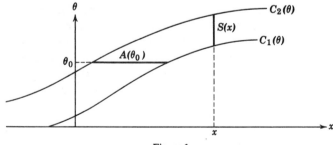

Figure 1

By Section 2 of Chapter 4, the conditions of the lemma are satisfied in particular for a one-parameter exponential family, provided the tests are nonrandomized. In cases such as that of binomial or Poisson distributions, where the family is exponential but X is integer-valued so that randomization is required, the intervals can be obtained by applying the lemma to the variable $X + V$ instead of X, where V is independent of X and uniformly distributed over $(0, 1)$.

In the binomial case, a table of the (randomized) uniformly most accurate unbiased confidence intervals is given by Blyth and Hutchinson (1960). The best choice of nonrandomized intervals and some large-sample approximations are discussed (and tables provided) by Blyth and Still (1983) and Blyth (1984). For additional discussion and references see Johnson and Kotz (1969, Section 3.7) and Ghosh (1979).

In Lemma 2, the distribution of X was assumed to depend only on θ. Consider now the exponential family (1) in which nuisance parameters are present in addition to θ. The UMP unbiased tests of $\theta = \theta_0$ are then performed as conditional tests given $T = t$, and the confidence intervals for θ will as a consequence also be obtained conditionally. If the conditional distributions are continuous, the acceptance regions will be of the form

$$C_1(\theta; t) \leq u \leq C_2(\theta; t),$$

where for each t the functions C_i are increasing by Lemma 2. The confidence intervals are then

$$C_2^{-1}(u; t) \leq \theta \leq C_1^{-1}(u; t).$$

If the conditional distributions are discrete, continuity can be obtained as before through addition of a uniform variable.

Example 8. Poisson ratio. Let X and Y be independent Poisson variables with means λ and μ, and let $\rho = \mu/\lambda$. The conditional distribution of Y given $X + Y = t$ is the binomial distribution $b(p, t)$ with

$$p = \frac{\rho}{1 + \rho}.$$

The UMP unbiased test $\phi(y, t)$ of the hypothesis $\rho = \rho_0$ is defined for each t as the UMP unbiased conditional test of the hypothesis $p = \rho_0/(1 + \rho_0)$. If

$$\underline{p}(t) \leq p \leq \bar{p}(t)$$

are the associated most accurate unbiased confidence intervals for p given t, it follows that the most accurate unbiased confidence intervals for μ/λ are

$$\frac{\underline{p}(t)}{1 - \underline{p}(t)} \leq \frac{\mu}{\lambda} \leq \frac{\bar{p}(t)}{1 - \bar{p}(t)}.$$

The binomial tests which determine the functions $\underline{p}(t)$ and $\bar{p}(t)$ are discussed in Example 1 of Chapter 4.

8.   REGRESSION

The relation between two variables X and Y can be studied by drawing an unrestricted sample and observing the two variables for each subject, obtaining n pairs of measurements $(X_1, Y_1), \ldots, (X_n, Y_n)$ (see Section 15 and Chapter, 5, Problem 10). Alternatively, it is frequently possible to control one of the variables such as the age of a subject, the temperature at which an experiment is performed, or the strength of the treatment that is being applied. Observations $Y_1, \ldots, Y_n$ of Y can then be obtained at a number of predetermined levels $x_1, \ldots, x_n$ of x. Suppose that for fixed x the distribution of Y is normal with constant variance σ^2 and a mean which is a function of x, *the regression of Y on x*, and which is assumed to be linear,

$$E[Y|x] = \alpha + \beta x.$$

If we put $v_i = (x_i - \bar{x})/\sqrt{\Sigma(x_j - \bar{x})^2}$ and $\gamma + \delta v_i = \alpha + \beta x_i$, so that $\Sigma v_i = 0$, $\Sigma v_i^2 = 1$, and

$$\alpha = \gamma - \delta \frac{\bar{x}}{\sqrt{\Sigma(x_j - \bar{x})^2}}, \qquad \beta = \frac{\delta}{\sqrt{\Sigma(x_j - \bar{x})^2}},$$

the joint density of $Y_1, \ldots, Y_n$ is

$$\frac{1}{(\sqrt{2\pi}\,\sigma)^n} \exp\left[-\frac{1}{2\sigma^2} \Sigma(y_i - \gamma - \delta v_i)^2 \right].$$

These densities constitute an exponential family (1) with

$$U = \Sigma v_i Y_i, \qquad T_1 = \Sigma Y_i^2, \qquad T_2 = \Sigma Y_i$$

$$\theta = \frac{\delta}{\sigma^2}, \qquad \vartheta_1 = -\frac{1}{2\sigma^2}, \qquad \vartheta_2 = \frac{\gamma}{\sigma^2}.$$

This representation implies the existence of UMP unbiased tests of the hypotheses $a\gamma + b\delta = c$ where a, b, and c are given constants, and therefore of most accurate unbiased confidence intervals for the parameter

$$\rho = a\gamma + b\delta.$$

To obtain these confidence intervals explicitly, one requires the UMP unbiased test of $H : \rho = \rho_0$, which is given by the acceptance region

(45)
$$\frac{\left|b\sum v_i Y_i + a\bar{Y} - \rho_0\right|\big/\sqrt{(a^2/n) + b^2}}{\sqrt{\left[\sum(Y_i - \bar{Y})^2 - \left(\sum v_i Y_i\right)^2\right]\big/(n-2)}} \leq C$$

where

$$\int_{-C}^{C} t_{n-2}(y)\, dy = 1 - \alpha.$$

(See Problem 33 and Chapter 7, Section 7, where there is also a discussion of the robustness of these procedures against nonnormality.) The resulting confidence intervals for ρ are centered at $b\sum v_i Y_i + a\bar{Y}$, and their length is

$$L = 2C\sqrt{\left[\frac{a^2}{n} + b^2\right]\frac{\sum(Y_i - \bar{Y})^2 - \left(\sum v_i Y_i\right)^2}{n-2}}.$$

It follows from the transformations given in Problem 33 that $[\sum(Y_i - \bar{Y})^2 - (\sum v_i Y_i)^2]/\sigma^2$ has a χ^2-distribution with $n - 2$ degrees of freedom and hence that the expected length of the intervals is

$$E(L) = 2C_n\sigma\sqrt{\frac{a^2}{n} + b^2}.$$

In particular applications, a and b typically are functions of the x's. If these are at the disposal of the experimenter and there is therefore some choice with respect to a and b, the expected length of L is minimized by minimizing $(a^2/n) + b^2$. Actually, it is not clear that the expected length is a good criterion for the accuracy of confidence intervals, since short intervals are desirable when they cover the true parameter value but not necessarily otherwise. However, the same result holds for other criteria such as the expected value of $(\bar{\rho} - \rho)^2 + (\rho - \underline{\rho})^2$ or more generally of $f_1(|\bar{\rho} - \rho|) + f_2(|\rho - \underline{\rho}|)$, where f_1 and f_2 are increasing functions of their arguments. (See Problem 33.) Furthermore, the same choice of a and b also minimizes the probability of the intervals covering any false value of the parameter. We shall therefore consider $(a^2/n) + b^2$ as an inverse measure of the accuracy of the intervals.

Example 9. Slope of regression line. Confidence levels for the slope $\beta = \delta/\sqrt{\Sigma(x_j - \bar{x})^2}$ are obtained from the above intervals by letting $a = 0$ and $b = 1/\sqrt{\Sigma(x_j - \bar{x})^2}$. Here the accuracy increases with $\Sigma(x_j - \bar{x})^2$, and if the x_j must be chosen from an interval $[C_0, C_1]$, it is maximized by putting half of the values at each end point. However, from a practical point of view, this is frequently not a good design, since it permits no check of the linearity of the regression.

Example 10. Ordinate of regression line. Another parameter of interest is the value $\alpha + \beta x_0$ to be expected from an observation Y at $x = x_0$. Since

$$\alpha + \beta x_0 = \gamma + \frac{\delta(x_0 - \bar{x})}{\sqrt{\Sigma(x_j - \bar{x})^2}},$$

the constants a and b are $a = 1$, $b = (x_0 - \bar{x})/\sqrt{\Sigma(x_j - \bar{x})^2}$. The maximum accuracy is obtained by minimizing $|\bar{x} - x_0|$ and, if $\bar{x} = x_0$ cannot be achieved exactly, also maximizing $\Sigma(x_j - \bar{x})^2$.

Example 11. Intercept of regression line. Frequently it is of interest to estimate the point x at which $\alpha + \beta x$ has a preassigned value. One may for example wish to find the dosage $x = -\alpha/\beta$ at which $E(Y|x) = 0$, or equivalently the value $v = (x - \bar{x})/\sqrt{\Sigma(x_j - \bar{x})^2}$ at which $\gamma + \delta v = 0$. Most accurate unbiased confidence sets for the solution $-\gamma/\delta$ of this equation can be obtained from the UMP unbiased tests of the hypotheses $-\gamma/\delta = v_0$. The acceptance regions of these tests are given by (45) with $a = 1$, $b = v_0$, and $\rho_0 = 0$, and the resulting confidence sets for v are the sets of values v satisfying

$$v^2\left[C^2S^2 - (\Sigma v_i Y_i)^2\right] - 2v\bar{Y}(\Sigma v_i Y_i) + \frac{1}{n}(C^2S^2 - n\bar{Y}^2) \geq 0,$$

where $S^2 = [\Sigma(Y_i - \bar{Y})^2 - (\Sigma v_i Y_i)^2]/(n - 2)$. If the associated quadratic equation in v has roots $\underline{v}, \bar{v}$, the confidence statement becomes

$$\underline{v} \leq v \leq \bar{v} \quad \text{when} \quad \frac{\left|\Sigma v_i Y_i\right|}{S} > C$$

and

$$v \leq \underline{v} \text{ or } v \geq \bar{v} \quad \text{when} \quad \frac{\left|\Sigma v_i Y_i\right|}{S} < C.$$

The somewhat surprising possibility that the confidence sets may be the outside of an interval actually is quite appropriate here. When the line $y = \gamma + \delta v$ is nearly parallel to the v-axis, the intercept with the v-axis will be large in absolute value, but its sign can be changed by a very small change in angle. There is the further possibility that the discriminant of the quadratic polynomial is negative,

$$n\overline{Y}^2 + \left(\sum v_i Y_i\right)^2 < C^2 S^2,$$

in which case the associated quadratic equation has no solutions. This condition implies that the leading coefficient of the quadratic polynomial is positive, so that the confidence set in this case becomes the whole real axis. The fact that the confidence sets are not necessarily finite intervals has led to the suggestion that their use be restricted to the cases in which they do have this form. Such usage will however affect the probability with which the sets cover the true value and hence the validity of the reported confidence coefficient.*

9. BAYESIAN CONFIDENCE SETS

The left side of the confidence statement (35) denotes the probability that the random set $S(X)$ will contain the constant point θ. The interpretation of this probability statement, before X is observed, is clear: it refers to the frequency with which this random event will occur. Suppose for example that X is distributed as $N(\theta, 1)$, and consider the confidence interval

$$X - 1.96 < \theta < X + 1.96$$

corresponding to confidence coefficient $\gamma = .95$. Then the random interval $(X - 1.96, X + 1.96)$ will contain θ with probability .95. Suppose now that X is observed to be 2.14. At this point, the earlier statement reduces to the inequality $0.18 < \theta < 4.10$, which no longer involves any random element. Since the only unknown quantity is θ, it is tempting (but not justified) to say that θ lies between 0.18 and 4.10 with probability .95.

To attach a meaningful probability to the event $\theta \in S(x)$ when x is fixed requires that θ be random. Inferences made under the assumption that the parameter θ is itself a random (though unobservable) quantity with a known distribution are called *Bayesian*, and the distribution Λ of θ before any observations are taken its *prior distribution*. After $X = x$ has been observed, inferences concerning θ can be based on its conditional distribution given x, the *posterior distribution*. In particular, any set $S(x)$ with the property

$$P\big[\theta \in S(x) \,\big|\, X = x\big] \geq \gamma \qquad \text{for all } x$$

---

*A method for obtaining the size of this effect was developed by Neyman, and tables have been computed on its basis by Fix. This work is reported by Bennett (1957).

is a $100\gamma\%$ Bayesian confidence set or *credible region* for θ. In the rest of this section, the random variable with prior distribution Λ will be denoted by Θ, with θ being the value taken on by Θ in the experiment at hand.

Example 12. Normal mean. Suppose that Θ has a normal prior distribution $N(\mu, b^2)$ and that given $\Theta = \theta$, the variables $X_1, \ldots, X_n$ are independent $N(\theta, \sigma^2)$, σ known. Then the posterior distribution of Θ given $x_1, \ldots, x_n$ is normal with mean (Problem 34)

$$\eta_x = E[\Theta|x] = \frac{n\bar{x}/\sigma^2 + \mu/b^2}{n/\sigma^2 + 1/b^2}$$

and variance

$$\tau_x^2 = \text{Var}[\Theta|x] = \frac{1}{n/\sigma^2 + 1/b^2}.$$

Since $[\Theta - \eta_x]/\tau_x$ then has a standard normal distribution, the interval $I(x)$ with endpoints

$$\frac{n\bar{x}/\sigma^2 + \mu/b^2}{n/\sigma^2 + 1/b^2} \pm \frac{1.96}{\sqrt{n/\sigma^2 + 1/b^2}}$$

satisfies $P[\Theta \in I(x)|X = x] = .95$ and is thus a 95% credible region.

For $n = 1$, $\mu = 0$, $\sigma = 1$, the interval reduces to

$$\frac{x}{1 + \dfrac{1}{b^2}} \pm \frac{1.96}{\sqrt{1 + \dfrac{1}{b^2}}}$$

which for large b is very close to the confidence interval for θ stated at the beginning of the section. But now the statement that θ lies between these limits with probability .95 is justified, since it is a probability statement concerning the random variable Θ.

The distribution $N(\mu, b^2)$ assigns higher probability to θ-values near μ than to those further away. Suppose instead that no information whatever about θ is available, so that one wishes to model a state of complete ignorance. This could be done by assigning the same probability density to all values of θ, that is, by assigning to Θ the probability density $\pi(\theta) \equiv c$, $-\infty < \theta < \infty$. Unfortunately, the resulting π is not a probability density, since $\int_{-\infty}^{\infty} \pi(\theta)\, d\theta = \infty$. However, if this fact is ignored and the posterior distribution of Θ given x is calculated in the usual way, it turns out (Problem 35) that $\pi(\theta|x)$ is the density of a genuine probability distribution, namely $N(\mu, \sigma^2/n)$, the limit of the earlier posterior distribution as $b \to \infty$. The *improper* (since it integrates to infinity), *noninformative* prior density $\pi(\theta) \equiv c$ thus leads approximately to the same results as the normal prior $N(\mu, b^2)$ for large b, and can be viewed as an approximation to the latter.

Unlike confidence sets, Bayesian credible regions provide exactly the desired kind of probability statement even after the observations are known. They do so, however, at the cost of an additional assumption: that θ is random and has a known prior distribution. Interpretations of such prior distributions as ways of utilizing past experience or as descriptions of a state of mind are discussed briefly in Chapter 4, Section 1 of *TPE*. Detailed accounts of the Bayesian approach and its application to credible regions can be found for example in Lindley (1965), Box and Tiao (1973), and Berger (1985); some frequency properties of such regions are discussed in Rubin (1984). The following examples provide a few illustrations and additional comments.

Example 13. Let X be binomial $b(p, n)$, and suppose that the prior distribution for p is the beta distribution* $B(a, b)$ with density $Cp^{a-1}(1 - p)^{b-1}, 0 < p < 1$, $0 < a, b$. Then the posterior distribution of p given $X = x$ is the beta distribution $B(a + x, b + n - x)$ (Problem 36). There are of course many sets $S(x)$ whose probability under this distribution is equal to the prescribed coefficient γ. A choice that is frequently recommended is the HPD (highest probability density) region, defined by the requirement that the posterior density of p given x be $\geq k$.

With a beta prior, only the following possibilities can occur: for fixed x,

(a) $\pi(p|x)$ is decreasing,
(b) $\pi(p|x)$ is increasing,
(c) $\pi(p|x)$ is increasing in $(0, p_0)$ and decreasing in $(p_0, 1)$ for some p_0,
(d) $\pi(p|x)$ is U-shaped, i.e. decreasing in $(0, p_0)$ and increasing in $(p_0, 1)$ for some p_0.

The HPD region then is of the form

(a) $p < K(x)$,
(b) $p > K(x)$,
(c) $K_1(x) < p < K_2(x)$,
(d) $p < K_1(x)$ or $p > K_2(x)$,

where the K's are determined by the requirement that the posterior probability of the region, given x, be γ; in cases (c) and (d) this condition must be supplemented by

$$\pi[K_1(x)|x] = \pi[K_2(x)|x].$$

In general, if $\pi(\theta|x)$ denotes the posterior density of θ, the HPD region is defined by

$$\pi(\theta|x) \geq k$$

---

*This is the so-called conjugate of the binomial distribution; for a more general discussion of conjugate distributions, see *TPE*, Chapter 4, Section 1.

with C determined by the size condition

$$P[\pi(\Theta|x) \geq k] = \gamma.$$

Example 14. Two-parameter normal: estimating the mean. Let $X_1, \ldots, X_n$ be independent $N(\xi, \sigma^2)$, and for the sake of simplicity suppose that (ξ, σ) has the joint improper prior density given by

$$\pi(\xi, \sigma) \, d\xi \, d\sigma = d\xi \frac{1}{\sigma} \, d\sigma \qquad \text{for all} \quad -\infty < \xi < \infty, \quad 0 < \sigma,$$

which is frequently used to model absence of information concerning the parameters. Then the joint posterior density of (ξ, σ) given $x = (x_1, \ldots, x_n)$ is of the form

$$\pi(\xi, \sigma|x) \, d\xi \, d\sigma = C(x) \frac{1}{\sigma^{n+1}} \exp\left(-\frac{1}{2\sigma^2} \sum_{i=1}^{n} (\xi - x_i)^2\right) d\xi \, d\sigma.$$

Determination of a credible region for ξ requires the marginal posterior density of ξ given x, which is obtained by integrating the joint posterior density with respect to σ. These densities depend only on the sufficient statistics $\bar{x}$ and $S^2 = \Sigma(x_i - \bar{x})^2$, and the posterior density of ξ is of the form (Problem 37)

$$A(x) \left[\frac{1}{1 + \dfrac{n(\xi - \bar{x})^2}{S^2}} \right]^{n/2}.$$

Here $\bar{x}$ and S enter only as location and scale parameters, and the linear function

$$t = \frac{\sqrt{n}(\xi - \bar{x})}{S/\sqrt{n-1}}$$

of ξ has the t-distribution with $n - 1$ degrees of freedom. Since this agrees with the distribution of t for fixed ξ and σ given in Section 2, the credible $100(1 - \alpha)\%$ region

$$\left| \frac{\sqrt{n}(\xi - \bar{x})}{S/\sqrt{n-1}} \right| \leq C$$

is formally identical with the confidence intervals (37). However, they are derived under different assumptions, and their interpretation differs accordingly.

Example 15. Two-parameter normal: estimating σ. Under the assumptions of the preceding example, credible regions for σ are based on the posterior distribution of σ given x, obtained by integrating the joint posterior density of (ξ, σ) with respect to ξ. Using the fact that $\Sigma(\xi - x_i)^2 = n(\xi - \bar{x})^2 + \Sigma(x_i - \bar{x})^2$, it is seen (Problem 38) that given x, the conditional (posterior) distribution of $\Sigma(x_i - \bar{x})^2/\sigma^2$ is χ^2 with $n - 1$ degrees of freedom. As in the case of the mean, this agrees with the sampling distribution of the same quantity when σ is a (constant) parameter, given in Section 2. (The agreement in both cases of two distributions derived under such different assumptions is a consequence of the particular choice of the prior distribution and the fact that it is invariant in the sense of *TPE*, Section 4.4.) A change of variables now gives the posterior density of σ and shows that $\pi(\sigma|x)$ is of the form (c) of Example 13, so that the HPD region is of the form $K_1(x) < \sigma < K_2(x)$ with $0 < K_1(x) < K_2(x) < \infty$.

Suppose that a credible region is required, not for σ, but for σ^r for some $r > 0$. For consistency, this should then be given by $[K_1(x)]^r < \sigma^r < [K_2(x)]^r$, but this is not the case, since the relative height of the density of a random variable at two points is not invariant under monotone transformations of the variable. In fact, in the present case, the HPD region for σ^r will become one-sided for sufficiently large r although it is two-sided for $r = 1$ (Problem 38).

Such inconsistencies do not occur if the HPD region is replaced by the equal-tails interval $(C_1(x), C_2(x))$ for which $P[\Theta < C_1(x) \mid X = x] = P[\Theta > C_2(x) \mid X = x] = (1 - \gamma)/2$.* More generally inconsistencies under transformations of Θ are avoided when the posterior distribution of Θ is summarized by a number of its percentiles corresponding to the standard confidence points mentioned in Chapter 3, Section 5. Such a set is a compromise between providing the complete posterior distribution and providing a single interval corresponding to only two percentiles.

Both the confidence and the Bayes approach present difficulties: the first, the problem of postdata interpretation; the second, the choice of a prior distribution and the interpretation of the posterior coverage probabilities if there is no clear basis for this choice. It is therefore not surprising that efforts have been made to find an approach without these drawbacks. The first such attempt, from which most later ones derive, is due to Fisher [1930; for his final account see Fisher (1973)].

To discuss Fisher's concept of fiducial probability, consider once more the example at the beginning of the section, in which X is distributed as $N(\theta, 1)$. Since then $X - \theta$ is distributed as $N(0, 1)$, so is $\theta - X$, and hence

$$P(\theta - X \le y) = \Phi(y) \qquad \text{for all } y.$$

For fixed $X = x$, this is the formal statement that a random variable θ has distribution $N(x, 1)$. Without assuming θ to be random, Fisher calls $N(x, 1)$ the *fiducial distribution* of θ. Since this distribution is to embody the

---

*They also do not occur when the posterior distribution of Θ is discrete.

information about θ provided by the data, it should be unique, and Fisher imposes conditions which he hopes will ensure uniqueness. This leads to some technical difficulties, but more basic is the question of how to interpret fiducial probability. In a series of independent repetitions of the experiment with arbitrarily varying θ_i, the quantities $\theta_1 - X_1, \theta_2 - X_2, \ldots$ will constitute a sequence of independent standard normal variables. From this fact, Fisher attempts to derive the fiducial distribution $N(x, 1)$ of θ as a frequency distribution with respect to an appropriate reference set. However, this argument is difficult to follow and unconvincing. For summaries of the fiducial literature and of later related developments by Dempster, Fraser, and others, see Pedersen (1978), Buehler (1980), Dawid and Stone (1982), and the encyclopedia articles by Fraser (1978), Edwards (1983), Buehler (1983), and Stone (1983).

Fisher's effort to define a suitable frame of reference led him to the important concept of *relevant subsets*, which will be discussed in Chapter 10.

10. PERMUTATION TESTS

For the comparison of a treatment with a control situation in which no treatment is given, it was shown in Section 3 that the one-sided t-test is UMP unbiased for testing $H : \eta = \xi$ against $\eta - \xi = \Delta > 0$ when the measurements $X_1, \ldots, X_m$ and $Y_1, \ldots, Y_n$ are samples from normal populations $N(\xi, \sigma^2)$ and $N(\eta, \sigma^2)$. It was further shown in Section 4 that the level of this test is (asymptotically) robust against nonnormality—that is, that except for small m or n the level of the test is approximately equal to the nominal level α when the X's and Y's are samples from any distributions with densities $f(x)$ and $f(y - \Delta)$ with finite variance. If such an approximate level is not satisfactory, one may prefer to try to obtain an exact level-α unbiased test (valid for all f) by replacing the original normal model with the nonparametric model for which the joint density of the variables is

$$(46) \qquad f(x_1) \ldots f(x_m) f(y_1 - \Delta) \ldots f(y_n - \Delta), \qquad f \in \mathscr{F},$$

where we shall take $\mathscr{F}$ to be the family of all probability densities that are continuous a.e.

If there is much variation in the population being sampled, the sensitivity of the experiment can frequently be increased by dividing the population into more homogeneous subgroups, defined for example by some characteristic such as age or sex. A sample of size N_i $(i = 1, \ldots, c)$ is then taken from the ith subpopulation: m_i to serve as controls, and the other $n_i = N_i - m_i$ to receive the treatment. If the observations in the ith subgroup of such a *stratified sample* are denoted by

$$\left(X_{i1}, \ldots, X_{im_i}; Y_{i1}, \ldots, Y_{in_i} \right) = \left(Z_{i1}, \ldots, Z_{iN_i} \right),$$

the density of $Z = (Z_{11}, \ldots, Z_{cN_c})$ is

$$(47) \quad p_\Delta(z) = \prod_{i=1}^{c} \left[f_i(x_{i1}) \ldots f_i(x_{im_i}) f_i(y_{i1} - \Delta) \ldots f_i(y_{in_i} - \Delta) \right].$$

Unbiasedness of a test ϕ for testing $\Delta = 0$ against $\Delta > 0$ implies that for all $f_1, \ldots, f_c$

$$(48) \quad \int \phi(z) p_0(z) \, dz = \alpha \quad \left(dz = dz_{11} \ldots dz_{cN_c} \right).$$

Theorem 6. *If $\mathscr{F}$ is the family of all probability densities f that are continuous a.e., then (48) holds for all $f_1, \ldots, f_c \in \mathscr{F}$ if and only if*

$$(49) \quad \frac{1}{N_1! \ldots N_c!} \sum_{z' \in S(z)} \phi(z') = \alpha \quad \text{a.e.,}$$

where $S(z)$ is the set of points obtained from z by permuting for each $i = 1, \ldots, c$ the coordinates z_{ij} $(j = 1, \ldots, N_i)$ within the ith subgroup in all $N_1! \ldots N_c!$ possible ways.

Proof. To prove the result for the case $c = 1$, note that the set of order statistics $T(Z) = (Z_{(1)}, \ldots, Z_{(N)})$ is a complete sufficient statistic for $\mathscr{F}$ (Chapter 4, Example 6). A necessary and sufficient condition for (48) is therefore

$$(50) \quad E\left[\phi(Z) | T(z) \right] = \alpha \quad \text{a.e.}$$

The set $S(z)$ in the present case $(c = 1)$ consists of the $N!$ points obtained from z through permutation of coordinates, so that $S(z) = \{ z' : T(z') = T(z) \}$. It follows from Section 4 of Chapter 2 that the conditional distribution of Z given $T(z)$ assigns probability $1/N!$ to each of the $N!$ points of $S(z)$. Thus (50) is equivalent to

$$(51) \quad \frac{1}{N!} \sum_{z' \in S(z)} \phi(z') = \alpha \quad \text{a.e.,}$$

as was to be proved. The proof for general c is completely analogous and is left as an exercise (Problem 44.)

The tests satisfying (49) are called *permutation tests*. An extension of this definition is given in Problem 54.

11. MOST POWERFUL PERMUTATION TESTS

For the problem of testing the hypothesis $H: \Delta = 0$ of no treatment effect on the basis of a stratified sample with density (47) it was shown in the preceding section that unbiasedness implies (49). We shall now determine the test which, subject to (49), maximizes the power against a fixed alternative (47) or more generally against an alternative with arbitrary fixed density $h(z)$.

The power of a test ϕ against an alternative h is

$$\int \phi(z)h(z)\, dz = \int E[\phi(Z)|t]\, dP^T(t).$$

Let $t = T(z) = (z_{(1)}, \ldots, z_{(N)})$, so that $S(z) = S(t)$. As was seen in Example 7 and Problem 5 of Chapter 2, the conditional expectation of $\phi(Z)$ given $T(Z) = t$ is

$$\psi(t) = \frac{\displaystyle\sum_{z \in S(t)} \phi(z)h(z)}{\displaystyle\sum_{z \in S(t)} h(z)}.$$

To maximize the power of ϕ subject to (49) it is therefore necessary to maximize $\psi(t)$ for each t subject to this condition. The problem thus reduces to the determination of a function ϕ which, subject to

$$\sum_{z \in S(t)} \phi(z) \frac{1}{N_1! \ldots N_c!} = \alpha,$$

maximizes

$$\sum_{z \in S(t)} \phi(z) \frac{h(z)}{\displaystyle\sum_{z' \in S(t)} h(z')}.$$

By the Neyman–Pearson fundamental lemma, this is achieved by rejecting H for those points z of $S(t)$ for which the ratio

$$\frac{h(z)N_1! \ldots N_c!}{\displaystyle\sum_{z' \in S(t)} h(z')}$$

is too large. Thus the most powerful test is given by the critical function

(52)
$$\phi(z) = \begin{cases} 1 & \text{when} & h(z) > C[T(z)], \\ \gamma & \text{when} & h(z) = C[T(z)], \\ 0 & \text{when} & h(z) < C[T(z)]. \end{cases}$$

To carry out the test, the $N_1! \ldots N_c!$ points of each set $S(z)$ are ordered according to the values of the density h. The hypothesis is rejected for the k largest values and with probability γ for the $(k+1)$st value, where k and γ are defined by

$$k + \gamma = \alpha N_1! \ldots N_c!.$$

Consider now in particular the alternatives (47). The most powerful permutation test is seen to depend on Δ and the f_i, and is therefore not UMP.

Of special interest is the class of normal alternatives with common variance:

$$f_i = N(\xi_i, \sigma^2).$$

The most powerful test against these alternatives, which turns out to be independent of the ξ_i, σ^2, and Δ, is appropriate when approximate normality is suspected but the assumption is not felt to be reliable. It may then be desirable to control the size of the test at level α regardless of the form of the densities f_i and to have the test unbiased against all alternatives (47). However, among the class of tests satisfying these broad restrictions it is natural to make the selection so as to maximize the power against the type of alternative one expects to encounter, that is, against the normal alternatives.

With the above choice of f_i, (47) becomes

(53) $$h(z) = (\sqrt{2\pi}\sigma)^{-N} \exp\left[-\frac{1}{2\sigma^2} \sum_{i=1}^{c} \left(\sum_{j=1}^{m_i} (z_{ij} - \xi_i)^2 \right.\right.$$
$$\left.\left. + \sum_{j=m_i+1}^{N_i} (z_{ij} - \xi_i - \Delta)^2 \right) \right].$$

Since the factor $\exp[-\sum_i\sum_{j=1}^{N_i}(z_{ij} - \xi_i)^2/2\sigma^2]$ is constant over $S(t)$, the test (52) therefore rejects H when $\exp(\Delta\sum_i\sum_{j=m_i+1}^{N_i}z_{ij}) > C[T(z)]$ and hence when

(54)
$$\sum_{i=1}^{c} \sum_{j=1}^{n_i} y_{ij} = \sum_{i=1}^{c} \sum_{j=m_i+1}^{N_i} z_{ij} > C[T(z)].$$

Of the $N_1! \ldots N_c!$ values that the test statistic takes on over $S(t)$, only

$$\binom{N_1}{n_1} \cdots \binom{N_c}{n_c}$$

are distinct, since the value of the statistic is the same for any two points z' and z'' for which $(z'_{i1}, \ldots, z'_{im_i})$ and $(z''_{i1}, \ldots, z''_{im_i})$ are permutations of each other for each i. It is therefore enough to compare these distinct values, and to reject H for the k' largest ones and with probability γ' for the $(k' + 1)$st, where

$$k' + \gamma' = \alpha \binom{N_1}{n_1} \cdots \binom{N_c}{n_c}.$$

The test (54) is most powerful against the normal alternatives under consideration among all tests which are unbiased and of level α for testing $H : \Delta = 0$ in the original family (47) with $f_1, \ldots, f_c \in \mathcal{F}$.* To complete the proof of this statement it is still necessary to prove the test unbiased against the alternatives (47). We shall show more generally that it is unbiased against all alternatives for which X_{ij} ($j = 1, \ldots, m_i$), Y_{ik} ($k = 1, \ldots, n_i$) are independently distributed with cumulative distribution functions F_i, G_i respectively such that Y_{ik} is stochastically larger than X_{ij}, that is, such that $G_i(z) \leq F_i(z)$ for all z. This is a consequence of the following lemma.

Lemma 3. *Let $X_1, \ldots, X_m, Y_1, \ldots, Y_n$ be samples from continuous distributions F, G, and let $\phi(x_1, \ldots, x_m; y_1, \ldots, y_n)$ be a critical function such that (a) its expectation is α whenever $G = F$, and (b) $y_i \leq y'_i$ for $i = 1, \ldots, n$ implies*

$$\phi(x_1, \ldots, x_m; y_1, \ldots, y_n) \leq \phi(x_1, \ldots, x_m; y'_1, \ldots, y'_n).$$

Then the expectation $\beta = \beta(F, G)$ of ϕ is $\geq \alpha$ for all pairs of distributions for which Y is stochastically larger than X; it is $\leq \alpha$ if X is stochastically larger than Y.

Proof. By Lemma 1 of Chapter 3 there exist functions f, g and independent random variables $V_1, \ldots, V_{m+n}$ such that the distributions of $f(V_i)$

---

*For a closely related result, see Odén and Wedel (1975).

and $g(V_i)$ are F and G respectively and that $f(z) \leq g(z)$ for all z. Then

$$E\phi[f(V_1),\ldots,f(V_m); f(V_{m+1}),\ldots,f(V_{m+n})] = \alpha$$

and

$$E\phi[f(V_1),\ldots,f(V_m); g(V_{m+1}),\ldots,g(V_{m+n})] = \beta.$$

Since for all $(v_1,\ldots,v_{m+n})$,

$$\phi[f(v_1),\ldots,f(v_m); f(v_{m+1}),\ldots,f(v_{m+n})]$$

$$\leq \phi[f(v_1),\ldots,f(v_m); g(v_{m+1}),\ldots,g(v_{m+n})],$$

the same inequality holds for the expectations of both sides, and hence $\alpha \leq \beta$.

The proof for the case that X is stochastically larger than Y is completely analogous.

The lemma also generalizes to the case of c vectors $(X_{i1},\ldots,X_{im_i}; Y_{i1},\ldots,Y_{in_i})$ with distributions (F_i, G_i). If the expectation of a function ϕ is then α when $F_i = G_i$ and ϕ is nondecreasing in each y_{ij} when all other variables are held fixed, it follows as before that the expectation of ϕ is $\geq \alpha$ when the random variables with distribution G_i are stochastically larger than those with distribution F_i.

In applying the lemma to the permutation test (54) it is enough to consider the case $c = 1$, the argument in the more general case being completely analogous. Since the rejection probability of the test (54) is α whenever $F = G$, it is only necessary to show that the critical function ϕ of the test satisfies (b). Now $\phi = 1$ if $\sum_{i=m+1}^{m+n} z_i$ exceeds sufficiently many of the sums $\sum_{i=m+1}^{m+n} z_{j_i}$, and hence if sufficiently many of the differences

$$\sum_{i=m+1}^{m+n} z_i - \sum_{i=m+1}^{m+n} z_{j_i}$$

are positive. For a particular permutation $(j_1,\ldots,j_{m+n})$

$$\sum_{i=m+1}^{m+n} z_i - \sum_{i=m+1}^{m+n} z_{j_i} = \sum_{i=1}^{p} z_{s_i} - \sum_{i=1}^{p} z_{r_i},$$

where $r_1 < \cdots < r_p$ denote those of the integers $j_{m+1},\ldots,j_{m+n}$ that are $\leq m$, and $s_1 < \cdots < s_p$ those of the integers $m+1,\ldots,m+n$ not included in the set $(j_{m+1},\ldots,j_{m+n})$. If $\Sigma z_{s_i} - \Sigma z_{r_i}$ is positive and $y_i \leq y_i'$, that is, $z_i \leq z_i'$ for $i = m + 1, \ldots, m + n$, then the difference $\Sigma z_{s_i}' - \Sigma z_{r_i}$ is also positive and hence ϕ satisfies (b).

The same argument also shows that the rejection probability of the test is $\leq \alpha$ when the density of the variables is given by (47) with $\Delta \leq 0$. The test is therefore equally appropriate if the hypothesis $\Delta = 0$ is replaced by $\Delta \leq 0$.

Except for small values of the sample sizes N_i, the amount of computation required to carry out the permutation test (54) is very large. Computational methods are discussed by Green (1977) and John and Robinson (1983b). Alternatively, several large-sample approximations for the critical value are available; see, for example, Robinson (1982).

A particularly simple approximation relates the permutation test to the corresponding t-test. On multiplying both sides of the inequality

$$\Sigma y_j > C[T(z)]$$

by $(1/m) + (1/n)$ and subtracting $(\Sigma x_i + \Sigma y_j)/m$, the rejection region for $c = 1$ becomes $\bar{y} - \bar{x} > C[T(z)]$ or $W = (\bar{y} - \bar{x})/\sqrt{\Sigma_{i=1}^N (z_i - \bar{z})^2} > C[T(z)]$, since the denominator of W is constant over $S(z)$ and hence depends only on $T(z)$. As was seen at the end of Section 3, this is equivalent to

$$(55) \quad \frac{(\bar{y} - \bar{x})/\sqrt{\dfrac{1}{m} + \dfrac{1}{n}}}{\sqrt{\left[\Sigma(x_i - \bar{x})^2 + \Sigma(y_j - \bar{y})^2\right]/(m + n - 2)}} > C[T(z)].$$

The rejection region therefore has the form of a t-test in which the constant cutoff point C_0 of (27) has been replaced by a random one. It turns out that when the hypothesis is true, so that the Z's are identically and independently distributed, and if $E|Z|^3 < \infty$ and m/n is bounded away from zero and infinity as m and n tend to infinity, the difference between the random cutoff point $C[T(Z)]$ and C_0 tends to zero in probability. In the limit, the permutation test therefore becomes equivalent to the t-test given by (27)–(29).* It follows that the *permutation test can be approximated for large samples by the standard t-test.* Exactly analogous results hold for $c > 1$; the appropriate t-test is provided in Chapter 7, Problem 9.

*This equivalence is not limited to the behavior under the hypothesis. For large samples, it is shown by Hoeffding (1952) and Bickel and van Zwet (1978, Theorem 7.2) that also the power of the permutation test is approximately equal to that of the t-test. For some implications and further references see Lambert (1985).

12. RANDOMIZATION AS A BASIS FOR INFERENCE

The problem of testing for the effect of a treatment was considered in Section 3 under the assumption that the treatment and control measurements $X_1, \ldots, X_m$ and $Y_1, \ldots, Y_n$ constitute samples from normal distributions, and in Sections 10 and 11 without relying on the assumption of normality. We shall now consider in somewhat more detail the structure of the experiment from which the data are obtained, resuming for the moment the assumption that the distributions involved are normal.

Suppose that the experimental material consists of $m + n$ patients, plants, pieces of material, or the like, drawn at random from the population to which the treatment could be applied. The treatment is given to n of these while the other m serve as controls. The characteristic that is to be influenced by the treatment is then measured in each case, leading to observations $X_1, \ldots, X_m; Y_1, \ldots, Y_n$.

To be specific, suppose that the treatment is carried out by injecting a drug and that $m + n$ ampules are assigned to the $m + n$ patients. The ith measurement can be considered as the sum of two components. One, say U_i, is associated with the ith patient; the other, V_i, with the ith ampule and the circumstances under which it is administered and under which the measurements are taken. The variables U_i and V_i are assumed to be independently distributed, the V's with normal distribution $N(\eta, \sigma^2)$ or $N(\xi, \sigma^2)$ as the ampule contains the drug or is one of those used for control. If in addition the U's are assumed to constitute a random sample from $N(\mu, \sigma_1^2)$, it follows that the X's and Y's are independently normally distributed with common variance $\sigma^2 + \sigma_1^2$ and means

$$E(X) = \mu + \xi, \qquad E(Y) = \mu + \eta.$$

Except for a change of notation their joint distribution is then given by (26), and the hypothesis $\eta = \xi$ can be tested by the standard t-test.

Unfortunately, under actual experimental conditions, it is frequently not possible to ensure that the patients or other experimental units constitute a random sample from the population of such units. They may be patients in a certain hospital at a given time, or volunteers for an experiment, and may constitute a haphazard rather than a random sample. In this case the U's would have to be considered as unknown constants, since they are not obtained by any definite sampling procedure. This assumption is appropriate also in a different context. Suppose that the experimental units are all the machines in a shop or fields on a farm. If the experiment is performed only to determine the best method for this particular shop or farm, these experimental units are the only relevant ones; that is, a replication of the experiment would consist in comparing the two treatments again for the same machines or fields rather than for a new batch drawn at random from a large population. In this case the units themselves, and therefore the u's, are constant.

Under the above assumptions the joint density of the $m + n$ measurements is

$$\frac{1}{(\sqrt{2\pi}\,\sigma)^{m+n}}\exp\left[-\frac{1}{2\sigma^2}\left(\sum_{i=1}^{m}(x_i - u_i - \xi)^2 + \sum_{j=1}^{n}(y_j - u_{m+j} - \eta)^2\right)\right].$$

Since the u's are completely arbitrary, it is clearly impossible to distinguish between $H: \eta = \xi$ and the alternatives $K: \eta > \xi$. In fact, every distribution of K also belongs to H and vice versa, and the most powerful level-α test for testing H against any simple alternative specifying ξ, η, σ, and the u's rejects H with probability α regardless of the observations.

Data which could serve as a basis for testing whether or not the treatment has an effect can be obtained through the fundamental device of *randomization*. Suppose that the $N = m + n$ patients are assigned to the N ampules at random, that is, in such a way that each of the $N!$ possible assignments has probability $1/N!$ of being chosen. Then for a given assignment the N measurements are independently normally distributed with variance σ^2 and means $\xi + u_{j_i}$ $(i = 1, \ldots, m)$ and $\eta + u_{j_i}$ $(i = m + 1, \ldots, m + n)$. The overall joint density of the variables

$$(Z_1, \ldots, Z_N) = (X_1, \ldots, X_m; Y_1, \ldots, Y_n)$$

is therefore

(56) $$\frac{1}{N!} \sum_{(j_1, \ldots, j_N)} \frac{1}{(\sqrt{2\pi}\,\sigma)^N}$$

$$\times \exp\left[-\frac{1}{2\sigma^2}\left(\sum_{i=1}^{m}(x_i - u_{j_i} - \xi)^2 + \sum_{i=1}^{n}(y_i - u_{j_{m+i}} - \eta)^2\right)\right]$$

where the outer summation extends over all $N!$ permutations $(j_1, \ldots, j_N)$ of $(1, \ldots, N)$. Under the hypothesis $\eta = \xi$ this density can be written as

(57) $$\frac{1}{N!} \sum_{(j_1, \ldots, j_N)} \frac{1}{(\sqrt{2\pi}\,\sigma)^N}\exp\left[-\frac{1}{2\sigma^2}\sum_{i=1}^{N}(z_i - \zeta_{j_i})^2\right],$$

where $\zeta_{j_i} = u_{j_i} + \xi = u_{j_i} + \eta$.

Without randomization, a set of y's which is large relative to the x-values could be explained entirely in terms of the unit effects u_i. However, if these are assigned to the y's at random, they will on the average balance those assigned to the x's. As a consequence, a marked superiority of the second sample becomes very unlikely under the hypothesis, and must therefore be attributed to the effectiveness of the treatment.

The method of assigning the treatments to the experimental units completely at random permits the construction of a level-α test of the hypothesis $\eta = \xi$, whose power exceeds α against all alternatives $\eta - \xi > 0$. The actual power of such a test will however depend not only on the alternative value of $\eta - \xi$, which measures the effect of the treatment, but also on the unit effects u_i. In particular, if there is excessive variation among the u's, this will swamp the treatment effect (much in the same way as an increase in the variance σ^2 would), and the test will accordingly have little power to detect any given alternative $\eta - \xi$.

In such cases the sensitivity of the experiment can be increased by an approach exactly analogous to the method of stratified sampling discussed in Section 10. In the present case this means replacing the process of complete randomization described above by a more restricted randomization procedure. The experimental material is divided into subgroups, which are more homogeneous than the material as a whole, so that within each group the differences among the u's are small. In animal experiments, for example, this can frequently be achieved by a division into litters. Randomization is then applied only within each group. If the ith group contains N_i units, n_i of these are selected at random to receive the treatment, and the remaining $m_i = N_i - n_i$ serve as controls ($\Sigma N_i = N, \Sigma m_i = m, \Sigma n_i = n$).

An example of this approach is the method of *matched pairs*. Here the experimental units are divided into pairs, which are as like each other as possible with respect to all relevant properties, so that within each pair the difference of the u's will be as small as possible. Suppose that the material consists of n such pairs, and denote the associated unit effects (the U's of the previous discussion) by $U_1, U_1'; \ldots; U_n, U_n'$. Let the first and second member of each pair receive the treatment or serve as control respectively, and let the observations for the ith pair be X_i and Y_i. If the matching is completely successful, as may be the case, for example, when the same patient is used twice in the investigation of a sleeping drug, or when identical twins are used, then $U_i' = U_i$ for all i, and the density of the X's and Y's is

$$(58) \quad \frac{1}{(\sqrt{2\pi}\sigma)^{2n}} \exp\left[-\frac{1}{2\sigma^2}\left[\sum(x_i - \xi - u_i)^2 + \sum(y_i - \eta - u_i)^2\right]\right].$$

The UMP unbiased test for testing $H: \eta = \xi$ against $\eta > \xi$ is then given in terms of the differences $W_i = Y_i - X_i$ by the rejection region

$$(59) \qquad \sqrt{n}\,\bar{w}\,\Big/\sqrt{\frac{1}{n-1}\sum(w_i - \bar{w})^2} > C.$$

(See Problem 48.)

However, usually one is not willing to trust the assumption $u_i' = u_i$ even after matching, and it again becomes necessary to randomize. Since as a result of the matching the variability of the u's within each pair is presumably considerably smaller than the overall variation, randomization is carried out only within each pair. For each pair, one of the units is selected with probability $\frac{1}{2}$ to receive the treatment, while the other serves as control. The density of the X's and Y's is then

$$(60) \qquad \frac{1}{2^n}\,\frac{1}{(\sqrt{2\pi}\,\sigma)^{2n}}\prod_{i=1}^{n}\left\{\exp\left[-\frac{1}{2\sigma^2}\big[(x_i - \xi - u_i)^2 + (y_i - \eta - u_i')^2\big]\right]\right.$$

$$\left. + \exp\left[-\frac{1}{2\sigma^2}\big[(x_i - \xi - u_i')^2 + (y_i - \eta - u_i)^2\big]\right]\right\}.$$

Under the hypothesis $\eta = \xi$, and writing

$$z_{i1} = x_i, \quad z_{i2} = y_i, \quad \zeta_{i1} = \xi + u_i, \quad \zeta_{i2} = \eta + u_i' \qquad (i = 1,\dots, n),$$

this becomes

$$(61) \qquad \frac{1}{2^n}\sum \frac{1}{(\sqrt{2\pi}\,\sigma)^{2n}}\exp\left[-\frac{1}{2\sigma^2}\sum_{i=1}^{n}\sum_{j=1}^{2}\big(z_{ij} - \zeta_{ij}'\big)^2\right].$$

Here the outer summation extends over the 2^n points $\zeta' = (\zeta_{11}', \dots, \zeta_{n2}')$ for which $(\zeta_{i1}', \zeta_{i2}')$ is either (ζ_{i1}, ζ_{i2}) or (ζ_{i2}, ζ_{i1}).

13. PERMUTATION TESTS AND RANDOMIZATION

It was shown in the preceding section that randomization provides a basis for testing the hypothesis $\eta = \xi$ of no treatment effect, without any assumptions concerning the experimental units. In the present section, a specific test will be derived for this problem. When the experimental units are treated as constants, the probability density of the observations is given by (56) in the case of complete randomization and by (60) in the case of matched pairs. More generally, let the experimental material be divided into c subgroups, let the randomization be applied within each subgroup, and let the observations in the ith subgroup be

$$(Z_{i1}, \dots, Z_{iN_i}) = (X_{i1}, \dots, X_{im_i}; Y_{i1}, \dots, Y_{in_i}).$$

For any point $u = (u_{11}, \dots, u_{cN_c})$, let $S(u)$ denote as before the set of $N_1! \dots N_c!$ points obtained from u by permuting the coordinates within each subgroup in all $N_1! \dots N_c!$ possible ways. Then the joint density of the Z's given u is

$$(62) \quad \frac{1}{N_1! \dots N_c!} \sum_{u' \in S(u)} \frac{1}{(\sqrt{2\pi}\,\sigma)^N}$$

$$\times \exp\left[-\frac{1}{2\sigma^2} \sum_{i=1}^{c} \left(\sum_{j=1}^{m_i} (z_{ij} - \xi - u'_{ij})^2 + \sum_{j=m_i+1}^{N_i} (z_{ij} - \eta - u'_{ij})^2 \right) \right],$$

and under the hypothesis of no treatment effect

$$(63)$$
$$p_{\sigma,\zeta}(z) = \frac{1}{N_1! \dots N_c!} \sum_{\zeta' \in S(\zeta)} \frac{1}{(\sqrt{2\pi}\,\sigma)^N} \exp\left[-\frac{1}{2\sigma^2} \sum_{i=1}^{c} \sum_{j=1}^{N_i} (z_{ij} - \zeta'_{ij})^2 \right].$$

It may happen that the coordinates of u or ζ are not distinct. If then some of the points of $S(u)$ or $S(\zeta)$ also coincide, each should be counted with its proper multiplicity. More precisely, if the $N_1! \dots N_c!$ relevant permutations of $N_1 + \cdots + N_c$ coordinates are denoted by g_k, $k = 1, \dots, N_1! \dots N_c!$, then $S(\zeta)$ can be taken to be the ordered set of points $g_k\zeta$, $k = 1, \dots, N_1! \dots N_c!$, and (63), for example, becomes

$$p_{\sigma,\zeta}(z) = \frac{1}{N_1! \dots N_c!} \sum_{k=1}^{N_1! \dots N_c!} \frac{1}{(\sqrt{2\pi}\,\sigma)^N} \exp\left(-\frac{1}{2\sigma^2} |z - g_k\zeta|^2 \right)$$

where $|u|^2$ stands for $\sum_{i=1}^{c}\sum_{j=1}^{N_i} u_{ij}^2$.

Theorem 7. *A necessary and sufficient condition for a critical function ϕ to satisfy*

$$(64) \qquad \int \phi(z) p_{\sigma,\zeta}(z)\, dz \le \alpha \qquad (dz = dz_{11} \dots dz_{cN_c})$$

for all $\sigma > 0$ and all vectors ζ is that

(65) $$\frac{1}{N_1! \dots N_c!} \sum_{z' \in S(z)} \phi(z') \le \alpha \qquad \text{a.e.}$$

The proof will be based on the following lemma.

Lemma 4. *Let A be a set in N-space with positive Lebesgue measure $\mu(A)$. Then for any $\epsilon > 0$ there exist real numbers $\sigma > 0$ and $\xi_1, \dots, \xi_N$ such that*

$$P\{(X_1, \dots, X_N) \in A\} \ge 1 - \epsilon,$$

where the X's are independently normally distributed with means $E(X_i) = \xi_i$ and variance $\sigma^2_{X_i} = \sigma^2$.

Proof. Suppose without loss of generality that $\mu(A) < \infty$. Given any $\eta > 0$, there exists a square Q such that

$$\mu(Q \cap \tilde{A}) \le \eta \mu(Q).$$

This follows from the fact that almost every point of A has metric density 1,* or from the more elementary fact that a measurable set can be approximated in measure by unions of disjoint squares. Let a be such that

$$\frac{1}{\sqrt{2\pi}} \int_{-a}^{a} \exp\left(-\frac{t^2}{2}\right) dt = \left(1 - \frac{\epsilon}{2}\right)^{1/N},$$

and let

$$\eta = \frac{\epsilon}{2} \left(\frac{\sqrt{2\pi}}{2a}\right)^N.$$

If $(\xi_1, \dots, \xi_N)$ is the center of Q, and if $\sigma = b/a = (1/2a)[\mu(Q)]^{1/N}$, where $2b$ is the length of the side of Q, then

$$\frac{1}{(\sqrt{2\pi}\,\sigma)^N} \int_{A \cap Q} \exp\left[-\frac{1}{2\sigma^2} \sum (x_i - \xi_i)^2\right] dx_1 \dots dx_N$$

$$\le \frac{1}{(\sqrt{2\pi}\,\sigma)^N} \int_Q \exp\left[-\frac{1}{2\sigma^2} \sum (x_i - \xi_i)^2\right] dx_1 \dots dx_N$$

$$= 1 - \left[\frac{1}{\sqrt{2\pi}} \int_{-a}^{a} \exp\left(-\frac{t^2}{2}\right) dt\right]^N = \frac{\epsilon}{2}.$$

*See for example Hobson (1927).

On the other hand,

$$\frac{1}{(\sqrt{2\pi}\,\sigma)^N} \int_{\tilde{A} \cap Q} \exp\left[-\frac{1}{2\sigma^2}\sum(x_i - \xi_i)^2\right] dx_1 \ldots dx_N$$

$$\leq \frac{1}{(\sqrt{2\pi}\,\sigma)^N}\mu(\tilde{A} \cap Q) < \frac{\epsilon}{2},$$

and by adding the two inequalities one obtains the desired result.

Proof of the theorem. Let ϕ be any critical function, and let

$$\psi(z) = \frac{1}{N_1! \ldots N_c!} \sum_{z' \in S(z)} \phi(z').$$

If (65) does not hold, there exists $\eta > 0$ such that $\psi(z) > \alpha + \eta$ on a set A of positive measure. By the Lemma there exists $\sigma > 0$ and $\zeta = (\zeta_{11}, \ldots, \zeta_{cN_c})$ such that $P\{Z \in A\} > 1 - \eta$ when $Z_{11}, \ldots, Z_{cN_c}$ are independently normally distributed with common variance σ^2 and means $E(Z_{ij}) = \zeta_{ij}$. It follows that

(66)

$$\int \phi(z) p_{\sigma,\zeta}(z)\, dz = \int \psi(z) p_{\sigma,\zeta}(z)\, dz$$

$$\geq \int_A \psi(z) \frac{1}{(\sqrt{2\pi}\,\sigma)^N} \exp\left[-\frac{1}{2\sigma^2}\sum\sum(z_{ij} - \zeta_{ij})^2\right] dz$$

$$> (\alpha + \eta)(1 - \eta),$$

which is $> \alpha$, since $\alpha + \eta < 1$. This proves that (64) implies (65). The converse follows from the first equality in (66).

Corollary 3. *Let H be the class of densities*

$$\{p_{\sigma,\zeta}(z): \sigma > 0, -\infty < \zeta_{ij} < \infty\}.$$

A complete family of tests for H at level of significance α is the class of tests $\mathscr{C}$ satisfying

(67)
$$\frac{1}{N_1! \ldots N_c!} \sum_{z' \in S(z)} \phi(z') = \alpha \qquad \text{a.e.}$$

Proof. The corollary states that for any given level-α test ϕ_0 there exists an element ϕ of $\mathscr{C}$ which is uniformly at least as powerful as ϕ_0. By the preceding theorem the average value of ϕ_0 over each set $S(z)$ is $\leq \alpha$. On the sets for which this inequality is strict, one can increase ϕ_0 to obtain a critical function ϕ satisfying (67), and such that $\phi_0(z) \leq \phi(z)$ for all z. Since against all alternatives the power of ϕ is at least that of ϕ_0, this establishes the result. An explicit construction of ϕ, which shows that it can be chosen to be measurable, is given in Problem 51.

This corollary shows that the normal randomization model (62) leads exactly to the class of tests that was previously found to be relevant when the U's constituted a sample but the assumption of normality was not imposed. It therefore follows from Section 11 that the most powerful level-α test for testing (63) against a simple alternative (62) is given by (52) with $h(z)$ equal to the probability density (62). If $\eta - \xi = \Delta$, the rejection region of this test reduces to

$$(68) \quad \sum_{u' \in S(u)} \exp\left[\frac{1}{\sigma^2} \sum_{i=1}^{c} \left(\sum_{j=1}^{N_i} z_{ij} u'_{ij} + \Delta \sum_{j=m_i+1}^{N_i} (z_{ij} - u'_{ij}) \right) \right] > C[T(z)],$$

since both $\Sigma\Sigma z_{ij}$ and $\Sigma\Sigma z_{ij}^2$ are constant on $S(z)$ and therefore functions only of $T(z)$. It is seen that this test depends on Δ and the unit effects u_{ij}, so that a UMP test does not exist.

Among the alternatives (62) a subclass occupies a central position and is of particular interest. This is the class of alternatives specified by the assumption that the unit effects u_i constitute a sample from a normal distribution. Although this assumption cannot be expected to hold exactly—in fact, it was just as a safeguard against the possibility of its breakdown that randomization was introduced—it is in many cases reasonable to suppose that it holds at least approximately. The resulting subclass of alternatives is given by the probability densities

$$(69) \quad \frac{1}{(\sqrt{2\pi}\,\sigma)^N}$$

$$\times \exp\left[-\frac{1}{2\sigma^2} \sum_{i=1}^{c} \left(\sum_{j=1}^{m_i} (z_{ij} - u_i - \xi)^2 + \sum_{j=m_i+1}^{N_i} (z_{ij} - u_i - \eta)^2 \right) \right].$$

These alternatives are suggestive also from a slightly different point of view. The procedure of assigning the experimental units to the treatments at random within each subgroup was seen to be appropriate when the variation of the u's is small within these groups and is employed when this is believed to be the case. This suggests, at least as an approximation, the assumption of constant $u_{ij} = u_i$, which is the limiting case of a normal distribution as the variance tends to zero, and for which the density is also given by (69).

Since the alternatives (69) are the same as the alternatives (53) of Section 11 with $u_i - \xi = \xi_i$, $u_i - \eta = \xi_i - \Delta$, *the permutation test (54) is seen to be most powerful for testing the hypothesis $\eta = \xi$ in the normal randomization model (62) against the alternatives (69) with $\eta - \xi > 0$.* The test retains this property in the still more general setting in which neither normality nor the sample property of the U's is assumed to hold. Let the joint density of the variables be

$$(70) \qquad \sum_{u' \in S(u)} \prod_{i=1}^{c} \left[\prod_{j=1}^{m_i} f_i(z_{ij} - u'_{ij} - \xi) \prod_{j=m_i+1}^{N_i} f_i(z_{ij} - u'_{ij} - \eta) \right],$$

with f_i continuous a.e. but otherwise unspecified.* Under the hypothesis $H : \eta = \xi$, this density is symmetric in the variables $(z_{i1}, \dots, z_{iN_i})$ of the ith subgroup for each i, so that any permutation test (49) has rejection probability α for all distributions of H. By Corollary 3, these permutation tests therefore constitute a complete class, and the result follows.

14. RANDOMIZATION MODEL AND CONFIDENCE INTERVALS

In the preceding section, the unit responses u_i were unknown constants (parameters) which were observed with error, the latter represented by the random terms V_i. A limiting case assumes that the variation of the V's is so small compared with that of the u's that these error variables can be taken to be constant, i.e. that $V_i = v$. The constant v can then be absorbed into the u's, and can therefore be assumed to be zero. This leads to the following two-sample *randomization model*:

N subjects would give "true" responses $u_1, \dots, u_N$ if used as controls. The subjects are assigned at random, n to treatment and m to control. If the responses are denoted by $X_1, \dots, X_m$ and $Y_1, \dots, Y_n$ as before, then under the hypothesis H of no treatment effect, the X's and Y's are a random permutation of the u's. Under this model, in which the random

---

*Actually, all that is needed is that $f_1, \dots, f_c \in \mathcal{F}$, where $\mathcal{F}$ is any family containing all normal distributions.

assignment of the subjects to treatment and control constitutes the only random element, the probability of the rejection region (55) is the same as under the more elaborate models of the preceding sections.

The corresponding limiting model under the alternatives assumes that the treatment has the effect of adding a constant amount Δ to the unit response, so that the X's and Y's are given by $(u_{i_1}, \ldots, u_{i_m}; u_{i_{m+1}} + \Delta, \ldots, u_{i_{m+n}} + \Delta)$ for some parmutation $(i_1, \ldots, i_N)$ of $(1, \ldots, N)$.

These models generalize in the obvious way to stratified samples. In particular, for paired comparisons it is assumed under H that the unit effects (u_i, u_i') are constants, of which one is assigned at random to treatment and the other to control. Thus the pair (X_i, Y_i) is equal to (u_i, u_i') or (u_i', u_i) with probability $\frac{1}{2}$ each, and the assignments in the n pairs are independent; the sample space consists of 2^n points each of which has probability $(\frac{1}{2})^n$. Under the alternative, it is assumed as before that Δ is added to each treated subject, so that $P(X_i = u_i, Y_i = u_i' + \Delta) = P(X_i = u_i', Y_i = u_i + \Delta) = \frac{1}{2}$. The distribution generated for the observations by such a randomization model is exactly the conditional distribution given $T(z)$ of the preceding sections. In the two-sample case, for example, this common distribution is specified by the fact that all permutations of $(X_1, \ldots, X_m; Y_1 - \Delta, \ldots, Y_n - \Delta)$ are equally likely. As a consequence, the power of the test (55) in the randomization model is also the conditional power in the two-sample model (46). As was pointed out in Chapter 4, Section 4, the conditional power $\beta(\Delta|T(z))$ can be interpreted as an unbiased estimate of the unconditional power $\beta_F(\Delta)$ in the two-sample model. The advantage of $\beta(\Delta|T(z))$ is that it depends only on Δ, not on the unknown F. Approximations to $\beta(\Delta|T(z))$ are discussed by Robinson (1973, 1982), John and Robinson (1983a), and Gabriel and Hsu (1983).

The tests (54), which apply to all three models—the sampling model (47), the randomization model, and the intermediate model (70)—can be inverted in the usual way to produce confidence sets for Δ. We shall now determine these sets explicitly for the paired comparisons and the two-sample case. The derivations will be carried out in the randomization model. However, they apply equally in the other two models, since the tests, and therefore the associated confidence sets, are identical for the three models.

Consider first the case of paired observations (x_i, y_i), $i = 1, \ldots, n$. The one-sided test rejects $H: \Delta = 0$ in favor of $\Delta > 0$ when $\sum_{i=1}^{n} y_i$ is among the K largest of the 2^n sums obtained by replacing y_i by x_i for all, some, or none of the values $i = 1, \ldots, n$. (It is assumed here for the sake of simplicity that $\alpha = K/2^n$, so that the test requires no randomization to achieve the exact level α.) Let $d_i = y_i - x_i = 2y_i - t_i$, where $t_i = x_i + y_i$ is fixed. Then the test is equivalent to rejecting when Σd_i is one of the K largest of the 2^n values $\Sigma \pm d_i$, since an interchange of y_i with x_i is equivalent to replacing

d_i by $-d_i$. Consider now testing $H: \Delta = \Delta_0$ against $\Delta > \Delta_0$. The test then accepts when $\Sigma(d_i - \Delta_0)$ is one of the $l = 2^n - K$ smallest of the 2^n sums $\Sigma \pm (d_i - \Delta_0)$, since it is now $y_i - \Delta_0$ that is being interchanged with x_i. We shall next invert this statement, replacing Δ_0 by Δ, and see that it is equivalent to a lower confidence bound for Δ.

In the inequality

$$(71) \qquad \Sigma(d_i - \Delta) < \Sigma[\pm(d_i - \Delta)],$$

suppose that on the right side the minus sign attaches to the $(d_i - \Delta)$ with $i = i_1, \ldots, i_r$ and the plus sign to the remaining terms. Then (71) is equivalent to

$$d_{i_1} + \cdots + d_{i_r} - r\Delta < 0, \qquad \text{or} \qquad \frac{d_{i_1} + \cdots + d_{i_r}}{r} < \Delta.$$

Thus, $\Sigma(d_i - \Delta)$ is among the l smallest of the $\Sigma \pm (d_i - \Delta)$ if and only if at least $2^n - l$ of the $M = 2^n - 1$ averages $(d_{i_1} + \cdots + d_{i_r})/r$ are $< \Delta$, i.e. if and only if $\delta_{(K)} < \Delta$, where $\delta_{(1)} < \cdots < \delta_{(M)}$ is the ordered set of averages $(d_{i_1} + \cdots + d_{i_r})/r$, $r = 1, \ldots, M$. This establishes $\delta_{(K)}$ as a lower confidence bound for Δ at confidence level $\gamma = K/2^n$. [Among all confidence sets that are unbiased in the model (47) with $m_i = n_i = 1$ and $c = n$, these bounds minimize the probability of falling below any value $\Delta' < \Delta$ for the normal model (53).]

By putting successively $K = 1, 2, \ldots, 2^n$, it is seen that the $M + 1$ intervals

$$(72) \qquad (-\infty, \delta_{(1)}), (\delta_{(1)}, \delta_{(2)}), \ldots, (\delta_{(M-1)}, \delta_{(M)}), (\delta_M, \infty)$$

each have probability $1/(M + 1) = 1/2^n$ of containing the unknown Δ. The two-sided confidence intervals $(\delta_{(K)}, \delta_{(2^n - K)})$ with $\gamma = (2^{n-1} - K)/2^{n-1}$ correspond to the two-sided version of the test (54) with error probability $(1 - \gamma)/2$ in each tail. A suitable subset of the points $\delta_{(1)}, \ldots, \delta_{(M)}$ constitutes a set of confidence points in the sense of Chapter 3, Section 5.

The inversion procedure for the two-group case is quite analogous. Let $(x_1, \ldots, x_m, y_1, \ldots, y_n)$ denote the m control and n treatment observations, and suppose without loss of generality that $m \leq n$. Then the hypothesis $\Delta = \Delta_0$ is accepted against $\Delta > \Delta_0$ if $\Sigma_{j=1}^n(y_j - \Delta_0)$ is among the l smallest of the $\binom{m+n}{n}$ sums obtained by replacing a subset of the $(y_j - \Delta_0)$'s with

x's. The inequality

$$\sum (y_j - \Delta_0) < (x_{i_1} + \cdots + x_{i_r}) + \left[y_{j_1} + \cdots + y_{j_{n-r}} - (n - r)\Delta \right],$$

with $(i_1, \ldots, i_r, j_1, \ldots, j_{n-r})$ a permutation of $(1, \ldots, n)$, is equivalent to $y_{i_1} + \cdots + y_{i_r} - r\Delta_0 < x_{i_1} + \cdots + x_{i_r}$, or

$$(73) \qquad\qquad \bar{y}_{i_1, \ldots, i_r} - \bar{x}_{i_1, \ldots, i_r} < \Delta_0.$$

Note that the number of such averages with $r \geq 1$ (i.e. omitting the empty set of subscripts) is equal to

$$\sum_{K=1}^{m} \binom{m}{K}\binom{n}{K} = \binom{m+n}{n} - 1 = M$$

(Problem 57). Thus, $H: \Delta = \Delta_0$ is accepted against $\Delta > \Delta_0$ at level $\alpha = 1 - l/(M + 1)$ if and only if at least K of the M differences (73) are less than Δ_0, and hence if and only if $\delta_{(K)} < \Delta_0$, where $\delta_{(1)} < \cdots < \delta_{(M)}$ denote the ordered set of differences (73). This establishes $\delta_{(K)}$ as a lower confidence bound for Δ with confidence coefficient $\gamma = 1 - \alpha$.

As in the paired comparisons case, it is seen that the intervals (72) each have probability $1/(M + 1)$ of containing Δ. Thus, two-sided confidence intervals and standard confidence points can be derived as before. For the generalization to stratified samples, see Problem 58.

Algorithms for computing the order statistics $\delta_{(1)}, \ldots, \delta_{(M)}$ in the paired-comparison and two-sample cases are discussed by Tritchler (1984). If M is too large for the computations to be practicable, reduced analyses based on either a fixed or random subset of the set of all $M + 1$ permutations are discussed, for example, by Gabriel and Hall (1983) and Vadiveloo (1983). [See also Problem 60(i).] Different such methods are compared by Forsythe and Hartigan (1970). For some generalizations, and relations to other subsampling plans, see Efron (1982, Chapter 9).

15. TESTING FOR INDEPENDENCE IN A BIVARIATE NORMAL DISTRIBUTION

So far, the methods of the present chapter have been illustrated mainly by the two-sample problem. As a further example, we shall now apply two of the formulations that have been discussed, the normal model of Section 3 and the nonparametric one of Section 10, to the hypothesis of independence in a bivariate distribution.

The probability density of a sample $(X_1, Y_1), \ldots, (X_n, Y_n)$ from a bivariate normal distribution is

$$(74) \quad \frac{1}{\left(2\pi\sigma\tau\sqrt{1-\rho^2}\right)^n} \exp\left[-\frac{1}{2(1-\rho^2)}\left(\frac{1}{\sigma^2}\sum(x_i - \xi)^2\right.\right.$$

$$\left.\left. -\frac{2\rho}{\sigma\tau}\sum(x_i - \xi)(y_i - \eta) + \frac{1}{\tau^2}\sum(y_i - \eta)^2\right)\right].$$

Here (ξ, σ^2) and (η, τ^2) are the mean and variance of X and Y respectively, and ρ is the correlation coefficient between X and Y. The hypotheses $\rho \leq \rho_0$ and $\rho = \rho_0$ for arbitrary ρ_0 cannot be treated by the methods of the present chapter, and will be taken up in Chapter 6. For the present, we shall consider only the hypothesis $\rho = 0$ that X and Y are independent, and the corresponding one-sided hypothesis $\rho \leq 0$.

The family of densities (74) is of the exponential form (1) with

$$U = \sum X_i Y_i, \quad T_1 = \sum X_i^2, \quad T_2 = \sum Y_i^2, \quad T_3 = \sum X_i, \quad T_4 = \sum Y_i$$

and

$$\theta = \frac{\rho}{\sigma\tau(1-\rho^2)}, \quad \vartheta_1 = \frac{-1}{2\sigma^2(1-\rho^2)}, \quad \vartheta_2 = \frac{-1}{2\tau^2(1-\rho^2)},$$

$$\vartheta_3 = \frac{1}{1-\rho^2}\left(\frac{\xi}{\sigma^2} - \frac{\eta\rho}{\sigma\tau}\right), \quad \vartheta_4 = \frac{1}{1-\rho^2}\left(\frac{\eta}{\tau^2} - \frac{\xi\rho}{\sigma\tau}\right).$$

The hypothesis $H: \rho \leq 0$ is equivalent to $\theta \leq 0$. Since the sample correlation coefficient

$$R = \frac{\sum(X_i - \bar{X})(Y_i - \bar{Y})}{\sqrt{\sum(X_i - \bar{X})^2 \sum(Y_i - \bar{Y})^2}}$$

is unchanged when the X_i and Y_i are replaced by $(X_i - \xi)/\sigma$ and $(Y_i - \eta)/\tau$, the distribution of R does not depend on ξ, η, σ, or τ, but only on ρ. For $\theta = 0$ it therefore does not depend on $\vartheta_1, \ldots, \vartheta_4$, and hence by Theorem 2, R is independent of $(T_1, \ldots, T_4)$ when $\theta = 0$. It follows from Theorem 1 that the UMP unbiased test of H rejects when

$$(75) \quad R \geq C_0,$$

or equivalently when

(76)
$$\frac{R}{\sqrt{(1 - R^2)/(n - 2)}} > K_0.$$

The statistic R is linear in U, and its distribution for $\rho = 0$ is symmetric about 0. The UMP unbiased test of the hypothesis $\rho = 0$ against the alternative $\rho \neq 0$ therefore rejects when

(77)
$$\frac{|R|}{\sqrt{(1 - R^2)/(n - 2)}} > K_1.$$

Since $\sqrt{n - 2}\, R / \sqrt{1 - R^2}$ has the t-distribution with $n - 2$ degrees of freedom when $\rho = 0$ (Problem 64), the constants K_0 and K_1 in the above tests are given by

(78)
$$\int_{K_0}^{\infty} t_{n-2}(y)\, dy = \alpha \quad \text{and} \quad \int_{K_1}^{\infty} t_{n-2}(y)\, dy = \frac{\alpha}{2}.$$

Since the distribution of R depends only on the correlation coefficient ρ, the same is true of the power of these tests.

Paralleling the work of Section 4, let us ask how sensitive the level of the test (76) is to the assumption of normality. Suppose that $(X_1, Y_1), \ldots,$ (X_n, Y_n) are a sample from some bivariate distribution F with finite second moment and correlation coefficient ρ. In the normal case, the condition $\rho = 0$ is equivalent to the independence of X and Y. This is not true in general, and it then becomes necessary to distinguish between

$$H_1 : X \text{ and } Y \text{ are independent}$$

and the broader hypothesis that X and Y are uncorrelated,

$$H_2 : \rho = 0.$$

Assuming H_1 to hold, consider the distribution of

$$\sqrt{n}\, R = \frac{\sqrt{n} \left[\dfrac{\sum X_i Y_i}{n} - \overline{XY} \right]}{\sqrt{\dfrac{\sum (X_i - \overline{X})^2}{n} \cdot \dfrac{\sum (Y_i - \overline{Y})^2}{n}}}.$$

Since the distribution of R is independent of $\xi = E(X_i)$ and $\eta = E(Y_i)$, suppose without loss of generality that $\xi = \eta = 0$. Then the limit distribution of $\sqrt{n}\,(\Sigma X_i Y_i/n)$ is normal with mean zero and variance

$$\text{Var}(X_i Y_i) = E\big(X_i^2\big)E\big(Y_i^2\big) = \sigma^2 \tau^2.$$

The term $(\sqrt{n}\,\overline{X})\overline{Y}$ tends to zero in probability, since $\sqrt{n}\,\overline{X}$ is bounded in probability and $\overline{Y}$ tends to zero in probability. Finally, the denominator tends in probability to $\sigma\tau$. It follows that $\sqrt{n}\,R$ tends in law to the standard normal distribution for all F with finite second moments. If $\alpha_n(F)$ is the rejection probability of the one- or two-sided test (76) or (77) when F is the true distribution, it follows that $\alpha_n(F)$ tends to the nominal level α as $n \to \infty$. For studies of how close $\alpha_n(F)$ is to α for different F and n, see for example Kowalski (1972) and Edgell and Noon (1984).

Consider now the distribution of $\sqrt{n}\,R$ under H_2. The limit argument is the same as under H_1 with the only difference that $\text{Var}(X_i Y_i)$ need no longer be equal to $\text{Var}\,X_i \cdot \text{Var}\,Y_i = \sigma^2 \tau^2$. The limit distribution of $\sqrt{n}\,R$ is therefore normal with mean zero and variance $\text{Var}(X_i Y_i)/[\text{Var}\,X_i \cdot \text{Var}\,Y_i]$, which can take on any value between 0 and ∞ (Problem 79). Even asymptotically, the size of the tests (76) and (77) is thus completely uncontrolled under H_2. [It can of course be brought under control by appropriate Studentization; see Problem 72 and the papers by Hsu (1949), Steiger and Hakstian (1982, 1983), and Beran and Srivastava (1985).]

Let us now return to H_1. Instead of relying on the robustness of R, one can obtain an exact level-α unbiased test of independence for a nonparametric model, in analogy to the permutation test of Section 10. For any bivariate distribution of (X, Y), let Y_x denote a random variable whose distribution is the conditional distribution of Y given x. We shall say that there is *positive regression dependence* between X and Y if for any $x < x'$ the variable $Y_{x'}$ is stochastically larger than Y_x. Generally speaking, larger values of Y will then correspond to larger values of X; this is the intuitive meaning of positive dependence. An example is furnished by any normal bivariate distribution with $\rho > 0$. (See Problem 68.) Regression dependence is a stronger requirement than positive quadrant dependence, which was defined in Chapter 4, Problem 19. However, both reflect the intuitive meaning that large (small) values of Y will tend to correspond to large (small) values of X.

As alternatives to H_1 consider positive regression dependence in a general bivariate distribution possessing a probability density with respect to Lebesgue measure. To see that unbiasedness implies similarity, let F_1, F_2 be any two univariate distributions with densities f_1, f_2 and consider the one-parameter family of distribution functions

$$(79) \quad F_1(x)F_2(y)\{1 + \Delta[1 - F_1(x)][1 - F_2(y)]\}, \qquad 0 \le \Delta \le 1.$$

This is positively regression dependent (Problem 69), and by letting $\Delta \to 0$ one sees that unbiasedness of ϕ against these distributions implies that the rejection probability is α when X and Y are independent, and hence that

$$\int \phi(x_1, \ldots, x_n; y_1, \ldots, y_n) f_1(x_1) \cdots f_1(x_n) f_2(y_1) \cdots f_2(y_n) \, dx \, dy = \alpha$$

for all probability densities f_1 and f_2. By Theorem 6 this in turn implies

$$\frac{1}{(n!)^2} \sum \phi(x_{i_1}, \ldots, x_{i_n}; y_{j_1}, \ldots, y_{j_n}) = \alpha.$$

Here the summation extends over the $(n!)^2$ points of the set $S(x, y)$, which is obtained from a fixed point (x, y) with $x = (x_1, \ldots, x_n)$, $y = (y_1, \ldots, y_n)$ by permuting the x-coordinates and the y-coordinates, each among themselves in all possible ways.

Among all tests satisfying this condition, the most powerful one against the normal alternatives (74) with $\rho > 0$ rejects for the k' largest values of (74) in each set $S(x, y)$, where $k'/(n!)^2 = \alpha$. Since $\Sigma x_i^2, \Sigma y_i^2, \Sigma x_i, \Sigma y_i$ are all constant on $S(x, y)$, the test equivalently rejects for the k' largest values of $\Sigma x_i y_i$ in each $S(x, y)$.

Of the $(n!)^2$ values that the statistic $\Sigma X_i Y_i$ takes on over $S(x, y)$, only $n!$ are distinct, since the statistic remains unchanged if the X's and Y's are subjected to the same permutation. A simpler form of the test is therefore obtained, for example by rejecting H for the k largest values of $\Sigma x_{(i)} y_{j_i}$ of each set $S(x, y)$, where $x_{(1)} < \cdots < x_{(n)}$ and $k/n! = \alpha$. The test can be shown to be unbiased against all alternatives with positive regression dependence. (See Problem 48 of Chapter 6.)

In order to obtain a comparison of the permutation test with the standard normal test based on the sample correlation coefficient R, let $T(X, Y)$ denote the set of ordered X's and Y's,

$$T(X, Y) = (X_{(1)}, \ldots, X_{(n)}; Y_{(1)}, \ldots, Y_{(n)}).$$

The rejection region of the permutation test can then be written as

$$\sum X_i Y_i > C[T(X, Y)],$$

or equivalently as

$$R > K[T(X, Y)].$$

It again turns out* that the difference between $K[T(X, Y)]$ and the cutoff point C_0 of the corresponding normal test (75) tends to zero, and that the two tests become equivalent in the limit as n tends to infinity. Sufficient conditions for this are that $\sigma_X^2, \sigma_Y^2 > 0$ and $E(|X|^3), E(|Y|^3) < \infty$. For large n, the standard normal test (75) therefore serves as an approximation for the permutation test, which is impractical except for small sample sizes.

16. PROBLEMS

Section 2

1. Let $X_1, \ldots, X_n$ be a sample from $N(\xi, \sigma^2)$. The power of Student's t-test is an increasing function of ξ/σ in the one-sided case $H: \xi \leq 0$, $K: \xi > 0$, and of $|\xi|/\sigma$ in the two-sided case $H: \xi = 0$, $K: \xi \neq 0$.
   [If

   $$S = \sqrt{\frac{1}{n-1} \sum (X_i - \bar{X})^2},$$

   the power in the two-sided case is given by

   $$1 - P\left\{ -\frac{CS}{\sigma} - \frac{\sqrt{n}\,\xi}{\sigma} \leq \frac{\sqrt{n}\,(\bar{X} - \xi)}{\sigma} \leq \frac{CS}{\sigma} - \frac{\sqrt{n}\,\xi}{\sigma} \right\}$$

   and the result follows from the fact that it holds conditionally for each fixed value of S/σ.]

2. In the situation of the previous problem there exists no test for testing $H: \xi = 0$ at level α, which for all σ has power $\geq \beta > \alpha$ against the alternatives (ξ, σ) with $\xi = \xi_1 > 0$.
   [Let $\beta(\xi_1, \sigma)$ be the power of any level α test of H, and let $\beta(\sigma)$ denote the power of the most powerful test for testing $\xi = 0$ against $\xi = \xi_1$ when σ is known. Then $\inf_\sigma \beta(\xi_1, \sigma) \leq \inf_\sigma \beta(\sigma) = \alpha$.]

3. (i) Let Z and V be independently distributed as $N(\delta, 1)$ and χ^2 with f degrees of freedom respectively. Then the ratio $Z \div \sqrt{V/f}$ has the noncentral t-distribution with f degrees of freedom and noncentrality

*For a proof see Fraser (1957).

parameter δ, the probability density of which is*

$$(80) \quad p_\delta(t) = \frac{1}{2^{\frac{1}{2}(f-1)}\Gamma(\frac{1}{2}f)\sqrt{\pi f}} \int_0^\infty y^{\frac{1}{2}(f-1)}$$

$$\times \exp(-\tfrac{1}{2}y)\exp\left[-\frac{1}{2}\left(t\sqrt{\frac{y}{f}} - \delta\right)^2\right] dy$$

or equivalently

$$p_\delta(t) = \frac{1}{2^{\frac{1}{2}(f-1)}\Gamma(\frac{1}{2}f)\sqrt{\pi f}} \exp\left(-\frac{1}{2}\frac{f\delta^2}{f+t^2}\right)$$

$$\times \left(\frac{f}{f+t^2}\right)^{\frac{1}{2}(f+1)} \int_0^\infty v^f \exp\left[-\frac{1}{2}\left(v - \frac{\delta t}{\sqrt{f+t^2}}\right)^2\right] dv.$$

Another form is obtained by making the substitution $w = t\sqrt{y}/\sqrt{f}$ in (80).

(ii) If $X_1,\ldots, X_n$ are independently distributed as $N(\xi, \sigma^2)$, then $\sqrt{n}\,\overline{X} \div \sqrt{\Sigma(X_i - \overline{X})^2/(n-1)}$ has the noncentral t-distribution with $n-1$ degrees of freedom and noncentrality parameter $\delta = \sqrt{n}\,\xi/\sigma$.

[(i): The first expression is obtained from the joint density of Z and V by transforming to $t = z \div \sqrt{v/f}$ and v.]

4. Let $X_1,\ldots, X_n$ be a sample from $N(\xi, \sigma^2)$. Denote the power of the one-sided t-test of $H: \xi \le 0$ against the alternative ξ/σ by $\beta(\xi/\sigma)$, and by $\beta^*(\xi/\sigma)$ the power of the test appropriate when σ is known. Determine $\beta(\xi/\sigma)$ for $n = 5, 10, 15$, $\alpha = .05$, $\xi/\sigma = 0.7, 0.8, 0.9, 1.0, 1.1, 1.2$, and in each case compare it with $\beta^*(\xi/\sigma)$. Do the same for the two-sided case.

5. Let $Z_1,\ldots, Z_n$ be independently normally distributed with common variance σ^2 and means $E(Z_i) = \zeta_i (i = 1,\ldots, s)$, $E(Z_i) = 0$ $(i = s+1,\ldots, n)$. There exist UMP unbiased tests for testing $\zeta_1 \le \zeta_1^0$ and $\zeta_1 = \zeta_1^0$ given by the rejection regions

$$\frac{Z_1 - \zeta_1^0}{\sqrt{\displaystyle\sum_{i=s+1}^{n} Z_i^2/(n-s)}} > C_0 \quad \text{and} \quad \frac{|Z_1 - \zeta_1^0|}{\sqrt{\displaystyle\sum_{i=s+1}^{n} Z_i^2/(n-s)}} > C.$$

When $\zeta_1 = \zeta_1^0$, the test statistic has the t-distribution with $n-s$ degrees of freedom.

*A systematic account of this distribution can be found in Johnson and Kotz (1970, Vol. 2, Chapter 31) and in Owen (1985).

6. Let $X_1, \ldots, X_n$ be independently normally distributed with common variance σ^2 and means $\xi_1, \ldots, \xi_n$, and let $Z_i = \sum_{j=1}^{n} a_{ij} X_j$ be an orthogonal transformation (that is, $\sum_{i=1}^{n} a_{ij} a_{ik} = 1$ or 0 as $j = k$ or $j \neq k$). The Z's are normally distributed with common variance σ^2 and means $\zeta_i = \sum a_{ij} \xi_j$.
   [The density of the Z's is obtained from that of the X's by substituting $x_i = \sum b_{ij} z_j$, where (b_{ij}) is the inverse of the matrix (a_{ij}), and multiplying by the Jacobian, which is 1.]

7. If $X_1, \ldots, X_n$ is a sample from $N(\xi, \sigma^2)$, the UMP unbiased tests of $\xi \leq 0$ and $\xi = 0$ can be obtained from Problems 5 and 6 by making an orthogonal transformation to variables $Z_1, \ldots, Z_n$ such that $Z_1 = \sqrt{n}\, \overline{X}$.
   [Then

$$\sum_{i=2}^{n} Z_i^2 = \sum_{i=1}^{n} Z_i^2 - Z_1^2 = \sum_{i=1}^{n} X_i^2 - n\overline{X}^2 = \sum_{i=1}^{n} (X_i - \overline{X})^2.]$$

8. Let $X_1, X_2, \ldots$ be a sequence of independent variables distributed as $N(\xi, \sigma^2)$, and let $Y_n = [nX_{n+1} - (X_1 + \cdots + X_n)]/\sqrt{n(n+1)}$. Then the variables $Y_1, Y_2, \ldots$ are independently distributed as $N(0, \sigma^2)$.

Section 3

9. Let $X_1, \ldots, X_n$ and $Y_1, \ldots, Y_n$ be independent samples from $N(\xi, \sigma^2)$ and $N(\eta, \tau^2)$ respectively. Determine the sample size necessary to obtain power $\geq \beta$ against the alternatives $\tau/\sigma > \Delta$ when $\alpha = .05$, $\beta = .9$, $\Delta = 1.5, 2, 3$, and the hypothesis being tested is $H : \tau/\sigma \leq 1$.

10. If $m = n$, the acceptance region (23) can be written as

$$\max\left(\frac{S_Y^2}{\Delta_0 S_x^2}, \frac{\Delta_0 S_X^2}{S_Y^2} \right) \leq \frac{1 - C}{C},$$

where $S_X^2 = \Sigma(X_i - \overline{X})^2$, $S_Y^2 = \Sigma(Y_i - \overline{Y})^2$ and where C is determined by

$$\int_0^C B_{n-1, n-1}(w)\, dw = \frac{\alpha}{2}.$$

11. Let $X_1, \ldots, X_m$ and $Y_1, \ldots, Y_n$ be samples from $N(\xi, \sigma^2)$ and $N(\eta, \sigma^2)$. The UMP unbiased test for testing $\eta - \xi = 0$ can be obtained through Problems 5 and 6 by making an orthogonal transformation from $(X_1, \ldots, X_m, Y_1, \ldots, Y_n)$ to $(Z_1, \ldots, Z_{m+n})$ such that $Z_1 = (\overline{Y} - \overline{X})/\sqrt{(1/m) + (1/n)}$, $Z_2 = (\Sigma X_i + \Sigma Y_i)/\sqrt{m + n}$.

12. *Exponential densities.* Let $X_1, \ldots, X_n$ be a sample from a distribution with exponential density $a^{-1} e^{-(x-b)/a}$ for $x \geq b$.

(i)   For testing $a = 1$ there exists a UMP unbiased test given by the acceptance region

$$C_1 \leq 2\sum [x_i - \min(x_1, \ldots, x_n)] \leq C_2,$$

where the test statistic has a χ^2-distribution with $2n - 2$ degrees of freedom when $a = 1$, and C_1, C_2 are determined by

$$\int_{C_1}^{C_2} \chi^2_{2n-2}(y)\, dy = \int_{C_1}^{C_2} \chi^2_{2n}(y)\, dy = 1 - \alpha.$$

(ii) For testing $b = 0$ there exists a UMP unbiased test given by the acceptance region

$$0 \leq \frac{n\min(x_1, \ldots, x_n)}{\sum [x_i - \min(x_1, \ldots, x_n)]} \leq C.$$

When $b = 0$, the test statistic has probability density

$$p(u) = \frac{n-1}{(1+u)^n}, \qquad u \geq 0.$$

[These distributions for varying b do not constitute an exponential family, and Theorem 3 of Chapter 4 is therefore not directly applicable.

(i): One can restrict attention to the ordered variables $X_{(1)} < \cdots < X_{(n)}$, since these are sufficient for a and b, and transform to new variables $Z_1 = nX_{(1)}$, $Z_i = (n - i + 1)[X_{(i)} - X_{(i-1)}]$ for $i = 2, \ldots, n$, as in Problem 14 of Chapter 2. When $a = 1$, Z_1 is a complete sufficient statistic for b, and the test is therefore obtained by considering the conditional problem given z_1. Since $\sum_{i=2}^{n} Z_i$ is independent of Z_1, the conditional UMP unbiased test has the acceptance region $C_1 \leq \sum_{i=2}^{n} Z_i \leq C_2$ for each z_1, and the result follows.

(ii): When $b = 0$, $\sum_{i=1}^{n} Z_i$ is a complete sufficient statistic for a, and the test is therefore obtained by considering the conditional problem given $\sum_{i=1}^{n} z_i$. The remainder of the argument uses the fact that $Z_1 / \sum_{i=1}^{n} Z_i$ is independent of $\sum_{i=1}^{n} Z_i$ when $b = 0$, and otherwise is similar to that used to prove Theorem 1.]

13.   Extend the results of the preceding problem to the case, considered in Problem 10, Chapter 3, that observation is continued only until $X_{(1)}, \ldots, X_{(r)}$ have been observed.

Section 4

14.   Corollary 2 remains valid if c_n is replaced by a sequence of random variables C_n tending to c in probability.

15.   (i)   Let $X_1, \ldots, X_n$ be a sample from $N(\xi, \sigma^2)$. The power of the one-sided one-sample t-test against a sequence of alternatives (ξ_n, σ) for which $\sqrt{n}\,\xi_n/\sigma \to \delta$ tends to $\Phi(\delta - u_\alpha)$.

(ii)   The result of (i) remains valid if $X_1, \ldots, X_n$ are a sample from any distribution with mean ξ and finite variance σ^2.

16.   Generalize Problem 15(i) and (ii) to the two-sample t-test.

17.   (i)   Given ρ, find the smallest and largest value of (31) as σ^2/τ^2 varies from 0 to ∞.

      (ii)   For nominal level $\alpha = .05$ and $\rho = 1, .2, .3, .4$, determine the smallest and the largest asymptotic level of the t-test as σ^2/τ^2 varies from 0 to ∞.

Section 5

18.   *The Chebyshev inequality.* For any random variable Y and constants $a > 0$ and c,

$$E(Y - c)^2 \geq a^2 P(|Y - c| \geq a).$$

19.   If Y_n is a sequence of random variables and c a constant such that $E(Y_n - c)^2 \to 0$, then for any $a > 0$,

$$P(|Y_n - c| \geq a) \to 0,$$

that is, Y_n *tends to c in probability.*

20.   Verify the formula for $\mathrm{Var}(\overline{X})$ in Model A.

21.   In Model A, suppose that the number of observations in group i is n_i. If $n_i \leq M$ and $s \to \infty$, show that the assumptions of Lemma 1 are satisfied and determine γ.

22.   Show that the conditions of Lemma 1 are satisfied and γ has the stated value: (i) in Model B; (ii) in Model C.

23.   Determine the maximum asymptotic level of the one-sided t-test when $\alpha = .05$ and $m = 2, 4, 6$: (i) in Model A; (ii) in Model B.

24.   Let $X_i = \xi + U_i$, and suppose that the joint density of the U's is *spherically symmetric*, that is, a function of ΣU_i^2 only,

$$f(u_1, \ldots, u_n) = q\left(\sum u_i^2\right).$$

Then the null distribution of the one-sample t-statistic is independent of q and hence the same as in the normal case, namely Student's t with $n - 1$ degrees of freedom.
[Write t as

$$\frac{\sqrt{n}\,\overline{X} \Big/ \sqrt{\Sigma X_j^2}}{\sqrt{\Sigma(X_i - \overline{X})^2 \Big/ (n-1)\Sigma X_j^2}},$$

and use the fact that when $\xi = 0$, the density of $X_1, \ldots, X_n$ is constant over the spheres $\Sigma x_i^2 = c$ and hence the conditional distribution of the variables $X_i / \sqrt{\Sigma X_j^2}$ given $\Sigma X_j^2 = c$ is uniform over the conditioning sphere and hence independent of q.]

Note. This model represents one departure from the normal-theory assumption, which does not affect the level of the test. The effect of a much weaker symmetry condition more likely to arise in practice is investigated by Efron (1969).

Section 6

25. On the basis of a sample $X = (X_1, \ldots, X_n)$ of fixed size from $N(\xi, \sigma^2)$ there do not exist confidence intervals for ξ with positive confidence coefficient and of bounded length.
    [Consider any family of confidence intervals $\delta(X) \pm L/2$ of constant length L. Let $\xi_1, \ldots, \xi_{2N}$ be such that $|\xi_i - \xi_j| > L$ whenever $i \neq j$. Then the sets $S_i = \{x : |\delta(x) - \xi_i| \leq L/2\}$ $(i = 1, \ldots, 2N)$ are mutually exclusive. Also, there exists $\sigma_0 > 0$ such that

$$\left| P_{\xi_i, \sigma} \{ X \in S_i \} - P_{\xi_1, \sigma} \{ X \in S_i \} \right| \leq \frac{1}{2N} \qquad \text{for} \quad \sigma > \sigma_0,$$

as is seen by transforming to new variables $Y_j = (X_j - \xi_1)/\sigma$ and applying Lemmas 2 and 4 of the Appendix. Since $\min_i P_{\xi_1, \sigma} \{ X \in S_i \} \leq 1/2N$, it follows for $\sigma > \sigma_0$ that $\min_i P_{\xi_i, \sigma} \{ X \in S_i \} \leq 1/N$, and hence that

$$\inf_{\xi, \sigma} P_{\xi, \sigma} \left\{ |\delta(X) - \xi| \leq \frac{L}{2} \right\} \leq \frac{1}{N}.$$

The confidence coefficient associated with the intervals $\delta(X) \pm L/2$ is therefore zero, and the same must be true a fortiori of any set of confidence intervals of length $\leq L$.]

26. Stein's two-stage procedure.

    (i) If mS^2/σ^2 has a $\chi^2 =$ distribution with m degrees of freedom, and if the conditional distribution of Y given $S = s$ is $N(0, \sigma^2/S^2)$, then Y has Student's t-distribution with m degrees of freedom.

    (ii) Let $X_1, X_2, \ldots$ be independently distributed as $N(\xi, \sigma^2)$. Let $\bar{X}_0 = \Sigma_{i=1}^{n_0} X_i/n_0$, $S^2 = \Sigma_{i=1}^{n_0} (X_i - \bar{X}_0)^2/(n_0 - 1)$, and let $a_1 = \cdots = a_{n_0} = a$, $a_{n_0+1} = \cdots = a_n = b$, and $n \geq n_0$ be measurable functions of S. Then

$$Y = \frac{\displaystyle\sum_{i=1}^{n} a_i (X_i - \xi)}{\sqrt{S^2 \displaystyle\sum_{i=1}^{n} a_i^2}}$$

has Student's distribution with $n_0 - 1$ degrees of freedom.

(iii)   Consider a two-stage sampling scheme Π_1, in which S^2 is computed from an initial sample of size n_0, and then $n - n_0$ additional observations are taken. The size of the second sample is such that

$$n = \max\left\{ n_0 + 1, \left[\frac{S^2}{c}\right] + 1\right\}$$

where c is any given constant and where $[y]$ denotes the largest integer $\leq y$. There then exist numbers $a_1, \ldots, a_n$ such that $a_1 = \cdots = a_{n_0}, a_{n_0+1} = \cdots = a_n, \Sigma_{i=1}^n a_i = 1, \Sigma_{i=1}^n a_i^2 = c/S^2$. It follows from (ii) that $\Sigma_{i=1}^n a_i(X_i - \xi)/\sqrt{c}$ has Student's t-distribution with $n_0 - 1$ degrees of freedom.

(iv)   The following sampling scheme Π_2, which does not require that the second sample contain at least one observation, is slightly more efficient than Π_1 for the applications to be made in Problems 27 and 28. Let n_0, S^2, and c be defined as before; let

$$n = \max\left\{ n_0, \left[\frac{S^2}{c}\right] + 1\right\},$$

$a_i = 1/n$ $(i = 1, \ldots, n)$, and $\overline{X} = \Sigma_{i=1}^n a_i X_i$. Then $\sqrt{n}\,(\overline{X} - \xi)/S$ has again the t-distribution with $n_0 - 1$ degrees of freedom.

[(ii): Given $S = s$, the quantities a, b, and n are constants, $\Sigma_{i=1}^{n_0} a_i(X_i - \xi) = n_0 a(\overline{X}_0 - \xi)$ is distributed as $N(0, n_0 a^2 \sigma^2)$, and the numerator of Y is therefore normally distributed with zero mean and variance $\sigma^2 \Sigma_{i=1}^n a_i^2$. The result now follows from (i).]

27.   *Confidence intervals of fixed length for a normal mean.*

(i)   In the two-stage procedure Π_1 defined in part (iii) of the preceding problem, let the number c be determined for any given $L > 0$ and $0 < \gamma < 1$ by

$$\int_{-L/2\sqrt{c}}^{L/2\sqrt{c}} t_{n_0-1}(y)\,dy = \gamma,$$

where t_{n_0-1} denotes the density of the t-distribution with $n_0 - 1$ degrees of freedom. Then the intervals $\Sigma_{i=1}^n a_i X_i \pm L/2$ are confidence intervals for ξ of length L and with confidence coefficient γ.

(ii)   Let c be defined as in (i), and let the sampling procedure be Π_2 as defined in part (iv) of Problem 26. The intervals $\overline{X} \pm L/2$ are then confidence intervals of length L for ξ with confidence coefficient $\geq \gamma$, while the expected number of observations required is slightly lower than under Π_1.

[(i): The probability that the intervals cover ξ equals

$$P_{\xi,\sigma}\left\{-\frac{L}{2\sqrt{c}} \le \frac{\sum_{i=1}^{n} a_i(X_i - \xi)}{\sqrt{c}} \le \frac{L}{2\sqrt{c}}\right\} = \gamma.$$

(ii): The probability that the intervals cover ξ equals

$$P_{\xi,\sigma}\left\{\frac{\sqrt{n}\,|\overline{X} - \xi|}{S} \le \frac{\sqrt{n}\,L}{2S}\right\} \ge P_{\xi,\sigma}\left\{\frac{\sqrt{n}\,|\overline{X} - \xi|}{S} \le \frac{L}{2\sqrt{c}}\right\} = \gamma.]$$

28. *Two-stage t-tests with power independent of σ.*

   (i) For the procedure Π_1 with any given c, let C be defined by

$$\int_{C}^{\infty} t_{n_0-1}(y)\,dy = \alpha.$$

   Then the rejection region $(\sum_{i=1}^{n} a_i X_i - \xi_0)/\sqrt{c} > C$ defines a level-α test of $H: \xi \le \xi_0$ with strictly increasing power function $\beta_c(\xi)$ depending only on ξ.

   (ii) Given any alternative ξ_1 and any $\alpha < \beta < 1$, the number c can be chosen so that $\beta_c(\xi_1) = \beta$.

   (iii) The test with rejection region $\sqrt{n}(\overline{X} - \xi_0)/S > C$ based on Π_2 and the same c as in (i) is a level-α test of H which is uniformly more powerful than the test given in (i).

   (iv) Extend parts (i)–(iii) to the problem of testing $\xi = \xi_0$ against $\xi \ne \xi_0$.

[(i) and (ii): The power of the test is

$$\beta_c(\xi) = \int_{C-(\xi-\xi_0)/\sqrt{c}}^{\infty} t_{n_0-1}(y)\,dy.$$

(iii): This follows from the inequality $\sqrt{n}|\xi - \xi_0|/S \ge |\xi - \xi_0|/\sqrt{c}$.]

29. Let $S(x)$ be a family of confidence sets for a real-valued parameter θ, and let $\mu[S(x)]$ denote its Lebesgue measure. Then for every fixed distribution Q of X (and hence in particular for $Q = P_{\theta_0}$ where θ_0 is the true value of θ)

$$E_Q\{\mu[S(X)]\} = \int_{\theta \ne \theta_0} Q\{\theta \in S(X)\}\,d\theta$$

provided the necessary measurability conditions hold.

[Write the expectation on the left side as a double integral, apply Fubini's theorem, and note that the integral on the right side is unchanged if the point $\theta = \theta_0$ is added to the region of integration.]

30. Use the preceding problem to show that uniformly most accurate confidence sets also uniformly minimize the expected Lebesgue measure (length in the case of intervals) of the confidence sets.*

Section 7

31. Let $X_1, \ldots, X_n$ be distributed as in Problem 12. Then the most accurate unbiased confidence intervals for the scale parameter a are

$$\frac{2}{C_2} \sum [x_i - \min(x_1, \ldots, x_n)] \le a \le \frac{2}{C_1} \sum [x_i - \min(x_1, \ldots, x_n)].$$

32. Most accurate unbiased confidence intervals exist in the following situations:

(i) If X, Y are independent with binomial distributions $b(p_1, m)$ and $b(p_2, n)$, for the parameter $p_1 q_2 / p_2 q_1$.

(ii) In a 2×2 table, for the parameter Δ of Chapter 4, Section 6.

Section 8

33. (i) Under the assumptions made at the beginning of Section 8, the UMP unbiased test of $H: \rho = \rho_0$ is given by (45).

(ii) Let $(\underline{\rho}, \bar{\rho})$ be the associated most accurate unbiased confidence intervals for $\rho = a\gamma + b\delta$, where $\underline{\rho} = \underline{\rho}(a, b)$, $\bar{\rho} = \bar{\rho}(a, b)$. Then if f_1 and f_2 are increasing functions, the expected value of $f_1(|\bar{\rho} - \rho|) + f_2(|\rho - \underline{\rho}|)$ is an increasing function of $a^2/n + b^2$.

[(i): Make any orthogonal transformation from $y_1, \ldots, y_n$ to new variables $z_1, \ldots, z_n$ such that $z_1 = \Sigma_i [bv_i + (a/n)] y_i / \sqrt{(a^2/n) + b^2}$, $z_2 = \Sigma_i (av_i - b) y_i / \sqrt{a^2 + nb^2}$, and apply Problems 5 and 6.
(ii): If $a_1^2/n + b_1^2 < a_2^2/n + b_2^2$, the random variable $|\bar{\rho}(a_2, b_2) - \rho|$ is stochastically larger than $|\bar{\rho}(a_1, b_1) - \rho|$, and analogously for $\underline{\rho}$.]

Section 9

34. Verify the posterior distribution of Θ given x in Example 12.

35. If $X_1, \ldots, X_n$ are independent $N(\theta, 1)$ and θ has the improper prior $\pi(\theta) \equiv 1$, determine the posterior distribution of θ given the X's.

36. Verify the posterior distribution of p given x in Example 13.

*For the corresponding result concerning one-sided confidence bounds, see Madansky (1962).

37. In Example 14, verify the marginal posterior distribution of ξ given x.

38. In Example 15, show that

    (i)   the posterior density $\pi(\sigma|x)$ is of type (c) of Example 13;
    (ii) for sufficiently large r, the posterior density of σ^r given x is no longer of type (c).

39. If X is normal $N(\theta,1)$ and θ has a Cauchy density $b/\{\pi[b^2 + (\theta - \mu)^2]\}$, determine the possible shapes of the HPD regions for varying μ and b.

40. Let $\theta = (\theta_1,\ldots,\theta_s)$ with θ_i real-valued, X have density $p_\theta(x)$, and Θ a prior density $\pi(\theta)$. Then the $100\gamma\%$ HPD region is the $100\gamma\%$ credible region R that has minimum volume.
    [Apply the Neyman–Pearson fundamental lemma to the problem of minimizing the volume of R.]

41. Let $X_1,\ldots,X_m$ and $Y_1,\ldots,Y_n$ be independently distributed as $N(\xi,\sigma^2)$ and $N(\eta,\sigma^2)$ respectively, and let (ξ,η,σ) have the joint improper prior density given by

$$\pi(\xi,\eta,\sigma)\, d\xi\, d\eta\, d\sigma = d\xi\, d\eta \cdot \frac{1}{\sigma}\, d\sigma \qquad \text{for all} \quad -\infty < \xi,\eta < \infty, \quad 0 < \sigma.$$

Under these assumptions, extend the results of Examples 14 and 15 to inferences concerning (i) $\eta - \xi$ and (ii) σ.

42. Let $X_1,\ldots,X_m$ and $Y_1,\ldots,Y_n$ be independently distributed as $N(\xi,\sigma^2)$ and $N(\eta,\tau^2)$, respectively and let (ξ,η,σ,τ) have the joint improper prior density $\pi(\xi,\eta,\sigma,\tau)\, d\xi\, d\eta\, d\sigma\, d\tau = d\xi\, d\eta\, (1/\sigma)\, d\sigma\, (1/\tau)\, d\tau$. Extend the result of Example 15 to inferences concerning τ^2/σ^2.

    Note. The posterior distribution of $\eta - \xi$ in this case is the so-called Behrens–Fisher distribution. The credible regions for $\eta - \xi$ obtained from this distribution do not correspond to confidence intervals with fixed coverage probability, and the associated tests of $H: \eta = \xi$ thus do not have fixed size (which instead depends on τ/σ). From numerical evidence [see Robinson (1976) for a summary of his and earlier results] it appears that the confidence intervals are conservative, that is, the actual coverage probability always exceeds the nominal one.

43. Let $T_1,\ldots,T_{s-1}$ have the multinomial distribution (34) of Chapter 2, and suppose that $(p_1,\ldots,p_{s-1})$ has the Dirichlet prior density $D(a_1,\ldots,a_s)$ with density proportional to $p_1^{a_1-1} \ldots p_s^{a_s-1}$, where $p_s = 1 - (p_1 + \cdots + p_{s-1})$. Determine the posterior distribution of $(p_1,\ldots,p_{s-1})$ given the T's.

Section 10

44. Prove Theorem 6 for arbitrary values of c.

Section 11

45. If $c = 1$, $m = n = 4$, $\alpha = .1$ and the ordered coordinates $z_{(1)}, \ldots, z_{(N)}$ of a point z are $1.97, 2.19, 2.61, 2.79, 2.88, 3.02, 3.28, 3.41$, determine the points of $S(z)$ belonging to the rejection region (54).

46. *Confidence intervals for a shift.*

   (i) Let $X_1, \ldots, X_m$; $Y_1, \ldots, Y_n$ be independently distributed according to continuous distributions $F(x)$ and $G(y) = F(y - \Delta)$ respectively. Without any further assumptions concerning F, confidence intervals for Δ can be obtained from permutation tests of the hypotheses $H(\Delta_0)$: $\Delta = \Delta_0$. Specifically, consider the point $(z_1, \ldots, z_{m+n}) = (x_1, \ldots, x_m, y_1 - \Delta, \ldots, y_n - \Delta)$ and the $\binom{m+n}{m}$ permutations $i_1 < \cdots < i_m$; $i_{m+1} < \cdots < i_{m+n}$ of the integers $1, \ldots, m + n$. Suppose that the hypothesis $H(\Delta)$ is accepted for the k of these permutations which lead to the smallest values of

$$\left| \sum_{j=m+1}^{m+n} z_{i_j}/n - \sum_{j=1}^{m} z_{i_j}/m \right|$$

   where

$$k = (1 - \alpha)\binom{m+n}{m}.$$

   Then the totality of values Δ for which $H(\Delta)$ is accepted constitute an interval, and these intervals are confidence intervals for Δ at confidence level $1 - \alpha$.

   (ii) Let $Z_1, \ldots, Z_N$ be independently distributed, symmetric about θ, with distribution $F(z - \theta)$, where $F(z)$ is continuous and symmetric about 0. Without any further assumptions about F, confidence intervals for θ can be obtained by considering the 2^N points $Z'_1, \ldots, Z'_N$, where $Z'_i = \pm(Z_i - \theta_0)$, and accepting $H(\theta_0)$: $\theta = \theta_0$ for the k of these points which lead to the smallest values of $\Sigma|Z'_i|$, where $k = (1 - \alpha)2^N$.

   [(i): A point is in the acceptance region for $H(\Delta)$ if

$$\left| \frac{\Sigma(y_j - \Delta)}{n} - \frac{\Sigma x_i}{m} \right| = |\bar{y} - \bar{x} - \Delta|$$

   is exceeded by at least $\binom{m+n}{m} - k$ of the quantities $|\bar{y}' - \bar{x}' - \gamma\Delta|$, where $(x'_1, \ldots, x'_m, y'_1, \ldots, y'_n)$ is a permutation of $(x_1, \ldots, x_m, y_1, \ldots, y_n)$, the quantity γ is determined by this permutation, and $|\gamma| \leq 1$. The desired result now follows from the following facts (for an alternative proof, see Section 14): (a) The set of Δ's for which $(\bar{y} - \bar{x} - \Delta)^2 \leq (\bar{y}' - \bar{x}' - \gamma\Delta)^2$ is, with probability one, an interval containing $\bar{y} - \bar{x}$. (b) The set of Δ's for which $(\bar{y} - \bar{x} - \Delta)^2$

is exceeded by a particular set of at least $\binom{m+n}{m} - k$ of the quantities $(\bar{y}' - \bar{x}' - \gamma\Delta)^2$ is the intersection of the corresponding intervals (a) and hence is an interval containing $\bar{y} - \bar{x}$. (c) The set of Δ's of interest is the union of the intervals (b) and, since they have a nonempty intersection, also an interval.]

Section 12

47. In the matched-pairs experiment for testing the effect of a treatment, suppose that only the differences $Z_i = Y_i - X_i$ are observable. The Z's are assumed to be a sample from an unknown continuous distribution, which under the hypothesis of no treatment effect is symmetric with respect to the origin. Under the alternatives it is symmetric with respect to a point $\zeta > 0$. Determine the test which among all unbiased tests maximizes the power against the alternatives that the Z's are a sample from $N(\zeta, \sigma^2)$ with $\zeta > 0$.
[Under the hypothesis, the set of statistics $(\Sigma_{i=1}^n Z_i^2, \ldots, \Sigma_{i=1}^n Z_i^{2n})$ is sufficient; that it is complete is shown as the corresponding result in Theorem 6. The remainder of the argument follows the lines of Section 11.]

48. (i) If $X_1, \ldots, X_n; Y_1, \ldots, Y_n$ are independent normal variables with common variance σ^2 and means $E(X_i) = \xi_i$, $E(Y_i) = \xi_i + \Delta$, the UMP unbiased test of $\Delta = 0$ against $\Delta > 0$ is given by (59).

    (ii) Determine the most accurate unbiased confidence intervals for Δ.

    [(i): The structure of the problem becomes clear if one makes the orthogonal transformation $X_i' = (Y_i - X_i)/\sqrt{2}$, $Y_i' = (X_i + Y_i)/\sqrt{2}$.]

49. *Comparison of two designs.* Under the assumptions made at the beginning of Section 12, one has the following comparison of the methods of complete randomization and matched pairs. The unit effects and experimental effects U_i and V_i are independently normally distributed with variances σ_1^2, σ^2 and means $E(U_i) = \mu$ and $E(V_i) = \xi$ or η as V_i corresponds to a control or treatment. With complete randomization, the observations are $X_i = U_i + V_i$ $(i = 1, \ldots, n)$ for the controls and $Y_i = U_{n+i} + V_{n+i}$ $(i = 1, \ldots, n)$ for the treated cases, with $E(X_i) = \mu + \xi$, $E(Y_i) = \mu + \eta$. For the matched pairs, if the matching is assumed to be perfect, the X's are as before, but $Y_i = U_i + V_{n+i}$. UMP unbiased tests are given by (27) for complete randomization and by (59) for matched pairs. The distribution of the test statistic under an alternative $\Delta = \eta - \xi$ is the noncentral t-distribution with noncentrality parameter $\sqrt{n}\,\Delta/\sqrt{2(\sigma^2 + \sigma_1^2)}$ and $2n - 2$ degrees of freedom in the first case, and with noncentrality parameter $\sqrt{n}\,\Delta/\sqrt{2}\,\sigma$ and $n - 1$ degrees of freedom in the second. Thus the method of matched pairs has the disadvantage of a smaller number of degrees of freedom and the advantage of a larger noncentrality parameter. For $\alpha = .05$ and $\Delta = 4$, compare the power of the two methods as a function of n when $\sigma_1 = 1$, $\sigma = 2$ and when $\sigma_1 = 2$, $\sigma = 1$.

50. *Continuation.* An alternative comparison of the two designs is obtained by considering the expected length of the most accurate unbiased confidence

intervals for $\Delta = \eta - \xi$ in each case. Carry this out for varying n and confidence coefficient $1 - \alpha = .95$ when $\sigma_1 = 1$, $\sigma = 2$ and when $\sigma_1 = 2$, $\sigma = 1$.

Section 13

51. Suppose that a critical function ϕ_0 satisfies (65) but not (67), and let $\alpha < \frac{1}{2}$. Then the following construction provides a measurable critical function ϕ satisfying (67) and such that $\phi_0(z) \le \phi(z)$ for all z. Inductively, sequences of functions $\phi_1, \phi_2, \ldots$ and $\psi_0, \psi_1, \ldots$ are defined through the relations

$$\psi_m(z) = \sum_{z' \in S(z)} \frac{\phi_m(z')}{N_1! \ldots N_c!}, \qquad m = 0, 1, \ldots,$$

and

$$\phi_m(z) = \begin{cases} \phi_{m-1}(z) + [\alpha - \psi_{m-1}(z)] \\ \qquad \text{if both } \phi_{m-1}(z) \text{ and } \psi_{m-1}(z) \text{ are } < \alpha, \\ \phi_{m-1}(z) \qquad \text{otherwise.} \end{cases}$$

The function $\phi(z) = \lim \phi_m(z)$ then satisfies the required conditions.
[The functions ϕ_m are nondecreasing and between 0 and 1. It is further seen by induction that $0 \le \alpha - \psi_m(z) \le (1 - \gamma)^m[\alpha - \psi_0(z)]$, where $\gamma = 1/N_1! \ldots N_c!$.]

52. Consider the problem of testing $H: \eta = \xi$ in the family of densities (62) when it is given that $\sigma > c > 0$ and that the point $(\xi_{11}, \ldots, \xi_{cN_c})$ of (63) lies in a bounded region R containing a rectangle, where c and R are known. Then Theorem 7 is no longer applicable. However, unbiasedness of a test ϕ of H implies (67), and therefore reduces the problem to the class of permutation tests.
[Unbiasedness implies $\int \phi(z) p_{\sigma, \xi}(z) \, dz = \alpha$ and hence

$$\alpha = \int \psi(z) p_{\sigma, \xi}(z) \, dz = \int \psi(z) \frac{1}{(\sqrt{2\pi} \sigma)^N} \exp\left[-\frac{1}{2\sigma^2} \sum \sum (z_{ij} - \xi_{ij})^2\right]$$

for all $\sigma > c$ and ξ in R. The result follows from completeness of this last family.]

53. To generalize Theorem 7 to other designs, let $Z = (Z_1, \ldots, Z_N)$ and let $G = \{g_1, \ldots, g_r\}$ be a group of permutations of N coordinates or more generally a group of orthogonal transformations of N-space. If

$$(81) \qquad p_{\sigma, \xi}(z) = \frac{1}{r} \sum_{k=1}^{r} \frac{1}{(\sqrt{2\pi} \sigma)^N} \exp\left(-\frac{1}{2\sigma^2}|z - g_k \xi|^2\right),$$

where $|z|^2 = \Sigma z_i^2$, then $\int \phi(z) p_{\sigma, \zeta}(z)\, dz \le \alpha$ for all $\sigma > 0$ and all ζ implies

(82)
$$\frac{1}{r} \sum_{z' \in S(z)} \phi(z') \le \alpha \qquad \text{a.e.,}$$

where $S(z)$ is the set of points in N-space obtained from z by applying to it all the transformations g_k, $k = 1, \dots, r$.

54. *Generalization of Corollary 3.* Let H be the class of densities (81) with $\sigma > 0$ and $-\infty < \zeta_i < \infty$ $(i = 1, \dots, N)$. A complete family of tests of H at level of significance α is the class of *permutation tests* satisfying

(83)
$$\frac{1}{r} \sum_{z' \in S(z)} \phi(z') = \alpha \qquad \text{a.e.}$$

Section 14

55. If $c = 1$, $m = n = 3$, and if the ordered x's and y's are respectively $1.97, 2.19, 2.61$ and $3.02, 3.28, 3.41$, determine the points $\delta_{(1)}, \dots, \delta_{(19)}$ defined as the ordered values of (73).

56. If $c = 4$, $m_i = n_i = 1$, and the pairs (x_i, y_i) are $(1.56, 2.01)$, $(1.87, 2.22)$, $(2.17, 2.73)$, and $(2.31, 2.60)$, determine the points $\delta_{(1)}, \dots, \delta_{(15)}$ which define the intervals (72).

57. If m, n are positive integers with $m \le n$, then

$$\sum_{K=1}^{m} \binom{m}{K}\binom{n}{K} = \binom{m+n}{m} - 1.$$

58. (i) Generalize the randomization models of Section 14 for paired comparisons $(n_1 = \cdots = n_c = 2)$ and the case of two groups $(c = 1)$ to an arbitrary number c of groups of sizes $n_1, \dots, n_c$.

   (ii) Generalize the confidence intervals (72) and (73) to the randomization model of part (i).

59. Let $Z_1, \dots, Z_n$ be i.i.d. according to a continuous distribution symmetric about θ, and let $T_{(1)} < \cdots < T_{(M)}$ be the ordered set of $M = 2^n - 1$ subsamples $(Z_{i_1} + \cdots + Z_{i_r})/r$, $r \ge 1$. If $T_{(0)} = -\infty$, $T_{(M+1)} = \infty$, then

$$P_\theta\left[T_{(i)} < \theta < T_{(i+1)} \right] = \frac{1}{M+1} \qquad \text{for all} \quad i = 0, 1, \dots, M.$$

[Hartigan (1969).]

60. (i) Given n pairs $(x_1, y_1), \dots, (x_n, y_n)$, let G be the group of 2^n permutations of the 2^n variables which interchange x_i and y_i in all, some, or none

of the n pairs. Let G_0 be any subgroup of G, and let e be the number of elements in G_0. Any element $g \in G_0$ (except the identity) is characterized by the numbers $i_1, \ldots, i_r$ $(r \geq 1)$ of the pairs in which x_i and y_i have been switched. Let $d_i = y_i - x_i$, and let $\delta_{(1)} < \cdots < \delta_{(e-1)}$ denote the ordered values $(d_{i_1} + \cdots + d_{i_r})/r$ corresponding to G_0. Then (72) continues to hold with $e - 1$ in place of M.

(ii) State the generalization of Problem 59 to the situation of part (i).

[Hartigan (1969).]

61. The preceding problem establishes a $1:1$ correspondence between $e - 1$ permutations T of G_0 which are not the identity and $e - 1$ nonempty subsets $\{i_1, \ldots, i_r\}$ of the set $\{1, \ldots, n\}$. If the permutations T and T' correspond respectively to the subsets $R = \{i_1, \ldots, i_r\}$ and $R' = \{j_1, \ldots, j_s\}$, then the group product $T'T$ corresponds to the subset $(R \cap \tilde{S}) \cup (\tilde{R} \cap S) = (R \cup S) - (R \cap S)$. [Hartigan (1969).]

62. Determine for each of the following classes of subsets of $\{1, \ldots, n\}$ whether (together with the empty subset) it forms a group under the group operation of the preceding problem: All subsets $\{i_1, \ldots, i_r\}$ with

(i) $r = 2$;
(ii) $r =$ even;
(iii) r divisible by 3.
(iv) Give two other examples of subgroups G_0 of G.

Note. A class of such subgroups is discussed by Forsythe and Hartigan (1970).

63. Generalize Problems 60(i) and 61 to the case of two groups of sizes m and n $(c = 1)$.

Section 15

64. (i)   If the joint distribution of X and Y is the bivariate normal distribution (70), then the conditional distribution of Y given x is the normal distribution with variance $\tau^2(1 - \rho^2)$ and mean $\eta + (\rho\tau/\sigma)(x - \xi)$.

(ii) Let $(X_1, Y_1), \ldots, (X_n, Y_n)$ be a sample from a bivariate normal distribution, let R be the sample correlation coefficient, and suppose that $\rho = 0$. Then the conditional distribution of $\sqrt{n - 2} R / \sqrt{1 - R^2}$ given $x_1, \ldots, x_n$ is Student's t-distribution with $n - 2$ degrees of freedom provided $\Sigma(x_i - \bar{x})^2 > 0$. This is therefore also the unconditional distribution of this statistic.

(iii) The probability density of R itself is then

(84) $$p(r) = \frac{1}{\sqrt{\pi}} \frac{\Gamma[\frac{1}{2}(n - 1)]}{\Gamma[\frac{1}{2}(n - 2)]} (1 - r^2)^{\frac{1}{2}n - 2}.$$

[(ii): If $v_i = (x_i - \bar{x})/\sqrt{\Sigma(x_j - \bar{x})^2}$ so that $\Sigma v_i = 0$, $\Sigma v_i^2 = 1$, the statistic can be written as

$$\frac{\sum v_i Y_i}{\sqrt{\left[\sum Y_i^2 - n\bar{Y}^2 - \left(\sum v_i Y_i\right)^2\right]/(n-2)}}.$$

Since its distribution depends only on ρ one can assume $\eta = 0$, $\tau = 1$. The desired result follows from Problem 6 by making an orthogonal transformation from $(Y_1, \ldots, Y_n)$ to $(Z_1, \ldots, Z_n)$ such that $Z_1 = \sqrt{n}\,\bar{Y}$, $Z_2 = \Sigma v_i Y_i$.]

65. (i) Let $(X_1, Y_1), \ldots, (X_n, Y_n)$ be a sample from the bivariate normal distribution (70), and let $S_1^2 = \Sigma(X_i - \bar{X})^2$, $S_2^2 = \Sigma(Y_i - \bar{Y})^2$, $S_{12} = \Sigma(X_i - \bar{X})(Y_i - \bar{Y})$. There exists a UMP unbiased test for testing the hypothesis $\tau/\sigma = \Delta$. Its acceptance region is

$$\frac{|\Delta^2 S_1^2 - S_2^2|}{\sqrt{\left(\Delta^2 S_1^2 + S_2^2\right)^2 - 4\Delta^2 S_{12}^2}} \le C,$$

and the probability density of the test statistic is given by (84) when the hypothesis is true.

(ii) Under the assumption $\tau = \sigma$, there exists a UMP unbiased test for testing $\eta = \xi$, with acceptance region $|\bar{Y} - \bar{X}|/\sqrt{S_1^2 + S_2^2 - 2S_{12}} \le C$. On multiplication by a suitable constant the test statistic has Student's t-distribution with $n - 1$ degrees of freedom when $\eta = \xi$. (Without the assumption $\tau = \sigma$, this hypothesis is a special case of the one considered in Chapter 8, Example 2.)

[(i): The transformation $U = \Delta X + Y$, $V = X - (1/\Delta)Y$ reduces the problem to that of testing that the correlation coefficient in a bivariate normal distribution is zero.
(ii): Transform to new variables $V_i = Y_i - X_i$, $U_i = Y_i + X_i$.]

66. (i) Let $(X_1, Y_1), \ldots, (X_n, Y_n)$ be a sample from the bivariate normal distribution (74), and let $S_1^2 = \Sigma(X_i - \bar{X})^2$, $S_{12} = \Sigma(X_i - \bar{X})(Y_i - \bar{Y})$, $S_2^2 = \Sigma(Y_i - \bar{Y})^2$.

Then (S_1^2, S_{12}, S_2^2) are independently distributed of $(\bar{X}, \bar{Y})$, and their joint distribution is the same as that of $(\Sigma_{i=1}^{n-1} X_i'^2, \Sigma_{i=1}^{n-1} X_i' Y_i', \Sigma_{i=1}^{n-1} Y_i'^2)$, where (X_i', Y_i'), $i = 1, \ldots, n - 1$, are a sample from the distribution (74) with $\xi = \eta = 0$.

(ii) Let $X_1, \ldots, X_m$ and $Y_1, \ldots, Y_m$ be two samples from $N(0, 1)$. Then the joint density of $S_1^2 = \Sigma X_i^2$, $S_{12} = \Sigma X_i Y_i$, $S_2^2 = \Sigma Y_i^2$ is

$$\frac{1}{4\pi\Gamma(m-1)}\left(s_1^2 s_2^2 - s_{12}^2\right)^{\frac{1}{2}(m-3)}\exp\left[-\tfrac{1}{2}\left(s_1^2 + s_2^2\right)\right]$$

for $s_{12}^2 \le s_1^2 s_2^2$, and zero elsewhere.

(iii)   The joint density of the statistics (S_1^2, S_{12}, S_2^2) of part (i) is

(85) $$\frac{\left(s_1^2 s_2^2 - s_{12}^2\right)^{\frac{1}{2}(n-4)}}{4\pi\Gamma(n-2)\left(\sigma\tau\sqrt{1-\rho^2}\right)^{n-1}} \exp\left[-\frac{1}{2(1-\rho^2)}\left(\frac{s_1^2}{\sigma^2} - \frac{2\rho s_{12}}{\sigma\tau} + \frac{s_2^2}{\tau^2}\right)\right]$$

for $s_{12}^2 \leq s_1^2 s_2^2$, and zero elsewhere.

[(i): Make an orthogonal transformation from $X_1, \ldots, X_n$ to $X_1', \ldots, X_n'$ such that $X_n' = \sqrt{n}\,\bar{X}$, and apply the same orthogonal transformation also to $Y_1, \ldots, Y_n$. Then

$$Y_n' = \sqrt{n}\,\bar{Y}, \qquad \sum_{i=1}^{n-1} X_i' Y_i' = \sum_{i=1}^{n} (X_i - \bar{X})(Y_i - \bar{Y}),$$

$$\sum_{i=1}^{n-1} X_i'^2 = \sum_{i=1}^{n} (X_i - \bar{X})^2, \qquad \sum_{i=1}^{n-1} Y_i'^2 = \sum_{i=1}^{n} (Y_i - \bar{Y})^2.$$

The pairs of variables $(X_1', Y_1'), \ldots, (X_n', Y_n')$ are independent, each with a bivariate normal distribution with the same variances and correlation as those of (X, Y) and with means $E(X_i') = E(Y_i') = 0$ for $i = 1, \ldots, n-1$.
(ii): Consider first the joint distribution of $S_{12} = \Sigma x_i Y_i$ and $S_2^2 = \Sigma Y_i^2$ given $x_1, \ldots, x_m$. Letting $Z_1 = S_{12}/\sqrt{\Sigma x_i^2}$ and making an orthogonal transformation from $Y_1, \ldots, Y_m$ to $Z_1, \ldots, Z_m$ so that $S_2^2 = \Sigma_{i=1}^m Z_i^2$, the variables Z_1 and $\Sigma_{i=2}^m Z_i^2 = S_2^2 - Z_1^2$ are independently distributed as $N(0,1)$ and χ_{m-1}^2 respectively. From this the joint conditional density of $S_{12} = s_1 Z_1$ and S_2^2 is obtained by a simple transformation of variables. Since the conditional distribution depends on the x's only through s_1^2, the joint density of S_1^2, S_{12}, S_2^2 is found by multiplying the above conditional density by the marginal one of S_1^2, which is χ_m^2. The proof is completed through use of the identity

$$\Gamma\left[\tfrac{1}{2}(m-1)\right]\Gamma\left(\tfrac{1}{2}m\right) = \frac{\sqrt{\pi}\,\Gamma(m-1)}{2^{m-2}}.$$

(iii): If $(X', Y') = (X_1', Y_1'; \ldots; X_m', Y_m')$ is a sample from a bivariate normal distribution with $\xi = \eta = 0$, then $T = (\Sigma X_i'^2, \Sigma X_i' Y_i', \Sigma Y_i'^2)$ is sufficient for $\theta = (\sigma, \rho, \tau)$, and the density of T is obtained from that given in part (ii) for $\theta_0 = (1, 0, 1)$ through the identity [Chapter 3, Problem 14 (i)]

$$p_\theta^T(t) = p_{\theta_0}^T(t)\frac{p_\theta^{X', Y'}(x', y')}{p_{\theta_0}^{X', Y'}(x', y')}.$$

The result now follows from part (i) with $m = n - 1$.]

67. If $(X_1, Y_1), \ldots, (X_n, Y_n)$ is a sample from a bivariate normal distribution, the probability density of the sample correlation coefficient R is*

(86) $$p_\rho(r) = \frac{2^{n-3}}{\pi(n-3)!}(1-\rho^2)^{\frac{1}{2}(n-1)}(1-r^2)^{\frac{1}{2}(n-4)}$$

$$\times \sum_{k=0}^{\infty} \Gamma^2[\tfrac{1}{2}(n+k-1)]\frac{(2\rho r)^k}{k!}$$

or alternatively

(87) $$p_\rho(r) = \frac{n-2}{\pi}(1-\rho^2)^{\frac{1}{2}(n-1)}(1-r^2)^{\frac{1}{2}(n-4)}$$

$$\times \int_0^1 \frac{t^{n-2}}{(1-\rho rt)^{n-1}}\frac{1}{\sqrt{1-t^2}}\, dt.$$

Another form is obtained by making the transformation $t = (1-v)/(1-\rho rv)$ in the integral on the right-hand side of (87). The integral then becomes

(88) $$\frac{1}{(1-\rho r)^{\frac{1}{2}(2n-3)}}\int_0^1 \frac{(1-v)^{n-2}}{\sqrt{2v}}\left[1-\tfrac{1}{2}v(1+\rho r)\right]^{-1/2}dv.$$

Expanding the last factor in powers of v, the density becomes

(89) $$\frac{n-2}{\sqrt{2\pi}}\frac{\Gamma(n-1)}{\Gamma(n-\frac{1}{2})}(1-\rho^2)^{\frac{1}{2}(n-1)}(1-r^2)^{\frac{1}{2}(n-4)}(1-\rho r)^{-n+\frac{3}{2}}$$

$$\times F\left(\tfrac{1}{2};\tfrac{1}{2}; n-\tfrac{1}{2}; \frac{1+\rho r}{2}\right),$$

where

(90) $$F(a, b, c, x) = \sum_{j=0}^{\infty} \frac{\Gamma(a+j)}{\Gamma(a)}\frac{\Gamma(b+j)}{\Gamma(b)}\frac{\Gamma(c)}{\Gamma(c+j)}\frac{x^j}{j!}$$

is a hypergeometric function.

[To obtain the first expression make a transformation from (S_1^2, S_2^2, S_{12}) with density (85) to (S_1^2, S_2^2, R) and expand the factor $\exp\{\rho s_{12}/(1-\rho^2)\sigma\tau\} =$

*The distribution of R is reviewed by Johnson and Kotz (1970, Vol. 2, Section 32) and Patel and Read (1982).

$\exp\{\rho r s_1 s_2 / (1 - \rho^2)\sigma\tau\}$ into a power series. The resulting series can be integrated term by term with respect to s_1^2 and s_2^2. The equivalence with the second expression is seen by expanding the factor $(1 - \rho r t)^{-(n-1)}$ under the integral in (87) and integrating term by term.]

68. If X and Y have a bivariate normal distribution with correlation coefficient $\rho > 0$, they are positively regression-dependent.
[The conditional distribution of Y given x is normal with mean $\eta + \rho\tau\sigma^{-1}(x - \xi)$ and variance $\tau^2(1 - \rho^2)$. Through addition to such a variable of the positive quantity $\rho\tau\sigma^{-1}(x' - x)$ it is transformed into one with the conditional distribution of Y given $x' > x$.]

69. (i) The functions (79) are bivariate cumulative distributions functions.

    (ii) A pair of random variables with distribution (79) is positively regression-dependent.

70. If X, Y are positively regression dependent, they are positively quadrant dependent.
[Positive regression dependence implies that

$$(91) \quad P[Y \le y \mid X \le x] \ge P[Y \le y \mid X \le x'] \qquad \text{for all} \quad x < x' \text{ and } y,$$

and (91) implies positive quadrant dependence.]

71. There exist bivariate distributions F of (X, Y) for which $\rho = 0$ and $\mathrm{Var}(XY)/[\mathrm{Var}(X)\mathrm{Var}(Y)]$ takes on any given positive value.

Additional Problems

72. Let (X_i, Y_i), $i = 1, \ldots, n$, be i.i.d. according to a bivariate distribution F with $E(X_i^2)$, $E(Y_i^2) < \infty$.

    (i)   If R is the sample correlation coefficient, then $\sqrt{n}\, R$ is asymptotically normal with mean 0 and variance $\mathrm{Var}(X_i Y_i)/\mathrm{Var}\, X_i \,\mathrm{Var}\, Y_i$.

    (ii) The variance of part (i) can take on any value between 0 and ∞.

    (iii) For testing $H_2 : \rho = 0$ against $\rho > 0$, define a denominator D_n and critical value c_n such that the rejection region $R/D_n \ge c_n$ has probability $\alpha_n(F) \to \alpha$ for all F satisfying H_2.

73. *Shape parameter of a gamma distribution.* Let $X_1, \ldots, X_n$ be a sample from the gamma distribution $\Gamma(g, b)$ defined in Problem 43 of Chapter 3.

    (i) There exist UMP unbiased tests of $H: g \le g_0$ against $g > g_0$ and of $H' : g = g_0$ against $g \ne g_0$, and their rejection regions are based on $W = \Pi(X_i/\overline{X})$.

    (ii) There exist uniformly most accurate confidence intervals for g based on W.

[Shorack (1972).]

Notes.
(1) The null distribution of W is discussed in Bain and Engelhardt (1975), Glaser (1976), and Engelhardt and Bain (1978).
(2) For $g = 1$, $\Gamma(g, b)$ reduces to an exponential distribution, and (i) becomes the UMP unbiased test for testing that a distribution is exponential against the alternative that it is gamma with $g > 1$ or with $g \neq 1$.
(3) An alternative treatment of this and some of the following problems is given by Bar-Lev and Reiser (1982).

74. *Scale parameter of a gamma distribution.* Under the assumptions of the preceding problem, there exists

   (i) A UMP unbiased test of $H: b \leq b_0$ against $b > b_0$ which rejects when $\Sigma X_i > C(\Pi X_i)$.

   (ii) Most accurate unbiased confidence intervals for b.

[The conditional distribution of ΣX_i given ΠX_i, which is required for carrying out this test, is discussed by Engelhardt and Bain (1977).]

75. *Gamma two-sample problem.* Let $X_1, \ldots, X_m$; $Y_1, \ldots, Y_n$ be independent samples from gamma distributions $\Gamma(g_1, b_1)$, $\Gamma(g_2, b_2)$ respectively.

   (i) If g_1, g_2 are known, there exists a UMP unbiased test of $H: b_2 = b_1$ against one- and two-sided alternatives, which can be based on a beta distribution.
[Some applications and generalizations are discussed in Lentner and Buehler (1963).]

   (ii) If g_1, g_2 are unknown, show that a UMP unbiased test of H continues to exist, and describe its general form.

   (iii) If $b_2 = b_1 = b$ (unknown), there exists a UMP unbiased test of $g_2 = g_1$ against one- and two-sided alternatives; describe its general form.

[(i): If Y_i $(i = 1, 2)$ are independent $\Gamma(g_i, b)$, then $Y_1 + Y_2$ is $\Gamma(g_1 + g_2, b)$ and $Y_1/(Y_1 + Y_2)$ has a beta distribution.]

76. Let $X_1, \ldots, X_n$ be a sample from the Pareto distribution $P(c, \tau)$, both parameters unknown. Obtain UMP unbiased tests for the parameters c and τ.

[Problem 12, and Problem 44 of Chapter 3.]

77. *Inverse Gaussian distribution.** Let $X_1, \ldots, X_n$ be a sample from the inverse Gaussian distribution $I(\mu, \tau)$, both parameters unknown.

   (i) There exists a UMP unbiased test of $\mu \leq \mu_0$ against $\mu > \mu_0$, which rejects when $\overline{X} > C[\Sigma(X_i + 1/X_i)]$, and a corresponding UMP unbiased

---

*For additional information concerning inference in inverse Gaussian distributions, see Folks and Chhikara (1978).

test of $\mu = \mu_0$ against $\mu \neq \mu_0$.
[The conditional distribution needed to carry out this test is given by Chhikara and Folks (1976).]

(ii) There exist UMP unbiased tests of $H : \tau = \tau_0$ against both one- and two-sided hypotheses based on the statistic $V = \Sigma(1/X_i - 1/\overline{X})$.

(iii) When $\tau = \tau_0$, the distribution of $\tau_0 V$ is χ^2_{n-1}.

[Tweedie (1957).]

78. Let $X_1, \ldots, X_m$ and $Y_1, \ldots, Y_n$ be independent samples from $I(\mu, \sigma)$ and $I(\nu, \tau)$ respectively.

(i) There exist UMP unbiased tests of τ_2/τ_1 against one- and two-sided alternatives.

(ii) If $\tau = \sigma$, there exist UMP unbiased tests of ν/μ against one- and two-sided alternatives.

[Chhikara (1975).]

79. Consider a one-sided, one-sample, level-α t-test with rejection region $t(X) \geq c_n$, where $X = (X_1, \ldots, X_n)$ and $t(X)$ is given by (16). Let $\alpha_n(F)$ be the rejection probability when $X_1, \ldots, X_n$ are i.i.d. according to a distribution $F \in \mathscr{F}$, with $\mathscr{F}$ the class of all distributions with mean zero and finite variance. Then for any fixed n, no matter how large, $\sup_{F \in \mathscr{F}} \alpha_n(F) = 1$.
[Let F be a mixture of two normals, $F = \gamma N(1, \sigma^2) + (1 - \gamma)N(\mu, \sigma^2)$ with $\gamma + (1 - \gamma)\mu = 0$. By taking γ sufficiently close to 1, one can be virtually certain that all n observations are from $N(1, \sigma^2)$. By taking σ sufficiently small, one can make the power of the t-test against the alternative $N(1, \sigma^2)$ arbitrarily close to 1. The result follows.]

Note. This is a special case of results of Bahadur and Savage (1956); for further discussion, see Loh (1985).

17. REFERENCES

The optimal properties of the one- and two-sample normal-theory tests were obtained by Neyman and Pearson (1933) as some of the principal applications of their general theory. Concern about the robustness of these tests began to be voiced in the 1920s [Neyman and Pearson (1928), Shewhart and Winters (1928), Sophister (1928), and Pearson (1929)] and has been an important topic ever since. Particularly influential were Box (1953), which introduced the term "robustness", Scheffé (1959, Chapter 10), Tukey (1960), and Hotelling (1961). Permutation tests, as alternatives to the standard tests having fixed significance levels, were initiated by Fisher (1935) and further developed, among others, by Pitman (1937, 1938), Lehmann and Stein (1949), Hoeffding (1952), and Box and Andersen (1955). Some aspects of these tests are reviewed in Bell and Sen (1984). Explicit confidence intervals based on subsampling were given by Hartigan (1969). The theory of unbiased confidence sets and its relation to that of unbiased tests is due to Neyman (1937).

Albers, W.
(1978). "Testing the mean of a normal population under dependence." *Ann. Statist.* **6**, 1337–1344.

Anderson, T. W.
(1971). *The Statistical Analysis of Time Series*, Wiley, New York.

Bahadur, R. R.
(1979). "A note on UMV estimates and ancillary statistics." In *Contributions to Statistics*, J. Hájek Memorial Volume (Jureckova, ed.), Academia, Prague.

Bahadur, R. R. and Savage, L. J.
(1956). "The nonexistence of certain statistical procedures in nonparametric problems." *Ann. Math. Statist.* **27**, 1115–1122.

Bain, L. J. and Engelhardt, M. E.
(1975). "A two-moment chi-square approximation for the statistic $\log(\overline{X}/\tilde{X})$." *J. Amer. Statist. Assoc.* **70**, 948–950.

Bar-Lev, S. K. and Reiser, B.
(1982). "An exponential subfamily which admits UMPU tests based on a single test statistic." *Ann. Statist.* **10**, 979–989.

Basu, D.
(1955). "On statistics independent of a complete sufficient statistic." *Sankhyā* **15**, 377–380. [Theorem 2.]
(1958). "On statistics independent of a sufficient statistic." *Sankhyā* **20**, 223–226.

Bell, C. B. and Sen, P. K.
(1984). "Randomization procedures." In *Handbook of Statistics 4* (Krishnaiah and Sen, eds., Elsevier.

Benjamini, Y.
(1983). "Is the *t*-test really conservative when the parent distribution is long-tailed?" *J. Amer. Statist. Assoc.* **78**, 645–654.

Bennett, B.
(1957). "On the performance characteristic of certain methods of determining confidence limits." *Sankhyā* **18**, 1–12.

Beran, R. and Srivastava, M. S.
(1985). "Bootstrap tests and confidence regions for functions of a covariance matrix." *Ann. Statist.* **13**, 95–115.

Berger, J. O.
(1985). *Statistical Decision Theory and Bayesian Analysis*, 2nd ed., Springer, New York.

Bickel, P. J. and Van Zwet, W. R.
(1978). "Asymptotic expansions for the power of distribution free tests in the two-sample problem." *Ann. Statist.* **6**, 937–1004.

Billingsley, P.
(1979). *Probability and Measure*, Wiley, New York.

Blyth, C. R.
(1984). "Approximate binomial confidence limits." Queen's Math. Preprint 1984-6, Queens' Univ., Kingston, Ontario.

Blyth, C. R. and Hutchinson, D. W.
(1960). "Tables of Neyman—shortest confidence intervals for the binomial parameter." *Biometrika* **47**, 481–491.

Blyth, C. R. and Still, H. A.
(1983). "Binomial confidence intervals." *J. Amer. Statist. Assoc.* **78**, 108–116.

Box, G. E. P.
(1953). "Non-normality and tests for variances." *Biometrika* **40**, 318–335.

Box, G. E. P. and Andersen, S. L.
(1955). "Permutation theory in the derivation of robust criteria and the study of departures from assumptions." *J. Roy. Statist. Soc. (B)* **17**, 1–34.

Box, G. E. P. and Jenkins, G.
(1970). *Time Series Analysis*, Holden Day, San Francisco.

Box, G. E. P. and Tiao, G. C.
(1964). "A note on criterion robustness and inference robustness." *Biometrika* **51**, 169–173.
(1973) *Bayesian Inference in Statistical Analysis*, Addison-Wesley, Reading, Mass.

Buehler, R.
(1980). "Fiducial inference." In *R. A. Fisher: An Appreciation* (Fienberg and Hinkley, eds.), Springer, New York.
(1983). "Fiducial inference." In *Encycl. Statist. Sciences*, Vol. 3, Wiley, New York.

Chhikara, R. S.
(1975). "Optimum tests for the comparison of two inverse Gaussian distribution means." *Austral. J. Statist.* **17**, 77–83.

Chhikara, R. S. and Folks, J. L.
(1976). "Optimum test procedures for the mean of first passage time distribution in Brownian motion with positive drift (inverse Gaussian distribution)." *Technometrics* **18**, 189–193.

Cochran, W. G.
(1968). "Errors of measurement in statistics." *Technometrics* **10**, 637–666.

Conover, W. J., Johnson, M. E., and Johnson, M. M.
(1981). "A comparative study of tests for homogeneity of variances, with applications to the outer continental shelf bidding data." *Technometrics* **23**, 351–361.

Cramér, H.
(1946). *Mathematical Methods of Statistics*, Princeton U.P.

Cressie, N.
(1980). "Relaxing assumptions in the one-sample *t*-test." *Austral. J. Statist.* **22**, 143–153.

Dawid, A. P. and Stone, M.
(1982). "The functional model basis of fiducial inference" (with discussion). *Ann. Statist.* **10**, 1040–1074.

Edgell, S. E. and Noon, S. M.
(1984). "Effect of violation of normality on the *t*-test of the correlation coefficient." *Psych. Bull.* **95**, 576–583.

Edwards, A. W. F.
(1983). "Fiducial distributions." In *Encycl. of Statist. Sci.*, Vol. 3, Wiley, New York.

Efron, B.

(1969). "Student's *t*-test under symmetry conditions." *J. Amer. Statist. Assoc.* **64**, 1278–1302.

(1982). *The Jackknife, the Bootstrap and Other Resampling Plans*, SIAM, Philadelphia.

Engelhardt, M. and Bain, L. J.

(1977). "Uniformly most powerful unbiased tests on the scale parameter of a gamma distribution with a nuisance shape parameter." *Technometrics* **19**, 77–81.

(1978). "Construction of optimal unbiased inference proceures for the parameters of the gamma distribution." *Technometrics* **20**, 485–489.

Falk, M. and Kohne, W.

(1984). "A robustification of the sign test under mixing conditions." *Ann. Statist.* **12**, 716–729.

Fenstad, G. U.

(1983). "A comparison between the *U* and *V* tests in the Behrens–Fisher problem." *Biometrika* **70**, 300–302.

Fisher, R. A.

(1915). "Frequency distribution of the values of the correlation coefficient in samples from an indefinitely large population." *Biometrika* **10**, 507–521.

[Derives the distribution of the sample correlation coefficient from a bivariate normal distribution.]

(1930). "Inverse probability." *Proc. Cambridge Philos. Soc.* **26**, 528–535.

(1931). "Properties of the [*Hh*] functions." In *Brit. Assoc. Math. Tables* 1; 3rd ed., 1951, xxviii–xxxvii.

[Derivation of noncentral *t*-distribution.]

(1935). *The Design of Experiments*. Oliver and Boyd, Edinburgh.

[Contains the basic ideas concerning permutation tests. In particular, points out how randomization provides a basis for inference and proposes the permutation version of the *t*-test as not requiring the assumption of normality.]

(1973). *Statistical Methods and Scientific Inference*, 3rd ed., Hafner, New York.

Folks, J. L. and Chhikara, R. S.

(1978). "The inverse Gaussian distribution and its statistical applications—a review" (with discussion). *J. Roy. Statist. Soc.* (*B*) **40**, 263–289.

Forsythe, A. and Hartigan, J. A.

(1970). "Efficiency of confidence intervals generated by repeated subsample calculations." *Biometrika* **57**, 629–639.

Fraser, D. A. S.

(1957). *Nonparametric Methods in Statistics*. Wiley, New York.

(1978). "Fiducial inference." In *Internat. Encycl. of Statistics*, Vol. 1, Free Press, New York.

Gabriel, K. R. and Hall, W. J.

(1983). "Rerandomization inference on regression and shift effects: Computationally feasible methods." *J. Amer. Statist. Assoc.* **78**, 827–836.

Gabriel, K. R. and Hsu, C. F.

(1983). "Evaluation of the power of rerandomization tests, with application to weather modification experiments." *J. Amer. Statist. Assoc.* **78**, 766–775.

Gastwirth, J. L. and Rubin, H.

(1971). "Effect of dependence on the level of some one-sample tests." *J. Amer. Statist. Assoc.* **66**, 816–820.

Ghosh, B. K.

(1979). "A comparison of some approximate confidence intervals for the binomial parameter." *Multiv. Anal.* **9**, 116–129.

Glaser, R. E.

(1976). "The ratio of the geometric mean to the arithmetic mean for a random sample from a gamma distribution." *J. Amer. Statist. Assoc.* **71**, 481–487.

Green, B. F.

(1977). "A practical interactive program for randomization tests of location." *Amer. Statist.* **31**, 37–39.

Hartigan, J. A.

(1969). "Using subsample values as typical values." *J. Amer. Statist. Assoc.* **64**, 1303–1317.

Helmert, F. R.

(1876). "Die Genauigkeit der Formel von Peters zur Berechnung des wahrscheinlichen Beobachtungsfehlers direkter Beobachtungen gleicher Genauigkeit." *Astron. Nachr.* **88**, 113–132.

[Obtains the distribution of $\Sigma(X_i - \overline{X})^2$ when the X's are independently, normally distributed.]

Hobson, E. W.

(1927). *Theory of Functions of a Real Variable*, 3rd ed., Vol. 1, Cambridge Univ. Press, p. 194.

Hodges, J. L., Jr. and Lehmann, E. L.

(1954). "Testing the approximate validity of statistical hypotheses." *J. Roy. Statist. Soc. (B)* **16**, 261–268.

(1970). "Deficiency." *Ann. Math. Statist.* **41**, 783–801.

Hoeffding, W.

(1952). "The large-sample power of tests based on permutations of observations." *Ann. Math. Statist.* **23**, 169–192.

Hotelling, H.

(1961). "The behavior of some standard statistical tests under nonstandard conditions." In *Proc. 4th Berkeley Symp. Math. Statist. and Probab.*, Univ. of Calif. Press, Berkeley.

Hsu, C. T.

(1940). "On samples from a normal bivariate population." *Ann. Math. Statist.* **11**, 410–426. [Problem 65 (ii)]

Hsu, P. L.

(1949). "The limiting distribution of functions of sample means and application to testing hypotheses." In *Proc. [First] Berkeley Symp. Math. Statist. and Probab.*, Univ. of Calif. Press, Berkeley.

John, R. D. and Robinson, J.

(1983a). "Edgeworth expansions for the power of permutation tests." *Ann. Statist.* **11**, 625–631.

(1983b). "Significance levels and confidence intervals for permutation tests." *J. Statist. Comput. and Simul.* **16**, 161–173.

Johnson, N. L. and Kotz, S.

(1969, 1970). *Distributions in Statistics: Discrete Distributions; Continuous Distributions*, Houghton Mifflin, New York.

Kendall, M. G. and Stuart A.

(1979). *The Advanced Theory of Statistics*, 4th ed., Vol. 2, Macmillan, New York.

Koehn, U. and Thomas, D. L.

(1975). "On statistics independent of a sufficient statistic: Basu's Lemma." *Amer. Statist.* **29**, 40–41.

Kowalski, C. J.

(1972). "On the effects of non-normality on the distribution of the sample product–moment correlation coefficient." *Appl. Statist.* **21**, 1–12.

Lambert, D.
(1985). Robust two-sample permutation tests. *Ann. Statist.* **13**, 606–625.

Lehmann, E. L.
(1947). "On optimum tests of composite hypotheses with one constraint." *Ann. Math. Statist.* **18**, 473–494.
[Determines best similar regions for a number of problems, including Problem 12.]
(1966). "Some concepts of dependence." *Ann. Math. Statist.* **37**, 1137–1153.
[Problem 70.]
(1980). "An interpretation of completeness and Basu's theorem." *J. Amer. Statist. Assoc.* **76**, 335–340.

Lehmann, E. L. and Stein, C.
(1949). "On the theory of some non-parameteric hypotheses." *Ann. Math. Statist.* **20**, 28–45.
[Develops the theory of optimum permutation tests.]

Lentner, M. M. and Buehler, R. J.
(1963). "Some inferences about gamma parameters with an application to a reliability problem." *J. Amer. Statist. Assoc.* **58**, 670–677.

Lindley, D. V.
(1965). *Introduction to Probability and Statistics from a Bayesian Viewpoint. Part 2, Inference*, Cambridge U. P.

Loh, W.-Y.
(1985). "A new method for testing separate families of hypotheses." *J. Amer. Statist. Assoc.* **80**, 362–368.

Madansky, A.
(1962). "More on length of confidence intervals." *J. Amer. Statist. Assoc.* **57**, 586–589.

Mardia, K. V. and Zamroch, P. J.
(1978). *Tables of the F and Related Distributions with Algorithms*, Academic, New York.

Maritz, J. S.
(1979). "A note on exact robust confidence intervals for location." *Biometrika* **66**, 163–166.
[Problem 46(ii).]

Morgan, W. A.
(1939). "A test for the significance of the difference between the two variances in a sample from a normal bivariate population." *Biometrika* **31**, 13–19.
[Problem 65(i).]

Morgenstern, D.
(1956). "Einfache Beispiele zweidimensionaler Verteilungen." *Mitteil. Math. Statistik* **8**, 234–235.
[Introduces the distributions (79).]

Neyman, J.
(1937). "Outline of a theory of statistical estimation based on the classical theory of probability." *Phil. Trans. Roy. Soc., Ser. A* **236**, 333–380.
(1938). "On statistics the distribution of which is independent of the parameters involved in the original probability law of the observed variables," *Statist. Res. Mem.* **2**, 58–89.
[Essentially Theorem 2 under regularity assumptions.]

Neyman, J. and Pearson, E. S.
(1928). "On the use and interpretation of certain test criteria." *Biometrika* **20A**, 175–240.
(1933). "On the problem of the most efficient tests of statistical hypotheses." *Phil. Trans. Roy. Soc., Ser. A* **231**, 289–337.

Odén, A. and Wedel, H.
(1975). "Arguments for Fisher's permutation test." *Ann. Statist.* **3**, 518–520.

Owen, D. B.
(1985). "Noncentral *t*-distribution." *Encycl. Statist. Sci.* **6**, 286–290.

Patel, J. K. and Read, C. B.
(1982). *Handbook of the Normal Distribution*, Dekker, New York.

Paulson, E.
(1941). "On certain likelihood ratio tests associated with the exponential distribution." *Ann. Math. Statist.* **12**, 301–306.
[Discusses the power of the tests of Problem 12.]

Pearson, E. S.
(1929). "Some notes on sampling tests with two variables." *Biometrika* **21**, 337–360.

Pearson, E. S. and Adyanthaya, N. K.
(1929). "The distribution of frequency constants in small samples from non-normal symmetric and skew populations." *Biometrika* **21**, 259–286.

Pearson, E. S. and Please, N. W.
(1975). "Relation between the shape of population distribution and the robustness of four simple test statistics." *Biometrika* **62**, 223–242.

Pedersen, J. G.
(1978). "Fiducial inference." *Internat. Statist. Rev.* **46**, 147–170.

Pfanzagl, J.
(1979). "On optimal median unbiased estimators in the presence of nuisance parameters." *Ann. Statist.* **7**, 187–193.

Pitman, E. J. G.
(1937, 1938). "Significance tests which may be applied to samples from any population." *J. Roy. Statist. Soc. Suppl.* **4**, 119–130, 225–232; *Biometrika* **29**, 322–335.
[Develops the theory of randomization tests with many applications.]
(1939). "A note on normal correlation." *Biometrika* **31**, 9–12.
[Problem 39(i).]

Posten, H. O., Yeh, H. C., and Owen, D. B.
(1982). "Robustness of the two-sample *t*-test under violations of the homogeneity of variance assumption." *Commun. Statist.* **11**, 109–126.

Pratt, J. W.
(1961). "Length of confidence intervals." *J. Amer. Statist. Assoc.* **56**, 549–567.
[Problem 29.]
(1962). "A note on unbiased tests." *Ann. Math. Statist.* **33**, 292–294.
[Proposes and illustrates approach (ii) of Section 1.]
(1964). "Robustness of some procedures for the two-sample location problem." *J. Amer. Statist. Assoc.* **59**, 665–680.

Ramachandran, K. V.
(1958). "A test of variances." *J. Amer. Statist. Assoc.* **53**, 741–747.

Ramsey, P. H.
(1980). "Exact type 1 error rates for robustness of Student's *t*-test with unequal variances." *J. Ed. Statist.* **5**, 337–349.

Ratcliffe, J. F.
(1968). "The effect on the *t*-distribution of non-normality in the sampled population." *Appl. Statist.* **17**, 42–48.

Robinson, G. K.
(1976). "Properties of Student's t and of the Behrens–Fisher solution to the two means problem." *Ann. Statist.* 4, 963–971.

Robinson, J.
(1973). "The large-sample power of permutation tests for randomization models." *Ann. Statist.* 1, 291–296.
(1982). "Saddle point approximations to permutation tests and confidence intervals." *J. Roy. Statist. Soc.* 44, 91–101.

Rothenberg, T. J.
(1984). "Approximating the distributions of econometric estimators and test statistics." In *Handbook of Econometrics II* (Griliches and Intriligator, eds.), Elsevier, Chapter 15.

Rubin, D. B.
(1984). "Bayesianly justifiable and relevant frequency calculations for the applied statistician." *Ann. Statist.* 12, 1151–1172.

Scheffé, H.
(1942). "On the ratio of the variances of two normal populations." *Ann. Math. Statist.* 13, 371–388.
(1959). *The Analysis of Variance*, Wiley, New York.
(1970). "Practical solutions of the Behrens–Fisher problem." *J. Amer. Statist. Assoc.* 65, 1501–1508.

Serfling, R. J.
(1980). *Approximation Theorems of Mathematical Statistics*, Wiley, New York.

Shewhart, W. A. and Winters, F. W.
(1928). "Small samples—new experimental results." *J. Amer. Statist. Assoc.* 23, 144–153.

Shorack, G.
(1972). "The best test of exponentiality against gamma alternatives." *J. Amer. Statist. Assoc.* 67, 213–214.

Sophister (G. F. E. Story)
(1928). "Discussion of small samples drawn from an infinite skew population." *Biometrika* 20A, 389–423.

Steiger, J. H. and Hakstian, A. R.
(1982). "The asymptotic distribution of elements of a correlation matrix: Theory and application," *British J. Math. Statist. Psych.* 35, 208–215.
(1983). "A historical note on the asymptotic distribution of correlations." *British J. Math. Statist. Psych.* 36, 157.

Stein, C.
(1945). "A two-sample test for a linear hypothesis whose power is independent of the variance." *Ann. Math. Statist.* 16, 243–258.
[Problems 26–28.]

Stone, M.
(1983). "Fiducial probability." In *Encycl. Statist. Sci.*, Vol. 3, Wiley, New York.

Student (W. S. Gosset)
(1908). "On the probable error of a mean." *Biometrika* 6, 1–25.
[Obtains the distribution of the t-statistic when the X's are a sample from $N(0, \sigma^2)$. A rigorous proof was given by R. A. Fisher, "Note on Dr. Burnside's recent paper on error of observation," *Proc. Cambridge Phil. Soc.* 21 (1923), 655–658.]
(1927). "Errors of routine analysis." *Biometrika* 19, 151–164.

Tan, W. Y.

(1982). "Sampling distributions and robustness of t, F and variance-ratio in two samples and ANOVA models with respect to departure from normality." *Commun. Statist.—Theor. Meth.* **11**, 2485–2511.

Tate, R. F. and Klett, G. W.

(1959). "Optimal confidence intervals for the variance of a normal distribution." *J. Amer. Statist. Assoc.* **54**, 674–682.

Tiku, M. L. and Balakrishnan, N.

(1984). "Testing equality of population variances the robust way." *Commun. Statist.—Theor. Meth.* **13**, 2143–2159.

Tiku, M. L. and Singh, M.

(1981). "Robust test for means when population variances are unequal." *Commun. Statist.—Theor. Meth.* **A10**, 2057–2071.

Tritchler, D.

(1984). "On inverting permutation tests." *J. Amer. Statist. Assoc.* **79**, 200–207.

Tukey, J. W.

(1960). "A survey of sampling from contaminated distributions." In *Contributions to Probability and Statistics* (Olkin, ed.), Stanford U.P.

Tukey, J. W. and McLaughlin, D. H.

(1963). "Less vulnerable confidence and significance procedures for location based on a single sample: Trimming/Winsorization 1." *Sankhyā* **25**, 331–352.

Tweedie, M. C. K.

(1957). "Statistical properties of inverse Gaussian distributions I, II." *Ann. Math. Statist.* **28**, 362–377, 696–705.

Vadiveloo, J.

(1983). "On the theory of modified randomization tests for nonparametric hypotheses." *Commun. Statist.* **A12**, 1581–1596.

Wang, Y. Y.

(1971). "Probabilities of the type I errors of the Welch tests for the Behrens–Fisher problem." *J. Amer. Statist. Assoc.* **66**, 605–608.

Yuen, K. K.

(1974). "The two-sample trimmed t for unequal population variances." *Biometrika* **61**, 165–170.

CHAPTER 6

Invariance

1. SYMMETRY AND INVARIANCE

Many statistical problems exhibit symmetries, which provide natural restrictions to impose on the statistical procedures that are to be employed. Suppose, for example, that $X_1, \ldots, X_n$ are independently distributed with probability densities $p_{\theta_1}(x_1), \ldots, p_{\theta_n}(x_n)$. For testing the hypothesis $H : \theta_1 = \cdots = \theta_n$ against the alternative that the θ's are not all equal, the test should be symmetric in $x_1, \ldots, x_n$, since otherwise the acceptance or rejection of the hypothesis would depend on the (presumably quite irrelevant) numbering of these variables.

As another example consider a circular target with center O, on which are marked the impacts of a number of shots. Suppose that the points of impact are independent observations on a bivariate normal distribution centered on O. In testing this distribution for circular symmetry with respect to O, it seems reasonable to require that the test itself exhibit such symmetry. For if it lacks this feature, a two-dimensional (for example, Cartesian) coordinate system is required to describe the test, and acceptance or rejection will depend on the choice of this system, which under the assumptions made is quite arbitrary and has no bearing on the problem.

The mathematical expression of symmetry is invariance under a suitable group of transformations. In the first of the two examples above the group is that of all permutations of the variables $x_1, \ldots, x_n$ since a function of n variables is symmetric if and only if it remains invariant under all permutations of these variables. In the second example, circular symmetry with respect to the center O is equivalent to invariance under all rotations about O.

In general, let X be distributed according to a probability distribution P_θ, $\theta \in \Omega$, and let g be a transformation of the sample space $\mathscr{X}$. All such transformations considered in connection with invariance will be assumed

282

to be $1:1$ transformations of $\mathscr{X}$ onto itself. Denote by gX the random variable that takes on the value gx when $X = x$, and suppose that when the distribution of X is P_θ, $\theta \in \Omega$, the distribution of gX is $P_{\theta'}$ with θ' also in Ω. The element θ' of Ω which is associated with θ in this manner will be denoted by $\bar{g}\theta$, so that

(1) $$P_\theta\{gX \in A\} = P_{\bar{g}\theta}\{X \in A\}.$$

Here the subscript θ on the left member indicates the distribution of X, not that of gX. Equation (1) can also be written as $P_\theta(g^{-1}A) = P_{\bar{g}\theta}(A)$ and hence as

(2) $$P_{\bar{g}\theta}(gA) = P_\theta(A).$$

The parameter set Ω remains invariant under g (or is preserved by g) if $\bar{g}\theta \in \Omega$ for all $\theta \in \Omega$, and if in addition for any $\theta' \in \Omega$ there exists $\theta \in \Omega$ such that $\bar{g}\theta = \theta'$. These two conditions can be expressed by the equation

(3) $$\bar{g}\Omega = \Omega.$$

The transformation $\bar{g}$ of Ω onto itself defined in this way is $1:1$ provided the distributions P_θ corresponding to different values of θ are distinct. To see this let $\bar{g}\theta_1 = \bar{g}\theta_2$. Then $P_{\bar{g}\theta_1}(gA) = P_{\bar{g}\theta_2}(gA)$ and therefore $P_{\theta_1}(A) = P_{\theta_2}(A)$ for all A, so that $\theta_1 = \theta_2$.

Lemma 1. *Let g, g' be two transformations preserving Ω. Then the transformations $g'g$ and g^{-1} defined by*

$$(g'g)x = g'(gx) \quad and \quad g(g^{-1}x) = x \qquad for\ all \quad x \in \mathscr{X}$$

also preserve Ω and satisfy

(4) $$\overline{g'g} = \overline{g'} \cdot \bar{g} \quad and \quad \left(\overline{g^{-1}}\right) = (\bar{g})^{-1}.$$

Proof. If the distribution of X is P_θ, then that of gX is $P_{\bar{g}\theta}$ and that of $g'gX = g'(gX)$ is therefore $P_{\bar{g}'\bar{g}\theta}$. This establishes the first equation of (4); the proof of the second one is analogous.

We shall say that *the problem of testing $H: \theta \in \Omega_H$ against $K: \theta \in \Omega_K$ remains invariant* under a transformation g if $\bar{g}$ preserves both Ω_H and Ω_K, so that the equation

(5) $$\bar{g}\Omega_H = \Omega_H$$

holds in addition to (3). Let $\mathscr{C}$ be a class of transformations satisfying these two conditions, and let G be the smallest class of transformations containing $\mathscr{C}$ and such that $g, g' \in G$ implies that $g'g$ and g^{-1} belong to G. Then G is a group of transformations, all of which by Lemma 1 preserve both Ω and Ω_H. Any class $\mathscr{C}$ of transformations leaving the problem invariant can therefore be extended to a group G. It follows further from Lemma 1 that the class of induced transformations $\bar{g}$ form a group $\bar{G}$. The two equations (4) express the fact that $\bar{G}$ is a homomorphism of G.

In the presence of symmetries in both sample and parameter space represented by the groups G and $\bar{G}$, it is natural to restrict attention to tests ϕ which are also symmetric, that is, which satisfy

(6) $\qquad \phi(gx) = \phi(x) \qquad$ for all $\quad x \in X$ and $g \in G$.

A test ϕ satisfying (6) is said to be *invariant under* G. The restriction to invariant tests is a particular case of the principle of invariance formulated in Section 5 of Chapter 1. As was indicated there and in the examples above, a transformation g can be interpreted as a change of coordinates. From this point of view, a test is invariant if it is independent of the particular coordinate system in which the data are expressed.

A transformation g, in order to leave a problem invariant, must in particular preserve the class $\mathscr{A}$ of measurable sets over which the distributions P_θ are defined. This means that any set $A \in \mathscr{A}$ is transformed into a set of $\mathscr{A}$ and is the image of such a set, so that gA and $g^{-1}A$ both belong to $\mathscr{A}$. Any transformation satisfying this condition is said to be *bimeasurable*. Since a group with each element g also contains g^{-1}, its elements are automatically bimeasurable if all of them are measurable. If g' and g are bimeasurable, so are $g'g$ and g^{-1}. The transformations of the group G above generated by a class $\mathscr{C}$ are therefore all bimeasurable provided this is the case for the transformations of $\mathscr{C}$.

2. MAXIMAL INVARIANTS

If a problem is invariant under a group of transformations, the *principle of invariance* restricts attention to invariant tests. In order to obtain the best of these, it is convenient first to characterize the totality of invariant tests.

Let two points x_1, x_2 be considered equivalent under G,

$$x_1 \sim x_2 \,(\mathrm{mod}\, G),$$

if there exists a transformation $g \in G$ for which $x_2 = gx_1$. This is a true equivalence relation, since G is a group and the sets of equivalent points, the *orbits* of G, therefore constitute a partition of the sample space. (Cf. Appendix, Section 1.) A point x traces out an orbit as all transformations g of G are applied to it; this means that the orbit containing x consists of the totality of points gx with $g \in G$. It follows from the definition of invariance that a function is invariant if and only if it is constant on each orbit.

A function M is said to be *maximal invariant* if it is invariant and if

$$(7) \qquad M(x_1) = M(x_2) \quad \text{implies} \quad x_2 = gx_1 \quad \text{for some } g \in G,$$

that is, if it is constant on the orbits but for each orbit takes on a different value. All maximal invariants are equivalent in the sense that their sets of constancy coincide.

Theorem 1. *Let $M(x)$ be a maximal invariant with respect to G. Then a necessary and sufficient condition for ϕ to be invariant is that it depends on x only through $M(x)$, that is that there exists a function h for which $\phi(x) = h[M(x)]$ for all x.*

Proof. If $\phi(x) = h[M(x)]$ for all x, then $\phi(gx) = h[M(gx)] = h[M(x)] = \phi(x)$ so that ϕ is invariant. On the other hand, if ϕ is invariant and if $M(x_1) = M(x_2)$, then $x_2 = gx_1$ for some g and therefore $\phi(x_2) = \phi(x_1)$.

Example 1. (i) Let $x = (x_1, \ldots, x_n)$, and let G be the group of translations

$$gx = (x_1 + c, \ldots, x_n + c), \qquad -\infty < c < \infty.$$

Then the set of differences $y = (x_1 - x_n, \ldots, x_{n-1} - x_n)$ is invariant under G. To see that it is maximal invariant suppose that $x_i - x_n = x_i' - x_n'$ for $i = 1, \ldots, n - 1$. Putting $x_n' - x_n = c$, one has $x_i' = x_i + c$ for all i, as was to be shown. The function y is of course only one representation of the maximal invariant. Others are for example $(x_1 - x_2, x_2 - x_3, \ldots, x_{n-1} - x_n)$ or the redundant $(x_1 - \bar{x}, \ldots, x_n - \bar{x})$. In the particular case that $n = 1$, there are no invariants. The whole space is a single orbit, so that for any two points there exists a transformation of G taking one into the other. In such a case the transformation group G is said to be *transitive*. The only invariant functions are then the constant functions $\phi(x) \equiv c$.

(ii) if G is the group of transformations

$$gx = (cx_1, \ldots, cx_n), \qquad c \neq 0,$$

a special role is played by any zero coordinates. However, in statistical applications the set of points for which none of the coordinates is zero typically has probability 1; attention can then be restricted to this part of the sample space, and the set of ratios $x_1/x_n, \ldots, x_{n-1}/x_n$ is a maximal invariant. Without this restriction, two points x, x' are equivalent with respect to the maximal invariant partition if among their coordinates there are the same number of zeros (if any), if these occur at the same places, and if for any two nonzero coordinates x_i, x_j the ratios x_j/x_i and x_j'/x_i' are equal.

(iii)   Let $x = (x_1, \ldots, x_n)$, and let G be the group of all orthogonal transformations $x' = \Gamma x$ of n-space. Then Σx_i^2 is maximal invariant, that is, two points x and x^* can be transformed into each other by an orthogonal transformation if and only if they have the same distance from the origin. The proof of this is immediate if one restricts attention to the plane containing the points x, x^* and the origin.

Example 2.   (i)   Let $x = (x_1, \ldots, x_n)$, and let G be the set of $n!$ permutations of the coordinates of x. Then the set of ordered coordinates (*order statistics*) $x_{(1)} \leq \cdots \leq x_{(n)}$ is maximal invariant. A permutation of the x_i obviously does not change the set of values of the coordinates and therefore not the $x_{(i)}$. On the other hand, two points with the same set of ordered coordinates can be obtained from each other through a permutation of coordinates.

(ii)   Let G be the totality of transformations $x_i' = f(x_i)$, $i = 1, \ldots, n$, such that f is continuous and strictly increasing, and suppose that attention can be restricted to the points all of whose n coordinates are distinct. If the x_i are considered as n points on the real line, any such transformation preserves their order. Conversely, if $x_1, \ldots, x_n$ and $x_1', \ldots, x_n'$ are two sets of points in the same order, say $x_{i_1} < \cdots < x_{i_n}$ and $x_{i_1}' < \cdots < x_{i_n}'$, there exists a transformation f satisfying the required conditions and such that $x_i' = f(x_i)$ for all i. It can be defined for example as $f(x) = x + (x_{i_1}' - x_{i_1})$ for $x \leq x_{i_1}$, $f(x) = x + (x_{i_n}' - x_{i_n})$ for $x \geq x_{i_n}$, and to be linear between x_{i_k} and $x_{i_{k+1}}$ for $k = 1, \ldots, n - 1$. A formal expression for the maximal invariant in this case is the set of *ranks* $(r_1, \ldots, r_n)$ of $(x_1, \ldots, x_n)$. Here the rank r_i of x_i is defined through

$$x_i = x_{(r_i)}$$

so that r_i is the number of x's $\leq x_i$. In particular $r_i = 1$ if x_i is the smallest x, $r_i = 2$ if it is the second smallest, and so on.

Example 3.   Let x be an $n \times s$ matrix ($s \leq n$) of rank s, and let G be the group of linear transformations $gx = xB$, where B is any nonsingular $s \times s$ matrix. Then a maximal invariant under G is the matrix $t(x) = x(x'x)^{-1}x'$, where x' denotes the transpose of x. Here $(x'x)^{-1}$ is meaningful because the $s \times s$ matrix $x'x$ is nonsingular; in fact, it will be shown in Lemma 1 of Chapter 8 that $x'x$ is positive definite.

That $t(x)$ is invariant is clear, since

$$t(gx) = xB(B'x'xB)^{-1}B'x' = x(x'x)^{-1}x' = t(x).$$

To see that $t(x)$ is maximal invariant, suppose that

$$x_1(x_1'x_1)^{-1}x_1' = x_2(x_2'x_2)^{-1}x_2'.$$

Since $(x_i'x_i)^{-1}$ is positive definite, there exist nonsingular matrices C_i such that

$(x_i'x_i)^{-1} = C_iC_i'$ and hence

$$(x_1C_1)(x_1C_1)' = (x_2C_2)(x_2C_2)'.$$

As will be shown in Chapter 8, Section 2, this implies the existence of an orthogonal matrix Q such that $x_2C_2 = x_1C_1Q$ and thus $x_2 = x_1B$ with $B = C_1QC_2^{-1}$, as was to be shown.

In the special case $s = n$, we have $t(x) = I$, so that there are no nontrivial invariants. This corresponds to the fact that in this case G is transitive, since any two nonsingular $n \times n$ matrices x_1 and x_2 satisfy $x_2 = x_1B$ with $B = x_1^{-1}x_2$.

This result can be made more intuitive through a geometric interpretation. Consider the s-dimensional subspace S of R^n spanned by the s columns of x. Then $P = x(x'x)^{-1}x'$ has the property that for any y in R^n, the vector Py is the projection of y onto S. (This will be proved in Chapter 7, Section 2.) The invariance of P expresses the fact that the projection of y onto S is independent of the choice of vectors spanning S. To see that it is maximal invariant, suppose that the projection of every y onto the spaces S_1 and S_2 spanned by two different sets of s vectors is the same. Then $S_1 = S_2$, so that the two sets of vectors span the same space. There then exists a nonsingular transformation taking one of these sets into the other.

A somewhat more systematic way of determining maximal invariants is obtained by selecting, by means of a specified rule, a unique point $M(x)$ on each orbit. Then clearly $M(X)$ is maximal invariant. To illustrate this method, consider once more two of the earlier examples.

Example 1(i) (continued). The orbit containing the point $(a_1, \ldots, a_n)$ under the group of translations is the set $\{(a_1 + c, \ldots, a_n + c), -\infty < c < \infty\}$, which is a line in E_n.

(a) As representative point $M(x)$ on this line, take its intersection with the hyperplane $x_n = 0$. Since then $a_n + c = 0$, this point corresponds to the value $c = -a_n$ and thus has coordinates $(a_1 - a_n, \ldots, a_{n-1} - a_n, 0)$. This leads to the maximal invariant $(x_1 - x_n, \ldots, x_{n-1} - x_n)$.

(b) An alternative point on the line is its intersection with the hyperplane $\Sigma x_i = 0$. Then $c = -\bar{a}$, and $M(a) = (a_1 - \bar{a}, \ldots, a_n - \bar{a})$.

(c) The point need not be specified by an intersection property. It can for instance be taken as the point on the line that is closest to the origin. Since the value of c minimizing $\Sigma(a_i + c)^2$ is $c = -\bar{a}$, this leads to the same point as (b).

Example 1(iii) (continued). The orbit containing the point $(a_1, \ldots, a_n)$ under the group of orthogonal transformations is the hypersphere containing $(a_1, \ldots, a_n)$ and with center at the origin. As representative point on this sphere, take its north pole, i.e. the point with $a_1 = \cdots = a_{n-1} = 0$. The coordinates of this point are $\left(0, \ldots, 0, \sqrt{\Sigma a_i^2}\right)$ and hence lead to the maximal invariant Σx_i^2. (Note that in this example, the determination of the orbit is essentially equivalent to the determination of the maximal invariant.)

Frequently, it is convenient to obtain a maximal invariant in a number of steps, each corresponding to a subgroup of G. To illustrate the process and a difficulty that may arise in its application, let $x = (x_1, \ldots, x_n)$, suppose that the coordinates are distinct, and consider the group of transformations

$$gx = (ax_1 + b, \ldots, ax_n + b), \qquad a \neq 0, \quad -\infty < b < \infty.$$

Applying first the subgroup of translations $x_i' = x_i + b$, a maximal invariant is $y = (y_1, \ldots, y_{n-1})$ with $y_i = x_i - x_n$. Another subgroup consists of the scale changes $x_i'' = ax_i$. This induces a corresponding change of scale in the y's: $y_i'' = ay_i$, and a maximal invariant with respect to this group acting on the y-space is $z = (z_1, \ldots, z_{n-2})$ with $z_i = y_i/y_{n-1}$. Expressing this in terms of the x's, we get $z_i = (x_i - x_n)/(x_{n-1} - x_n)$, which is maximal invariant with respect to G.

Suppose now the process is carried out in the reverse order. Application first of the subgroup $x_i'' = ax_i$ yields as maximal invariant $u = (u_1, \ldots, u_{n-1})$ with $u_i = x_i/x_n$. However, the translations $x_i' = x_i + b$ do not induce transformations in u-space, since $(x_i + b)/(x_n + b)$ is not a function of x_i/x_n.

Quite generally, let a transformation group G be *generated* by two subgroups D and E in the sense that it is the smallest group containing D and E. Then G consists of the totality of products $e_m d_m \ldots e_1 d_1$ for $m = 1, 2, \ldots$, with $d_i \in D$, $e_i \in E(i = 1, \ldots, m)$.[†] The following theorem shows that whenever the process of determining a maximal invariant in steps can be carried out at all, it leads to a maximal invariant with respect to G.

Theorem 2.   *Let G be a group of transformations, and let D and E be two subgroups generating G. Suppose that $y = s(x)$ is maximal invariant with respect to D, and that for any $e \in E$*

(8) $$s(x_1) = s(x_2) \quad implies \quad s(ex_1) = s(ex_2).$$

If $z = t(y)$ is maximal invariant under the group E^* of transformations e^* defined by

$$e^*y = s(ex) \qquad when \quad y = s(x),$$

then $z = t[s(x)]$ is maximal invariant with respect to G.

Proof.   To show that $t[s(x)]$ is invariant, let $x' = gx$, $g = e_m d_m \ldots e_1 d_1$. Then

$$t[s(x')] = t[s(e_m d_m \ldots e_1 d_1 x)] = t[e_m^* s(d_m \ldots e_1 d_1 x)]$$
$$= t[s(e_{m-1} d_{m-1} \ldots e_1 d_1 x)],$$

[†]See Section 1 of the Appendix.

and the last expression can be reduced by induction to $t[s(x)]$. To see that $t[s(x)]$ is in fact maximal invariant, suppose that $t[s(x')] = t[s(x)]$. Setting $y' = s(x')$, $y = s(x)$, one has $t(y') = t(y)$, and since $t(y)$ is maximal invariant with respect to E^*, there exists e^* such that $y' = e^*y$. Then $s(x') = e^*s(x) = s(ex)$, and by the maximal invariance of $s(x)$ with respect to D there exists $d \in D$ such that $x' = dex$. Since de is an element of G this completes the proof.

Techniques for obtaining the distribution of maximal invariants are discussed by Andersson (1982), Eaton (1983), Farrell (1985), and Wijsman (1985).

3. MOST POWERFUL INVARIANT TESTS

The class of all invariant functions can be obtained as the totality of functions of a maximal invariant $M(x)$. Therefore, in particular the class of all invariant tests is the totality of tests depending only on the maximal invariant statistic M. The latter statement, while correct for all the usual situations, actually requires certain qualifications regarding the class of measurable sets in M-space. These conditions will be discussed at the end of the section; they are satisfied in the examples below.

Example 4. Let $X = (X_1, \ldots, X_n)$, and suppose that the density of X is $f_i(x_1 - \theta, \ldots, x_n - \theta)$ under H_i $(i = 0, 1)$, where θ ranges from $-\infty$ to ∞. The problem of testing H_0 against H_1 is invariant under the group G of transformations

$$gx = (x_1 + c, \ldots, x_n + c), \quad -\infty < c < \infty,$$

which in the parameter space induces the transformations

$$\bar{g}\theta = \theta + c.$$

By Example 1, a maximal invariant under G is $Y = (X_1 - X_n, \ldots, X_{n-1} - X_n)$. The distribution of Y is independent of θ and under H_i has the density

$$\int_{-\infty}^{\infty} f_i(y_1 + z, \ldots, y_{n-1} + z, z) \, dz.$$

When referred to Y, the problem of testing H_0 against H_1 therefore becomes one of testing a simple hypothesis against a simple alternative. The most powerful test is then independent of θ, and therefore UMP among all invariant tests. Its rejection region by the Neyman–Pearson lemma is

$$\frac{\int_{-\infty}^{\infty} f_1(y_1 + z, \ldots, y_{n-1} + z, z) \, dz}{\int_{-\infty}^{\infty} f_0(y_1 + z, \ldots, y_{n-1} + z, z) \, dz} = \frac{\int_{-\infty}^{\infty} f_1(x_1 + u, \ldots, x_n + u) \, du}{\int_{-\infty}^{\infty} f_0(x_1 + u, \ldots, x_n + u) \, du} > C.$$

A general theory of *separate families of hypotheses* (in which the family K of alternatives does not adjoin the hypothesis H but, as in Example 4, is separated from it) was initiated by Cox (1961, 1962). A bibliography of the subject is given in Pereira (1977); see also Loh (1985).

Before applying invariance, it is frequently convenient first to reduce the data to a sufficient statistic T. If there exists a test $\phi_0(T)$ that is UMP among all invariant tests depending only on T, one would like to be able to conclude that $\phi_0(T)$ is also UMP among all invariant tests based on the original X. Unfortunately, this does not follow, since it is not clear that for any invariant test based on X there exists an equivalent test based on T, which is also invariant. Sufficient conditions for $\phi_0(T)$ to have this property are provided by Hall, Wijsman, and Ghosh (1965) and Hooper (1982a), and a simple version of such a result (applicable to Examples 5 and 6 below) will be given by Theorem 6 in Section 5. The relationship between sufficiency and invariance is discussed further in Berk (1972) and Landers and Rogge (1973).

Example 5. If $X_1, \ldots, X_n$ is a sample from $N(\xi, \sigma^2)$, the hypothesis $H: \sigma \geq \sigma_0$ remains invariant under the transformations $X_i' = X_i + c$, $-\infty < c < \infty$. In terms of the sufficient statistics $Y = \overline{X}$, $S^2 = \Sigma(X_i - \overline{X})^2$ these transformations become $Y' = Y + c$, $(S^2)' = S^2$, and a maximal invariant is S^2. The class of invariant tests is therefore the class of tests depending on S^2. It follows from Theorem 2 of Chapter 3 that there exists a UMP invariant test, with rejection region $\Sigma(X_i - \overline{X})^2 \leq C$. This coincides with the UMP unbiased test (9) of Chapter 5.

Example 6. If $X_1, \ldots, X_m$ and $Y_1, \ldots, Y_n$ are samples from $N(\xi, \sigma^2)$ and $N(\eta, \tau^2)$, a set of sufficient statistics is $T_1 = \overline{X}$, $T_2 = \overline{Y}$, $T_3 = \sqrt{\Sigma(X_i - \overline{X})^2}$, and $T_4 = \sqrt{\Sigma(Y_j - \overline{Y})^2}$. The problem of testing $H: \tau^2/\sigma^2 \leq \Delta_0$ remains invariant under the transformations $T_1' = T_1 + c_1$, $T_2' = T_2 + c_2$, $T_3' = T_3$, $T_4' = T_4$, $-\infty < c_1, c_2 < \infty$, and also under a common change of scale of all four variables. A maximal invariant with respect to the first group is (T_3, T_4). In the space of this maximal invariant, the group of scale changes induces the transformations $T_3'' = cT_3$, $T_4'' = cT_4$, $0 < c$, which has as maximal invariant the ratio T_4/T_3. The statistic $Z = [T_4^2/(n-1)] \div [T_3^2/(m-1)]$ on division by $\Delta = \tau^2/\sigma^2$ has an F-distribution with density given by (21) of Chapter 5, so that the density of Z is

$$\frac{C(\Delta)z^{\frac{1}{2}(n-3)}}{\left(\Delta + \dfrac{n-1}{m-1}z\right)^{\frac{1}{2}(m+n-2)}}, \quad z > 0.$$

For varying Δ, these densities constitute a family with monotone likelihood ratio, so that among all tests of H based on Z, and therefore among all invariant tests, there exists a UMP one given by the rejection region $Z > C$. This coincides with the UMP unbiased test (20) of Chapter 5.

Example 7. In the method of *paired comparisons* for testing whether a treatment has a beneficial effect, the experimental material consists of n pairs of subjects. From each pair, a subject is selected at random for treatment while the other serves as control. Let X_i be 1 or 0 as for the ith pair the experiment turns out in favor of the treated subject or the control, and let $p_i = P\{X_i = 1\}$. The hypothesis of no effect, $H: p_i = \frac{1}{2}$ for $i = 1, \ldots, n$, is to be tested against the alternatives that $p_i > \frac{1}{2}$ for all i.

The problem remains invariant under all permutations of the n variables $X_1, \ldots, X_n$, and a maximal invariant under this group is the total number of successes $X = X_1 + \cdots + X_n$. The distribution of X is

$$P\{X = k\} = q_1 \ldots q_n \sum \frac{p_{i_1}}{q_{i_1}} \cdots \frac{p_{i_k}}{q_{i_k}},$$

where $q_i = 1 - p_i$ and where the summation extends over all $\binom{n}{k}$ choices of subscripts $i_1 < \cdots < i_k$. The most powerful invariant test against an alternative $(p'_1, \ldots, p'_n)$ rejects H when

$$f(k) = \frac{1}{\binom{n}{k}} \sum \frac{p'_{i_1}}{q'_{i_1}} \cdots \frac{p'_{i_k}}{q'_{i_k}} > C.$$

To see that f is an increasing function of k, note that $a_i = p'_i / q'_i > 1$, and that

$$\sum_j \sum a_j a_{i_1} \ldots a_{i_k} = (k + 1) \sum a_{i_1} \ldots a_{i_{k+1}}$$

and

$$\sum_j \sum a_{i_1} \ldots a_{i_k} = (n - k) \sum a_{i_1} \ldots a_{i_k}.$$

Here, in both equations, the second summation on the left-hand side extends over all subscripts $i_1 < \cdots < i_k$ of which none is equal to j, and the summation on the right-hand side extends over all subscripts $i_1 < \cdots < i_{k+1}$ and $i_1 < \cdots < i_k$ respectively without restriction. Then

$$f(k + 1) = \frac{1}{\binom{n}{k+1}} \sum a_{i_1} \ldots a_{i_{k+1}} = \frac{1}{(n-k)\binom{n}{k}} \sum_j \sum a_j a_{i_1} \ldots a_{i_k}$$

$$> \frac{1}{\binom{n}{k}} \sum a_{i_1} \ldots a_{i_k} = f(k),$$

as was to be shown. Regardless of the alternative chosen, the test therefore rejects when $k > C$, and hence is UMP invariant. If the ith comparison is considered plus or minus as X_i is 1 or 0, this is seen to be another example of the sign test. (Cf. Chapter 3, Example 8, and Chapter 4, Section 9.)

Sufficient statistics provide a simplification of a problem by reducing the sample space; this process involves no change in the parameter space. Invariance, on the other hand, by reducing the data to a maximal invariant statistic M, whose distribution may depend only on a function of the parameter, typically also shrinks the parameter space. The details are given in the following theorem.

Theorem 3. *If $M(x)$ is invariant under G, and if $v(\theta)$ is maximal invariant under the induced group $\overline{G}$, then the distribution of $M(X)$ depends only on $v(\theta)$.*

Proof. Let $v(\theta_1) = v(\theta_2)$. Then $\theta_2 = \overline{g}\theta_1$, and hence

$$P_{\theta_2}\{M(X) \in B\} = P_{\overline{g}\theta_1}\{M(X) \in B\} = P_{\theta_1}\{M(gX) \in B\}$$

$$= P_{\theta_1}\{M(X) \in B\}.$$

This result can be paraphrased by saying that the principle of invariance identifies all parameter points that are equivalent with respect to $\overline{G}$.

In application, for instance in Examples 5 and 6, the maximal invariants $M(x)$ and $\delta = v(\theta)$ under G and $\overline{G}$ are frequently real-valued, and the family of probability densities $p_\delta(m)$ of M has monotone likelihood ratio. For testing the hypothesis $H: \delta \leq \delta_0$ there exists then a UMP test among those depending only on M, and hence a UMP invariant test. Its rejection region is $M \geq C$, where

$$(9) \qquad \int_C^\infty p_{\delta_0}(m) \, dm = \alpha.$$

Consider this problem now as a two-decision problem with decisions d_0 and d_1 of accepting or rejecting H, and a loss function $L(\theta, d_i) = L_i(\theta)$. Suppose that $L_i(\theta)$ depends only on the parameter δ, $L_i(\theta) = L_i'(\delta)$ say, and satisfies

$$(10) \qquad L_1'(\delta) - L_0'(\delta) \gtrless 0 \qquad \text{as} \quad \delta \lessgtr \delta_0.$$

It then follows from Theorem 3 of Chapter 3 that the family of rejection regions $M \geq C(\alpha)$, as α varies from 0 to 1, forms a complete family of decision procedures among those depending only on M, and hence a complete family of invariant procedures. As before, the choice of a particular significance level α can be considered as a convenient way of specifying a test from this family.

At the beginning of the section it was stated that the class of invariant tests coincides with the class of tests based on a maximal invariant statistic $M = M(X)$. However, a statistic is not completely specified by a function, but requires also specification of a class $\mathscr{B}$ of measurable sets. If in the present case $\mathscr{B}$ is the class of all sets B for which $M^{-1}(B) \in \mathscr{A}$, the desired statement is correct. For let $\phi(x) = \psi[M(x)]$ and ϕ by $\mathscr{A}$-measurable, and let C be a Borel set on the line. Then $\phi^{-1}(C) = M^{-1}[\psi^{-1}(C)] \in \mathscr{A}$ and hence $\psi^{-1}(C) \in \mathscr{B}$, so that ψ is $\mathscr{B}$-measurable and $\phi(x) = \psi[M(x)]$ is a test based on the statistic M.

In most applications, $M(x)$ is a measurable function taking on values in a Euclidean space and it is convenient to take $\mathscr{B}$ as the class of Borel sets. If $\phi(x) = \psi[M(x)]$ is then an arbitrary measurable function depending only on $M(x)$, it is not clear that $\psi(m)$ is necessarily $\mathscr{B}$-measurable. This measurability can be concluded if $\mathscr{X}$ is also Euclidean with $\mathscr{A}$ the class of Borel sets, and if the range of M is a Borel set. We shall prove it here only under the additional assumption (which in applications is usually obvious, and which will not be verified explicitly in each case) that there exists a vector-valued Borel-measurable function $Y(x)$ such that $[M(x), Y(x)]$ maps $\mathscr{X}$ onto a Borel subset of the product space $\mathscr{M} \times \mathscr{Y}$, that this mapping is $1:1$, and that the inverse mapping is also Borel-measurable. Given any measurable function ϕ of x, there exists then a measurable function ϕ' of (m, y) such that $\phi(x) \equiv \phi'[M(x), Y(x)]$. If ϕ depends only on $M(x)$, then ϕ' depends only on m, so that $\phi'(m, y) = \psi(m)$ say, and ψ is a measurable function of m.* In Example 1(i) for instance, where $x = (x_1, \ldots, x_n)$ and $M(x) = (x_1 - x_n, \ldots, x_{n-1} - x_n)$, the function $Y(x)$ can be taken as $Y(x) = x_n$.

4. SAMPLE INSPECTION BY VARIABLES

A sample is drawn from a lot of some manufactured product in order to decide whether the lot is of acceptable quality. In the simplest case, each sample item is classified directly as satisfactory or defective (*inspection by attributes*), and the decision is based on the total number of defectives. More generally, the quality of an item is characterized by a variable Y (*inspection by variables*), and an item is considered satisfactory if Y exceeds a given constant u. The probability of a defective is then

$$p = P\{Y \le u\}$$

and the problem becomes that of testing the hypothesis $H : p \ge p_0$.

---

*The last statement is an immediate consequence, for example, of Theorem B, Section 34, of Halmos (1974).

As was seen in Example 8 of Chapter 3, no use can be made of the actual value of Y unless something is known concerning the distribution of Y. In the absence of such information, the decision will be based, as before, simply on the number of defectives in the sample. We shall consider the problem now under the assumption that the measurements $Y_1, \ldots, Y_n$ constitute a sample from $N(\eta, \sigma^2)$. Then

$$p = \int_{-\infty}^{u} \frac{1}{\sqrt{2\pi}\,\sigma} \exp\left[-\frac{1}{2\sigma^2}(y - \eta)^2\right] dy = \Phi\left(\frac{u - \eta}{\sigma}\right),$$

where

$$\Phi(y) = \int_{-\infty}^{y} \frac{1}{\sqrt{2\pi}} \exp\left(-\tfrac{1}{2}t^2\right) dt$$

denotes the cumulative distribution function of a standard normal distribution, and the hypothesis H becomes $(u - \eta)/\sigma \geq \Phi^{-1}(p_0)$. In terms of the variables $X_i = Y_i - u$, which have mean $\xi = \eta - u$ and variance σ^2, this reduces to

$$H : \frac{\xi}{\sigma} \leq \theta_0$$

with $\theta_0 = -\Phi^{-1}(p_0)$. This hypothesis, which was considered in Chapter 5, Section 2, for $\theta_0 = 0$, occurs also in other contexts. It is appropriate when one is interested in the mean ξ of a normal distribution, expressed in σ-units rather than on a fixed scale.

For testing H, attention can be restricted to the pair of variables $\overline{X}$ and $S = \sqrt{\Sigma(X_i - \overline{X})^2}$, since they form a set of sufficient statistics for (ξ, σ), which satisfy the conditions of Theorem 6 of the next section. These variables are independent, the distribution of $\overline{X}$ being $N(\xi, \sigma^2/n)$ and that of S/σ being χ_{n-1}. Multiplication of $\overline{X}$ and S by a common constant $c > 0$ transforms the parameters into $\xi' = c\xi$, $\sigma' = c\sigma$, so that ξ/σ and hence the problem of testing H remain invariant. A maximal invariant under these transformations is $\overline{x}/s$ or

$$t = \frac{\sqrt{n}\,\overline{x}}{s/\sqrt{n - 1}},$$

the distribution of which depends only on the maximal invariant in the parameter space $\theta = \xi/\sigma$ (cf. Chapter 5, Section 2). Thus, the invariant tests are those depending only on t, and it remains to find the most powerful test of $H : \theta \leq \theta_0$ within this class.

The probability density of t is (Chapter 5, Problem 3)

$$p_\delta(t) = C \int_0^\infty \exp\left[-\frac{1}{2}\left(t\sqrt{\frac{w}{n-1}} - \delta\right)^2\right] w^{\frac{1}{2}(n-2)} \exp\left(-\tfrac{1}{2}w\right) dw,$$

where $\delta = \sqrt{n}\,\theta$ is the noncentrality parameter, and this will now be shown to constitute a family with monotone likelihood ratio. To see that the ratio

$$r(t) = \frac{\displaystyle\int_0^\infty \exp\left[-\frac{1}{2}\left(t\sqrt{\frac{w}{n-1}} - \delta_1\right)^2\right] w^{\frac{1}{2}(n-2)} \exp\left(-\tfrac{1}{2}w\right) dw}{\displaystyle\int_0^\infty \exp\left[-\frac{1}{2}\left(t\sqrt{\frac{w}{n-1}} - \delta_0\right)^2\right] w^{\frac{1}{2}(n-2)} \exp\left(-\tfrac{1}{2}w\right) dw}$$

is an increasing function of t for $\delta_0 < \delta_1$, suppose first that $t < 0$ and let $v = -t\sqrt{w/(n-1)}$. The ratio then becomes proportional to

$$\frac{\displaystyle\int_0^\infty f(v)\exp\left[-(\delta_1 - \delta_0)v - \frac{(n-1)v^2}{2t^2}\right] dv}{\displaystyle\int_0^\infty f(v)\exp\left[-\frac{(n-1)v^2}{2t^2}\right] dv}$$

$$= \int \exp[-(\delta_1 - \delta_0)v]\, g_{t^2}(v)\, dv$$

where

$$f(v) = \exp(-\delta_0 v)\, v^{n-1} \exp(-v^2/2)$$

and

$$g_{t^2}(v) = \frac{f(v)\exp\left[-\dfrac{(n-1)v^2}{2t^2}\right]}{\displaystyle\int_0^\infty f(z)\exp\left[-\dfrac{(n-1)z^2}{2t^2}\right] dz}.$$

Since the family of probability densities $g_{t^2}(v)$ is a family with monotone likelihood ratio, the integral of $\exp[-(\delta_1 - \delta_0)v]$ with respect to this density is a decreasing function of t^2 (Problem 14 of Chapter 3), and hence an increasing function of t for $t < 0$. Similarly one finds that $r(t)$ is an increasing function of t for $t > 0$ by making the transformation $v = t\sqrt{w/(n-1)}$. By continuity it is then an increasing function of t for all t.

There exists therefore a UMP invariant test of $H: \xi/\sigma \le \theta_0$, which rejects when $t > C$, where C is determined by (9). In terms of the original variables Y_i the rejection region of the UMP invariant test of $H: p \ge p_0$ becomes

$$(11) \qquad \frac{\sqrt{n}\,(\bar{y} - u)}{\sqrt{\Sigma(y_i - \bar{y})^2/(n-1)}} > C.$$

If the problem is considered as a two-decision problem with losses $L_0(p)$ and $L_1(p)$ for accepting or rejecting $p \ge p_0$, which depend only on p and satisfy the condition corresponding to (10), the class of tests (11) constitutes a complete family of invariant procedures as C varies from $-\infty$ to ∞.

Consider next the comparison of two products on the basis of samples $X_1, \ldots, X_m; Y_1, \ldots, Y_n$ from $N(\xi, \sigma^2)$ and $N(\eta, \sigma^2)$. If

$$p = \Phi\left(\frac{u - \xi}{\sigma}\right), \qquad \pi = \Phi\left(\frac{u - \eta}{\sigma}\right),$$

one wishes to test the hypothesis $p \le \pi$, which is equivalent to

$$H: \eta \le \xi.$$

The statistics $\bar{X}$, $\bar{Y}$, and $S = \sqrt{\Sigma(X_i - \bar{X})^2 + \Sigma(Y_j - \bar{Y})^2}$ are a set of sufficient statistics for ξ, η, σ. The problem remains invariant under the addition of an arbitrary common constant to $\bar{X}$ and $\bar{Y}$, which leaves $\bar{Y} - \bar{X}$ and S as maximal invariants. It is also invariant under multiplication of $\bar{X}$, $\bar{Y}$, and S, and hence of $\bar{Y} - \bar{X}$ and S, by a common positive constant, which reduces the data to the maximal invariant $(\bar{Y} - \bar{X})/S$. Since

$$t = \frac{(\bar{y} - \bar{x})/\sqrt{\dfrac{1}{m} + \dfrac{1}{n}}}{s/\sqrt{m+n-2}}$$

has a noncentral t-distribution with noncentrality parameter $\delta = \sqrt{mn}\,(\eta - \xi)/\sqrt{m+n}\,\sigma$, the UMP invariant test of $H: \eta - \xi \le 0$ rejects when $t > C$. This coincides with the UMP unbiased test (27) of Chapter 5, Section 3. Analogously, the corresponding two-sided test (30) of Chapter 5, with rejection region $|t| \ge C$, is UMP invariant for testing the hypothesis $p = \pi$ against the alternatives $p \ne \pi$ (Problem 9).

5. ALMOST INVARIANCE

Let G be a group of transformations leaving a family $\mathscr{P} = \{ P_\theta, \theta \in \Omega \}$ of distributions of X invariant. A test ϕ is said to be *equivalent to an invariant test* if there exists an invariant test ψ such that $\phi(x) = \psi(x)$ for all x except possibly on a $\mathscr{P}$-null set N; ϕ is said to be *almost invariant with respect to G if*

$$(12) \qquad \phi(gx) = \phi(x) \qquad \text{for all} \quad x \in \mathscr{X} - N_g, \quad g \in G$$

where the exceptional null set N_g is permitted to depend on g. This concept is required for investigating the relationship of invariance to unbiasedness and to certain other desirable properties. In this connection it is important to know whether a UMP invariant test is also UMP among almost invariant tests. This turns out to be the case under assumptions which are made precise in Theorem 4 below and which are satisfied in all the usual applications.

If ϕ is equivalent to an invariant test, then $\phi(gx) = \phi(x)$ for all $x \notin N \cup g^{-1}N$. Since $P_\theta(g^{-1}N) = P_{\bar{g}\theta}(N) = 0$, it follows that ϕ is then almost invariant. The following theorem gives conditions under which conversely any almost invariant test is equivalent to an invariant one.

Theorem 4. *Let G be a group of transformations of $\mathscr{X}$, and let $\mathscr{A}$ and $\mathscr{B}$ be σ-fields of subsets of $\mathscr{X}$ and G such that for any set $A \in \mathscr{A}$ the set of pairs (x, g) for which $gx \in A$ is measurable $\mathscr{A} \times \mathscr{B}$. Suppose further that there exists a σ-finite measure ν over G such that $\nu(B) = 0$ implies $\nu(Bg) = 0$ for all $g \in G$. Then any measurable function that is almost invariant under G (where "almost" refers to some σ-finite measure μ) is equivalent to an invariant function.*

Proof. Because of the measurability assumptions, the function $\phi(gx)$ considered as a function of the two variables x and g is measurable $\mathscr{A} \times \mathscr{B}$. It follows that $\phi(gx) - \phi(x)$ is measurable $\mathscr{A} \times \mathscr{B}$, and so therefore is the set S of points (x, g) with $\phi(gx) \neq \phi(x)$. If ϕ is almost invariant, any section of S with fixed g is a μ-null set. By Fubini's theorem (Theorem 3 of Chapter 2) there exists therefore a μ-null set N such that for all $x \in \mathscr{X} - N$

$$\phi(gx) = \phi(x) \qquad \text{a.e. } \nu.$$

Without loss of generality suppose that $\nu(G) = 1$, and let A be the set of points x for which

$$\int \phi(g'x) \, d\nu(g') = \phi(gx) \qquad \text{a.e. } \nu.$$

If

$$f(x, g) = \left| \int \phi(g'x) \, d\nu(g') - \phi(gx) \right|,$$

then A is the set of points x for which

$$\int f(x, g) \, d\nu(g) = 0.$$

Since this integral is a measurable function of x, it follows that A is measurable. Let

$$\psi(x) = \begin{cases} \int \phi(gx) \, d\nu(g) & \text{if } x \in A, \\ 0 & \text{if } x \notin A. \end{cases}$$

Then ψ is measurable and $\psi(x) = \phi(x)$ for $x \notin N$, since $\phi(gx) = \phi(x)$ a.e. ν implies that $\int \phi(g'x) \, d\nu(g') = \phi(x)$ and that $x \in A$. To show that ψ is invariant it is enough to prove that the set A is invariant. For any point $x \in A$, the function $\phi(gx)$ is constant except on a null subset N_x of G. Then $\phi(ghx)$ has the same constant value for all $g \notin N_x h^{-1}$, which by assumption is again a ν-null set; and hence $hx \in A$, which completes the proof.

Additional results concerning the relation of invariance and almost invariance are given by Berk and Bickel (1968) and Berk (1970). In particular, the basic idea of the following example is due to Berk (1970).

Example 8. Counterexample. Let $Z, Y_1, \ldots, Y_n$ be independently distributed as $N(\theta, 1)$, and consider the $1:1$ transformations $y_i' = y_i$ $(i = 1, \ldots, n)$ and

$z' = z$ except for a finite number of points $a_1, \ldots, a_k$ for which $a_i' = a_{j_i}$ for some permutation $(j_1, \ldots, j_k)$ of $(1, \ldots, k)$.

If the group G is generated by taking for $(a_1, \ldots, a_k)$, $k = 1, 2, \ldots$, all finite sets and for $(j_1, \ldots, j_k)$ all permutations of $(1, \ldots, k)$, then $(z, y_1, \ldots, y_n)$ is almost invariant. It is however not equivalent to an invariant function, since $(y_1, \ldots, y_n)$ is maximal invariant.

Corollary 1. *Suppose that the problem of testing $H : \theta \in \omega$ against $K : \theta \in \Omega - \omega$ remains invariant under G and that the assumptions of Theorem 4 hold. Then if ϕ_0 is UMP invariant, it is also UMP within the class of almost invariant tests.*

Proof. If ϕ is almost invariant, it is equivalent to an invariant test ψ by Theorem 4. The tests ϕ and ψ have the same power function, and hence ϕ_0 is uniformly at least as powerful as ϕ.

In applications, $\mathscr{P}$ is usually a dominated family, and μ any σ-finite measure equivalent to $\mathscr{P}$ (which exists by Theorem 2 of the Appendix). If ϕ is almost invariant with respect to $\mathscr{P}$, it is then almost invariant with respect to μ and hence equivalent to an invariant test. Typically, the sample space $\mathscr{X}$ is an n-dimensional Euclidean space, $\mathscr{A}$ is the class of Borel sets, and the elements of G are transformations of the form $y = f(x, \tau)$, where τ ranges over a set of positive measure in an m-dimensional space and f is a Borel-measureable vector-valued function of $m + n$ variables. If $\mathscr{B}$ is taken as the class of Borel sets in m-space, the measurability conditions of the theorem are satisfied.

The requirement that for all $g \in G$ and $B \in \mathscr{B}$

$$(13) \qquad \nu(B) = 0 \quad \text{implies} \quad \nu(Bg) = 0$$

is satisfied in particular when

$$(14) \qquad \nu(Bg) = \nu(B) \qquad \text{for all} \quad g \in G, \quad B \in \mathscr{B}.$$

The existence of such a *right invariant measure* is guaranteed for a large class of groups by the theory of Haar measure. Alternatively, it is usually not difficult to check the condition (13) directly.

Example 9. Let G be the group of all nonsingular linear transformations of n-space. Relative to a fixed coordinate system the elements of G can be represented by nonsingular $n \times n$ matrices $A = (a_{ij})$, $A' = (a'_{ij}), \ldots$ with the matrix product serving as the group product of two such elements. The σ-field $\mathscr{B}$ can be taken to be the class of Borel sets in the space of the n^2 elements of the matrices, and the measure ν can be taken as Lebesgue measure over $\mathscr{B}$. Consider now a set S of matrices with $\nu(S) = 0$, and the set S^* of matrices $A'A$ with $A' \in S$ and A fixed. If $a = \max|a_{ij}|$, $C' = A'A$, and $C'' = A''A$, the inequalities $|a''_{ij} - a'_{ij}| \le \epsilon$ for all i, j imply $|c''_{ij} - c'_{ij}| \le na\epsilon$. Since a set has ν-measure zero if and only if it can be covered by a union of rectangles whose total measure does not exceed any given $\epsilon > 0$, it follows that $\nu(S^*) = 0$, as was to be proved.

In the preceding chapters, tests were compared purely in terms of their power functions (possibly weighted according to the seriousness of the losses involved). Since the restriction to invariant tests is a departure from this point of view, it is of interest to consider the implications of applying invariance to the power functions rather than to the tests themselves. Any test that is invariant or almost invariant under a group G has a power function which is invariant under the group $\overline{G}$ induced by G in the parameter space.

To see that the converse is in general not true, let X_1, X_2, X_3 be independently, normally distributed with mean ξ and variance σ^2, and consider the hypothesis $\sigma \geq \sigma_0$. The test with rejection region

$$|X_2 - X_1| > k \quad \text{when} \quad \overline{X} < 0,$$

$$|X_3 - X_2| > k \quad \text{when} \quad \overline{X} \geq 0$$

is not invariant under the group G of transformations $X_i' = X_i + c$, but its power function is invariant under the associated group $\overline{G}$.

The two properties, almost invariance of a test ϕ and invariance of its power function, become equivalent if before the application of invariance considerations the problem is reduced to a sufficient statistic whose distributions constitute a boundedly complete family.

Lemma 2. *Let the family $\mathscr{P}^T = \{P_\theta^T, \theta \in \Omega\}$ of distributions of T be boundedly complete, and let the problem of testing $H : \theta \in \Omega_H$ remain invariant under a group G of transformations of T. Then a necessary and sufficient condition for the power function of a test $\psi(t)$ to be invariant under the induced group $\overline{G}$ over Ω is that $\psi(t)$ is almost invariant under G.*

Proof. For all $\theta \in \Omega$ we have $E_{\overline{g}\theta}\psi(T) = E_\theta\psi(gT)$. If ψ is almost invariant, $E_\theta\psi(T) = E_\theta\psi(gT)$ and hence $E_{\overline{g}\theta}\psi(T) = E_\theta\psi(T)$, so that the power function of ψ is invariant. Conversely, if $E_\theta\psi(T) = E_{\overline{g}\theta}\psi(T)$, then $E_\theta\psi(T) = E_\theta\psi(gT)$, and it follows from the bounded completeness of $\mathscr{P}^T$ that $\psi(gt) = \psi(t)$ a.e. $\mathscr{P}^T$.

As a consequence, it is seen that UMP almost invariant tests also possess the following optimum property.

Theorem 5. *Under the assumptions of Lemma 2, let $v(\theta)$ be maximal invariant with respect to $\overline{G}$, and suppose that among the tests of H based on the sufficient statistic T there exists a UMP almost invariant one, say $\psi_0(t)$. Then $\psi_0(t)$ is UMP in the class of all tests based on the original observations X, whose power function depends only on $v(\theta)$.*

Proof. Let $\phi(x)$ be any such test, and let $\psi(t) = E[\phi(X)|t]$. The power function of $\psi(t)$, being identical with that of $\phi(x)$, depends then only on $v(\theta)$, and hence is invariant under $\overline{G}$. It follows from Lemma 2 that $\psi(t)$ is almost invariant under G, and $\psi_0(t)$ is uniformly at least as powerful as $\psi(t)$ and therefore as $\phi(x)$.

Example 10. For the hypothesis $\tau^2 \leq \sigma^2$ concerning the variances of two normal distributions, the statistics $(\overline{X}, \overline{Y}, S_X^2, S_Y^2)$ constitute a complete set of sufficient statistics. It was shown in Example 6 that there exists a UMP invariant test with respect to a suitable group G, which has rejection region $S_Y^2/S_X^2 > C_0$.

Since in the present case almost invariance of a test with respect to G implies that it is equivalent to an invariant one (Problem 12), Theorem 5 is applicable with $v(\theta) = \Delta = \tau^2/\sigma^2$, and the test is therefore UMP among all tests whose power function depends only on Δ.

Theorem 4 makes it possible to establish a simple condition under which reduction to sufficiency before the application of invariance is legitimate.

Theorem 6. *Let X be distributed according to P_θ, $\theta \in \Omega$, and let T be sufficient for θ. Suppose G leaves invariant the problem of testing $H : \theta \in \Omega_H$ and that T satisfies*

$$T(x_1) = T(x_2) \quad \text{implies} \quad T(gx_1) = T(gx_2) \quad \text{for all } g \in G,$$

so that G induces a group $\tilde{G}$ of transformations of T-space through

$$\tilde{g}T(x) = T(gx).$$

(i)   *If $\varphi(x)$ is any invariant test of H, there exists an almost invariant test ψ based on T, which has the same power function as φ.*

(ii)   *If in addition the assumptions of Theorem 4 are satisfied, the test ψ of (i) can be taken to be invariant.*

(iii)   *If there exists a test $\psi_0(T)$ which is UMP among all $\tilde{G}$-invariant tests based on T, then under the assumptions of (ii), ψ_0 is also UMP among all G-invariant tests based on X.*

This theorem justifies the derivation of the UMP invariant tests of Examples 5 and 6.

Proof.   (i): Let $\psi(t) = E[\varphi(X)|t]$. Then ψ has the same power function as φ. To complete the proof, it suffices to show that $\psi(t)$ is almost invariant, i.e. that

$$\psi(\tilde{g}t) = \psi(t) \quad \left(\text{a.e. } \mathscr{P}^T\right).$$

It follows from (1) that

$$E_\theta\big[\varphi(gX)|\tilde{g}t\big] = E_{\tilde{g}\theta}\big[\varphi(X)|t\big] \quad \left(\text{a.e. } P_\theta\right).$$

Since T is sufficient, both sides of this equation are independent of θ. Furthermore $\varphi(gx) = \varphi(x)$ for all x and g, and this completes the proof.

Part (ii) follows immediately from (i) and Theorem 4, and part (iii) from (ii).

6. UNBIASEDNESS AND INVARIANCE

The principles of unbiasedness and invariance complement each other in that each is successful in cases where the other is not. For example, there exist UMP unbiased tests for the comparison of two binomial or Poisson distributions, problems to which invariance considerations are not applicable. UMP unbiased tests also exist for testing the hypothesis $\sigma = \sigma_0$ against $\sigma \neq \sigma_0$ in a normal distribution, while invariance does not reduce this problem sufficiently far. Conversely, there exist UMP invariant tests of hypotheses specifying the values of more than one parameter (to be considered in Chapter 7) but for which the class of unbiased tests has no UMP member. There are also hypotheses, for example the one-sided hypothesis $\xi/\sigma \leq \theta_0$ in a univariate normal distribution or $\rho \leq \rho_0$ in a bivariate one (Problem 10) with $\theta_0, \rho_0 \neq 0$, where a UMP invariant test exists but the existence of a UMP unbiased test does not follow by the methods of Chapter 5 and is an open question.

On the other hand, to some problems both principles have been applied successfully. These include Student's hypotheses $\xi \leq \xi_0$ and $\xi = \xi_0$ concerning the mean of a normal distribution, and the corresponding two-sample problems $\eta - \xi \leq \Delta_0$ and $\eta - \xi = \Delta_0$ when the variances of the two samples are assumed equal. Other examples are the one-sided hypotheses $\sigma^2 \geq \sigma_0^2$ and $\tau^2/\sigma^2 \geq \Delta_0$ concerning the variances of one or two normal distributions. The hypothesis of independence $\rho = 0$ in a bivariate normal distribution is still another case in point (Problem 10). In all these examples the two optimum procedures coincide. We shall now show that this is not accidental but is the case whenever the UMP invariant test is UMP also among all almost invariant tests and the UMP unbiased test is unique. In this sense, the principles of unbiasedness and of almost invariance are consistent.

Theorem 7. *Suppose that for a given testing problem there exists a UMP unbiased test ϕ^* which is unique (up to sets of measure zero), and that there also exists a UMP almost invariant test with respect to some group G. Then the latter is also unique (up to sets of measure zero), and the two tests coincide a.e.*

Proof. If $U(\alpha)$ is the class of unbiased level-α tests, and if $g \in G$, then $\phi \in U(\alpha)$ if and only if $\phi g \in U(\alpha)$.[†] Denoting the power function of the

---

[†] ϕg denotes the critical function which assigns to x the value $\phi(gx)$.

test ϕ by $\beta_\phi(\theta)$, we thus have

$$\beta_{\phi^* g}(\theta) = \beta_{\phi^*}(\bar{g}\theta) = \sup_{\phi \in U(\alpha)} \beta_\phi(\bar{g}\theta) = \sup_{\phi \in U(\alpha)} \beta_{\phi g}(\theta)$$

$$= \sup_{\phi g \in U(\alpha)} \beta_{\phi g}(\theta) = \beta_{\phi^*}(\theta).$$

It follows that ϕ^* and $\phi^* g$ have the same power function, and, because of the uniqueness assumption, that ϕ^* is almost invariant. Therefore, if ϕ' is UMP almost invariant, we have $\beta_{\phi'}(\theta) \geq \beta_{\phi^*}(\theta)$ for all θ. On the other hand, ϕ' is unbiased, as is seen by comparing it with the invariant test $\phi(x) \equiv \alpha$, and hence $\beta_{\phi'}(\theta) \leq \beta_{\phi^*}(\theta)$ for all θ. Since ϕ' and ϕ^* therefore have the same power function, they are equal a.e. because of the uniqueness of ϕ^*, as was to be proved.

This theorem provides an alternative derivation for some of the tests of Chapter 5. In Theorem 3 of Chapter 4, the existence of UMP unbiased tests was established for one- and two-sided hypotheses concerning the parameter θ of the exponential family (10) of Chapter 4. For this family, the statistics (U, T) are sufficient and complete, and in terms of these statistics the UMP unbiased test is therefore unique. Convenient explicit expressions for some of these tests, which were derived in Chapter 5, can instead be obtained by noting that when a UMP almost invariant test exists, the same test by Theorem 7 must also be UMP unbiased. This proves for example that the tests of Examples 5 and 6 of the present chapter are UMP unbiased.

The principles of unbiasedness and invariance can be used to supplement each other in cases where neither principle alone leads to a solution but where they do so when applied in conjunction. As an example consider a sample $X_1, \ldots, X_n$ from $N(\xi, \sigma^2)$ and the problem of testing $H: \xi/\sigma = \theta_0 \neq 0$ against the two-sided alternatives that $\xi/\sigma \neq \theta_0$. Here sufficiency and invariance reduce the problem to the consideration of $t = \sqrt{n}\,\bar{x}/\sqrt{\Sigma(x_i - \bar{x})^2/(n-1)}$. The distribution of this statistic is the noncentral t-distribution with noncentrality parameter $\delta = \sqrt{n}\,\xi/\sigma$ and $n-1$ degrees of freedom. For varying δ, the family of these distributions can be shown to be STP_∞ [Karlin (1968, pp. 118–119; see Chapter 3, Problem 27] and hence in particular STP_3. It follows by Problem 29 of Chapter 3 that among all tests of H based on t, there exists a UMP unbiased one with acceptance region $C_1 \leq t \leq C_2$, where C_1, C_2 are determined by the conditions

$$P_{\delta_0}\{C_1 \leq t \leq C_2\} = 1 - \alpha \quad \text{and} \quad \frac{\partial P_\delta\{C_1 \leq t \leq C_2\}}{\partial \delta}\bigg|_{\delta = \delta_0} = 0.$$

In terms of the original observations, this test then has the property of being UMP among all tests that are unbiased and invariant. Whether it is also UMP unbiased without the restriction to invariant tests is an open problem.

An analogous example occurs in the testing of the hypotheses $H : \rho = \rho_0$ and $H' : \rho_1 \leq \rho \leq \rho_2$ against two-sided alternatives on the basis of a sample from a bivariate normal distribution with correlation coefficient ρ. (The testing of $\rho \leq \rho_0$ against $\rho > \rho_0$ is treated in Problem 10.) The distribution of the sample correlation coefficient has not only monotone likelihood ratio as shown in Problem 10, but is in fact STP_∞ [Karlin (1968, Section 3.4)]. Hence there exist tests of both H and H' which are UMP among all tests that are both invariant and unbiased.

Another case in which the combination of invariance and unbiasedness appears to offer a promising approach is the *Behrens-Fisher problem*. Let $X_1, \ldots, X_m$ and $Y_1, \ldots, Y_n$ be samples from normal distributions $N(\xi, \sigma^2)$ and $N(\eta, \tau^2)$ respectively. The problem is that of testing $H : \eta \leq \xi$ (or $\eta = \xi$) without assuming equality of the variances σ^2 and τ^2. A set of sufficient statistics for $(\xi, \eta, \sigma, \tau)$ is then $(\overline{X}, \overline{Y}, S_X^2, S_Y^2)$, where $S_X^2 = \Sigma(X_i - \overline{X})^2/(m - 1)$ and $S_Y^2 = \Sigma(Y_j - \overline{Y})^2/(n - 1)$. Adding the same constant to $\overline{X}$ and $\overline{Y}$ reduces the problem to $\overline{Y} - \overline{X}$, S_X^2, S_Y^2, and multiplication of all variables by a common positive constant to $(\overline{Y} - \overline{X})/\sqrt{S_X^2 + S_Y^2}$ and S_Y^2/S_X^2. One would expect any reasonable invariant rejection region to be of the form

$$(15) \qquad \frac{\overline{Y} - \overline{X}}{\sqrt{S_X^2 + S_Y^2}} \geq g\left(\frac{S_Y^2}{S_X^2}\right)$$

for some suitable function g. If this test is also to be unbiased, the probability of (15) must equal α when $\eta = \xi$ for all values of τ/σ. It has been shown by Linnik and others that only pathological functions g with this property can exist. [This work is reviewed by Pfanzagl (1974).] However, approximate solutions are available which provide tests that are satisfactory for all practical purposes. These are the Welch approximate t-solution described in Chapter 5, Section 4, and the Welch–Aspin test. Both are discussed, and evaluated, in Scheffé (1970) and Wang (1971); see also Chernoff (1949), Wallace (1958), and Davenport and Webster (1975).

The property of a test ϕ_1 being UMP invariant is relative to a particular group G_1, and does not exclude the possibility that there might exist another test ϕ_2 which is UMP invariant with respect to a different group G_2. Simple instances can be obtained from Examples 8 and 11.

Example 8. (continued). If G_1 is the group G of Example 8, a UMP invariant test of $H : \theta \leq \theta_0$ against $\theta > \theta_0$ rejects when $Y_1 + \cdots + Y_n > C$. Let G_2 be the group obtained by interchanging the role of Z and Y_1. Then a UMP invariant test with respect to G_2 rejects when $Z + Y_2 + \cdots + Y_n > C$. Analogous UMP invariant tests are obtained by interchanging the role of Z and any one of the other Y's, and further examples by applying the transformations of G in Example 8 to more than one variable. In particular, if it is applied independently to all $n + 1$ variables, only the constants remain invariant, and the test $\phi \equiv \alpha$ is UMP invariant.

Example 11.* For another example, let (X_{11}, X_{12}) and (X_{21}, X_{22}) be independent and have bivariate normal distributions with zero means and covariance matrices

$$\begin{pmatrix} \sigma_1^2 & \rho\sigma_1\sigma_2 \\ \rho\sigma_1\sigma_2 & \sigma_2^2 \end{pmatrix} \quad \text{and} \quad \begin{pmatrix} \Delta\sigma_1^2 & \Delta\rho\sigma_1\sigma_2 \\ \Delta\rho\sigma_1\sigma_2 & \Delta\sigma_2^2 \end{pmatrix}.$$

Suppose that these matrices are nonsingular, or equivalently that $|\rho| \neq 1$, but that σ_1, σ_2, ρ, and Δ are otherwise unknown. The problem of testing $\Delta = 1$ against $\Delta > 1$ remains invariant under the group G_1 of all nonsingular transformations

$$\begin{aligned} X'_{i1} &= bX_{i1} \\ X'_{i2} &= a_1 X_{i1} + a_2 X_{i2} \end{aligned}, \quad (a_2, b > 0).$$

Since the probability is 0 that $X_{11}X_{22} = X_{12}X_{21}$, the 2×2 matrix (X_{ij}) is nonsingular with probability 1, and the sample space can therefore be restricted to be the set of all nonsingular such matrices. A maximal invariant under the subgroup corresponding to $b = 1$ is the pair (X_{11}, X_{21}). The argument of Example 6 then shows that there exists a UMP invariant test under G_1 which rejects when $X_{21}^2/X_{11}^2 > C$.

By interchanging 1 and 2 in the second subscript of the X's one sees that under the corresponding group G_2 the UMP invariant test rejects when $X_{22}^2/X_{12}^2 > C$.

A third group leaving the problem invariant is the smallest group containing both G_1 and G_2, namely the group G of all common nonsingular transformations

$$\begin{aligned} X'_{i1} &= a_{11} X_{i1} + a_{12} X_{i2} \\ X'_{i2} &= a_{21} X_{i1} + a_{22} X_{i2} \end{aligned}, \quad (i = 1, 2).$$

Given any two nonsingular sample points $Z = (X_{ij})$ and $Z' = (X'_{ij})$, there exists a nonsingular linear transformation A such that $Z' = AZ$. There are therefore no invariants under G, and the only invariant size-α test is $\phi \equiv \alpha$. It follows vacuously that this is UMP invariant under G.

7. ADMISSIBILITY

Any UMP unbiased test has the important property of admissibility (Problem 1 of Chapter 4), in the sense that there cannot exist another test which is uniformly at least as powerful and against some alternatives actually more powerful than the given one. The corresponding property does not necessarily hold for UMP invariant tests, as is shown by the following example.

---

*Due to Charles Stein.

Example 11. (continued). Under the assumptions of Example 11 it was seen that the UMP invariant test under G is the test $\varphi \equiv \alpha$ which has power $\beta(\Delta) \equiv \alpha$. On the other hand, X_{11} and X_{21} are independently distributed as $N(0, \sigma_1^2)$ and $N(0, \Delta\sigma_1^2)$. On the basis of these observations there exists a UMP test for testing $\Delta = 1$ against $\Delta > 1$ with rejection region $X_{21}^2/X_{11}^2 > C$ (Chapter 3 Problem 38). The power function of this test is strictly increasing in Δ and hence $> \alpha$ for all $\Delta > 1$.

Admissibility of optimum invariant tests therefore cannot be taken for granted but must be established separately for each case.

We shall distinguish two slightly different concepts of admissibility. A test φ_0 will be called *α-admissible* for testing $H : \theta \in \Omega_H$ against a class of alternatives $\theta \in \Omega'$ if for any other level-α test φ

$$(16) \qquad E_\theta \varphi(X) \geq E_\theta \varphi_0(X) \qquad \text{for all} \quad \theta \in \Omega'$$

implies $E_\theta \varphi(X) = E_\theta \varphi_0(X)$ for all $\theta \in \Omega'$. This definition takes no account of the relationship of $E_\theta \varphi(X)$ and $E_\theta \varphi_0(X)$ for $\theta \in \Omega_H$ beyond the requirement that both tests are of level α. A concept closer to the decision-theoretic notion of admissibility discussed in Chapter 1, Section 8, defines φ_0 to be *d-admissible* for testing H against Ω' if (16) and

$$(17) \qquad E_\theta \varphi(X) \leq E_\theta \varphi_0(X) \qquad \text{for all} \quad \theta \in \Omega_H$$

jointly imply $E_\theta \varphi(X) = E_\theta \varphi_0(X)$ for all $\theta \in \Omega_H \cup \Omega'$ (see Problem 20).

Any level-α test φ_0 that is α-admissible is also d-admissible provided no other test φ exists with $E_\theta \varphi(X) = E_\theta \varphi_0(X)$ for all $\theta \in \Omega'$ but $E_\theta \varphi(X) \neq E_\theta \varphi_0(X)$ for some $\theta \in \Omega_H$. That the converse does not hold is shown by the following example.

Example 12. Let X be normally distributed with mean ξ and known variance σ^2. For testing $H : \xi \leq -1$ or ≥ 1 against $\Omega' : \xi = 0$, there exists a level-α test φ_0, which rejects when $C_1 \leq X \leq C_2$ and accepts otherwise, such that (Problem 21)

$$E_\xi \varphi_0(X) \leq E_{\xi = -1} \varphi_0(X) = \alpha \qquad \text{for} \quad \xi \leq -1$$

and

$$E_\xi \varphi_0(X) \leq E_{\xi = +1} \varphi_0(X) = \alpha' < \alpha \qquad \text{for} \quad \xi \geq +1.$$

A slight modification of the proof of Theorem 6 of Chapter 3 shows that φ_0 is the unique test maximizing the power at $\xi = 0$ subject to

$$E_\xi \varphi(X) \leq \alpha \quad \text{for} \quad \xi \leq -1 \quad \text{and} \quad E_\xi \varphi(X) \leq \alpha' \quad \text{for} \quad \xi \geq 1,$$

and hence that φ_0 is d-admissible.

On the other hand, the test φ with rejection region $|X| \leq C$, where $E_{\xi=-1}\varphi(X)$ $= E_{\xi=1}\varphi(X) = \alpha$, is the unique test maximizing the power at $\xi = 0$ subject to $E_{\xi}\varphi(X) \leq \alpha$ for $\xi \leq -1$ or ≥ 1, and hence is more powerful against Ω' than φ_0, so that φ_0 is not α-admissible.

A test that is admissible under either definition against Ω' is also admissible against any Ω'' containing Ω' and hence in particular against the class of all alternatives $\Omega_K = \Omega - \Omega_H$. The terms α- and d-admissible without qualification will be reserved for admissibility against Ω_K. Unless a UMP test exists, any α-admissible test will be admissible against some $\Omega' \subset \Omega_K$ and inadmissible against others. Both the strength of an admissibility result and the method of proof will depend on the set Ω'.

Consider in particular the admissibility of a UMP unbiased test mentioned at the beginning of the section. This does not rule out the existence of a test with greater power for all alternatives of practical importance and smaller power only for alternatives so close to H that the value of the power there is immaterial. In the present section, we shall discuss two methods for proving admissibility against various classes of alternatives.

Theorem 8. *Let X be distributed according to an exponential family with density*

$$p_\theta(x) = C(\theta)\exp\left(\sum_{j=1}^{s} \theta_j T_j(x)\right)$$

with respect to a σ-finite measure μ over a Euclidean sample space $(\mathcal{X}, \mathcal{A})$, and let Ω be the natural parameter space of this family. Let Ω_H and Ω' be disjoint nonempty subsets of Ω, and suppose that φ_0 is a test of $H : \theta \in \Omega_H$ based on $T = (T_1, \ldots, T_s)$ with acceptance region A_0 which is a closed convex subset of R^s possessing the following property: If $A_0 \cap \{\sum a_i t_i > c\}$ is empty for some c, there exists a point $\theta^* \in \Omega$ and a sequence $\lambda_n \to \infty$ such that $\theta^* + \lambda_n a \in \Omega'$ [where λ_n is a scalar and $a = (a_1, \ldots, a_s)$]. Then if A is any other acceptance region for H satisfying

$$P_\theta(X \in A) \leq P_\theta(X \in A_0) \qquad \text{for all} \quad \theta \in \Omega',$$

A is contained in A_0, except for a subset of measure 0, i.e. $\mu(A \cap \tilde{A}_0) = 0$.

Proof. Suppose to the contrary that $\mu(A \cap \tilde{A}_0) > 0$. Then it follows from the closure and convexity of A_0 that there exist $a \in R^s$ and a real number c such that

(18) $$A_0 \cap \left\{t : \sum a_i t_i > c\right\} \text{ is empty}$$

and

(19) $A \cap \{t : \sum a_i t_i > c\}$ has positive μ-measure,

that is, the set A protrudes in some direction from the convex set A_0. We shall show that this fact and the exponential nature of the densities imply that

(20) $P_\theta(A) > P_\theta(A_0)$ for some $\theta \in \Omega'$,

which provides the required contradiction. Let φ_0 and φ denote the indicators of $\tilde{A}_0$ and $\tilde{A}$ respectively, so that (20) is equivalent to

$$\int [\varphi_0(t) - \varphi(t)] \, dP_\theta(t) > 0   \text{ for some }   \theta \in \Omega'.$$

If $\theta = \theta^* + \lambda_n a \in \Omega'$, the left side becomes

$$\frac{C(\theta^* + \lambda_n a)}{C(\theta^*)} e^{c\lambda_n} \int [\varphi_0(t) - \varphi(t)] e^{\lambda_n(\sum a_i t_i - c)} \, dP_{\theta^*}(t).$$

Let this integral be $I_n^+ + I_n^-$, where I_n^+ and I_n^- denote the contributions over the regions of integration $\{t : \sum a_i t_i > c\}$ and $\{t : \sum a_i t_i \le c\}$ respectively. Since I_n^- is bounded, it is enough to show that $I_n^+ \to \infty$ as $n \to \infty$. By (18), $\varphi_0(t) = 1$ and hence $\varphi_0(t) - \varphi(t) \ge 0$ when $\sum a_i t_i > c$, and by (19)

$$\mu \{\varphi_0(t) - \varphi(t) > 0 \text{ and } \sum a_i t_i > c\} > 0.$$

This shows that $I_n^+ \to \infty$ as $\lambda_n \to \infty$ and therefore completes the proof.

Corollary 2. *Under the assumptions of Theorem 8, the test with accep-tance region A_0 is d-admissible. If its size is α and there exists a finite point θ_0 in the closure $\bar{\Omega}_H$ of Ω_H for which $E_{\theta_0}\varphi_0(X) = \alpha$, then φ_0 is also α-admissi-ble.*

Proof.

(i) Suppose φ satisfies (16). Then by Theorem 8, $\varphi_0(x) \le \varphi(x)$ (a.e. μ). If $\varphi_0(x) < \varphi(x)$ on a set of positive measure, then $E_\theta\varphi_0(X) < E_\theta\varphi(X)$ for all θ and hence (17) cannot hold.

(ii) By the argument of part (i), (16) implies $\alpha = E_{\theta_0}\varphi_0(X) < E_{\theta_0}\varphi(X)$, and hence by the continuity of $E_\theta\varphi(X)$ there exists a point $\theta \in \Omega_H$ for which $\alpha < E_\theta\varphi(X)$. Thus φ is not a level-α test.

Theorem 8 and the corollary easily extend to the case where the competitors φ of φ_0 are permitted to be randomized, but the assumption that φ_0 is nonrandomized is essential. Thus, the main applications of these results are to the case that μ is absolutely continuous with respect to Lebesgue measure. The boundary of A_0 will then typically have measure zero, so that the closure requirement for A_0 can be dropped.

Example 13. Normal mean. If $X_1, \ldots, X_n$ is a sample from the normal distribution $N(\xi, \sigma^2)$, the family of distributions is exponential with $T_1 = \overline{X}$, $T_2 = \Sigma X_i^2$, $\theta_1 = n\xi/\sigma^2$, $\theta_2 = -1/2\sigma^2$. Consider first the one-sided problem $H: \theta_1 \le 0$, $K: \theta_1 > 0$ with $\alpha < \frac{1}{2}$. Then the acceptance region of the t-test is $A: T_1/\sqrt{T_2} \le C$ ($C > 0$), which is convex [Problem 22(i)]. The alternatives $\theta \in \Omega' \subset K$ will satisfy the conditions of Theorem 8 if for any half plane $a_1 t_1 + a_2 t_2 > c$ that does not intersect the set $t_1 \le C\sqrt{t_2}$ there exists a ray $(\theta_1^* + \lambda a_1, \theta_2^* + \lambda a_2)$ in the direction of the vector (a_1, a_2) for which $(\theta_1^* + \lambda a_1, \theta_2^* + \lambda a_2) \in \Omega'$ for all sufficiently large λ. In the present case, this condition must hold for all $a_1 > 0 > a_2$. Examples of sets Ω' satisfying this requirement (and against which the t-test is therefore admissible) are

$$\Omega_1' : \theta_1 > k_1 \text{ or } \frac{\xi}{\sigma^2} > k_1'$$

and

$$\Omega_2' : \frac{\theta_1}{\sqrt{-\theta_2}} > k_2 \text{ or } \frac{\xi}{\sigma} > k_2'.$$

On the other hand, the condition is not satisfied for $\Omega' : \xi > k$ (Problem 22).

Analogously, the acceptance region $A: T_1^2 \le CT_2$ of the two-sided t-test for testing $H: \theta_1 = 0$ against $\theta_1 \ne 0$ is convex, and the test is admissible against $\Omega_1' : |\xi/\sigma^2| > k_1$ and $\Omega_2' : |\xi/\sigma| > k_2$.

In decision theory, a quite general method for proving admissibility consists in exhibiting a procedure as a unique Bayes solution. In the present case, this is justified by the following result, which is closely related to Theorem 7 of Chapter 3.

Theorem 9. *Suppose the set $\{x : f_\theta(x) > 0\}$ is independent of θ, and let a σ-field be defined over the parameter space Ω, containing both Ω_H and Ω_K and such that the densities $f_\theta(x)$ (with respect to μ) of X are jointly measurable in θ and x. Let Λ_0 and Λ_1 be probability distributions over this σ-field with $\Lambda_0(\Omega_H) = \Lambda_1(\Omega_K) = 1$, and let*

$$h_i(x) = \int f_\theta(x) \, d\Lambda_i(\theta).$$

Suppose φ_0 is a nonrandomized test of H against K defined by

$$\varphi_0(x) = \begin{cases} 1 \\ 0 \end{cases} \quad \text{if} \quad \frac{h_1(x)}{h_0(x)} \gtrless k,$$

and that $\mu\{x : h_1(x)/h_0(x) = k\} = 0$.

(i)   *Then φ_0 is d-admissible for testing H against K.*

(ii)   *Let $\sup_{\Omega_H} E_\theta \varphi_0(X) = \alpha$ and $\omega = \{\theta : E_\theta \varphi_0(X) = \alpha\}$. If $\omega \subset \Omega_H$ and $\Lambda_0(\omega) = 1$, then φ_0 is also α-admissible.*

(iii)   *If Λ_1 assigns probability 1 to $\Omega' \subset \Omega_K$, the conclusions of (i) and (ii) apply with Ω' in place of Ω_K.*

Proof.   (i): Suppose φ is any other test, satisfying (16) and (17) with $\Omega' = \Omega_K$. Then also

$$\int E_\theta \varphi(X) \, d\Lambda_0(\theta) \leq \int E_\theta \varphi_0(X) \, d\Lambda_0(\theta)$$

and

$$\int E_\theta \varphi(X) \, d\Lambda_1(\theta) \geq \int E_\theta \varphi_0(X) \, d\Lambda_1(\theta).$$

By the argument of Theorem 7 of Chapter 3, these inequalities are equivalent to

$$\int \varphi(x) h_0(x) \, d\mu(x) \leq \int \varphi_0(x) h_0(x) \, d\mu(x)$$

and

$$\int \varphi(x) h_1(x) \, d\mu(x) \geq \int \varphi_0(x) h_1(x) \, d\mu(x),$$

and the $h_i(x)$ $(i = 0, 1)$ are probability densities with respect to μ. This contradicts the uniqueness of the most powerful test of h_0 against h_1 at level $\int \varphi_0(x) h_0(x) \, d\mu(x)$.

(ii): By assumption, $\int E_\theta \varphi_0(x) \, d\Lambda_0(\theta) = \alpha$, so that φ_0 is a level-α test of h_0. If φ is any other level-α test of H satisfying (16) with $\Omega' = \Omega_K$, it is also a level-α test of h_0 and the argument of part (i) can be applied as before.

(iii): This follows immediately from the proofs of (i) and (ii).

Example 13. (continued).   In the two-sided normal problem of Example 13 with $H : \xi = 0$, $K : \xi \neq 0$ consider the class $\Omega'_{a,b}$ of alternatives (ξ, σ) satisfying

$$(21) \qquad \sigma^2 = \frac{1}{a + \eta^2}, \quad \xi = \frac{b\eta}{a + \eta^2}, \quad -\infty < \eta < \infty$$

for some fixed $a, b > 0$, and the subset ω of Ω_H of points $(0, \sigma^2)$ with $\sigma^2 < 1/a$. Let Λ_0, Λ_1 be distributions over ω and $\Omega'_{a,b}$ defined by the densities [Problem 23(i)]

$$\lambda_0(\eta) = \frac{C_0}{\left(a + \eta^2\right)^{n/2}}$$

and

$$\lambda_1(\eta) = \frac{C_1 e^{(n/2)b^2\eta^2/(a+\eta^2)}}{\left(a + \eta^2\right)^{n/2}}.$$

Straightforward calculation then shows [Problem 23(ii)] that the densities h_0 and h_1 of Theorem 9 become

$$h_0(x) = \frac{C_0 e^{-(a/2)\Sigma x_i^2}}{\sqrt{\Sigma x_i^2}}$$

and

$$h_1(x) = \frac{C_1 \exp\left(-\frac{a}{2}\sum x_i^2 + \frac{b^2\left(\sum x_i\right)^2}{2\sum x_i^2}\right)}{\sqrt{\Sigma x_i^2}},$$

so that the Bayes test φ_0 of Theorem 9 rejects when $\bar{x}^2/\Sigma x_i^2 > k$ and hence reduces to the two-sided t-test.

The condition of part (ii) of the theorem is clearly satisfied so that the t-test is both d- and α-admissible against $\Omega'_{a,b}$.

When dealing with invariant tests, it is of particular interest to consider admissibility against invariant classes of alternatives. In the case of the two-sided test φ_0, this means sets Ω' depending only on $|\xi/\sigma|$. It was seen in Example 13 that φ_0 is admissible against $\Omega' : |\xi/\sigma| \geq B$ for any B, that is, against distant alternatives, and it follows from the test being UMP unbiased or from Example 13 (continued) that φ_0 is admissible against $\Omega' : |\xi/\sigma| \leq A$ for any $A > 0$, that is, against alternatives close to H. This leaves open the question whether φ_0 is admissible against sets $\Omega' : 0 < A < |\xi/\sigma| < B < \infty$, which include neither nearby nor distant alternatives. It was in fact shown by Lehmann and Stein (1953) that φ_0 is admissible for testing H against $|\xi|/\sigma = \delta$ for any $\delta > 0$ and hence that it is admissible against any invariant Ω'. It was also shown there that the one-sided t-test of $H : \xi = 0$ is admissible against $\xi/\sigma = \delta'$ for any $\delta' > 0$. These results will not be proved here. The proof is based on assigning to $\log \sigma$ the uniform density on $(-N, N)$ and letting $N \to \infty$, thereby approximating the "improper" prior distribution which assigns to $\log \sigma$ the uniform distribution on $(-\infty, \infty)$, that is, Lebesgue measure.

That the one-sided t-test φ_1 of $H : \xi < 0$ is not admissible against all Ω' is shown by Brown and Sackrowitz (1984), who exhibit a test φ satisfying

$$E_{\xi,\sigma}\varphi(X) < E_{\xi,\sigma}\varphi_1(X) \qquad \text{for all} \quad \xi < 0, \ 0 < \sigma < \infty$$

and

$$E_{\xi,\sigma}\varphi(X) > E_{\xi,\sigma}\varphi_1(X) \qquad \text{for all} \quad 0 < \xi_1 < \xi < \xi_2 < \infty, \quad 0 < \sigma < \infty.$$

Example 14. Normal variance. For testing the variance σ^2 of a normal distribution on the basis of a sample $X_1, \ldots, X_n$ from $N(\xi, \sigma^2)$, the Bayes approach of Theorem 9 easily proves α-admissibility of the standard test against any location invariant set of alternatives Ω', that is, any set Ω' depending only on σ^2. Consider first the one-sided hypothesis $H: \sigma \le \sigma_0$ and the alternatives $\Omega': \sigma = \sigma_1$ for any $\sigma_1 > \sigma_0$. Admissibility of the UMP invariant (and unbiased) rejection region $\Sigma(X_i - \bar{X})^2 > C$ follows immediately from Chapter 3, Section 9, where it was shown that this test is Bayes for a pair of prior distributions (Λ_0, Λ_1): namely, Λ_1 assigning probability 1 to any point (ξ_1, σ_1), and Λ_0 putting $\sigma = \sigma_0$ and assigning to ξ the normal distribution $N(\xi_1, (\sigma_1^2 - \sigma_0^2)/n)$. Admissibility of $\Sigma(X_i - \bar{X})^2 \le C$ when the hypothesis is $H: \sigma \ge \sigma_0$ and $\Omega' = \{(\xi, \sigma): \sigma = \sigma_1\}$, $\sigma_1 < \sigma_0$, is seen by interchanging Λ_0 and Λ_1, σ_0 and σ_1.

A similar approach proves α-admissibility of any size-α rejection region

$$(22) \qquad \sum (X_i - \bar{X})^2 \le C_1 \text{ or } \ge C_2$$

for testing $H: \sigma = \sigma_0$ against $\Omega': \{\sigma = \sigma_1\} \cup \{\sigma = \sigma_2\}$ $(\sigma_1 < \sigma_0 < \sigma_2)$. On Ω_H, where the only variable is ξ, the distribution Λ_0 for ξ can be taken as the normal distribution with an arbitrary mean ξ_1 and variance $(\sigma_2^2 - \sigma_0^2)/n$. On Ω', let the conditional distribution of ξ given $\sigma = \sigma_2$ assign probability 1 to the value ξ_1, and let the conditional distribution of ξ given $\sigma = \sigma_1$ be $N(\xi_1, (\sigma_2^2 - \sigma_1^2)/n)$. Finally, let Λ_1 assign probabilities p and $1 - p$ to $\sigma = \sigma_1$ and $\sigma = \sigma_2$, respectively. Then the rejection region satisfies (22), and any constants C_1 and C_2 for which the test has size α can be attained by proper choice of p [Problem 24(i)].

The results of Examples 13 and 14 can be used as the basis for proving admissibility results in many other situations involving normal distributions. The main new difficulty tends to be the presence of additional (nuisance) means. These can often be eliminated by use of the following lemma.

Lemma 3. *For any given σ^2 and $M^2 > \sigma^2$ there exists a distribution Λ_σ such that*

$$I(z) = \int \frac{1}{\sqrt{2\pi}\,\sigma} e^{-(1/2\sigma^2)(z-\zeta)^2} \, d\Lambda_\sigma(\zeta)$$

is the normal density with mean zero and variance M^2.

Proof. Let $\theta = \zeta/\sigma$, and let θ be normally distributed with zero mean and variance τ^2. Then it is seen [Problem 24(ii)] that

$$I(z) = \frac{1}{\sqrt{2\pi}\,\sigma\sqrt{1 + \tau^2}} \exp\left[-\frac{1}{2\sigma^2(1 + \tau^2)} z^2 \right].$$

The result now follows by letting $\tau^2 = (M^2/\sigma^2) - 1$, so that $\sigma^2(1 + \tau^2) = M^2$.

Example 15. Let $X_1, \ldots, X_m; Y_1, \ldots, Y_n$ be samples from $N(\xi, \sigma^2)$ and $N(\eta, \tau^2)$ respectively, and consider the problem of testing $H: \tau/\sigma = 1$ against $\tau/\sigma = \Delta > 1$.

(i) Suppose first that $\xi = \eta = 0$. If Λ_0 and Λ_1 assign probability 1 to the points $(\sigma_0, \tau_0 = \sigma_0)$ and $(\sigma_1, \tau_1 = \Delta\sigma_1)$ respectively, the ratio h_1/h_0 of Theorem 9 is proportional to

$$\exp\left\{ -\frac{1}{2}\left[\left(\frac{1}{\Delta^2\sigma_1^2} - \frac{1}{\sigma_0^2} \right)\sum y_j^2 - \left(\frac{1}{\sigma_0^2} - \frac{1}{\sigma_1^2} \right)\sum x_i^2 \right]\right\},$$

and for suitable choice of critical value and $\sigma_1 < \sigma_0$, the rejection region of the Bayes test reduces to

$$\frac{\sum y_j^2}{\sum x_i^2} > \frac{\Delta^2\sigma_1^2 - \sigma_0^2}{\sigma_0^2 - \sigma_1^2}.$$

The values σ_0^2 and σ_1^2 can then be chosen to give this test any preassigned size α.

(ii) If ξ and η are unknown, then $\bar{X}, \bar{Y}, S_X^2 = \Sigma(X_i - \bar{X})^2$, $S_Y^2 = \Sigma(Y_j - \bar{Y})^2$ are sufficient statistics, and S_X^2 and S_Y^2 can be represented as $S_X^2 = \Sigma_{i=1}^{m-1}U_i^2$, $S_Y^2 = \Sigma_{j=1}^{n-1}V_j^2$, with the U_i, V_j independent normal with means 0 and variances σ^2 and τ^2 respectively.

To σ and τ assign the distributions Λ_0 and Λ_1 of part (i) and conditionally, given σ and τ, let ξ and η be independently distributed according to $\Lambda_{0\sigma}, \Lambda_{0\tau}$ over Ω_H and $\Lambda_{1\sigma}, \Lambda_{1\tau}$ over Ω_K, with these four conditional distributions determined from Lemma 3 in such a way that

$$\int \frac{\sqrt{m}}{\sqrt{2\pi}\,\sigma_0} e^{-(m/2\sigma_0^2)(\bar{x}-\xi)^2}\, d\Lambda_{0\sigma_0}(\xi) = \int \frac{\sqrt{m}}{\sqrt{2\pi}\,\sigma_1} e^{-(m/2\sigma_1^2)(\bar{x}-\xi)^2}\, d\Lambda_{0\sigma_1}(\xi),$$

and analogously for η. This is possible by choosing the constant M^2 of Lemma 3 greater than both σ_0^2 and σ_1^2. With this choice of priors, the contribution from $\bar{x}$ and $\bar{y}$ to the ratio h_1/h_0 of Theorem 9 disappears, so that h_1/h_0 reduces to the expression for this ratio in part (i), with Σx_i^2 and Σy_j^2 replaced by $\Sigma(x_i - \bar{x})^2$ and $\Sigma(y_j - \bar{y})^2$ respectively.

This approach applies quite generally in normal problems with nuisance means, provided the prior distribution of the variances $\sigma^2, \tau^2, \ldots$ assigns probability 1 to a bounded set, so that M^2 can be chosen to exceed all possible values of these variances.

Admissibility questions have been considered not only for tests but also for confidence sets. These will not be treated here (but see Chapter 9, Example 10); a convenient entry to the literature is Cohen and Strawderman (1973). For additional results, see Hooper (1982b) and Arnold (1984).

8. RANK TESTS

One of the basic problems of statistics is the two-sample problem of testing the equality of two distributions. A typical example is the comparison of a treatment with a control, where the hypothesis of no treatment effect is tested against the alternatives of a beneficial effect. This was considered in Chapter 5 under the assumption of normality, and the appropriate test was seen to be based on Student's t. It was also shown that when approximate normality is suspected but the assumption cannot be trusted, one is led to replacing the t-test by its permutation analogue, which in turn can be approximated by the original t-test.

We shall consider the same problem below without, at least for the moment, making any assumptions concerning even the approximate form of the underlying distributions, assuming only that they are continuous. The observations then consist of samples $X_1, \ldots, X_m$ and $Y_1, \ldots, Y_n$ from two distributions with continuous cumulative distribution functions F and G, and the problem becomes that of testing the hypothesis

$$H_1 : G = F.$$

If the treatment effect is assumed to be additive, the alternatives are $G(y) = F(y - \Delta)$. We shall here consider the more general possibility that the size of the effect may depend on the value of y (so that Δ becomes a nonnegative function of y) and therefore test H_1 against the one-sided alternatives that the Y's are stochastically larger than the X's,

$$K_1 : G(z) \leq F(z) \quad \text{for all } z, \quad \text{and} \quad G \neq F.$$

An alternative experiment that can be performed to test the effect of a treatment consists of the comparison of N pairs of subjects, which have been matched so as to eliminate as far as possible any differences not due to the treatment. One member of each pair is chosen at random to receive the treatment while the other serves as control. If the normality assumption of Chapter 5, Section 12, is dropped and the pairs of subjects can be considered to constitute a sample, the observations $(X_1, Y_1), \ldots, (X_N, Y_N)$ are a sample from a continuous bivariate distribution F. The hypothesis of no effect is then equivalent to the assumption that F is symmetric with respect to the line $y = x$:

$$H_2 : F(x, y) = F(y, x).$$

Another basic problem, which occurs in many different contexts, concerns the dependence or independence of two variables. In particular, if

$(X_1, Y_1), \ldots, (X_N, Y_N)$ is a sample from a bivariate distribution F, one will be interested in the hypothesis

$$H_3 : F(x, y) = G_1(x)G_2(y)$$

that X and Y are independent, which was considered for normal distributions in Section 15 of Chapter 5. The alternatives of interest may, for example, be that X and Y are positively dependent. An alternative formulation results when x, instead of being random, can be selected for the experiment. If the chosen values are $x_1 < \cdots < x_N$ and F_i denotes the distribution of Y given x_i, the Y's are independently distributed with continuous cumulative distribution functions $F_1, \ldots, F_N$. The hypothesis of independence of Y from x becomes

$$H_4 : F_1 = \cdots = F_N,$$

while under the alternatives of positive regression dependence the variables Y_i are stochastically increasing with i.

In these and other similar problems, invariance reduces the data so completely that the actual values of the observations are discarded and only certain order relations between different groups of variables are retained. It is nevertheless possible on this basis to test the various hypotheses in question, and the resulting tests frequently are nearly as powerful as the standard normal tests. We shall now carry out this reduction for the four problems above.

The two-sample problem of testing H_1 against K_1 remains invariant under the group G of all transformations

$$x_i' = \rho(x_i), \quad y_j' = \rho(y_j) \qquad (i = 1, \ldots, m, \quad j = 1, \ldots, n)$$

such that ρ is continuous and strictly increasing. This follows from the fact that these transformations preserve both the continuity of a distribution and the property of two variables being either identically distributed or one being stochastically larger than the other. As was seen (with a different notation) in Example 3, a maximal invariant under G is the set of ranks

$$(R'; S') = (R_1', \ldots, R_m'; S_1', \ldots, S_n')$$

of $X_1, \ldots, X_m; Y_1, \ldots, Y_n$ in the combined sample. Since the distribution of $(R_1', \ldots, R_m'; S_1', \ldots, S_n')$ is symmetric in the first m and in the last n variables for all distributions F and G, a set of sufficient statistics for (R', S') is the set of the X-ranks and that of the Y-ranks without regard to the subscripts of the X's and Y's. This can be represented by the ordered X-ranks and Y-ranks

$$R_1 < \cdots < R_m \quad \text{and} \quad S_1 < \cdots < S_n,$$

and therefore by one of these sets alone since each of them determines the other. Any invariant test is thus a *rank test*, that is, it depends only on the ranks of the observations, for example on $(S_1, \ldots, S_n)$.

That almost invariant tests are equivalent to invariant ones in the present context was shown first by Bell (1964). A streamlined and generalized version of his approach is given by Berk and Bickel (1968) and Berk (1970), who also show that the conclusion of Theorem 6 remains valid in this case.

To obtain a similar reduction for H_2, it is convenient first to make the transformation $Z_i = Y_i - X_i$, $W_i = X_i + Y_i$. The pairs of variables (Z_i, W_i) are then again a sample from a continuous bivariate distribution. Under the hypothesis this distribution is symmetric with respect to the w-axis, while under the alternatives the distribution is shifted in the direction of the positive z-axis. The problem is unchanged if all the w's are subjected to the same transformation $w_i' = \lambda(w_i)$, where λ is $1:1$ and has at most a finite number of discontinuities, and $(Z_1, \ldots, Z_N)$ constitutes a maximal invariant under this group. [Cf. Problem 2(ii).]

The Z's are a sample from a continuous univariate distribution D, for which the hypothesis of symmetry with respect to the origin,

$$H_2': D(z) + D(-z) = 1 \quad \text{for all } z,$$

is to be tested against the alternatives that the distribution is shifted toward positive z-values. This problem is invariant under the group G of all transformations

$$z_i' = \rho(z_i) \quad (i = 1, \ldots, N)$$

such that ρ is continuous, odd, and strictly increasing. If $z_{i_1}, \ldots, z_{i_m} < 0 < z_{j_1}, \ldots, z_{j_n}$ where $i_1 < \cdots < i_m$ and $j_1 < \cdots < j_n$, let $s_1', \ldots, s_n'$ denote the ranks of $z_{j_1}, \ldots, z_{j_n}$ among the absolute values $|z_1|, \ldots, |z_N|$, and $r_1', \ldots, r_m'$ the ranks of $|z_{i_1}|, \ldots, |z_{i_m}|$ among $|z_1|, \ldots, |z_N|$. The transformations ρ preserve the sign of each observation, and hence in particular also the numbers m and n. Since ρ is a continuous, strictly increasing function of $|z|$, it leaves the order of the absolute values invariant and therefore the ranks r_j' and s_j'. To see that the latter are maximal invariant, let $(z_1, \ldots, z_N)$ and $(z_1', \ldots, z_N')$ be two sets of points with $m' = m$, $n' = n$, and the same r_i' and s_j'. There exists a continuous, strictly increasing function on the positive real axis such that $|z_i'| = \rho(|z_i|)$ and $\rho(0) = 0$. If ρ is defined for negative z by $\rho(-z) = -\rho(z)$, it belongs to G and $z_i' = \rho(z_i)$ for all i, as was to be proved. As in the preceding problem, sufficiency permits the further reduction to the ordered ranks $r_1 < \cdots < r_m$ and $s_1 < \cdots < s_n$. This retains the information for the rank of each absolute value whether it belongs to a positive or negative observation, but not with which positive or negative observation it is associated.

The situation is very similar for the hypotheses H_3 and H_4. The problem of testing for independence in a bivariate distribution against the alternatives of positive dependence is unchanged if the X_i and Y_i are subjected to transformations $X_i' = \rho(X_i)$, $Y_i' = \lambda(Y_i)$ such that ρ and λ are continuous and strictly increasing. This leaves as maximal invariant the ranks $(R_1', \ldots, R_N')$ of $(X_1, \ldots, X_N)$ among the X's and the ranks $(S_1', \ldots, S_N')$ of $(Y_1, \ldots, Y_N)$ among the Y's. The distribution of $(R_1', S_1'), \ldots, (R_N', S_N')$ is symmetric in these N pairs for all distributions of (X, Y). It follows that a sufficient statistic is $(S_1, \ldots, S_N)$ where $(1, S_1), \ldots, (N, S_N)$ is a permutation of $(R_1', S_1'), \ldots, (R_N', S_N')$ and where therefore S_i is the rank of the variable Y associated with the ith smallest X.

The hypothesis H_4 that $Y_1, \ldots, Y_n$ constitutes a sample is to be tested against the alternatives K_4 that the Y_i are stochastically increasing with i. This problem is invariant under the group of transformations $y_i' = \rho(y_i)$ where ρ is continuous and strictly increasing. A maximal invariant under this group is the set of ranks $S_1, \ldots, S_N$ of $Y_1, \ldots, Y_N$.

Some invariant tests of the hypotheses H_1 and H_2 will be considered in the next two sections. Corresponding results concerning H_3 and H_4 are given in Problems 46–48.

9. THE TWO-SAMPLE PROBLEM

The problem of testing the two-sample hypothesis $H : G = F$ against the one-sided alternatives K that the Y's are stochastically larger than the X's is reduced by the principle of invariance to the consideration of tests based on the ranks $S_1 < \cdots < S_n$ of the Y's. The specification of the S_i is equivalent to specifying for each of the $N = m + n$ positions within the combined sample (the smallest, the next smallest, etc.) whether it is occupied by an x or a y. Since for any set of observations n of the N positions are occupied by y's and since the $\binom{N}{n}$ possible assignments of n positions to the y's are all equally likely when $G = F$, the joint distribution of the S_i under H is

(23) $$P\{S_1 = s_1, \ldots, S_n = s_n\} = 1 \Big/ \binom{N}{n}$$

for each set $1 \le s_1 < s_2 < \cdots < s_n \le N$. Any rank test of H of size

$$\alpha = k \bigg/ \binom{N}{n}$$

therefore has a rejection region consisting of exactly k points $(s_1, \ldots, s_n)$.

For testing H against K there exists no UMP rank test, and hence no UMP invariant test. This follows for example from a consideration of two of the standard tests for this problem, since each is most powerful among all rank tests against some alternative. The two tests in question have rejection regions of the form

$$(24) \qquad\qquad h(s_1) + \cdots + h(s_n) > C.$$

One, the Wilcoxon *two-sample test*, is obtained from (24) by letting $h(s) = s$, so that it rejects H when the sum of the y-ranks is too large. We shall show below that for sufficiently small Δ, this is most powerful against the alternatives that F is the logistic distribution $F(x) = 1/(1 + e^{-x})$, and that $G(y) = F(y - \Delta)$. The other test, the *normal-scores test*, has the rejection region (24) with $h(s) = E(W_{(s)})$, where $W_{(1)} < \cdots < W_{(N)}$ is an ordered sample of size N from a standard normal distribution.[†] This is most powerful against the alternatives that F and G are normal distributions with common variance and means ξ and $\eta = \xi + \Delta$, when Δ is sufficiently small.

To prove that these tests have the stated properties it is necessary to know the distribution of $(S_1, \ldots, S_n)$ under the alternatives. If F and G have densities f and g such that f is positive whenever g is, the joint distribution of the S_i is given by

$$(25) \quad P\{S_1 = s_1, \ldots, S_n = s_n\} = E\left[\frac{g(V_{(s_1)})}{f(V_{(s_1)})} \cdots \frac{g(V_{(s_n)})}{f(V_{(s_n)})} \right] \bigg/ \binom{N}{n},$$

where $V_{(1)} < \cdots < V_{(N)}$ is an ordered sample of size N from the distribution F. (See Problem 29.) Consider in particular the translation (or shift) alternatives

$$g(y) = f(y - \Delta),$$

and the problem of maximizing the power for small values of Δ. Suppose

---

[†] Tables of the expected order statistics from a normal distribution are given in *Biometrika Tables for Statisticians*, Vol. 2, Cambridge U. P., 1972, Table 9. For additional references, see David (1981, Appendix, Section 3.2).

that f is differentiable and that the probability (25), which is now a function of Δ, can be differentiated with respect to Δ under the expectation sign. The derivative of (25) at $\Delta = 0$ is then

$$\frac{\partial}{\partial \Delta} P_\Delta\{S_1 = s_1, \ldots, S_n = s_n\}\bigg|_{\Delta=0} = -E\left[\frac{f'(V_{(s_1)})}{f(V_{(s_1)})} \cdots \frac{f'(V_{(s_n)})}{f(V_{(s_n)})}\right]\bigg/\binom{N}{n}.$$

Since under the hypothesis the probability of any ranking is given by (23), it follows from the Neyman–Pearson lemma in the extended form of Theorem 5, Chapter 3, that the derivative of the power function at $\Delta = 0$ is maximized by the rejection region

(26)
$$- \sum_{i=1}^{n} E\left[\frac{f'(V_{(s_i)})}{f(V_{(s_i)})}\right] > C.$$

The same test maximizes the power itself for sufficiently small Δ. To see this let s denote a general rank point $(s_1, \ldots, s_n)$, and denote by $s^{(j)}$ the rank point giving the jth largest value to the left-hand side of (26). If

$$\alpha = k\bigg/\binom{N}{n},$$

the power of the test is then

$$\beta(\Delta) = \sum_{j=1}^{k} P_\Delta(s^{(j)}) = \sum_{j=1}^{k}\left[\frac{1}{\binom{N}{n}} + \Delta\frac{\partial}{\partial \Delta}P_\Delta(s^{(j)})\bigg|_{\Delta=0} + \cdots\right].$$

Since there is only a finite number of points s, there exists for each j a number $\Delta_j > 0$ such that the point $s^{(j)}$ also gives the jth largest value to $P_\Delta(s)$ for all $\Delta < \Delta_j$. If Δ is less than the smallest of the numbers

$$\Delta_j, \qquad j = 1, \ldots, \binom{N}{n},$$

the test also maximizes $\beta(\Delta)$.

If $f(x)$ is the normal density $N(\xi, \sigma^2)$, then

$$-\frac{f'(x)}{f(x)} = -\frac{d}{dx}\log f(x) = \frac{x - \xi}{\sigma^2},$$

and the left-hand side of (26) becomes

$$\sum E \frac{V_{(s_i)} - \xi}{\sigma^2} = \frac{1}{\sigma} \sum E\left(W_{(s_i)}\right)$$

where $W_{(1)} < \cdots < W_{(N)}$ is an ordered sample from $N(0, 1)$. The test that maximizes the power against these alternatives (for sufficiently small Δ) is therefore the normal-scores test.

In the case of the logistic distribution,

$$F(x) = \frac{1}{1 + e^{-x}}, \qquad f(x) = \frac{e^{-x}}{(1 + e^{-x})^2},$$

and hence

$$-\frac{f'(x)}{f(x)} = 2F(x) - 1.$$

The locally most powerful rank test therefore rejects when $\sum E[F(V_{(s_i)})] > C$. If V has the distribution F and $0 \le y \le 1$,

$$P\{F(V) \le y\} = P\{V \le F^{-1}(y)\} = F[F^{-1}(y)] = y,$$

so that $U = F(V)$ is uniformly distributed over $(0, 1)$.* The rejection region can therefore be written as $\sum E(U_{(s_i)}) > C$, where $U_{(1)} < \cdots < U_{(N)}$ is an ordered sample of size N from the uniform distribution $U(0, 1)$. Since $E(U_{(s_i)}) = s_i/(N + 1)$, the test is seen to be the Wilcoxon test.

Both the normal-scores test and the Wilcoxon test are unbiased against the one-sided alternatives K. In fact, let ϕ be the critical function of any test determined by (24) with h nondecreasing. Then ϕ is nondecreasing in the y's, and the probability of rejection is α for all $F = G$. By Lemma 3 of Chapter 5 the test is therefore unbiased against all alternatives of K.

It follows from the unbiasedness properties of these tests that the most powerful invariant tests in the two cases considered are also most powerful against their respective alternatives among all tests that are invariant and unbiased. The nonexistence of a UMP test is thus not relieved by restricting the tests to be unbiased as well as invariant. Nor does the application of the unbiasedness principle alone lead to a solution, as was seen in the discussion of permutation tests in Chapter 5, Section 11. With the failure of these two

*This transformation, which takes a random variable with continuous distribution F into a uniformly distributed variable, is known as the *probability integral transformation*.

principles, both singly and in conjunction, the problem is left not only without a solution but even without a formulation. A possible formulation (stringency) will be discussed in Chapter 9. However, the determination of a most stringent test for the two-sample hypothesis is an open problem.

Both tests mentioned above appear to be very satisfactory in practice. Even when F and G are normal with common variance, they are nearly as powerful as the t-test. To obtain a numerical comparison, suppose that the two samples are of equal size, and consider the ratio n^*/n of the number of observations required by two tests to obtain the same power β against the same alternative. Let $m = n$ and $m^* = n^* = g(n)$ be the sample sizes required by one of the rank tests and the t-test respectively, and suppose (as is the case for the tests under consideration) that the ratio n^*/n tends to a limit e independent of α and β as $n \to \infty$. Then e is called the *asymptotic efficiency* of the rank test relative to the t-test. Thus, if in a particular case $e = \frac{1}{2}$, then the rank test requires approximately twice as many observations as the t-test to achieve the same power.

In the particular case of the Wilcoxon test, e turns out to be equal to $3/\pi \sim 0.95$ when F and G are normal distributions with equal variance. When F and G are not necessarily normal but differ only in location, e depends on the form of the distribution. It is always ≥ 0.864, but may exceed 1 and can in fact be infinite.[†] The situation is even more favorable for the normal-scores test. Its asymptotic efficiency relative to the t-test is always ≥ 1 when F and G differ only in location; it is 1 in the particular case that F is normal (and only then).

The above results do not depend on the assumption of equal sample sizes; they are also valid if m/n and m^*/n^* tend to a common limit ρ as $n \to \infty$ where $0 < \rho < \infty$. At least in the case that F is normal, the asymptotic results agree well with those found for very small samples. For a more detailed discussion of these and related efficiency results, see for example, Lehmann (1975), Randles and Wolfe (1979), and Blair and Higgins (1980).

It was seen in Chapter 5, Sections 4 and 11, that both the size and the power of the t-test and its permutation version are robust against nonnormality, that is, that the actual size and power, at least for large m and n, are approximately equal to the values asserted by the normal theory even when F is not normal. The two tests are thus *performance-robust*: under mild assumptions on F, their actual performance is, asymptotically, independent of F. However, as was pointed out in Chapter 5, Section 4, the insensitivity of the power to the shape of F is not as advantageous as may appear at first sight, since the optimality of the t-test is tied to the assumption of normal-

---

[†] Upper bounds for certain classes of distributions are given by Loh (1984).

ity. The above results concerning the efficiency of the Wilcoxon and normal-scores tests show in fact that for many distributions F the t-test is far from optimal, so that the efficiency and optimality properties of t are quite nonrobust.

The most ambitious goal in the nonparametric two-sample shift model (46) of Chapter 5 would be to find a test which asymptotically preserves the optimality for arbitrary F which the t-test possesses exactly in the normal case. Such a test should have asymptotic efficiency 1 not with respect to a fixed test, but for each possible true F with respect to the tests which are asymptotically most powerful for that F. Such *adaptive* tests (which achieve simultaneous optimality by adapting themselves to the unknown F) do in fact exist if F is sufficiently smooth, although they are not yet practical. Their possibility was first suggested by Stein (1956b), whose program has been implemented for point-estimation problems [see for example Beran (1974), Stone (1975), and Bickel (1982)], but not yet for testing problems.

For testing $H: G = F$ against the two-sided alternatives that the Y's are either stochastically smaller or larger than the X's, two-sided versions of the rank tests of this section can be used. In particular, suppose that h is increasing and that $h(s) + h(N + 1 - s)$ is independent of s, as is the case for the Wilcoxon and normal-scores statistics. Then under H, the statistic $\Sigma h(s_j)$ is symmetrically distributed about $n\Sigma_{i=1}^{N}h(i)/N = \mu$, and (24) suggests the rejection region

$$\left|\sum h(s_j) - \mu\right| = \frac{1}{N}\left|m \sum_{j=1}^{n} h(s_j) - n \sum_{i=1}^{m} h(r_i)\right| > C.$$

The theory here is still less satisfactory than in the one-sided case. These tests need not even be unbiased [Sugiura (1965)], and it is not known whether they are admissible within the class of all rank tests. On the other hand, the relative asymptotic efficiencies are the same as in the one-sided case.

The two-sample hypothesis $G = F$ can also be tested against the general alternatives $G \neq F$. This problem arises in deciding whether two products, two sets of data, or the like can be pooled when nothing is known about the underlying distributions. Since the alternatives are now unrestricted, the problem remains invariant under all transformations $x_i' = f(x_i)$, $y_j' = f(y_j)$, $i = 1, \ldots, m$, $j = 1, \ldots, n$, such that f has only a finite number of discontinuities. There are no invariants under this group, so that the only invariant test is $\phi(x, y) \equiv \alpha$. This is however not admissible, since there do exist tests of H that are strictly unbiased against all alternatives $G \neq F$ (Problem 41). One of the tests most commonly employed for this problem is the *Smirnov test*. Let the *empirical distribution functions* of the two samples be defined by

$$S_{x_1,\ldots,x_m}(z) = \frac{a}{m}, \qquad S_{y_1,\ldots,y_n}(z) = \frac{b}{n},$$

where a and b are the numbers of x's and y's less or equal to z respectively. Then H is rejected according to this test when

$$\sup_z \left| S_{x_1,\ldots,x_m}(z) - S_{y_1,\ldots,y_n}(z) \right| > C.$$

Accounts of the theory of this and related tests are given, for example, in Hájek and Šidák (1967), Durbin (1973), and Serfling (1980).

Two-sample rank tests are distribution-free for testing $H: G = F$ but not for the nonparametric Behrens–Fisher situation of testing $H: \eta = \xi$ when the X's and Y's are samples from $F((x - \xi)/\sigma)$ and $F((y - \eta)/\tau)$ with σ, τ unknown. A detailed study of the effect of the difference in scales on the levels of the Wilcoxon and normal-scores tests is provided by Pratt (1964).

10. THE HYPOTHESIS OF SYMMETRY

When the method of paired comparisons is used to test the hypothesis of no treatment effect, the problem was seen in Section 8 to reduce through invariance to that of testing the hypothesis

$$H_2': D(z) + D(-z) = 1 \text{ for all } z,$$

which states that the distribution D of the differences $Z_i = Y_i - X_i$ ($i = 1,\ldots, N$) is symmetric with respect to the origin. The distribution D can be specified by the triple (ρ, F, G) where

$$\rho = P\{Z \le 0\}, \qquad F(z) = P\{|Z| \le z | Z < 0\},$$

$$G(z) = P\{Z \le z | Z > 0\},$$

and the hypothesis of symmetry with respect to the origin then becomes

$$H: \rho = \tfrac{1}{2}, G = F.$$

Invariance and sufficiency were shown to reduce the data to the ranks $S_1 < \cdots < S_n$ of the positive Z's among the absolute values $|Z_1|, \ldots, |Z_N|$. The probability of $S_1 = s_1, \ldots, S_n = s_n$ is the probability of this event given that there are n positive observations multiplied by the probability that the number of positive observations is n. Hence

$$P\{S_1 = s_1, \ldots, S_n = s_n\}$$

$$= \binom{N}{n}(1 - \rho)^n \rho^{N-n} P_{F, G}\{S_1 = s_1, \ldots, S_n = s_n \mid n\}$$

where the second factor is given by (25). Under H, this becomes

$$P\{S_1 = s_1, \ldots, S_n = s_n\} = \frac{1}{2^N}$$

for each of the

$$\sum_{n=0}^{N} \binom{N}{n} = 2^N$$

n-tuples $(s_1, \ldots, s_n)$ satisfying $1 \leq s_1 < \cdots < s_n \leq N$. Any rank test of size $\alpha = k/2^N$ therefore has a rejection region containing exactly k such points $(s_1, \ldots, s_n)$.

The alternatives K of a beneficial treatment effect are characterized by the fact that the variable Z being sampled is stochastically larger than some random variable which is symmetrically distributed about 0. It is again suggestive to use rejection regions of the form $h(s_1) + \cdots + h(s_n) > C$, where however n is no longer a constant as it was in the two-sample problem, but depends on the observations. Two particular cases are the *Wilcoxon one-sample test*, which is obtained by putting $h(s) = s$, and the analogue of the normal-scores test with $h(s) = E(W_{(s)})$ where $W_{(1)} < \cdots < W_{(N)}$ are the ordered values of $|V_1|, \ldots, |V_N|$, the V's being a sample from $N(0, 1)$. The W's are therefore an ordered sample of size N from a distribution with density $\sqrt{2/\pi}\, e^{-w^2/2}$ for $w \geq 0$.

As in the two-sample problem, it can be shown that each of these tests is most powerful (among all invariant tests) against certain alternatives, and that they are both unbiased against the class K. Their asymptotic efficiencies relative to the t-test for testing that the mean of Z is zero have the same values $3/\pi$ and 1 as the corresponding two-sample tests, when the distribution of Z is normal.

In certain applications, for example when the various comparisons are made under different experimental conditions or by different methods, it may be unrealistic to assume that the variables $Z_1, \ldots, Z_N$ have a common distribution. Suppose instead that the Z_i are still independently distributed but with arbitrary continuous distributions D_i. The hypothesis to be tested is that each of these distributions is symmetric with respect to the origin.

This problem remains invariant under all transformations $z_i' = f_i(z_i)$ $i = 1, \ldots, N$, such that each f_i is continuous, odd, and strictly increasing. A maximal invariant is then the number n of positive observations, and it follows from Example 8 that there exists a UMP invariant test, the *sign test*, which rejects when n is too large. This test reflects the fact that the magnitude of the observations or of their absolute values can be explained entirely in terms of the spread of the distributions D_i, so that only the signs of the Z's are relevant.

Frequently, it seems reasonable to assume that the Z's are identically distributed, but the assumption cannot be trusted. One would then prefer to use the information provided by the ranks s_i but require a test which controls the probability of false rejection even when the assumption fails. As is shown by the following lemma, this requirement is in fact satisfied for every (symmetric) rank test. Actually, the lemma will not require even the independence of the Z's; it will show that any symmetric rank test continues to correspond to the stated level of significance provided only the treatment is assigned at random within each pair.

Lemma 4. *Let* $\phi(z_1, \ldots, z_N)$ *be symmetric in its* N *variables and such that*

$$(27) \qquad\qquad E_D \phi(Z_1, \ldots, Z_N) = \alpha$$

when the Z's *are a sample from any continuous distribution* D *which is symmetric with respect to the origin. Then*

$$(28) \qquad\qquad E\phi(Z_1, \ldots, Z_N) = \alpha$$

if the joint distribution of the Z's *is unchanged under the* 2^N *transformations* $Z_1' = \pm Z_1, \ldots, Z_N' = \pm Z_N$.

Proof. The condition (27) implies

$$(29) \qquad \sum_{(j_1, \ldots, j_N)} \sum \frac{\phi(\pm z_{j_1}, \ldots, \pm z_{j_N})}{2^N \cdot N!} = \alpha \qquad \text{a.e.,}$$

where the outer summation extends over all $N!$ permutations $(j_1, \ldots, j_N)$ and the inner one over all 2^N possible choices of the signs $+$ and $-$. This is proved exactly as was Theorem 6 of Chapter 5. If in addition ϕ is symmetric, (29) implies

$$(30) \qquad \sum \frac{\phi(\pm z_1, \ldots, \pm z_N)}{2^N} = \alpha.$$

Suppose that the distribution of the Z's is invariant under the 2^N transformations in question. Then the conditional probability of any sign combination of $Z_1, \ldots, Z_N$ given $|Z_1|, \ldots, |Z_N|$ is $1/2^N$. Hence (30) is equivalent to

$$(31) \qquad E\left[\phi(Z_1, \ldots, Z_N) \middle| |Z_1|, \ldots, |Z_N|\right] = \alpha \qquad \text{a.e.,}$$

and this implies (28) which was to be proved.

The tests discussed above can be used to test symmetry about any known value θ_0 by applying them to the variables $Z_i - \theta_0$. The more difficult problem of testing for symmetry about an unknown point θ will not be considered here. Tests of this hypothesis are discussed, among others, by Antille, Kersting, and Zucchini (1982), Bhattacharya, Gastwirth, and Wright (1982), Boos (1982), and Koziol (1983).

As was pointed out in Section 5 of Chapter 5, the one-sample t-test is not robust against dependence. Unfortunately, this is also true—although to a somewhat lesser extent—of the sign and one-sample Wilcoxon tests [Gastwirth and Rubin (1971)].

11. EQUIVARIANT CONFIDENCE SETS

Confidence sets for a parameter θ in the presence of nuisance parameters ϑ were discussed in Chapter 5 (Sections 6 and 7) under the assumption that θ is real-valued. The correspondence between acceptance regions $A(\theta_0)$ of the hypotheses $H(\theta_0): \theta = \theta_0$ and confidence sets $S(x)$ for θ given by (34) and (35) of Chapter 5 is, however, independent of this assumption; it is valid regardless of whether θ is real-valued, vector-valued, or possibly a label for a completely unknown distribution function (in the latter case, confidence intervals become confidence bands for the distribution function). This correspondence, which can be summarized by the relationship

$$(32) \qquad \theta \in S(x) \quad \text{if and only if} \quad x \in A(\theta),$$

was the basis for deriving uniformly most accurate and uniformly most accurate unbiased confidence sets. In the present section, it will be used to obtain uniformly most accurate equivariant confidence sets.

We begin by defining equivariance for confidence sets. Let G be a group of transformations of the variable X preserving the family of distributions $\{P_{\theta, \vartheta}, (\theta, \vartheta) \in \Omega\}$ and let $\bar{G}$ be the induced group of transformations of Ω. If $\bar{g}(\theta, \vartheta) = (\theta', \vartheta')$, we shall suppose that θ' depends only on $\bar{g}$ and θ and not on ϑ, so that $\bar{g}$ induces a transformation in the space of θ. In order to keep the notation from becoming unnecessarily complex, it will then be convenient to write also $\theta' = \bar{g}\theta$. For each transformation $g \in G$, denote by g^* the transformation acting on sets S in θ-space and defined by

$$(33) \qquad\qquad g^*S = \{ \bar{g}\theta : \theta \in S \},$$

so that g^*S is the set obtained by applying the transformation $\bar{g}$ to each point θ of S. The invariance argument of Chapter 1, Section 5, then suggests restricting consideration to confidence sets satisfying

$$(34) \qquad g^*S(x) = S(gx) \qquad \text{for all } \quad x \in \mathcal{X}, \quad g \in G.$$

We shall say that such confidence sets are *equivariant* under G. This terminology avoids the impression created by the term invariance (used by some authors and in the first edition of this book) that the confidence sets remain unchanged under the transformation $X' = gX$. If the transformation g is interpreted as a change of coordinates, (34) means that the confidence statement does not depend on the coordinate system used to express the data. The statement that the transformed parameter $\bar{g}\theta$ lies in $S(gx)$ is equivalent to stating that $\theta \in g^{*-1}S(gx)$, which is equivalent to the original statement $\theta \in S(x)$ provided (34) holds.

Example 16. Let X, Y be independently normally distributed with means ξ, η and unit variance, and let G be the group of all rigid motions of the plane, which is generated by all translations and orthogonal transformations. Here $\bar{g} = g$ for all $g \in G$. An example of an equivariant class of confidence sets is given by

$$S(x, y) = \left\{ (\xi, \eta) : (x - \xi)^2 + (y - \eta)^2 \leq C \right\},$$

the class of circles with radius $\sqrt{C}$ and center (x, y). The set $g^*S(x, y)$ is the set of all points $g(\xi, \eta)$ with $(\xi, \eta) \in S(x, y)$, and hence is obtained by subjecting $S(x, y)$ to the rigid motion g. The result is the circle with radius $\sqrt{C}$ and center $g(x, y)$, and (34) is therefore satisfied.

In accordance with the definitions given in Chapters 3 and 5, a class of confidence sets for θ will be said to be *uniformly most accurate equivariant* at confidence level $1 - \alpha$ if among all equivariant classes of sets $S(x)$ at that level it minimizes the probability

$$P_{\theta, \vartheta}\{ \theta' \in S(X) \} \qquad \text{for all } \quad \theta' \neq \theta.$$

In order to derive confidence sets with this property from families of UMP invariant tests, we shall now investigate the relationship between equivariance of confidence sets and invariance of the associated tests.

Suppose that for each θ_0 there exists a group of transformations G_{θ_0} which leaves invariant the problem of testing $H(\theta_0): \theta = \theta_0$, and denote by G the group of transformations generated by the totality of groups G_θ.

Lemma 5.

(i) *Let $S(x)$ be any class of confidence sets that is equivariant under G, and let $A(\theta) = \{x: \theta \in S(x)\}$; then the acceptance region $A(\theta)$ is invariant under G_θ for each θ.*

(ii) *If in addition, for each θ_0 the acceptance region $A(\theta_0)$ is UMP invariant for testing $H(\theta_0)$ at level α, the class of confidence sets $S(x)$ is uniformly most accurate among all equivariant confidence sets at confidence level $1 - \alpha$.*

Proof. (i): Consider any fixed θ, and let $g \in G_\theta$. Then

$$gA(\theta) = \{gx: \theta \in S(x)\} = \{x: \theta \in S(g^{-1}x)\} = \{x: \theta \in g^{*-1}S(x)\}$$

$$= \{x: \bar{g}\theta \in S(x)\} = \{x: \theta \in S(x)\} = A(\theta).$$

Here the third equality holds because $S(x)$ is equivariant, and the fifth one because $g \in G_\theta$ and therefore $\bar{g}\theta = \theta$.

(ii): If $S'(x)$ is any other equivariant class of confidence sets at the prescribed level, the associated acceptance regions $A'(\theta)$ by (i) define invariant tests of the hypotheses $H(\theta)$. It follows that these tests are uniformly at most as powerful as those with acceptance regions $A(\theta)$ and hence that

$$P_{\theta, \vartheta}\{\theta' \in S(X)\} \le P_{\theta, \vartheta}\{\theta' \in S'(X)\} \qquad \text{for all} \quad \theta' \ne \theta,$$

as was to be proved.

It is an immediate consequence of the lemma that if UMP invariant acceptance regions $A(\theta)$ have been found for each hypothesis $H(\theta)$ (invariant with respect to G_θ), and if the confidence sets $S(x) = \{\theta: x \in A(\theta)\}$ are equivariant under G, then they are uniformly most accurate equivariant.

Example 17. Under the assumptions of Example 16, the problem of testing $\xi = \xi_0$, $\eta = \eta_0$ is invariant under the group G_{ξ_0, η_0} of orthogonal transformations about the point (ξ_0, η_0):

$$X' - \xi_0 = a_{11}(X - \xi_0) + a_{12}(Y - \eta_0),$$

$$Y' - \eta_0 = a_{21}(X - \xi_0) + a_{22}(Y - \eta_0),$$

where the matrix (a_{ij}) is orthogonal. There exists under this group a UMP invariant test, which has the acceptance region (Problem 8 of Chapter 7)

$$(X - \xi_0)^2 + (Y - \eta_0)^2 \leq C.$$

Let G_0 be the smallest group containing the groups $G_{\xi, \eta}$ for all ξ, η. Since this is a subgroup of the group G of Example 16 (the two groups actually coincide, but this is immaterial for the argument), the confidence sets $(X - \xi)^2 + (Y - \eta)^2 \leq C$ are equivariant under G_0 and hence uniformly most accurate equivariant.

Example 18. Let $X_1, \ldots, X_n$ be independently normally distributed with mean ξ and variance σ^2. Confidence intervals for ξ are based on the hypotheses $H(\xi_0) : \xi = \xi_0$, which are invariant under the groups G_{ξ_0} of transformations $X_i' = a(X_i - \xi_0) + \xi_0 (a \neq 0)$. The UMP invariant test of $H(\xi_0)$ has acceptance region

$$\frac{\sqrt{(n-1)n}\,|\overline{X} - \xi_0|}{\sqrt{\sum (X_i - \overline{X})^2}} \leq C,$$

and the associated confidence intervals are

$$(35) \quad \overline{X} - \frac{C}{\sqrt{n(n-1)}} \sqrt{\sum (X_i - \overline{X})^2} \leq \xi \leq \overline{X} + \frac{C}{\sqrt{n(n-1)}} \sqrt{\sum (X_i - \overline{X})^2}.$$

The group G in the present case consists of all transformations $g : X_i' = aX_i + b$ $(a \neq 0)$, which on ξ induces the transformation $\overline{g} : \xi' = a\xi + b$. Application of the associated transformation g^* to the interval (35) takes it into the set of points $a\xi + b$ for which ξ satisfies (35), that is, into the interval with end points

$$a\overline{X} + b - \frac{|a|C}{\sqrt{n(n-1)}} \sqrt{\sum (X_i - \overline{X})^2}, \qquad a\overline{X} + b + \frac{|a|C}{\sqrt{n(n-1)}} \sqrt{\sum (X_i - \overline{X})^2}$$

Since this coincides with the interval obtained by replacing X_i in (35) with $aX_i + b$, the confidence intervals (35) are equivariant under G_0 and hence uniformly most accurate equivariant.

Example 19. In the two-sample problem of Section 9, assume the shift model in which the X's and Y's have densities $f(x)$ and $g(y) = f(y - \Delta)$ respectively, and consider the problem of obtaining confidence intervals for the shift parameter Δ which are distribution-free in the sense that the coverage probability is independent of the true f. The hypothesis $H(\Delta_0) : \Delta = \Delta_0$ can be tested, for example, by means of the Wilcoxon test applied to the observations $X_i, Y_j - \Delta_0$, and confidence sets for Δ can then be obtained by the usual inversion process. The resulting confidence intervals are of the form $D_{(k)} < \Delta < D_{(mn+1-k)}$ where $D_{(1)} < \cdots < D_{(mn)}$ are the mn ordered differences $Y_j - X_i$. [For details see Problem 39 and for fuller accounts nonparametric books such as Lehmann (1975) and Randles and Wolfe (1979).] By their construction, these intervals have a coverage probability $1 - \alpha$ which is independent of f. However, the invariance considerations of Sections 8 and 9 do not apply. The hypothesis $H(\Delta_0)$ is invariant under the transformations $X_i' = \rho(X_i)$, $Y_j' = \rho(Y_j - \Delta_0) + \Delta_0$ with ρ continuous and strictly increasing, but the shift model, and hence the problem under consideration, is not invariant under these transformations.

12. AVERAGE SMALLEST EQUIVARIANT CONFIDENCE SETS

In the examples considered so far, the invariance and equivariance properties of the confidence sets corresponded to invariant properties of the associated tests. In the following examples this is no longer the case.

Example 20. Let $X_1, \ldots, X_n$ be a sample from $N(\xi, \sigma^2)$, and consider the problem of estimating σ^2.

The model is invariant under translations $X_i' = X_i + a$, and sufficiency and invariance reduce the data to $S^2 = \Sigma(X_i - \overline{X})^2$. The problem of estimating σ^2 by confidence sets also remains invariant under scale changes $X_i' = bX_i$, $S' = bS$, $\sigma' = b\sigma$ $(0 < b)$, although these do not leave the corresponding problem of testing the hypothesis $\sigma = \sigma_0$ invariant. (Instead, they leave invariant the *family* of these testing problems, in the sense that they transform one such hypothesis into another.) The totality of equivariant confidence sets based on S is given by

$$(36) \qquad\qquad \frac{\sigma^2}{S^2} \in A,$$

where A is any fixed set on the line satisfying

$$(37) \qquad\qquad P_{\sigma=1}\left(\frac{1}{S^2} \in A \right) = 1 - \alpha.$$

That any set $\sigma^2 \in S^2 \cdot A$ is equivariant is obvious. Conversely, suppose that $\sigma^2 \in C(S^2)$ is an equivariant family of confidence sets for σ^2. Then $C(S^2)$ must satisfy $b^2 C(S^2) = C(b^2 S^2)$ and hence

$$\sigma^2 \in C(S^2) \quad \text{if and only if} \quad \frac{\sigma^2}{S^2} \in \frac{1}{S^2} C(S^2) = C(1),$$

which establishes (36) with $A = C(1)$.

Among the confidence sets (36) with A satisfying (37) there does not exist one that uniformly minimizes the probability of covering false values (Problem 55). Consider instead the problem of determining the confidence sets that are physically smallest in the sense of having minimum Lebesgue measure. This requires minimizing $\int_A dv$ subject to (37). It follows from the Neyman–Pearson lemma that the minimizing A^* is

$$(38) \qquad\qquad A^* = \{ v : p(v) > C \},$$

where $p(v)$ is the density of $V = 1/S^2$ when $\sigma = 1$, and where C is determined by (37). Since $p(v)$ is unimodal (Problem 56), these smallest confidence sets are intervals, $aS^2 < \sigma^2 < bS^2$. Values of a and b are tabled by Tate and Klett (1959), who also table the corresponding (different) values a', b' for the uniformly most accurate unbiased confidence intervals $a'S^2 < \sigma^2 < b'S^2$ (given in Example 5 of Chapter 5).

Instead of minimizing the Lebesgue measure $\int_A dv$ of the confidence sets A, one may prefer to minimize the scale-invariant measure

$$(39) \qquad \qquad \int_A \frac{1}{v}\, dv.$$

To an interval (a, b), (39) assigns, in place of its length $b - a$, its logarithmic length $\log b - \log a = \log(b/a)$. The optimum solution A^{**} with respect to this new measure is again obtained by applying the Neyman–Pearson lemma, and is given by

$$(40) \qquad \qquad A^{**} = \{ v : vp(v) > C \},$$

which coincides with the uniformly most accurate unbiased confidence sets [Problem 57(i)].

One advantage of minimizing (39) instead of Lebesgue measure is that it then does not matter whether one estimates σ or σ^2 (or σ^r for some other power of r), since under (39), if (a, b) is the best interval for σ, then (a^r, b^r) is the best interval for σ^r [Problem 57(ii)].

Example 21. Let X_i $(i = 1,\ldots, r)$ be independently normally distributed as $N(\xi_i, 1)$. A slight generalization of Example 17 shows that uniformly most accurate equivariant confidence sets for $(\xi_1,\ldots, \xi_r)$ exist with respect to the group G of all rigid transformations and are given by

$$(41) \qquad \qquad \sum (X_i - \xi_i)^2 \le C.$$

Suppose that the context of the problem does not possess the symmetry which would justify invoking invariance with respect to G, but does allow the weaker assumption of invariance under the group G_0 of translations $X_i' = X_i + a_i$. The totality of equivariant confidence sets with respect to G_0 is given by

$$(42) \qquad \qquad (X_1 - \xi_1,\ldots, X_r - \xi_r) \in A,$$

where A is any fixed set in r-space satisfying

$$(43) \qquad \qquad P_{\xi_1 = \,\cdots\, = \xi_r = 0}((X_1,\ldots, X_r) \in A) = 1 - \alpha.$$

Since uniformly most accurate equivariant confidence sets do not exist (Problem 55), let us consider instead the problem of determining the confidence sets of smallest Lebesgue measure. (This measure is invariant under G_0.) This is given by (38) with $v = (v_1,\ldots, v_r)$ and $p(v)$ the density of $(X_1,\ldots, X_r)$ when $\xi_1 = \cdots = \xi_r = 0$, and hence coincides with (41).

Example 22. In the preceding example, suppose that the X_i are distributed as $N(\xi_i, \sigma^2)$ with σ^2 unknown, and that a variable S^2 is available for estimating σ^2. Of S^2 assume that it is independent of the X's and that S^2/σ^2 has a χ^2-distribution with f degrees of freedom.

The estimation of $(\xi_1, \ldots, \xi_r)$ by confidence sets on the basis of X's and S^2 remains invariant under the group G_0 of transformations

$$X_i' = bX_i + a_i, \qquad S' = bS, \qquad \xi_i' = b\xi_i + a_i, \qquad \sigma' = b\sigma,$$

and the most general equivariant confidence set is of the form

(44)
$$\left(\frac{X_1 - \xi_1}{S}, \ldots, \frac{X_r - \xi_r}{S} \right) \in A,$$

where A is any fixed set in r-space satisfying

(45)
$$P_{\xi_1 = \cdots = \xi_r = 0}\left[\left(\frac{X_1}{S}, \ldots, \frac{X_r}{S} \right) \in A \right] = 1 - \alpha.$$

The confidence sets (44) can be written as

(46)
$$(\xi_1, \ldots, \xi_r) \in (X_1, \ldots, X_r) - SA,$$

where $-SA$ is the set obtained by multiplying each point of A by the scalar $-S$.

To see (46), suppose that $C(X_1, \ldots, X_r; S)$ is an equivariant confidence set for $(\xi_1, \ldots, \xi_r)$. Then the r-dimensional set C must satisfy

$$C(bX_1 + a_1, \ldots, bX_r + a_r; bS) = b[C(X_1, \ldots, X_r; S)] + (a_1, \ldots, a_r)$$

for all $a_1, \ldots, a_r$ and all $b > 0$. It follows that $(\xi_1, \ldots, \xi_r) \in C$ if and only if

$$\left(\frac{X_1 - \xi_1}{S}, \ldots, \frac{X_r - \xi_r}{S} \right) \in \frac{(X_1, \ldots, X_r) - C(X_1, \ldots, X_r; S)}{S} = C(0, \ldots, 0; 1) = A.$$

The equivariant confidence sets of smallest volume are obtained by choosing for A the set A^* given by (38) with $v = (v_1, \ldots, v_r)$ and $p(v)$ the joint density of $(X_1/S, \ldots, X_r/S)$ when $\xi_1 = \cdots = \xi_r = 0$. This density is a decreasing function of Σv_i^2 (Problem 58), and the smallest equivariant confidence sets are therefore given by

(47)
$$\Sigma(X_i - \xi_i)^2 \le CS^2.$$

[Under the larger group G generated by all rigid transformations of $(X_1, \ldots, X_r)$ together with the scale changes $X_i' = bX_i$, $S' = bS$, the same sets have the stronger property of being uniformly most accurate equivariant; see Problem 59.]

Examples 20–22 have the common feature that the equivariant confidence sets $S(X)$ for $\theta = (\theta_1, \ldots, \theta_r)$ are characterized by an r-valued *pivotal quantity*, that is, a function $h(X, \theta) = (h_1(X, \theta), \ldots, h_r(X, \theta))$ of the observations X and parameters θ being estimated that has a fixed distribution, and such that the most general equivariant confidence sets are of the form

$$(48) \qquad\qquad h(X, \theta) \in A$$

for some fixed set A.* When the functions h_i are linear in θ, the confidence sets $C(X)$ obtained by solving (48) for θ are linear transforms of A (with random coefficients), so that the volume or invariant measure of $C(X)$ is minimized by minimizing

$$(49) \qquad\qquad \int_A \rho(v_1, \ldots, v_r) \, dv_1 \ldots dv_r$$

for the appropriate ρ. The problem thus reduces to that of minimizing (49) subject to

$$(50) \quad P_{\theta_0}\{h(X, \theta_0) \in A\} = \int_A p(v_1, \ldots, v_r) \, dv_1 \ldots dv_r = 1 - \alpha,$$

where $p(v_1, \ldots, v_r)$ is the density of the pivotal quantity $h(X, \theta)$. The minimizing A is given by

$$(51) \qquad\qquad A^* = \left\{ v : \frac{p(v_1, \ldots, v_r)}{\rho(v_1, \ldots, v_r)} > C \right\},$$

with C determined by (50).

The following is one more illustration of this approach.

Example 23. Let $X_1, \ldots, X_m$ and $Y_1, \ldots, Y_n$ be samples from $N(\xi, \sigma^2)$ and $N(\eta, \tau^2)$ respectively, and consider the problem of estimating $\Delta = \tau^2/\sigma^2$. Sufficiency and invariance under translations $X_i' = X_i + a_1$, $Y_j' = Y_j + a_2$ reduce the data to $S_X^2 = \Sigma(X_i - \bar{X})^2$ and $S_Y^2 = \Sigma(Y_j - \bar{Y})^2$. The problem of estimating Δ also remains invariant under the scale changes

$$X_i' = b_1 X_i, \quad Y_j' = b_2 Y_j, \qquad 0 < b_1, b_2 < \infty,$$

which induce the transformations

$$(52) \qquad S_X' = b_1 S_X, \qquad S_Y' = b_2 S_Y, \qquad \sigma' = b_1 \sigma, \qquad \tau' = b_2 \tau.$$

*More general results concerning the relationship of equivariant confidence sets and pivotal quantities are given in Problems 78–81.

The totality of equivariant confidence sets for Δ is given by $\Delta/V \in A$, where $V = S_Y^2/S_X^2$ and A is any fixed set on the line satisfying

(53)
$$P_{\Delta=1}\left(\frac{1}{V} \in A\right) = 1 - \alpha.$$

To see this, suppose that $C(S_X, S_Y)$ are any equivariant confidence sets for Δ. Then C must satisfy

(54)
$$C(b_1 S_X, b_2 S_Y) = \frac{b_2^2}{b_1^2} C(S_X, S_Y),$$

and hence $\Delta \in C(S_X, S_Y)$ if and only if the pivotal quantity V/Δ satisfies

$$\frac{\Delta}{V} = \frac{S_X^2 \Delta}{S_Y^2} \in \frac{S_X^2}{S_Y^2} C(S_X, S_Y) = C(1,1) = A.$$

As in Example 20, one may now wish to choose A so as to minimize either its Lebesgue measure $\int_A dv$ or the invariant measure $\int_A (1/v)\, dv$. The resulting confidence sets are of the form

(55)
$$p(v) > C \quad \text{and} \quad vp(v) > C$$

respectively. In both cases, they are intervals $V/b < \Delta < V/a$ [Problem 60(i)]. The values of a and b minimizing Lebesgue measure are tabled by Levy and Narula (1974); those for the invariant measure coincide with the uniformly most accurate unbiased intervals [Problem 60(ii)].

13. CONFIDENCE BANDS FOR A DISTRIBUTION FUNCTION

Suppose that $X = (X_1, \ldots, X_n)$ is a sample from an unknown continuous cumulative distribution function F, and that lower and upper bounds L_X and M_X are to be determined such that with preassigned probability $1 - \alpha$ the inequalities

$$L_X(u) \le F(u) \le M_X(u) \qquad \text{for all } u$$

hold for all continuous cumulative distribution functions F. This problem is invariant under the group G of transformations

$$X_i' = g(X_i), \qquad i = 1, \ldots, n,$$

where g is any continuous strictly increasing function. The induced transformation in the parameter space is $\bar{g}F = F(g^{-1})$.

If $S(x)$ is the set of continuous cumulative distribution functions

$$S(x) = \{F: L_x(u) \le F(u) \le M_x(u) \text{ for all } u\},$$

then

$$g^*S(x) = \{\bar{g}F: L_x(u) \le F(u) \le M_x(u) \text{ for all } u\}$$

$$= \{F: L_x[g^{-1}(u)] \le F(u) \le M_x[g^{-1}(u)] \text{ for all } u\}.$$

For an equivariant procedure, this must coincide with the set

$$S(gx) = \left\{F: L_{g(x_1),\ldots,g(x_n)}(u) \le F(u) \le M_{g(x_1),\ldots,g(x_n)}(u) \text{ for all } u\right\}.$$

The condition of equivariance is therefore

$$L_{g(x_1),\ldots,g(x_n)}[g(u)] = L_x(u), \quad M_{g(x_1),\ldots,g(x_n)}[g(u)] = M_x(u)$$

for all x and u.

To characterize the totality of equivariant procedures, consider the *empirical distribution function* (EDF) T_x given by

$$T_x(u) = \frac{i}{n} \quad \text{for } x_{(i)} \le u < x_{(i+1)}, \quad i = 0,\ldots,n,$$

where $x_{(1)} < \cdots < x_{(n)}$ is the ordered sample and where $x_{(0)} = -\infty$, $x_{(n+1)} = \infty$. Then a necessary and sufficient condition for L and M to satisfy the above equivariance condition is the existence of numbers $a_0,\ldots,a_n; a'_0,\ldots,a'_n$ such that

$$L_x(u) = a_i, \quad M_x(u) = a'_i \quad \text{for } x_{(i)} < u < x_{(i+1)}.$$

That this condition is sufficient is immediate. To see that it is also necessary, let u, u' be any two points satisfying $x_{(i)} < u < u' < x_{(i+1)}$. Given any $y_1,\ldots,y_n$ and v with $y_{(i)} < v < y_{(i+1)}$, there exist $g, g' \in G$ such that

$$g(y_{(i)}) = g'(y_{(i)}) = x_{(i)}, \quad g(v) = u, \quad g'(v) = u'.$$

If L_x, M_x are equivariant, it then follows that $L_x(u') = L_y(v)$ and $L_x(u) = L_y(v)$, and hence that $L_x(u') = L_x(u)$ and similarly $M_x(u') = M_x(u)$, as was to be proved. This characterization shows L_x and M_x to be step functions whose discontinuity points are restricted to those of T_x.

Since any two continuous strictly increasing cumulative distribution functions can be transformed into one another through a transformation $\bar{g}$, it follows that all these distributions have the same probability of being covered by an equivariant confidence band. (See Problem 66.) Suppose now that F is continuous but no longer strictly increasing. If I is any interval of constancy of F, there are no observations in I, so that I is also an interval of constancy of the sample cumulative distribution function. It follows that the probability of the confidence band covering F is not affected by the presence of I and hence is the same for all continuous cumulative distribution functions F.

For any numbers a_i, a_i' let Δ_i, Δ_i' be determined by

$$a_i = \frac{i}{n} - \Delta_i, \qquad a_i' = \frac{i}{n} + \Delta_i'.$$

Then it was seen above that any numbers $\Delta_0, \ldots, \Delta_n; \Delta_0', \ldots, \Delta_n'$ define a confidence band for F, which is equivariant and hence has constant probability of covering the true F. From these confidence bands a test can be obtained of the hypothesis of *goodness of fit* $F = F_0$ that the unknown F equals a hypothetical distribution F_0. The hypothesis is accepted if F_0 lies entirely within the band, that is, if

$$-\Delta_i < F_0(u) - T_x(u) < \Delta_i'$$

for all $x_{(i)} < u < x_{(i+1)}$ and all $i = 1, \ldots, n$.

Within this class of tests there exists no UMP member, and the most common choice of the Δ's is $\Delta_i = \Delta_i' = \Delta$ for all i. The acceptance region of the resulting *Kolmogorov test* can be written as

$$(56) \qquad \sup_{-\infty < u < \infty} |F_0(u) - T_x(u)| < \Delta.$$

Tables of the null distribution of the Kolmogorov statistic are given by Birnbaum (1952). For large n, approximate critical values can be obtained from the limit distribution K of $\sqrt{n} \sup |F_0(u) - T_x(u)|$, due to Kolmogorov and tabled by Smirnov (1948). Derivations of K can be found, for example, in Feller (1948), Hájek and Sidák (1967), and Billingsley (1968).

Alternative goodness-of-fit tests are based on other measures of the distance between the cumulative distribution functions F_0 and T_x. Surveys dealing with properties of such tests, including tests for goodness of fit when the hypothesis specifies a parametric family rather than a single distribution, are provided by Durbin (1973), Kendall and Stuart (1979, Chapter 30), Neuhaus (1979), and Tallis (1983).

14. PROBLEMS

Section 1

1. Let G be a group of measurable transformations of $(\mathcal{X}, \mathcal{A})$ leaving $\mathcal{P} = \{ P_\theta, \theta \in \Omega \}$ invariant, and let $T(x)$ be a measurable transformation to $(\mathcal{T}, \mathcal{B})$. Suppose that $T(x_1) = T(x_2)$ implies $T(gx_1) = T(gx_2)$ for all $g \in G$, so that G induces a group G^* on $\mathcal{T}$ through $g^*T(x) = T(gx)$, and suppose further that the induced transformations g^* are measurable $\mathcal{B}$. Then G^* leaves the family $\mathcal{P}^T = \{ P_\theta^T, \theta \in \Omega \}$ of distributions of T invariant.

Section 2

2. (i)   Let $\mathcal{X}$ be the totality of points $x = (x_1, \ldots, x_n)$ for which all coordinates are different from zero, and let G be the group of transformations $x_i' = cx_i$, $c > 0$. Then a maximal invariant under G is $(\text{sgn } x_n, x_1/x_n, \ldots, x_{n-1}/x_n)$ where sgn x is 1 or -1 as x is positive or negative.

    (ii) Let $\mathcal{X}$ be the space of points $x = (x_1, \ldots, x_n)$ for which all coordinates are distinct, and let G be the group of all transformations $x_i' = f(x_i)$, $i = 1, \ldots, n$, such that f is a $1:1$ transformation of the real line onto itself with at most a finite number of discontinuities. Then G is transitive over $\mathcal{X}$.

    [(ii): Let $x = (x_1, \ldots, x_n)$ and $x' = (x_1', \ldots, x_n')$ be any two points of $\mathcal{X}$. Let $I_1, \ldots, I_n$ be a set of mutually exclusive open intervals which (together with their end points) cover the real line and such that $x_j \in I_j$. Let $I_1', \ldots, I_n'$ be a corresponding set of intervals for $x_1', \ldots, x_n'$. Then there exists a transformation f which maps each I_j continuously onto I_j', maps x_j into x_j', and maps the set of $n - 1$ end points of $I_1, \ldots, I_n$ onto the set of end points of $I_1', \ldots, I_n'$.]

3. (i)   A sufficient condition for (8) to hold is that D is a normal subgroup of G.

    (ii) If G is the group of transformations $x' = ax + b$, $a \neq 0$, $-\infty < b < \infty$, then the subgroup of translations $x' = x + b$ is normal but the subgroup $x' = ax$ is not.

    [The defining property of a normal subgroup is that given $d \in D$, $g \in G$, there exists $d' \in D$ such that $gd = d'g$. The equality $s(x_1) = s(x_2)$ implies $x_2 = dx_1$ for some $d \in D$, and hence $ex_2 = e\,dx_1 = d'ex_1$. The result (i) now follows, since s is invariant under D.]

Section 3

4. Let X, Y have the joint probability density $f(x, y)$. Then the integral $h(z) = \int_{-\infty}^{\infty} f(y - z, y) \, dy$ is finite for almost all z, and is the probability density of $Z = Y - X$.
    [Since $P\{Z \le b\} = \int_{-\infty}^{b} h(z) \, dz$, it is finite and hence h is finite almost everywhere.]

5. (i) Let $X = (X_1, \ldots, X_n)$ have probability density $(1/\theta^n) f[(x_1 - \xi)/\theta, \ldots, (x_n - \xi)/\theta]$, where $-\infty < \xi < \infty$, $0 < \theta$ are unknown, and where f is even. The problem of testing $f = f_0$ against $f = f_1$ remains invariant under the transformations $x_i' = a x_i + b$ $(i = 1, \ldots, n)$, $a \ne 0$, $-\infty < b < \infty$, and the most powerful invariant test is given by the rejection region

    $$\int_{-\infty}^{\infty} \int_{0}^{\infty} v^{n-2} f_1(v x_1 + u, \ldots, v x_n + u) \, dv \, du$$

    $$> C \int_{-\infty}^{\infty} \int_{0}^{\infty} v^{n-2} f_0(v x_1 + u, \ldots, v x_n + u) \, dv \, du.$$

    (ii) Let $X = (X_1, \ldots, X_n)$ have probability density $f(x_1 - \sum_{j=1}^{k} w_{1j} \beta_j, \ldots, x_n - \sum_{j=1}^{k} w_{nj} \beta_j)$, where $k < n$, the w's are given constants, the matrix (w_{ij}) is of rank k, the β's are unknown, and we wish to test $f = f_0$ against $f = f_1$. The problem remains invariant under the transformations $x_i' = x_i + \sum_{j=1}^{k} w_{ij} \gamma_j$, $-\infty < \gamma_1, \ldots, \gamma_k < \infty$, and the most powerful invariant test is given by the rejection region

    $$\frac{\int \cdots \int f_1(x_1 - \sum w_{1j}\beta_j, \ldots, x_n - \sum w_{nj}\beta_j) \, d\beta_1, \ldots, d\beta_k}{\int \cdots \int f_0(x_1 - \sum w_{1j}\beta_j, \ldots, x_n - \sum w_{nj}\beta_j) \, d\beta_1, \ldots, d\beta_k} > C.$$

    [A maximal invariant is given by $y =$

    $$\left(x_1 - \sum_{r=n-k+1}^{n} a_{1r} x_r, \ x_2 - \sum_{r=n-k+1}^{n} a_{2r} x_r, \ \ldots, x_{n-k} - \sum_{r=n-k+1}^{n} a_{n-k, r} x_r \right)$$

    for suitably chosen constants a_{ir}.]

6. Let $X_1, \ldots, X_m$; $Y_1, \ldots, Y_n$ be samples from exponential distributions with densities $\sigma^{-1} e^{-(x-\xi)/\sigma}$ for $x \ge \xi$, and $\tau^{-1} e^{-(y-\eta)/\tau}$ for $y \ge \eta$.

    (i) For testing $\tau/\sigma \le \Delta$ against $\tau/\sigma > \Delta$, there exists a UMP invariant test with respect to the group $G: X_i' = a X_i + b$, $Y_j' = a Y_j + c$, $a > 0$, $-\infty$

$< b, c < \infty$, and its rejection region is

$$\frac{\sum [y_j - \min(y_1, \ldots, y_n)]}{\sum [x_i - \min(x_1, \ldots, x_m)]} > C.$$

(ii)   This test is also UMP unbiased.

(iii) Extend these results to the case that only the r smallest X's and the s smallest Y's are observed.

[(ii): See Problem 12 of Chapter 5.]

7. If $X_1, \ldots, X_n$ and $Y_1, \ldots, Y_n$ are samples from $N(\xi, \sigma^2)$ and $N(\eta, \tau^2)$ respectively, the problem of testing $\tau^2 = \sigma^2$ against the two-sided alternatives $\tau^2 \neq \sigma^2$ remains invariant under the group G generated by the transformations $X_i' = aX_i + b$, $Y_i' = aY_i + c$, $a \neq 0$, and $X_i' = Y_i$, $Y_i' = X_i$. There exists a UMP invariant test under G with rejection region

$$W = \max \left\{ \frac{\sum (Y_i - \bar{Y})^2}{\sum (X_i - \bar{X})^2}, \frac{\sum (X_i - \bar{X})^2}{\sum (Y_i - \bar{Y})^2} \right\} \geq k.$$

[The ratio of the probability densities of W for $\tau^2/\sigma^2 = \Delta$ and $\tau^2/\sigma^2 = 1$ is proportional to $[(1 + w)/(\Delta + w)]^{n-1} + [(1 + w)/(1 + \Delta w)]^{n-1}$ for $w \geq 1$. The derivative of this expression is ≥ 0 for all Δ.]

Section 4

8. (i)   When testing $H: p \leq p_0$ against $K: p > p_0$ by means of the test corresponding to (11), determine the sample size required to obtain power β against $p = p_1$, $\alpha = .05$, $\beta = .9$ for the cases $p_0 = .1$, $p_1 = .15, .20, .25$; $p_0 = .05$, $p_1 = .10, .15, .20, .25$; $p_0 = .01$, $p_1 = .02, .05, .10, .15, .20$.

(ii) Compare this with the sample size required if the inspection is by attributes and the test is based on the total number of defectives.

9. *Two-sided t-test.*

(i)   Let $X_1, \ldots, X_n$ be a sample from $N(\xi, \sigma^2)$. For testing $\xi = 0$ against $\xi \neq 0$, there exists a UMP invariant test with respect to the group $X_i' = cX_i$, $c \neq 0$, given by the two-sided t-test (17) of Chapter 5.

(ii) Let $X_1, \ldots, X_m$ and $Y_1, \ldots, Y_n$ be samples from $N(\xi, \sigma^2)$ and $N(\eta, \sigma^2)$ respectively. For testing $\eta = \xi$ against $\eta \neq \xi$ there exists a UMP invariant test with respect to the group $X_i' = aX_i + b$, $Y_j' = aY_j + b$, $a \neq 0$, given by the two-sided t-test (30) of Chapter 5.

[(i): Sufficiency and invariance reduce the problem to $|t|$, which in the notation of Section 4 has the probability density $p_\delta(t) + p_\delta(-t)$ for $t > 0$. The ratio of this density for $\delta = \delta_1$ to its value for $\delta = 0$ is proportional to $\int_0^\infty (e^{\delta_1 v} + e^{-\delta_1 v}) g_{t^2}(v) \, dv$, which is an increasing function of t^2 and hence of $|t|$.]

10. *Testing a correlation coefficient.* Let $(X_1, Y_1), \ldots, (X_n, Y_n)$ be a sample from a bivariate normal distribution.

   (i) For testing $\rho \leq \rho_0$ against $\rho > \rho_0$ there exists a UMP invariant test with respect to the group of all transformations $X_i' = aX_i + b$, $Y_i' = cY_i + d$ for which $a, c > 0$. This test rejects when the sample correlation coefficient R is too large.

   (ii) The problem of testing $\rho = 0$ against $\rho \neq 0$ remains invariant in addition under the transformation $Y_i' = -Y_i$, $X_i' = X_i$. With respect to the group generated by this transformation and those of (i) there exists a UMP invariant test, with rejection region $|R| \geq C$.

[(i): To show that the probability density $p_\rho(r)$ of R has monotone likelihood ratio, apply the condition of Chapter 3, Problem 8(i), to the expression (88) given for this density in Chapter 5. Putting $t = \rho r + 1$, the second derivative $\partial^2 \log p_\rho(r) / \partial \rho \, \partial r$ up to a positive factor is

$$\frac{\displaystyle\sum_{i,j=0}^{\infty} c_i c_j t^{i+j-2}\left[(j-i)^2(t-1) + (i+j)\right]}{2\left[\displaystyle\sum_{i=0}^{\infty} c_i t^i\right]^2}.$$

To see that the numerator is positive for all $t > 0$, note that it is greater than

$$2\sum_{i=0}^{\infty} c_i t^{i-2} \sum_{j=i+1}^{\infty} c_j t^j \left[(j-i)^2(t-1) + (i+j)\right].$$

Holding i fixed and using the inequality $c_{j+1} < \frac{1}{2}c_j$, the coefficient of t^j in the interior sum is ≥ 0.]

11. For testing the hypothesis that the correlation coefficient ρ of a bivariate normal distribution is $\leq \rho_0$, determine the power against the alternative $\rho = \rho_1$ when the level of significance α is .05, $\rho_0 = .3$, $\rho_1 = .5$, and the sample size n is $50, 100, 200$.

Section 5

12. Almost invariance of a test ϕ with respect to the group G of either Problem 6(i) or Example 6 implies that ϕ is equivalent to an invariant test.

Section 6

13. Show that

   (i) G_1 of Example 11 is a group;

   (ii) the test which rejects when $X_{21}^2 / X_{11}^2 > C$ is UMP invariant under G_1;

   (iii) the smallest group containing G_1 and G_2 is the group G of Example 11.

14. Consider a testing problem which is invariant under a group G of transformations of the sample space, and let $\mathscr{C}$ be a class of tests which is closed under G, so that $\phi \in \mathscr{C}$ implies $\phi g \in \mathscr{C}$, where ϕg is the test defined by $\phi g(x) = \phi(gx)$. If there exists an a.e. unique UMP member ϕ_0 of $\mathscr{C}$, then ϕ_0 is almost invariant.

15. *Envelope power function.* Let $S(\alpha)$ be the class of all level-α tests of a hypothesis H, and let $\beta_\alpha^*(\theta)$ be the *envelope power function*, defined by

$$\beta_\alpha^*(\theta) = \sup_{\phi \in S(\alpha)} \beta_\phi(\theta),$$

where β_ϕ denotes the power function of ϕ. If the problem of testing H is invariant under a group G, then $\beta_\alpha^*(\theta)$ is invariant under the induced group $\overline{G}$.

16. (i)   A generalization of equation (1) is

$$\int_A f(x)\, dP_\theta(x) = \int_{gA} f(g^{-1}x)\, dP_{\bar{g}\theta}(x).$$

(ii) If P_{θ_1} is absolutely continuous with respect to P_{θ_0}, then $P_{\bar{g}\theta_1}$ is absolutely continuous with respect to $P_{\bar{g}\theta_0}$ and

$$\frac{dP_{\theta_1}}{dP_{\theta_0}}(x) = \frac{dP_{\bar{g}\theta_1}}{dP_{\bar{g}\theta_0}}(gx) \qquad \left(\text{a.e. } P_{\theta_0}\right).$$

(iii) The distribution of $dP_{\theta_1}/dP_{\theta_0}(X)$ when X is distributed as P_{θ_0} is the same as that of $dP_{\bar{g}\theta_1}/dP_{\bar{g}\theta_0}(X')$ when X' is distributed as $P_{\bar{g}\theta_0}$.

17. *Invariance of likelihood ratio.* Let the family of distributions $\mathscr{P} = \{P_\theta, \theta \in \Omega\}$ be dominated by μ, let $p_\theta = dP_\theta/d\mu$, let μg^{-1} be the measure defined by $\mu g^{-1}(A) = \mu[g^{-1}(A)]$, and suppose that μ is absolutely continuous with respect to μg^{-1} for all $g \in G$.

(i)   Then

$$p_\theta(x) = p_{\bar{g}\theta}(gx)\frac{d\mu}{d\mu g^{-1}}(gx) \qquad (\text{a.e. } \mu).$$

(ii) Let Ω and ω be invariant under $\overline{G}$, and countable. Then the likelihood ratio $\sup_\Omega p_\theta(x)/\sup_\omega p_\theta(x)$ is almost invariant under G.

(iii) Suppose that $p_\theta(x)$ is continuous in θ for all x, that Ω is a separable pseudometric space, and that Ω and ω are invariant. Then the likelihood ratio is almost invariant under G.

18. *Inadmissible likelihood-ratio test.* In many applications in which a UMP invariant test exists, it coincides with the likelihood-ratio test. That this is,

however, not always the case is seen from the following example. Let $P_1, \ldots, P_n$ be n equidistant points on the circle $x^2 + y^2 = 4$, and $Q_1, \ldots, Q_n$ on the circle $x^2 + y^2 = 1$. Denote the origin in the (x, y) plane by O, let $0 < \alpha \leq \frac{1}{2}$ be fixed, and let (X, Y) be distributed over the $2n + 1$ points $P_1, \ldots, P_n, Q_1, \ldots, Q_n, O$ with probabilities given by the following table:

	P_i	Q_i	O
H	α/n	$(1 - 2\alpha)/n$	α
K	p_i/n	0	$(n - 1)/n$

where $\Sigma p_i = 1$. The problem remains invariant under rotations of the plane by the angles $2k\pi/n$ $(k = 0, 1, \ldots, n - 1)$. The rejection region of the likelihood-ratio test consists of the points $P_1, \ldots, P_n$, and its power is $1/n$. On the other hand, the UMP invariant test rejects when $X = Y = 0$, and has power $(n - 1)/n$.

19. Let G be a group of transformations of $\mathscr{X}$, and let $\mathscr{A}$ be a σ-field of subsets of $\mathscr{X}$, and μ a measure over $(\mathscr{X}, \mathscr{A})$. Then a set $A \in \mathscr{A}$ is said to be almost invariant if its indicator function is almost invariant.

   (i) The totality of almost invariant sets forms a σ-field $\mathscr{A}_0$, and a critical function is almost invariant if and only if it is $\mathscr{A}_0$-measurable.

   (ii) Let $\mathscr{P} = \{P_\theta, \theta \in \Omega\}$ be a dominated family of probability distributions over $(\mathscr{X}, \mathscr{A})$, and suppose that $\bar{g}\theta = \theta$ for all $\bar{g} \in \bar{G}$, $\theta \in \Omega$. Then the σ-field $\mathscr{A}_0$ of almost invariant sets is sufficient for $\mathscr{P}$.

   [Let $\lambda = \Sigma c_i P_{\theta_i}$ be equivalent to $\mathscr{P}$. Then

   $$\frac{dP_\theta}{d\lambda}(gx) = \frac{dP_{g^{-1}\theta}}{\Sigma c_i dP_{g^{-1}\theta_i}}(x) = \frac{dP_\theta}{d\lambda}(x) \qquad (\text{a.e. } \lambda),$$

   so that $dP_\theta/d\lambda$ is almost invariant and hence $\mathscr{A}_0$-measurable.]

Section 7

20. The definition of d-admissibility of a test coincides with the admissibility definition given in Chapter 1, Section 8 when applied to a two-decision procedure with loss 0 or 1 as the decision taken is correct or false.

21. (i) The following example shows that α-admissibility does not always imply d-admissibility. Let X be distributed as $U(0, \theta)$, and consider the tests φ_1 and φ_2 which reject when respectively $X < 1$ and $X < \frac{3}{2}$ for testing $H: \theta = 2$ against $K: \theta = 1$. Then for $\alpha = \frac{3}{4}$, φ_1 and φ_2 are both α-admissible but φ_2 is not d-admissible.

   (ii) Verify the existence of the test φ_0 of Example 12.

22. (i) The acceptance region $T_1/\sqrt{T_2} \leq C$ of Example 13 is a convex set in the (T_1, T_2) plane.

(ii) In Example 13, the conditions of Theorem 8 are not satisfied for the sets $A: T_1/\sqrt{T_2} \leq C$ and $\Omega': \xi > k$.

23. (i) In Example 13 (continued) show that there exist C_0, C_1 such that $\lambda_0(\eta)$ and $\lambda_1(\eta)$ are probability densities (with respect to Lebesgue measure).

(ii) Verify the densities h_0 and h_1.

24. Verify

(i) the admissibility of the rejection region (22);

(ii) the expression for $I(Z)$ given in the proof of Lemma 3.

25. Let $X_1, \ldots, X_m; Y_1, \ldots, Y_n$ be independent $N(\xi, \sigma^2)$ and $N(\eta, \sigma^2)$ respectively. The one-sided t-test of $H: \delta = \xi/\sigma \leq 0$ is admissible against the alternatives (i) $0 < \delta < \delta_1$ for any $\delta_1 > 0$; (ii) $\delta > \delta_2$ for any $\delta_2 > 0$.

26. For the model of the preceding problem, generalize Example 13 (continued) to show that the two-sided t-test is a Bayes solution for an appropriate prior distribution.

Section 9

27. *Wilcoxon two-sample test.* Let $U_{ij} = 1$ or 0 as $X_i < Y_j$ or $X_i > Y_j$, and let $U = \Sigma\Sigma U_{ij}$ be the number of pairs X_i, Y_j with $X_i < Y_j$.

(i) Then $U = \Sigma S_i - \frac{1}{2}n(n + 1)$, where $S_1 < \cdots < S_n$ are the ranks of the Y's, so that the test with rejection region $U > C$ is equivalent to the Wilcoxon test.

(ii) Any given arrangement of x's and y's can be transformed into the arrangement $x \ldots xy \ldots y$ through a number of interchanges of neighboring elements. The smallest number of steps in which this can be done for the observed arrangement is $mn - U$.

28. *Expectation and variance of Wilcoxon statistic.* If the X's and Y's are samples from continuous distributions F and G respectively, the expectation and variance of the Wilcoxon statistic U defined in the preceding problem are given by

(57)
$$E\left(\frac{U}{mn}\right) = P\{X < Y\} = \int F\,dG$$

and

(58)
$$mn\,\text{Var}\left(\frac{U}{mn}\right) = \int F\,dG + (n - 1)\int(1 - G)^2\,dF$$

$$+ (m - 1)\int F^2\,dG - (m + n - 1)\left(\int F\,dG\right)^2.$$

Under the hypothesis $G = F$, these reduce to

(59) $$E\left(\frac{U}{mn}\right) = \frac{1}{2}, \qquad \text{Var}\left(\frac{U}{mn}\right) = \frac{m+n+1}{12\,mn}.$$

29. (i) Let $Z_1, \ldots, Z_N$ be independently distributed with densities $f_1, \ldots, f_N$, and let the rank of Z_i be denoted by T_i. If f is any probability density which is positive whenever at least one of the f_i is positive, then

(60) $$P\{T_1 = t_1, \ldots, T_N = t_N\} = \frac{1}{N!} E\left[\frac{f_1\left(V_{(t_1)}\right)}{f\left(V_{(t_1)}\right)} \cdots \frac{f_N\left(V_{(t_N)}\right)}{f\left(V_{(t_N)}\right)}\right],$$

where $V_{(1)} < \cdots < V_{(N)}$ is an ordered sample from a distribution with density f.

(ii) If $N = m + n$, $f_1 = \cdots = f_m = f$, $f_{m+1} = \cdots = f_{m+n} = g$, and $S_1 < \cdots < S_n$ denote the ordered ranks of $Z_{m+1}, \ldots, Z_{m+n}$ among all the Z's, the probability distribution of $S_1, \ldots, S_n$ is given by (25).

[(i): The probability in question is $\int \ldots \int f_1(z_1) \ldots f_N(z_N) \, dz_1 \ldots dz_N$ integrated over the set in which z_i is the t_ith smallest of the z's for $i = 1, \ldots, N$. Under the transformation $w_{t_i} = z_i$ the integral becomes $\int \ldots \int f_1(w_{t_1}) \ldots f_N(w_{t_N}) \, dw_1 \ldots dw_N$, integrated over the set $w_1 < \cdots < w_N$. The desired result now follows from the fact that the probability density of the order statistics $V_{(1)} < \cdots < V_{(N)}$ is $N! f(w_1) \cdots f(w_N)$ for $w_1 < \cdots < w_N$.]

30. (i) For any continuous cumulative distribution function F, define $F^{-1}(0) = -\infty$, $F^{-1}(y) = \inf\{x : F(x) = y\}$ for $0 < y < 1$, $F^{-1}(1) = \infty$ if $F(x) < 1$ for all finite x, and otherwise $\inf\{x : F(x) = 1\}$. Then $F[F^{-1}(y)] = y$ for all $0 \le y \le 1$, but $F^{-1}[F(y)]$ may be $< y$.

(ii) Let Z have a cumulative distribution function $G(z) = h[F(z)]$, where F and h are continuous cumulative distribution functions, the latter defined over $(0, 1)$. If $Y = F(Z)$, then $P\{Y < y\} = h(y)$ for all $0 \le y \le 1$.

(iii) If Z has the continuous cumulative distribution function F, then $F(Z)$ is uniformly distributed over $(0, 1)$.

[(ii): $P\{F(Z) < y\} = P\{Z < F^{-1}(y)\} = F[F^{-1}(y)] = y$.]

31. Let Z_i have a continuous cumulative distribution function F_i $(i = 1, \ldots, N)$, and let G be the group of all transformations $Z_i' = f(Z_i)$ such that f is continuous and strictly increasing.

(i) The transformation induced by f in the space of distributions is $F_i' = F_i(f^{-1})$.

(ii) Two N-tuples of distributions $(F_1, \ldots, F_N)$ and $(F_1', \ldots, F_N')$ belong to the same orbit with respect to $\bar{G}$ if and only if there exist continuous distribution functions $h_1, \ldots, h_N$ defined on $(0, 1)$ and strictly increasing continuous distribution functions F and F' such that $F_i = h_i(F)$ and $F_i' = h_i(F')$.

[(i): $P\{f(Z_i) \leq y\} = P\{Z_i \leq f^{-1}(y)\} = F_i[f^{-1}(y)]$.
(ii): If $F_i = h_i(F)$ and the F_i' are on the same orbit, so that $F_i' = F_i(f^{-1})$, then $F_i' = h_i(F')$ with $F' = F(f^{-1})$. Conversely, if $F_i = h_i(F)$, $F_i' = h_i(F')$, then $F_i' = F_i(f^{-1})$ with $f = F'^{-1}(F)$.]

32. Under the assumptions of the preceding problem, if $F_i = h_i(F)$, the distribution of the ranks $T_1, \ldots, T_N$ of $Z_1, \ldots, Z_N$ depends only on the h_i, not on F. If the h_i are differentiable, the distribution of the T_i is given by

$$(61) \qquad P\{T_1 = t_1, \ldots, T_N = t_N\} = \frac{E\left[h_1'\left(U_{(t_1)}\right) \ldots h_N'\left(U_{(t_N)}\right)\right]}{N!},$$

where $U_{(1)} < \cdots < U_{(N)}$ is an ordered sample of size N from the uniform distribution $U(0, 1)$.

[The left-hand side of (61) is the probability that of the quantities $F(Z_1), \ldots, F(Z_N)$, the ith one is the t_ith smallest for $i = 1, \ldots, N$. This is given by $\int \ldots \int h_1'(y_1) \ldots h_N'(y_N) \, dy$ integrated over the region in which y_i is the t_ith smallest of the y's for $i = 1, \ldots, N$. The proof is completed as in Problem 29.]

33. *Distribution of order statistics.*

(i) If $Z_1, \ldots, Z_N$ is a sample from a cumulative distribution function F with density f, the joint density of $Y_i = Z_{(s_i)}$, $i = 1, \ldots, n$, is

$$(62) \qquad \frac{N! f(y_1) \ldots f(y_n)}{(s_1 - 1)!(s_2 - s_1 - 1)! \ldots (N - s_n)!}$$

$$\times [F(y_1)]^{s_1 - 1}[F(y_2) - F(y_1)]^{s_2 - s_1 - 1} \ldots [1 - F(y_n)]^{N - s_n}$$

for $y_1 < \cdots < y_n$.

(ii) For the particular case that the Z's are a sample from the uniform distribution on $(0, 1)$, this reduces to

$$(63) \qquad \frac{N!}{(s_1 - 1)!(s_2 - s_1 - 1)! \ldots (N - s_n)!}$$

$$y_1^{s_1 - 1}(y_2 - y_1)^{s_2 - s_1 - 1} \ldots (1 - y_n)^{N - s_n}.$$

For $n = 1$, (63) is the density of the beta-distribution $B_{s, N-s+1}$, which therefore is the distribution of the single order statistic $Z_{(s)}$ from $U(0, 1)$.

(iii)   Let the distribution of $Y_1, \ldots, Y_n$ be given by (63), and let V_i be defined by $Y_i = V_i V_{i+1} \ldots V_n$ for $i = 1, \ldots, n$. Then the joint distribution of the V_i is

$$\frac{N!}{(s_1 - 1)! \ldots (N - s_n)!} \prod_{i=1}^{n} v_i^{s_i - 1} (1 - v_i)^{s_{i+1} - s_i - 1} \qquad (s_{n+1} = N + 1),$$

so that the V_i are independently distributed according to the beta-distribution $B_{s_i, s_{i+1} - s_i}$.

[(i): If $Y_1 = Z_{(s_1)}, \ldots, Y_n = Z_{(s_n)}$ and $Y_{n+1}, \ldots, Y_N$ are the remaining Z's in the original order of their subscripts, the joint density of $Y_1, \ldots, Y_n$ is $N(N - 1) \ldots (N - n + 1) \int \ldots \int f(y_{n+1}) \ldots f(y_N) \, dy_{n+1} \ldots dy_N$ integrated over the region in which $s_1 - 1$ of the y's are $< y_1$, $s_2 - s_1 - 1$ between y_1 and $y_2, \ldots$, and $N - s_n > y_n$. Consider any set where a particular $s_1 - 1$ of the y's is $< y_1$, a particular $s_2 - s_1 - 1$ of them is between y_1 and y_2, and so on. There are $N!/(s_1 - 1)! \ldots (N - s_n)!$ of these regions, and the integral has the same value over each of them, namely $[F(y_1)]^{s_1 - 1}[F(y_2) - F(y_1)]^{s_2 - s_1 - 1} \ldots [1 - F(y_n)]^{N - s_n}.]$

34.   (i)   If $X_1, \ldots, X_m$ and $Y_1, \ldots, Y_n$ are samples with continuous cumulative distribution functions F and $G = h(F)$ respectively, and if h is differentiable, the distribution of the ranks $S_1 < \cdots < S_n$ of the Y's is given by

$$(64) \quad P\{S_1 = s_1, \ldots, S_n = s_n\} = \frac{E\left[h'(U_{(s_1)}) \ldots h'(U_{(s_n)})\right]}{\binom{m+n}{m}}$$

where $U_{(1)} < \cdots < U_{(m+n)}$ is an ordered sample from the uniform distribution $U(0, 1)$.

(ii)   If in particular $G = F^k$, where k is a positive integer, (64) reduces to

$$(65) \quad P\{S_1 = s_1, \ldots, S_n = s_n\}$$

$$= \frac{k^n}{\binom{m+n}{m}} \prod_{j=1}^{n} \frac{\Gamma(s_j + jk - j)}{\Gamma(s_j)} \cdot \frac{\Gamma(s_{j+1})}{\Gamma(s_{j+1} + jk - j)}.$$

35.   For sufficiently small $\theta > 0$, the Wilcoxon test at level

$$\alpha = k \Big/ \binom{N}{n}, \qquad k \text{ a positive integer,}$$

maximizes the power (among rank tests) against the alternatives (F, G) with $G = (1 - \theta)F + \theta F^2$.

36. An alternative proof of the optimum property of the Wilcoxon test for detecting a shift in the logistic distribution is obtained from the preceding problem by equating $F(x - \theta)$ with $(1 - \theta)F(x) + \theta F^2(x)$, neglecting powers of θ higher than the first. This leads to the differential equation $F - \theta F' = (1 - \theta)F + \theta F^2$, the solution of which is the logistic distribution.

37. Let $\mathscr{F}_0$ be a family of probability measures over $(\mathscr{X}, \mathscr{A})$, and let $\mathscr{C}$ be a class of transformations of the space $\mathscr{X}$. Define a class $\mathscr{F}_1$ of distributions by $F_1 \in \mathscr{F}_1$ if there exists $F_0 \in \mathscr{F}_0$ and $f \in \mathscr{C}$ such that the distribution of $f(X)$ is F_1 when that of X is F_0. If ϕ is any test satisfying (a) $E_{F_0}\phi(X) = \alpha$ for all $F_0 \in \mathscr{F}_0$, and (b) $\phi(x) \leq \phi[f(x)]$ for all x and all $f \in \mathscr{C}$, then ϕ is unbiased for testing $\mathscr{F}_0$ against $\mathscr{F}_1$.

38. Let $X_1, \ldots, X_m; Y_1, \ldots, Y_n$ be samples from a common continuous distribution F. Then the Wilcoxon statistic U defined in Problem 27 is distributed symmetrically about $\frac{1}{2}mn$ even when $m \neq n$.

39. (i) If $X_1, \ldots, X_m$ and $Y_1, \ldots, Y_n$ are samples from $F(x)$ and $G(y) = F(y - \Delta)$ respectively (F continuous), and $D_{(1)} < \cdots < D_{(mn)}$ denote the ordered differences $Y_j - X_i$, then

$$P\left[D_{(k)} < \Delta < D_{(mn+1-k)} \right] = P_0[k \leq U \leq mn - k],$$

where U is the statistic defined in Problem 27 and the probability on the right side is calculated for $\Delta = 0$.

(ii) Determine the above confidence interval for Δ when $m = n = 6$, the confidence coefficient is $\frac{20}{21}$, and the observations are x: .113, .212, .249, .522, .709, .788, and y: .221, .433, .724, .913, .917, 1.58.

(iii) For the data of (ii) determine the confidence intervals based on Student's t for the case that F is normal.

[(i): $D_{(i)} \leq \Delta < D_{(i+1)}$ if and only if $U_\Delta = mn - i$, where U_Δ is the statistic U of Problem 27 calculated for the observations

$$X_1, \ldots, X_m; Y_1 - \Delta, \ldots, Y_n - \Delta.]$$

40. (i) Let X, X' and Y, Y' be independent samples of size 2 from continuous distributions F and G respectively. Then

$$p = P\{\max(X, X') < \min(Y, Y')\} + P\{\max(Y, Y') < \min(X, X')\}$$

$$= \tfrac{1}{3} + 2\Delta,$$

where $\Delta = \int (F - G)^2 \, d[(F + G)/2]$.

(ii) $\Delta = 0$ if and only if $F = G$.

[(i): $p = \int (1 - F)^2 \, dG^2 + \int (1 - G)^2 \, dF^2$, which after some computation reduces to the stated form.

(ii): $\Delta = 0$ implies $F(x) = G(x)$ except on a set N which has measure zero both under F and G. Suppose that $G(x_1) - F(x_1) = \eta > 0$. Then there exists x_0 such that $G(x_0) = F(x_0) + \frac{1}{2}\eta$ and $F(x) < G(x)$ for $x_0 \leq x \leq x_1$. Since $G(x_1) - G(x_0) > 0$, it follows that $\Delta > 0$.]

41. *Continuation.*

  (i) There exists at every significance level α a test of $H : G = F$ which has power $> \alpha$ against all continuous alternatives (F, G) with $F \neq G$.

  (ii) There does not exist a nonrandomized unbiased rank test of H against all $G \neq F$ at level

$$\alpha = 1 \Big/ \binom{m+n}{m}.$$

[(i): let $X_i, X_i'; Y_i, Y_i'$ $(i = 1, \ldots, n)$ be independently distributed, the X's with distribution F, the Y's with distribution G, and let $V_i = 1$ if $\max(X_i, X_i') < \min(Y_i, Y_i')$ or $\max(Y_i, Y_i') < \min(X_i, X_i')$, and $V_i = 0$ otherwise. Then ΣV_i has a binomial distribution with the probability p defined in Problem 40, and the problem reduces to that of testing $p = \frac{1}{3}$ against $p > \frac{1}{3}$.

(ii): Consider the particular alternatives for which $P\{X < Y\}$ is either 1 or 0.]

Section 10

42. (i) Let m and n be the numbers of negative and positive observations among $Z_1, \ldots, Z_N$, and let $S_1 < \cdots < S_n$ denote the ranks of the positive Z's among $|Z_1|, \ldots, |Z_N|$. Consider the $N + \frac{1}{2}N(N - 1)$ distinct sums $Z_i + Z_j$ with $i = j$ as well as $i \neq j$. The Wilcoxon signed rank statistic ΣS_j is equal to the number of these sums that are positive.

  (ii) If the common distribution of the Z's is D, then

$$E\Big(\sum S_j\Big) = \frac{1}{2}N(N + 1) - ND(0) - \frac{1}{2}N(N - 1)\int D(-z) \, dD(z).$$

[(i) Let K be the required number of postive sums. Since $Z_i + Z_j$ is positive if and only if the Z corresponding to the larger of $|Z_i|$ and $|Z_j|$ is positive, $K = \sum_{i=1}^{N}\sum_{j=1}^{N} U_{ij}$, where $U_{ij} = 1$ if $Z_j > 0$ and $|Z_i| \leq Z_j$ and $U_{ij} = 0$ otherwise.]

43. Let $Z_1, \ldots, Z_N$ be a sample from a distribution with density $f(z - \theta)$, where $f(z)$ is positive for all z and f is symmetric about 0, and let m, n, and the S_j be defined as in the preceding problem.

  (i) The distribution of n and the S_j is given by

  (66) $P\{$the number of positive Z's is n and $S_1 = s_1, \ldots, S_n = s_n\}$

$$= \frac{1}{2^N} E\left[\frac{f(V_{(r_1)} + \theta) \ldots f(V_{(r_m)} + \theta) f(V_{(s_1)} - \theta) \ldots f(V_{(s_n)} - \theta)}{f(V_{(1)}) \ldots f(V_{(N)})}\right],$$

where $V_{(1)} < \cdots < V_{(N)}$ is an ordered sample from a distribution with density $2f(v)$ for $v > 0$, and 0 otherwise.

(ii)   The rank test of the hypothesis of symmetry with respect to the origin, which maximizes the derivative of the power function at $\theta = 0$ and hence maximizes the power for sufficiently small $\theta > 0$, rejects, under suitable regularity conditions, when

$$-E\left[\sum_{j=1}^{n} \frac{f'\left(V_{(s_j)}\right)}{f\left(V_{(s_j)}\right)}\right] > C.$$

(iii) In the particular case that $f(z)$ is a normal density with zero mean, the rejection region of (ii) reduces to $\Sigma E(V_{(s_j)}) > C$, where $V_{(1)} < \cdots < V_{(N)}$ is an ordered sample from a χ-distribution with 1 degree of freedom.

(iv)   Determine a density f such that the one-sample Wilcoxon test is most powerful against the alternatives $f(z - \theta)$ for sufficiently small positive θ.

[(i): Apply Problem 29(i) to find an expression for $P\{S_1 = s_1, \ldots, S_n = s_n$ given that the number of positive Z's is $n\}$.]

44.   An alternative expression for (66) is obtained if the distribution of Z is characterized by (ρ, F, G). If then $G = h(F)$ and h is differentiable, the distribution of n and the S_j is given by

$$(67) \qquad \rho^m (1 - \rho)^n E\left[h'\left(U_{(s_1)}\right) \ldots h'\left(U_{(s_n)}\right)\right],$$

where $U_{(1)} < \cdots < U_{(N)}$ is an ordered sample from $U(0, 1)$.

45.   *Unbiased tests of symmetry.* Let $Z_1, \ldots, Z_N$ be a sample, and let ϕ be any rank test of the hypothesis of symmetry with respect to the origin such that $z_i \le z_i'$ for all i implies $\phi(z_1, \ldots, z_N) \le \phi(z_1', \ldots, z_N')$. Then ϕ is unbiased against the one-sided alternatives that the Z's are stochastically larger than some random variable that has a symmetric distribution with respect to the origin.

46.   *The hypothesis of randomness.* Let $Z_1, \ldots, Z_N$ be independently distributed with distributions $F_1, \ldots, F_N$, and let T_i denote the rank of Z_i among the Z's. For testing the *hypothesis of randomness* $F_1 = \cdots = F_N$ against the alternatives K of an *upward trend*, namely that Z_i is stochastically increasing with i, consider the rejection regions

$$(68) \qquad\qquad\qquad \sum i t_i > C$$

and

$$(69) \qquad\qquad\qquad \sum i E\left(V_{(t_i)}\right) > C,$$

where $V_{(1)} < \cdots < V_{(N)}$ is an ordered sample from a standard normal distribution and where t_i is the value taken on by T_i.

(i)   The second of these tests is most powerful among rank tests against the normal alternatives $F = N(\gamma + i\delta, \sigma^2)$ for sufficiently small δ.

(ii) Determine alternatives against which the first test is a most powerful rank test.

(iii) Both tests are unbiased against the alternatives of an upward trend; so is any rank test ϕ satisfying $\phi(z_1, \ldots, z_N) \le \phi(z_1', \ldots, z_N')$ for any two points for which $i < j$, $z_i < z_j$ implies $z_i' < z_j'$ for all i and j.

[(iii): Apply Problem 37 with $\mathscr{C}$ the class of transformations $z_1' = z_1$, $z_i' = f_i(z_i)$ for $i > 1$, where $z < f_2(z) < \cdots < f_N(z)$ and each f_i is nondecreasing. If $\mathscr{F}_0$ is the class of N-tuples $(F_1, \ldots, F_N)$ with $F_1 = \cdots = F_N$, then $\mathscr{F}_1$ coincides with the class K of alternatives.]

47. In the preceding problem let $U_{ij} = 1$ if $(j - i)(Z_j - Z_i) > 0$, and $= 0$ otherwise.

(i)   The test statistic $\Sigma i T_i$ can be expressed in terms of the U's through the relation

$$\sum_{i=1}^{N} iT_i = \sum_{i<j} (j - i)U_{ij} + \frac{N(N + 1)(N + 2)}{6}.$$

(ii) The smallest number of steps [in the sense of Problem 27(ii)] by which $(Z_1, \ldots, Z_N)$ can be transformed into the ordered sample $(Z_{(1)}, \ldots, Z_{(N)})$ is $[N(N - 1)/2] - U$, where $U = \sum_{i<j} U_{ij}$. This suggests $U > C$ as another rejection region for the preceding problem.

[(i): Let $V_{ij} = 1$ or 0 as $Z_i \le Z_j$ or $Z_i > Z_j$. Then $T_j = \sum_{i=1}^{N} V_{ij}$, and $V_{ij} = U_{ij}$ or $1 - U_{ij}$ as $i < j$ or $i \ge j$. Expressing $\sum_{j=1}^{N} jT_j = \sum_{j=1}^{N} j\sum_{i=1}^{N} V_{ij}$ in terms of the U's and using the fact that $U_{ij} = U_{ji}$, the result follows by a simple calculation.]

48. *The hypothesis of independence.* Let $(X_1, Y_1), \ldots, (X_N, Y_N)$ be a sample from a bivariate distribution, and $(X_{(1)}, Z_1), \ldots, (X_{(N)}, Z_N)$ be the same sample arranged according to increasing values of the X's, so that the Z's are a permutation of the Y's. Let R_i be the rank of X_i among the X's, S_i the rank of Y_i among the Y's, and T_i the rank of Z_i among the Z's, and consider the hypothesis of independence of X and Y against the alternatives of positive regression dependence.

(i)   Conditionally, given $(X_{(1)}, \ldots, X_{(N)})$, this problem is equivalent to testing the hypothesis of randomness of the Z's against the alternatives of an upward trend.

(ii)   The test (68) is equivalent to rejecting when the *rank correlation coefficient*

$$\frac{\sum(R_i - \bar{R})(S_i - \bar{S})}{\sqrt{\sum(R_i - \bar{R})^2 \sum(S_i - \bar{S})^2}} = \frac{12}{N^3 - N} \sum\left(R_i - \frac{N+1}{2}\right)\left(S_i - \frac{N+1}{2}\right)$$

is too large.

(iii) An alternative expression for the rank correlation coefficient* is

$$1 - \frac{6}{N^3 - N}\sum(S_i - R_i)^2 = 1 - \frac{6}{N^3 - N}\sum(T_i - i)^2.$$

(iv)   The test $U > C$ of Problem 47(ii) is equivalent to rejecting when Kendall's *t*-statistic* $\sum_{i<j} V_{ij}/N(N-1)$ is too large where V_{ij} is $+1$ or -1 as $(Y_j - Y_i)(X_j - X_i)$ is positive or negative.

(v)   The tests (ii) and (iv) are unbiased against the alternatives of positive regression dependence.

Section 11

49. In Example 16, a family of sets $S(x, y)$ is a class of equivariant confidence sets if and only if there exists a set $\mathscr{R}$ of real numbers such that

$$S(x, y) = \bigcup_{r \in \mathscr{R}}\left\{(\xi, \eta) : (x - \xi)^2 + (y - \eta)^2 = r^2\right\}.$$

50. Let $X_1, \ldots, X_n; Y_1, \ldots, Y_n$ be samples from $N(\xi, \sigma^2)$ and $N(\eta, \tau^2)$ respectively. Then the confidence intervals (43) of Chapter 5 for τ^2/σ^2, which can be written as

$$\frac{\sum(Y_j - \bar{Y})^2}{k\sum(X_i - \bar{X})^2} \leq \frac{\tau^2}{\sigma^2} \leq \frac{k\sum(Y_j - \bar{Y})^2}{\sum(X_i - \bar{X})^2},$$

are uniformly most accurate equivariant with respect to the smallest group G containing the transformations $X_i' = aX + b$, $Y_i' = aY + c$ for all $a \neq 0$, b, c and the transformation $X_i' = dY_i$, $Y_i' = X_i/d$ for all $d \neq 0$.
[Cf. Problem 7.]

51. (i)   *One-sided equivariant confidence limits.* Let θ be real-valued, and suppose that for each θ_0, the problem of testing $\theta \leq \theta_0$ against $\theta > \theta_0$ (in the presence of nuisance parameters ϑ) remains invariant under a group

---

*For further material on these statistics see Kendall (1970); Aiyar, Guillier, and Albers (1979); and books on nonparametric inference.

G_{θ_0} and that $A(\theta_0)$ is a UMP invariant acceptance region for this hypothesis at level α. Let the associated confidence sets $S(x) = \{\theta : x \in A(\theta)\}$ be one-sided intervals $S(x) = \{\theta : \underline{\theta}(x) \leq \theta\}$, and suppose they are equivariant under all G_θ and hence under the group G generated by these. Then the lower confidence limits $\underline{\theta}(X)$ are uniformly most accurate equivariant at confidence level $1 - \alpha$ in the sense of minimizing $P_{\theta,\vartheta}\{\underline{\theta}(X) \leq \theta'\}$ for all $\theta' < \theta$.

(ii) Let $X_1, \ldots, X_n$ be independently distributed as $N(\xi, \sigma^2)$. The upper confidence limits $\sigma^2 \leq \Sigma(X_i - \overline{X})^2/C_0$ of Example 5, Chapter 5, are uniformly most accurate equivariant under the group $X_i' = X_i + c$, $-\infty < c < \infty$. They are also equivariant (and hence uniformly most accurate equivariant) under the larger group $X_i' = aX_i + c$, $-\infty < a, c < \infty$.

52. *Counterexample.* The following example shows that the equivariance of $S(x)$ assumed in the paragraph following Lemma 5 does not follow from the other assumptions of this lemma. In Example 8, let $n = 1$, let $G^{(1)}$ be the group G of Example 8, and let $G^{(2)}$ be the corresponding group when the roles of Z and $Y = Y_1$ are reversed. For testing $H(\theta_0) : \theta = \theta_0$ against $\theta \neq \theta_0$ let G_{θ_0} be equal to $G^{(1)}$ augmented by the transformation $Y' = \theta_0 - (Y_1 - \theta_0)$ when $\theta \leq 0$, and let G_{θ_0} be equal to $G^{(2)}$ augmented by the transformation $Z' = \theta_0 - (Z - \theta_0)$ when $\theta > 0$. Then there exists a UMP invariant test of $H(\theta_0)$ under G_{θ_0} for each θ_0, but the associated confidence sets $S(x)$ are not equivariant under $G = \{G_\theta, -\infty < \theta < \infty\}$.

53. (i) Let $X_1, \ldots, X_n$ be independently distributed as $N(\xi, \sigma^2)$, and let $\theta = \xi/\sigma$. The lower confidence bounds $\underline{\theta}$ for θ, which at confidence level $1 - \alpha$ are uniformly most accurate invariant under the transformations $X_i' = aX_i$, are

$$\underline{\theta} = C^{-1}\left(\frac{\sqrt{n}\,\overline{X}}{\sqrt{\Sigma(X_i - \overline{X})^2/(n-1)}}\right)$$

where the function $C(\theta)$ is determined from a table of noncentral t so that

$$P_\theta\left\{\frac{\sqrt{n}\,\overline{X}}{\sqrt{\Sigma(X_i - \overline{X})^2/(n-1)}} \leq C(\theta)\right\} = 1 - \alpha.$$

(ii) Determine $\underline{\theta}$ when the x's are 7.6, 21.2, 15.1, 32.0, 19.7, 25.3, 29.1, 18.4 and the confidence level is $1 - \alpha = .95$.

54. (i) Let $(X_1, Y_1), \ldots, (X_n, Y_n)$ be a sample from a bivariate normal distribution, and let

$$\underline{\rho} = C^{-1}\left(\frac{\Sigma(X_i - \overline{X})(Y_i - \overline{Y})}{\sqrt{\Sigma(X_i - \overline{X})^2 \Sigma(Y_i - \overline{Y})^2}}\right),$$

where $C(\rho)$ is determined such that

$$P_\rho \left\{ \frac{\sum (X_i - \bar{X})(Y_i - \bar{Y})}{\sqrt{\sum (X_i - \bar{X})^2 \sum (Y_i - \bar{Y})^2}} \le C(\rho) \right\} = 1 - \alpha.$$

Then $\underline{\rho}$ is a lower confidence limit for the population correlation coefficient ρ at confidence level $1 - \alpha$; it is uniformly most accurate invariant with respect to the group of transformations $X_i' = aX_i + b$, $Y_i' = cY_i + d$, with $ac > 0$, $-\infty < b, d < \infty$.

(ii)   Determine $\underline{\rho}$ at level $1 - \alpha = .95$ when the observations are $(12.9, .56)$, $(9.8, .92)$, $(13.1, .42)$, $(12.5, 1.01)$, $(8.7, .63)$, $(10.7, .58)$, $(9.3, .72)$, $(11.4, .64)$.

Section 12

55.   In Examples 20 and 21 there do not exist equivariant sets that uniformly minimize the probability of covering false values.

56.   In Example 20, the density $p(v)$ of $V = 1/S^2$ is unimodal.

57.   Show that in Example 20,

(i)   the confidence sets $\sigma^2/S^2 \in A^{**}$ with A^{**} given by (40) coincide with the uniformly most accurate unbiased confidence sets for σ^2;

(ii)   if (a, b) is best with respect to (39) for σ, then (a', b') is best for σ^r $(r > 0)$.

58.   Let $X_1, \ldots, X_r$ be independent $N(0, 1)$, and let S^2 be independent of the X's and distributed as χ_ν^2. Then the distribution of $(X_1/S\sqrt{\nu}, \ldots, X_r/S\sqrt{\nu})$ is a central multivariate t-distribution, and its density is

$$p(v_1, \ldots, v_r) = \frac{\Gamma\left(\frac{1}{2}(\nu + r)\right)}{(\pi\nu)^{r/2}\Gamma(\nu/2)} \left(1 + \frac{1}{\nu}\sum v_i^2\right)^{-\frac{1}{2}(\nu+r)}.$$

59.   The confidence sets (47) are uniformly most accurate equivariant under the group G defined at the end of Example 22.

60.   In Example 23, show that

(i)   both sets (55) are intervals;

(ii)   the sets given by $vp(v) > C$ coincide with the intervals (42) of Chapter 5.

61.   Let $X_1, \ldots, X_m$; $Y_1, \ldots, Y_n$ be independently normally distributed as $N(\xi, \sigma^2)$ and $N(\eta, \sigma^2)$ respectively. Determine the equivariant confidence sets for $\eta - \xi$ that have smallest Lebesgue measure when

(i) σ is known;

(ii) σ is unknown.

62. Generalize the confidence sets of Example 18 to the case that the X_i are $N(\xi_i, d_i \sigma^2)$ where the d's are known constants.

63. Solve the problem corresponding to Example 20 when

(i) $X_1, \ldots, X_n$ is a sample from the exponential density $E(\xi, \sigma)$, and the parameter being estimated is σ;

(ii) $X_1, \ldots, X_n$ is a sample from the uniform density $U(\xi, \xi + \tau)$, and the parameter being estimated is τ.

64. Let $X_1, \ldots, X_n$ be a sample from the exponential distribution $E(\xi, \sigma)$. With respect to the transformations $X_i' = bX_i + a$ determine the smallest equivariant confidence sets

(i) for σ, both when size is defined by Lebesgue measure and by the equivariant measure (39);

(ii) for ξ.

65. Let X_{ij} ($j = 1, \ldots, n_i$; $i = 1, \ldots, s$) be samples from the exponential distribution $E(\xi_i, \sigma)$. Determine the smallest equivariant confidence sets for $(\xi_1, \ldots, \xi_r)$ with respect to the group $X_{ij}' = bX_{ij} + a_i$.

Section 13

66. If the confidence sets $S(x)$ are equivariant under the group G, then the probability $P_\theta\{\theta \in S(X)\}$ of their covering the true value is invariant under the induced group $\bar{G}$.

67. Consider the problem of obtaining a (two-sided) confidence band for an unknown continuous cumulative distribution function F.

(i) Show that this problem is invariant both under strictly increasing and strictly decreasing continuous transformations $X_i' = f(X_i)$, $i = 1, \ldots, n$, and determine a maximal invariant with respect to this group.

(ii) Show that the problem is not invariant under the transformation

$$X_i' = \begin{cases} X_i & \text{if } |X_i| \geq 1, \\ X_i - 1 & \text{if } 0 < X_i < 1, \\ X_i + 1 & \text{if } -1 < X_i < 0. \end{cases}$$

[(ii): For this transformation g, the set $g^*S(x)$ is no longer a band.]

Additional Problems

68. Let $X_1, \ldots, X_n$ be a sample from a distribution with density

$$\frac{1}{\tau^n} f\left(\frac{x_1}{\tau}\right) \cdots f\left(\frac{x_n}{\tau}\right),$$

where $f(x)$ is either zero for $x < 0$ or symmetric about zero. The most powerful scale-invariant test for testing $H : f = f_0$ against $K : f = f_1$ rejects when

$$\frac{\int_0^\infty v^{n-1} f_1(vx_1) \ldots f_1(vx_n) \, dv}{\int_0^\infty v^{n-1} f_0(vx_1) \ldots f_0(vx_n) \, dv} > C.$$

69. *Normal vs. double exponential.* For $f_0(x) = e^{-x^2/2}/\sqrt{2\pi}$, $f_1(x) = e^{-|x|}/2$, the test of the preceding problem reduces to rejecting when $\sqrt{\Sigma x_i^2}/\Sigma|x_i| < C$.

(Hogg, 1972.)

Note. The corresponding test when both location and scale are unknown is obtained in Uthoff (1973). Testing normality against Cauchy alternatives is discussed by Franck (1981).

70. *Uniform vs. triangular.*

   (i)   For $f_0(x) = 1$ $(0 < x < 1)$, $f_1(x) = 2x$ $(0 < x < 1)$, the test of Problem 68 reduces to rejecting when $T = x_{(n)}/\bar{x} < C$.
   (ii) Under f_0, the statistic $2n \log T$ is distributed as χ^2_{2n}.

(Quesenberry and Starbuck, 1976.)

71. Show that the test of Problem 5(i) reduces to

   (i) $[x_{(n)} - x_{(1)}]/S < c$ for normal vs. uniform;
   (ii) $[\bar{x} - x_{(1)}]/S < c$ for normal vs. exponential;
   (iii) $[\bar{x} - x_{(1)}]/[x_{(n)} - x_{(1)}] < c$ for uniform vs. exponential.

(Uthoff, 1970.)

Note. When testing for normality, one is typically not interested in distinguishing the normal from some other given shape but would like to know more generally whether the data are or are not consonant with a normal distribution. This is a special case of the problem of testing for goodness of fit, briefly referred to at the end of Section 13. Methods particularly suitable for testing normality are discussed for example in Shapiro, Wilk, and Chen (1968), Hegazy and Green (1975), D'Agostino (1982), Hall and Welsh (1983), and Spiegelhalter (1983), and for testing exponentiality in Galambos (1982), Brain and Shapiro (1983), Spiegelhalter (1983), Deshpande (1983), Doksum and Yandell (1984), and Spurrier (1984). See also Kent and Quesenberry (1982).

72. The UMP invariant test of Problem 69 is also UMP similar.
    [Consider the problem of testing $\alpha = 0$ vs. $\alpha > 0$ in the two-parameter exponential family with density

$$C(\alpha, \tau) \exp\left(-\frac{\alpha}{2\tau^2} \Sigma x_i^2 - \frac{1-\alpha}{\tau} \Sigma|x_i|\right), \qquad 0 \le \alpha < 1.]$$

Note. For the analogous result for the tests of Problem 70, 71, see Quesenberry and Starbuck (1976).

73. The following UMP unbiased tests of Chapter 5 are also UMP invariant under change in scale:

   (i) The test of $g \leq g_0$ in a gamma distribution (Problem 73 of Chapter 5).

   (ii) The test of $b_1 \leq b_2$ in Problem 75(i) of Chapter 5.

74. Let $X_1, \ldots, X_n$ be a sample from $N(\xi, \sigma^2)$, and consider the UMP invariant level-α test of $H: \xi/\sigma \leq \theta_0$ (Section 6.4). Let $\alpha_n(F)$ be the actual significance level of this test when $X_1, \ldots, X_n$ is a sample from a distribution F with $E(X_i) = \xi$, $\mathrm{Var}(X_i) = \sigma^2 < \infty$. Then the relation $\alpha_n(F) \to \alpha$ will not in general hold unless $\theta_0 = 0$.
   [Use the fact that the joint distribution of $\sqrt{n}(\overline{X} - \xi)$ and $\sqrt{n}(S^2 - \sigma^2)$ tends to the bivariate normal distribution with mean zero and covariance matrix

$$\begin{pmatrix} \sigma^2 & \mu_3 \\ \mu_3 & \mu_4 - \sigma^2 \end{pmatrix},$$

   where $S^2 = \Sigma(X_i - \overline{X})^2/n$ and $\mu_k = E(X_i - \xi)^k$. See for example Serfling (1980).]

75. The totality of permutations of K distinct numbers $a_1, \ldots, a_K$ for varying $a_1, \ldots, a_K$ can be represented as a subset C_K of Euclidean K-space R_K, and the group G of Example 8 as the union of $C_2, C_3, \ldots$. Let ν be the measure over G which assigns to a subset B of G the value $\Sigma_{k=2}^{\infty} \mu_K(B \cap C_K)$, where μ_K denotes Lebesgue measure in E_K. Give an example of a set $B \subset G$ and an element $g \in G$ such that $\nu(B) > 0$ but $\nu(Bg) = 0$.
   [If a, b, c, d are distinct numbers, the permutations g, g' taking (a, b) into (b, a) and (c, d) into (d, c) respectively are points in C_2, but gg' is a point in C_4.]

76. The Kolmogorov test (56) for testing $H: F = F_0$ (F_0 continuous) is consistent against any alternative $F_1 \neq F_0$, that is, its power against any fixed F_1 tends to 1 as $n \to \infty$.
   [The critical value $\Delta = \Delta_n$ of (56) corresponding to a given α satisfies $\sqrt{n}\Delta \to K$ for some $K > 0$ as $n \to \infty$. Let a be any value for which $F_1(a) \neq F_0(a)$, and use the facts that (a) $|F_0(a) - T_X(a)| \leq \sup|F_0(u) - T_X(u)|$ and (b) if $F = F_1$, the statistic $T_X(a)$ has a binomial distribution with success probability $p = F_1(a) \neq F_0(a)$.]   [Massey (1950).]

   Note. For exact power calculations in both the continuous and discrete case, see for example Niederhausen (1981) and Gleser (1985).

77. (i) Let $X_1, \ldots, X_m; Y_1, \ldots, Y_n$ be i.i.d. according to a continuous distribution F, let the ranks of the Y's be $S_1 < \cdots < S_n$, and let $T = h(S_1) + \cdots + h(S_n)$. Then if either $m = n$ or $h(s) + h(N + 1 - s)$ is independent of s, the distribution of T is symmetric about $n\Sigma_{i=1}^{N} h(i)/N$.

(ii) Show that the two-sample Wilcoxon and normal-scores statistics are symmetrically distributed under H, and determine their centers of symmetry.

[(i): Let $S_i' = N + 1 - S_i$, and use the fact that $T' = \Sigma h(S_j')$ has the same distribution under H as T.]

Note. The following problems explore the relationship between pivotal quantities and equivariant confidence sets. For more details see Arnold (1984). Let X be distributed according $P_{\theta, \vartheta}$, and consider confidence sets for θ that are equivariant under a group G^*, as in Section 11. If w is the set of possible θ-values, define a group $\tilde{G}$ on $\mathscr{X} \times w$ by $\tilde{g}(\theta, x) = (gx, \bar{g}\theta)$.

78. Let $V(X, \theta)$ be any pivotal quantity [i.e. have a fixed probability distribution independent of (θ, ϑ)], and let B be any set in the range space of V with probability $P(V \in B) = 1 - \alpha$. Then the sets $S(x)$ defined by

(70) $\theta \in S(x)$ if and only if $V(\theta, x) \in B$

are confidence sets for θ with confidence coefficient $1 - \alpha$.

79. (i) If $\tilde{G}$ is transitive over $\mathscr{X} \times w$ and $V(X, \theta)$ is maximal invariant under $\tilde{G}$, then $V(X, \theta)$ is pivotal.

   (ii) By (i), any quantity $W(X, \theta)$ which is invariant under $\tilde{G}$ is pivotal; give an example showing that the converse need not be true.

80. Under the assumptions of the preceding problem, the confidence set $S(x)$ is equivariant under G^*.

81. Under the assumptions of Problem 79, suppose that a family of confidence sets $S(x)$ is equivariant under G^*. Then there exists a set B in the range space of the pivotal V such that (70) holds. In this sense, all equivariant confidence sets can be obtained from pivotals.
[Let A be the subset of $\mathscr{X} \times w$ given by $A = \{(x, \theta) : \theta \in S(x)\}$. Show that $\tilde{g}A = A$, so that any orbit of $\tilde{G}$ is either in A or in the complement of A. Let the maximal invariant $V(x, \theta)$ be represented as in Section 2 by a uniquely defined point on each orbit, and let B be the set of these points whose orbits are in A. Then $V(x, \theta) \in B$ if and only if $(x, \theta) \in A$.]

Note. Problem 80 provides a simple check of the equivariance of confidence sets. In Example 21, for instance, the confidence sets (41) are based on the pivotal vector $(X_1 - \xi_1, \ldots, X_r - \xi_r)$, and hence are equivariant.

15. REFERENCES

Invariance considerations were introduced for particular classes of problems by Hotelling and Pitman. (See the references to Chapter 1.) The general theory of invariant and almost invariant tests, together with its principal parametric applications, was developed by Hunt and Stein (1946) in an unpublished paper. In their paper, invariance was not proposed as a desirable property in itself but as a tool for deriving most stringent tests (cf. Chapter 9). Apart from this difference in point of view, the present account is based on the ideas of Hunt and Stein, about which I learned through conversations with Charles Stein during the years 1947–1950.

Of the admissibility results of Section 7, Theorem 8 is due to Birnbaum (1955) and Stein (1956a); Example 13 (continued) and Lemma 3, to Kiefer and Schwartz (1965).

Aiyar, R. J., Guillier, C. L., and Albers, W.
   (1979). "Asymptotic relative efficiencies of rank tests for trend alternatives." *J. Amer. Statist. Assoc.* **74**, 226–231.

Anderson, T. W.
   (1984). *An Introduction to Multivariate Statistical Analysis*, 2nd ed., Wiley, New York. [Problem 10.]

Andersson, S.
   (1982). "Distributions of maximal invariants using quotient measures." *Ann. Statist.* **10**, 955–961.

Antille, A., Kersting, G., and Zucchini, W.
   (1982). "Testing symmetry." *J. Amer. Statist. Assoc.* **77**, 639–651.

Arnold, S. F.
   (1984). "Pivotal quantities and invariant confidence regions." *Statist. and Decisions* **2**, 257–280.

Barnard, G.
   (1950). "The Behrens–Fisher test." *Biometrika* **37**, 203–207.

Bell, C. B.
   (1964). "A characterization of multisample distribution-free statistics." *Ann. Math. Statist.* **35**, 735–738.

Beran, R.
   (1974). "Asymptotically efficient adaptive rank estimates in location models." *Ann. Statist.* **2**, 63–74.

Berk, R. H.
   (1967). "A special group structure and equivariant estimation." *Ann. Math. Statist.* **38**, 1436–1445.
   (1970). "A remark on almost invariance." *Ann. Math. Statist.* **41**, 733–735.
   (1972). "A note on sufficiency and invariance." *Ann. Math. Statist.* **43**, 647–650.

Berk, R. J. and Bickel, P. J.
   (1968). "On invariance and almost invariance." *Ann. Math. Statist.* **39**, 1573–1576.

Bhattacharya, P. K., Gastwirth, J. L., and Wright, A. L.
   (1982). "Two modified Wilcoxon tests for symmetry about an unknown location parameter." *Biometrika* **69**, 377–382.

Bickel, P. J.
   (1982). "On adaptive estimation." *Ann. Statist.* **10**, 647–671.

Billingsley, P.
   (1968). *Convergence of Probability Measures*. Wiley, New York.

Birnbaum, A.
(1955). "Characterization of complete classes of tests of some multiparameter hypotheses, with applications to likelihood ratio tests." *Ann. Math. Statist.* **26**, 31–36.

Birnbaum, Z. W.
(1952). "Numerical tabulation of the distribution of Kolmogorov's statistic for finite sample size." *J. Amer. Statist. Assoc.* **47**, 425–441.

Blair, R. C. and Higgins, J. J.
(1980), "A comparison of the power of Wilcoxon's rank-sum statistic to that of Student's *t*-statistic under various nonnormal distributions." *J. Ed. Statist.* **5**, 309–335.

Bondar, J. V.
(1976). "Borel cross sections and maximal invariants." *Ann. Statist.* **4**, 866–877.

Boos, D. D.
(1982). "A test for asymmetry associated with the Hodges–Lehmann estimator." *J. Amer. Statist. Assoc.* **77**, 647–651.

Brain, C. W. and Shapiro, S. S.
(1983). "A regression test for exponentiality: Censored and complete samples." *Technometrics* **25**, 69–76.

Brown, L. D. and Sackrowitz, H.
(1984). "An alternative to Student's *t*-test for problems with indifference zones." *Ann. Statist.* **12**, 451–469.

Chernoff, H.
(1949). "Asymptotic studentization in testing of hypotheses." *Ann. Math. Statist.* **20**, 268–278.

Cohen, A. and Strawderman, W. E.
(1973). "Admissibility implications for different criteria in confidence estimation." *Ann. Statist.* **1**, 363–366.

Cox, D. R.
(1961). "Tests of separate families of hypotheses." In *Proc. 4th Berkeley Symp.*, Vol. 1, 105–123.
(1962). "Further results on tests of separate families of hypotheses." *J. Roy. Statist. Soc. (B)* **24**, 406–423.

D'Agostino, R. B.
(1982). "Departures from normality, tests for." In *Encycl. Statist. Sci.* Vol. 2, Wiley, New York.

Davenport, J. M. and Webster, J. T.
(1975). "The Behrens–Fisher problem. An old solution revisited." *Metrika* **22**, 47–54.

David, H. A.
(1981). *Order Statistics*, 2nd ed., Wiley, New York.

Davis, C. E. and Quade, D.
(1978). "*U*-statistics for skewness or symmetry." *Comm. Statist.* **A7**, 413–418.

Deshpande, J. V.
(1983). "A class of tests for exponentiality against increasing failure rate average alternatives." *Biometrika* **70**, 514–518.

Deuchler, G.
(1914). "Ueber die Methoden der Korrelationsrechnung in der Paedagogik und Psychologie." *Z. Pädag. Psychol.* **15**, 114–131, 145–159, 229–242.

[Appears to contain the first proposal of the two-sample procedure known as the Wilcoxon test, which was later discovered independently by many different authors. A history of this test is given by W. H. Kruskal, "Historical notes on the Wilcoxon unpaired two-sample test." *J. Amer. Statist. Assoc.* **52** (1957), 356–360.]

Doksum, K. A., Fenstad, G., and Aaberge, R.
(1977). "Plots and tests for symmetry." *Biometrika* **64**, 473–487.

Doksum, K. A. and Yandell, B. S.
(1984). "Tests for exponentiality." In *Handbook of Statistics* (Krishnaiah and Sen, eds.), Vol. 4, 579–611.

Durbin, J.
(1973). *Distribution Theory Based on the Sample Distribution Function*, SIAM, Philadelphia.

Eaton, M. L.
(1983). *Multivariate Statistics*, Wiley, New York.

Epps, T. W., Singleton, K. J., and Pulley, L. B.
(1982). "A test of separate families of distributions based on the empirical moment generating function." *Biometrika* **69**, 391–399.

Epstein, B. and Tsao, C. K.
(1953). "Some tests based on ordered observations from two exponential populations." *Ann. Math. Statist.* **24**, 458–466.

Farrell, R. H.
(1968). "Towards a theory of generalized Bayes tests." *Ann. Math. Statist.* **39**, 1–22.
(1985). *Multivariate Calculation: Use of the Continuous Groups*, Springer, Berlin.

Feller, W.
(1948). "On the Kolmogorov–Smirnov limit theorems for empirical distributions." *Ann. Math. Statist.* **19**, 177–189.

Fisher, R. A.
(1956, 1959, 1973). *Statistical Methods and Scientific Inference*, Oliver and Boyd, Edinburgh (1956, 1959); Hafner, New York (1973).
[In Chapter IV the author gives his views on hypothesis testing and in particular discusses his ideas on the Behrens–Fisher problem.]

Fisher, R. A. and Yates, F.
(1948). *Statistical Tables for Biological, Agricultural and Medical Research*, 3rd ed., Oliver and Boyd, London.
[Implicit in the introduction to Tables XX and XXI is a consideration of rank-order tests such as (19).]

Franck, W. E.
(1981). "The most powerful invariant test of normal versus Cauchy with applications to stable alternatives." *J. Amer. Statist. Assoc.* **76**, 1002–1005.

Fraser, D. A. S.
(1957). *Nonparametric Methods in Statistics*, Wiley, New York.

Galambos, J.
(1982). "Exponential distribution." In *Encycl. Statist. Sci.*, Vol. 2, Wiley, New York.

Gastwirth, J. L. and Rubin, H.
(1971). "Effect of dependence on the level of some one-sample tests." *J. Amer. Statist. Assoc.* **66**, 816–820.

Gleser, L. J.
(1985). "Exact power of goodness-of-fit tests of Kolmogorov type for discontinuous distributions." *J. Amer. Statist. Assoc.* **80**, 954–958.

Hájek, J. and Šidák, Z.
(1967). *Theory of Rank Tests*, Academia, Prague.

Hall, P. and Welsh, A. H.
(1983). "A test for normality based on the empirical characteristic function." *Biometrika* **70**, 485–489.

Hall, W. J., Wijsman, R. A., and Ghosh, J. K.
(1965). "The relationship between sufficiency and invariance with applications in sequential analysis." *Ann. Math. Statist.* **36**, 575–614.

Halmos, P. R.
(1974). *Measure Theory*, Springer, New York.

Hegazy, Y. A. S. and Green, J. R.
(1975). "Some new goodness-of-fit tests using order statistics." *Appl. Statist.* **24**, 299–308.

Hemelrijk, J.
(1950). "A family of parameter-free tests for symmetry with respect to a given point." *Proc. Koninkl. Ned. Akad. Wetenschap.* **53**, 945–955, 1186–1198.
[Discusses the relationship of the hypothesis of symmetry with the two-sample problem.]

Hill, D. L. and Rao, P. V.
(1977). "Tests of symmetry based on Cramér–von Mises statistics." *Biometrika* **64**, 489–494.

Hoeffding, W.
(1951). "'Optimum' nonparametric tests." in *Proc. 2nd Berkeley Symposium on Mathematical Statistics and Probability*, Univ. of Calif. Press., Berkeley, 83–92.
[Derives a basic rank distribution of which (18) is a special case, and from it obtains locally optimum tests of the type (19). His results are specialized to the two-sample problem by Milton E. Terry, "Some rank order tests which are most powerful against specific parametric alternatives," *Ann. Math. Statist.* **23** (1952), 346–366.]

Hogg, R. V.
(1972). "More light on the kurtosis and related statistics," *J. Amer. Statist. Assoc.* **67**,422–424.

Hooper, P. M.
(1982a). "Sufficiency and invariance in confidence set estimation." *Ann. Statist.* **10**, 549–555.
(1982b). "Invariant confidence sets with smallest expected measure." *Ann. Statist.* **10**, 1283–1294.

Hoyle, M. H.
(1973). "Transformations—an introduction and a bibliography." *Int. Statist. Rev.* **41**, 203–223.

Hsu, P. L.
(1938). "Contributions to the theory of Student's *t*-test as applied to the problem of two samples." *Statist. Res. Mem.* **II**, 1–24.
[Shows that the two-sample *t*-test, in the case of equal and not very small sample sizes, is approximately unbiased even when the variances are unequal, and that for this case the *t*-test therefore constitutes an approximate solution to the Behrens–Fisher problem.]

Hunt, G. and Stein, C.
(1946). "Most stringent tests of statistical hypotheses." Unpublished.

Karlin, S.
(1968). *Total Positivity*. Stanford U.P.

Kendall, M. G.
(1970). *Rank Correlation Methods*, 4th ed., Griffin, London.

Kendall, M. G. and Stuart, A.
(1979). *The Advanced Theory of Statistics*. 4th ed., Vol. 2, MacMillan, New York.

Kent, J. and Quesenberry, C. P.
(1982). "Selecting among probability distributions used in reliability." *Technometrics* **24**, 59–65.

Kiefer, J. and Schwartz, R.
(1965). "Admissible Bayes character of T^2-, R^2-, and other fully invariant tests for classical multivariate normal problems." *Ann. Math. Statist.* **36**, 747–770.

Koziol, J. A.
(1983). "Tests for symmetry about an unknown value based on the empirical distribution function." *Comm. Statist.* **12**, 2823–2846.

Kruskal, W.
(1954). "The monotonicity of the ratio of two non-central t density functions." *Ann. Math. Statist.* **25**, 162–165.

Landers, D. and Rogge, L.
(1973). "On sufficiency and invariance." *Ann. Statist.* **1**, 543–544.

Lehmann, E. L.
(1950). "Some principles of the theory of testing hypotheses." *Ann. Math. Statist.* **21**, 1–26.
[Lemma 2; Theorem 7; presents an example of Stein on which Problem 18 is patterned.]
(1951). "Consistency and unbiasedness of certain nonparametric tests." *Ann. Math. Statist.* **22**, 165–179.
[Problems 33, 34.]
(1953). "The power of rank tests." *Ann. Math. Statist.* **24**, 28–43.
[Applies invariance considerations to nonparametric problems.]
(1975). *Nonparametrics: Statistical Methods Based on Ranks*. Holden Day, San Francisco.

Lehmann, E. L. and Stein, C. M.
(1953). "The admissibility of certain invariant statistical tests involving a translation parameter." *Ann. Math. Statist.* **24**, 473–479.

Levy, K. J. and Narula, S. C.
(1974). "Shortest confidence intervals for the ratio of two normal variances." *Canad. J. Statist.* **2**, 83–87.

Loh, W.-Y.
(1984). "Bounds on ARE's for restricted classes of distributions defined via tail-orderings." *Ann. Statist.* **12**, 685–701.
(1985). "A new method for testing separate families of hypotheses." *J. Amer. Statist. Assoc.* **80**, 362–368.

Marden, J. L.
(1982). "Minimal complete classes of tests of hypotheses with multivariate one-sided alternatives." *Ann. Statist.* **10**, 962–970.

Massey, F. J.
(1950). "A note on the power of a non-parametric test." *Ann. Math. Statist.* **21**, 440–443.

Moses, L. E.
(1953). "Nonparametric inference." In *Statistical Inference* (Walker and Lev), Henry Holt, New York, Chapter 18.
[Proposes the confidence intervals for Δ of Example 15.]

Neuhaus, G.
(1979). "Asymptotic theory of goodness of fit tests when parameters are present: A survey." *Statistics* **10**, 479–494.

Niederhausen, H.
(1981). Scheffer polynomials for computing exact Kolmogorov–Smirnov and Renyi type distributions." *Ann. Statist.* **9**, 923–944.

Pereira, B. De B.
(1977). "Discriminating among separate models: A bibliography." *Internat. Statist. Rev.* **45**, 163–172.

Pfanzagl, J.
(1974). "On the Behrens–Fisher problem." *Biometrika* **61**, 39–47.

Pitman, E. J. G.
(1939). "Tests of hypotheses concerning location and scale parameters." *Biometrika* **31**, 200–215.
[Invariance considerations are introduced, and are applied to problems similar to that treated in Example 4.]
(1949). "Lecture notes on nonparametric statistical inference," unpublished.
[Develops the concept of relative asymptotic efficiency and applies it to several examples including the Wilcoxon test.]

Pratt, J. W.
(1964). "Robustness of some procedures for the two-sample location problem." *J. Amer. Statist. Assoc.* **59**, 665–680.

Quesenberry, C. P. and Starbuck, R. R.
(1976). "On optimal tests for separate hypotheses and conditional probability integral transformations." *Comm. Statist. A* **1**, 507–524.

Randles, R. H., Fligner, M. A., Policello, G. E., II, and Wolfe, D. A.
(1980). "An asymptotically distribution-free test for symmetry versus asymmetry." *J. Amer. Statist. Assoc.* **75**, 168–172.

Randles, R. H. and Wolfe, D. A.
(1979). *Introduction to the Theory of Nonparametric Statistics.* Wiley, New York.

Scheffé, H.
(1942). "On the ratio of the variances of two normal populations." *Ann. Math. Statist.* **13**, 371–388.
[Introduces the idea of logarithmically shortest confidence intervals for ratios of scale parameters.]
(1970). "Practical solutions of the Behrens–Fisher problem." *J. Amer. Statist. Assoc.* **65**, 1501–1504.

Serfling, R. H.
(1980). *Approximation Theorems of Mathematical Statistics.* Wiley, New York.

Shapiro, S. S., Wilk, M. B., and Chen, H. J.
(1968). "A comparative study of various tests of normality." *J. Amer. Statist. Assoc.* **63**, 1343–1372.

Smirnov, N. V.
(1948). "Tables for estimating the goodness of fit of empirical distributions." *Ann. Math. Statist.* **19**, 279–281.

Spiegelhalter, D. J.
(1983). "Diagnostic tests of distributional shape." *Biometrika* **70**, 401–409.

Spurrier, J. D.
(1984). "An overview of tests for exponentiality." *Commun. Statist.—Theor. Meth.* **13**, 1635–1654.

Stein, C. M.
(1956a). "The admissibility of Hotelling's T^2-test." *Ann. Math. Statist.* **27**, 616–623.
(1956b). "Efficient nonparametric testing and estimation." in *Proc. 3rd Berkeley Symp. Math. Statist. and Probab.* Univ. of Calif. Press, Berkeley.

Stone, C. J.
(1975). "Adaptive maximum likelihood estimators of a location parameter." *Ann. Statist.* **3**, 267–284.

Sugiura, N.
(1965). "An example of the two-sided Wilcoxon test which is not unbiased." *Ann. Inst. Statist. Math.* **17**, 261–263.

Sukhatme, P. V.
(1936). "On the analysis of *k* samples from exponential distributions with especial reference to the problem of random intervals." *Statist. Res. Mem.* **1**, 94–112.

Tallis, G. M.
(1983). "Goodness of fit." In *Encycl. Statist. Sci.*, Vol. 3, Wiley, New York.

Tate, R. F. and Klett, G. W.
(1959). "Optimal confidence intervals for the variance of a normal distribution." *J. Amer. Statist. Assoc.* **54**, 674–682.

Uthoff, V. A.
(1970). "An optimum test property of two well-known statistics." *J. Amer. Statist. Assoc.* **65**, 1597–1600.
(1973). "The most powerful scale and location invariant test of normal versus double exponential." *Ann. Statist.* **1**, 170–174.

Wallace, D. L.
(1958). "Asymptotic approximations to distributions." *Ann. Math. Statist.* **29**, 635–654.

Walsh, J. E.
(1949). "Some significance tests for the median which are valid under very general conditions." *Ann. Math. Statist.* **20**, 64–81.
[Lemma 3; proposes the Wilcoxon one-sample test in the form given in Problem 35. The equivalence of the two tests was shown by Tukey in an unpublished mimeographed report dated 1949.]

Wang, Y. Y.
(1971). "Probabilities of type I errors of Welch tests for the Behrens–Fisher problem." *J. Amer. Statist. Assoc.* **66**, 605–608.

Wijsman, R. A.
(1985). "Proper action in steps, with application to density ratios of maximal invariants." *Ann. Statist.* **13**, 395–402.

Wilcoxon, F.
(1945). "Individual comparisons by ranking methods." *Biometrics* **1**, 80–83.
[Proposes the two tests bearing his name. (See also Deuchler, 1914.)]

Wolfowitz, J.
(1949). "The power of the classical tests associated with the normal distribution." *Ann. Math. Statist.* **20**, 540–551.
[Proves Lemma 2 for a number of special cases.]

CHAPTER 7

Linear Hypotheses

1. A CANONICAL FORM

Many testing problems concern the means of normal distributions and are special cases of the following *general univariate linear hypothesis.* Let $X_1, \ldots, X_n$ be independently normally distributed with means $\xi_1, \ldots, \xi_n$ and common variance σ^2. The vector of means* $\underline{\xi}$ is known to lie in a given s-dimensional linear subspace Π_Ω ($s < n$), and the hypothesis H is to be tested that $\underline{\xi}$ lies in a given $(s - r)$-dimensional subspace Π_ω of Π_Ω ($r \leq s$).

Example 1. In the two-sample problem of testing equality of two normal means (considered with a different notation in Chapter 5, Section 3), it is given that $\xi_i = \xi$ for $i = 1, \ldots, n_1$ and $\xi_i = \eta$ for $i = n_1 + 1, \ldots, n_1 + n_2$, and the hypothesis to be tested is $\eta = \xi$. The space Π_Ω is then the space of vectors

$$(\xi, \ldots, \xi, \eta, \ldots, \eta) = \xi(1, \ldots, 1, 0, \ldots, 0) + \eta(0, \ldots, 0, 1, \ldots, 1)$$

spanned by $(1, \ldots, 1, 0, \ldots, 0)$ and $(0, \ldots, 0, 1, \ldots, 1)$, so that $s = 2$. Similarly, Π_ω is the set of all vectors $(\xi, \ldots, \xi) = \xi(1, \ldots, 1)$, and hence $r = 1$.

Another hypothesis that can be tested in this situation is $\eta = \xi = 0$. The space Π_ω is then the origin, $s - r = 0$ and hence $r = 2$. The more general hypothesis $\xi = \xi_0$, $\eta = \eta_0$ is not a linear hypothesis, since Π_ω does not contain the origin. However, it reduces to the previous case through the transformation $X_i' = X_i - \xi_0$ ($i = 1, \ldots, n_1$), $X_i' = X_i - \eta_0$ ($i = n_1 + 1, \ldots, n_1 + n_2$).

Example 2. The regression problem of Chapter 5, Section 8, is essentially a linear hypothesis. Changing the notation to make it conform with that of the present section, let $\xi_i = \alpha + \beta t_i$, where α, β are unknown, and the t_i known and not all equal. Since Π_Ω is the space of all vectors $\alpha(1, \ldots, 1) + \beta(t_1, \ldots, t_n)$, it has dimension $s = 2$. The hypothesis to be tested may be $\alpha = \beta = 0$ ($r = 2$) or it may

---

*Throughout this chapter, a fixed coordinate system is assumed given in n-space. A vector with components $\xi_1, \ldots, \xi_n$ is denoted by $\underline{\xi}$, and an $n \times 1$ column matrix with elements $\xi_1, \ldots, \xi_n$ by ξ.

only specify that one of the parameters is zero ($r = 1$). The more general hypotheses $\alpha = \alpha_0$, $\beta = \beta_0$ can be reduced to the previous case by letting $X_i' = X_i - \alpha_0 - \beta_0 t_i$, since then $E(X_i') = \alpha' + \beta' t_i$ with $\alpha' = \alpha - \alpha_0$, $\beta' = \beta - \beta_0$.

Higher polynomial regression and regression in several variables also fall under the linear-hypothesis scheme. Thus if $\xi_i = \alpha + \beta t_i + \gamma t_i^2$ or more generally $\xi_i = \alpha + \beta t_i + \gamma u_i$, where the t_i and u_i are known, it can be tested whether one or more of the regression coefficients α, β, γ are zero, and by transforming to the variables $X_i' - \alpha_0 - \beta_0 t_i - \gamma_0 u_i$ also whether these coefficients have specified values other than zero.

In the general case, the hypothesis can be given a simple form by making an orthogonal transformation to variables $Y_1, \ldots, Y_n$

$$(1) \qquad\qquad Y = CX, \qquad C = (c_{ij}) \quad i, j = 1, \ldots, n,$$

such that the first s row vectors $\underline{c}_1, \ldots, \underline{c}_s$ of the matrix C span Π_Ω, with $\underline{c}_{r+1}, \ldots, \underline{c}_s$ spanning Π_ω. Then $Y_{s+1} = \cdots = Y_n = 0$ if and only if $\underline{X}$ is in Π_Ω, and $Y_1 = \cdots = Y_r = Y_{s+1} = \cdots = Y_n = 0$ if and only if $\underline{X}$ is in Π_ω. Let $\eta_i = E(Y_i)$, so that $\eta = C\xi$. Then since ξ lies in Π_Ω a priori and in Π_ω under H, it follows that $\eta_i = 0$ for $i = s + 1, \ldots, n$ in both cases, and $\eta_i = 0$ for $i = 1, \ldots, r$ when H is true. Finally, since the transformation is orthogonal, the variables $Y_1, \ldots, Y_n$ are again independently normally distributed with common variance σ^2, and the problem reduces to the following canonical form.

The variables $Y_1, \ldots, Y_n$ are independently, normally distributed with common variance σ^2 and means $E(Y_i) = \eta_i$ for $i = 1, \ldots, s$ and $E(Y_i) = 0$ for $i = s + 1, \ldots, n$, so that their joint density is

$$(2) \qquad\qquad \frac{1}{(\sqrt{2\pi}\,\sigma)^n} \exp\left[-\frac{1}{2\sigma^2}\left(\sum_{i=1}^{s} (y_i - \eta_i)^2 + \sum_{i=s+1}^{n} y_i^2 \right) \right].$$

The η's and σ^2 are unknown, and the hypothesis to be tested is

$$(3) \qquad\qquad H : \eta_1 = \cdots = \eta_r = 0 \qquad (r \le s < n).$$

Example 3. To illustrate the determination of the transformation (1), consider once more the regression model $\xi_i = \alpha + \beta t_i$ of Example 2. It was seen there that Π_Ω is spanned by $(1, \ldots, 1)$ and $(t_1, \ldots, t_n)$. If the hypothesis being tested is $\beta = 0$, Π_ω is the one-dimensional space spanned by the first of these vectors. The row vector $\underline{c}_2$ is in Π_ω and of length 1, and hence $\underline{c}_2 = (1/\sqrt{n}, \ldots, 1/\sqrt{n})$. Since $\underline{c}_1$ is in Π_Ω, of length 1, and orthogonal to $\underline{c}_2$, its coordinates are of the form $a + bt_i$, $i = 1, \ldots, n$, where a and b are determined by the conditions $\Sigma(a + bt_i) = 0$ and $\Sigma(a + bt_i)^2 = 1$. The solutions of these equations are $a = -b\bar{t}$, $b =$

$1/\sqrt{\Sigma(t_j - \bar{t})^2}$, and therefore $a + bt_i = (t_i - \bar{t})/\sqrt{\Sigma(t_j - \bar{t})^2}$, and

$$Y_1 = \frac{\Sigma X_i(t_i - \bar{t})}{\sqrt{\Sigma(t_j - \bar{t})^2}} = \frac{\Sigma(X_i - \bar{X})(t_i - \bar{t})}{\sqrt{\Sigma(t_j - \bar{t})^2}}.$$

The remaining row vectors of C can be taken to be any set of orthogonal unit vectors which are orthogonal to Π_Ω; it turns out not to be necessary to determine them explicitly.

If the hypothesis to be tested is $\alpha = 0$, Π_ω is spanned by $(t_1, \ldots, t_n)$, so that the ith coordinate of c_2 is $t_i/\sqrt{\Sigma t_j^2}$. The coordinates of c_1 are again of the form $a + bt_i$ with a and b now determined by the equations $\Sigma(a + bt_i)t_i = 0$ and $\Sigma(a + bt_i)^2 = 1$. The solutions are $b = -a n\bar{t}/\Sigma t_j^2$, $a = \sqrt{\Sigma t_j^2/n\Sigma(t_j - \bar{t})^2}$, and therefore

$$Y_1 = \sqrt{\frac{n\Sigma t_j^2}{\Sigma(t_j - \bar{t})^2}} \left(\bar{X} - \frac{\bar{t}}{\Sigma t_j^2}\Sigma t_i X_i \right).$$

In the case of the hypothesis $\alpha = \beta = 0$, Π_ω is the origin and c_1, c_2 can be taken as any two orthogonal unit vectors in Π_Ω. One possible choice is that appropriate to the hypothesis $\beta = 0$, in which case Y_1 is the linear function given there and $Y_2 = \sqrt{n}\,\bar{X}$.

The general linear-hypothesis problem in terms of the Y's remains invariant under the group G_1 of transformations $Y_i' = Y_i + c_i$ for $i = r + 1, \ldots, s$; $Y_i' = Y_i$ for $i = 1, \ldots, r$; $s + 1, \ldots, n$. This leaves $Y_1, \ldots, Y_r$ and $Y_{s+1}, \ldots, Y_n$ as maximal invariants. Another group of transformations leaving the problem invariant is the group G_2 of all orthogonal transformations of $Y_1, \ldots, Y_r$. The middle set of variables having been eliminated, it follows from Chapter 6, Example 1(iii), that a maximal invariant under G_2 is $U = \Sigma_{i=1}^r Y_i^2, Y_{s+1}, \ldots, Y_n$. This can be reduced to U and $V = \Sigma_{i=s+1}^n Y_i^2$ by sufficiency. Finally, the problem also remains invariant under the group G_3 of scale changes $Y_i' = cY_i$, $c \neq 0$, for $i = 1, \ldots, n$. In the space of U and V this induces the transformation $U^* = c^2 U$, $V^* = c^2 V$, under which $W = U/V$ is maximal invariant. Thus the principle of invariance reduces the data to the single statistic*

(4)
$$W = \frac{\sum_{i=1}^r Y_i^2}{\sum_{i=s+1}^n Y_i^2}.$$

*A corresponding reduction without assuming normality is discussed by Jagers (1980).

Each of the three transformation groups G_i ($i = 1, 2, 3$) which lead to the above reduction induces a corresponding group $\overline{G}_i$ in the parameter space. The group $\overline{G}_1$ consists of the translations $\eta_i' = \eta_i + c_i$ ($i = r + 1, \ldots, s$), $\eta_i' = \eta_i$ ($i = 1, \ldots, r$), $\sigma' = \sigma$, which leaves $(\eta_1, \ldots, \eta_r, \sigma)$ as maximal invariants. Since any orthogonal transformation of $Y_1, \ldots, Y_r$ induces the same transformation on $\eta_1, \ldots, \eta_r$ and leaves σ^2 unchanged, a maximal invariant under $\overline{G}_2$ is $(\Sigma_{i=1}^r \eta_i^2, \sigma^2)$. Finally the elements of $\overline{G}_3$ are the transformations $\eta_i' = c\eta_i$, $\sigma' = |c|\sigma$, and hence a maximal invariant with respect to the totality of these transformations is

$$(5) \qquad \psi^2 = \frac{\displaystyle\sum_{i=1}^r \eta_i^2}{\sigma^2}.$$

It follows from Theorem 3 of Chapter 6 that the distribution of W depends only on ψ^2, so that the principle of invariance reduces the problem to that of testing the simple hypothesis $H : \psi = 0$. More precisely, the probability density of W is (cf. Problems 2 and 3)

$$(6) \qquad p_\psi(w) = e^{-\frac{1}{2}\psi^2} \sum_{k=0}^{\infty} c_k \frac{\left(\frac{1}{2}\psi^2\right)^k}{k!} \frac{w^{\frac{1}{2}r-1+k}}{(1+w)^{\frac{1}{2}(r+n-s)+k}},$$

where

$$c_k = \frac{\Gamma\left[\frac{1}{2}(r + n - s) + k\right]}{\Gamma\left(\frac{1}{2}r + k\right)\Gamma\left[\frac{1}{2}(n - s)\right]}.$$

For any ψ_1 the ratio $p_{\psi_1}(w)/p_0(w)$ is an increasing function of w, and it follows from the Neyman–Pearson fundamental lemma that the most powerful invariant test for testing $\psi = 0$ against $\psi = \psi_1$ rejects when W is too large, or equivalently when

$$(7) \qquad W^* = \frac{\displaystyle\sum_{i=1}^r Y_i^2/r}{\displaystyle\sum_{i=s+1}^n Y_i^2/(n-s)} > C.$$

The cutoff point C is determined so that the probability of rejection is α when $\psi = 0$. Since in this case W^* is the ratio of two independent χ^2 variables, each divided by the number of its degrees of freedom, the distribution of W^* is the F-distribution with r and $n - s$ degrees of freedom, and hence C is determined by

(8)
$$\int_C^\infty F_{r,\,n-s}(y)\,dy = \alpha.$$

The test is independent of ψ_1, and hence is UMP among all invariant tests. By Theorem 5 of Chapter 6, it is also UMP among all tests whose power function depends only on ψ^2.

The rejection region (7) can also be expressed in the form

(9)
$$\frac{\displaystyle\sum_{i=1}^{r} Y_i^2}{\displaystyle\sum_{i=1}^{r} Y_i^2 + \sum_{i=s+1}^{n} Y_i^2} > C'.$$

When $\psi = 0$, the left-hand side is distributed according to the beta-distribution with r and $n - s$ degrees of freedom [defined through (24) of Chapter 5], so that C' is determined by

(10)
$$\int_{C'}^1 B_{\frac{1}{2}r,\,\frac{1}{2}(n-s)}(y)\,dy = \alpha.$$

For an alternative value of ψ, the left-hand side of (9) is distributed according to the *noncentral beta-distribution* with noncentrality parameter ψ, the density of which is (Problem 3)

(11)
$$g_\psi(y) = e^{-\frac{1}{2}\psi^2} \sum_{k=0}^{\infty} \frac{\left(\frac{1}{2}\psi^2\right)^k}{k!} B_{\frac{1}{2}r+k,\,\frac{1}{2}(n-s)}(y).$$

The power of the test against an alternative ψ is therefore*

$$\beta(\psi) = \int_{C'}^1 g_\psi(y)\,dy.$$

In the particular case $r = 1$, the rejection region (7) reduces to

(12)
$$\frac{|Y_1|}{\sqrt{\displaystyle\sum_{i=s+1}^{n} Y_i^2/(n - s)}} > C_0.$$

---

*Tables of the power of the F-test are provided by Tiku (1967, 1972) [reprinted in Graybill (1976)] and Cohen (1977); charts are given in Pearson and Hartley (1972). Various approximations are discussed by Johnson and Kotz (1970).

This is a two-sided t-test, which by the theory of Chapter 5 (see for example Problem 5 of that chapter) is UMP unbiased. On the other hand, no UMP unbiased test exists for $r > 1$.

The F-test (7) shares the admissibility properties of the two-sided t-test discussed in Chapter 6, Section 7. In particular, the test is admissible against distant alternatives $\psi^2 \geq \psi_1^2$ (Problem 6) and against nearby alternatives $\psi^2 \leq \psi_2^2$ (Problem 7). It was shown by Lehmann and Stein (1953) that the test is in fact admissible against the alternatives $\psi^2 = \psi_1^2$ for any ψ_1 and hence against all invariant alternatives.

2. LINEAR HYPOTHESES AND LEAST SQUARES

In applications to specific problems it is usually not convenient to carry out the reduction to canonical form explicitly. The test statistic W can be expressed in terms of the original variables by noting that $\sum_{i=s+1}^{n} Y_i^2$ is the minimum value of

$$\sum_{i=1}^{s} (Y_i - \eta_i)^2 + \sum_{i=s+1}^{n} Y_i^2 = \sum_{i=1}^{n} [Y_i - E(Y_i)]^2$$

under unrestricted variation of the η's. Also, since the transformation $Y = CX$ is orthogonal and orthogonal transformations leave distances unchanged,

$$\sum_{i=1}^{n} [Y_i - E(Y_i)]^2 = \sum_{i=1}^{n} (X_i - \xi_i)^2.$$

Furthermore, there is a $1:1$ correspondence between the totality of s-tuples $(\eta_1, \ldots, \eta_s)$ and the totality of vectors $\underline{\xi}$ in Π_Ω. Hence

(13)
$$\sum_{i=s+1}^{n} Y_i^2 = \sum_{i=1}^{n} (X_i - \hat{\xi}_i)^2,$$

where the $\hat{\xi}$'s are the least-squares estimates of the ξ's under Ω, that is, the values that minimize $\sum_{i=1}^{n} (X_i - \xi_i)^2$ subject to $\underline{\xi}$ in Π_Ω.

In the same way it is seen that

$$\sum_{i=1}^{r} Y_i^2 + \sum_{i=s+1}^{n} Y_i^2 = \sum_{i=1}^{n} (X_i - \hat{\hat{\xi}}_i)^2 .$$

where the $\hat{\hat{\xi}}$'s are the values that minimize $\sum (X_i - \xi_i)^2$ subject to $\underline{\xi}$ in Π_ω.

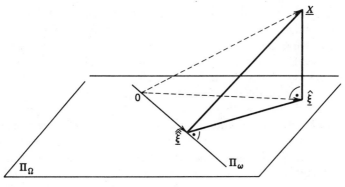

Figure 1

The test (7) therefore becomes

(14)
$$W^* = \frac{\left[\sum\limits_{i=1}^{n}(X_i - \hat{\hat{\xi}}_i)^2 - \sum\limits_{i=1}^{n}(X_i - \hat{\xi}_i)^2\right]/r}{\sum\limits_{i=1}^{n}(X_i - \hat{\xi}_i)^2/(n-s)} > C,$$

where C is determined by (8). Geometrically the vectors $\hat{\xi}$ and $\hat{\hat{\xi}}$ are the projections of $\underline{X}$ on Π_Ω and Π_ω, so that the triangle formed by $\underline{X}$, $\hat{\xi}$, and $\hat{\hat{\xi}}$ has a right angle at $\hat{\xi}$. (Figure 1.) Thus the denominator and numerator of W^*, except for the factors $1/(n-s)$ and $1/r$, are the squares of the distances between $\underline{X}$ and $\hat{\xi}$ and between $\hat{\xi}$ and $\hat{\hat{\xi}}$ respectively. An alternative expression for W^* is therefore

(15)
$$W^* = \frac{\sum\limits_{i=1}^{n}(\hat{\xi}_i - \hat{\hat{\xi}}_i)^2/r}{\sum\limits_{i=1}^{n}(X_i - \hat{\xi}_i)^2/(n-s)}.$$

It is desirable to express also the *noncentrality parameter* $\psi^2 = \sum_{i=1}^{r}\eta_i^2/\sigma^2$ in terms of the ξ's. Now $X = C^{-1}Y$, $\xi = C^{-1}\eta$, and

(16)
$$\sum\limits_{i=1}^{r}Y_i^2 = \sum\limits_{i=1}^{n}(X_i - \hat{\hat{\xi}}_i)^2 - \sum\limits_{i=1}^{n}(X_i - \hat{\xi}_i)^2.$$

If the right-hand side of (16) is denoted by $f(X)$, it follows that $\sum_{i=1}^{r}\eta_i^2 = f(\xi)$.

A slight generalization of a linear hypothesis is the inhomogeneous hypothesis which specifies for the vector of means ξ a subhyperplane Π'_ω of Π_Ω not passing through the origin. Let Π_ω denote the subspace of Π_Ω which passes through the origin and is parallel to Π'_ω. If ξ^0 is any point of Π'_ω, the set Π'_ω consists of the totality of points $\xi = \xi^* + \xi^0$ as ξ^* ranges over Π_ω. Applying the transformation (1) with respect to Π_ω, the vector of means η for $\xi \in \Pi'_\omega$ is then given by $\eta = C\xi = C\xi^* + C\xi^0$ in the canonical form (2), and the totality of these vectors is therefore characterized by the equations $\eta_1 = \eta_1^0, \ldots, \eta_r = \eta_r^0$, $\eta_{s+1} = \cdots = \eta_n = .0$, where η_i^0 is the ith coordinate of $C\xi^0$. In the canonical form, the inhomogeneous hypothesis $\xi \in \Pi'_\omega$ therefore becomes $\eta_i = \eta_i^0$ ($i = 1, \ldots, r$). This reduces to the homogeneous case on replacing Y_i with $Y_i - \eta_i^0$, and it follows from (7) that the UMP invariant test has the rejection region

$$
(17) \qquad \frac{\sum\limits_{i=1}^{r} \left(Y_i - \eta_i^0 \right)^2 / r}{\sum\limits_{i=s+1}^{n} Y_i^2 / (n - s)} > C,
$$

and that the noncentrality parameter is $\psi^2 = \sum_{i=1}^{r}(\eta_i - \eta_i^0)^2/\sigma^2$.

In applications it is usually most convenient to apply the transformation $X_i - \xi_i^0$ directly to (14) or (15). It follows from (17) that such a transformation always leaves the denominator unchanged. This can also be seen geometrically, since the transformation is a translation of n-space parallel to Π_Ω and therefore leaves the distance $\sum(X_i - \hat{\xi}_i)^2$ from $\underline{X}$ to Π_Ω unchanged. The noncentrality parameter can be computed as before by replacing X with ξ in the transformed numerator (16).

Some examples of linear hypotheses, all with $r = 1$, were already discussed in Chapter 5. The following treats two of these from the present point of view.

Example 4. Let $X_1, \ldots, X_n$ be independently, normally distributed with common mean μ and variance σ^2, and consider the hypothesis $H: \mu = 0$. Here Π_Ω is the line $\xi_1 = \cdots = \xi_n$, Π_ω is the origin, and s and r are both equal to 1. From the identity

$$
\sum (X_i - \mu)^2 = \sum (X_i - \bar{X})^2 + n(\bar{X} - \mu)^2, \qquad \left(\bar{X} = \sum \frac{X_i}{n} \right)
$$

it is seen that $\hat{\xi}_i = \bar{X}$, while $\hat{\hat{\xi}}_i = 0$. The test statistic and ψ^2 are therefore given by

$$
W = \frac{n\bar{X}^2}{\sum (X_i - \bar{X})^2} \quad \text{and} \quad \psi^2 = \frac{n\mu^2}{\sigma^2}.
$$

Under the hypothesis, the distribution of $(n - 1)W$ is that of the square of a variable having Student's t-distribution with $n - 1$ degrees of freedom.

Example 5. In the two-sample problem considered in Example 1, the sum of squares

$$\sum_{i=1}^{n_1} (X_i - \xi)^2 + \sum_{i=n_1+1}^{n} (X_i - \eta)^2$$

is minimized by

$$\hat{\xi} = X^{(1)}_{\cdot} = \sum_{i=1}^{n_1} \frac{X_i}{n_1}, \qquad \hat{\eta} = X^{(2)}_{\cdot} = \sum_{i=n_1+1}^{n} \frac{X_i}{n_2},$$

while under the hypothesis $\eta - \xi = 0$

$$\hat{\hat{\xi}} = \hat{\hat{\eta}} = \overline{X} = \frac{n_1 X^{(1)}_{\cdot} + n_2 X^{(2)}_{\cdot}}{n}.$$

The numerator of the test statistic (15), is therefore

$$n_1\left(X^{(1)}_{\cdot} - \overline{X}\right)^2 + n_2\left(X^{(2)}_{\cdot} - \overline{X}\right)^2 = \frac{n_1 n_2}{n_1 + n_2}\left[X^{(2)}_{\cdot} - X^{(1)}_{\cdot}\right]^2.$$

The more general hypothesis $\eta - \xi = \theta_0$ reduces to the previous case on replacing X_i with $X_i - \theta_0$ for $i = n_1 + 1, \ldots, n$, and is therefore rejected when

$$\frac{\left(X^{(2)}_{\cdot} - X^{(1)}_{\cdot} - \theta_0\right)^2 / \left(\dfrac{1}{n_1} + \dfrac{1}{n_2}\right)}{\left[\displaystyle\sum_{i=1}^{n_1}\left(X_i - X^{(1)}_{\cdot}\right)^2 + \sum_{i=n_1+1}^{n}\left(X_i - X^{(2)}_{\cdot}\right)^2\right] / (n_1 + n_2 - 2)} > C.$$

The noncentrality parameter is $\psi^2 = (\eta - \xi - \theta_0)^2/(1/n_1 + 1/n_2)\sigma^2$. Under the hypothesis, the square root of the test statistic has the t-distribution with $n_1 + n_2 - 2$ degrees of freedom.

Explicit formulae for the $\hat{\xi}_i$ and $\hat{\hat{\xi}}_i$ can be obtained by introducing a coordinate system into the parameter space. Suppose in such a system, Π_Ω is defined by the equations

$$\xi_i = \sum_{j=1}^{s} a_{ij}\beta_j, \qquad i = 1, \ldots, n,$$

or, in matrix notation,

(18) $$\underset{n\times 1}{\xi} = \underset{n\times s}{A}\ \underset{s\times 1}{B},$$

where A is known and of rank s, and $\beta_1, \ldots, \beta_s$ are unknown parameters. If $\hat{\beta}_1, \ldots, \hat{\beta}_s$ are the least-squares estimators minimizing $\Sigma_i(X_i - \Sigma_j a_{ij}\beta_j)^2$, it is seen by differentiation that the $\hat{\beta}_j$ are the solutions of the equations

$$A'A\beta = A'X$$

and hence are given by

$$\hat{\beta} = (A'A)^{-1}A'X.$$

(That $A'A$ is nonsingular is shown in Lemma 1 of Chapter 8.) Thus, we obtain

$$\hat{\xi} = A(A'A)^{-1}A'X.$$

Since $\hat{\xi} = \hat{\xi}(X)$ is the projection of X into the space Π_Ω spanned by the s columns of A, the formula $\hat{\xi} = A(A'A)^{-1}A'X$ shows that $P = A(A'A)^{-1}A'$ has the property claimed for it in Example 3 of Chapter 6, that for any X in R^n, PX is the projection of X into Π_Ω.

3. TESTS OF HOMOGENEITY

The UMP invariant test obtained in the preceding section for testing the equality of the means of two normal distributions with common variance is also UMP unbiased (Section 3 of Chapter 5). However, when a number of populations greater than 2 is to be tested for homogeneity of means, a UMP unbiased test no longer exists, so that invariance considerations lead to a new result. Let X_{ij} $(j = 1, \ldots, n_i; i = 1, \ldots, s)$ be independently distributed as $N(\mu_i, \sigma^2)$, and consider the hypothesis

$$H: \mu_1 = \cdots = \mu_s.$$

This arises, for example, in the comparison of a number of different treatments, processes, varieties, or locations, when one wishes to test whether these differences have any effect on the outcome X. It may arise more generally in any situation involving a *one-way classification* of the outcomes, that is, in which the outcomes are classified according to a single factor.

The hypothesis H is a linear hypothesis with $r = s - 1$, with Π_Ω given by the equations $\xi_{ij} = \xi_{ik}$ for $j, k = 1, \ldots, n$, $i = 1, \ldots, s$ and with Π_ω the line on which all $n = \Sigma n_i$ coordinates ξ_{ij} are equal. We have

$$\Sigma\Sigma(X_{ij} - \mu_i)^2 = \Sigma\Sigma(X_{ij} - X_{i.})^2 + \Sigma n_i(X_{i.} - \mu_i)^2$$

with $X_{i.} = \Sigma_{j=1}^{n_i} X_{ij}/n_i$, and hence $\hat{\xi}_{ij} = X_{i.}$. Also,

$$\Sigma\Sigma(X_{ij} - \mu)^2 = \Sigma\Sigma(X_{ij} - X_{..})^2 + n(X_{..} - \mu)^2$$

with $X_{..} = \Sigma\Sigma X_{ij}/n$, so that $\hat{\hat{\xi}}_{ij} = X_{..}$. Using the form (15) of W^*, the test therefore becomes

(19)
$$W^* = \frac{\Sigma n_i(X_{i.} - X_{..})^2/(s - 1)}{\Sigma\Sigma(X_{ij} - X_{i.})^2/(n - s)} > C.$$

The noncentrality parameter is

$$\psi^2 = \frac{\Sigma n_i(\mu_i - \mu_.)^2}{\sigma^2}$$

with

$$\mu_. = \frac{\Sigma n_i \mu_i}{n}.$$

The sum of squares in both numerator and denominator of (19) admits three interpretations, which are closely related: (i) as the two components in the decomposition of the total variation

$$\Sigma\Sigma(X_{ij} - X_{..})^2 = \Sigma\Sigma(X_{ij} - X_{i.})^2 + \Sigma n_i(X_{i.} - X_{..})^2,$$

of which the first represents the variation within, and the second the variation between populations; (ii) as a basis, through the test (19), for comparing these two sources of variation; (iii) as estimates of their expected values, $(n - s)\sigma^2$ and $(s - 1)\sigma^2 + \Sigma n_i(\mu_i - \mu_.)^2$ (Problem 13). This breakdown of the total variation, together with the various interpretations of the components, is an example of an *analysis of variance*,* which will be applied to more complex problems in the succeeding sections.

*For conditions under which such a breakdown is possible, see Albert (1976).

We shall now digress for a moment from the linear hypothesis scheme to consider the hypothesis of equality of variances when the variables X_{ij} are distributed as $N(\mu_i, \sigma_i^2)$, $i = 1, \ldots, s$. A UMP unbiased test of this hypothesis was obtained in Chapter 5, Section 3, for the case $s = 2$, but does not exist for $s > 2$ (see, for example, Problem 6 of Chapter 4). Unfortunately, neither is there available for this problem a group for which there exists a UMP invariant test. To obtain a test, we shall now give a large-sample approximation, which for sufficiently large n essentially reduces the problem to that of testing the equality of s means.

It is convenient first to reduce the observations to the set of sufficient statistics $X_{i\cdot} = \sum_j X_{ij}/n_i$ and $S_i^2 = \sum_j (X_{ij} - X_{i\cdot})^2$, $i = 1, \ldots, s$. The hypothesis

$$H: \sigma_1 = \cdots = \sigma_s$$

remains invariant under the transformations $X'_{ij} = X_{ij} + c_i$, which in the space of sufficient statistics induce the transformations $S_i'^2 = S_i^2$, $X'_{i\cdot} = X_{i\cdot} + c_i$. A set of maximal invariants under this group are $S_1^2, \ldots, S_s^2$. Each statistic S_i^2 is the sum of squares of $n_i - 1$ independent normal variables with zero mean and variance σ_i^2, and it follows from the central limit theorem that for large n_i

$$\sqrt{n_i - 1} \left(\frac{S_i^2}{n_i - 1} - \sigma_i^2 \right)$$

is approximately distributed as $N(0, 2\sigma_i^4)$. This approximation is inconvenient for the present purpose, since the unknown parameters σ_i enter not only into the mean but also the variance of the limiting distribution.

The difficulty can be avoided through the use of a suitable *variance-stabilizing* transformation. Such transformations can be obtained with the help of Theorem 5 of Chapter 5, which shows that if $\sqrt{n}(T_n - \theta)$ is asymptotically normal with variance $\tau^2(\theta)$, then $\sqrt{n}[f(T_n) - f(\theta)]$ is asymptotically normal with variance $\tau^2(\theta)[f'(\theta)]^2$. Thus f is variance-stabilizing [i.e., the distribution of $f(T_n)$ has approximately constant variance] if $f'(\theta)$ is proportional to $1/\tau(\theta)$.

This applies to the present case with $n = n_i - 1$, $T_n = S_i^2/(n_i - 1)$, $\theta = \sigma_i^2$, and $\tau^2 = 2\theta^2$, and leads to the transformation $f(\theta) = \log \theta$ for which the derivative is proportional to $1/\theta$. The limiting distribution of $\sqrt{n_i - 1} \{\log[S_i^2/(n_i - 1)] - \log \sigma_i^2\}$ is the normal distribution with zero mean and variance 2, so that for large n_i the variable $Z_i = \log[S_i^2/(n_i - 1)]$ has the approximate distribution $N(\zeta_i, a_i^2)$ with $\zeta_i = \log \sigma_i^2$, $a_i^2 = 2/(n_i - 1)$.

The problem is now reduced to that of testing the equality of means of s independent variables Z_i distributed as $N(\zeta_i, a_i^2)$ where the a_i are known. In the particular case that the n_i are equal, the variances a_i^2 are equal and the asymptotic problem is a simpler version (in that the variance is known) of the problem considered at the beginning of the section. The hypothesis $\zeta_1 = \cdots = \zeta_s$ is invariant under addition of a common constant to each of the Z's and under orthogonal transformations of the hyperplanes which are perpendicular to the line $Z_1 = \cdots = Z_s$. The UMP invariant rejection region is then

$$\frac{\sum(Z_i - \bar{Z})^2}{a^2} > C$$

where a^2 is the common variance of the Z_i and where C is determined by

$$(20) \qquad \int_C^\infty \chi_{s-1}^2(y)\, dy = \alpha.$$

In the more general case of unequal a_i, the problem reduces to a linear hypothesis with known variance through the transformation $Z_i' = Z_i/a_i$, and the UMP invariant test under a suitable group of linear transformations rejects when

$$(21) \qquad \sum \frac{1}{a_i^2}\left(Z_i - \frac{\sum Z_j/a_j^2}{\sum 1/a_j^2}\right)^2 = \sum\left(\frac{Z_i}{a_i}\right)^2 - \frac{\left(\sum Z_j/a_j^2\right)^2}{\sum\left(1/a_j^2\right)} > C$$

(see Problem 14), where C is again determined by (20). This rejection region, which is UMP invariant for testing $\zeta_1 = \cdots = \zeta_s$ in the limiting distribution, can then be said to have this property asymptotically for testing the original hypothesis $H: \sigma_1 = \cdots = \sigma_s$.

When applying the principle of invariance, it is important to make sure that the underlying symmetry assumptions really are satisfied. In the problem of testing the equality of a number of normal means $\mu_1, \ldots, \mu_s$ for example, all parameter points, which have the same value of $\psi^2 = \sum n_i(\mu_i - \mu_.)^2/\sigma^2$, are identified under the principle of invariance. This is appropriate only when these alternatives can be considered as being equidistant from the hypothesis. In particular, it should then be immaterial whether the given value of ψ^2 is built up by a number of small contributions or a single large one. Situations where instead the main emphasis is on the detection of large individual deviations do not possess the required symmetry, and the test based on (19) need no longer be optimum.

The robustness properties against nonnormality of the t-test, and the nonrobustness of the F-test for variances, found in Chapter 5, Section 4 for the two-sample problem, carry over to the comparison of more than two means or variances. Specifically, the size and power of the F-test (19) of $H: \mu_1 = \cdots = \mu_s$ is robust for large n_i if the X_{ij} $(j = 1, \ldots, n_i)$ are samples from distributions $F(x - \mu_i)$ where F is an arbitrary distribution with finite variance. [A discussion of the corresponding permutation test with references to the literature can be found for example in Robinson (1983). For an elementary treatment see Edgington (1980).] On the other hand, the test for equality of variances described above (or Bartlett's test,[†] which is the classical test for this problem) is highly sensitive to the assumption of normality, and therefore is rarely appropriate. More robust tests for this latter hypothesis are reviewed in Conover, Johnson, and Johnson (1981).

That the size of the test (19) is robust against nonnormality follows from the fact that if the X_{ij}, $j = 1, \ldots, n_i$, are independent samples from $F(x - \mu_i)$, then under $H: \mu_1 = \cdots = \mu_s$

(i)   the distribution of the numerator of W^*, multiplied by $(s - 1)/\sigma^2$, tends to the χ^2_{s-1} distribution provided $n_i/n \to \rho_i > 0$ for all i and

(ii)   the denominator of W^* tends in probability to σ^2.

To see (i), assume without loss of generality that $\mu_1 = \cdots = \mu_s = 0$. Then the variables $\sqrt{n_i}\, X_i.$ are independent, each with a distribution which by the central limit theorem tends to $N(0, \sigma^2)$ as $n_i \to \infty$ for any F with finite variance. It follows (see Section 5.1, Theorem 7 of *TPE*) that for any function h, the limit distribution of $h(\sqrt{n_1}\, X_1., \ldots, \sqrt{n_s}\, X_s.)$ is the distribution of $h(U_1, \ldots, U_s)$ where $U_1, \ldots, U_s$ are independent $N(0, \sigma^2)$, provided

$$\{ (u_1, \ldots, u_s) : h(u_1, \ldots, u_s) = c \}$$

has Lebesgue measure 0 for any c. Suppose that $n_i/n = \rho_i$ as $n_1, \ldots, n_s$ tend to infinity. This condition is satisfied for

$$h(\sqrt{n_1}\, X_1., \ldots, \sqrt{n_s}\, X_s.) = \sum n_i (X_i. - X..)^2,$$

and the limit distribution of the numerator of W^* is (for all F with finite variance) what it is when F is normal, namely σ^2 times χ^2_{s-1}. A slight modification shows the result to remain true if $n_i/n \to \rho_i$.

[†]For a discussion of this test, see for example Cyr and Manoukian (1982) and Glaser (1982).

Part (ii) is a special case of the following more general result: Let $X_1, \ldots, X_n$ be independently distributed, X_i according to $F(x_i - \mu_i)$ with $E(X_i) = \mu_i$ and $\mathrm{Var}(X_i) = \sigma^2 < \infty$, and suppose that for each n the vector $(\mu_1, \ldots, \mu_n)$ is known to lie in an s-dimensional space Π_{Ω_n} with s fixed. Then the denominator D of (14) tends to σ^2 in probability as $n \to \infty$.

This can be seen from the canonical form (7) of W^*, in which

$$D = \frac{1}{n-s} \sum_{i=s+1}^{n} Y_i^2 = \frac{n}{n-s} \left[\frac{1}{n} \sum_{i=1}^{n} Y_i^2 \right] - \frac{1}{n-s} \sum_{i=1}^{s} Y_i^2$$

and the fact that $\Sigma Y_i^2 / n = \Sigma X_i^2 / n$. Since $E(Y_i) = 0$ for $i = s+1, \ldots, n$, assume, without loss of generality for the distribution of $\Sigma_{i=s+1}^n Y_i^2$, that $E(X_i) = E(Y_i) = 0$ for all i. Then by the law of large numbers $\Sigma X_i^2 / n$ tends in probability to $E(X_i^2) = \sigma^2$. On the other hand, we shall now show that the second term on the right side of D tends in probability to zero. The result then follows.

To see this, it is enough to show that each of $Y_1^2, \ldots, Y_s^2$ is bounded in probability. Now $Y_i = \Sigma c_{ij}^{(n)} X_j$, where the vectors $(c_{i1}^{(n)}, \ldots, c_{in}^{(n)})$ are orthogonal and of length 1. Therefore, by the Chebyshev inequality

$$P\left(Y_i^2 \geq a^2\right) < \frac{1}{a^2} E\left(\Sigma c_{ij}^{(n)} X_j\right)^2 = \frac{\sigma^2}{a^2}$$

and this completes the proof.

Another robustness aspect of the s-sample F-test concerns the assumption of a common variance. Here the situation is even worse than in the two-sample case. If the X_{ij} are independently distributed as $N(\mu_i, \sigma_i^2)$ and if $s > 2$, the size of the F-test (19) of $H: \mu_1 = \cdots = \mu_s$ is not asymptotically robust as $n_i \to \infty$, $n_i / \Sigma n_j \to \rho_i$, regardless of the values of the ρ_i [Scheffé (1959)]. More appropriate tests for this generalized Behrens–Fisher problem have been proposed by Welch (1951), James (1951), and Brown and Forsythe (1974a), and are further discussed by Clinch and Kesselman (1982). The corresponding robustness problem for more general linear hypotheses is treated by James (1954) and Johansen (1980); see also Rothenberg (1984).

The linear model F-test—as was seen to be the case for the t-test—is highly nonrobust against dependence of the observations. Tests of the hypothesis that the covariance matrix is proportional to the identity against various specified forms of dependence are considered in King and Hillier (1985).

The test (19), although its level and power are asymptotically independent of the distribution F, tends to be inefficient if F has heavier tails than the normal distribution. More efficient tests are obtained by generalizing the considerations of Sections 8 and 9 of Chapter 6. Suppose the X_{ij} are samples of size n_i from continuous distributions F_i $(i = 1, \ldots, s)$ and that we wish to test $H : F_1 = \cdots = F_s$. Invariance, by the argument of Chapter 6, Section 8, then reduces the data to the ranks R_{ij} of the X_{ij} in the combined sample of $n = \Sigma n_i$ observations. A natural analogue of the two-sample Wilcoxon test is the *Kruskal–Wallis test*, which rejects H when $\Sigma n_i (R_i. - R..)^2$ is too large. For the shift model $F_i(y) = F(y - \mu_i)$, the asymptotic efficiency of this test relative to (19) is the same as that of the Wilcoxon to the t-test in the case $s = 2$. The theory of this and related rank tests is developed in books on nonparametric statistics such as Hájek and Šidák (1967), Lehmann (1975), Randles and Wolfe (1979), and Hettmansperger (1984).

Unfortunately, such rank tests are available only for the very simplest linear models. An alternative approach capable of achieving similar efficiencies for much wider classes of linear models can be obtained through large-sample theory. It replaces the least-squares estimators by estimators with better efficiency properties for nonnormal distributions and obtains an asymptotically valid significance level through "Studentization",* that is, by dividing the statistic by a suitable estimator of its standard deviation. Different ways of implementing such a program are reviewed, for example, by Draper (1981, 1983), McKean and Schrader (1982), and Ronchetti (1982). [For a simple alternative of this kind to Student's t-test, see Prescott (1975).]

Sometimes, it is of interest to test the hypothesis $H : \mu_1 = \cdots = \mu_s$ considered at the beginning of the section, against only the ordered alternatives $\mu_1 \leq \cdots \leq \mu_s$ rather than against the general alternatives of any inequalities among the μ's. Then the F-test (19) is no longer reasonable; more powerful alternative tests for this and other problems involving ordered alternatives are discussed in Barlow et al. (1972).

4. MULTIPLE COMPARISONS

Testing equality of a number of means as a simple choice between acceptance and rejection usually leaves many questions unanswered. In particular, when the hypothesis is rejected one would like to obtain more detailed

---

*This term (after Student, the pseudonym of W. S. Gosset) is a misnomer. The procedure of dividing the sample mean $\bar{X}$ by its estimated standard deviation and referring the resulting statistic to the standard normal distribution (without regard to the distribution of the X's) was used already by Laplace. Student's contribution consisted in pointing out that if the X's are normal, the approximate normal distribution of the t-statistic can be replaced by its exact distribution—Student's t.

information about the relative positions of the means. In order to determine just where the differences in the μ's occur, one may want to begin by testing the hypothesis $H_s: \mu_1 = \cdots = \mu_s$, as before, with the F-test (19). If this test accepts, the means are judged to exhibit no significant differences, the set $\{\mu_1, \ldots, \mu_s\}$ is declared homogeneous, and the procedure terminates. If H_s is rejected, a search for the source of the differences can be initiated by proceeding to a second stage, which consists in testing the s hypotheses

$$H_{s-1, i}: \mu_1 = \cdots = \mu_{i-1} = \mu_{i+1} = \cdots = \mu_s$$

by means of the appropriate F-test for each. This requires the obvious modification of the numerator of (19), while the denominator is being retained at all the steps. This is justified by the assumption of a common variance σ^2 of which the denominator is an estimate. For any hypothesis that is accepted, the associated set of means and all its subsets are judged not to have shown any significant differences and are not tested further. For any rejected hypothesis the $s - 1$ subsets of size $s - 2$ are tested [except those that are subsets of an $(s - 1)$-set whose homogeneity has been accepted], and the procedure is continued in this way until nothing is left to be tested.

It is clear from this description that a particular set of μ's is declared heterogeneous if and only if the hypothesis of homogeneity is rejected for it and all sets containing it.

Instead of the F-tests, other tests of homogeneity could be used at the various stages. When the sample sizes $n_i = n$ are equal, as we shall assume throughout the remainder of this section, the most common alternative is based on the *Studentized range* statistic

$$(22) \qquad \frac{\max|X_{j\cdot} - X_{i\cdot}|}{\sqrt{\sum\sum (X_{ij} - X_{i\cdot})^2 / sn(n - 1)}}$$

where the maximum is taken over all pairs (i, j) within the set being tested. We shall here restrict attention to procedures where the test statistics are either F or Studentized range, not necessarily the same at all stages.

To complete the description of the procedure, once the test statistics have been chosen, it is necessary to specify the critical values which they must exceed for rejection, or equivalently, the significance levels at which the various tests are to be performed. Suppose all tests at a given stage are performed at the same level, and denote this level by α_k when the equality of k means is being tested, and the associated critical values by C_k, $k = 2, \ldots, s$.

Before discussing the best choice of α's let us consider some specific methods that have been proposed in the literature. Additional properties and uses of some of these will be mentioned at the end of the section.

(i) *Tukey's T-method.* This procedure employs the Studentized range test at each stage with a common critical value $C_k = C$ for all k. The method has an unusual feature which makes it particularly simple to apply. In general, in order to determine whether a particular subset S_0 of means should be called nonhomogeneous, it is necessary to proceed stagewise since the homogeneity of S_0 itself is not tested unless homogeneity has been rejected for all sets containing S_0. However, with Tukey's T-method it is only necessary to test S_0 itself. If the Studentized range of S_0 exceeds C, so will that of any set containing S_0, and S_0 is declared nonhomogeneous. In the contrary case, homogeneity of S_0 is accepted. The two facts which jointly eliminate the need for a stagewise procedure in this case are (a) that the range, and hence the Studentized range, of S_0 cannot exceed that of any set S containing S_0, and (b) the constancy of the critical value. The next method applies this idea to a procedure based on F-tests.

(ii) *Gabriel's simultaneous test procedure.* F-statistics do not have property (a) above. However, this property is possessed by the statistics νF, where ν is the number of numerator degrees of freedom (Problem 16). Hence a procedure based on F-statistics with critical values $C_k = C/(k-1)$ satisfies both (a) and (b), since $k - 1$ is the number of numerator degrees of freedom when k means are being tested, that is, at the $(s - k + 1)$st stage. This procedure, which in this form was proposed by Gabriel (1964), permits the testing of many additional hypotheses and when these are included becomes Scheffé's S-method, which will be discussed in Sections 9 and 10.

(iii) *Fisher's least-significant-difference method* employs an F-test at the first stage, and Studentized range tests with a common critical value $C_{s-1} = \cdots = C_2$ at all succeeding stages. The constants C_s and C_2 are related by the fact that the first stage F-test and the pairwise t-test of the last stage have the same level.

The usual descriptions of (iii) and (i) consider only the first and last stage of these procedures, and omit the conclusions which can be drawn from the intermediate stages.

Several classes of procedures have been defined by prescribing the significance levels α_k, which can then be applied to the chosen test statistic at each stage. Examples are:

(iv) *The Newman–Keuls levels:*

$$\alpha_k = \alpha.$$

(v)   *The Duncan levels:*

$$\alpha_k = 1 - \gamma^{k-1}.$$

(vi)   *The Tukey levels:*

$$\alpha_k = \begin{cases} 1 - \gamma^{k/2}, & 1 < k < s - 1, \\ 1 - \gamma^{s/2}, & k = s - 1, s. \end{cases}$$

In both (v) and (vi), $\gamma = 1 - \alpha_2$.

Most of the above methods and some others are reviewed and their justification discussed by Spjøtvoll (1974); comparisons of different methods are provided, for example, by Einot and Gabriel (1975).

Let us now consider the choice of the levels α_k more systematically. In generalizing the usual significance level α for a single test, it is desirable to control some overall measure of the extent to which a procedure leads to false rejections. One such measure is the maximum probability α_0 of at least one false rejection, that is, of rejecting homogeneity of at least one set of μ's which is in fact homogeneous. The probability of at least one false rejection for a given $(\mu_1, \ldots, \mu_s)$ will be denoted by $\alpha(\mu_1, \ldots, \mu_s)$, so that $\alpha_0 = \sup \alpha(\mu_1, \ldots, \mu_s)$, where the supremum is taken over all s-tuples $(\mu_1, \ldots, \mu_s)$.

In order to study the best choice of $\alpha_2, \ldots, \alpha_s$ subject to

$$(23) \qquad\qquad\qquad \alpha_0 \leq \alpha_0^*$$

for a given level α_0^*, let us simplify the problem by assuming σ^2 to be known, say $\sigma^2 = 1$. Then the F-tests (19) are replaced by the χ^2-tests with rejection region $\Sigma n_i (X_{i.} - X_{..})^2 > C$, and the Studentized range tests are replaced by the range tests which reject when the range of the subgroup being tested is too large.

Theorem 1.   *Suppose that at each stage either a χ^2- or a range test is used (not necessarily the same at all stages) and that the μ's fall into r distinct groups of sizes $v_1, \ldots, v_r$ $(\Sigma v_i = s)$, say*

$$(24) \qquad \mu_{i_1} = \cdots = \mu_{i_{v_1}}, \qquad \mu_{i_{v_1+1}} = \cdots = \mu_{i_{v_1+v_2}}, \ldots,$$

where $(i_1, \ldots, i_s)$ is a permutation of $(1, \ldots, s)$. Then

$$(25) \qquad \sup \alpha(\mu_1, \ldots, \mu_s) = 1 - \prod_{i=1}^{r} (1 - \alpha_{v_i}),$$

where $\alpha_1 = 0$ and the supremum is taken over all $(\mu_1, \ldots, \mu_s)$ satisfying (24).

Proof. Since false rejection can occur only when at least one of the hypotheses

$$(26) \qquad H_1' : \mu_{i_1} = \cdots = \mu_{i_{v_1}}, \qquad H_2' : \mu_{i_{v_1+1}} = \cdots = \mu_{i_{v_1+v_2}}, \ldots$$

is rejected,

$$\alpha(\mu_1, \ldots, \mu_s) \le P \text{ (rejecting at least one } H_i')$$

$$= 1 - P \text{ (accepting all the } H_i')$$

$$= 1 - \prod_{i=1}^{r} (1 - \alpha_{v_i}).$$

Here the last equality follows from the fact that the test statistics for testing the hypotheses $H_1', \ldots, H_r'$ are independent.

To see that the upper bound is sharp, let the distances between the different groups of means (24) all tend to infinity. Then the probability of accepting homogeneity of any set containing $\{\mu_{i_1}, \ldots, \mu_{i_{v_1}}\}$ as a proper subset, and therefore not reaching the stage at which H_1' is tested, tends to zero. The same is true for $H_2', \ldots, H_r'$, and hence $\alpha(\mu_1, \ldots, \mu_s)$ tends to the right side of (25).

It is interesting to note that sup $\alpha(\mu_1, \ldots, \mu_s)$ depends only on $\alpha_2, \ldots, \alpha_s$ and not on whether χ^2- or range statistics are used at the various stages. In fact, Theorem 1 remains true for many other statistics (Problem 17).

It follows from Theorem 1 that a procedure with levels $(\alpha_2, \ldots, \alpha_s)$ satisfies (23) if and only if

$$(27) \qquad \prod_{i=1}^{r} (1 - \alpha_{v_i}) \ge 1 - \alpha_0^* \qquad \text{for all} \quad (v_1, \ldots, v_r) \quad \text{with } \sum v_i = s.$$

To see how to choose $\alpha_2, \ldots, \alpha_s$ subject to (23) or (27), let us say that $(\alpha_2, \ldots, \alpha_s)$ is *inadmissible* if there exists another set of levels $(\alpha_2', \ldots, \alpha_s')$ satisfying (27) and such that

$$(28) \qquad \alpha_i \le \alpha_i' \qquad \text{for all } i, \text{ with strict inequality for some } i.$$

These inequalities imply that the procedure with the levels α_i' has uniformly better chance of detecting existing inhomogeneities than the procedure based on the α_i. The definition is thus in the spirit of α-admissibility discussed in Chapter 6, Section 7.

Lemma 1. *Under the assumptions of Theorem* 1, *necessary conditions for* $(\alpha_2, \ldots, \alpha_s)$ *to be admissible are*

(i)　$\alpha_2 \leq \cdots \leq \alpha_s$, *and*

(ii)　$\alpha_s = \alpha_{s-1} = \alpha_0^*$.

Proof. (i): Suppose to the contrary that there exists k such that $\alpha_{k+1} < \alpha_k$, and consider the procedure in which $\alpha_i' = \alpha_i$ for $i \neq k + 1$ and $\alpha_{k+1}' = \alpha_k$. To show that $\alpha_0' \leq \alpha_0^*$, we need only show that $\Pi(1 - \alpha_{v_i}') \geq 1 - \alpha_0^*$ for all $(v_1, \ldots, v_r)$. If none of the v's is equal to $k + 1$, then $\alpha_{v_i}' = \alpha_{v_i}$ for all i, and the result follows. Otherwise replace each v that is equal to $k + 1$ by two v's—one equal to k and one equal to 1—and denote the resulting set of v's by $\omega_1, \ldots, \omega_{r'}$. Then

$$\prod_{i=1}^{r} \left(1 - \alpha_{v_i}'\right) = \prod_{i=1}^{r'} \left(1 - \alpha_{\omega_i}\right) \geq 1 - \alpha_0^*.$$

(ii): The left side of (27) involves α_s if and only if $r = 1$, $v_1 = s$. Thus the only restriction on α_s is $\alpha_s \leq \alpha_0^*$, and the only admissible choice is $\alpha_s = \alpha_0^*$. The argument for α_{s-1} is analogous.

Part (ii) of this lemma shows that procedures (i) and (ii) are inadmissible since in both $\alpha_{s-1} < \alpha_s$. The same argument shows Duncan's set of levels to be inadmissible. [However, choices (i), (ii), and (v) can be justified from other points of view; see for example Spjøtvoll (1974) and comment 5 at the end of the section.] It also follows from the lemma that for $s = 3$ there is a unique best choice of levels, namely $\alpha_2 = \alpha_3 = \alpha_0^*$.

Having fixed $\alpha_0 = \alpha_s = \alpha_{s-1} = \alpha_0^*$, how should we choose the remaining α's? In order to have a reasonable chance of detecting existing inhomogeneities for all patterns, we should like to have none of the α's too small. In view of part (i) of Lemma 1, this aim is perhaps best achieved by maximizing α_2.

Lemma 2. *Under the assumptions of Theorem* 1, *the maximum value of* α_2 *subject to* (23) *is*

$$(29) \qquad \alpha_2 = 1 - \left(1 - \alpha_0^*\right)^{[s/2]^{-1}}$$

where $[A]$ *denotes the largest integer* $\leq A$.

Proof. Instead of fixing α_0 at α_0^* and maximizing α_2, it is more convenient instead to fix α_2, at, say α, and then to minimize α_0. The lemma will be proved by showing that the resulting minimum value of α_0 is

$$\alpha_0^* = 1 - \left(1 - \alpha\right)^{[s/2]}.$$

Suppose first that s is even. Since α_2 is fixed at α, it follows from Theorem 1 that the right side of (25) can be made arbitrarily close to α_0^*. This is seen by letting $v_1 = \cdots = v_{s/2} = 2$. When s is odd, the same argument applies if we put an additional v equal to 1.

Lemmas 1 and 2 show that any procedure with $\alpha_2 = \alpha_s$, and hence Fisher's least-significant-difference procedure and the Newman–Keuls choice of levels, is admissible for $s = 3$ but inadmissible for $s \geq 4$. The second of these statements is seen from the fact that $\alpha_0 \leq \alpha_0^*$ implies $\alpha_2 \leq 1 - (1 - \alpha_0^*)^{[s/2]^{-1}} < \alpha_0^*$ when $s \geq 4$. The choice $\alpha_s = \alpha_2$ thus violates Lemma 1(ii).

Once α_2 has been fixed at the value given by Lemma 2, it turns out that subject to (23) there exists a unique optimal choice of the remaining α's when s is odd, and a narrow range of choices when s is even.

Theorem 2. *When s is odd, then $\alpha_3, \ldots, \alpha_s$ are maximized, subject to (23) and (29), by*

$$(30) \qquad \alpha_i^* = 1 - (1 - \alpha_2)^{[i/2]},$$

and these values can be attained simultaneously.

Proof. If we put $\gamma_i = 1 - \alpha_i$ and $\gamma = \gamma_2$, then by (27) and (29) any procedure satisfying the conditions of the theorem must satisfy

$$\prod \gamma_{v_i} \geq \gamma^{[s/2]} = \gamma^{(s-1)/2}.$$

Let i be odd, and consider any configuration in which $v_1 = i$ and all the remaining v's are equal to 2. Then

$$\gamma_i \gamma^{(s-i)/2} \geq \gamma^{(s-1)/2},$$

and hence

$$(31) \qquad \gamma_i \geq \gamma_i^* = 1 - \alpha_i^*.$$

An analogous argument proves (31) for even i.

Consider now the procedure defined by $\gamma_i = \gamma_i^*$. This clearly satisfies (29), and it only remains to show that it also satisfies (23) or equivalently (27), and hence that

$$\prod \gamma^{[v_i/2]} \geq \gamma^{(s-1)/2}$$

or that

$$\sum_{i=1}^{r} \left[\frac{v_i}{2} \right] \leq \frac{s-1}{2}.$$

Now $\Sigma[v_i/2] = (s - b)/2$, where b is the number of odd v's (including ones). Since s is odd, $b \geq 1$, and this completes the proof.

Note that the levels (30) are close to the Tukey levels (vi), which are admissible but do not satisfy (29).

When s is even, a uniformly best choice is not available. In this case, the Tukey levels (vi) satisfy (29), are admissible, and constitute a reasonable choice. [See Lehmann and Shaffer (1979).]

Even in the simplified version with known variance the multiple testing problem considered in the present section is clearly much more difficult than the testing of a single hypothesis; the solution presented above still ignores many important aspects of the problem.

1. *Choice of test statistic.* The most obvious feature that has not been dealt with is the choice of test statistics. Unfortunately it does not appear that the invariance considerations which were so helpful in the case of a single hypothesis play a similar role here.

2. *Order relation of significant means.* Whenever two means $X_{i.}$, $X_{j.}$ are judged to differ, we should like to state not only that $\mu_i \neq \mu_j$, but that if $X_{i.} < X_{j.}$ then also $\mu_i < \mu_j$. Such additional statements introduce the possibility of additional errors (stating $\mu_i < \mu_j$ when in fact $\mu_i > \mu_j$), and it is not obvious that when these are included, the probability of at least one error is still bounded by α_0^*. [This problem of directional errors has been solved in a simpler situation in Shaffer (1980).]

3. *Nominal versus true levels.* The levels $\alpha_2, \ldots, \alpha_s$, sometimes called *nominal levels*, are the levels at which the hypotheses $\mu_i = \mu_j$, $\mu_i = \mu_j = \mu_k, \ldots$ are tested. They are however not the true probabilities of falsely rejecting the homogeneity of these sets, but only the upper bounds of these probabilities with respect to variation of the remaining μ's. The true probabilities tend to be much smaller (particularly when s is large), since they take into account that homogeneity of a set S_0 is rejected only if it is also rejected for all sets S containing S_0.

4. *Interpretability.* The totality of acceptance and rejection statements resulting from a multiple comparison procedure typically does not lead to a simple pattern of means. This is illustrated by the possibility that the hypothesis of homogeneity is rejected for a set S but for none of its subsets. As another example, consider the case $s = 3$, where it may happen that the hypotheses $\mu_i = \mu_j$ and $\mu_j = \mu_k$ are accepted but $\mu_i = \mu_k$ is rejected. The number of such "inconsistencies" and the corresponding difficulty of interpreting the results may be formidable. Measures of the *complexity* of the totality of statements as a third criterion (besides level and power) are discussed by Shaffer (1981).

5. Procedures (i) and (ii) can be inverted to provide simultaneous confidence intervals for all differences $\mu_j - \mu_i$. The T-method (discussed in Problems 65–68) was designed to give simultaneous intervals for all differences $\mu_j - \mu_i$; it can be extended to cover also all *contrasts* in the μ's, that is, all linear functions $\Sigma c_i \mu_i$ with $\Sigma c_i = 0$, but against more complex contrasts the intervals tend to be longer than those of Scheffé's S-method, which was intended for the simultaneous consideration of all contrasts. [For a comparison of the two methods, see for example Scheffé (1959, Section 3.7) and Arnold (1981, Chapter 12).] It is a disadvantage of the remaining (truly stagewise) procedures of this section that they do not permit such an inversion.

6. To control the rate of false rejections, we have restricted attention to procedures controlling the probability of at least one error. This is sometimes called the *error rate per experiment*, since it counts any experiment as faulty in which even one false rejection occurs. Instead, one might wish to control the expected proportion or number of false rejections. An optimality theory based on the latter criterion is given in Spjøtvoll (1972).

7. The optimal choice of the α_k discussed in this section can be further improved, at the cost of considerable additional complication, by permitting the α's to depend on the outcomes of the other tests. This possibility is discussed, for example, in Marcus, Peritz, and Gabriel (1976); see also Holm (1979) and Shaffer (1984).

8. If the variance σ^2 is unknown, the dependence introduced by the common denominator S when X_i is replaced by X_i/S invalidates Theorems 1 and 2, and no analogous results are available in this case.

5. TWO-WAY LAYOUT: ONE OBSERVATION PER CELL

The hypothesis of equality of several means arises when a number of different treatments, procedures, varieties, or manifestations of some other factors are to be compared. Frequently one is interested in studying the effects of more than one factor, or the effects of one factor as certain other conditions of the experiment vary, which then play the role of additional factors. In the present section we shall consider the case that the number of factors affecting the outcomes of the experiment is two.

Suppose that one observation is obtained at each of a number of levels of these factors, and denote by X_{ij} $(i = 1, \ldots, a;\ j = 1, \ldots, b)$ the value observed when the first factor is at the ith and the second at the jth level. It is assumed that the X_{ij} are independently normally distributed with constant variance σ^2, and for the moment also that the two factors act independently (they are then said to be *additive*), so that ξ_{ij} is of the form

$\alpha_i' + \beta_j'$. Putting $\mu = \alpha'. + \beta'.$ and $\alpha_i = \alpha_i' - \alpha'.$, $\beta_j = \beta_j' - \beta'.$, this can be written as

(32) $$\xi_{ij} = \mu + \alpha_i + \beta_j, \qquad \sum \alpha_i = \sum \beta_j = 0,$$

where the α's and β's (the *main effects* of A and B) and μ are uniquely determined by (32) as*

(33) $$\alpha_i = \xi_{i.} - \xi.., \qquad \beta_j = \xi._{j} - \xi.., \qquad \mu = \xi...$$

Consider the hypothesis

(34) $$H : \alpha_1 = \cdots = \alpha_a = 0$$

that the first factor has no effect on the outcome being observed. This arises in two quite different contexts. The factor of interest, corresponding say to a number of treatments, may be β, while α corresponds to a classification according to, for example, the site on which the observations are obtained (farm, laboratory, city, etc.). The hypothesis then represents the possibility that this subsidiary classification has no effect on the experiment so that it need not be controlled. Alternatively, α may be the (or a) factor of primary interest. In this case, the formulation of the problem as one of hypothesis testing would usually be an oversimplification, since in case of rejection of H, one would require estimates of the α's or at least a grouping according to high and low values.

The hypothesis H is a linear hypothesis with $r = a - 1$, $s = 1 + (a - 1) + (b - 1) = a + b - 1$, and $n - s = (a - 1)(b - 1)$. The least-squares estimates of the parameters under Ω can be obtained from the identity

$$\sum\sum (X_{ij} - \xi_{ij})^2 = \sum\sum (X_{ij} - \mu - \alpha_i - \beta_j)^2$$

$$= \sum\sum \left[(X_{ij} - X_{i.} - X_{.j} + X..) + (X_{i.} - X.. - \alpha_i) \right.$$

$$\left. + (X_{.j} - X.. - \beta_j) + (X.. - \mu) \right]^2$$

$$= \sum\sum (X_{ij} - X_{i.} - X_{.j} + X..)^2 + b\sum (X_{i.} - X.. - \alpha_i)^2$$

$$+ a\sum (X_{.j} - X.. - \beta_j)^2 + ab(X.. - \mu)^2,$$

---

*The replacing of a subscript by a dot indicates that the variable has been averaged with respect to that subscript.

which is valid because in the expansion of the third sum of squares the cross-product terms vanish. It follows that

$$\hat{\alpha}_i = X_{i.} - X_{..}, \qquad \hat{\beta}_j = X_{.j} - X_{..}, \qquad \hat{\mu} = X_{..},$$

and that

$$\sum\sum(X_{ij} - \hat{\xi}_{ij})^2 = \sum\sum(X_{ij} - X_{i.} - X_{.j} + X_{..})^2.$$

Under the hypothesis H we still have $\hat{\hat{\beta}}_j = X_{.j} - X_{..}$ and $\hat{\hat{\mu}} = X_{..}$, and hence $\hat{\hat{\xi}}_{ij} - \hat{\xi}_{ij} = X_{i.} - X_{..}$. The best invariant test therefore rejects when

$$(35) \qquad W^* = \frac{b\sum(X_{i.} - X_{..})^2/(a-1)}{\sum\sum(X_{ij} - X_{i.} - X_{.j} + X_{..})^2/(a-1)(b-1)} > C.$$

The noncentrality parameter, on which the power of the test depends, is given by

$$(36) \qquad \psi^2 = \frac{b\sum(\xi_{i.} - \xi_{..})^2}{\sigma^2} = \frac{b\sum\alpha_i^2}{\sigma^2}.$$

This problem provides another example of an analysis of variance. The total variation can be broken into three components,

$$\sum\sum(X_{ij} - X_{..})^2 = b\sum(X_{i.} - X_{..})^2 + a\sum(X_{.j} - X_{..})^2$$

$$+ \sum\sum(X_{ij} - X_{i.} - X_{.j} + X_{..})^2.$$

Of these, the first contains the variation due to the α's, the second that due to the β's. The last component, in the canonical form of Section 1, is equal to $\sum_{i=s+1}^{n} Y_i^2$. It is therefore the sum of squares of those variables whose means are zero even under Ω. Since this residual part of the variation, which on division by $n - s$ is an estimate of σ^2, cannot be put down to any effects such as the α's or β's, it is frequently labeled "error," as an indication that it is due solely to the randomness of the observations, not to any differences of the means. Actually, the breakdown is not quite as sharp as is suggested by the above description. Any component such as that attributed to the α's always also contains some "error," as is seen for example from its expectation, which is

$$E\sum(X_{i\cdot}-X_{\cdot\cdot})^2 = (a-1)\sigma^2 + b\sum\alpha_i^2.$$

Instead of testing whether a certain factor has any effect, one may wish to estimate the size of the effect at the various levels of the factor. Other parameters, which it is sometimes interesting to estimate, are the average outcomes (for example yields) $\xi_1\cdot,\ldots,\xi_a\cdot$ when the factor is at the various levels. If $\theta_i = \mu + \alpha_i = \xi_i\cdot$, confidence sets for $(\theta_1,\ldots,\theta_a)$ are obtained by considering the hypotheses $H(\theta^0): \theta_i = \theta_i^0$ $(i = 1,\ldots, a)$. For testing $\theta_1 = \cdots = \theta_a = 0$, the least-squares estimates of the ξ_{ij} are $\hat{\xi}_{ij} = X_{i\cdot}+ X_{\cdot j} - X_{\cdot\cdot}$ and $\hat{\hat{\xi}}_{ij} = X_{\cdot j} - X_{\cdot\cdot}$. The denominator sum of squares is therefore $\sum\sum(X_{ij} - X_{i\cdot}- X_{\cdot j} + X_{\cdot\cdot})^2$ as before, while the numerator sum of squares is

$$\sum\sum\left(\hat{\xi}_{ij} - \hat{\hat{\xi}}_{ij}\right)^2 = b\sum X_{i\cdot}^2.$$

The general hypothesis reduces to this special case on replacing X_{ij} with the variable $X_{ij} - \theta_i^0$. Since $s = a + b - 1$ and $r = a$, the hypothesis $H(\theta^0)$ is rejected when

$$\frac{b\sum(X_{i\cdot}- \theta_i^0)^2/a}{\sum\sum(X_{ij} - X_{i\cdot}- X_{\cdot j} + X_{\cdot\cdot})^2/(a-1)(b-1)} > C.$$

The associated confidence sets for $(\theta_1,\ldots,\theta_a)$ are the spheres

$$\sum(\theta_i - X_{i\cdot})^2 \le \frac{aC\sum\sum(X_{ij} - X_{i\cdot}- X_{\cdot j} + X_{\cdot\cdot})^2}{(a-1)(b-1)b}.$$

When considering confidence sets for the effects $\alpha_1,\ldots,\alpha_a$ one must take account of the fact that the α's are not independent. Since they add up to zero, it would be enough to restrict attention to $\alpha_1,\ldots,\alpha_{a-1}$. However, an easier and more symmetric solution is found by retaining all the α's. The rejection region of $H: \alpha_i = \alpha_i^0$ for $i = 1,\ldots, a$ (with $\sum\alpha_i^0 = 0$) is obtained from (35) by letting $X'_{ij} = X_{ij} - \alpha_i^0$, and hence is given by

$$b\sum(X_{i\cdot}- X_{\cdot\cdot}- \alpha_i^0)^2 > \frac{C\sum\sum(X_{ij} - X_{i\cdot}- X_{\cdot j} + X_{\cdot\cdot})^2}{b - 1}.$$

The associated confidence set consists of the totality of points $(\alpha_1,\ldots,\alpha_a)$

satisfying $\Sigma \alpha_i = 0$ and

$$\sum [\alpha_i - (X_i.- X..)]^2 \leq \frac{c\sum\sum(X_{ij} - X_i.- X._j + X..)^2}{b(b-1)}.$$

In the space of $(\alpha_1, \ldots, \alpha_a)$, this inequality defines a sphere whose center $(X_1.- X.., \ldots, X_a.- X..)$ lies on the hyperplane $\Sigma \alpha_i = 0$. The confidence sets for the α's therefore consist of the interior and surface of the great hyperspheres obtained by cutting the a-dimensional spheres with the hyperplane $\Sigma \alpha_i = 0$.

In both this and the previous case, the usual method shows the class of confidence sets to be invariant under the appropriate group of linear transformations, and the sets are therefore uniformly most accurate invariant.

A rank test of (34) analogous to the Kruskal–Wallis test for the one-way layout is Friedman's test, obtained by ranking the s observations $X_{1j}, \ldots, X_{sj}$ separately from 1 to s at each level j of the second factor. If these ranks are denoted by $R_{1j}, \ldots, R_{sj}$, Friedman's test rejects for large values of $\Sigma(R_i.- R..)^2$. Unless s is large, this test suffers from the fact that comparisons are restricted to observations at the same level of factor 2. The test can be improved by "aligning" the observations from different levels, for example, by subtracting from each observation at the jth level its mean $X._j$ for that level, and then ranking the aligned observations from 1 to ab. For a discussion of these tests and their efficiency see Lehmann (1975, Chapter 6), and for an extension to tests of (34) in the model (32) when there are several observations per cell, Mack and Skillings (1980). Further discussion is provided by Hettmansperger (1984).

That in the experiment described at the beginning of the section there is only one observation per cell, and that as a consequence hypotheses about the α's and β's cannot be tested without some restrictions on the means ξ_{ij}, does not of course justify the assumption of additivity. Rather, the other way around, the experiment should not be performed with just one observation per cell unless the factors can safely be assumed to be additive. Faced with such an experiment without prior assurance that the assumption holds, one should test the hypothesis of additivity. A number of tests for this purpose are discussed, for example, in Hegemann and Johnson (1976) and in Marasinghe and Johnson (1981).

6. TWO-WAY LAYOUT: m OBSERVATIONS PER CELL

In the preceding section it was assumed that the effects of the two factors α and β are independent and hence additive. The factors may, however, interact in the sense that the effect of one depends on the level of the other.

Thus the effectiveness of a teacher depends for example on the quality or the age of the students, and the benefit derived by a crop from various amounts of irrigation depends on the type of soil as well as on the variety being planted. If the additivity assumption is dropped, the means ξ_{ij} of X_{ij} are no longer given by (32) under Ω but are completely arbitrary. More than ab observations, one for each combination of levels, are then required, since otherwise $s = n$. We shall here consider only the simple case in which the number of observations is the same at each combination of levels.

Let X_{ijk} ($i = 1, \ldots, a$; $j = 1, \ldots, b$; $k = 1, \ldots, m$) be independent normal with common variance σ^2 and mean $E(X_{ijk}) = \xi_{ij}$. In analogy with the previous notation we write

$$\xi_{ij} = \xi.. + (\xi_{i\cdot} - \xi..) + (\xi_{\cdot j} - \xi..) + (\xi_{ij} - \xi_{i\cdot} - \xi_{\cdot j} + \xi..)$$

$$= \mu + \alpha_i + \beta_j + \gamma_{ij}$$

with $\Sigma_i \alpha_i = \Sigma_j \beta_j = \Sigma_i \gamma_{ij} = \Sigma_j \gamma_{ij} = 0$. Then α_i is the average effect of factor 1 at level i, averaged over the b levels of factor 2, and a similar interpretation holds for the β's. The γ's are called *interactions*, since γ_{ij} measures the extent to which the joint effect $\xi_{ij} - \xi..$ of factors 1 and 2 at levels i and j exceeds the sum $(\xi_{i\cdot} - \xi..) + (\xi_{\cdot j} - \xi..)$ of the individual effects. Consider again the hypothesis that the α's are zero. Then $r = a - 1$, $s = ab$, and $n - s = (m - 1)ab$. From the decomposition

$$\Sigma\Sigma\Sigma(X_{ijk} - \xi_{ij})^2 = \Sigma\Sigma\Sigma(X_{ijk} - X_{ij\cdot})^2 + m\Sigma\Sigma(X_{ij\cdot} - \xi_{ij})^2$$

and

$$\Sigma\Sigma(X_{ij\cdot} - \xi_{ij})^2 = \Sigma\Sigma(X_{ij\cdot} - X_{i\cdot\cdot} - X_{\cdot j\cdot} + X_{\cdots} - \gamma_{ij})^2$$

$$+ b\Sigma(X_{i\cdot\cdot} - X_{\cdots} - \alpha_i)^2 + a\Sigma(X_{\cdot j\cdot} - X_{\cdots} - \beta_j)^2$$

$$+ ab(X_{\cdots} - \mu)^2$$

it follows that

$$\hat{\mu} = \hat{\hat{\mu}} = \hat{\xi}.. = X_{\cdots}, \qquad \hat{\alpha}_i = \hat{\xi}_{i\cdot} - \hat{\xi}.. = X_{i\cdot\cdot} - X_{\cdots},$$

$$\hat{\beta}_j = \hat{\hat{\beta}}_j = \hat{\xi}_{\cdot j} - \hat{\xi}.. = X_{\cdot j\cdot} - X_{\cdots},$$

$$\hat{\gamma}_{ij} = \hat{\hat{\gamma}}_{ij} = X_{ij\cdot} - X_{i\cdot\cdot} - X_{\cdot j\cdot} + X_{\cdots},$$

and hence that

$$\sum\sum\sum \left(X_{ijk} - \hat{\xi}_{ij} \right)^2 = \sum\sum\sum \left(X_{ijk} - X_{ij.} \right)^2,$$

$$\sum\sum\sum \left(\hat{\xi}_{ij} - \hat{\hat{\xi}}_{ij} \right)^2 = mb\sum \left(X_{i..} - X_{...} \right)^2.$$

The most powerful invariant test therefore rejects when

$$(37) \qquad W^* = \frac{mb\sum \left(X_{i..} - X_{...} \right)^2/(a-1)}{\sum\sum\sum \left(X_{ijk} - X_{ij.} \right)^2/(m-1)ab} > C,$$

and the noncentrality parameter in the distribution of W^* is

$$(38) \qquad \frac{mb\sum \left(\xi_{i.} - \xi_{..} \right)^2}{\sigma^2} = \frac{mb\sum \alpha_i^2}{\sigma^2}.$$

Another hypothesis of interest is the hypothesis H' that the two factors are additive,[†]

$$H' : \gamma_{ij} = 0 \qquad \text{for all } i, j.$$

The least-squares estimates of the parameters are easily derived as before, and the UMP invariant test is seen to have the rejection region (Problem 22)

$$(39) \quad W^* = \frac{m\sum\sum \left(X_{ij.} - X_{i..} - X_{.j.} + X_{...} \right)^2/(a-1)(b-1)}{\sum\sum\sum \left(X_{ijk} - X_{ij.} \right)^2/(m-1)ab} > C.$$

Under H', the statistic W^* has the F-distribution with $(a-1)(b-1)$ and $(m-1)ab$ degrees of freedom; the noncentrality parameter for any alternative set of γ's is

$$(40) \qquad \psi^2 = \frac{m\sum\sum \gamma_{ij}^2}{\sigma^2}.$$

[†]A test of H' against certain restricted alternatives has been proposed for the case of one observation per cell by Tukey (1949); see Hegemann and Johnson (1976) for further discussion.

The decomposition of the total variation into its various components, in the present case, is given by

$$\sum\sum\sum(X_{ijk} - X...)^2 = mb\sum(X_{i..} - X...)^2 + ma\sum(X_{.j.} - X...)^2$$

$$+ m\sum\sum(X_{ij.} - X_{i..} - X_{.j.} + X...)^2$$

$$+ \sum\sum\sum(X_{ijk} - X_{ij.})^2.$$

Here the first three terms contain the variation due to the α's, β's and γ's respectively, and the last component corresponds to error. The tests for the hypotheses that the α's, β's, or γ's are zero, the first and third of which have the rejection regions (37) and (39), are then obtained by comparing the α, β, or γ sum of squares with that for error.

An analogous decomposition is possible when the γ's are assumed a priori to be equal to zero. In that case, the third component which previously was associated with γ represents an additional contribution to error, and the breakdown becomes

$$\sum\sum\sum(X_{ijk} - X...)^2 = mb\sum(X_{i..} - X...)^2 + ma\sum(X_{.j.} - X...)^2$$

$$+ \sum\sum\sum(X_{ijk} - X_{i..} - X_{.j.} + X...)^2,$$

with the last term corresponding to error. The hypothesis $H: \alpha_1 = \cdots = \alpha_a = 0$ is then rejected when

$$\frac{mb\sum(X_{i..} - X...)^2/(a-1)}{\sum\sum\sum(X_{ijk} - X_{i..} - X_{.j.} + X...)^2/(abm - a - b + 1)} > C.$$

Suppose now that the assumption of no interaction, under which this test was derived, is not justified. The denominator sum of squares then has a noncentral χ^2-distribution instead of a central one; and is therefore stochastically larger than was assumed (Problem 25). It follows that the actual rejection probability is less than it would be for $\sum\sum\gamma_{ij}^2 = 0$. This shows that the probability of an error of the first kind will not exceed the nominal level of significance, regardless of the values of the γ's. However, the power also decreases with increasing $\sum\sum\gamma_{ij}^2/\sigma^2$ and tends to zero as this ratio tends to infinity.

The analysis of variance and the associated tests derived in this section for two factors extend in a straightforward manner to a larger number of factors (see for example Problem 26). On the other hand, if the number of observations is not the same for each combination of levels (each *cell*), explicit formulae for the least-squares estimators may no longer be available, but there is no difficulty in computing these estimators and the associated UMP invariant tests numerically. However, in applications it is then not always clear how to define main effects, interactions, and other parameters of interest, and hence what hypothesis to test. These issues are discussed, for example, in Hocking and Speed (1975) and Speed, Hocking, and Hackney (1978). See also *TPE*, Chapter 3, Example 4.4, and Arnold (1981, Section 7.4).

Of great importance are arrangements in which only certain combinations of levels occur, since they permit reducing the size of the experiment. Thus for example three independent factors, at m levels each, can be analyzed with only m^2 observations, instead of the m^3 required if 1 observation were taken at each combination of levels, by adopting a Latin-square design (Problem 27).

The class of problems considered here contains as a special case the two-sample problem treated in Chapter 5, which concerns a single factor with only two levels. The questions discussed in that connection regarding possible inhomogeneities of the experimental material and the randomization required to offset it are of equal importance in the present, more complex situations. If inhomogeneous material is subdivided into more homogeneous groups, this classification can be treated as constituting one or more additional factors. The choice of these groups is an important aspect in the determination of a suitable experimental design.† A very simple example of this is discussed in Problems 49 and 50 of Chapter 5.

Multiple comparison procedures for two-way (and higher) layouts are discussed by Spjøtvoll (1974); additional references can be obtained from the bibliography of R. G. Miller (1977).

7. REGRESSION

Hypotheses specifying one or both of the regression coefficients α, β when $X_1, \ldots, X_n$ are independently normally distributed with common variance σ^2 and means

$$(41) \qquad\qquad \xi_i = \alpha + \beta t_i$$

---

† For a discussion of various designs and the conditions under which they are appropriate see, for example, Cox (1958), John (1971), John and Quenouille (1977), and Box, Hunter, and Hunter (1978). Optimum properties of certain designs, proved by Wald, Ehrenfeld, Kiefer, and others, are discussed by Kiefer (1958, 1980) and Silvey (1980). The role of randomization, treated for the two-sample problem in Chapter 5, Section 12, is studied by Kempthorne (1955), Wilk and Kempthorne (1955), Scheffé (1959), and others; see, for example, Lorenzen (1984).

are essentially linear hypotheses, as was pointed out in Example 2. The hypotheses $H: \alpha = \alpha_0$ and $H_2: \beta = \beta_0$ were treated in Chapter 5, Section 8, where they were shown to possess UMP unbiased tests. We shall now consider H_1 and H_2, as well as the hypothesis $H_3: \alpha = \alpha_0$, $\beta = \beta_0$, from the present point of view. By the general theory of Section 1 the resulting tests will be UMP invariant under suitable groups of linear transformations. For the first two cases, in which $r = 1$, this also provides, by the argument of Chapter 6, Section 6, an alternative proof of their being UMP unbiased.

The space Π_Ω is the same for all three hypotheses. It is spanned by the vectors $(1,\ldots,1)$ and $(t_1,\ldots,t_n)$ and has therefore dimension $s = 2$ unless the t_i are all equal, which we shall assume not to be the case. The least-squares estimates α and β under Ω are obtained by minimizing $\Sigma(X_i - \alpha - \beta t_i)^2$. For any fixed value of β, this is achieved by the value $\alpha = \bar{X} - \beta \bar{t}$, for which the sum of squares reduces to $\Sigma[(X_i - \bar{X}) - \beta(t_i - \bar{t})]^2$. By minimizing this with respect to β one finds

$$(42) \qquad \hat{\beta} = \frac{\Sigma(X_i - \bar{X})(t_i - \bar{t})}{\Sigma(t_j - \bar{t})^2}, \qquad \hat{\alpha} = \bar{X} - \hat{\beta}\bar{t};$$

and

$$\Sigma(X_i - \hat{\alpha} - \hat{\beta}t_i)^2 = \Sigma(X_i - \bar{X})^2 - \hat{\beta}^2\Sigma(t_i - \bar{t})^2$$

is the denominator sum of squares for all three hypotheses. The numerator of the test statistic (7) for testing the two hypotheses $\alpha = 0$ and $\beta = 0$ is Y_1^2, and for testing $\alpha = \beta = 0$ is $Y_1^2 + Y_2^2$.

For the hypothesis $\alpha = 0$, the statistic Y_1 was shown in Example 3 to be equal to

$$\left(\bar{X} - \bar{t}\frac{\Sigma t_i X_i}{\Sigma t_j^2}\right)\sqrt{n\frac{\Sigma t_j^2}{\Sigma(t_j - \bar{t})^2}} = \hat{\alpha}\sqrt{n\frac{\Sigma(t_j - \bar{t})^2}{\Sigma t_j^2}}.$$

Since then

$$E(Y_1) = \alpha\sqrt{n\frac{\Sigma(t_j - \bar{t})^2}{\Sigma t_j^2}},$$

the hypothesis $\alpha = \alpha_0$ is equivalent to the hypothesis $E(Y_1) = \eta_1^0 = \alpha_0\sqrt{n\Sigma(t_j - \bar{t})^2/\Sigma t_j^2}$, for which the rejection region (17) is $(n - s)(Y_1 - $

$\eta_1^0)^2/\sum_{i=s+1}^n Y_i^2 > C_0$ and hence

$$(43) \qquad \frac{|\hat{\alpha} - \alpha_0|\sqrt{n\sum(t_j - \bar{t})^2/\sum t_j^2}}{\sqrt{\sum(X_i - \hat{\alpha} - \hat{\beta}t_i)^2/(n-2)}} > C_0.$$

For the hypothesis $\beta = 0$, Y_1 was shown to be equal to

$$\frac{\sum(X_i - \bar{X})(t_i - \bar{t})}{\sqrt{\sum(t_j - \bar{t})^2}} = \hat{\beta}\sqrt{\sum(t_j - \bar{t})^2}.$$

Since then $E(Y_1) = \beta\sqrt{\sum(t_j - \bar{t})^2}$, the hypothesis $\beta = \beta_0$ is equivalent to $E(Y_1) = \eta_1^0 = \beta_0\sqrt{\sum(t_j - \bar{t})^2}$ and the rejection region is

$$(44) \qquad \frac{|\hat{\beta} - \beta_0|\sqrt{\sum(t_j - \bar{t})^2}}{\sqrt{\sum(X_i - \hat{\alpha} - \hat{\beta}t_i)^2/(n-2)}} > C_0.$$

For testing $\alpha = \beta = 0$, it was shown in Example 3 that

$$Y_1 = \hat{\beta}\sqrt{\sum(t_j - \bar{t})^2}, \qquad Y_2 - \sqrt{n}\,\bar{X} = \sqrt{n}\,(\hat{\alpha} + \hat{\beta}\bar{t});$$

and the numerator of (7) is therefore

$$\frac{Y_1^2 + Y_2^2}{2} = \frac{n(\hat{\alpha} + \hat{\beta}\bar{t})^2 + \hat{\beta}^2\sum(t_j - \bar{t})^2}{2}.$$

The more general hypothesis $\alpha = \alpha_0$, $\beta = \beta_0$ is equivalent to $E(Y_1) = \eta_1^0$, $E(Y_2) = \eta_2^0$, where $\eta_1^0 = \beta_0\sqrt{\sum(t_j - \bar{t})^2}$, $\eta_2^0 = \sqrt{n}\,(\alpha_0 + \beta_0\bar{t})$; and the rejection region (17) can therefore be written as

$$(45) \qquad \frac{\left[n(\hat{\alpha} - \alpha_0)^2 + 2n\bar{t}(\hat{\alpha} - \alpha_0)(\hat{\beta} - \beta_0) + \sum t_i^2(\hat{\beta} - \beta_0)^2\right]/2}{\sum(X_i - \hat{\alpha} - \hat{\beta}t_i)^2/(n-2)} > C.$$

The associated confidence sets for (α, β) are obtained by reversing this inequality and replacing α_0 and β_0 by α and β. The resulting sets are ellipses centered at $(\hat{\alpha}, \hat{\beta})$.

The simple regression model (41) can be generalized in many directions; the means ξ_i may for example be polynomials in t_i of higher than the first degree (see Problem 30), or more complex functions such as trigonometric polynomials; or they may be functions of several variables, t_i, u_i, v_i. Some further extensions will now be illustrated by a number of examples.

Example 6. A variety of problems arise when there is more than one regression-line. Suppose that the variables X_{ij} are independently normally distributed with common variance and means

$$(46) \qquad \xi_{ij} = \alpha_i + \beta_i t_{ij} \qquad (j = 1, \ldots, n_i; \quad i = 1, \ldots, b).$$

The hypothesis that these regression lines have equal slopes

$$H : \beta_1 = \cdots = \beta_b$$

may occur for example when the equality of a number of growth rates is to be tested. The parameter space Π_Ω has dimension $s = 2b$ provided none of the sums $\sum_j (t_{ij} - t_{i.})^2$ is zero; the number of constraints imposed by the hypothesis is $r = b - 1$. The minimum value of $\sum\sum(X_{ij} - \xi_{ij})^2$ under Ω is obtained by minimizing $\sum_j (X_{ij} - \alpha_i - \beta_i t_{ij})^2$ for each i, so that by (42),

$$\hat{\beta}_i = \frac{\sum\limits_j (X_{ij} - X_{i.})(t_{ij} - t_{i.})}{\sum\limits_j (t_{ij} - t_{i.})^2}, \qquad \hat{\alpha}_i = X_{i.} - \hat{\beta}_i t_{i.}.$$

Under H, one must minimize $\sum\sum(X_{ij} - \alpha_i - \beta t_{ij})^2$, which for any fixed β leads to $\alpha_i = X_{i.} - \beta t_{i.}$ and reduces the sum of squares to $\sum\sum[(X_{ij} - X_{i.}) - \beta(t_{ij} - t_{i.})]^2$. Minimizing this with respect to β, one finds

$$\hat{\hat{\beta}} = \frac{\sum\sum(X_{ij} - X_{i.})(t_{ij} - t_{i.})}{\sum\sum(t_{ij} - t_{i.})^2}, \qquad \hat{\hat{\alpha}}_i = X_{i.} - \hat{\hat{\beta}} t_{i.}.$$

Since

$$X_{ij} - \hat{\xi}_{ij} = X_{ij} - \hat{\alpha}_i - \hat{\beta}_i t_{ij} = (X_{ij} - X_{i.}) - \hat{\beta}_i(t_{ij} - t_{i.})$$

and

$$\hat{\xi}_{ij} - \hat{\hat{\xi}}_{ij} = (\hat{\alpha}_i - \hat{\hat{\alpha}}_i) + t_{ij}(\hat{\beta}_i - \hat{\hat{\beta}}) = (\hat{\beta}_i - \hat{\hat{\beta}})(t_{ij} - t_{i.}),$$

the rejection region (15) is

$$(47) \qquad \frac{\sum_i \left(\hat{\beta}_i - \hat{\hat{\beta}} \right)^2 \sum_j \left(t_{ij} - t_{i.} \right)^2 / (b - 1)}{\sum\sum \left[\left(X_{ij} - X_{i.} \right) - \hat{\beta}_i \left(t_{ij} - t_{i.} \right) \right]^2 / (n - 2b)} > C,$$

where the left-hand side under H has the F-distribution with $b - 1$ and $n - 2b$ degrees of freedom.

Since

$$E(\hat{\beta}_i) = \beta_i \quad \text{and} \quad E(\hat{\hat{\beta}}) = \frac{\sum_i \beta_i \sum_j \left(t_{ij} - t_{i.} \right)^2}{\sum\sum \left(t_{ij} - t_{i.} \right)^2},$$

the noncentrality parameter of the distribution for an alternative set of β's is $\psi^2 = \sum_i (\beta_i - \bar{\beta})^2 \sum_j (t_{ij} - t_{i.})^2 / \sigma^2$, where $\bar{\beta} = E(\hat{\hat{\beta}})$. In the particular case that the n_i and the t_{ij} are independent of i, $\bar{\beta}$ reduces to $\bar{\beta} = \sum \beta_j / b$.

Example 7. The regression model (46) arises in the comparison of a number of treatments when the experimental units are treated as fixed and the unit effects u_{ij} (defined in Chapter 5, Section 11) are proportional to known constants t_{ij}. Here t_{ij} might for example be a measure of the fertility of the i, jth piece of land or the weight of the i, jth experimental animal prior to the experiment. It is then frequently possible to assume that the proportionality factor β_i does not depend on the treatment, in which case (46) reduces to

$$(48) \qquad \xi_{ij} = \alpha_i + \beta t_{ij}$$

and the hypothesis of no treatment effect becomes

$$H: \alpha_1 = \cdots = \alpha_b.$$

The space Π_Ω coincides with Π_ω of the previous example, so that $s = b + 1$ and

$$\hat{\beta} = \frac{\sum\sum \left(X_{ij} - X_{i.} \right)\left(t_{ij} - t_{i.} \right)}{\sum\sum \left(t_{ij} - t_{i.} \right)^2}, \qquad \hat{\alpha}_i = X_{i.} - \hat{\beta} t_{i.}.$$

Minimization of $\sum\sum (X_{ij} - \alpha - \beta t_{ij})^2$ gives

$$\hat{\hat{\beta}} = \frac{\sum\sum \left(X_{ij} - X_{..} \right)\left(t_{ij} - t_{..} \right)}{\sum\sum \left(t_{ij} - t_{..} \right)^2}, \qquad \hat{\hat{\alpha}} = X_{..} - \hat{\hat{\beta}} t_{..},$$

where $X_{..} = \sum\sum X_{ij}/n$, $t_{..} = \sum\sum t_{ij}/n$, $n = \sum n_i$. The sum of squares in the numerator of W^* in (15) is thus

$$\sum\sum\left(\hat{\xi}_{ij} - \hat{\hat{\xi}}_{ij}\right)^2 = \sum\sum\left[(X_{i\cdot} - X_{\cdot\cdot}) + \hat{\beta}(t_{ij} - t_{i\cdot}) - \hat{\hat{\beta}}(t_{ij} - t_{\cdot\cdot})\right]^2.$$

The hypothesis H is therefore rejected when

$$(49) \qquad \frac{\sum\sum\left[(X_{i\cdot} - X_{\cdot\cdot}) + \hat{\beta}(t_{ij} - t_{i\cdot}) - \hat{\hat{\beta}}(t_{ij} - t_{\cdot\cdot})\right]^2/(b-1)}{\sum\sum\left[(X_{ij} - X_{i\cdot}) - \hat{\beta}(t_{ij} - t_{i\cdot})\right]^2/(n-b-1)} > C,$$

where under H the left-hand side has the F-distribution with $b-1$ and $n-b-1$ degrees of freedom.

The hypothesis H can be tested without first ascertaining the values of the t_{ij}; it is then the hypothesis of no effect in a one-way classification considered in Section 3, and the test is given by (19). Actually, since the unit effects u_{ij} are assumed to be constants, which are now completely unknown, the treatments are assigned to the units either completely at random or at random within subgroups. The appropriate test is then a randomization test for which (19) is an approximation.

Example 7 illustrates the important class of situations in which an analysis of variance (in the present case concerning a one-way classification) is combined with a regression problem (in the present case linear regression on the single "concomitant variable" t). Both parts of the problem may of course be considerably more complex than was assumed here. Quite generally, in such combined problems one can test (or estimate) the treatment effects as was done above, and a similar analysis can be given for the regression coefficients. The breakdown of the variation into its various treatment and regression components is the so-called *analysis of covariance*.

8. ROBUSTNESS AGAINST NONNORMALITY

The F-test for the equality of a set of means was shown to be robust against nonnormal errors in Section 3. The proof given there extends without much change to the analysis of variance tests of Sections 5 and 6, but the situation is more complicated for regression tests.

As an example, consider the simple linear-regression situation (41). More specifically, let $U_1, U_2, \ldots$ be a sequence of independent random variables with common distribution F, which has mean 0 and finite variance σ^2, and let

$$X_i = \alpha + \beta t_i + U_i.$$

If F is normal, the distribution of $\hat{\beta}$ given by (42) is $N(0, \sigma^2/\Sigma(t_i - \bar{t})^2)$ for all sample sizes and therefore also asymptotically. However, for nonnormal

F, the exact distribution of $\hat{\beta}$ will depend on the t's in a more complicated way. An asymptotic theory requires a sequence of constants $t_1, t_2, \dots$. A sufficient condition on this sequence for asymptotic normality of $\hat{\beta}$ can be obtained from the following lemma, which we shall not prove here but which is an easy consequence of the Lindeberg form of the central limit theorem. [See for example Arnold (1981, Theorem 10.3).]

Lemma 3. *Let $Y_1, Y_2, \dots$ be independently identically distributed with mean zero and finite variance σ^2, and let $c_1, c_2, \dots$ be a sequence of constants. Then a sufficient condition for $\sum_{i=1}^{n} c_i Y_i / \sqrt{\sum c_i^2}$ to tend in law to $N(0, \sigma^2)$ is that*

$$(50) \qquad \frac{\max\limits_{i=1,\dots,n} c_i^2}{\sum\limits_{j=1}^{n} c_j^2} \to 0 \qquad as \quad n \to \infty.$$

The condition (50) prevents the c's from increasing so fast that the last term essentially dominates the sum, in which case there is no reason to expect asymptotic normality. Applying the lemma to the estimator $\hat{\beta}$ of β, we see that

$$\hat{\beta} - \beta = \frac{\sum (X_i - \alpha - \beta t_i)(t_i - \bar{t})}{\sum (t_i - \bar{t})^2},$$

and it follows that

$$\frac{(\hat{\beta} - \beta)\sqrt{\sum (t_i - \bar{t})^2}}{\sigma}$$

tends in law to $N(0, 1)$ provided

$$(51) \qquad \frac{\max (t_i - \bar{t})^2}{\sum (t_j - \bar{t})^2} \to 0.$$

Example 8. The condition (51) holds in the case of equal spacing $t_i = a + i\Delta$, but not when the t's grow exponentially, for example, when $t_i = 2^i$ (Problem 31).

In case of doubt about normality we may, instead of relying on the above result, prefer to utilize tests based on the ranks of the X's, which are exactly distribution-free and which tend to be more efficient when F is heavy-tailed. Such tests are discussed in the nonparametric books cited in Section 3; see also Aiyar, Guillier, and Albers (1979).

Lemma 3 holds not only for a single sequence $c_1, c_2, \ldots$, but also when the c's are allowed to change with n so that they form a triangular array c_{in}, $i = 1, \ldots, n$, $n = 1, 2, \ldots$, and the condition (51) generalizes analogously.

Let us next extend (51) to arbitrary linear hypotheses with $r = 1$. The model will be taken to be in the parametric form (18) where the elements a_{ij} may depend on n, but s remains fixed. Throughout, the notation will suppress the dependence on n. Without loss of generality suppose that $A'A = I$, so that the columns of A are mutually orthogonal and of length 1. Consider the hypothesis

$$H: \theta = \sum_{j=1}^{s} b_j \beta_j = 0$$

where the b's are constants with $\Sigma b_j^2 = 1$. Then

$$\hat{\theta} = \hat{\theta}_b = \Sigma b_j \hat{\beta}_j = \Sigma d_i X_i,$$

where by (18)

(52) $$d_i = \sum a_{ij} b_j.$$

By the orthogonality of A, $\Sigma d_i^2 = \Sigma b_j^2 = 1$, so that under H,

$$E(\hat{\theta}) = 0 \quad \text{and} \quad \text{Var}(\hat{\theta}) = \sigma^2.$$

Thus, H is rejected when the t-statistic

(53) $$\frac{|\hat{\theta}|}{\sqrt{\Sigma(X_i - \hat{\xi}_i)^2/(n-s)}} \geq C.$$

It was shown in Section 3 that the denominator tends to σ^2 in probability, and it follows from Lemma 3 that $\hat{\theta}$ tends in law to $N(0, \sigma^2)$ provided

(54) $$\max d_i^2 \to 0 \quad \text{as} \quad n \to \infty.$$

Under this condition, the level of the t-test is therefore robust against nonnormality.

So far, $b = (b_1, \ldots, b_s)$ has been fixed. To determine when the level of (53) is robust for all b with $\Sigma b_j^2 = 1$, it is only necessary to find the maximum value of d_i as b varies. By the Schwarz inequality

$$d_i^2 = \left(\sum_j a_{ij} b_j \right)^2 \le \sum_{j=1}^{s} a_{ij}^2,$$

with equality holding when $b_j = a_{ij}/\sqrt{\Sigma_k a_{ik}^2}$. The desired maximum of d_i^2 is therefore $\Sigma_j a_{ij}^2$, and

$$(55) \qquad \max_i \sum_{j=1}^{s} a_{ij}^2 \to 0 \qquad \text{as} \quad n \to \infty$$

is a sufficient condition for the asymptotic normality of every $\hat{\theta}_b$.

The condition (55) depends on the choice of coordinate system in the parameter space, and in particular on the assumed orthogonality of A. To obtain a condition that is coordinate-free, consider an arbitrary change of coordinates $\beta^* = B^{-1}\beta$, where B is nonsingular. Then $\xi = A\beta = AB\beta^* = A^*\beta^*$ with $A^* = AB$. To be independent of the coordinate system, the condition on A must therefore be invariant under the group G of transformations $A \to AB$ for all nonsingular B. It was seen in Example 3 of Chapter 6 that the maximal invariant under G is $P_A = A(A'A)^{-1}A'$, so that the condition must depend only on P_A. We are therefore looking for a function of P_A which reduces to $\Sigma_j a_{ij}^2$ when the columns of A are orthogonal. In this case $P_A = AA'$, and $\Sigma_j a_{ij}^2$ is the ith diagonal element of P_A. If Π_{ij} denotes the ijth element of P_A, (55) is thus equivalent to the *Huber condition*

$$(56) \qquad \max_i \Pi_{ii} \to 0 \qquad \text{as} \quad n \to \infty,$$

which is coordinate-free.

If $\Pi_{ii} \le M_n$ for all $i = 1, \ldots, n$, then also $\Pi_{ij} \le M_n$ for all i and j. This follows from the fact (see Example 3 of Chapter 6) that there exists a nonsingular E with $P = EE'$, on applying the Schwarz inequality to the ijth element of EE'. It follows that (56) is equivalent to

$$(57) \qquad \max_{i,j} \Pi_{ij} \to 0 \qquad \text{as} \quad n \to \infty.$$

Theorem 3. *Let $X_i = \xi_i + U_i$ $(i = 1, \ldots, n)$, where the U's are iid according to a distribution F with $E(U_i) = 0$, $\mathrm{Var}(U_i) = \sigma^2$, and where for* *each n the vector* $\xi = (\xi_1, \ldots, \xi_n)$ *is known to lie in an s-dimensional linear subspace* $\Pi_\Omega^{(n)}$ *of* R^n *given by* (18) *and satisfying* (56). *Then the size* $\alpha_n(F)$ *of the normal theory test given by* (7) *and* (8) *for testing* $H: \xi \in \Pi_\omega^{(n)}$, *where* $\Pi_\omega^{(n)}$ *is any subspace of* $\Pi_\Omega^{(n)}$ *of fixed dimension* $s - r$ ($0 < r \leq s$), *satisfies* $\alpha_n(F) \to \alpha$ *as* $n \to \infty$.

Proof. It was seen earlier that when (56) holds, the distribution of $\hat{\theta}_b = \Sigma b_j \hat{\beta}_j$ tends to $N(0, \sigma^2)$ for any b with $\Sigma b_j^2 = 1$. By the Cramér–Wold theorem [see for example Billingsley (1979), Theorem 29.4)], this implies that $\hat{\beta}_1, \ldots, \hat{\beta}_s$ have a joint s-variate normal limit distribution with mean 0 (under H) and covariance matrix $\sigma^2 I$. Without loss of generality suppose that $\hat{\beta}_i = \eta_i$, where the η's are given by the canonical form of Section 1. Then the columns of A are orthogonal and of length 1, and $\hat{\beta}_i = Y_i$. By standard multivariate asymptotic theory (Theorem 1.7 of *TPE*), the limit distribution of $\Sigma_{i=1}^r Y_i^2 = \Sigma_{i=1}^r \hat{\beta}_i^2$ under H is then that of a sum of squares of independent normal variables with means zero and variance σ^2, that is, $\sigma^2 \chi_r^2$, independent of F. The robustness of the level of (7) now follows from the fact, shown in Section 3, that the denominator of W^* tends to σ^2 in probability.

For evaluating Π_{ii}, it is helpful to note that $\hat{\xi}_i = \Sigma_{j=1}^n \Pi_{ij} X_j$ ($i = 1, \ldots, n$), so that Π_{ii} is simply the coefficient of X_i in $\hat{\xi}_i$, which must be calculated in any case to carry out the test.

As an example, consider once more the regression example that opened the section. From (42), it is seen that the coefficient of X_i in $\hat{\xi}_i = \hat{\alpha} + \hat{\beta} t_i$ is $\Pi_{ii} = 1/n + (t_i - \bar{t})^2 / \Sigma (t_j - \bar{t})^2$, and (56) is thus equivalent to the condition (51) found earlier for this example.

As a second example, consider a two-way layout with m observations per cell, and the additive model $\xi_{ijk} = E(X_{ijk}) = \mu + \alpha_i + \beta_j$ ($i = 1, \ldots, a$; $j = 1, \ldots, b$), $\Sigma \alpha_i = \Sigma \beta_j = 0$. Then $\hat{\xi}_{ijk} = X_{i\cdot\cdot} + X_{\cdot j\cdot} - X_{\cdots}$, and it is seen that for fixed a and b, (56) holds as $m \to \infty$.

The condition (56) guarantees asymptotic robustness for all linear hypotheses $\Pi_\omega \subset \Pi_\Omega$. If one is concerned only with a particular hypothesis, a weaker condition will suffice (Problem 40).

9.   SCHEFFÉ'S S-METHOD: A SPECIAL CASE

If $X_1, \ldots, X_r$ are independent normal with common variance σ^2 and expectations $E(X_i) = \alpha + \beta t_i$, confidence sets for (α, β) were obtained in the preceding section. A related problem is that of determining confidence bands for the whole regression line $\xi = \alpha + \beta t$, that is, functions $L'(t; X)$, $M'(t; X)$ such that

(58) $P\{ L'(t; X) \leq \alpha + \beta t \leq M'(t; X) \text{ for all } t \} = \gamma.$

The problem of obtaining simultaneous confidence intervals for a continuum of parametric functions arises also in other contexts. In the present section, a general problem of this kind will be considered for linear models. Confidence bands for an unknown distribution function were treated in Section 13 of Chapter 6.

Suppose first that $X_1, \ldots, X_r$ are independent normal with variance $\sigma^2 = 1$ and with means $E(X_i) = \xi_i$, and that simultaneous confidence intervals are required for all linear functions $\sum u_i \xi_i$. No generality is lost by dividing $\sum u_i \xi_i$ and its lower and upper bound by $\sqrt{\sum u_i^2}$, so that attention can be restricted to confidence sets

$$(59) \quad S(x): L(u; x) \le \sum u_i \xi_i \le M(u; x) \quad \text{for all} \quad u \in U,$$

where x, u denote both the vectors with coordinates x_i, u_i and the $r \times 1$ column matrices with these elements, and where U is the set of all u with $\sum u_i^2 = 1$. The sets $S(x)$ are to satisfy

$$(60) \quad P_\xi[S(X)] = \gamma \quad \text{for all} \quad \xi = (\xi_1, \ldots, \xi_r).$$

Since $u = (u_1, \ldots, u_r) \in U$ if and only if $-u = (-u_1, \ldots, -u_r) \in U$, the simultaneous inequalities (59) imply $L(-u; x) \le -\sum u_i \xi_i \le M(-u; x)$, and hence

$$-M(-u; x) \le \sum u_i \xi_i \le -L(-u; x)$$

and

$$\max\left(L(u; x), -M(-u; x)\right) \le \sum u_i \xi_i \le \min\left(M(u; x), -L(-u; x)\right).$$

Nothing is therefore lost by assuming that L and M satisfy

$$(61) \quad L(u; x) = -M(-u; x).$$

The problem of determining suitable confidence bounds $L(u; x)$ and $M(u; x)$ is invariant under the group G_1 of orthogonal transformations

$$G_1: gx = Qx, \quad \bar{g}\xi = Q\xi \quad (Q \text{ an orthogonal } r \times r \text{ matrix}).$$

Writing $\sum u_i \xi_i = u'\xi$, we have

$$g^*S(x) = \{Q\xi: L(u; x) \le u'\xi \le M(u; x) \text{ for all } u \in U\}$$

$$= \{\xi: L(u; x) \le u'(Q^{-1}\xi) \le M(u; x) \text{ for all } u \in U\}$$

$$= \{\xi: L(Q^{-1}u; x) \le u'\xi \le M(Q^{-1}u; x) \text{ for all } u \in U\},$$

where the last equality uses the fact that U is invariant under orthogonal transformations of u.

Since

$$S(gx) = \{\xi\colon L(u; Qx) \le u'\xi \le M(u; Qx) \text{ for all } u \in U\},$$

the confidence sets $S(x)$ are equivariant under G_1 if and only if

$$L(u; Qx) = L(Q^{-1}u; x), \qquad M(u, Qx) = M(Q^{-1}u; x),$$

or equivalently if

(62) $\qquad L(Qu; Qx) = L(u; x), \quad M(Qu; Qx) = M(u; x)$

$$\text{for all} \quad x, Q \text{ and } u \in U,$$

that is, if L and M are invariant under common orthogonal transformations of u and x.

A function L of u and x is invariant under these transformations if and only if it depends on u and x only through $u'x$, $x'x$, and $u'u$ [Problem 42(i)] and hence (since $u'u = 1$) if there exists h such that

(63) $\qquad L(u; x) = h(u'x, x'x).$

A second group of transformations leaving the problem invariant is the group of translations

$$G_2\colon gx = x + a, \ \bar{g}\xi = \xi + a$$

where $x + a = (x_1 + a_1, \dots, x_r + a_r)$. An argument paralleling that leading to (62) shows that $L(u; x)$ is equivariant under G_2 if and only if [Problem 42(ii)]

(64) $\qquad L(u; x + a) = L(u; x) + \sum a_i u_i \qquad$ for all x, a, and u.

The function h of (63) must therefore satisfy

$$h[u'(x + a), (x + a)'(x + a)] = h(u'x, x'x) + a'u$$

$$\text{for all} \quad a, x \text{ and } u \in U,$$

and hence, putting $x = 0$,

$$h(u'a, a'a) = a'u + h(0,0).$$

A necessary condition (which clearly is also sufficient) for $S(x)$ to be equivariant under both G_1 and G_2 is therefore the existence of constants c and d such that

$$S(x) = \left\{ \xi : \sum u_i x_i - c \leq \sum u_i \xi_i \leq \sum u_i x_i + d \quad \text{for all } u \in U \right\}.$$

From (61) it follows that $c = d$, so that the only equivariant families $S(x)$ are given by

(65) $$S(x) = \left\{ \xi : \left| \sum u_i(x_i - \xi_i) \right| \leq c \quad \text{for all } u \in U \right\}.$$

The constant c is determined by (60), which now reduces to

(66) $$P_0 \left\{ \left| \sum u_i X_i \right| \leq c \quad \text{for all } u \in U \right\} = \gamma.$$

By the Schwarz inequality $(\sum u_i X_i)^2 \leq \sum X_i^2$, since $\sum u_i^2 = 1$, and hence

(67) $$\left| \sum u_i X_i \right| \leq c \quad \text{for all } u \in U \quad \text{if and only if} \quad \sum X_i^2 \leq c^2.$$

The constant c in (65) is therefore given by

(68) $$P\left(\chi_r^2 \leq c^2 \right) = \gamma.$$

In (65), it is of course possible to drop the restriction $u \in U$ by writing (65) in the equivalent form

(69) $$S(x) = \left\{ \xi : \left| \sum u_i(x_i - \xi_i) \right| \leq c \sqrt{\sum u_i^2} \quad \text{for all } u \right\}.$$

So far attention has been restricted to the confidence bands (59). However, confidence sets do not have to be intervals, and it may be of interest to consider more general simultaneous confidence sets

(70) $$S(x): \sum u_i \xi_i \in A(u, x) \text{ for all } u \in U.$$

For these sets, the equivariance conditions (62) and (64) become respectively (Problem 43)

(71) $$A(Qu, Qx) = A(u, x) \quad \text{for all} \quad x, Q \text{ and } u \in U$$

and

(72) $$A(u, x + a) = A(u, x) + u'a \quad \text{for all} \quad u, x, \text{ and } a.$$

The first of these is equivalent to the condition that the set $A(u, x)$ depends on $u \in U$ and x only through $u'x$ and $x'x$. On the other hand putting $x = 0$ in (72) gives

$$A(u, a) = A(u, 0) + u'a.$$

It follows from (71) that $A(u, 0)$ is a fixed set A_1 independent of u, so that

(73) $$A(u, x) = A_1 + u'x.$$

The most general equivariant sets (under G_1 and G_2) are therefore of the form

(74) $$\sum u_i(x_i - \xi_i) \in A \qquad \text{for all} \quad u \in U,$$

where $A = -A_1$.

We shall now suppose that $r > 1$ and then show that among all A which define confidence sets (74) with confidence coefficient $\geq \gamma$, the sets (65) are smallest[†] in the very strong sense that if $A_0 = [-c_0, c_0]$ denotes the set (65) with confidence coefficient γ, then A_0 is a subset of A.

To see this, note that if $Y_i = X_i - \xi_i$, the sets A are those satisfying

(75) $$P\left(\sum u_i Y_i \in A \quad \text{for all } u \in U\right) \geq \gamma.$$

Now the set of values taken on by $\sum u_i y_i$ for a fixed $y = (y_1, \ldots, y_r)$ as u ranges over U is the interval (Problem 43)

$$I(y) = \left[-\sqrt{\sum y_i^2}, +\sqrt{\sum y_i^2}\right].$$

Let c^* be the largest value of c for which the interval $[-c, c]$ is contained in A. Then the probability (75) is equal to

$$P\{I(Y) \subset A\} = P\{I(Y) \subset [-c^*, c^*]\}.$$

Since $P\{I(Y) \subset A\} \geq \gamma$, it follows that $c^* \geq c_0$, and this completes the proof.

It is of interest to compare the simultaneous confidence intervals (65) for all $\sum u_i \xi_i$, $u \in U$, with the joint confidence spheres for $(\xi_1, \ldots, \xi_r)$ given by (41) of Chapter 6. These two sets of confidence statements are equivalent in the following sense.

[†]A more general definition of smallness is due to Wijsman (1979). It has been pointed out to me by Professor Wijsman that his concept is equivalent to that of tautness defined by Wynn and Bloomfield (1971).

Theorem 4. *The parameter vector* $(\xi_1, \ldots, \xi_r)$ *satisfies* $\Sigma(X_i - \xi_i)^2 \le c^2$ *if and only if it satisfies* (65).

Proof. The result follows immediately from (67) with X_i replaced by $X_i - \xi_i$.

Another comparison of interest is that of the simultaneous confidence intervals (69) for all u with the corresponding interval

$$(76) \qquad S'(x) = \left\{ \xi: \left| \Sigma u_i(x_i - \xi_i) \right| \le c' \sqrt{\Sigma u_i^2} \right\}$$

for a single given u. Since $\Sigma u_i(X_i - \xi_i)/\sqrt{\Sigma u_i^2}$ has a standard normal distribution, the constant c' is determined by $P(\chi_1^2 \le c'^2) = \gamma$ instead of by (68). If $r > 1$, the constant $c^2 = c_r^2$ is clearly larger than $c'^2 = c_1^2$. The lengthening of the confidence intervals by the factor c_r/c_1 in going from (76) to (69) is the price one must pay for asserting confidence γ for all $\Sigma u_i \xi_i$ instead of a single one.

In (76), it is assumed that the vector u defines the linear combination of interest and is given before any observations are available. However, it often happens that an interesting linear combination $\Sigma \hat{u}_i \xi_i$ to be estimated is suggested by the data. The intervals

$$(77) \qquad \left| \Sigma \hat{u}_i(x_i - \xi_i) \right| \le c \sqrt{\Sigma \hat{u}_i^2}$$

with c given by (68) then provide confidence limits for $\Sigma \hat{u}_i \xi_i$ at confidence level γ, since they are included in the set of intervals (69). [The notation $\hat{u}_i$ in (77) indicates that the u's were suggested by the data rather than fixed in advance.]

Example 9. Two groups. *Suppose the data exhibit a natural split into a lower and upper group, say* $\xi_{i_1}, \ldots, \xi_{i_k}$ *and* $\xi_{j_1}, \ldots, \xi_{j_{r-k}}$, *with averages* $\bar{\xi}_-$ *and* $\bar{\xi}_+$, *and that confidence limits are required for* $\bar{\xi}_+ - \bar{\xi}_-$. *Letting* $\bar{X}_- = (X_{i_1} + \cdots + X_{i_k})/k$ *and* $\bar{X}_+ = (X_{j_1} + \cdots + X_{j_{r-k}})/(r-k)$ *denote the associated averages of the X's, we see that*

$$(78) \qquad \bar{X}_+ - \bar{X}_- - c\sqrt{\frac{1}{k} + \frac{1}{r-k}} \le \bar{\xi}_+ - \bar{\xi}_- \le \bar{X}_+ - \bar{X}_- + c\sqrt{\frac{1}{k} + \frac{1}{r-k}}$$

with c given by (68) *provide the desired limits. Similarly*

$$(79) \qquad \bar{X}_- - \frac{c}{\sqrt{k}} \le \bar{\xi}_- \le \bar{X}_- + \frac{c}{\sqrt{k}}, \qquad \bar{X}_+ - \frac{c}{\sqrt{r-k}} \le \bar{\xi}_+ \le \bar{X}_+ + \frac{c}{\sqrt{r-k}}$$

provide simultaneous confidence intervals for the two group means separately, with c
again given by (68). [For a discussion of related examples and issues see Peritz (1965).]

Instead of estimating a data-based function $\Sigma \hat{u}_i \xi_i$, one may be interested
in testing it. At level $\alpha = 1 - \gamma$, the hypothesis $\Sigma \hat{u}_i \xi_i = 0$ is rejected when
the confidence intervals (77) do not cover the origin, i.e. when

$$\left| \sum \hat{u}_i x_i \right| \geq c \sqrt{\sum \hat{u}_i^2}.$$

Equivariance with respect to the group G_1 of orthogonal transformations
assumed at the beginning of this section is appropriate only when all linear
combinations $\Sigma u_i \xi_i$ with $u \in U$ are of equal importance. Suppose instead
that interest focuses on the individual means, so that simultaneous confidence intervals are required for $\xi_1, \ldots, \xi_r$. This problem remains invariant
under the translation group G_2. However, it is no longer invariant under G_1,
but only under the much smaller subgroup G_0 generated by the $n!$ permutations and the 2^n changes of sign of the X's. The only simultaneous intervals
that are equivariant under G_0 and G_2 are given by [Problem 44(i)]

$$(80) \qquad S(x) = \{ \xi : x_i - \Delta \leq \xi_i \leq x_i + \Delta \text{ for all } i \},$$

where Δ is determined by

$$(81) \qquad P[S(X)] = P(\max |Y_i| \leq \Delta) = \gamma$$

with $Y_1, \ldots, Y_r$ being independent $N(0, 1)$.

These *maximum-modulus* intervals for the ξ's can be extended to all
linear combinations $\Sigma u_i \xi_i$ of the ξ's by noting that the right side of (80) is
equal to the set [Problem 45(ii)]

$$(82) \qquad \left\{ \xi : \left| \sum u_i (X_i - \xi_i) \right| \leq \Delta \sum |u_i| \text{ for all } u \right\},$$

which therefore also has probability γ, but which is not equivariant under
G_1. A comparison of the intervals (82) with the Scheffé intervals (69) shows
[Problem 44(iii)] that the intervals (82) are shorter when $\Sigma u_j \xi_j = \xi_i$ (i.e.
when $u_j = 1$ for $j = i$, and $u_j = 0$ otherwise), but that they are longer for
example when $u_1 = \cdots = u_r$.

10. SCHEFFÉ'S S-METHOD FOR GENERAL LINEAR MODELS

The results obtained in the preceding section for the simultaneous estimation of all linear functions $\Sigma u_i \xi_i$ when the common variance of the variables
X_i is known easily extend to the general linear model of Section 1. In the

canonical form (2), the observations are n independent normal random variables with common unknown variance σ^2 and with means $E(Y_i) = \eta_i$ for $i = 1, \ldots, r, r + 1, \ldots, s$ and $E(Y_i) = 0$ for $i = s + 1, \ldots, n$. Simultaneous confidence intervals are required for all linear functions $\sum_{i=1}^{r} u_i \eta_i$ with $u \in U$, where U is the set of all $u = (u_1, \ldots, u_r)$ with $\sum_{i=1}^{r} u_i^2 = 1$. Invariance under the translation group $Y_i' = Y_i + a_i$, $i = r + 1, \ldots, s$, leaves $Y_1, \ldots, Y_r; Y_{s+1}, \ldots, Y_n$ as maximal invariants, and sufficiency justifies restricting attention to $Y = (Y_1, \ldots, Y_r)$ and $S^2 = \sum_{j=s+1}^{n} Y_j^2$. The confidence intervals corresponding to (59) are therefore of the form

$$(83) \qquad L(u; y, S) \leq \sum_{i=1}^{r} u_i \eta_i \leq M(u; y, S) \qquad \text{for all} \quad u \in U,$$

and in analogy to (61) may be assumed to satisfy

$$(84) \qquad\qquad L(u; y, S) = -M(-u; y, S).$$

By the argument leading to (63), it is seen in the present case that equivariance of $L(u; y, S)$ under G_1 requires that

$$L(u; y, S) = h(u'y, y'y, S),$$

and equivariance under G_2 requires that L be of the form

$$L(u; y, S) = \sum_{i=1}^{r} u_i y_i - c(S).$$

Since σ^2 is unknown, the problem is now also invariant under the group of scale changes

$$G_3 : y_i' = b y_i \; (i = 1, \ldots, r), \; S' = bS \; (b > 0).$$

Equivariance of the confidence intervals under G_3 leads to the condition [Problem 45(i)]

$$L(u; by, bS) = bL(u; y, S) \qquad \text{for all} \quad b > 0,$$

and hence to

$$b\sum u_i y_i - c(bS) = b\left[\sum u_i y_i - c(S)\right],$$

or $c(bS) = bc(S)$. Putting $S = 1$ shows that $c(S)$ is proportional to S.

Thus

$$L(u; y, S) = \sum u_i y_i - cS, \qquad M(u; y, S) = \sum u_i y_i + dS,$$

and by (84), $c = d$, so that the equivariant simultaneous intervals are given by

$$(85) \qquad \sum u_i y_i - cS \le \sum u_i \eta_i \le \sum u_i y_i + cS \qquad \text{for all} \quad u \in U.$$

Since (85) is equivalent to

$$\frac{\sum (y_i - \eta_i)^2}{S^2} \le c^2,$$

the constant c is determined from the F-distribution by

$$(86) \qquad P_0\left[\frac{\sum Y_i^2/r}{S^2/(n-s)} \le \frac{n-s}{r}c^2\right] = P_0\left(F_{r, n-s} \le \frac{n-s}{r}c^2\right) = \gamma.$$

As in (69), the restriction $u \in U$ can be dropped; this only requires replacing c in (85) and (86) by $c\sqrt{\sum u_i^2} = c\sqrt{\operatorname{Var}\sum u_i Y_i/\sigma^2}$.

As in the case of known variance, instead of restricting attention to the confidence bands (85), one may wish to permit more general simultaneous confidence sets

$$(87) \qquad \sum u_i \eta_i \in A(u; y, S).$$

The most general equivariant confidence sets are then of the form [Problem 45(ii)]

$$(88) \qquad \frac{\sum u_i(y_i - \eta_i)}{S} \in A \qquad \text{for all} \quad u \in U,$$

and for a given confidence coefficient, the set A is minimized by $A_0 = [-c, c]$, so that (88) reduces to (85).

For applications, it is convenient to express the intervals (85) in terms of the original variables X_i and ξ_i. Suppose as in Section 1 that $X_1, \ldots, X_n$ are independently distributed as $N(\xi_i, \sigma^2)$, where $\xi = (\xi_1, \ldots, \xi_n)$ is assumed to lie in a given s-dimensional linear subspace Π_Ω ($s < n$). Let V be an r-dimensional subspace of Π_Ω ($r < s$), let $\hat\xi_i$ be the least squares estimates of the ξ's under Π_Ω, and let $S^2 = \Sigma(X_i - \hat\xi_i)^2$. Then the inequalities

(89)

$$\sum v_i \hat\xi_i - cS\sqrt{\frac{\mathrm{Var}\left(\sum v_i \hat\xi_i\right)}{\sigma^2}} \le \sum v_i \xi_i \le \sum v_i \hat\xi_i + cS\sqrt{\frac{\mathrm{Var}\left(\sum v_i \hat\xi_i\right)}{\sigma^2}}$$

$$\text{for all} \quad v \in V,$$

with c given by (86), provide simultaneous confidence intervals for $\sum v_i \xi_i$ for all $v \in V$ with confidence coefficient γ.

This result is an immediate consequence of (85) and (86) together with the following three facts, which will be proved below:

(i)   If $\Sigma_{i=1}^s u_i \eta_i = \Sigma_{j=1}^n v_j \xi_j$, then $\Sigma_{i=1}^s u_i Y_i = \Sigma_{j=1}^n v_j \hat\xi_j$;

(ii) $\Sigma_{i=s+1}^n Y_i^2 = \Sigma_{j=1}^n (X_j - \hat\xi_j)^2$.

To state (iii), note that the η's are obtained as linear functions of the ξ's through the relationship

(90) $(\eta_1, \ldots, \eta_r, \eta_{r+1}, \ldots, \eta_s, 0, \ldots, 0)' = C(\xi_1, \ldots, \xi_n)'$

where C is defined by (1) and the prime indicates a transpose. This is seen by taking the expectation of both sides of (1). For each vector $u = (u_1, \ldots, u_r)$, (90) expresses $\Sigma u_i \eta_i$ as a linear function $\Sigma v_j^{(u)} \xi_j$ of the ξ's.

(iii) As u ranges over r-space, $v^{(u)} = (v_1^{(u)}, \ldots, v_n^{(u)})$ ranges over V.

Proof of (i).   Recall from Section 2 that

$$\sum_{j=1}^n (X_j - \xi_j)^2 = \sum_{i=1}^s (Y_i - \eta_i)^2 + \sum_{j=s+1}^n Y_j^2.$$

Since the right side is minimized by $\eta_i = Y_i$ and the left side by $\xi_j = \hat\xi_j$, this shows that

$$(Y_1 \; \cdots \; Y_s\, 0 \cdots 0)' = C(\hat\xi_1 \; \cdots \; \hat\xi_n)',$$

and the result now follows from comparison with (90).

Proof of (ii).   This is just equation (13).

Proof of (*iii*). Since $\eta_i = \sum_{j=1}^n c_{ij}\xi_j$, we have $\sum u_i \eta_i = \sum v_j^{(u)}\xi_j$ with $v_j^{(u)} = \sum_{i=1}^r u_i c_{ij}$. Thus the vectors $v^{(u)} = (v_1^{(u)}, \ldots, v_n^{(u)})$ are linear combinations, with weights $u_1, \ldots, u_r$, of the first r row vectors of C. Since the space spanned by these row vectors is V, the result follows.

The set of linear functions $\sum v_i \xi_i$, $v \in V$, for which the interval (89) does not cover the origin—that is, for which v satisfies

$$(91) \qquad \left|\sum v_i \hat{\xi}_i\right| > cS\sqrt{\frac{\mathrm{Var}\left(\sum v_i \hat{\xi}_i\right)}{\sigma^2}}$$

—is declared significantly different from 0 by the intervals (89). Thus (91) is a rejection region at level $\alpha = 1 - \gamma$ of the hypothesis $H: \sum v_i \xi_i = 0$ for all $v \in V$ in the sense that H is rejected if and only if at least one $v \in V$ satisfies (91). If Π_ω denotes the $(s - r)$-dimensional space of vectors $v \in \Pi_\Omega$ which are orthogonal to V, then H states that $\xi \in \Pi_\omega$, and the rejection region (91) is in fact equivalent to the F-test of $H: \xi \in \Pi_\omega$ of Section 1. In canonical form, this was seen in the sentence following (85).

To implement the intervals (89) in specific situations in which the corresponding intervals for a single given function $\sum v_i \xi_i$ are known, it is only necessary to designate the space V and to obtain its dimension r, the constant c then being determined by (86).

Example 10. All contrasts. Let X_{ij} ($j = 1, \ldots, n_i$; $i = 1, \ldots, s$) be independently distributed as $N(\xi_i, \sigma^2)$, and suppose V is the space of all vectors $v = (v_1, \ldots, v_n)$ satisfying

$$(92) \qquad \sum v_i = 0.$$

Any function $\sum v_i \xi_i$ with $v \in V$ is called a *contrast* among the ξ_i. The set of contrasts includes in particular the differences $\bar{\xi}_+ - \bar{\xi}_-$ discussed in Example 9. The space Π_Ω is the set of all vectors $(\xi_1, \ldots, \xi_1; \xi_2, \ldots, \xi_2; \xi_s, \ldots, \xi_s)$ and has dimension s, while V is the subspace of vectors Π_Ω that are orthogonal to $(1, \ldots, 1)$ and hence has dimension $r = s - 1$. It was seen in Section 3 that $\hat{\xi}_i = X_{i\cdot}$, and if the vectors of V are denoted by

$$\left(\frac{w_1}{n_1}, \ldots, \frac{w_1}{n_1}; \frac{w_2}{n_2}, \ldots, \frac{w_2}{n_2}; \frac{w_s}{n_s}, \ldots, \frac{w_s}{n_s}\right),$$

the simultaneous confidence intervals (89) become (Problem 47)

$$(93) \quad \sum w_i X_{i\cdot} - cS\sqrt{\frac{\sum w_i^2}{n_i}} \leq \sum w_i \xi_i \leq \sum w_i X_{i\cdot} + cS\sqrt{\frac{\sum w_i^2}{n_i}}$$

$$\text{for all } (w_1, \ldots, w_s) \text{ satisfying } \sum w_i = 0,$$

with $S^2 = \sum\sum(X_{ij} - X_{i\cdot})^2$.

In the present case the space Π_ω is the set of vectors with all coordinates equal, so that the associated hypothesis is $H : \xi_1 = \cdots = \xi_s$. The rejection region (91) is thus equivalent to that given by (19).

Instead of testing the overall homogeneity hypothesis H, we may be interested in testing one or more subhypotheses suggested by the data. In the situation corresponding to that of Example 9 (but with replications), for instance, interest may focus on the hypotheses $H_1 : \xi_{i_1} = \cdots = \xi_{i_k}$ and $H_2 : \xi_{j_1} = \cdots = \xi_{j_{s-k}}$. A level α simultaneous test of H_1 and H_2 is given by the rejection region

$$\frac{\Sigma^{(1)} n_i \left(X_{i\cdot} - X^{(1)}_{\cdot\cdot} \right)^2 / (k - 1)}{S^2 / (n - s)} > C, \qquad \frac{\Sigma^{(2)} n_i \left(X_{i\cdot} - X^{(2)}_{\cdot\cdot} \right)^2 / (s - k - 1)}{S^2 / (n - s)} > C,$$

where $\Sigma^{(1)}, \Sigma^{(2)}, X^{(1)}_{\cdot\cdot}, X^{(2)}_{\cdot\cdot}$ indicate that the summation or averaging extends over the sets $(i_1, \ldots, i_k)$ and $(j_1, \ldots, j_{s-k})$ respectively, $S^2 = \Sigma\Sigma(X_{ij} - X_{i\cdot})^2$, $\alpha = 1 - \gamma$, and the constant C is given by (86) with $r = s$ and is therefore the same as in (19), rather than being determined by the $F_{k-1, n-s}$ and $F_{s-k-1, n-s}$ distributions. The reason for this larger critical value is, of course, the fact the H_1 and H_2 were suggested by the data. The present procedure is an example of Gabriel's simultaneous test procedure mentioned in Section 4.

Example 11. Two-way layout. As a second example, consider first the additive model in the two-way classification of Section 5 or 6, and then the more general interaction model of Section 6.

Suppose X_{ij} are independent $N(\xi_{ij}, \sigma^2)$ $(i = 1, \ldots, a; \; j = 1, \ldots, b)$, with ξ_{ij} given by (32), and let V be the space of all linear functions $\Sigma w_i \alpha_i = \Sigma w_i (\xi_{i\cdot} - \xi..)$. As was seen in Section 5, $s = a + b - 1$. To determine r, note that V can also be represented as $\Sigma_{i-1} w_i \xi_i$ with $\Sigma w_i = 0$ [Problem 46(i)], which shows that $r = a - 1$. The least-squares estimators $\hat{\xi}_i$ were found in Section 5 to be $\hat{\xi}_{ij} = X_{i\cdot} + X_{\cdot j} - X..$, so that $\hat{\xi}_{i\cdot} = X_{i\cdot}$ and $S^2 = \Sigma\Sigma(X_{ij} - X_{i\cdot} - X_{\cdot j} + X..)^2$. The simultaneous confidence intervals (89) therefore can be written as

$$\Sigma w_i X_{i\cdot} - cS\sqrt{\frac{\Sigma w_i^2}{b}} \le \Sigma w_i \xi_{i\cdot} \le \Sigma w_i X_{i\cdot} + cS\sqrt{\frac{\Sigma w_i^2}{b}}$$

$$\text{for all } w \text{ with } \sum_{i=1}^{a} w_i = 0.$$

If there are m observations in each cell, and the model is additive as before, the only changes required are to replace $X_{i\cdot}$ by $X_{i\cdot\cdot}$, S^2 by $\Sigma\Sigma\Sigma(X_{ijk} - X_{i\cdot\cdot} - X_{\cdot j\cdot} + X...)^2$, and the expression under the square root by $\Sigma w_i^2 / bm$.

Let us now drop the assumption of additivity and consider the general linear model $\xi_{ijk} = \mu + \alpha_i + \beta_j + \gamma_{ij}$, with μ and the α's, β's, and γ's defined as in Section 6. The dimension s of Π_Ω is then ab, and the least-squares estimators of the parameters were seen in Section 6 to be

$$\hat{\mu} = X..., \qquad \hat{\alpha}_i = X_{i\cdot\cdot} - X..., \qquad \hat{\beta}_j = X_{\cdot j\cdot} - X...,$$

$$\hat{\gamma}_{ij} = X_{ij\cdot} - X_{i\cdot\cdot} - X_{\cdot j\cdot} + X... .$$

The simultaneous intervals for all $\Sigma w_i \alpha_i$, or for all $\Sigma w_i \xi_i$. with $\Sigma w_i = 0$, are therefore unchanged except for the replacement of $S^2 = \Sigma(X_{ijk} - X_{i\cdot\cdot} - X_{\cdot j\cdot} + X_{\cdots})^2$ by $S^2 = \Sigma(X_{ijk} - X_{ij\cdot})^2$ and of $n - s = n - a - b + 1$ by $n - s = n - ab = (m - 1)ab$ in (86).

Analogously, one can obtain simultaneous confidence intervals for the totality of linear functions $\Sigma w_{ij} \gamma_{ij}$, or equivalently the set of functions $\Sigma w_{ij} \xi_{ij}$. for the totality of w's satisfying $\Sigma_i w_{ij} = \Sigma_j w_{ij} = 0$ [Problem 46(ii), (iii)].

Example 12. Regression line. As a last example consider the problem of obtaining confidence bands for a regression line, mentioned at the beginning of the section. The problem was treated for a single value t_0 in Chapter 5, Section 8 (with a different notation) and in Section 7 of the present chapter. The simultaneous confidence intervals in the present case become

$$(94) \quad \hat{\alpha} + \hat{\beta}t - cS\left[\frac{1}{n} + \frac{(t - \bar{t})^2}{\Sigma(t_i - \bar{t})^2}\right]^{1/2} \leq \alpha + \beta t$$

$$\leq \hat{\alpha} + \hat{\beta}t + cS\left[\frac{1}{n} + \frac{(t - \bar{t})^2}{\Sigma(t_i - \bar{t})^2}\right]^{1/2},$$

where $\hat{\alpha}$ and $\hat{\beta}$ are given by (33),

$$S^2 = \Sigma(X_i - \hat{\alpha} - \hat{\beta}t_i)^2 = \Sigma(X_i - \bar{X})^2 - \hat{\beta}^2\Sigma(t_i - \bar{t})^2$$

and c is determined by (86) with $r = s = 2$. This is the Working–Hotelling confidence band for a regression line.

At the beginning of the section, the Scheffé intervals were derived as the only confidence bands that are equivariant under the indicated groups. If the requirement of equivariance (particular under orthogonal transformations) is dropped, other bounds exist which are narrower for certain sets of vectors u at the cost of being wider for others [Problems 45(iii) and 68]. A general method that gives special emphasis to a given subset is described by Richmond (1982). Some optimality results not requiring equivariance but instead permitting bands which are narrower for some values of t at the expense of being wider for others are provided, among others, by Bohrer (1973), Cima and Hochberg (1976), Richmond (1982), Naiman (1984a, b), and Piegorsch (1985a, b). If bounds are required only for a subset, it may be possible that intervals exist at the prescribed confidence level, which are uniformly narrower than the Scheffé intervals. This is the case for example for the intervals (94) when t is restricted to a given finite interval. For a discussion of this and related problems, and references to the literature, see for example Wynn and Bloomfield (1971) and Wynn (1984).

11. RANDOM-EFFECTS MODEL: ONE-WAY
CLASSIFICATION

In the factorial experiments discussed in Sections 3, 5, and 6, the factor levels were considered fixed, and the associated effects (the μ's in Section 3, the α's, β's and γ's in Sections 5 and 6) to be unknown constants. However, in many applications, these levels and their effects instead are (unobservable) random variables. If all the effects are constant or all random, one speaks of *fixed-effects model* (*model I*) or *random-effects model* (*model II*) respectively, and the term *mixed model* refers to situations in which both types occur. Of course, only the model I case constitutes a linear hypothesis according to the definition given at the beginning of the chapter. In the present section we shall treat as model II the case of a single factor (one-way classification), which was analyzed under the model I assumption in Section 3.

As an illustration of this problem, consider a material such as steel, which is manufactured or processed in batches. Suppose that a sample of size n is taken from each of s batches and that the resulting measurements X_{ij} ($j = 1, \ldots, n$; $i = 1, \ldots, s$) are independently normally distributed with variance σ^2 and mean ξ_i. If the factor corresponding to i were constant, with the same effect α_i in each replication of the experiment, we would have

$$\xi_i = \mu + \alpha_i \qquad \left(\sum \alpha_i = 0\right)$$

and

$$X_{ij} = \mu + \alpha_i + U_{ij}$$

where the U_{ij} are independently distributed as $N(0, \sigma^2)$. The hypothesis of no effect is $\xi_1 = \cdots = \xi_s$ or equivalently $\alpha_1 = \cdots = \alpha_s = 0$. However, the effect is associated with the batches, of which a new set will be involved in each replication of the experiment; and the effect therefore does not remain constant. Instead, we shall suppose that the batch effects constitute a sample from a normal distribution, and to indicate their random nature we shall write A_i for α_i, so that

$$(95) \qquad\qquad X_{ij} = \mu + A_i + U_{ij}.$$

The assumption of additivity (lack of interaction) of batch and unit effect, in the present model, implies that the A's and U's are independent. If the expectation of A_i is absorbed into μ, it follows that the A's and U's are independently normally distributed with zero means and variances σ_A^2 and σ^2 respectively. The X's of course are no longer independent.

The hypothesis of no batch effect, that the A's are zero and hence constant, takes the form

$$H : \sigma_A^2 = 0.$$

This is not realistic in the present situation, but is the limiting case of the hypothesis

$$H(\Delta_0) : \frac{\sigma_A^2}{\sigma^2} \leq \Delta_0$$

that the batch effect is small relative to the variation of the material within a batch. These two hypotheses correspond respectively to the model I hypotheses $\Sigma \alpha_i^2 = 0$ and $\Sigma \alpha_i^2 / \sigma^2 \leq \Delta_0$.

To obtain a test of $H(\Delta_0)$ it is convenient to begin with the same transformation of variables that reduced the corresponding model I problem to canonical form. Each set $(X_{i1}, \ldots, X_{in})$ is subjected to an orthogonal transformation $Y_{ij} = \Sigma_{k=1}^{n} c_{jk} X_{ik}$ such that $Y_{i1} = \sqrt{n}\, X_{i.}$. Since $c_{1k} = 1/\sqrt{n}$ for $k = 1, \ldots, n$ (see Example 3), it follows from the assumption of orthogonality that $\Sigma_{k=1}^{n} c_{jk} = 0$ for $j = 2, \ldots, n$ and hence that $Y_{ij} = \Sigma_{k=1}^{n} c_{jk} U_{ik}$ for $j > 1$. The Y_{ij} with $j > 1$ are therefore independently normally distributed with zero mean and variance σ^2. They are also independent of $U_{i.}$ since $(\sqrt{n}\, U_{i.}\, Y_{i2} \ldots Y_{in})' = C(U_{i1}\, U_{i2} \ldots U_{in})'$ (a prime indicates the transpose of a matrix). On the other hand, the variables $Y_{i1} = \sqrt{n}\, X_{i.} = \sqrt{n}\,(\mu + A_i + U_{i.})$ are also independently normally distributed but with mean $\sqrt{n}\,\mu$ and variance $\sigma^2 + n\sigma_A^2$. If an additional orthogonal transformation is made from $(Y_{11}, \ldots, Y_{s1})$ to $(Z_{11}, \ldots, Z_{s1})$ such that $Z_{11} = \sqrt{s}\, Y_{.1}$, the Z's are independently normally distributed with common variance $\sigma^2 + n\sigma_A^2$ and means $E(Z_{11}) = \sqrt{sn}\,\mu$ and $E(Z_{i1}) = 0$ for $i > 1$. Putting $Z_{ij} = Y_{ij}$ for $j > 1$ for the sake of conformity, the joint density of the Z's is then

$$(96) \quad (2\pi)^{-ns/2} \sigma^{-(n-1)s} \left(\sigma^2 + n\sigma_A^2\right)^{-s/2}$$

$$\times \exp\left[-\frac{1}{2(\sigma^2 + n\sigma_A^2)}\left((z_{11} - \sqrt{sn}\,\mu)^2 + \sum_{i=2}^{s} z_{i1}^2 \right) - \frac{1}{2\sigma^2} \sum_{i=1}^{s} \sum_{j=2}^{n} z_{ij}^2 \right].$$

The problem of testing $H(\Delta_0)$ is invariant under addition of an arbitrary constant to Z_{11}, which leaves the remaining Z's as a maximal set of invariants. These constitute samples of size $s(n - 1)$ and $s - 1$ from two normal distributions with means zero and variances σ^2 and $\tau^2 = \sigma^2 + n\sigma_A^2$.

The hypothesis $H(\Delta_0)$ is equivalent to $\tau^2/\sigma^2 \leq 1 + \Delta_0 n$, and the problem reduces to that of comparing two normal variances, which was considered in Example 6 of Chapter 6 without the restriction to zero means. The UMP invariant test, under multiplication of all Z_{ij} by a common positive constant, has the rejection region

$$(97) \qquad W^* = \frac{1}{1 + \Delta_0 n} \cdot \frac{S_A^2/(s-1)}{S^2/(n-1)s} > C,$$

where

$$S_A^2 = \sum_{i=2}^{s} Z_{i1}^2 \quad \text{and} \quad S^2 = \sum_{i=1}^{s} \sum_{j=2}^{n} Z_{ij}^2 = \sum_{i=1}^{s} \sum_{j=2}^{n} Y_{ij}^2.$$

The constant C is determined by

$$\int_C^\infty F_{s-1,(n-1)s}(y)\, dy = \alpha.$$

Since

$$\sum_{j=1}^{n} Y_{ij}^2 - Y_{i1}^2 = \sum_{j=1}^{n} U_{ij}^2 - nU_{i\cdot}^2$$

and

$$\sum_{i=1}^{s} Z_{i1}^2 - Z_{11}^2 = \sum_{i=1}^{s} Y_{i1}^2 - sY_{\cdot1}^2,$$

the numerator and denominator sums of squares of W^*, expressed in terms of the X's, become

$$S_A^2 = n \sum_{i=1}^{s} (X_{i\cdot} - X_{\cdot\cdot})^2 \quad \text{and} \quad S^2 = \sum_{i=1}^{s} \sum_{j=1}^{n} (X_{ij} - X_{i\cdot})^2.$$

In the particular case $\Delta_0 = 0$, the test (97) is equivalent to the corresponding model I test (19), but they are of course solutions of different problems, and also have different power functions. Instead of being distributed according to a noncentral χ^2-distribution as in model I, the numerator sum of squares of W^* is proportional to a central χ^2-variable even when the hypothesis is false, and the power of the test (97) against an alternative value of Δ is obtained from the F-distribution through

$$\beta(\Delta) = P_\Delta\{W^* > C\} = \int_{\frac{1+\Delta_0 n}{1+\Delta n}C}^{\infty} F_{s-1,(n-1)s}(y) \, dy.$$

The family of tests (97) for varying Δ_0 is equivalent to the confidence statements

(98)
$$\underline{\Delta} = \frac{1}{n}\left[\frac{S_A^2/(s-1)}{CS^2/(n-1)s} - 1\right] \leq \Delta.$$

The corresponding upper confidence bounds for Δ are obtained from the tests of the hypotheses $\Delta \geq \Delta_0$. These have the acceptance regions $W^* \geq C'$, where W^* is given by (97) and C' is determined by

$$\int_{C'}^{\infty} F_{s-1,(n-1)s} = 1 - \alpha,$$

and the resulting confidence bounds are

(99)
$$\Delta \leq \frac{1}{n}\left[\frac{S_A^2/(s-1)}{C'S^2/(n-1)s} - 1\right] = \overline{\Delta}.$$

Both the confidence sets (98) and (99) are equivariant with respect to the group of transformations generated by those considered for the testing problems, and hence are uniformly most accurate equivariant.

When $\underline{\Delta}$ is negative, the confidence set $(\underline{\Delta}, \infty)$ contains all possible values of the parameter Δ. For small Δ, this will happen with high probability $(1 - \alpha$ for $\Delta = 0)$, as must be the case, since $\underline{\Delta}$ is then required to be a safe lower bound for a quantity which is equal to or near zero. Even more awkward is the possibility that $\overline{\Delta}$ is negative, so that the confidence set $(-\infty, \overline{\Delta})$ is empty.[*] An interpretation is suggested by the fact that this occurs if and only if the hypothesis $\Delta \geq \Delta_0$ is rejected for all positive values of Δ_0. This may be taken as an indication that the assumed model is not appropriate,[†] although it must be realized that for small Δ the probability of the event $\overline{\Delta} < 0$ is near α even when the assumptions are satisfied, so that this outcome will occasionally be observed.

The tests of $\Delta \leq \Delta_0$ and $\Delta \geq \Delta_0$ are not only UMP invariant but also UMP unbiased, and UMP unbiased tests also exist for testing $\Delta = \Delta_0$

---

[*] Such awkward confidence sets are discussed further at the end of Chapter 10, Section 4.

[†] For a discussion of possibly more appropriate alternative models, see Smith and Murray (1984).

against the two-sided alternatives $\Delta \neq \Delta_0$. This follows from the fact that the joint density of the Z's constitutes an exponential family. The confidence sets associated with these three families of tests are then uniformly most accurate unbiased (Problem 48). That optimum unbiased procedures exist in the model II case but not in the corresponding model I problem is explained by the different structure of the two hypotheses. The model II hypothesis $\sigma_A^2 = 0$ imposes one constraint, since it concerns the single parameter σ_A^2. On the other hand, the corresponding model I hypothesis $\sum_{i=1}^{s} \alpha_i^2 = 0$ specifies the values of the s parameters $\alpha_1, \ldots, \alpha_s$, and since $s - 1$ of these are independent, imposes $s - 1$ constraints.

A UMP invariant test of $\Delta \leq \Delta_0$ does not exist if the sample sizes n_i are unequal. An invariant test with a weaker optimum property for this case is obtained by Spjøtvoll (1967).

Since Δ is a ratio of variances, it is not surprising that the test statistic W^* shares the great sensitivity to the assumption of normality found in Chapter 5, Section 4 for the corresponding two-sample problem. More robust alternatives are discussed, for example, by Arvesen and Layard (1975).

12. NESTED CLASSIFICATIONS

The theory of the preceding section does not carry over even to so simple a situation as the general one-way classification with unequal numbers in the different classes (Problem 51). However, the unbiasedness approach does extend to the important case of a *nested* (hierarchical) classification with equal numbers in each class. This extension is sufficiently well indicated by carrying it through for the case of two factors; it follows for the general case by induction with respect to the number of factors.

Returning to the illustration of a batch process, suppose that a single batch of raw material suffices for several batches of the finished product. Let the experimental material consist of ab batches, b coming from each of a batches of raw material, and let a sample of size n be taken from each. Then (95) becomes

$$(100) \qquad X_{ijk} = \mu + A_i + B_{ij} + U_{ijk}$$

$$(i = 1, \ldots, a; \quad j = 1, \ldots, b; \quad k = 1, \ldots, n)$$

where A_i denotes the effect of the ith batch of raw material, B_{ij} that of the jth batch of finished product obtained from this material, and U_{ijk} the effect of the kth unit taken from this batch. All these variables are assumed to be independently normally distributed with zero means and with variances σ_A^2,

σ_B^2, and σ^2 respectively. The main part of the induction argument consists in proving the existence of an orthogonal transformation to variables Z_{ijk} the joint density of which, except for a constant, is

$$(101) \quad \exp\left[-\frac{1}{2(\sigma^2 + n\sigma_B^2 + bn\sigma_A^2)}\left((z_{111} - \sqrt{abn}\,\mu)^2 + \sum_{i=2}^{a} z_{i11}^2 \right) \right.$$

$$\left. -\frac{1}{2(\sigma^2 + n\sigma_B^2)}\sum_{i=1}^{a}\sum_{j=2}^{b} z_{ij1}^2 - \frac{1}{2\sigma^2}\sum_{i=1}^{a}\sum_{j=1}^{b}\sum_{k=2}^{n} z_{ijk}^2 \right].$$

As a first step, there exists for each fixed i, j an orthogonal transformation from $(X_{ij1}, \ldots, X_{ijn})$ to $(Y_{ij1}, \ldots, Y_{ijn})$ such that

$$Y_{ij1} = \sqrt{n}\,X_{ij\cdot} = \sqrt{n}\,\mu + \sqrt{n}\,(A_i + B_{ij} + U_{ij\cdot}).$$

As in the case of a single classification, the variables Y_{ijk} with $k > 1$ depend only on the U's, are independently normally distributed with zero mean and variance σ^2, and are independent of the $U_{ij\cdot}$. On the other hand, the variables Y_{ij1} have exactly the structure of the Y_{ij} in the one-way classification,

$$Y_{ij1} = \mu' + A_i' + U_{ij}',$$

where $\mu' = \sqrt{n}\,\mu$, $A_i' = \sqrt{n}\,A_i$, $U_{ij}' = \sqrt{n}\,(B_{ij} + U_{ij\cdot})$, and where the variances of A_i' and U_{ij}' are $\sigma_A'^2 = n\sigma_A^2$ and $\sigma'^2 = \sigma^2 + n_B^2$ respectively. These variables can therefore be transformed to variables Z_{ij1} whose density is given by (96) with Z_{ij1} in place of $Z_{ij\cdot}$. Putting $Z_{ijk} = Y_{ijk}$ for $k > 1$, the joint density of all Z_{ijk} is then given by (101).

Two hypotheses of interest can be tested on the basis of (101)— $H_1 : \sigma_A^2/(\sigma^2 + n\sigma_B^2) \leq \Delta_0$ and $H_2 : \sigma_B^2/\sigma^2 \leq \Delta_0$, which state that one or the other of the classifications has little effect on the outcome. Let

$$S_A^2 = \sum_{i=2}^{a} z_{i11}^2, \qquad S_B^2 = \sum_{i=1}^{a}\sum_{j=2}^{b} z_{ij1}^2, \qquad S^2 = \sum_{i=1}^{a}\sum_{j=1}^{b}\sum_{k=2}^{n} z_{ijk}^2.$$

To obtain a test of H_1, one is tempted to eliminate S^2 through invariance under multiplication of Z_{ijk} for $k > 1$ by an arbitrary constant. However, these transformations do not leave (101) invariant, since they do not always preserve the fact that σ^2 is the smallest of the three variances σ^2, $\sigma^2 + n\sigma_B^2$, and $\sigma^2 + n\sigma_B^2 + bn\sigma_A^2$. We shall instead consider the problem from the point of view of unbiasedness. For any unbiased test of H_1, the probability of rejection is α whenever $\sigma_A^2/(\sigma^2 + n\sigma_B^2) = \Delta_0$, and hence in particular when the three variances are σ^2, τ_0^2, and $(1 + bn\Delta_0)\tau_0^2$ for any fixed τ_0^2 and all $\sigma^2 < \tau_0^2$. It follows by the techniques of Chapter 4 that the conditional probability of rejection given $S^2 = s^2$ must be equal to α for almost all values of s^2. With S^2 fixed, the joint distribution of the remaining variables is of the same type as (101) after the elimination of Z_{111}, and a UMP unbiased conditional test given $S^2 = s^2$ has the rejection region

$$(102) \qquad W_1^* = \frac{1}{1 + bn\Delta_0} \cdot \frac{S_A^2/(a-1)}{S_B^2/(b-1)a} \geq C_1.$$

Since S_A^2 and S_B^2 are independent of S^2, the constant C_1 is determined by the fact that when $\sigma_A^2/(\sigma^2 + n\sigma_B^2) = \Delta_0$, the statistic W_1^* is distributed as $F_{a-1,(b-1)a}$ and hence in particular does not depend on s. The test (102) is clearly unbiased and hence UMP unbiased.

An alternative proof of this optimality property can be obtained using Theorem 7 of Chapter 6. The existence of a UMP unbiased test follows from the exponential family structure of the density (101), and the test is the same whether τ^2 is equal to $\sigma^2 + n\sigma_B^2$ and hence $\geq \sigma^2$, or whether it is unrestricted. However, in the latter case, the test (102) is UMP invariant and therefore is UMP unbiased even when $\tau^2 \geq \sigma^2$.

The argument with respect to H_2 is completely analogous and shows the UMP unbiased test to have the rejection region

$$(103) \qquad W_2^* = \frac{1}{1 + n\Delta_0} \cdot \frac{S_B^2/(b-1)a}{S^2/(n-1)ab} \geq C_2,$$

where C_2 is determined by the fact that for $\sigma_B^2/\sigma^2 = \Delta_0$, the statistic W_2^* is distributed as $F_{(b-1)a,(n-1)ab}$.

It remains to express the statistics S_A^2, S_B^2, and S^2 in terms of the X's. From the corresponding expressions in the one-way classification, it follows that

$$S_A^2 = \sum_{i=1}^{a} Z_{i11}^2 - Z_{111}^2 = b\sum (Y_{i\cdot1} - Y_{\cdot\cdot1})^2,$$

$$S_B^2 = \sum_{i=1}^{a} \left[\sum_{j=1}^{b} Z_{ij1}^2 - Z_{i11}^2 \right] = \sum\sum (Y_{ij1} - Y_{i\cdot1})^2,$$

$$S^2 = \sum_{i=1}^{a} \sum_{j=1}^{b} \left[\sum_{k=1}^{n} Y_{ijk}^2 - Y_{ij1}^2 \right] = \sum_i \sum_j \left[\sum_{k=1}^{n} U_{ijk}^2 - nU_{ij.}^2 \right]$$

$$= \sum_i \sum_j \sum_k (U_{ijk} - U_{ij.})^2.$$

Hence

(104) $$S_A^2 = bn \sum (X_{i..} - X_{...})^2, \qquad S_B^2 = n \sum \sum (X_{ij.} - X_{i..})^2,$$

$$S^2 = \sum \sum \sum (X_{ijk} - X_{ij.})^2.$$

It is seen from the expression of the statistics in terms of the Z's that their expectations are $E[S_A^2/(a-1)] = \sigma^2 + n\sigma_B^2 + bn\sigma_A^2$, $E[S_B^2/(b-1)a] = \sigma^2 + n\sigma_B^2$, and $E[S^2/(n-1)ab] = \sigma^2$. The decomposition

$$\sum \sum \sum (X_{ijk} - X_{...})^2 = S_A^2 + S_B^2 + S^2$$

therefore forms a basis for the analysis of the variance of X_{ijk},

$$\text{Var}(X_{ijk}) = \sigma_A^2 + \sigma_B^2 + \sigma^2$$

by providing estimates of the *components of variance* σ_A^2, σ_B^2, and σ^2, and tests of certain ratios of these components.

Nested two-way classifications also occur as mixed models. Suppose for example that a firm produces the material of the previous illustrations in different plants. If α_i denotes the effect of the ith plant (which is fixed, since the plants do not change in the replication of the experiment), B_{ij} the batch effect, and U_{ijk} the unit effect, the observations have the structure

(105) $$X_{ijk} = \mu + \alpha_i + B_{ij} + U_{ijk}.$$

Instead of reducing the X's to the fully canonical form in terms of the Z's as before, it is convenient to carry out only the reduction to the Y's (such that $Y_{ij1} = \sqrt{n} X_{ij.}$) and the first of the two transformations which take the Y's into the Z's. If the resulting variables are denoted by W_{ijk}, they satisfy $W_{i11} = \sqrt{b}\, Y_{i\cdot 1}$, $W_{ijk} = Y_{ijk}$ for $k > 1$ and

$$\sum_{i=1}^{a} (W_{i11} - W_{\cdot 11})^2 = S_A^2, \qquad \sum_{i=1}^{a}\sum_{j=2}^{b} W_{ij1}^2 = S_B^2, \qquad \sum_{i=1}^{a}\sum_{j=1}^{b}\sum_{k=2}^{n} W_{ijk}^2 = S^2$$

where S_A^2, S_B^2, and S^2 are given by (104). The joint density of the W's is, except for a constant,

$$(106) \quad \exp\left[-\frac{1}{2(\sigma^2 + n\sigma_B^2)} \left(\sum_{i=1}^{a} (w_{i11} - \mu - \alpha_i)^2 + \sum_{i=1}^{a}\sum_{j=2}^{b} w_{ij1}^2 \right) \right.$$
$$\left. -\frac{1}{2\sigma^2} \sum_{i=1}^{a}\sum_{j=1}^{b}\sum_{k=2}^{n} w_{ijk}^2 \right].$$

This shows clearly the different nature of the problem of testing that the plant effect is small,

$$H : \alpha_1 = \cdots = \alpha_a = 0 \quad \text{or} \quad H' : \frac{\sum \alpha_i^2}{\sigma^2 + n\sigma_B^2} \leq \Delta_0$$

and testing the corresponding hypothesis for the batch effect: $\sigma_B^2/\sigma^2 \leq \Delta_0$. The first of these is essentially a model I problem (linear hypothesis). As before, unbiasedness implies that the conditional rejection probability given $S^2 = s^2$ is equal to α a.e. With S^2 fixed, the problem of testing H is a linear hypothesis, and the rejection region of the UMP invariant conditional test given $S^2 = s^2$ has the rejection region (102) with $\Delta_0 = 0$. The constant C_1 is again independent of S^2, and the test is UMP among all tests that are both unbiased and invariant. A test with the same property also exists for testing H'. Its rejection region is

$$\frac{S_A^2/(a-1)}{S_B^2/(b-1)a} \geq C',$$

where C' is determined from the noncentral F-distribution instead of, as before, the (central) F-distribution.

On the other hand, the hypothesis $\sigma_B^2/\sigma^2 \leq \Delta_0$ is essentially model II. It is invariant under addition of an arbitrary constant to each of the variables W_{i11}, which leaves $\sum_{i=1}^{a}\sum_{j=2}^{b} W_{ij1}^2$ and $\sum_{i=1}^{a}\sum_{j=1}^{b}\sum_{k=2}^{n} W_{ijk}^2$ as maximal invariants, and hence reduces the structure to pure model II with one classification. The test is then given by (103) as before. It is both UMP invariant and UMP unbiased.

A two-factor mixed model in which there is interaction between the two factors will be considered in Example 2 of Chapter 8. Very general mixed models (containing general type II models as special cases) are discussed, for example, by Harville (1978), J. Miller (1977), and Brown (1984), but see the note following Problem 63.

The different one- and two-factor models are discussed from a Bayesian point of view, for example, in Box and Tiao (1973) and Broemeling (1985). In distinction to the approach presented here, the Bayesian treatment also includes inferences concerning the values of the individual random components such as the batch means ξ_i of Section 11.

13. PROBLEMS

1. *Expected sums of squares.* The expected values of the numerator and denominator of the statistic W^* defined by (7) are

$$E\left(\sum_{i=1}^r \frac{Y_i^2}{r}\right) = \sigma^2 + \frac{1}{r}\sum_{i=1}^r \eta_i^2 \quad \text{and} \quad E\left[\sum_{i=s+1}^n \frac{Y_i^2}{n-s}\right] = \sigma^2.$$

2. *Noncentral χ^2-distribution*.*

   (i) If X is distributed as $N(\psi,1)$, the probability density of $V = X^2$ is $p_\psi^V(v) = \sum_{k=0}^\infty P_k(\psi)f_{2k+1}(v)$, where $P_k(\psi) = (\psi^2/2)^k e^{-(1/2)\psi^2}/k!$ and where f_{2k+1} is the probability density of a χ^2-variable with $2k+1$ degrees of freedom.

   (ii) Let $Y_1,\ldots,Y_r$ be independently normally distributed with unit variance and means $\eta_1,\ldots,\eta_r$. Then $U = \sum Y_i^2$ is distributed according to the noncentral χ^2-distribution with r degrees of freedom and noncentrality parameter $\psi^2 = \sum_{i=1}^r \eta_i^2$, which has probability density

   (107) $$p_\psi^U(u) = \sum_{k=0}^\infty P_k(\psi)f_{r+2k}(u).$$

   Here $P_k(\psi)$ and $f_{r+2k}(u)$ have the same meaning as in (i), so that the distribution is a mixture of χ^2-distributions with Poisson weights.

   [(i): This is seen from

   $$p_\psi^V(v) = \frac{e^{-\frac{1}{2}(\psi^2+v)}\left(e^{\psi\sqrt{v}} + e^{-\psi\sqrt{v}}\right)}{2\sqrt{2\pi v}}$$

---

*The literature on noncentral χ^2, including tables, is reviewed in Chapter 28 of Johnson and Kotz (1970, Vol. 2), in Chou, Arthur, Rosenstein, and Owen (1984), and in Tiku (1985a).

by expanding the expression in parentheses into a power series, and using the fact that $\Gamma(2k) = 2^{2k-1}\Gamma(k)\Gamma(k + \tfrac{1}{2})/\sqrt{\pi}$.

(ii): Consider an orthogonal transformation to $Z_1, \ldots, Z_r$ such that $Z_1 = \Sigma \eta_i Y_i/\psi$. Then the Z's are independent normal with unit variance and means $E(Z_1) = \psi$ and $E(Z_i) = 0$ for $i > 1$.]

3. *Noncentral F- and beta-distribution.*[†] Let $Y_1, \ldots, Y_r$; $Y_{s+1}, \ldots, Y_n$ be independently normally distributed with common variance σ^2 and means $E(Y_i) = \eta_i (i = 1, \ldots, r)$; $E(Y_i) = 0$ ($i = s + 1, \ldots, n$).

(i) The probability density of $W = \Sigma_{i=1}^r Y_i^2 / \Sigma_{i=s+1}^n Y_i^2$ is given by (6). The distribution of the constant multiple $(n - s)W/r$ of W is the *noncentral F-distribution*.

(ii) The distribution of the statistic $B = \Sigma_{i=1}^r Y_i^2 / (\Sigma_{i=1}^r Y_i^2 + \Sigma_{i=s+1}^n Y_i^2)$ is the *noncentral beta-distribution*, which has probability density

(108)
$$\sum_{k=0}^{\infty} P_k(\psi) g_{\frac{1}{2}r+k, \frac{1}{2}(n-s)}(b),$$

where

(109)
$$g_{p,q}(b) = \frac{\Gamma(p + q)}{\Gamma(p)\Gamma(q)} b^{p-1}(1 - b)^{q-1}, \qquad 0 \le b \le 1$$

is the probability density of the (central) beta-distribution.

4. (i) The noncentral χ^2 and F distributions have strictly monotone likelihood ratio.

(ii) Under the assumptions of Section 1, the hypothesis $H' : \psi^2 \le \psi_0^2$ ($\psi_0 > 0$ given) remains invariant under the transformations G_i ($i = 1, 2, 3$) that were used to reduce $H : \psi = 0$, and there exists a UMP invariant test with rejection region $W > C'$. The constant C' is determined by $P_{\psi_0}\{W > C'\} = \alpha$, with the density of W given by (6).

[(i): Let $f(z) = \Sigma_{k=0}^{\infty} b_k z^k / \Sigma_{k=0}^{\infty} a_k z^k$ where the constants a_k, b_k are > 0 and $\Sigma a_k z^k$ and $\Sigma b_k z^k$ converge for all $z > 0$, and suppose that $b_k/a_k < b_{k+1}/a_{k+1}$ for all k. Then

$$f'(z) = \frac{\displaystyle\sum_{k<n}\sum (n - k)(a_k b_n - a_n b_k) z^{k+n-1}}{\left(\displaystyle\sum_{k=0}^{\infty} a_k z^k\right)^2}$$

is positive, since $(n - k)(a_k b_n - a_n b_k) > 0$ for $k < n$, and hence f is increasing.]

[†] For literature on noncentral F, see Johnson and Kotz (1970, Vol. 2) and Tiku (1985b).

Note. The noncentral χ^2- and F-distributions are in fact STP_∞ [see for example Marshall and Olkin (1979) and Brown, Johnstone and MacGibbon (1981)], and there thus exists a test of $H : \psi = \psi_0$ against $\psi \neq \psi_0$ which is UMP among all tests that are both invariant and unbiased.

5. *Best average power.*

   (i) Consider the general linear hypothesis H in the canonical form given by (2) and (3) of Section 1, and for any $\eta_{r+1}, \ldots, \eta_s$, σ, and ρ let $S = S(\eta_{r+1}, \ldots, \eta_s, \sigma; \rho)$ denote the sphere $\{(\eta_1, \ldots, \eta_r) : \Sigma_{i=1}^r \eta_i^2 / \sigma^2 = \rho^2\}$. If $\beta_\phi(\eta_1, \ldots, \eta_s, \sigma)$ denotes the power of a test ϕ of H, then the test (9) maximizes the average power

   $$\frac{\int_S \beta_\phi(\eta_1, \ldots, \eta_s, \sigma)\, dA}{\int_S dA}$$

   for every $\eta_{r+1}, \ldots, \eta_s$, σ, and ρ among all unbiased (or similar) tests. Here dA denotes the differential of area on the surface of the sphere.

   (ii) The result (i) provides an alternative proof of the fact that the test (9) is UMP among all tests whose power function depends only on $\Sigma_{i=1}^r \eta_i^2 / \sigma^2$.

[(i): if $U = \Sigma_{i=1}^r Y_i^2$, $V = \Sigma_{i=s+1}^n Y_i^2$, unbiasedness (or similarity) implies that the conditional probability of rejection given $Y_{r+1}, \ldots, Y_s$, and $U + V$ equals α a.e. Hence for any given $\eta_{r+1}, \ldots, \eta_s$, σ, and ρ, the average power is maximized by rejecting when the ratio of the average density to the density under H is larger than a suitable constant $C(y_{r+1}, \ldots, y_s, u + v)$, and hence when

$$g(y_1, \ldots, y_r; \eta_1, \ldots, \eta_r) = \int_S \exp\left(\sum_{i=1}^r \frac{\eta_i y_i}{\sigma^2} \right) dA > C(y_{r+1}, \ldots, y_s, u + v).$$

As will be indicated below, the function g depends on $y_1, \ldots, y_r$ only through u and is an increasing function of u. Since under the hypothesis $U/(U + V)$ is independent of $Y_{r+1}, \ldots, Y_s$ and $U + V$, it follows that the test is given by (9). The exponent in the integral defining g can be written as $\Sigma_{i=1}^r \eta_i y_i / \sigma^2 = (\rho \sqrt{u} \cos \beta)/\sigma$, where β is the angle ($0 \leq \beta \leq \pi$) between $(\eta_1, \ldots, \eta_r)$ and $(y_1, \ldots, y_r)$. Because of the symmetry of the sphere, this is unchanged if β is replaced by the angle γ between $(\eta_1, \ldots, \eta_r)$ and an arbitrary fixed vector. This shows that g depends on the y's only through u; for fixed $\eta_1, \ldots, \eta_r$, σ denote it by $h(u)$. Let S' be the subset of S in which $0 \leq \gamma \leq \pi/2$. Then

$$h(u) = \int_{S'} \left[\exp\left(\frac{\rho \sqrt{u} \cos \gamma}{\sigma} \right) + \exp\left(\frac{-\rho \sqrt{u} \cos \gamma}{\sigma} \right) \right] dA,$$

which proves the desired result.]

6. Use Theorem 8 of Chapter 6 to show that the F-test (7) is α-admissible against $\Omega': \psi \geq \psi_1$ for any $\psi_1 > 0$.

7. Given any $\psi_2 > 0$, apply Theorem 9 and Lemma 3 of Chapter 6 to obtain the F-test (7) as a Bayes test against a set Ω' of alternatives contained in the set $0 < \psi \leq \psi_2$.

Section 2

8. Under the assumptions of Section 1 suppose that the means ξ_i are given by

$$\xi_i = \sum_{j=1}^{s} a_{ij}\beta_j,$$

where the constants a_{ij} are known and the matrix $A = (a_{ij})$ has full rank, and where the β_j are unknown parameters. Let $\theta = \sum_{j=1}^{s} e_j \beta_j$ be a given linear combination of the β_j.

(i) If $\hat{\beta}_j$ denotes the values of the β_j minimizing $\sum(X_i - \xi_i)^2$ and if $\hat{\theta} = \sum_{j=1}^{s} e_j \hat{\beta}_j = \sum_{j=1}^{n} d_i X_i$, the rejection region of the hypothesis $H: \theta = \theta_0$ is

(110)
$$\frac{|\hat{\theta} - \theta_0|/\sqrt{\sum d_i^2}}{\sqrt{\sum(X_i - \hat{\xi}_i)^2/(n-s)}} > C_0$$

where the left-hand side under H has the distribution of the absolute value of Student's t with $n - s$ degrees of freedom.

(ii) The associated confidence intervals for θ are

(111)
$$\hat{\theta} - k\sqrt{\frac{\sum(X_i - \hat{\xi}_i)^2}{n-s}} \leq \theta \leq \hat{\theta} + k\sqrt{\frac{\sum(X_i - \hat{\xi}_i)^2}{n-s}}$$

with $k = C_0\sqrt{\sum d_i^2}$. These intervals are uniformly most accurate equivariant under a suitable group of transformations.

[(i): Consider first the hypothesis $\theta = 0$, and suppose without loss of generality that $\theta = \beta_1$; the general case can be reduced to this by making a linear transformation in the space of the β's. If $\underline{a}_1, \ldots, \underline{a}_s$ denote the column vectors of the matrix A which by assumption span Π_Ω, then $\underline{\xi} = \beta_1 \underline{a}_1 + \cdots + \beta_s \underline{a}_s$, and since $\hat{\underline{\xi}}$ is in Π_Ω, also $\hat{\underline{\xi}} = \hat{\beta}_1 \underline{a}_1 + \cdots + \hat{\beta}_s \underline{a}_s$. The space Π_ω defined by the hypothesis $\beta_1 = 0$ is spanned by the vectors $\underline{a}_2, \ldots, \underline{a}_s$ and also by the row vectors $\underline{c}_2, \ldots, \underline{c}_s$ of the matrix C of (1), while $\underline{c}_1$ is orthogonal to Π_ω. By (1), the vector $\underline{X}$ is given by $\underline{X} = \sum_{i=1}^{n} Y_i \underline{c}_i$, and its projection $\hat{\underline{\xi}}$ on Π_Ω therefore satisfies $\hat{\underline{\xi}} = \sum_{i=1}^{s} Y_i \underline{c}_i$. Equating the two expressions for $\hat{\xi}$ and taking the inner product of both sides of this equation with $\underline{c}_1$ gives $Y_1 = \hat{\beta}_1\Sigma_{i=1}^n a_{i1}c_{1i}$, since the $\underline{c}$'s are an orthogonal set of unit vectors. This shows that Y_1 is proportional to $\hat{\beta}_1$ and, since the variance of Y_1 is the same as that of the X's, that $|Y_1| = |\hat{\beta}_1|/\sqrt{\Sigma d_i^2}$. The result for testing $\beta_1 = 0$ now follows from (12) and (13). The test for $\beta_1 = \beta_1^0$ is obtained by making the transformation $X_i^* = X_i - a_{i1}\beta_1^0$.

(ii): The invariance properties of the intervals (111) can again be discussed without loss of generality by letting θ be the parameter β_1. In the canonical form of Section 1, one then has $E(Y_1) = \eta_1 = \lambda\beta_1$ with $|\lambda| = 1/\sqrt{\Sigma d_i^2}$ while $\eta_2, \ldots, \eta_s$ do not involve β_1. The hypothesis $\beta_1 = \beta_1^0$ is therefore equivalent to $\eta_1 = \eta_1^0$ with $\eta_1^0 = \lambda\beta_1^0$. This is invariant (a) under addition of arbitrary constants to $Y_2, \ldots, Y_s$; (b) under the transformations $Y_1^* = -(Y_1 - \eta_1^0) + \eta_1^0$; (c) under the scale changes $Y_i^* = cY_i$ $(i = 2, \ldots, n)$, $Y_1^* - \eta_1^{0*} = c(Y_1 - \eta_1^0)$. The confidence intervals for $\theta = \beta_1$ are then uniformly most accurate equivariant under the group obtained from (a), (b), and (c) by varying η_1^0.]

9. Let X_{ij} $(j = 1, \ldots, m_i)$ and Y_{ik} $(k = 1, \ldots, n_i)$ be independently normally distributed with common variance σ^2 and means $E(X_{ij}) = \xi_i$ and $E(Y_{ij}) = \xi_i + \Delta$. Then the UMP invariant test of $H: \Delta = 0$ is given by (110) with $\theta = \Delta$, $\theta_0 = 0$ and

$$\hat{\theta} = \frac{\sum_i \dfrac{m_i n_i}{N_i}(Y_{i\cdot} - X_{i\cdot})}{\sum_i \dfrac{m_i n_i}{N_i}}, \qquad \hat{\xi}_i = \frac{\sum_{j=1}^{m_i} X_{ij} + \sum_{k=1}^{n_i}(Y_{ik} - \hat{\theta})}{N_i},$$

where $N_i = m_i + n_i$.

10. Let $X_1, \ldots, X_n$ be independently normally distributed with known variance σ_0^2 and means $E(X_i) = \xi_i$, and consider any linear hypothesis with $s \leq n$ (instead of $s < n$ which is required when the variance is unknown). This remains invariant under a subgroup of that employed when the variance was unknown, and the UMP invariant test has rejection region

(112) $$\sum(X_i - \hat{\hat{\xi}}_i)^2 - \sum(X_i - \hat{\xi}_i)^2 = \sum(\hat{\xi}_i - \hat{\hat{\xi}}_i)^2 > C\sigma_0^2$$

with C determined by

(113) $$\int_C^\infty \chi_r^2(y)\, dy = \alpha.$$

11. Consider two experiments with observations $(X_1, \ldots, X_n)$ and $(Y_1, \ldots, Y_n)$ respectively, where the X_i and Y_i are independent normal with variance $\sigma^2 = 1$ and means $E(X_i) = c_i\theta_i$, $E(Y_i) = \theta_i$. Then the experiment based on the Y_i is more informative than that based on the X_i if and only if $|c_i| \leq 1$ for all i.

[If $1/c_i^2 = 1 + d_i$ with $d_i > 0$, let $Y_i' = Y_i + V_i$, where V_i is $N(0, d_i)$ and independent of Y_i. Then $c_i Y_i'$ has the same distribution as X_i. Conversely, if $c_i > 1$, the UMP unbiased test of $H: \theta_i = \theta$ against $\theta_i > 0$ based on $(X_1, \ldots, X_n)$ is more powerful than the corresponding test based on $(Y_1, \ldots, Y_n)$.]

12. Under the assumptions of the preceding problem suppose that $E(X_i) = \xi_i = \sum_{j=1}^s a_{ij}\theta_j$, $E(Y_i) = \eta_i = \sum_{j=1}^s b_{ij}\theta_j$ with the $n \times s$ matrices $A = (a_{ij})$ and $B = (b_{ij})$ of rank s. Then the experiment based on the Y_i is more informative than that based on the X_i if and only if $B'B - A'A$ is nonnegative definite. [There exists a nonsingular matrix F such that $F'A'AF = I$ and $F'B'BF = \Lambda$, where I is the identity and Λ is diagonal. The transformation $X' = FX$, $Y' = FY$ reduces the situation to that of Problem 11.]

Note. The results of Problems 11 and 12 no longer hold when σ^2 is unknown. See Hansen and Torgersen (1974).

Section 3

13. If the variables X_{ij} ($j = 1, \ldots, n_i$; $i = 1, \ldots, s$) are independently distributed as $N(\mu_i, \sigma^2)$, then

$$E\left[\sum n_i(X_i. - X..)^2\right] = (s - 1)\sigma^2 + \sum n_i(\mu_i - \mu.)^2,$$

$$E\left[\sum \sum (X_{ij} - X_i.)^2\right] = (n - s)\sigma^2.$$

14. Let $Z_1, \ldots, Z_s$ be independently distributed as $N(\zeta_i, a_i^2)$, $i = 1, \ldots, s$, where the a_i are known constants.

   (i) With respect to a suitable group of linear transformations there exists a UMP invariant test of $H: \zeta_1 = \cdots = \zeta_s$ given by the rejection region (21).

   (ii) The power of this test is the integral from C to ∞ of the noncentral χ^2-density with $s - 1$ degrees of freedom and noncentrality parameter λ^2 obtained by substituting ζ_i for Z_i in the left-hand side of (21).

15. (i) If $X_1, \ldots, X_n$ is a sample from a Poisson distribution with mean $E(X_i)$ $= \lambda$, then $\sqrt{n}(\sqrt{X} - \sqrt{\lambda})$ tends in law to $N(0, \frac{1}{4})$ as $n \to \infty$.

   (ii) If X has the binomial distribution $b(p, n)$, then $\sqrt{n}[\arcsin\sqrt{X/n} - \arcsin\sqrt{p}]$ tends in law to $N(0, \frac{1}{4})$ as $n \to \infty$.

   (iii) If $(X_1, Y_1), \ldots, (X_n, Y_n)$ is a sample from a bivariate normal distribution, then as $n \to \infty$ (in the notation of Chapter 5, Section 15)

$$\sqrt{n}\left[\log\frac{1 + R}{1 - R} - \log\frac{1 + \rho}{1 - \rho}\right] \to N(0, 4).$$

Note. Certain refinements of these transformations are discussed by Anscombe (1948), Freeman and Tukey (1950), and Hotelling (1953). Transformations of data to achieve approximately a normal linear model are considered by Box and Cox (1964); for later developments stemming from this work see Bickel and Doksum (1981), Box and Cox (1982), and Hinkley and Runger (1984).

Section 4

16. Show that

$$\sum_{i=1}^{r+1}\left(Y_i - \frac{Y_1 + \cdots + Y_{r+1}}{r+1}\right)^2 - \sum_{i=1}^{r}\left(Y_i - \frac{Y_1 + \cdots + Y_r}{r}\right)^2 \geq 0.$$

17. (i) For the validity of Theorem 1 it is only required that the probability of rejecting homogeneity of any set containing $\{\mu_{i_1}, \ldots, \mu_{i_{v_1}}\}$ as a proper subset tends to 1 as the distance between the different groups (26) all $\rightarrow \infty$, with the analogous condition holding for $H'_2, \ldots, H'_r$.

　　(ii) The condition of part (i) is satisfied for example if homogeneity of a set S is rejected for large values of $\Sigma|X_i. - X..|$, where the sum extends over the subscripts i for which $\mu_i \in S$.

18. In Lemma 1, show that $\alpha_{s-1} = \alpha_0^*$ is necessary for admissibility.

19. Prove Lemma 2 when s is odd.

20. Show that the Tukey levels (vi) satisfy (29) when s is even but not when s is odd.

21. The Tukey T-method leads to the simultaneous confidence intervals

$$(114) \quad \left|(X_j. - X_i.) - (\mu_j - \mu_i)\right| \leq \frac{CS}{\sqrt{sn(n-1)}} \quad \text{for all } i, j.$$

[The probability of (114) is independent of the μ's and hence equal to $1 - \alpha_s$.]

Section 6

22. The linear-hypothesis test of the hypothesis of no interaction in a two-way layout with m observations per cell is given by (39).

23. In the two-way layout of Section 6 with $a = b = 2$, denote the first three terms in the partition of $\Sigma\Sigma\Sigma(X_{ijk} - X_{ij.})^2$ by S_A^2, S_B^2, and S_{AB}^2, corresponding to the A, B, and AB effects (i.e. the α's, β's, and γ's), and denote by H_A, H_B, and H_{AB} the hypotheses of these effects being zero. Define a new two-level factor B' which is at level 1 when A and B are both at level 1 or both at level

2, and which is at level 2 when A and B are at different levels. Then

$$H_{B'} = H_{AB}, \qquad S_{B'} = S_{AB}, \qquad H_{AB'} = H_B, \qquad S_{AB'} = S_B,$$

so that the B-effect has become an interaction, and the AB-interaction the effect of the factor B'. [Shaffer (1977b).]

24. The size of each of the following tests is robust against nonnormality:

   (i)   the test (35) as $b \to \infty$,
   (ii) the test (37) as $mb \to \infty$,
   (iii) the test (39) as $m \to \infty$.

   Note. Nonrobustness against inequality of variances is discussed in Brown and Forsythe (1974a).

25. Let X_λ denote a random variable distributed as noncentral χ^2 with f degrees of freedom and noncentrality parameter λ^2. Then X_λ is stochastically larger than X_λ if $\lambda < \lambda'$.
   [It is enough to show that if Y is distributed as $N(0,1)$, then $(Y + \lambda')^2$ is stochastically larger than $(Y + \lambda)^2$. The equivalent fact that for any $z > 0$,

$$P\{|Y + \lambda'| \leq z\} \leq P\{|Y + \lambda| \leq z\},$$

   is an immediate consequence of the shape of the normal density function. An alternative proof is obtained by combining Problem 4 with Lemma 2 of Chapter 3.]

26. Let X_{ijk} $(i = 1,\ldots,a;\ j = 1,\ldots,b;\ k = 1,\ldots,m)$ be independently normally distributed with common variance σ^2 and mean

$$E(X_{ijk}) = \mu + \alpha_i + \beta_j + \gamma_k \qquad \left(\sum \alpha_i = \sum \beta_j = \sum \gamma_k = 0\right).$$

   Determine the linear hypothesis test for testing $H : \alpha_1 = \cdots = \alpha_a = 0$.

27. In the three-factor situation of the preceding problem, suppose that $a = b = m$. The hypothesis H can then be tested on the basis of m^2 observations as follows. At each pair of levels (i, j) of the first two factors one observation is taken, to which we refer as being in the ith row and the jth column. If the levels of the third factor are chosen in such a way that each of them occurs once and only once in each row and column, the experimental design is a *Latin square*. The m^2 observations are denoted by $X_{ij(k)}$, where the third subscript indicates the level of the third factor when the first two are at levels i and j. It is assumed that $E(X_{ij(k)}) = \xi_{ij(k)} = \mu + \alpha_i + \beta_j + \gamma_k$, with $\sum \alpha_i = \sum \beta_j = \sum \gamma_k = 0$.

   (i) The parameters are determined from the ξ's through the equations

$$\xi_{i\cdot(\cdot)} = \mu + \alpha_i, \qquad \xi_{\cdot j(\cdot)} = \mu + \beta_j, \qquad \xi_{\cdot\cdot(k)} = \mu + \gamma_k, \qquad \xi_{\cdot\cdot(\cdot)} = \mu.$$

(Summation over j with i held fixed automatically causes summation also over k.)

(ii) The least-squares estimates of the parameters may be obtained from the identity

$$\sum_i \sum_j \left[x_{ij(k)} - \xi_{ij(k)} \right]^2$$

$$= m \sum \left[x_{i \cdot (\cdot)} - x_{\cdot \cdot (\cdot)} - \alpha_i \right]^2 + m \sum \left[x_{\cdot j(\cdot)} - x_{\cdot \cdot (\cdot)} - \beta_j \right]^2$$

$$+ m \sum \left[x_{\cdot \cdot (k)} - x_{\cdot \cdot (\cdot)} - \gamma_k \right]^2 + m^2 \left[x_{\cdot \cdot (\cdot)} - \mu \right]^2$$

$$+ \sum_i \sum_k \left[x_{ij(k)} - x_{i \cdot (\cdot)} - x_{\cdot j(\cdot)} - x_{\cdot \cdot (k)} + 2 x_{\cdot \cdot (\cdot)} \right]^2.$$

(iii) For testing the hypothesis $H: \alpha_1 = \cdots = \alpha_m = 0$, the test statistic W^* of (15) is

$$\frac{m \sum \left[X_{i \cdot (\cdot)} - X_{\cdot \cdot (\cdot)} \right]^2}{\sum \sum \left[X_{ij(k)} - X_{i \cdot (\cdot)} - X_{\cdot j(\cdot)} - X_{\cdot \cdot (k)} + 2 X_{\cdot \cdot (\cdot)} \right]^2 / (m - 2)}.$$

The degrees of freedom are $m - 1$ for the numerator and $(m - 1)(m - 2)$ for the denominator, and the noncentrality parameter is $\psi^2 = m \sum \alpha_i^2 / \sigma^2$.

Section 7

28. In a regression situation, suppose that the observed values X_j and Y_j of the independent and dependent variable differ from certain true values X_j' and Y_j' by errors U_j, V_j which are independently normally distributed with zero means and variances σ_U^2 and σ_V^2. The true values are assumed to satisfy a linear relation: $Y_j' = \alpha + \beta X_j'$. However, the variables which are being controlled, and which are therefore constants, are the X_j rather than the X_j'. Writing x_j for X_j, we have $x_j = X_j' + U_j$, $Y_j = Y_j' + V_j$, and hence $Y_j = \alpha + \beta x_j + W_j$, where $W_j = V_j - \beta U_j$. The results of Section 7 can now be applied to test that β or $\alpha + \beta x_0$ have a specified value.

29. Let $X_1, \ldots, X_m; Y_1, \ldots, Y_n$ be independently normally distributed with common variance σ^2 and means $E(X_i) = \alpha + \beta(u_i - \bar{u})$, $E(Y_j) = \gamma + \delta(v_j - \bar{v})$, where the u's and v's are known numbers. Determine the UMP invariant tests of the linear hypotheses $H: \beta = \delta$ and $H: \alpha = \gamma$, $\beta = \delta$.

30. Let $X_1, \ldots, X_n$ be independently normally distributed with common variance σ^2 and means $\xi_i = \alpha + \beta t_i + \gamma t_i^2$, where the t_i are known. If the coefficient vectors $(t_1^k, \ldots, t_n^k)$, $k = 0, 1, 2$, are linearly independent, the parameter space Π_Ω has dimension $s = 3$, and the least-squares estimates $\hat{\alpha}, \hat{\beta}, \hat{\gamma}$ are the

unique solutions of the system of equations

$$\alpha \sum t_i^k + \beta \sum t_i^{k+1} + \gamma \sum t_i^{k+2} = \sum t_i^k X_i \quad (k = 0, 1, 2).$$

The solutions are linear functions of the X's, and if $\hat{\gamma} = \sum c_i X_i$, the hypothesis $\gamma = 0$ is rejected when

$$\frac{|\hat{\gamma}|/\sqrt{\sum c_i^2}}{\sqrt{\sum (X_i - \hat{\alpha} - \hat{\beta} t_i - \hat{\gamma} t_i^2)^2/(n-3)}} > C_0.$$

Section 8

31. Verify the claims made in Example 8.

32. Let X_{ijk} $(k = 1, \ldots, n_{ij};\ i = 1, \ldots, a;\ j = 1, \ldots, b)$ be independently normally distributed with mean $E(X_{ijk}) = \xi_{ij}$ and variance σ^2. Then the test of any linear hypothesis concerning the ξ_{ij} has a robust level provided $n_{ij} \to \infty$ for all i and j.

33. In the two-way layout of the preceding problem give examples of submodels $\Pi_\Omega^{(1)}$ and $\Pi_\Omega^{(2)}$ of dimensions s_1 and s_2, both less than ab, such that in one case the condition (56) continues to require $n_{ij} \to \infty$ for all i and j but becomes a weaker requirement in the other case.

34. Suppose (56) holds for some particular sequence $\Pi_\Omega^{(n)}$ with fixed s. Then it holds for any sequence $\Pi_\Omega'^{(n)} \subset \Pi_\Omega^{(n)}$ of dimension $s' < s$.
     [If Π_Ω is spanned by the s columns of A, let Π_Ω' be spanned by the first s' columns of A.]

35. Let $\{c_n\}$ and $\{c_n'\}$ be two increasing sequences of constants such that $c_n'/c_n \to 1$ as $n \to \infty$. Then $\{c_n\}$ satisfies (56) if and only if $\{c_n'\}$ does.

36. Let $c_n = u_0 + u_1 n + \cdots + u_k n^k$, $u_i \geq 0$ for all i. Then c_n satisfies (56).
     [Apply Problem 35 with $c_n' = n^k$.]

37. (i) Under the assumptions of Problem 30, express the condition (56) in terms of the t's.

     (ii) Determine whether the condition of part (i) is equivalent to (51).

38. If $\xi_i = \alpha + \beta t_i + \gamma u_i$, express the condition (56) in terms of the t's and u's.

39. Show that $\sum_{i=1}^n \Pi_{ii} = s$.
     [Since the Π_{ii} are independent of A, take A to be orthogonal.]

40. Show how to weaken (56) if a robustness condition is required only for testing a particular subspace Π_ω of Π_Ω.
     [Suppose that Π_ω is given by $\beta_1 = \cdots = \beta_r = 0$, and use (54).]

41. Give an example of an analysis of covariance (46) in which (56) does not hold but the level of the F-test of $H: \alpha_1 = \cdots = \alpha_b$ is robust against nonnormality.

Section 9

42. (i) A function L satisfies the first equation of (62) for all u, x, and orthogonal transformations Q if and only if it depends on u and x only through $u'x$, $x'x$, and $u'u$.

   (ii) A function L is equivariant under G_2 if and only if it satisfies (64).

43. (i) For the confidence sets (70), equivariance under G_1 and G_2 reduces to (71) and (72) respectively.

   (ii) For fixed $(y_1, \ldots, y_r)$, the statements $\Sigma u_i y_i \in A$ hold for all $(u_1, \ldots, u_r)$ with $\Sigma u_i^2 = 1$ if and only if A contains the interval $I(y) = [-\sqrt{\Sigma y_i^2}, + \sqrt{\Sigma y_i^2}]$.

   (iii) Show that the statement following (74) ceases to hold when $r = 1$.

44. Let X_i ($i = 1, \ldots, r$) be independent $N(\xi_i, 1)$.

   (i) The only simultaneous confidence intervals equivariant under G_0 are those given by (80).

   (ii) The inequalities (80) and (82) are equivalent.

   (iii) Compared with the Scheffé intervals (69), the intervals (82) for $\Sigma u_j \xi_j$ are shorter when $\Sigma u_j \xi_j = \xi_i$ and longer when $u_1 = \cdots = u_r$.

   [(ii): For a fixed $u = (u_1, \ldots, u_r)$, $\Sigma u_i y_i$ is maximized subject to $|y_i| \leq \Delta$ for all i, by $y_i = \Delta$ when $u_i > 0$ and $y_i = -\Delta$ when $u_i < 0$.]

Section 10

45. (i) The confidence intervals $L(u; y, S) = \Sigma u_i y_i - c(S)$ are equivariant under G_3 if and only if $L(u; by, bS) = bL(u; y, S)$ for all $b > 0$.

   (ii) The most general confidence sets (87) which are equivariant under G_1, G_2, and G_3 are of the form (88).

46. (i) In Example 11, the set of linear functions $\Sigma w_i \alpha_i = \Sigma w_i(\xi_i. - \xi..)$ for all w can also be represented as the set of functions $\Sigma w_i \xi_i.$ for all w satisfying $\Sigma w_i = 0$.

   (ii) The set of linear functions $\Sigma\Sigma w_{ij} \gamma_{ij} = \Sigma\Sigma w_{ij}(\xi_{ij} - \xi_i. - \xi_{.j} + \xi..)$ for all w is equivalent to the set $\Sigma\Sigma w_{ij} \xi_{ij}.$ for all w satisfying $\Sigma_i w_{ij} = \Sigma_j w_{ij} = 0$.

   (iii) Determine the simultaneous confidence intervals (89) for the set of linear functions of part (ii).

47. (i) In Example 10, the simultaneous confidence intervals (89) reduce to (93).

   (ii) What change is needed in the confidence intervals of Example 10 if the v's are not required to satisfy (92), i.e. if simultaneous confidence intervals are desired for all linear functions $\Sigma v_i \xi_i$ instead of all contrasts? Make a table showing the effect of this change for $s = 2, 3, 4, 5$; $n_i = n = 3, 5, 10$.

Section 11

48. (i) The test (97) of $H : \Delta \leq \Delta_0$ is UMP unbiased.

    (ii) Determine the UMP unbiased test of $H : \Delta = \Delta_0$ and the associated uniformly most accurate unbiased confidence sets for Δ.

49. In the model (95), the correlation coefficient ρ between two observations X_{ij}, X_{ik} belonging to the same class, the so-called *intraclass correlation coefficient*, is given by $\rho = \sigma_A^2 / (\sigma_A^2 + \sigma^2)$.

Section 12

50. The tests (102) and (103) are UMP unbiased.

51. If X_{ij} is given by (95) but the number n_i of observations per batch is not constant, obtain a canonical form corresponding to (96) by letting $Y_{i1} = \sqrt{n_i} \, X_{i.}$. Note that the set of sufficient statistics has more components than when n_i is constant.

52. The general nested classification with a constant number of observations per cell, under model II, has the structure

$$X_{ijk\ldots} = \mu + A_i + B_{ij} + C_{ijk} + \cdots + U_{ijk\ldots},$$

$$i = 1, \ldots, a; \; j = 1, \ldots, b; \; k = 1, \ldots, c; \ldots.$$

    (i) This can be reduced to a canonical form generalizing (101).

    (ii) There exist UMP unbiased tests of the hypotheses

$$H_A : \frac{\sigma_A^2}{cd \ldots \sigma_B^2 + d \ldots \sigma_C^2 + \cdots + \sigma^2} \leq \Delta_0,$$

$$H_B : \frac{\sigma_B^2}{d \ldots \sigma_C^2 + \cdots + \sigma^2} \leq \Delta_0.$$

53. Consider the model II analogue of the two-way layout of Section 6, according to which

$$(115) \qquad X_{ijk} = \mu + A_i + B_j + C_{ij} + E_{ijk}$$

$$(i = 1, \ldots, a; \quad j = 1, \ldots, b; \quad k = 1, \ldots, n),$$

where the A_i, B_j, C_{ij}, and E_{ijk} are independently normally distributed with mean zero and with variances σ_a^2; σ_B^2, σ_C^2, and σ^2 respectively. Determine tests which are UMP among all tests that are invariant (under a suitable group) and unbiased of the hypotheses that the following ratios do not exceed a given constant (which may be zero):

  (i) σ_C^2/σ^2;
  (ii) $\sigma_A^2/(n\sigma_C^2 + \sigma^2)$;
  (iii) $\sigma_B^2/(n\sigma_C^2 + \sigma^2)$.

Note that the test of (i) requires $n > 1$, but those of (ii) and (iii) do not.
[Let $S_A^2 = nb\Sigma(X_{i..} - X_{...})^2$, $S_B^2 = na\Sigma(X_{.j.} - X_{...})^2$, $S_C^2 = n\Sigma\Sigma(X_{ij.} - X_{i..} - X_{.j.} + X_{...})^2$, $S^2 = \Sigma\Sigma\Sigma(X_{ijk} - X_{ij.})^2$, and make a transformation to new variables Z_{ijk} (independent, normal, and with mean zero except when $i = j = k = 1$) such that

$$S_A^2 = \sum_{i=2}^{a} Z_{i11}^2, \qquad S_B^2 = \sum_{j=2}^{b} Z_{1j1}^2, \qquad S_C^2 = \sum_{i=2}^{a} \sum_{j=2}^{b} Z_{ij1}^2,$$

$$S^2 = \sum_{i=1}^{a} \sum_{j=1}^{b} \sum_{k=2}^{n} Z_{ijk}^2.]$$

54. Consider the mixed model obtained from (115) by replacing the random variables A_i by unknown constants α_i satisfying $\Sigma\alpha_i = 0$. With (ii) replaced by (ii') $\Sigma\alpha_i^2/(n\sigma_C^2 + \sigma^2)$, there again exist tests which are UMP among all tests that are invariant and unbiased, and in cases (i) and (iii) these coincide with the corresponding tests of Problem 53.

55. Consider the following generalization of the univariate linear model of Section 1. The variables X_i $(i = 1, \ldots, n)$ are given by $X_i = \xi_i + U_i$, where $(U_1, \ldots, U_n)$ have a joint density which is *spherical*, that is, a function of $\Sigma_{i-1}^n u_i^2$, say

$$f(U_1, \ldots, U_n) = q\left(\sum U_i^2\right).$$

The parameter spaces Π_Ω and Π_ω and the hypothesis H are as in Section 1.

  (i) The orthogonal transformation (1) reduces $(X_1, \ldots, X_n)$ to canonical variables $(Y_1, \ldots, Y_n)$ with $Y_i = \eta_i + V_i$, where $\eta_i = 0$ for $i = s + 1, \ldots, n$, H reduces to (3), and the V's have joint density $q(v_1, \ldots, v_n)$.

  (ii) In the canonical form of (i), the problem is invariant under the groups G_1, G_2, and G_3 of Section 1, and the statistic W^* given by (7) is maximal invariant.

56. Under the assumptions of the preceding problem, the null distribution of W^* is independent of q and hence the same as in the normal case, namely, F with r and $n - s$ degrees of freedom.
[See Chapter 5, Problem 24].

Note. The analogous multivariate problem is treated by Kariya (1981), who also shows that the test (9) of Chapter 8 continues to be UMP invariant provided q is a nonincreasing convex function. The same method shows that this conclusion holds under the same conditions also in the present case. For a review of work on spherically and elliptically symmetric distributions, see Chmielewski (1981).

Additional Problems

57. Consider the additive random-effects model

$$X_{ijk} = \mu + A_i + B_j + U_{ijk} \qquad (i = 1, \ldots, a; \quad j = 1, \ldots, b; \quad k = 1, \ldots, n),$$

where the A's, B's, and U's are independent normal with zero means and variances σ_A^2, σ_B^2, and σ^2 respectively. Determine

(i) the joint density of the X's,
(ii) the UMP unbiased test of $H : \sigma_B^2/\sigma^2 \le \delta$.

58. For the mixed model

$$X_{ij} = \mu + \alpha_i + B_j + U_{ij} \qquad (i = 1, \ldots, a; \quad j = 1, \ldots, n),$$

where the B's and u's are as in Problem 57 and the α's are constants adding to zero, determine (with respect to a suitable group leaving the problem invariant)

(i)   a UMP invariant test of $H : \alpha_1 = \cdots = \alpha_a$;
(ii) a UMP invariant test of $H : \xi_1 = \cdots = \xi_a = 0$ $(\xi_i = \mu + \alpha_i)$;
(iii) a test of $H : \sigma_B^2/\sigma^2 \le \delta$ which is both UMP invariant and UMP unbiased.

59. Let $(X_{1j}, \ldots, X_{pj})$, $j = 1, \ldots, n$, be a sample from a p-variate normal distribution with mean $(\xi_1, \ldots, \xi_p)$ and covariance matrix $\Sigma = (\sigma_{ij})$ where $\sigma_{ij} = \sigma^2$ when $j = i$, and $\sigma_{ij} = \rho\sigma^2$ when $j \ne i$. Show that the covariance matrix is positive definite if and only if $\rho > -1/(p - 1)$.
    [For fixed σ and $\rho < 0$, the quadratic form $(1/\sigma^2)\Sigma\Sigma\sigma_{ij}y_iy_j = \Sigma y_i^2 + \rho\Sigma\Sigma y_iy_j$ takes on its minimum value over $\Sigma y_i^2 = 1$ when all the y's are equal.]

60. Under the assumptions of the preceding problem, determine the UMP invariant test (with respect to a suitable G) of $H : \xi_1 = \cdots = \xi_p$.
    [Show that this model agrees with that of Problem 58 if $\rho = \sigma_b^2/(\sigma_b^2 + \sigma^2)$, except that instead of being positive, ρ now only needs to satisfy $\rho > -1/(p - 1)$.]

61. Permitting interactions in the model of Problem 57 leads to the model

$$X_{ijk} = \mu + A_i + B_j + C_{ij} + U_{ijk} \qquad (i = 1, \ldots, a; j = 1, \ldots, b; k = 1, \ldots, n).$$

where the A's, B's, C's, and U's are independent normal with mean zero and variances σ_A^2, σ_B^2, σ_C^2, and σ^2.

(i) Give an example of a situation in which such a model might be appropriate.

(ii) Reduce the model to a convenient canonical form along the lines of Sections 5 and 8.

(iii) Determine UMP unbiased tests of (a) $H_1 : \sigma_B^2 = 0$; (b) $H_2 : \sigma_C^2 = 0$.

62. Formal analogy with the model of Problem 61 suggests the mixed model

$$X_{ijk} = \mu + \alpha_i + B_j + C_{ij} + U_{ijk}$$

with the B's, C's, and U's as in Problem 61. Reduce this model to a canonical form involving $X_{...}$ and the sums of squares

$$\frac{\sum(X_{i..} - X_{...} - \alpha_i)^2}{n\sigma_C^2 + \sigma^2}, \quad \frac{\sum\sum(X_{.j.} - X_{...})^2}{an\sigma_B^2 + n\sigma_C^2 + \sigma^2},$$

$$\frac{\sum\sum(X_{ij.} - X_{i..} - X_{.j.} + X_{...})^2}{n\sigma_C^2 + \sigma^2}, \quad \frac{\sum\sum\sum(X_{ijk} - X_{i..} - X_{.j.} + X_{...})^2}{\sigma^2}.$$

63. Among all tests that are both unbiased and invariant under suitable groups under the assumptions of Problem 62, there exist UMP tests of

(i) $H_1 : \alpha_1 = \cdots = \alpha_a = 0$;

(ii) $H_2 : \sigma_B^2/(n\sigma_C^2 + \sigma^2) \leq C$;

(iii) $H_3 : \sigma_C^2/\sigma^2 \leq C$.

Note. The independence assumptions of Problems 62 and 63 often are not realistic. For alternative models, derived from more basic assumptions, see Scheffé (1956, 1959). Relations between the two types of models are discussed in Hocking (1973), Cohen and Miller (1976), and Kendall, Stuart, and Ord (1983).

64. Let $(X_{1j1}, \ldots, X_{1jn}; X_{2j1}, \ldots, X_{2jn}; \ldots; X_{aj1}, \ldots, X_{ajn})$, $j = 1, \ldots, b$, be a sample from an an-variate normal distribution. Let $E(X_{ijk}) = \xi_i$, and denote by $\Sigma_{ii'}$ the matrix of covariances of $(X_{ij1}, \ldots, X_{ijn})$ with $(X_{i'j1}, \ldots, X_{i'jn})$. Suppose that for all i, the diagonal elements of Σ_{ii} are $= \tau^2$ and the off-diagonal elements $= \rho_1 \tau^2$, and that for $i \neq i'$ all n^2 elements of $\Sigma_{ii'}$ are $= \rho_2 \tau^2$.

(i) Find necessary and sufficient conditions on ρ_1 and ρ_2 for the overall $abn \times abn$ covariance matrix to be positive definite.

(ii) Show that this model agrees with that of Problem 62 for suitable values of ρ_1 and ρ_2.

65. *Tukey's T-Method.* Let X_i $(i = 1, \ldots, r)$ be independent $N(\xi_i, 1)$, and consider simultaneous confidence intervals

(116) $L[(i, j); x] \leq \xi_j - \xi_i \leq M[(i, j); x]$ for all $i \neq j$.

The problem of determining such confidence intervals remains invariant under the group G_0' of all permutations of the X's and under the group G_2 of translations $gx = x + a$.

(i) In analogy with (61), attention can be restricted to confidence bounds satisfying

(117) $L[(i, j); x] = -M[(j, i); x]$.

(ii) The only simultaneous confidence intervals satisfying (117) and equivariant under G_0' and G_2 are those of the form

(118) $S(x) = \left\{ \xi : x_j - x_i - \Delta < \xi_j - \xi_i < x_j - x_i + \Delta \text{ for all } i \neq j \right\}$.

(iii) The constant Δ for which (118) has probability γ is determined by

(119) $P_0\left\{ \max |X_j - X_i| < \Delta \right\} = P_0\left\{ X_{(n)} - X_{(1)} < \Delta \right\} = \gamma$,

where the probability P_0 is calculated under the assumption that $\xi_1 = \cdots = \xi_r$.

66. In the preceding problem consider arbitrary contrasts $\Sigma c_i \xi_i$, $\Sigma c_i = 0$. The event

(120) $\left| (X_j - X_i) - (\xi_j - \xi_i) \right| \leq \Delta$ for all $i \neq j$

is equivalent to the event

(121) $\left| \Sigma c_i X_i - \Sigma c_i \xi_i \right| \leq \dfrac{\Delta}{2} \Sigma |c_i|$ for all c with $\Sigma c_i = 0$,

which therefore also has probability γ. This shows how to extend the Tukey intervals for all pairs to all contrasts.

[That (121) implies (120) is obvious. To see that (120) implies (121), let $y_i = x_i - \xi_i$ and maximize $|\Sigma c_i y_i|$ subject to $|y_j - y_i| \leq \Delta$ for all i and j. Let P and N denote the sets $\{i : c_i > 0\}$ and $\{i : c_i < 0\}$, so that

$$\Sigma c_i y_i = \sum_{i \in P} c_i y_i - \sum_{i \in N} |c_i| y_i.$$

Then for fixed c, the sum $\Sigma c_i y_i$ is maximized by maximizing the y_i's for $i \in P$ and minimizing those for $i \in N$. Since $|y_j - y_i| \leq \Delta$, it is seen that $\Sigma c_i y_i$ is maximized by $y_i = \Delta/2$ for $i \in P$, $y_i = -\Delta/2$ for $i \in N$. The minimization of $\Sigma c_i y_i$ is handled analogously.]

67. (i) Let X_{ij} ($j = 1, \dots, n$; $i = 1, \dots, s$) be independent $N(\xi_i, \sigma^2)$, σ^2 unknown. Then the problem of obtaining simultaneous confidence intervals for all differences $\xi_j - \xi_i$ is invariant under G_0', G_2, and the scale changes G_3.

   (ii) The only equivariant confidence bounds based on the sufficient statistics X_i. and $S^2 = \Sigma\Sigma(X_{ij} - X_i.)^2$ and satisfying the condition corresponding to (117) are those given by

(122)
$$S(x) = \left\{ x : x_j. - x_i. - \frac{\Delta'}{\sqrt{n-s}} S \le \xi_j - \xi_i \right.$$

$$\left. \le x_j. - x_i. + \frac{\Delta'}{\sqrt{n-s}} S \right.$$

$$\left. \text{for all } i \ne j \right\}$$

with Δ' determined by the null distribution of the *Studentized range*

(123)
$$P_0 \left\{ \frac{\max |X_j. - X_i.|}{S/\sqrt{n-s}} < \Delta' \right\} = \gamma.$$

   (iii) Extend the results of Problem 66 to the present situation.

68. Construct an example [i.e., choose values $n_1 = \cdots = n_s = n$ and α and a particular contrast $(c_1, \dots, c_s)$] for which the Tukey confidence intervals (121) are shorter than the Scheffé intervals (93), and an example in which the situation is reversed.

69. *Dunnett's method.* Let X_{0j} ($j = 1, \dots, m$) and X_{ik} ($i = 1, \dots, s$; $k = 1, \dots, n$) represent measurements on a standard and s competing new treatments, and suppose the X's are independently distributed as $N(\xi_0, \sigma^2)$ and $N(\xi_i, \sigma^2)$ respectively. Generalize Problems 65 and 67 to the problem of obtaining simultaneous confidence intervals for the s differences $\xi_i - \xi_0$ ($i = 1, \dots, s$).

70. In generalization of Problem 66, show how to extend the Dunnett intervals of Problem 69 to the set of all contrasts.
   [Use the fact that the event $|y_i - y_0| \le \Delta$ for $i = 1, \dots, s$ is equivalent to the event $|\Sigma_{i=0}^{s} c_i y_i| \le \Delta \Sigma_{i=1}^{s} |c_i|$ for all $(c_0, \dots, c_s)$ satisfying $\Sigma_{i=0}^{s} c_i = 0$.]
   Note. As is pointed out in Problems 45(iii) and 68, the intervals resulting from the extension of the Tukey (and Dunnett) methods to all contrasts are shorter than the Scheffé intervals for the differences for which these methods

were designed and for contrasts close to them, and longer for some other contrasts. For details and generalizations, see for example Miller (1981), Richmond (1982), and Shaffer (1977a).

71. In the regression model of Problem 8, generalize the confidence bands of Example 12 to the regression surfaces

(i) $h_1(e_1, \ldots, e_s) = \sum_{j=1}^{s} e_j \beta_j$;

(ii) $h_2(e_2, \ldots, e_s) = \beta_1 + \sum_{j=2}^{s} e_j \beta_j$.

14. REFERENCES

The general linear model [in the parametric form (18)] was formulated at the beginning of the 19th century by Legendre and Gauss, who were concerned with estimating the unknown parameters. [For an account of its history, see Seal (1967).] The canonical form (2) of the model is due to Kolodziejczyk (1935). The analysis of variance, including the concept of interaction, was developed by Fisher in the 1920s and 1930s, and a systematic account is provided by Scheffé (1959) in a book that includes a careful treatment of alternative models and of robustness questions.

The first simultaneous confidence intervals (for a regression line) were obtained by Working and Hotelling (1929). The optimal property of the Scheffé intervals presented in Section 9 is a special case of results of Wijsman (1979, 1980). A review of the literature on the relationship of tests and confidence sets for a parameter vector with the associated simultaneous confidence intervals for functions of its components can be found in Kanoh and Kusunoki (1984).

Aiyar, R. J., Guillier, C. L., and Albers, W.
(1979). "Asymptotic relative efficiencies of rank tests for trend alternatives." *J. Amer. Statist. Assoc.* **74**, 226–231.

Albert, A.
(1976). "When is a sum of squares an analysis of variance?" *Ann. Statist.* **4**, 775–778.

Anscombe, F.
(1948). "Transformations of Poisson, binomial and negative binomial data." *Biometrika* **35**, 246–254.

Arnold, S. F.
(1980). "Asymptotic validity of F-tests for the ordinary linear model and the multiple correlation model." *J. Amer. Statist. Assoc.* **75**, 890–894.
(1981). *The Theory of Linear Models and Multivariate Analysis*, Wiley, New York.

Arvesen, J. N. and Layard, M. W. J.
(1975). "Asymptotically robust tests in unbalanced variance component models." *Ann. Statist.* **3**, 1122–1134.

Barlow, R. E., Bartholomew, D. J., Bremner, J. M., and Brunk, H. D.
(1972). *Statistical Inference under Order Restrictions*, Wiley, New York.

Bartlett, M. S.
(1947). "The use of transformations." *Biometrics* **3**, 39–52.
[Discussion of, among others, the logarithmic, square-root, and arcsine transformations.]

Bickel, P. J. and Doksum, K. A.
(1981). "An analysis of transformations revisited." *J. Amer. Statist. Assoc.* **76**, 296–311.

Billingsley, P.
(1979). *Probability and Measure*, Wiley, New York.

Bohrer, R.
(1973). "An optimality property of Scheffé bounds." *Ann. Statist.* **1**, 766–772.

Box, G. E. P.
(1949). "A general distribution theory for a class of likelihood ratio criteria." *Biometrika* **36**, 317–346.

Box, G. E. P. and Cox, D. R.
(1964). "An analysis of transformations." *J. Roy. Statist. Soc. (B)* **26**, 211–252.
(1982). "An analysis of transformations revisited, rebutted." *J. Amer. Statist. Assoc.* **77**, 209–210.

Box, G. E. P., Hunter, W. G., and Hunter, J. S.
(1978). *Statistics for Experimenters*, Wiley, New York.

Box, G. E. P. and Tiao, G. C.
(1973). *Bayesian Inference in Statistical Analysis* Addison-Wesley, Reading, MA.

Broemeling, L. D.
(1985). *Bayesian Analysis of Linear Models*. Dekker, New York.

Brown, K. G.
(1984). "On analysis of variance in the mixed model." *Ann. Statist.* **12**, 1488–1499.

Brown, L. D., Johnstone, I. M. and MacGibbon, K. B.
(1981). "Variation diminishing transformations: A direct approach to total positivity and its statistical applications." *J. Amer. Statist. Assoc.* **76**, 824–832.

Brown, M. B. and Forsythe, A. B.
(1974a). "The small sample behavior of some statistics which test the equality of several means." *Technometrics* **16**, 129–132.
(1974b). "Robust tests for the equality of variances." *J. Amer. Statist. Assoc.* **69**, 364–367.

Chmielewski, M. A.
(1981). "Elliptically symmetric distributions: A review and bibliography." *Int. Statist. Rev.* **49**, 67–74.

Chou, Y.-M., Arthur, K. H., Rosenstein, R. B., and Owen, D. B.
(1984). "New representations of the noncentral chi-square density and cumulative." *Comm. Statist.—Theor. Meth.* **13**, 2673–2678.

Cima, J. A. and Hochberg, Y.
(1976). "On optimality criteria in simultaneous interval estimation." *Comm. Statist.—Theor. Meth.* **A5**(9), 875–882.

Clinch, J. C. and Kesselman, H. J.
(1982). "Parametric alternatives to the analysis of variance." *J. Educ. Statist.* **7**, 207–214.

Cochran, W. G.
(1957). "Analysis of covariance: Its nature and uses." *Biometrics* **13**, 261–281.

Cochran, W. G. and Cox, G. H.
(1957). *Experimental Designs*, 2nd ed., Wiley, New York.

Cohen, A. and Miller, J. J.

(1976). "Some remarks on Scheffé's two-way mixed model." *Amer. Statistician* **30**, 36–37.

Cohen, J.

(1977). *Statistical Power Analysis for the Behavioral Sciences*, revised ed., Academic, New York.

[Advocates the consideration of power attainable against the alternatives of interest, and provides the tables needed for this purpose for some of the most common tests.]

Conover, W. J., Johnson, M. E., and Johnson, M. M.

(1981). "A comparative study of tests for homogeneity of variances, with applications to the outer continental shelf bidding data." *Technometrics* **23**, 351–361.

Cox, D. R.

(1958). *Planning of Experiments*. Wiley, New York.

Cyr, J. L. and Manoukian, E. B.

(1982). "Approximate critical values with error bounds for Bartlett's test of homogeneity of variances for unequal sample sizes." *Comm. Statist.—Theor. Meth.* **11**, 1671–1680.

Darroch, J. N. and Speed, T. P.

(1983). "Additive and multiplicative models and interactions." *Ann. Statist.* **11**, 724–738.

Das Gupta, S. and Perlman, M. D.

(1974). "Power of the noncentral *F*-test: Effect of additional variates on Hotelling's *T*-test." *J. Amer. Statist. Assoc.* **69**, 174–180.

Draper, D.

(1981). *Rank-Based Robust Analysis of Linear Models*, Ph.D. Thesis, Dept. of Statistics, Univ. of California, Berkeley.

(1983). *Rank-Based Robust Analysis of Linear Models. I. Exposition and Background*, Tech. Report No. 17, Dept. of Statistics, Univ. of California, Berkeley.

Duncan, D. B.

(1955). "Multiple range and multiple *F*-tests." *Biometrics* **11**, 1–42.

[An exposition of the ideas of one of the early workers in the area of multiple comparisons.]

Edgington, E. S.

(1980). *Randomization Tests*, Marcel Dekker, New York.

Eicker, F.

(1963). "Asymptotic normality and consistency of the least squares estimators for families of linear regressions." *Ann. Math. Statist.* **34**, 447–456.

Einot, I. and Gabriel, K. R.

(1975). "A study of the powers of several methods of multiple comparisons." *J. Amer. Statist. Assoc.* **70**, 574–583.

Eisenhart, C.

(1947). "The assumptions underlying the analysis of variance." *Biometrics* **3**, 1–21.

[Discusses the distinction between model I and model II.]

Fisher, R. A.

(1924). "On a distribution yielding the error functions of several well-known statistics." In *Proc. Int. Math. Congress*, Toronto, 805–813.

[Discusses the use of the *z*-distribution (which is equivalent to the *F*-distribution) in analysis of variance (model I) and regression analysis.]

(1925). *Statistical Methods for Research Workers*, 1st ed., Oliver and Boyd, Edinburgh.

(1928). "The general sampling distribution of the multiple correlation coefficient." *Proc. Roy. Soc.* (*A*) **121**, 654–673.

[Derives the noncentral χ^2- and noncentral beta-distributions and the distribution of the sample multiple correlation coefficient for arbitrary values of the population multiple correlation coefficient.]

(1935). *The Design of Experiments*. 1st ed., Oliver and Boyd, Edinburgh.

Freeman, M. F. and Tukey, J. W.

(1950). "Transformations related to the angular and the square root." *Ann. Math. Statist.* **21** 607–611.

Gabriel, K. R.

(1964). "A procedure for testing the homogeneity of all sets of means in analysis of variance." *Biometrics* **20**, 459–477.

Glaser, R. E.

(1982). "Bartlett's test of homogeneity of variances." *Encycl. Statist. Sci.* **1**, 189–191.

Graybill, F. A.

(1976). *Theory and Application of the Linear Model*. Duxbury Press, North Scituate, Mass.

Hahn, G. J.

(1982). "Design of experiments: An annotated bibliography." In *Encycl. Statist. Sci.*, Vol. 2, Wiley, New York.

Hájek, J. and Šidák, Z.

(1967). *Theory of Rank Tests*, Academia, Prague.

Hansen, O. H. and Torgersen, E. N.

(1974). "Comparison of linear normal experiments." *Ann. Statist.* **2**, 367–373.

[Problems 11, 12].

Harville, D. A.

(1978). "Alternative formulations and procedures for the two-way mixed model." *Biometrics* **34**, 441–454.

Hegemann, V. and Johnson, D. E.

(1976). "The power of two tests for nonadditivity." *J. Amer. Statist. Assoc.* **71**, 945–948.

Herbach, L. H.

(1959). "Properties of Model II–type analysis of variance tests." *Ann. Math. Statist.* **30**, 939–959.

Hettmansperger, T. P.

(1984). *Statistical Inference Based on Ranks*, Wiley, New York.

Hinkley, D. V. and Runger, G.

(1984). "The analysis of transformed data." (with discussion). *J. Amer. Statist. Assoc.* **79**, 302–320.

Hocking, R. R.

(1973). "A discussion of the two-way mixed model." *Amer. Statistician* **27**, 148–152.

Hocking, R. R. and Speed, F. M.

(1975). "A full rank analysis of some linear model problems." *J. Amer. Statist. Assoc.* **70**, 706–712.

Holm, S.

(1979). "A simple sequentially rejective multiple test procedure." *Scand. J. Statist.* **6**, 65–70.

Hotelling, H.

(1953). "New light on the correlation coefficient and its transforms." *J. Roy. Statist. Soc.* (*B*) **15**, 193–224.

Houtman, A. M. and Speed, T. P.

(1983). "Balance in designed experiments with orthogonal block structure." *Ann. Statist.* **11**, 1069–1085.

Hsu, P. L.
(1941). "Analysis of variance from the power function stand-point." *Biometrika* **32**, 62–69. [Shows that the test (7) is UMP among all tests whose power function depends only on the noncentrality parameter.]

Huber, P. J.
(1973). "Robust regression: Asymptotics, conjectures and Monte Carlo." *Ann. Statist.* **1**, 799–821.
[Obtains the robustness conditions (55) and (56); related results are given by Eicker (1963).]

Hunt, G. and Stein, C.M.
(1946). "Most stringent tests of statistical hypotheses." Unpublished.
[Proves the test (7) to be UMP almost invariant.]

Jagers, P.
(1980). "Invariance in the linear model—an argument for χ^2 and F in nonnormal situations." *Statistics* **11**, 455–464.

James, G. S.
(1951). "The comparison of several groups of observations when the ratios of the population variances are unknown." *Biometrika* **38**, 324–329.
(1954). "Tests of linear hypotheses in univariate and multivariate analysis when the ratios of the population variances are unknown." *Biometrika* **41**, 19–43.

Johansen, S.
(1980). "The Welch–James approximation to the distribution of the residual sum of squares in a weighted linear regression." *Biometrika* **67**, 85–92.

John, A. and Quenouille, M. H.
(1977). *Experiments: Design and Analysis*, 2nd ed., Hafner, New York.

John, P. W.
(1971). *Statistical Design and Analysis of Experiments*, Macmillan, New York.

Johnson, N. L. and Kotz, S.
(1970). *Distributions in Statistics: Continuous Univariate Distributions* (2 vols.), Houghton Mifflin, New York.

Kanoh, S. and Kusunoki, U.
(1984). "One sided simultaneous bounds in linear regression." *J. Amer. Statist. Assoc.* **79**, 715–719.

Kariya, T.
(1981). "Robustness of multivariate tests." *Ann. Statist.* **9**, 1267–1275.

Kempthorne, O.
(1952). *The Design and Analysis of Experiments*, Wiley, New York.
(1955). "The randomization theory of experimental inference." *J. Amer. Statist. Assoc.* **50**, 946–967.

Kendall, M. G., Stuart, A., and Ord, J. K.
(1983). *The Advanced Theory of Statistics* 4th ed., Vol. 3, Hafner, New York.

Kiefer, J.
(1958). "On the nonrandomized optimality and randomized nonoptimality of symmetrical designs." *Ann. Math. Statist.* **29**, 675–699.
(1980). "Optimal design theory in relation to combinatorial design." In *Combinatorial Mathematics, Optimal Designs, and Their Applications* (Shrivastava, ed.), North Holland.

King, M. L. and Hillier, G. H.
(1985) "Locally best invariant tests of the error covariance matrix of the linear regression model. *J. Roy. Statist. Soc.* **47**, 98–102.

Kolodziejczyk, S.
(1935). "An important class of statistical hypotheses." *Biometrika* **37**, 161–190.
[Discussion of the general linear univariate hypothesis from the likelihood-ratio point of view.]

Lehmann, E. L.
(1975). *Nonparametrics: Statistical Methods Based on Ranks*, Holden-Day, San Francisco.

Lehmann, E. L. and Shaffer, J. P.
(1979). "Optimal significance levels for multistage comparison procedures." *Ann. Statist.* **7**, 27–45.

Lehmann, E. L. and Stein, C.M.
(1953). "The admissibility of certain invariant statistical tests involving a translation parameter." *Ann. Math. Statist.* **24**, 473–479.

Lorenzen, T. J.
(1984). "Randomization and blocking in the design of experiments." *Comm. Statist.—Theor. Meth.* **13**, 2601–2623.

Mack, G. A. and Skillings, J. H.
(1980). "A Friedman type rank test for main effects in a two-factor ANOVA." *J. Amer. Statist. Assoc.* **75**, 947–951.

Marasinghe, M. C. and Johnson, D. E.
(1981). "Testing subhypotheses in the multiplicative interaction model." *Technometrics* **23**, 385–393.

Marcus, R., Peritz, E. and Gabriel, K. R.
(1976). "On closed testing procedures with special reference to ordered analysis of variance." *Biometrika* **63**, 655–660.

Mardia, K. V. and Zemroch, P. J.
(1978). *Tables of the F- and Related Distributions with Algorithms*, Academic, London.
[Extensive tables of critical values for the central F- and related distributions.]

Marshall, A. W. and Olkin, I.
(1979). *Inequalities: Theory of Majorization and Its Applications*, Academic, New York.

McKean, J. W. and Schrader, R. M.
(1982). "The use and interpretation of robust analysis of variance." In *Modern Data Analysis* (Launer and Siegel, eds.), Academic, New York.

Miller, J.
(1977). "Asymptotic properties of maximum likelihood estimates in the mixed model of the analysis of variance." *Ann. Statist.* **5**, 746–762.

Miller, R. G.
(1977). "Developments in multiple comparisons 1966–1976." *J. Amer. Statist. Assoc.* **72**, 779–788.
(1981). *Simultaneous Statistical Inference*, 2nd ed., Springer, Berlin–New York.

Naiman, D. Q.
(1984a). "Average width optimality of simultaneous confidence bounds." *Ann. Statist.* **12**, 1199–1214.
(1984b). "Optimal simultaneous confidence bounds." *Ann. Statist.* **12**, 702–715.

Olshen, R. A.
(1973). "The conditional level of the F-test." *J. Amer. Statist. Assoc.* **68**, 692–698.

Pearson, E. S. and Hartley, H. O.
(1972). *Biometrika Tables for Statisticians*. Cambridge U.P., Cambridge.

Peritz, E.
(1965). "On inferring order relations in analysis of variance." *Biometrics* **21**, 337–344.

Piegorsch, W. W.
(1985a). "Admissible and optimal confidence bounds in simple linear regression." *Ann. Statist.* **13**, 801–817.
(1985b). "Average width optimality for confidence bands in simple linear regression." *J. Amer. Statist. Assoc.* **80**, 692–697.

Prescott, P.
(1975). "A simple alternative to Student's *t*." *Appl. Statist.* **24**, 210–217.

Randles, R. H. and Wolfe, D. A.
(1979). *Introduction to the Theory of Nonparametric Statistics*, Wiley, New York.

Richmond, J.
(1982). "A general method for constructing simultaneous confidence intervals." *J. Amer. Statist. Assoc.* **77**, 455–460.

Robinson, J.
(1973). "The large-sample power of permutation tests for randomization models." *Ann. Statist.* **1**, 291–296.
[Discusses the asymptotic performance of the permutation version of the *F*-test in randomized block experiments.]
(1983). "Approximations to some test statistics for permutation tests in a completely randomized design." *Austr. J. Statist.* **25**, 358–369.

Ronchetti, E.
(1982). "Robust alternatives to the *F*-test for the linear model." In *Probability and Statistical Inference* (Grossman, Pflug, and Wertz, eds.), D. Reidel, Dordrecht.

Rothenberg, T. J.
(1984). "Hypothesis testing in linear models when the error covariance matrix is nonscalar." *Econometrica* **52**, 827–842.

Scheffé, H.
(1953). "A method for judging all contrasts in the analysis of variance." *Biometrika* **40** 87–104.
[Develops Scheffé's *S*-method. See also Olshen (1973) and Scheffé (1977).]
(1956). "A 'mixed model' for the analysis of variance." *Ann. Math. Statist.* **27**, 23–36 and 251–271.
[Example 12.]
(1958). "Fitting straight lines when one variable is controlled." *J. Amer. Statist. Assoc.* **53**, 106–117.
[Problem 28.]
(1959). *Analysis of Variance*, Wiley, New York.
(1977). "A note on a reformulation of the *S*-method of multiple comparison (with discussion)." *J. Amer. Statist. Assoc.* **72**, 143–146.

Seal, H. L.
(1967). "Studies in the history of probability and statistics XV. The historical development of the Gauss linear model." *Biometrika* **54**, 1–24.

Seber, G. A. F.
(1977). *Linear Regression Analysis*, Wiley, New York.

Serfling, R. J.
(1980). *Approximation Theorems of Mathematical Statistics*, Wiley, New York.

Shaffer, J. P.

(1977a), "Multiple comparisons emphasizing selected contrasts: An extension and generalization of Dunnett's procedure." *Biometrics* **33**, 293–303.

(1977b). "Reorganization of variables in analysis of variance and multidimensional contingency tables." *Psych. Bull.* **84**, 220–228.

(1980). "Control of directional errors with stagewise multiple test procedures." *Ann. Statist.* **8**, 1342–1347.

(1981). "Complexity: An interpretability criterion for multiple comparisons." *J. Amer. Statist. Assoc.* **76**, 395–401.

(1984). "Issues arising in multiple comparisons among populations." In *Proc. Seventh Conference on Probab. Theory* (Iosifescu, ed.). Edit. Acad. Republ. Soc. Romania. Bucharest.

Silvey, S. D.

(1980). *Optimal Design*, Chapman and Hall, London.

Smith, D. W. and Murray, L. W.

(1984). "An alternative to Eisenhart's Model II and mixed model in the case of negative variance estimates." *J. Amer. Statist. Assoc.* **79**, 145–151.

Speed, F. M., Hocking, R. R. and Hackney, O. P.

(1978). "Methods of analysis of linear models with unbalanced data." *J. Amer. Statist. Assoc.* **73**, 105–112.

Spjøtvoll, E.

(1967). "Optimum invariant tests in unbalanced variance components models." *Ann. Math. Statist.* **38**, 422–428.

(1972). "On the optimality of some multiple comparison procedures." *Ann. Math. Statist.* **43**, 398–411.

(1974). "Multiple testing in analysis of variance." *Scand. J. Statist.* **1**, 97–114.

Tiku, M. L.

(1967). "Tables of the power of the F-test." *J. Amer. Statist. Assoc.* **62**, 525–539.

(1972). "More tables of the power of the F-test." *J. Amer. Statist. Assoc.* **67**, 709–710.

(1985a). "Noncentral chi-square distribution." *Encycl. Statist. Sci.*, **6**, 276–280.

(1985b). "Noncentral F-distribution." *Encycl. Statist. Sci.* **6**, 280–284.

Tukey, J. W.

(1949). "One degree of freedom for non-additivity." *Biometrics* **5**, 232–242.

(1953). "The problem of multiple comparisons."

[This unpublished MS* was widely distributed and exerted a strong influence on the development and acceptance of multiple comparison procedures. It pioneered many of the basic ideas, including the T-method and a first version of Theorem 1.]

Wald, A.

(1942). "On the power function of the analysis of variance test." *Ann. Math. Statist.* **13**, 434–439.

[Problem 5. This problem is also treated by Hsu, "On the power function of the E^2-test and the T^2-test", *Ann. Math. Statist.* **16** (1945), 278–286.]

Welch, B. L.

(1951). "On the comparison of several mean values: An alternative approach." *Biometrika* **38**, 330–336.

Wijsman, R. A.

(1979). "Constructing all smallest simultaneous confidence sets in a given class, with applications to MANOVA." *Ann. Statist.* **7**, 1003–1018.

*To be published as part of Tukey's collected papers by Wadsworth.

(1980). "Smallest simultaneous confidence sets with applications in multivariate analysis." *Multivariate Anal. V*, 483–489.

[Optimality results for simultaneous confidence sets including those of Section 9.]

Wilk, M. B. and Kempthorne, O.

(1955). "Fixed, mixed, and random models." *J. Amer. Statist. Assoc.* **50**, 1144–1167.

Working, H. and Hotelling, H.

(1929). "Application of the theory of error to the interpretation of trends." *J. Amer. Statist. Assoc.* **24**, Mar. Suppl., 73–85.

Wynn, H. P.

(1984). "An exact confidence band for one-dimensional polynomial regression." *Biometrika* **71**, 375–379.

Wynn, H. P. and Bloomfield, P.

(1971). "Simultaneous confidence bands in regression analysis" (with discussion). *J. Roy. Statist. Soc. (B)* **33**, 202–217.

CHAPTER 8

Multivariate Linear Hypotheses

1. A CANONICAL FORM

The univariate linear models of the preceding chapter arise in the study of the effects of various experimental conditions (factors) on a single characteristic such as yield, weight, length of life, or blood pressure. This characteristic is assumed to be normally distributed with a mean which depends on the various factors under investigation, and a variance which is independent of these factors. We shall now consider the multivariate analogue of this model, which is appropriate when one is concerned with the effect of one or more factors simultaneously on several characteristics, for example the effect of a change in the diet of dairy cows on both fat content and quantity of milk.

The multivariate generalization of a real-valued normally distributed random variable is a random vector $(X_1, \ldots, X_p)$ with the *multivariate normal probability density*

$$(1) \qquad \frac{\sqrt{|A|}}{(2\pi)^{\frac{1}{2}p}} \exp\left[-\tfrac{1}{2} \sum\sum a_{ij}(x_i - \xi_i)(x_j - \xi_j)\right],$$

where the matrix $A = (a_{ij})$ is positive definite, and $|A|$ denotes its determinant. The means and covariance matrix of the X's are given by

$$(2) \qquad E(X_i) = \xi_i, \qquad E(X_i - \xi_i)(X_j - \xi_j) = \sigma_{ij}, \qquad (\sigma_{ij}) = A^{-1}.$$

Consider now n independent multivariate normal vectors $X_\alpha = (X_{\alpha 1}, \ldots, X_{\alpha p})$, $\alpha = 1, \ldots, n$, with means $E(X_{\alpha i}) = \xi_{\alpha i}$ and common covariance matrix A^{-1}. As in the univariate case, a *multivariate linear hypothesis* is defined in terms of two linear subspaces Π_Ω and Π_ω of n-dimensional space having dimensions $s < n$ and $0 \leq s - r < s$. It is assumed known that for all $i = 1, \ldots, p$, the vectors $(\xi_{1i}, \ldots, \xi_{ni})$ lie in Π_Ω; the hypothesis to be tested specifies that they lie in Π_ω. This problem is reduced to canonical form by applying to each of the p vectors $(X_{1i}, \ldots, X_{ni})$ the orthogonal transformation (1) of Chapter 7. If

$$
X = \begin{pmatrix} X_{11} & \cdots & X_{1p} \\ \vdots & & \vdots \\ X_{n1} & \cdots & X_{np} \end{pmatrix}
$$

and the transformed variables are denoted by $X_{\alpha i}^*$, the transformation may be written in matrix form as

$$
X^* = CX,
$$

where $C = (c_{\alpha\beta})$ is an orthogonal matrix.

To obtain the joint distribution of the $X_{\alpha i}^*$ consider first the covariance of any two of them, say $X_{\alpha i}^* = \sum_{\gamma=1}^n c_{\alpha\gamma} X_{\gamma i}$ and $X_{\beta j}^* = \sum_{\delta=1}^n c_{\beta\delta} X_{\delta j}$. Using the fact that the covariance of $X_{\gamma i}$ and $X_{\delta j}$ is zero when $\gamma \neq \delta$ and σ_{ij} when $\gamma = \delta$, we have

$$
\mathrm{Cov}(X_{\alpha i}^*, X_{\beta j}^*) = \sum_{\gamma=1}^n \sum_{\delta=1}^n c_{\alpha\gamma} c_{\beta\delta} \mathrm{Cov}(X_{\gamma i}, X_{\delta j})
$$

$$
= \sigma_{ij} \sum_{\gamma=1}^n c_{\alpha\gamma} c_{\beta\gamma} = \begin{cases} \sigma_{ij} & \text{when } \alpha = \beta, \\ 0 & \text{when } \alpha \neq \beta. \end{cases}
$$

The rows of X^* are therefore again independent multivariate normal vectors with common covariance matrix A^{-1}. It follows as in the univariate case that the vectors of means satisfy

$$
\xi_{s+1, i}^* = \cdots = \xi_{ni}^* = 0 \qquad (i = 1, \ldots, p)
$$

under Ω, and that the hypothesis becomes

$$
H: \xi_{1i}^* = \cdots = \xi_{ri}^* = 0 \qquad (1 = 1, \ldots, p).
$$

Changing notation so that Y's, U's, and Z's denote the first r, the next $s - r$, and the last $m = n - s$ sample vectors, we therefore arrive at the following *canonical form*. The vectors Y_α, U_β, Z_γ ($\alpha = 1, \ldots, r$; $\beta = 1, \ldots, s - r$; $\gamma = 1, \ldots, m$) are independently distributed according to p-variate normal distributions with common covariance matrix A^{-1}. The means of the Z's are given to be zero, and the hypothesis H is to be tested that the means of the Y's are zero. If

$$Y = \begin{pmatrix} Y_{11} & \cdots & Y_{1p} \\ \vdots & & \vdots \\ Y_{r1} & \cdots & Y_{rp} \end{pmatrix} \quad \text{and} \quad Z = \begin{pmatrix} Z_{11} & \cdots & Z_{1p} \\ \vdots & & \vdots \\ Z_{m1} & \cdots & Z_{mp} \end{pmatrix},$$

invariance and sufficiency will be shown below to reduce the observations to the $p \times p$ matrices $Y'Y$ and $Z'Z$. It will then be convenient to have an expression of these statistics in terms of the original observations.

As in the univariate case, let $(\hat{\xi}_{1i}, \ldots, \hat{\xi}_{ni})$ and $(\hat{\hat{\xi}}_{1i}, \ldots, \hat{\hat{\xi}}_{ni})$ denote the projections of the vector $(X_{1i}, \ldots, X_{ni})$ on Π_Ω and Π_ω. Then

$$\sum_{\alpha=1}^{n} (X_{\alpha i} - \hat{\xi}_{\alpha i})(X_{\alpha j} - \hat{\xi}_{\alpha j})$$

is the inner product of two vectors, each of which is the difference between a given vector and its projection on Π_Ω. It follows that this quantity is unchanged under orthogonal transformations of the coordinate system in which the variables are expressed. Now the transformation

$$C \begin{pmatrix} X_{1i} \\ \vdots \\ X_{ni} \end{pmatrix}$$

may be interpreted as expressing the vector $(X_{1i}, \ldots, X_{ni})$ in a new coordinate system, the first s coordinate axes of which lie in Π_Ω. The projection on Π_Ω of the transformed vector $(Y_{1i}, \ldots, Y_{ri}, U_{1i}, \ldots, U_{s-r,i}, Z_{1i}, \ldots, Z_{mi})$ is $(Y_{1i}, \ldots, Y_{ri}, U_{1i}, \ldots, U_{s-r,i}, 0, \ldots, 0)$, so that the difference between the vector and its projection is $(0, \ldots, 0, Z_{1i}, \ldots, Z_{mi})$. The ijth element of $Z'Z$ is therefore given by

$$(3) \qquad \sum_{\gamma=1}^{m} Z_{\gamma i} Z_{\gamma j} = \sum_{\alpha=1}^{n} (X_{\alpha i} - \hat{\xi}_{\alpha i})(X_{\alpha j} - \hat{\xi}_{\alpha j}).$$

Analogously, the projection of the transformed vector $(Y_{1i}, \ldots, Y_{ri}, U_{1i}, \ldots, U_{s-r,i}, 0, \ldots, 0)$ on Π_ω is $(0, \ldots, 0, U_{1i}, \ldots, U_{s-r,i}, 0, \ldots, 0)$, and the difference between the projections on Π_Ω and Π_ω is therefore $(Y_{1i}, \ldots, Y_{ri}, 0, \ldots, 0, \ldots, 0)$. It follows that the sum $\sum_{\beta=1}^{r} Y_{\beta i} Y_{\beta j}$ is equal to the inner product (for the ith and jth vector) of the difference of these projections. On comparing this sum with the expression of the same inner product in the original coordinate system, it is seen that the ijth element of $Y'Y$ is given by

$$(4) \qquad \sum_{\beta=1}^{r} Y_{\beta i} Y_{\beta j} = \sum_{\alpha=1}^{n} \left(\hat{\xi}_{\alpha i} - \hat{\hat{\xi}}_{\alpha i} \right) \left(\hat{\xi}_{\alpha j} - \hat{\hat{\xi}}_{\alpha j} \right).$$

2. REDUCTION BY INVARIANCE

The multivariate linear hypothesis, described in the preceding section in canonical form, remains invariant under certain groups of transformations. To obtain maximal invariants under these groups we require, in addition to some of the standard theorems concerning quadratic forms, the following lemma.

Lemma 1.　*If M is any $m \times p$ matrix, then*

(i)　*$M'M$ is positive semidefinite,*

(ii)　*the rank of $M'M$ equals the rank of M, so that in particular $M'M$ is nonsingular if and only if $m \geq p$ and M is of rank p.*

Proof.　(i): Consider the quadratic form $Q = u'(M'M)u$. If $w = Mu$, then

$$Q = w'w \geq 0.$$

(ii): The sum of squares $w'w$ is zero if and only if the vector w is zero, and the result follows from the fact that the solutions u of the system of equations $Mu = 0$ form a linear space of dimension $p - \rho$, where ρ is the rank of M.

We shall now consider three groups under which the problem remains invariant.

G_1.　Addition of an arbitrary constant $d_{\beta i}$ to each of the variables $U_{\beta i}$ leaves the problem invariant, and this eliminates the U's, since the Y's and Z's are maximal invariant under this group.

G_2.　In the process of reducing the problem to canonical form it was seen that an orthogonal transformation

$$Y^* = CY$$

affects neither the independence of the row vectors of Y nor the covariance matrix of these vectors. The means of the Y^*'s are zero if and only if those of the Y's are, and hence the problem remains invariant under these transformations.

The matrix $Y'Y$ of inner products of the column vectors of Y is invariant under G_2, since $Y^{*\prime}Y^* = Y'C'CY = Y'Y$. The matrix $Y'Y$ will be proved to be maximal invariant by showing that $Y'Y = Y^{*\prime}Y^*$ implies the existence of an orthogonal matrix C such that $Y^* = CY$. Consider first the case $r = p$. Without loss of generality the p column vectors of Y can be assumed to be linearly independent, since the exceptional set of Y's for which this does not hold has measure zero. The equality $Y'Y = Y^{*\prime}Y^*$ implies that $C = Y^*Y^{-1}$ is orthogonal and that $Y^* = CY$, as was to be proved. Suppose next that $r > p$. There is again no loss of generality in assuming the p column vectors of Y to be linearly independent. Since for any two p-dimensional subspaces of r-space there exists an orthogonal transformation taking one into the other, it can be assumed that (after a suitable orthogonal transformation) the p column vectors of Y and Y^* lie in the same p-space, and the problem is therefore reduced to the case $r = p$. If finally $r < p$, the first r column vectors of Y can be assumed to be linearly independent. Denoting the matrices formed by the first r and last $p - r$ columns of Y by Y_1 and Y_2, so that

$$Y = \begin{pmatrix} Y_1 & Y_2 \end{pmatrix},$$

one has $Y_1^{*\prime}Y_1^* = Y_1'Y_1$, and by the previous argument there exists an orthogonal matrix B such that $Y_1^* = BY_1$. From the relation $Y_1^{*\prime}Y_2^* = Y_1'Y_2$ it now follows that $Y_2^* = (Y_1^{*\prime})^{-1}Y_1'Y_2 = BY_2$, and this completes the proof.

Similarly the problem remains invariant under the orthogonal transformations

$$Z^* = DZ,$$

which leave $Z'Z$ as maximal invariant. Alternatively the reduction to $Z'Z$ can be argued from the fact that $Z'Z$ together with the Y's and U's form a set of sufficient statistics. In either case the problem under the groups G_1 and G_2 reduces to the two matrices $V = Y'Y$ and $S = Z'Z$.

G_3. We now impose the restriction $m \geq p$ (see Problem 1), which assures that there are enough degrees of freedom to provide a reasonable estimate of the covariance matrix, and consider the transformations

$$Y^* = YB, \qquad Z^* = ZB,$$

where B is any nonsingular $p \times p$ matrix. These transformations act separately on each of the independent multivariate normal vectors $(Y_{\beta 1}, \ldots, Y_{\beta p})$, $(Z_{\gamma 1}, \ldots, Z_{\gamma p})$, and clearly leave the problem invariant. The induced transformation in the space of $V = Y'Y$ and $S = Z'Z$ is

$$V^* = B'VB, \qquad S^* = B'SB.$$

Since $|B'(V - \lambda S)B| = |B|^2 |V - \lambda S|$, the roots of the determinantal equation

(5) $$|V - \lambda S| = 0$$

are invariant under this group. To see that they are maximal invariant, suppose that the equations $|V - \lambda S| = 0$ and $|V^* - \lambda S^*| = 0$ have the same roots. One may again without loss of generality restrict attention to the case that p of the row vectors of Z are linearly independent, so that the matrix Z has rank p, and that the same is true of Z^*. The matrix S is then positive definite by Lemma 1, and it follows from the theory of the simultaneous reduction to diagonal form of two quadratic forms[†] that there exists a nonsingular matrix B_1 such that

$$B_1'VB_1 = \Lambda, \qquad B_1'SB_1 = I,$$

where Λ is a diagonal matrix whose elements are the roots of (5) and I is the identity matrix. There also exists B_2 such that

$$B_2'V^*B_2 = \Lambda, \qquad B_2'S^*B_2 = I,$$

and thus $B = B_1 B_2^{-1}$ transforms V into V^* and S into S^*.

Of the roots of (5), which constitute a maximal set of invariants, some may be zero. In fact, since these roots are the diagonal elements of Λ, the number of nonzero roots is equal to the rank of Λ and hence to the rank of $V = B_1'^{-1}\Lambda B_1^{-1}$, which by Lemma 1 is $\min(p, r)$. When this number is > 1, a UMP invariant test does not exist. The case $p = 1$ is that of a univariate linear hypothesis treated in Section 1 of Chapter 7. We shall now consider the remaining possibility that $r = 1$.

When $r = 1$, the equation (5), and hence the equivalent equation

$$|VS^{-1} - \lambda I| = 0,$$

has only one nonzero root. All coefficients of powers of λ of degree $< p - 1$ therefore vanish in the expression of the determinant as a polynomial in λ, and the equation becomes

$$(-\lambda)^p + W(-\lambda)^{p-1} = 0,$$

[†] See for example Anderson (1984, Appendix A, Theorem A.2.2).

where W is the sum of the diagonal elements (trace) of VS^{-1}. If S^{ij} denotes the ijth element of S^{-1} and the single Y-vector is $(Y_1, \ldots, Y_p)$, an easy computation shows that

$$(6) \qquad W = \sum_{i=1}^{p} \sum_{j=1}^{p} S^{ij} Y_i Y_j.$$

A necessary and sufficient condition for a test to be invariant under G_1, G_2, and G_3 is therefore that it depends only on W.

The distribution of W depends only on the maximal invariant in the parameter space; this is found to be

$$(7) \qquad \psi^2 = \sum_{i=1}^{p} \sum_{j=1}^{p} a_{ij} \eta_i \eta_j,$$

where $\eta_i = E(Y_i)$, and the probability density of W is given by (Problems 5–7)

$$(8) \qquad p_\psi(w) = e^{-\frac{1}{2}\psi^2} \sum_{k=0}^{\infty} \frac{\left(\frac{1}{2}\psi^2\right)^k}{k!} C_k \frac{w^{\frac{1}{2}p-1+k}}{(1+w)^{\frac{1}{2}(m+1)+k}}.$$

This is the same as the density of the test statistic in the univariate case, given as (6) of Chapter 7, with $r = p$ and $n - s = m + 1 - p$. For any $\psi_0 < \psi_1$ the ratio $p_{\psi_1}(w)/p_{\psi_0}(w)$ is an increasing function of w, and it follows from the Neyman–Pearson lemma that the most powerful invariant test for testing $H: \eta_1 = \cdots = \eta_p = 0$ rejects when W is too large, or equivalently when

$$(9) \qquad \frac{m+1-p}{p} W > C.$$

The quantity mW, which for $p = 1$ reduces to the square of Student's t, is essentially Hotelling's T^2-statistic, to which it specializes in the one-sample test to be considered in the next section. The constant C is determined from the fact that for $\psi = 0$ the statistic $(m + 1 - p)W/p$ has the F-distribution with p and $m + 1 - p$ degrees of freedom. As in the univariate case, there also exists a UMP invariant test of the more general hypothesis $H': \psi^2 \leq \psi_0^2$, with rejection region $W > C'$.

3. THE ONE- AND TWO-SAMPLE PROBLEMS

The simplest special case of a linear hypothesis with $r = 1$ is the hypothesis $H: \xi_1 = \cdots = \xi_p = 0$, where $(X_{\alpha 1}, \ldots, X_{\alpha p})$, $\alpha = 1, \ldots, n$, is a sample from a p-variate normal distribution (1) with unknown mean $(\xi_1, \ldots, \xi_p)$, covariance matrix $\Sigma = A^{-1}$, and $p \leq n - 1$. It is seen from Example 4 of Chapter 7 that

$$\hat{\xi}_{\alpha i} = \sum_{\beta=1}^{n} \frac{X_{\beta i}}{n} = X_{\cdot i}, \qquad \hat{\hat{\xi}}_{\alpha i} = 0.$$

By (3), the ijth element S_{ij} of $S = Z'Z$ is therefore

$$S_{ij} = \sum_{\alpha=1}^{n} (X_{\alpha i} - X_{\cdot i})(X_{\alpha j} - X_{\cdot j}),$$

and by (4)

$$Y_i Y_j = n X_{\cdot i} X_{\cdot j}.$$

With these expressions the test statistic is the quantity W of (6), and the test is given by (9) with $s = 1$ and hence with $m = n - s = n - 1$. The statistic $T^2 = (n - 1)W$ is known as *Hotelling's T^2*. The noncentrality parameter (7) in the present case reduces to $\psi^2 = \Sigma\Sigma a_{ij} \xi_i \xi_j$.

The test shares the robustness properties of the corresponding univariate t-test discussed in Chapter 5, Section 4. Suppose that $(X_{\alpha 1}, \ldots, X_{\alpha p})$ is a sample from any p-variate distribution F with vector mean zero and finite, nonsingular covariance matrix Σ, and write

(10) $$T^2 = \Sigma\Sigma \sqrt{n} X_{\cdot i}(n - 1)S^{ij}\sqrt{n} X_{\cdot j}.$$

Using the fact that $S_{ij}/(n - 1)$ tends in probability to σ_{ij} and that $(\sqrt{n} X_{\cdot 1}, \ldots, \sqrt{n} X_{\cdot p})$ has a p-variate normal limit distribution with covariance matrix Σ, it is seen (Problem 8) that the null distribution of T^2 tends to the χ_p^2-distribution as $n \to \infty$. Thus, asymptotically the significance level of the T^2-test is independent of F. However, for small n, the differences may be substantial. For details see for example Everitt (1979), Davis (1982), Srivastava and Awan (1982), and Seber (1984).

The T^2-test was shown by Stein (1956) to be admissible against the class of alternatives $\psi^2 \geq c$ for any $c > 0$ by the method of Theorem 8 of Chapter 6. Against the class of alternatives $\psi^2 \leq c$ admissibility was proved by Kiefer and Schwartz (1965) [see Problem 47, and also Schwartz (1967) and (1969)].

The problem of testing H against one-sided alternatives such as $K : \xi_i \geq 0$ for all i, with at least one inequality strict, is treated by Perlman (1969) and in Barlow et al. (1972), which gives a survey of the literature. Minimal complete classes for this and related problems are discussed by Marden (1982).

Most accurate equivariant confidence sets for the unknown mean vector $(\xi_1, \ldots, \xi_p)$ are obtained from the UMP invariant test of $H: \xi_i = \xi_{i0}$ $(i = 1, \ldots, p)$, which has acceptance region

$$n \sum\sum (X_{\cdot i} - \xi_{i0})(n-1)S^{ij}(X_{\cdot j} - \xi_{j0}) \leq C.$$

The associated confidence sets are therefore ellipsoids

(11) $$n \sum\sum (\xi_i - X_{\cdot i})(n-1)S^{ij}(\xi_j - X_{\cdot j}) \leq C$$

centered at $(X_{\cdot 1}, \ldots, X_{\cdot p})$. These confidence sets are equivariant under the groups G_1–G_3 of Section 2 (Problem 9), and by Lemma 4 of Chapter 6 are therefore uniformly most accurate among all equivariant confidence sets at the specified level.

Consider next the two-sample problem in which $(X_{\alpha 1}^{(1)}, \ldots, X_{\alpha p}^{(1)})$, $\alpha = 1, \ldots, n_1$, and $(X_{\beta 1}^{(2)}, \ldots, X_{\beta p}^{(2)})$, $\beta = 1, \ldots, n_2$, are independent samples from multivariate normal distributions with common covariance matrix A^{-1} and means $(\xi_1^{(1)}, \ldots, \xi_p^{(1)})$ and $(\xi_1^{(2)}, \ldots, \xi_p^{(2)})$. Suppose that $p \leq n_1 + n_2 - 2$,* and consider the hypothesis $H: \xi_i^{(1)} = \xi_i^{(2)}$ for $i = 1, \ldots, p$. Then $s = 2$, and it follows from Example 5 of Chapter 7 that for all α and β

$$\hat{\xi}_{\alpha i}^{(1)} = X_{\cdot i}^{(1)}, \qquad \hat{\xi}_{\beta_i}^{(2)} = X_{\cdot i}^{(2)}$$

and

$$\hat{\hat{\xi}}_{\alpha i}^{(1)} = \hat{\hat{\xi}}_{\beta_i}^{(2)} = \frac{\sum_{\alpha=1}^{n_1} X_{\alpha i}^{(1)} + \sum_{\beta=1}^{n_2} X_{\beta i}^{(2)}}{n_1 + n_2} = \overline{X}_i.$$

Hence

$$S_{ij} = \sum_{\alpha=1}^{n_1} \left(X_{\alpha i}^{(1)} - X_{\cdot i}^{(1)} \right)\left(X_{\alpha j}^{(1)} - X_{\cdot j}^{(1)} \right) + \sum_{\beta=1}^{n_2} \left(X_{\beta i}^{(2)} - X_{\cdot i}^{(2)} \right)\left(X_{\beta j}^{(2)} - X_{\cdot j}^{(2)} \right),$$

and the expression for $Y_i Y_j$ can be simplified to

$$Y_i Y_j = n_1 \left(X_{\cdot i}^{(1)} - \overline{X}_i \right)\left(X_{\cdot j}^{(1)} - \overline{X}_j \right) + n_2 \left(X_{\cdot i}^{(2)} - \overline{X}_i \right)\left(X_{\cdot j}^{(2)} - \overline{X}_j \right).$$

*A test of H for the case that $p > n_1 + n_2 - 2$ is discussed by Dempster (1958).

Since $m = n - 2$, $T^2 = mW$ is given by

$$(12) \qquad T^2 = n(n - 2)\big(X_{\cdot}^{(1)} - X_{\cdot}^{(2)} \big)' S^{-1}\big(X_{\cdot}^{(1)} - X_{\cdot}^{(2)} \big),$$

where $n = n_1 + n_2$ and $X_{\cdot}^{(k)} = (X_{\cdot 1}^{(k)} \cdots X_{\cdot p}^{(k)})'$, $k = 1, 2$.

As in the one-sample problem, this test is robust against nonnormality for large n_1 and n_2 (Problem 10). In the two-sample case, the robustness question arises also with respect to the assumption of equal covariances for the two samples. The result here parallels that for the corresponding univariate situation: if $n_1/n_2 \to 1$, the asymptotic distribution of T^2 is the same when Σ_1 and Σ_2 are unequal as when they are equal; if $n_1/n_2 \to \rho \neq 1$, the limit distribution of T^2 derived for $\Sigma_1 = \Sigma_2$ no longer applies when the covariances differ (Problem 11).

Tests of the hypothesis $\xi_i^{(1)} = \xi_i^{(2)}$ ($i = 1, \ldots, p$) when the covariance matrices are not assumed to be equal (i.e. for the multivariate Behrens–Fisher problem) have been proposed by James (1954) and Yao (1965) and are studied further in Subrahmaniam and Subrahmaniam (1973, 1975) and Johansen (1980). Their results are summarized in Seber (1984). For related work, see Dalal (1978), Dalal and Fortini (1982), and Anderson (1984). The effect of outliers is studied by Bauer (1981).

Both the one- and the two-sample problem are examples of multivariate linear hypotheses with r equal to 1, so that a UMP invariant test exists and is of the T^2 type (9). Other problems with $r = 1$ arise in multivariate regression (Problem 13) and in some repeated-measurement problems (Section 5).

Instead of testing the value of a mean vector or the equality of two mean vectors in the one- and the two-sample problem respectively, it may be of interest to test the corresponding hypotheses $\Sigma = \Sigma_0$ or $\Sigma_1 = \Sigma_2$ concerning the covariance matrices. Since the resulting tests, as in the univariate case, are extremely sensitive to the assumption of normality, they are not very useful and we shall not consider them here. They are treated from an invariance point of view by Arnold (1981) and by Anderson (1984), who also discusses more robust alternatives. In the one-sample case, another problem of interest is that of testing the hypothesis of independence of two sets of components from each other. For the case $p = 2$, this was considered in Chapter 5, Section 13. For general p, see Problem 45.

4. MULTIVARIATE ANALYSIS OF VARIANCE (MANOVA)

When the number r of vector constraints imposed by H on a multivariate linear model with $p > 1$ exceeds 1, a UMP invariant test no longer exists. Tests based on various functions of the roots λ_i of (5) have been proposed for this case, among them

   (i) the *Lawley–Hotelling trace test*, which rejects for large values of $\sum \lambda_i$;

  (ii) the likelihood-ratio test (*Wilks* Λ), which rejects for small values of $|V|/|V + S|$ or equivalently of $\prod 1/(1 + \lambda_i)$ (Problem 18);

 (iii) the *Pillai–Bartlett trace test*, which rejects for large values of $\sum \lambda_i/(1 + \lambda_i)$;

 (iv) *Roy's maximum-root test*, which rejects for large values of max λ_i.

Since these test statistics are all invariant under the groups G_1–G_3 of Section 1, their distribution depends only on the maximal invariants in the parameter space, which are the nonzero roots of the equation

$$(13) \qquad\qquad |B - \lambda \Sigma| = 0,$$

where Σ is the common covariance matrix of $(X_{\alpha 1}, \ldots, X_{\alpha p})$ and B is the $p \times p$ matrix with (i, j)th element

$$\sum_{\alpha=1}^{n} E\left(\hat{\xi}_{\alpha i} - \hat{\bar{\xi}}_{\alpha i}\right) E\left(\hat{\xi}_{\alpha j} - \hat{\bar{\xi}}_{\alpha j}\right).$$

Some comparisons of the power of the tests (i)–(iv) are given among others by Pillai and Jayachandran (1967), Olson (1976), and Stevens (1980), and suggest that there is little difference in the power of (i)–(iii), but considerable difference with (iv). This last test tends to be more powerful against alternatives that approximate the situation in which (13) has only one nonzero root, that is, alternatives in which all but one of the roots are close to zero and there is one (positive) root that is widely separated from the others (see Problem 19 for an example). On the other hand, the maximum-root test tends to be less powerful than the other three when (13) has several roots which differ considerably from zero.

The lack of difference among (i)–(iii) is supported by a corresponding asymptotic result. To motivate the asymptotics, consider first the s-sample problem in which $(X_{\alpha 1}^{(k)}, \ldots, X_{\alpha p}^{(k)})$, $\alpha = 1, \ldots, n_k$, $k = 1, \ldots, s$, are samples of size n_k from p-variate normal distributions with mean $(\xi_1^{(k)}, \ldots, \xi_p^{(k)})$ and common covariance matrix Σ. For testing $H: \xi_i^{(1)} = \cdots = \xi_i^{(s)}$ for all $i = 1, \ldots, p$, the matrices V and S have elements (Problem 16)

$$(14) \qquad\qquad V_{ij} = \sum_k n_k \left(X_{\cdot i}^{(k)} - \bar{X}_{\cdot i}\right)\left(X_{\cdot j}^{(k)} - \bar{X}_{\cdot j}\right)$$

and

$$(15) \qquad S_{ij} = \sum_{k=1}^{s} \sum_{\alpha=1}^{n_k} \left(X_{\alpha i}^{(k)} - X_{\cdot i}^{(k)} \right)\left(X_{\alpha j}^{(k)} - X_{\cdot j}^{(k)} \right),$$

where $\overline{X}_{\cdot i} = \sum n_k X_{\cdot i}^{(k)} / \sum n_k$. Under the hypothesis, the joint distribution of the V_{ij} is independent of $n_1, \ldots, n_s$, while $S_{ij}/(n - s)$ tends in probability to the (i, j)th element σ_{ij} of Σ.

Analogously, in other analysis-of-variance situations, as the cell sizes tend to infinity, the distribution of V under H remains constant while $S_{ij}/(n - s)$ tends in probability to σ_{ij}.

Let $\lambda_1, \ldots, \lambda_a$ denote the $a = \min(p, r)$ nonzero roots of

$$(16) \qquad |V - \lambda S| = \left| V - (n - s)\lambda \frac{S}{n - s} \right| = 0,$$

and $\lambda_1^*, \ldots, \lambda_a^*$ the nonzero roots of

$$(17) \qquad |V - \lambda \Sigma| = 0,$$

the null distribution of which we suppose to be independent of n. Then it is plausible and easy to show (Problem 21) that $((n - s)\lambda_1, \ldots, (n - s)\lambda_a)$ tends in law to $(\lambda_1^*, \ldots, \lambda_a^*)$ and hence that the distribution of $T_1 = (n - s)\sum \lambda_i$ tends to that of $\sum \lambda_i^*$ as $n \to \infty$. If

$$T_2 = (n - s)\sum \frac{\lambda_i}{1 + \lambda_i} \quad \text{and} \quad T_3 = (n - s)\log \prod (1 + \lambda_i),$$

we shall now show that $T_2 - T_1$ and $T_3 - T_1$ tend to zero in probability, so that T_1, T_2, and T_3 are asymptotically equivalent and in particular have the same limit distribution.

(a) The convergence of the distribution of $(n - s)\lambda_i$ implies that $\lambda_i \to 0$ in probability and hence that $T_2 - T_1$ tends to zero in probability.

(b) The expansion $\log(1 + x) = x[1 + o(1)]$ as $x \to 0$ gives

$$(n - s)\log \prod (1 + \lambda_i) = (n - s)\sum \log(1 + \lambda_i) = (n - s)\sum \lambda_i + R_n,$$

where $R_n \to 0$ in probability by (a).

Thus, the distributions of T_1, T_2, and T_3 all tend to that of $\sum \lambda_i^*$. On the other hand, the distribution of the normalized maximum-root statistic $(n - s)\max \lambda_i$ tends to the quite different distribution of $\max \lambda_i^*$.

The null distribution of $\sum \lambda_i^*$ is the limit distribution of T_1, T_2, and T_3 and therefore provides a first, crude approximation to the distribution of these statistics under H. We shall now show that this limit distribution is χ^2 with rp degrees of freedom.

To see this, consider the linear model in its canonical form of Section 1, in which the rows of the $r \times p$ matrix Y are independent p-variate normal with common covariance matrix Σ and mean $\eta = E(Y)$, but where Σ is now assumed to be known. Under H, the matrix η is the $r \times p$ zero matrix. There exists a nonsingular transformation $Y^* = YB$ such that the covariance matrix $B'\Sigma B$ of the rows of Y^* is the identity matrix. The variables $Y_{\alpha i}^*$ ($\alpha = 1, \ldots, r$; $i = 1, \ldots, p$) are then independent normal with means $\eta_{\alpha i}^* = E(Y_{\alpha i}^*)$ and unit variance. The hypothesis becomes $H : \eta_{\alpha i}^* = 0$ for all α and i, and the UMP invariant test (under orthogonal transformations of the pr-dimensional sample space) rejects when $\sum\sum Y_{\alpha i}^{*2} > C$. The test statistic $\sum\sum Y_{\alpha i}^{*2}$ is the trace of the matrix $V^* = Y^{*\prime} Y^* = B'VB$ and is therefore the sum of the roots of the equation $|B'VB - \lambda I| = 0$. Since $I = B'\Sigma B$, they are also the roots of $|V - \lambda \Sigma| = 0$ and hence $\sum\sum Y_{\alpha i}^{*2} = \sum \lambda_i^*$, and this completes the proof.

More accurate approximations, and tables of the null distributions of the four tests, are given in Anderson (1984) and Seber (1984). p-values are also provided by the standard computer packages.

The robustness against nonnormality of tests for univariate linear hypotheses extends to the joint distribution of the roots λ_i of (5) as it did for the single root in the case $r = 1$. This is seen by showing that, as before, $S_{ij}/(n - s)$ tends in probability to σ_{ij}, and that the joint distribution of the variables Y_{ij} ($i = 1, \ldots, r$; $j = 1, \ldots, p$) and hence of the elements of V tends to a limit which is independent of the underlying error distribution (see for example Problems 20 and 21). For more details, see Arnold (1981). Simulation studies by Olson (1974) suggest that of the four tests, the size of (iii) is the most and that of (iv) the least robust.

Discussion of multivariate linear models from a Bayesian point of view can be found, for example, in Box and Tiao (1973), in Press and Shigemasu (1985), and in the references cited there.

5. FURTHER APPLICATIONS

The invariant tests of multivariate linear hypotheses discussed in the preceding sections apply to the multivariate analogue of any univariate linear hypothesis, and the extension of the univariate to the corresponding multivariate test is routine. In addition, these tests have applications to some hypotheses that are not multivariate linear hypotheses as defined in Section 1 but which can be brought to this form, through suitable transformation and reduction.

In the linear hypotheses of Section 1, the parameter vectors being tested are linear combinations

$$\sum_{\gamma=1}^{n} c_{\nu\gamma}\underline{\xi}_{\gamma} = \sum_{\gamma=1}^{n} c_{\nu\gamma}E(\underline{X}_{\gamma}), \qquad \nu = 1,\ldots,r$$

where the $\underline{X}_{\gamma}$ are the n independent rows of the observation matrix X. We shall now instead consider linear combinations of the corresponding column vectors, and thus of the (dependent) components of the p-variate distribution.

Example 1. Let $(X_{\alpha 1},\ldots, X_{\alpha q}, X_{\alpha, q+1},\ldots, X_{\alpha,2q})$, $\alpha = 1,\ldots, n$, be a sample from a multivariate normal distribution, and consider the problem of testing $H: \xi_{q+i} = \xi_i$ for $i = 1,\ldots, q$. This might arise for example when $X_{\alpha 1},\ldots, X_{\alpha q}$ and $X_{\alpha, q+1},\ldots, X_{\alpha, 2q}$ are q measurements taken on the same subject before and after a certain treatment, or on the left and right sides of the subject.

Example 2. Let $(X_{\alpha 1},\ldots, X_{\alpha p})$, $\alpha = 1,\ldots, n$, be a sample from a p-variate normal distribution, and consider the problem of testing the hypothesis $H: \xi_1 = \cdots = \xi_p$. As an application suppose that a shop has p machines for manufacturing a certain product, the quality of which is measured by a random variable X. In an experiment involving n workers, each worker is put on all p machines, with $X_{\alpha i}$ being the result of the αth worker on the ith machine. If the n workers are considered as a random sample from a large population, the vectors $(X_{\alpha 1},\ldots, X_{\alpha p})$ may be assumed to be a sample from a p-variate normal distribution. Of the two factors involved in this experiment one is fixed (machines) and one random (workers), in the sense that a replication of the experiment would employ the same machines but a new sample of workers. The hypothesis being tested is that the fixed effect is absent. The test in this mixed model is quite different from the corresponding model I test where both effects are fixed, and which was treated in Section 5 of Chapter 7.

An important feature of such *repeated measurement designs* is that the p component measurements are measured on a common scale, so that it is meaningful to compare them. (This is not necessary in the general linear-hypothesis situations of the earlier sections, where the comparisons are made separately for each fixed component over different groups of subjects.) Although both Examples 1 and 2 are concerned with a single multivariate sample, this is not a requirement of such designs. Both examples extend for instance to the case of several groups of subjects (corresponding to different conditions or treatments) on all of which the same comparisons are made for each measurement.

Quite generally, consider the multivariate linear model of Section 1 in which each of the p column vectors of the matrix

$$\xi = \begin{pmatrix} \xi_{11} & \cdots & \xi_{1p} \\ \vdots & & \\ \xi_{n1} & \cdots & \xi_{np} \end{pmatrix}$$

is assumed to lie in a common s-dimensional linear subspace Π_Ω of n-dimensional space. However, the hypothesis H is now different. It specifies that each of the *row vectors* of ξ lies in a $(p - d)$-dimensional subspace Π'_ω of p-space. In Example 1, $s = 1$, $p - d = q$; in Example 2, $s = p - d = 1$.

As a first step toward a canonical form, make a transformation $\tilde{Y} = XE$, E nonsingular, such that under H the first d columns of $\tilde{\eta} = E(\tilde{Y})$ are equal to zero. This is achieved by any E the last $p - d$ columns of which span Π'_ω. The rows of $\tilde{Y}$ are then again independent, normally distributed with common covariance matrix, which is now $E'\Sigma E$. Also, since each column of $\tilde{\eta}$ is a linear combination of the columns of the matrix $\xi = E(X)$, the columns of $\tilde{\eta}$ lie in Π_Ω. If we write

$$\tilde{Y} = \begin{pmatrix} \tilde{Y}_1 & \tilde{Y}_2 \\ d & p-d \end{pmatrix}, \qquad \tilde{\eta} = \begin{pmatrix} \tilde{\eta}_1 & \tilde{\eta}_2 \\ d & p-d \end{pmatrix},$$

the matrix $\tilde{\eta}_1$ under H reduces to the $n \times d$ zero matrix.

Next, subject $\tilde{Y}$ to an orthogonal transformation $C\tilde{Y}$, with the first s rows of C spanning Π_Ω, and denote the resulting matrix by

$$(18) \qquad C\tilde{Y} = \begin{pmatrix} Y & U \\ Z & V \end{pmatrix} \begin{matrix} s \\ m \end{matrix} .$$
$$\qquad\qquad\qquad\quad d \quad p-d=l$$

Then it follows from Chapter 7, Section 1 that the rows of (18) are p-variate normal with common covariance matrix $E'\Sigma E$ and with means

$$E(Y) = \eta, \qquad E(Z) = 0, \qquad E(U) = \nu, \qquad E(V) = 0.$$

In this canonical form, the hypothesis becomes $H : \eta = 0$.

The problem of testing H remains invariant under the group G_1 of adding arbitrary constants to the ls elements of U, which leaves Y, Z, and V as maximal invariants. The next step is to show that invariance considerations also permit the discarding of V.

Let G_2 be the group of transformations

(19) $$V^* = ZB + VC, \qquad Z^* = Z, \qquad Y^* = Y,$$

where B is any $d \times l$ and C any nonsingular $l \times l$ matrix. Before applying the principle of invariance, it will be convenient to reduce the problem by sufficiency. The matrix Y together with the matrices of inner products $Z'Z$, $V'V$, and $Z'V$ form a set of sufficient statistics, and it follows from Theorem 6 of Chapter 6 that the search for a UMP invariant test can restrict attention to these sufficient statistics (Problem 24). We shall now show that under the transformations (19), the matrices Y and $Z'Z$ are maximal invariant on the basis of Y, $Z'Z$, $V'V$, and $Z'V$.

To prove this, it is necessary to show that for any given $m \times l$ matrix V^{**} there exist B and C such that $V^* = ZB + VC$ satisfies

$$Z'V^* = Z'V^{**} \qquad \text{and} \qquad V^{*'}V^* = V^{**'}V^{**}.$$

Geometrically, these equations state that there exist vectors $(V_{1i}^*, \ldots, V_{mi}^*)$, $i = 1, \ldots, l$ in the space S spanned by the columns of Z and V which have a preassigned set of inner products with each other and with the column vectors of Z.

Consider first the case $l = 1$. If $d + 1 \geq m$, one can assume that Z and the column of V span S, and one can then take $V^{**} = V^*$. If $d + 1 < m$, then Z and the column of V may be assumed to be linearly independent. There then exists a rotation about the columns of Z as axis, which takes V^{**} into a vector lying in S, and this vector has the properties required of V^*.

The proof is now completed by repeated application of the result for this special case. It can be applied first to the vector $(V_{11}, \ldots, V_{m1})$, to determine the first column of B and a number c_{11} to which one may add zeros to construct the first column of C. By adjoining the transformed vector $(V_{11}^*, \ldots, V_{m1}^*)$ to the columns of Z and applying the result to the vector $(V_{12}, \ldots, V_{m2})$, one obtains a vector $(V_{12}^*, \ldots, V_{m2}^*)$ which lies in the space spanned by $(V_{11}, \ldots, V_{m1})$, $(V_{12}, \ldots, V_{m2})$ and the column vectors of Z, and which in addition has the preassigned inner products with $(V_{11}^*, \ldots, V_{m1}^*)$, with the columns of Z and with itself. This second step determines the second column of B and two numbers c_{12}, c_{22} to which zeros can be added to provide the second column of C. Proceeding inductively in this way, one obtains for C a triangular matrix with zeros below the main diagonal, so that C is nonsingular. Since Z, V, and V^{**} can be assumed to have maximal rank, it follows from Lemma 1 and the equation $V^{*'}V^* = V^{**'}V^{**}$ that the rank of V^* is also maximal, and this completes the proof.

Thus invariance reduces consideration to the matrices Y and Z, the rows of which are independently distributed according to a d-variate normal distribution with common unknown covariance matrix. The expectations are $E(Y) = \eta$, $E(Z) = 0$, and the hypothesis being tested is $H : \eta = 0$, a multivariate linear hypothesis with $r = s$. In particular when $s = 1$, as was the case in Examples 1 and 2, there exists a UMP invariant test based on Hotelling's T^2. When $s > 1$, the tests of Section 4 become applicable. In either case, the tests require that $m \geq d$.

In the reduction to canonical form, the $p \times p$ matrix E could have been restricted to be orthogonal. However, since the covariance matrix of the rows is unknown (rather than being proportional to the identity matrix as was the case for the columns), this restriction is unnecessary, and for applications it is convenient not to impose it.

It is also worth noting that

$$\begin{pmatrix} Y \\ Z \end{pmatrix} = C\tilde{Y}_1,$$

so that (Y, Z) is equivalent to $\tilde{Y}_1$. In terms of $(\tilde{Y}_1, \tilde{Y}_2)$, the invariance argument thus reduces the data to the maximal invariant $\tilde{Y}_1$.

Example 1. (continued). For the transformation XE take

$$D_{\alpha i} = X_{\alpha, q+i} - X_{\alpha i}, \quad W_{\alpha i} = X_{\alpha i}, \quad \alpha = 1, \ldots, n, \quad i = 1, \ldots, q.$$

By the last remark preceding the example, invariance then reduces the data to the matrix $(D_{\alpha i})$, which was previously denoted by $\tilde{Y}_1$. The $(D_{\alpha 1}, \ldots, D_{\alpha q})$ constitute a sample from a q-variate normal distribution with mean $(\delta_1, \ldots, \delta_q)$, $\delta_i = \xi_{q+i} - \xi_i$. The hypothesis H reduces to $\delta_i = 0$ for all i, and the UMP invariant test is Hotelling's one-sample test discussed in Section 3 (with q in place of p).

To illustrate the case $s > 1$, suppose that the experimental subjects consist of two groups, and denote the $p = 2q$ measurements on each subject by

$$\left(X_{\alpha 1}, \ldots, X_{\alpha q}; X_{\alpha, q+1}, \ldots, X_{\alpha, 2q} \right), \quad \alpha = 1, \ldots, n_1$$

and

$$\left(X_{\beta 1}^*, \ldots, X_{\beta q}^*; X_{\beta, q+1}^*, \ldots, X_{\beta, 2q}^* \right), \quad \beta = 1, \ldots, n_2.$$

Consider the hypothesis $H : \xi_{q+i} = \xi_i$, $\xi_{q+i}^* = \xi_i^*$ for $i = 1, \ldots, q$, which might arise under the same circumstances as in the one-sample case. The same argument as before now reduces the data to the two samples

$$\left(D_{\alpha 1}, \ldots, D_{\alpha q} \right), \quad \alpha = 1, \ldots, n_1,$$

and

$$\left(D_{\beta 1}^*, \ldots, D_{\beta q}^* \right), \quad \beta = 1, \ldots, n_2,$$

with means $(\delta_1, \ldots, \delta_q)$ and $(\delta_1^*, \ldots, \delta_q^*)$, and the hypothesis being tested becomes $H: \delta_1 = \cdots = \delta_q = 0$, $\delta_1^* = \cdots = \delta_q^* = 0$. This is a multivariate linear hypothesis with $r = s = 2$ and $p = q$, which can be tested by the tests of Section 4.

A linear hypothesis concerning the row vectors $(\xi_{\alpha 1}, \ldots, \xi_{\alpha p})$ has been seen in this section to be reducible to the linear hypothesis $\eta = 0$ on the reduced variables Y and Z. To consider the robustness of the resulting tests against nonnormality in the original variables, suppose that $X_{\alpha i} = \xi_{\alpha i} + W_{\alpha i}$, where $(W_{\alpha 1}, \ldots, W_{\alpha p})$, $\alpha = 1, \ldots, n$, is a sample from a p-variate distribution F with mean zero, where the ξ' and H are as at the beginning of the section. As before, let $XE = \tilde{Y} = (\tilde{Y}_1 \tilde{Y}_2)$. Then the rows of $\tilde{Y} - E(\tilde{Y})$ will be independent and have a common distribution, and the n rows of $\tilde{Y}_1$ will therefore be independently distributed according to d-variate distributions $\tilde{F}(\tilde{y}_{\alpha 1} - \tilde{\eta}_{\alpha 1}, \ldots, \tilde{y}_{\alpha d} - \tilde{\eta}_{\alpha d})$. The vectors $(\tilde{\eta}_{1i}, \ldots, \tilde{\eta}_{ni})$, $i = 1, \ldots, d$, all lie in Π_Ω, and under H they are all equal to zero. It follows that if the size of the normal-theory test of this reduced problem is robust against nonnormality (in $\tilde{F}$), the test is also robust against nonnormality in the original distribution F. In particular, the tests of Examples 1 and 2 are therefore robust against nonnormality.

In some multivariate studies of the kind described in Section 1, observations are taken not only on the characteristics of interest but also on certain covariates.

Example 3. Consider the two-sample problem of Section 3, where $(X_{\alpha 1}^{(1)}, \ldots, X_{\alpha p}^{(1)})$ and $(X_{\beta 1}^{(2)}, \ldots, X_{\beta p}^{(2)})$ represent p measurements under treatments 1 and 2 on random samples of n_1 and n_2 subjects respectively, but suppose that in addition q control measurements $(X_{\alpha, p+1}^{(1)}, \ldots, X_{\alpha, p+q}^{(1)})$ and $(X_{\beta, p+1}^{(2)}, \ldots, X_{\beta, p+q}^{(2)})$ are available on each subject. The $n = n_1 + n_2$ $(p + q)$-vectors of X's are assumed to be independently distributed according to $(p + q)$-variate normal distributions with common covariance matrix and with expectations $E(X_{\alpha i}^{(1)}) = \xi_i$, $E(X_{\beta i}^{(2)}) = \eta_i$ for $i = 1, \ldots, p$ and $E(X_{\alpha i}^{(1)}) = E(X_{\beta i}^{(2)}) = \nu_i$ for $i = p + 1, \ldots, p + q$. The hypothesis being tested is $H: \xi_i = \eta_i$ for $i = 1, \ldots, p$. It is hoped that the control measurements through their correlations with the p treatment measurements will make it possible to obtain a test with increased power despite the fact that these auxiliary observations have no direct bearing on the hypothesis.

More generally, suppose that the total set of measurements on the αth subject is $\underline{X}_\alpha = (X_{\alpha 1}, \ldots, X_{\alpha p}, X_{\alpha, p+1}, \ldots, X_{\alpha, p+q})$, and that the vectors $\underline{X}_\alpha$, $\alpha = 1, \ldots, n$ are independent, $(p + q)$-variate normal with common covariance matrix. For $i = 1, \ldots, p$, the mean vectors $(\xi_{1i}, \ldots, \xi_{ni})$ are assumed as in Section 1 to lie in an s-dimensional subspace Π_Ω of n-space, the hypothesis specifying that $(\xi_{1i}, \ldots, \xi_{ni})$ lies in an $(s - r)$-dimensional subspace Π_ω of Π_Ω. For $i = p + 1, \ldots, p + q$, the vectors $(\xi_{1i}, \ldots, \xi_{ni})$ are assumed to lie in Π_ω under both the hypothesis and the alternatives. Application of the orthogonal transformation CX of Section 1 to the augmented data matrix and some of the invariance considerations of the present section result in the reduced canonical form

$$\begin{pmatrix} Y & U \\ Z & V \end{pmatrix} \begin{matrix} r \\ m = n - s \end{matrix}$$
$$ p q$$

where the $r + m$ rows are independent ($p + q$)-variate normal with common covariance matrix and means

$$E(Y) = \eta, \quad E(Z) = 0, \quad E(U) = 0, \quad E(V) = 0.$$

The hypothesis being tested is $H : \eta = 0$. This problem bears a close formal resemblance to that considered for the model (18), with the important difference that the expectations $E(U) = \nu$ are now assumed to be zero. A number of invariant tests making use of the auxiliary variables U and V have been proposed, and it is shown in Marden and Perlman (1980) for the case $r = 1$ that some of these are substantially more powerful than the corresponding T^2-test based on Y and Z alone. For reduction by invariance, comparative power, and admissibility of various tests in the case of general r, see Kariya (1978) and Marden (1983), where there is also a survey of the literature. A detailed theoretical treatment of this and related testing problems is given by Kariya (1985).

6. SIMULTANEOUS CONFIDENCE INTERVALS

In the preceding sections, the tests and confidence sets of Chapter 7 were generalized from the univariate to the multivariate linear model. The present section is concerned with the corresponding generalization of Scheffé's simultaneous confidence intervals (Chapter 7, Section 9). In the canonical form of Section 2, the means of interest are the expectations $\eta_{ij} = E(Y_{ij})$, $i = 1, \ldots, r$, $j = 1, \ldots, p$. We shall here consider simultaneous confidence intervals not for all linear functions $\Sigma\Sigma c_{ij}\eta_{ij}$, but only those of the form*

$$\sum_{i=1}^{r} \sum_{j=1}^{p} u_i v_j \eta_{ij} = \sum_{j=1}^{p} v_j \left(\sum_{i=1}^{r} u_i \eta_{ij} \right).$$

This is in line with the linear hypotheses of Section 1 in that the same linear function $\Sigma u_i \eta_{ij}$ is considered for each of the p components of the multivariate distribution. The objects of interest are linear combinations of these functions. [For a more general discussion, see Wijsman (1979, 1980).]

---

*Simultaneous confidence intervals for other linear functions (based on the Lawley-Hotelling trace test) are discussed by Anderson (1984, Section 8.7.3).

When $r = 1$, one is dealing with a single vector $(\eta_1, \ldots, \eta_p)$, and the simultaneous estimation of all linear functions $\sum_{j=1}^{p} v_j \eta_j$ is conceptually very similar to the univariate case treated in Chapter 7, Section 9.

Example 4. Contrasts in the s-sample problem. Consider the comparison of two products, of which p quality characteristics $(\xi_{11}, \ldots, \xi_{1p})$ and $(\xi_{21}, \ldots, \xi_{2p})$ are measured on two samples. The parametric functions of interest are the linear combinations $\sum v_j (\xi_{2j} - \xi_{1j})$. Since for fixed j only the difference $\xi_{2j} - \xi_{1j}$ is of interest, invariance permits restricting attention to the variables $Y_j = (X_{2j} - X_{1j})/\sqrt{2}$ and S, and hence $r = 1$. If instead there are $s > 2$ groups, one may be interested in all contrasts $\sum_{i=1}^{s} w_i \xi_{ij}$, $\sum w_i = 0$. One may wish to combine the same contrasts from the p different components into $\sum v_j (\sum w_i \xi_{ij})$, $\sum w_i = 0$, and is then dealing with the more general case in which $r = s - 1$.

As in the univariate case, it will be assumed without loss of generality that $\sum u_i^2 = 1$ so that $u \in U$, and the problem becomes that of determining simultaneous confidence intervals

$$(20) \quad L(u, v; y, S) \le u'\eta v \le M(u, v; y, S) \quad \text{for all} \quad u \in U \quad \text{and all } v$$

with confidence coefficient γ. The argument of the univariate case shows that attention may be restricted to L and M satisfying

$$(21) \qquad\qquad L(u, v; y, S) = -M(-u, v; y, S).$$

We shall show that there exists a unique set of such intervals that remain invariant under a number of groups, and begin by noticing that the problem remains invariant under the group G_1 of Section 2, which leaves the sample matrices Y and Z as maximal invariants to which attention may therefore be restricted.

Consider next the group G_2 of Section 2, that is, the group of orthogonal transformations $Y^* = QY$, $\eta^* = Q\eta$. The argument of Chapter 7, Section 9 with respect to the same group shows that L and M depend on u, y only through $u'y$ and $y'y$, so that

$$L(u, v; y, S) = L_1(u'y, y'y; v, S), \quad M(u, v; y, S) = M_1(u'y, y'y; v, S).$$

Apply next the group G_1' of translations $Y^* = Y + a$, $\eta^* = \eta + a$, where a is an arbitrary $r \times p$ matrix. Since $u'\eta^* v = u'\eta v + u'av$, equivariance requires that

$$L_1(u'(y + a), (y + a)'(y + a); v, S) = L_1(u'y, y'y; v, S) + u'av,$$

and hence, putting $y = 0$, $L_1(0, 0; v, S) = L_2(v, S)$, and replacing a by y,

$$L_1(u'y, y'y; v, S) = u'yv + L_2(v, S)$$

and the analogous condition for M.

In order to determine L_2, consider the group G_3 of Section 2, that is, the group of linear transformations $Y^* = YB$, $Z^* = ZB$, and thus $S^* = B'SB$. An argument paralleling that for G_2 shows that an equivariant L_2 and M_2 must satisfy

$$(22) \quad L_2(Bv, S) = L_2(v, B'SB), \qquad M_2(Bv, S) = M_2(v, B'SB)$$

for all nonsingular B, positive definite S, and all v. In particular, when $S = I$ one has

$$L_2(v, I) = L_2(Bv, I) \qquad \text{for all orthogonal } B$$

so that $L_2(v, I) = L_3(v'v)$. With $B = S^{-1/2}$, so that $B'SB = I$, and $w = S^{-1/2}v$, (22) then reduces to

$$L_2(w, S) = L_3(w'Sw).$$

Thus,

$$L(u, v; y, S) = u'yv + L_3(v'Sv), \qquad M(u, v; y, S) = u'yv + M_3(v'Sv),$$

and by (21), $L_3(v'Sv) = -M_3(v'Sv)$.

The derivation of the simultaneous confidence intervals will now be completed by an invariance argument that does not involve a transformation of the observations (Y, S) but only a reparametrization of the linear functions $u'\eta v$. If v is replaced by cv for some positive c, then $u'\eta v$ becomes $cu'\eta v$, and equivariance therefore requires that

$$L_3(cv'Svc) = cL_3(v'Sv) \qquad \text{for all} \quad v, S \text{ and } c > 0.$$

For $v'Sv = 1$, this gives $L_3(c^2) = cL_3(1) = kc$, say, and hence

$$L_3(v'Sv) = k\sqrt{v'Sv}.$$

The only confidence intervals satisfying all of the above equivariance conditions are therefore given by

$$(23) \quad |u'\eta v - u'yv| \le k\sqrt{v'Sv} \qquad \text{for all} \quad u \in U \text{ and all } v.$$

It remains to evaluate the constant k, for which the probability (23) equals the given confidence coefficient γ. This requires determining the maximum

$$(24) \qquad \max_{u \in U, v} \frac{[u'(\eta - y)v]^2}{v'Sv}.$$

For fixed v, it follows from the Schwarz inequality that the numerator of (24) is maximized for

$$u = \frac{(\eta - y)v}{\sqrt{v'(\eta - y)'(\eta - y)v}}$$

and that the maximum is equal to

(25) $$\max_{u \in U} [u'(\eta - y)v]^2 = v'(\eta - y)'(\eta - y)v.$$

Substitution of this maximum value into (24) leaves a maximization problem which is solved by the following lemma.

Lemma 2. *Let B and S be symmetric $p \times p$ matrices, and suppose that S is positive definite. Then the maximum of*

$$f(v) = \frac{v'Bv}{v'Sv}$$

is equal to the largest root $\lambda_{\max}$ of the equation

(26) $$|B - \lambda S| = 0,$$

and the maximum is attained for any vector v which is proportional to an eigenvector corresponding to this root, that is, any v satisfying $(B - \lambda_{max}S)v = 0$.

Proof. Since $f(cv) = f(v)$ for all $c \neq 0$, assume without loss of generality that $v'Sv = 1$, and subject to this condition, maximize $v'Bv$. There exists a nonsingular transformation $w = Av$ for which

$$v'Bv = \sum \lambda_i w_i^2, \qquad v'Sv = \sum w_i^2 = 1$$

where $\lambda_1 \geq \lambda_2 \geq \cdots \geq \lambda_p$ are the roots of (26). In terms of the w's it is clear that the maximum value of $f(v)$ is obtained by putting $w_1 = 1$ and the remaining w's equal to zero, and that the maximum value is λ_1. That the maximizing vector is an associated eigenvector is seen in terms of the w's by noting that $w' = (1, 0, \ldots, 0)$ satisfies $(\Lambda - \lambda_1 I)w = 0$, where Λ is the diagonal matrix whose diagonal entries are the λ's.

Application of this lemma, with $B = (\eta - Y)'(\eta - Y)$, shows that

$$\max_{u \in U, v} \frac{[u'(\eta - Y)v]^2}{v'Sv} = \lambda_1(Y - \eta, S),$$

where $\lambda_1 = \lambda_1(Y - \eta, S)$ is the maximum root of

(27) $|(Y - \eta)'(Y - \eta) - \lambda S| = 0.$

Since the distribution of $Y - \eta$ is independent of η, the constant k in (23) is thus determined by

$$P_{\eta=0}\big[\lambda_1(Y, S) \le k^2\big] = \gamma$$

and hence coincides with the critical value of Roy's maximum-root test at level $\alpha = 1 - \gamma$. In particular when $r = 1$, the statistic $(m + 1 - p)\lambda_1/p$ has the F-distribution with p and $m + 1 - p$ degrees of freedom.

As in the univariate case, one may wish to permit more general simultaneous confidence sets

$$u'\eta v \in A(u, v; y, s) \qquad \text{for all} \quad u \in U, \quad v.$$

If the restriction to intervals is dropped, equivariant confidence sets are no longer unique, and by essentially repeating the derivation of the intervals it is easy to show that (Problem 30) the most general equivariant confidence sets are of the form

(28) $$\frac{u'(\eta - y)v}{\sqrt{v'Sv}} \in A \qquad \text{for all} \quad u \in U \quad \text{and all } v,$$

where A is any fixed one-dimensional set. However, as in the univariate case, if the confidence coefficient of (28) is γ, the set A contains the interval $(-k, k)$ for which the probability of (23) is γ, and the intervals (23) are therefore the smallest confidence sets at the given level.

There are three confidence statements which, though less detailed, are essentially equivalent to (23):

(i)   It follows from (25) that (23) is equivalent to the statement

(29) $v'(\eta - y)'(\eta - y)v \le k^2 v'Sv \qquad \text{for all } v.$

These inequalities provide simultaneous confidence ellipsoids for all vectors ηv.

(ii)   Alternatively, one may be interested in simultaneous confidence sets for all vectors $u'\eta$, $u \in U$. For this purpose, write

$$\frac{[u'(\eta - y)v]^2}{v'Sv} = \frac{v'(\eta - y)'uu'(\eta - y)v}{v'Sv}$$

By Lemma 2, the maximum (with respect to v) of this ratio is the largest root of

$$(30) \qquad |(\eta - y)'uu'(\eta - y) - \lambda S| = 0.$$

As was seen in Section 2, with y in place of $u'(\eta - y)$, this equation has only one nonzero root, which is equal to

$$u'(\eta - y)S^{-1}(\eta - y)'u,$$

and (23) is therefore equivalent to

$$(31) \qquad u'(\eta - y)S^{-1}(\eta - y)'u \leq k^2 \qquad \text{for all} \quad u \in U.$$

This provides the desired simultaneous confidence ellipsoids for the vectors $u'\eta$, $u \in U$.

Both (29) and (31) can be shown to be smallest equivariant confidence sets under some of the transformation groups considered earlier in the section (Problem 31).

(iii)   Finally, it is seen from the definition of λ_1 that (23) is equivalent to the inequalities

$$(32) \qquad \lambda_1(Y - \eta, S) \leq k^2,$$

which constitute the confidence sets for η obtained from Roy's maximum-root test.

As in the univariate case, the simultaneous confidence intervals (23) for $u'\eta v$ for all $u \in U$ and all v have the same form as the uniformly most accurate unbiased confidence intervals

$$(33) \qquad |u'\eta v - u'yv| \leq k_0\sqrt{v'Sv}$$

for a single given $u \in U$ and v (Problem 32). Clearly, $k_0 < k$, since the probability of (33) equals that of (23). The increase from k_0 to k is the price paid for the stronger assertion, which permits making the confidence statements

$$|\hat{u}'\eta\hat{v} - \hat{u}'y\hat{v}| \leq k\sqrt{\hat{v}'S\hat{v}}$$

for any linear combinations $\hat{u}'\eta\hat{v}$ suggested by the data.

The simultaneous confidence intervals of the present section were derived for the model in canonical form. For particular applications, Y and S must be expressed in terms of the original variables X. (See for example, Problems 33, 34.)

7. χ^2-TESTS: SIMPLE HYPOTHESIS AND UNRESTRICTED ALTERNATIVES

UMP invariant tests exist only for rather restricted classes of problems, among which linear hypotheses are perhaps the most important. However, when the number of observations is large, there frequently exist tests which possess this property at least approximately. Although a detailed treatment of large-sample theory is outside the scope of this book, we shall indicate briefly some theory of two types of tests possessing such properties: χ^2-tests and likelihood-ratio tests. In both cases the approximate optimum property is a consequence of the asymptotic equivalence of the problem with one of testing a linear hypothesis. This relationship will be sketched in the next section. As preparation we discuss first a special class of χ^2 problems.

It will be convenient to begin by considering the linear hypothesis model with known covariance matrix. Let $Y = (Y_1, \ldots, Y_q)$ have the multivariate normal probability density

$$(34) \qquad \frac{\sqrt{|A|}}{(2\pi)^{\frac{1}{2}q}} \exp\left[-\frac{1}{2} \sum_{i=1}^{q} \sum_{j=1}^{q} a_{ij}(y_i - \eta_i)(y_j - \eta_j) \right]$$

with known covariance matrix A^{-1}. The point of means $\eta = (\eta_1, \ldots, \eta_q)$ is known to lie in a given s-dimensional linear space Π_Ω with $s \le q$; the hypothesis to be tested is that η lies in a given $(s - r)$-dimensional linear subspace Π_ω of Π_Ω ($r \le s$). This problem (which was considered in canonical form in Section 4) is invariant under a suitable group G of linear transformations, and there exists a UMP invariant test with respect to G, given by the rejection region

$$(35) \qquad \sum\sum a_{ij}(y_i - \hat{\hat{\eta}}_i)(y_j - \hat{\hat{\eta}}_j) - \sum\sum a_{ij}(y_i - \hat{\eta}_i)(y_j - \hat{\eta}_j)$$

$$= \sum\sum a_{ij}(\hat{\eta}_i - \hat{\hat{\eta}}_i)(\hat{\eta}_j - \hat{\hat{\eta}}_j)$$

$$\ge C.$$

Here $\hat{\eta}$ is the point of Π_Ω which is closest to the sample point y in the metric defined by the quadratic form $\sum\sum a_{ij}x_ix_j$, that is, which minimizes the quantity $\sum\sum a_{ij}(y_i - \eta_i)(y_j - \eta_j)$ for η in Π_Ω. Similarly $\hat{\hat{\eta}}$ is the point in Π_ω minimizing this quantity.

When the hypothesis is true, the left-hand side of (35) has a χ^2-distribution with r degrees of freedom, so that C is determined by

$$(36) \qquad \int_C^\infty \chi_r^2(z)\, dz = \alpha.$$

When η is not in Π_ω, the probability of rejection is

$$(37) \qquad \int_C^\infty p_\lambda(z)\, dz,$$

where $p_\lambda(z)$ is the noncentral χ^2-density of Chapter 7, Problem 2 with r degrees of freedom and noncentrality parameter λ^2, obtained by replacing $y_i, \hat{\eta}_i, \hat{\hat{\eta}}_i$ in (35) with their expectations, or equivalently, if (35) is considered as a function of y, by replacing y with η throughout. This expression for the power is valid even when the assumed model is not correct so that $E(Y) = \eta$ does not lie in Π_Ω. For the particular case that $\eta \in \Pi_\Omega$, the second term in this expression for λ^2 equals 0. A proof of the above statements is obtained by reducing the problem to a linear hypothesis through a suitable linear transformation. (See Problem 35).

Returning to the theory of χ^2-tests, which deals with hypotheses concerning multinomial distributions,* consider n multinomial trials with m possible outcomes. If $p = (p_1, \ldots, p_m)$ denotes the probabilities of these outcomes and X_i the number of trials resulting in the ith outcome, the distribution of $X = (X_1, \ldots, X_m)$ is

$$(38) \quad P(x_1, \ldots, x_m) = \frac{n!}{x_1! \ldots x_m!} p_1^{x_1} \cdots p_m^{x_m} \quad \left(\sum x_i = n, \quad \sum p_i = 1 \right).$$

The simplest χ^2 problems are those of testing a hypothesis $H: p = \pi$ where $\pi = (\pi_1, \ldots, \pi_m)$ is given, against the unrestricted alternatives $p \neq \pi$. As $n \to \infty$, the power of the tests to be considered will tend to one against any fixed alternative. (A sequence of tests with this property is called *consistent*.) In order to study the power function of such tests for large n, it is of interest to consider a sequence of alternatives $p^{(n)}$ tending to π as $n \to \infty$. If the rate of convergence is faster than $1/\sqrt{n}$, the power of even the most powerful test will tend to the level of significance α. The sequences reflecting the aspects of the power that are of greatest interest, and which are most likely to provide a useful approximation to the actual power for large but finite n, are the sequences for which $\sqrt{n}(p^{(n)} - \pi)$ tends to a nonzero limit, so that

$$(39) \qquad p_i^{(n)} = \pi_i + \frac{\Delta_i}{\sqrt{n}} + R_i^{(n)}$$

say, where $\sqrt{n}\, R_i^{(n)}$ tends to zero as n tends to infinity.

---

*For an alternative approach to such hypotheses see Hoeffding (1965).

Let

(40)
$$Y_i = \frac{X_i - n\pi_i}{\sqrt{n}}.$$

Then $\sum_{i=1}^m Y_i = 0$, and the mean of Y_i is zero under H and tends to Δ_i under the alternatives (39). The covariance matrix of the Y's is

(41)
$$\sigma_{ij} = -\pi_i\pi_j \quad \text{if } i \neq j, \qquad \sigma_{ii} = \pi_i(1 - \pi_i)$$

when H is true, and tends to these values for the alternatives (39). As $n \to \infty$, the distribution of $Y = (Y_1, \ldots, Y_{m-1})$ tends to the multivariate normal distribution with means $E(Y_i) = 0$ under H and $E(Y_i) = \Delta_i$ for the sequence of alternatives (39), and with covariance matrix (41) in both cases. [A proof assuming H is given for example by Cramér (1946, Section 30.1). It carries over with only the obvious changes to the case that H is not true.] The density of the limiting distribution is

(42)
$$c \exp\left[-\frac{1}{2} \left(\sum_{i=1}^{m-1} \frac{(y_i - \Delta_i)^2}{\pi_i} + \frac{\left(\sum_{j=1}^{m-1} (y_j - \Delta_j) \right)^2}{\pi_m} \right) \right]$$

and the hypothesis to be tested becomes $H: \Delta_1 = \cdots = \Delta_{m-1} = 0$.

According to (35), the UMP invariant test in this asymptotic model rejects when

$$\sum_{i=1}^{m-1} \frac{y_i^2}{\pi_i} + \frac{1}{\pi_m} \left(\sum_{j=1}^{m-1} y_j \right)^2 > C$$

and hence when

(43)
$$n \sum_{i=1}^m \frac{(\nu_i - \pi_i)^2}{\pi_i} > C$$

where $\nu_i = X_i/n$ and C is determined by (36) with $r = m - 1$. [The accuracy of the χ^2-approximation to the exact null distribution of the test statistic in this case is discussed for example by Radlow and Alf (1975). For more accurate approximations in this and related problems, see McCullagh (1985) and the literature cited there.] The limiting power of the test against the sequence of alternatives (39) is given by (37) with $\lambda^2 = \sum_{i=1}^{m} \Delta_i^2 / \pi_i$. This provides an approximation to the power for fixed n and a particular alternative p if one identifies p with $p^{(n)}$ for this value of n. From (39) one finds approximately $\Delta_i = \sqrt{n}\,(p_i - \pi_i)$, so that the noncentrality parameter becomes

$$(44) \qquad \lambda^2 = n \sum_{i=1}^{m} \frac{(p_i - \pi_i)^2}{\pi_i}.$$

Example 5. Suppose the hypothesis is to be tested that certain events (births, deaths, accidents) occur uniformly over a stated time interval such as a day or a year. If the time interval is divided into m equal parts and p_i denotes the probability of an occurrence in the ith subinterval, the hypothesis becomes $H: p_i = 1/m$ for $i = 1, \ldots, m$. The test statistic is then

$$mn \sum_{i=1}^{m} \left(\nu_i - \frac{1}{m} \right)^2,$$

where ν_i is the relative frequency of occurrence in the ith subinterval. The approximate power of the test is given by (37) with $r = m - 1$ and $\lambda^2 = mn\sum_{i=1}^{m}[p_i - (1/m)]^2$.

Unbiasedness of the test (43) and a local optimality property among tests based on the frequencies ν_i are established by Cohen and Sackrowitz (1975).

Example 5 illustrates the use of the χ^2-test (43) for providing a particularly simple alternative to goodness-of-fit tests such as that of Kolmogorov, mentioned at the end of Chapter 6, Section 13. However, when not only the frequencies ν_i but also the original observations X_i are available, reduction of the data through grouping results in tests that tend to be less efficient than those based on the Kolmogorov or related statistics. For further discussion of χ^2 and its many generalizations, comparison with other goodness-of-fit tests, and references to the extensive literature, see Kendall and Stuart (1979, Section 30.60). The choice of the number m of groups is considered, among others, by Quine and Robinson (1985) and by Kallenberg, Oosterhoff, and Schriever (1985).

8. χ^2-AND LIKELIHOOD-RATIO TESTS

It is both a strength and a weakness of the χ^2-test of the preceding section that its asymptotic power depends only on the weighted sum of squared deviations (44), not on the signs of these deviations and their distribution over the different values of i. This is an advantage if no knowledge is available concerning the alternatives, since the test then provides equal protection against all alternatives that are equally distant from $H: p = \pi$ in the metric (44). However, frequently one does know the type of deviations to be expected if the hypothesis is not true, and in such cases the test can be modified so as to increase its asymptotic power against the alternatives of interest by concentrating it on these alternatives.

To derive the modified test, suppose that a restricted class of alternatives to H has been defined

$$K: p \in \mathscr{S}, \; p \neq \pi.$$

Let the surface $\mathscr{S}$ have a parametric representation

$$p_i = f_i(\theta_1, \ldots, \theta_s), \qquad i = 1, \ldots, m,$$

and let

$$\pi_i = f_i(\theta_1^0, \ldots, \theta_s^0).$$

Suppose that the θ_j are real-valued, that the derivatives $\partial f_i / \partial \theta_j$ exist and are continuous at θ^0, and that the Jacobian matrix $(\partial f_i / \partial \theta_j)$ has rank s at θ^0. If $\theta^{(n)}$ is any sequence such that

$$(45) \qquad \sqrt{n}\left(\theta_j^{(n)} - \theta_j^0\right) \to \delta_j,$$

the limiting distribution of the variables $(Y_1, \ldots, Y_{m-1})$ of the preceding section is normal with mean

$$(46) \qquad E(Y_i) = \Delta_i = \sum_{j=1}^{s} \delta_j \frac{\partial f_i}{\partial \theta_j}\bigg|_{\theta^0}$$

and covariance matrix (41). This is seen by expanding f_i about the point θ^0 and applying the limiting distribution (42). The problem of testing H against all sequences of alternatives in K satisfying (45) is therefore asymptotically equivalent to testing the hypothesis

$$\Delta_1 = \cdots = \Delta_{m-1} = 0$$

in the family (42) against the alternatives $\overline{K}: (\Delta_1, \ldots, \Delta_{m-1}) \in \Pi_\Omega$ where Π_Ω, is the linear space formed by the totality of points with coordinates

$$(47) \qquad \Delta_i = \sum_{j=1}^{s} \delta_j \frac{\partial f_i}{\partial \theta_j}\bigg|_{\theta^0}.$$

We note for later use that for any fixed n, the totality of points

$$p_i = \pi_i + \frac{\Delta_i}{\sqrt{n}}, \qquad i = 1, \ldots, m,$$

with the Δ_i satisfying (47), constitute the tangent plane to $\mathscr{S}$ at π, which will be denoted by $\bar{\mathscr{S}}$.

Let $(\hat{\Delta}_1, \ldots, \hat{\Delta}_m)$ be the values minimizing $\sum_{i=1}^m (y_i - \Delta_i)^2/\pi_i$ subject to the conditions $(\Delta_1, \ldots, \Delta_{m-1}) \in \Pi_\Omega$ and $\Delta_m = -(\Delta_1 + \cdots + \Delta_{m-1})$. Then by (35), the asymptotically UMP invariant test rejects H in favor of $\bar{K}$ if

$$\frac{\sum_{i=1}^m y_i^2}{\pi_i} - \frac{\sum_{i=1}^m (y_i - \hat{\Delta}_i)^2}{\pi_i} = \frac{\sum_{i=1}^m \hat{\Delta}_i^2}{\pi_i} > C,$$

or equivalently if

$$(48) \qquad \frac{n \sum_{i=1}^m (\nu_i - \pi_i)^2}{\pi_i} - \frac{n \sum_{i=1}^m (\nu_i - \hat{p}_i)^2}{\pi_i} = \frac{n \sum_{i=1}^m (\hat{p}_i - \pi_i)^2}{\pi_i} > C,$$

where the $\hat{p}_i$ minimize $\sum(\nu_i - p_i)^2/\pi_i$ subject to $p \in \bar{\mathscr{S}}$. The constant C is determined by (36) with $r = s$. An asymptotically equivalent test, which, however, frequently is more difficult to compute explicitly, is obtained by letting the $\hat{p}_i$ be the minimizing values subject to $p \in \mathscr{S}$ instead of $p \in \bar{\mathscr{S}}$. An approximate expression for the power of the test against an alternative p is given by (37) with λ^2 obtained from (48) by substituting p_i for ν_i when the $\hat{p}_i$ are considered as functions of the ν_i.

Example 6. Suppose that in Example 5, where the hypothesis of a uniform distribution is being tested, the alternatives of interest are those of a cyclic movement, which may be represented at least approximately by a sine wave

$$p_i = \frac{1}{m} + \rho \int_{(i-1)2\pi/m}^{i2\pi/m} \sin(u - \theta) \, du, \qquad i = 1, \ldots, m.$$

Here ρ is the amplitude and θ the phase of the cyclic disturbance. Putting $\xi = \rho \cos \theta$, $\eta = \rho \sin \theta$, we get

$$p_i = \frac{1}{m}(1 + a_i \xi + b_i \eta),$$

where

$$a_i = 2m \sin \frac{\pi}{m} \sin (2i - 1) \frac{\pi}{m}, \qquad b_i = -2m \sin \frac{\pi}{m} \cos (2i - 1) \frac{\pi}{m}.$$

The equations for p_i define the surface $\mathcal{S}$, which in the present case is a plane, so that it coincides with $\bar{\mathcal{S}}$.

The quantities $\hat{\xi}, \hat{\eta}$ minimizing $\Sigma(\nu_i - p_i)^2/\pi_i$ subject to $p \in \mathcal{S}$ are

$$\hat{\xi} = \frac{\Sigma a_i \nu_i}{\Sigma a_i^2 \pi_i}, \qquad \hat{\eta} = \frac{\Sigma b_i \nu_i}{\Sigma b_i^2 \pi_i}$$

with $\pi_i = 1/m$. Let $m > 2$. Using the fact that $\Sigma a_i = \Sigma b_i = \Sigma a_i b_i = 0$ and that

$$\sum_{i=1}^{m} \sin^2 (2i - 1) \frac{\pi}{m} = \sum_{i=1}^{m} \cos^2 (2i - 1) \frac{\pi}{m} = \frac{m}{2},$$

the test becomes after some simplification

$$2n \left[\sum_{i=1}^{m} \nu_i \sin (2i - 1) \frac{\pi}{m} \right]^2 + 2n \left[\sum_{i=1}^{m} \nu_i \cos (2i - 1) \frac{\pi}{m} \right]^2 > C,$$

where the number of degrees of freedom of the left-hand side is $s = 2$. The noncentrality parameter determining the approximate power is

$$\lambda^2 = n \left(\xi m \sin \frac{\pi}{m} \right)^2 + n \left(\eta m \sin \frac{\pi}{m} \right)^2 = n \rho^2 m^2 \sin^2 \frac{\pi}{m}.$$

The χ^2-tests discussed so far were for simple hypotheses. Consider now the more general problem of testing $H: p \in \mathcal{T}$ against the alternatives $K: p \in \mathcal{S}$, $p \notin \mathcal{T}$ where $\mathcal{T} \subset \mathcal{S}$ and where $\mathcal{S}$ and $\mathcal{T}$ have parametric representations

$$\mathcal{S}: p_i = f_i(\theta_1, \ldots, \theta_s), \qquad \mathcal{T}: p_i = f_i(\theta_1^0, \ldots, \theta_r^0, \theta_{r+1}, \ldots, \theta_s).$$

The basis for a large-sample analysis of this problem is the fact that for large n a sphere of radius $\rho/\sqrt{n}$ can be located which for sufficiently large ρ contains the true point p with arbitrarily high probability. Attention can therefore be restricted to sequences of points $p^{(n)} \in \mathcal{S}$ which tend to some fixed point $\pi \in \mathcal{T}$ at the rate of $1/\sqrt{n}$. More specifically, let $\pi_i = f_i(\theta_1^0, \ldots, \theta_s^0)$, and let $\theta^{(n)}$ be a sequence satisfying (45). Then the variables $(Y_1, \ldots, Y_{m-1})$ have a normal limiting distribution with covariance matrix (41) and a vector of means given by (46). Let Π_Ω be defined as before, let

Π_ω be the linear space

$$\Pi_\omega : \Delta_i = \sum_{j=r+1}^{s} \delta_j \frac{\partial p_i}{\partial \theta_j}\bigg|_{\theta^0},$$

and consider the problem of testing that $p^{(n)}$ is a sequence in H for which $\theta^{(n)}$ satisfies (45) against all sequences in K satisfying this condition. This is asymptotically equivalent to the problem, discussed at the beginning of Section 7, of testing $(\Delta_1, \ldots, \Delta_{m-1}) \in \Pi_\omega$ in the family (42) when it is given that $(\Delta_1, \ldots, \Delta_{m-1}) \in \Pi_\Omega$. By (35), the rejection region for this problem is

$$\sum \frac{\left(y_i - \hat{\hat{\Delta}}_i\right)^2}{\pi_i} - \sum \frac{\left(y_i - \hat{\Delta}_i\right)^2}{\pi_i} > C,$$

where the $\hat{\Delta}_i$ and $\hat{\hat{\Delta}}_i$ minimize $\Sigma(y_i - \Delta_i)^2/\pi_i$ subject to $\Delta_m = -(\Delta_1 + \cdots + \Delta_{m-1})$ and $(\Delta_1, \ldots, \Delta_{m-1})$ in Π_Ω and Π_ω respectively. In terms of the original variables, the rejection region becomes

$$(49) \qquad \frac{n\Sigma(\nu_i - \hat{\hat{p}}_i)^2}{\pi_i} - \frac{n\Sigma(\nu_i - \hat{p}_i)^2}{\pi_i} > C.$$

Here the $\hat{p}_i$ and $\hat{\hat{p}}_i$ minimize

$$(50) \qquad \sum \frac{(\nu_i - p_i)^2}{\pi_i}$$

when p is restricted to lie in the tangent plane at π to $\mathscr{S}$ and $\mathscr{T}$ respectively, and the constant C is determined by (36).

The above solution of the problem depends on the point π, which is not given. A test which is asymptotically equivalent to (49) and does not depend on π is obtained if $\hat{p}_i$ and $\hat{\hat{p}}_i$ are replaced by p_i^* and p_i^{**} which minimize (50) for p restricted to $\mathscr{S}$ and $\mathscr{T}$ instead of to their tangents, and if further π_i is replaced in (49) and (50) by a suitable estimate, for example by ν_i. This leads to the rejection region

$$(51) \quad n\sum \frac{(\nu_i - p_i^{**})^2}{\nu_i} - n\sum \frac{(\nu_i - p_i^*)^2}{\nu_i} = n\sum \frac{(p_i^* - p_i^{**})^2}{\nu_i} > C,$$

where the p_i^{**} and p_i^* minimize

$$(52) \qquad\qquad \sum \frac{(\nu_i - p_i)^2}{\nu_i}$$

subject to $p \in \mathscr{T}$ and $p \in \mathscr{S}$ respectively, and where C is determined by (36) as before. An approximation to the power of the test for fixed n and a particular alternative p is given by (37) with λ^2 obtained from (51) by substituting p_i for ν_i when the p_i^* and p_i^{**} are considered as functions of the ν_i.[†]

A more general large-sample approach, which unlike χ^2 is not tied to the multinomial distribution, is based on the method of maximum likelihood. We shall here indicate this theory only briefly, and in particular shall state the main facts without the rather complex regularity assumptions required for their validity.[‡]

Let $p_\theta(x)$, $\theta = (\theta_1, \ldots, \theta_r)$, be a family of univariate probability densities, and consider the problem of testing, on the basis of a (large) sample $X_1, \ldots, X_n$, the simple hypothesis $H: \theta_i = \theta_i^0$, $i = 1, \ldots, r$. Let $\hat{\theta} = (\hat{\theta}_1, \ldots, \hat{\theta}_r)$ be the maximum-likelihood estimate of θ, that is, the parameter vector maximizing $p_\theta(x_1) \ldots p_\theta(x_n)$. Then asymptotically as $n \to \infty$, attention can be restricted to the $\hat{\theta}_i$, since they are "asymptotically sufficient".[§] The power of the tests to be considered will tend to one against any fixed alternative, and the alternatives of interest, as in the χ^2 case, are sequences $\theta_i^{(n)}$ satisfying

$$(53) \qquad\qquad \sqrt{n}\left(\theta_i^{(n)} - \theta_i^0\right) \to \Delta_i.$$

If $Y_i = \sqrt{n}(\hat{\theta}_i - \theta_i^0)$, the limiting distribution of $Y_1, \ldots, Y_r$ is the multivariate normal distribution (34) with

$$(54) \qquad a_{ij} = a_{ij}(\theta^0) = -E\left(\frac{\partial^2 \log p_\theta(X)}{\partial \theta_i \, \partial \theta_j}\right)\Bigg|_{\theta = \theta^0}$$

and with $\eta_i = 0$ under H and $\eta_i = \Delta_i$ for the alternatives satisfying (53).

[†]A proof of the above statements and a discussion of certain tests which are asymptotically equivalent to (48) and sometimes easier to determine explicitly are given, for example, in Fix, Hodges, and Lehmann (1959).

[‡]For a detailed treatment and references to the literature see Serfling (1980, Section 4.4).

[§]This was shown by Wald (1943); for a definition of asymptotic sufficiency and further results concerning this concept see LeCam (1956, 1960).

By (35), the UMP invariant test in this asymptotic model rejects when

$$(55) \qquad -\sum_{i=1}^{r}\sum_{j=1}^{r} a_{ij}n\big(\hat{\theta}_i - \theta_i^0\big)\big(\hat{\theta}_j - \theta_j^0\big) > C.$$

Under H, the left-hand side has a limiting χ^2-distribution with r degrees of freedom, while under the alternatives (53) the limiting distribution is noncentral χ^2 with noncentrality parameter

$$(56) \qquad \lambda^2 = \lim \sum_{i=1}^{r}\sum_{j=1}^{r} a_{ij}n\big(\theta_i^{(n)} - \theta_i^0\big)\big(\theta_j^{(n)} - \theta_j^0\big).$$

The approximate power against a specific alternative θ is therefore given by (37), with λ^2 obtained from (56) by substituting θ for $\theta^{(n)}$.

The test (55) is asymptotically equivalent to the likelihood-ratio test, which rejects when

$$(57) \qquad \Lambda_n = \frac{p_\theta(x_1)\dots p_\theta(x_n)}{p_{\theta^0}(x_1)\dots p_{\theta^0}(x_n)} \geq k.$$

This is seen by expanding $\sum_{v=1}^{n}\log p_{\theta^0}(x_v)$ about $\sum_{v=1}^{n}\log p_{\hat\theta}(x_v)$ and using the fact that at $\theta = \hat{\theta}$ the derivatives $\partial\sum\log p_\theta(x_v)/\partial\theta_i$ are zero. Application of the law of large numbers shows that $-2\log\Lambda_n$ differs from the left-hand side of (55) by a term tending to zero in probability as $n \to \infty$. In particular, the two statistics therefore have the same limiting distribution.

The extension of this method to composite hypotheses is quite analogous to the corresponding extension in the χ^2 case. Let $\theta = (\theta_1,\dots,\theta_s)$ and $H: \theta_i = \theta_i^0$ for $i = 1,\dots,r$ $(r < s)$. If attention is restricted to sequences $\theta^{(n)}$ satisfying (53) for $i = 1,\dots,s$ and some arbitrary $\theta_{r+1}^0,\dots,\theta_s^0$, the asymptotic problem becomes that of testing $\eta_1 = \cdots = \eta_r = 0$ against unrestricted alternatives $(\eta_1,\dots,\eta_s)$ for the distributions (34) with $a_{ij} = a_{ij}(\theta^0)$ given by (54). Then $\hat{\eta}_i = Y_i$ for all i, while $\hat{\hat{\eta}}_i = 0$ for $i = 1,\dots,r$ and $= Y_i$ for $i = r + 1,\dots,s$, so that the UMP invariant test is given by (55). The coefficients $a_{ij} = a_{ij}(\theta^0)$ depend on $\theta_{r+1}^0,\dots,\theta_s^0$ but as before an asymptotically equivalent test statistic is obtained by replacing $a_{ij}(\theta^0)$ with $a_{ij}(\hat{\theta})$. Again, the statistic is also asymptotically equivalent to minus twice the logarithm of the likelihood ratio, and the test is therefore asymptotically equivalent to the likelihood-ratio test,* which rejects when

$$(58) \qquad \Lambda_n = \frac{p_\theta(x_1)\dots p_\theta(x_n)}{p_{\hat\theta}(x_1)\dots p_{\hat\theta}(x_n)} \geq k$$

---

*The asymptotic theory of likelihood-ratio tests has been extended to more general types of problems, including in particular the case of restricted classes of alternatives, by Chernoff (1954). See also Serfling (1980).

where $\hat{\hat{\theta}}$ is the maximum-likelihood estimate of θ under H, and where $-2 \log \Lambda_n$ as before has a limiting χ^2-distribution with r degrees of freedom.

Example 7. *Independence in a two-dimensional contingency table.* In generalization of the multinomial model for a 2×2 table discussed in Chapter 4, Section 6, consider a twofold classification of n subjects, drawn at random from a large population, into classes $A_1, \ldots, A_a$ and $B_1, \ldots, B_b$ respectively. If n_{ij} denotes the number of subjects belonging to both A_i and B_j, the joint probability of the ab variables n_{ij} is

$$(59) \qquad \frac{N!}{\prod\limits_{i,j} n_{ij}!} \prod_{i,j} p_{ij}^{n_{ij}} \qquad \left(\sum n_{ij} = n, \quad \sum p_{ij} = 1 \right).$$

The hypothesis to be tested is that the two classifications are independent, that is, that p_{ij} is of the form

$$(60) \qquad\qquad\qquad H: p_{ij} = p_i p_j'$$

for some p_i, p_j' satisfying $\sum p_i = \sum p_j' = 1$.

Alternative, asymptotically equivalent tests are provided by (51) and the likelihood-ratio test. Since the minimization required by the former leads to a system of equations that cannot be solved explicitly, let us consider the likelihood-ratio approach. In the unrestricted multinomial model, the probability (59) is maximized by $\hat{p}_{ij} = n_{ij}/n$; under H, the maximizing probabilities are given by

$$\hat{\hat{p}}_i = \frac{n_{i\cdot}}{n}, \qquad \hat{\hat{p}}_j' = \frac{n_{\cdot j}}{n}$$

where $n_{i\cdot} = \sum_j n_{ij}/b$ and $n_{\cdot j} = \sum_i n_{ij}/a$ (Problem 39). Substitution in (58) gives

$$\Lambda = \frac{\prod\limits_{i,j} n_{ij}^{n_{ij}}}{\prod\limits_{i} n_{i\cdot}^{bn_{i\cdot}} \prod\limits_{j} n_{\cdot j}^{an_{\cdot j}}}.$$

Since under Ω the p_{ij} are subject only to the restriction $\sum\sum p_{ij} = 1$, it is seen that $s = ab - 1$. Similarly, $s - r = (a - 1) + (b - 1)$ and hence $-2 \log \Lambda$, under H, has a limiting χ^2-distribution with $r = (ab - 1) - (a + b - 2) = (a - 1)(b - 1)$ degrees of freedom. The accuracy of the χ^2-approximation, and possible improvements, in this and related problems are discussed by Lawal and Upton (1984) and Lewis, Saunders, and Westcott (1984), and in the literature cited in these papers.

For further work on two- and higher-dimensional contingency tables, see for example the books by Haberman (1974), Bishop, Fienberg, and Holland (1975), and Plackett (1981), and the paper by Goodman (1985).

9. PROBLEMS

Section 2

1. (i) If $m < p$, the matrix S, and hence the matrix S/m (which is an unbiased estimate of the unknown covariance matrix of the underlying p-variate distribution), is singular. If $m \geq p$, it is nonsingular with probability 1.

   (ii) If $r + m \leq p$, the test $\phi(y, u, z) \equiv \alpha$ is the only test that is invariant under the groups G_1 and G_3 of Section 2.

   [(ii): The U's are eliminated through G_1. Since the $r + m$ row vectors of the matrices Y and Z may be assumed to be linearly independent, any such set of vectors can be transformed into any other through an element of G_3.]

2. (i) If $p < r + m$, and $V = Y'Y$, $S = Z'Z$, the $p \times p$ matrix $V + S$ is nonsingular with probability 1, and the characteristic roots of the equation

   (61) $$|V - \lambda(V + S)| = 0$$

   constitute a maximal set of invariants under G_1, G_2, and G_3.

   (ii) Of the roots of (61), $p - \min(r, p)$ are zero and $p - \min(m, p)$ are equal to one. There are no other constant roots, so that the number of variable roots, which constitute a maximal invariant set, is $\min(r, p) + \min(m, p) - p$.

   [The multiplicity of the root $\lambda = 1$ is p minus the rank of S, and hence $p - \min(m, p)$. Equation (61) cannot hold for any constant $\lambda \neq 0, 1$ for almost all V, S, since for any $\mu \neq 0$, $V + \mu S$ is nonsingular with probability 1.]

3. (i) If A and B are $k \times m$ and $m \times k$ matrices respectively, then the product matrices AB and BA have the same nonzero characteristic roots.

   (ii) This provides an alternative derivation of the fact that W defined by (6) is the only nonzero characteristic root of the determinantal equation (5).

   [(i): If x is a nonzero solution of the equation $ABx = \lambda x$ with $\lambda \neq 0$, then $y = Bx$ is a nonzero solution of $BAy = \lambda y$.]

4. In the case $r = 1$, the statistic W given by (6) is maximal invariant under the group induced by G_1 and G_3 on the statistics Y_i, $U_{\alpha i}$ ($i = 1, \ldots, p$; $\alpha = 1, \ldots, s - 1$), and $S = Z'Z$.
   [There exists a nonsingular matrix B such that $B'SB = I$ and such that only the first coordinate of YB is nonzero. This is seen by first finding B_1 such that $B_1'SB_1 = I$ and then an orthogonal Q such that only the first coordinate of YB_1Q is nonzero.]

5. Let $Z_{\alpha i}$ ($\alpha = 1, \ldots, m$; $i = 1, \ldots, p$) be independently distributed as $N(0, 1)$, and let $Q = Q(Y)$ be an orthogonal $m \times m$ matrix depending on a random

variable Y that is independent of the Z's. If $Z_{\alpha i}^*$ is defined by

$$(Z_{1i}^* \ldots Z_{mi}^*) = (Z_{1i} \ldots Z_{mi})Q',$$

then the $Z_{\alpha i}^*$ are independently distributed as $N(0,1)$ and are independent of Y.

[For each y, the conditional distribution of the $(Z_{1i} \ldots Z_{mi})Q'(y)$, given $Y = y$, is as stated.]

6. Let Z be the $m \times p$ matrix $(Z_{\alpha i})$, where $p \leq m$ and the $Z_{\alpha i}$ are independently distributed as $N(0,1)$, let $S = Z'Z$, and let S_1 be the matrix obtained by omitting the last row and column of S. Then the ratio of determinants $|S|/|S_1|$ has a χ^2-distribution with $m - p + 1$ degrees of freedom.

[Let q be an orthogonal matrix (dependent on $Z_{11}, \ldots, Z_{m1}$) such that $(Z_{11} \ldots Z_{m1})Q' = (R\ 0\ \ldots\ 0)$, where $R^2 = \sum_{\alpha=1}^{m} Z_{\alpha 1}^2$. Then

$$S = Z'Q'QZ = \begin{vmatrix} R & 0 & \cdots & 0 \\ Z_{12}^* & Z_{22}^* & \cdots & Z_{m2}^* \\ \vdots & \vdots & \vdots & \vdots \\ Z_{1p}^* & Z_{2p}^* & \cdots & Z_{mp}^* \end{vmatrix} \begin{vmatrix} R & Z_{12}^* & \cdots & Z_{1p}^* \\ 0 & Z_{22}^* & & Z_{2p}^* \\ \vdots & \vdots & \vdots & \vdots \\ 0 & Z_{m2}^* & \cdots & Z_{mp}^* \end{vmatrix},$$

where the $Z_{\alpha i}^*$ denote the transforms under Q. The first of the matrices on the right-hand side is equal to the product

$$\left(\begin{array}{c|c} R & 0 \\ \hline Z_1^* & I \end{array} \right) \left(\begin{array}{c|c} I & 0 \\ \hline 0 & Z^* \end{array} \right),$$

where Z^* is the $(m - 1) \times (p - 1)$ matrix with elements $Z_{\alpha i}^*$ ($\alpha = 2, \ldots, m$; $i = 2, \ldots, p$), I is the $(p - 1) \times (p - 1)$ identity matrix, Z_1^* is the column vector $(Z_{12}^* \ldots Z_{1p}^*)'$, and 0 indicates a row or column of zeros. It follows that $|S|$ is equal to R^2 multiplied by the determinant of $Z^{*'}Z^*$. Since S_1 is the product of the $m \times (p - 1)$ matrix obtained by omitting the last column of Z multiplied on the left by the transpose of this $m \times (p - 1)$ matrix, $|S_1|$ is equal to R^2 multiplied by the determinant of the matrix obtained by omitting the last row and column of $Z^{*'}Z^*$. The ratio $|S|/|S_1|$ has therefore been reduced to the corresponding ratio in terms of the $Z_{\alpha i}^*$ with m and p replaced by $m - 1$ and $p - 1$, and by induction the problem is seen to be unchanged if m and p are replaced by $m - k$ and $p - k$ for any $k < p$. In particular, $|S|/|S_1|$ can be evaluated under the assumption that m and p have been replaced by $m - (p - 1)$ and $p - (p - 1) = 1$. In this case, the matrix Z' is a row matrix $(Z_{11} \ldots Z_{m-p+1,1})$; the determinant of S is $|S| = \sum_{\alpha=1}^{m-p+1} Z_{\alpha 1}^2$, which has a χ_{m-p+1}^2-distribution; and since S is a 1×1 matrix, $|S_1|$ is replaced by 1.]

7. *Null distribution of Hotelling's T^2.* The statistic $W = YS^{-1}Y'$ defined by (6), where Y is a row vector, has the distribution of a ratio, of which the numerator

and denominator are distributed independently, as noncentral χ^2 with non-centrality parameter ψ^2 and p degrees of freedom and as central χ^2 with $m + 1 - p$ degrees of freedom respectively.

[Since the distribution of W is unchanged if the same nonsingular transformation is applied to $(Y_1, \ldots, Y_p)$ and each of the m vectors $(Z_{\alpha 1}, \ldots, Z_{\alpha p})$, the common covariance matrix of these vectors can be assumed to be the identity matrix. Let Q be an orthogonal matrix (depending on the Y's) such that $(Y_1 \ \ldots \ Y_p)Q = (0 \ 0 \ \ldots \ T)$, where $T^2 = \Sigma Y_i^2$. Since QQ' is the identity matrix, one has

$$W = (YQ)(Q'S^{-1}Q)(Q'Y') = (0 \ \cdots \ 0 \ T)(Q'S^{-1}Q)(0 \ \cdots \ 0 \ T)'.$$

Hence W is the product of T^2, which has a noncentral χ^2-distribution with p degrees of freedom and noncentrality parameter ψ^2, and the element which lies in the pth row and the pth column of the matrix $Q'S^{-1}Q = (Q'SQ)^{-1} = (Q'Z'ZQ)^{-1}$. By Problems 5 and 6, this matrix is distributed independently of the Y's, and the reciprocal of the element in question is distributed as χ^2_{m-p+1}.]

Note. An alternative derivation of this distribution begins by obtaining the distribution of S, known as the *Wishart distribution*. This is essentially a p-variate analogue of χ^2 and plays a central role in tests concerning covariance matrices. [See for example Seber (1984).]

Section 3

8. Let $(X_{\alpha 1}, \ldots, X_{\alpha p})$, $\alpha = 1, \ldots, n$, be a sample from any p-variate distribution with zero mean and finite nonsingular covariance matrix Σ. Then the distribution of T^2 defined by (10) tends to χ^2 with p degrees of freedom.

9. The confidence ellipsoids (11) for $(\xi_1, \ldots, \xi_p)$ are equivariant under the groups G_1–G_3 of Section 2.

10. The two-sample test based on (12) is robust against nonnormality as n_1 and $n_2 \to \infty$.

11. The two-sample test based on (12) is robust against heterogeneity of covariances as n_1 and $n_2 \to \infty$ when $n_1/n_2 \to 1$, but not in general.

12. Inversion of the two-sample test based on (12) leads to confidence ellipsoids for the vector $(\xi_1^{(2)} - \xi_1^{(1)}, \ldots, \xi_p^{(2)} - \xi_p^{(1)})$ which are uniformly most accurate equivariant under the groups G_1–G_3 of Section 2.

13. *Simple multivariate regression.* In the model of Section 1 with

    (62) $\xi_{vi} = \alpha_i + \beta_i t_v$ $(v = 1, \ldots, n; \ i = 1, \ldots, s)$,

    the UMP invariant test of $H : \beta_1 = \cdots = \beta_p = 0$ is given by (6) and (9), with

    $$Y_i = \hat{\beta}_i, \ S_{ij} = \sum_{v=1}^{n} \left[X_{vi} - \hat{\alpha}_i - \hat{\beta}_i t_v \right]\left[X_{vj} - \hat{\alpha}_j - \hat{\beta}_j t_v \right]$$

    where $\hat{\beta}_i = \Sigma X_{vi}(t_v - \bar{t})/\sqrt{\Sigma(t_v - \bar{t})^2}$, $\hat{\alpha}_i = X_{\cdot i} - \hat{\beta}_i \bar{t}$.

14. Let $(Y_{v1}, \ldots, Y_{vp})$, $v = 1, \ldots, n$, be a sample from a p-variate distribution F with mean zero and covariance matrix Σ, and let $Z_i^{(n)} = \sum_{v=1}^{n} c_v Y_{vi} / \sqrt{\sum_{v=1}^{n} c_v^2}$ for some sequence of constants $c_1, c_2, \ldots$. Then $(Z_1^{(n)}, \ldots, Z_p^{(n)})$ tends in law to $N(0, \Sigma)$ provided the c's satisfy the condition (50) of Chapter 7.

    [By the Cramér–Wold theorem [see for example Serfling (1980)], it is enough to prove that $\sum a_i Z_i^{(n)} \to N(0, a'\Sigma a)$ for all $a = (a_1, \ldots, a_n)$ with $\sum a_i^2 = 1$, and this follows from Lemma 3 of Chapter 7.]

15. Suppose $X_{vi} = \xi_{vi} + U_{vi}$, where the ξ_{vi} are given by (62) and where $(U_{v1}, \ldots, U_{vp})$, $v = 1, \ldots, n$, is a sample from a p-variate distribution with mean 0 and covariance matrix Σ. The size of the test of Problem 13 is robust for this model as $n \to \infty$.

    [Apply Problem 14 and the univariate robustness result of Chapter 7, Section 8.]

    Note. This problem illustrates how the robustness of a univariate linear test carries over to its multivariate analogue. For a general result see Arnold (1981, Section 19.8).

Section 4

16. Verify the elements of V and S given by (14) and (15).

17. Let V and S be $p \times p$ matrices, V of rank $a \leq p$ and S nonsingular, and let $\lambda_1, \ldots, \lambda_a$ denote the nonzero roots of $|V - \lambda S| = 0$. Then

    (i) $\mu_i = 1/(1 + \lambda_i)$, $i = 1, \ldots, a$, are the a smallest roots of

    (63) $|S - \mu(V + S)| = 0$

    (the other $p - a$ being $= 1$);

    (ii) $v_i = 1 + \lambda_i$ are the a largest roots of

    (64) $|V + S - vS| = 0.$

18. Under the assumptions of Problem 17, show that

$$\prod \frac{1}{1 + \lambda_i} = \frac{|V|}{|V + S|}.$$

    [The determinant of a matrix is equal to the product of its characteristic roots.]

19. (i)   If (13) has only one nonzero root, then B is of rank 1. In canonical form $B = \eta'\eta$, and there then exists a vector $(a_1, \ldots, a_p)$ and constants $c_1, \ldots, c_v$ such that

    (65) $(\eta_{v1}, \ldots, \eta_{vp}) = c_v(a_1, \ldots, a_p)$ for $v = 1, \ldots, r.$

    (ii)   For the s-sample problem considered in Section 4, restate (65) in terms of the means $(\xi_1^{(k)}, \ldots, \xi_p^{(k)})$ of the text.

20. Let $(X_{\alpha 1}, \ldots, X_{\alpha p})$, $\alpha = 1, \ldots, n$, be independently distributed according to p-variate distributions $F(x_{\alpha 1} - \xi_{\alpha 1}, \ldots, x_{\alpha p} - \xi_{\alpha p})$ with finite covariance matrix Σ, and suppose the ξ's satisfy the linear model assumptions of Section 1. Then under H, $S_{ij}/(n - s)$ tends in probability to the (ij)th element σ_{ij} of Σ.
[See the corresponding univariate result of Chapter 7, Section 3.]

21. Let $(X_{\alpha 1}^{(k)}, \ldots, X_{\alpha p}^{(k)})$, $\alpha = 1, \ldots, n_k$, $k = 1, \ldots, s$, be samples from p-variate distributions $F(x_1 - \xi_1^{(k)}, \ldots, x_p - \xi_p^{(k)})$ with finite covariance matrix Σ, and let $\lambda_1, \ldots, \lambda_a$ be the nonzero roots of (16) and $(\lambda_1^*, \ldots, \lambda_a^*)$ those of (17), with V and S given by (14) and (15). Then the joint distribution of $((n - s)\lambda_1, \ldots, (n - s)\lambda_a)$ tends to that of $(\lambda_1^*, \ldots, \lambda_a^*)$ as $n \to \infty$.

22. Give explicit expressions for the elements of V and S in the multivariate analogues of the following situations:

   (i)   The hypothesis (34) in the two-way layout (32) of Chapter 7.

   (ii) The hypothesis (34) in the two-way layout of Section 6 of Chapter 7.

   (iii) The hypothesis $H' : \gamma_{ij} = 0$ for all i, j, in the two-way layout of Section 6 of Chapter 7.

23. The probability of a type-I error for each of the tests of the preceding problem is robust against nonnormality: in case (i) as $b \to \infty$; in case (ii) as $mb \to \infty$; in case (iii) as $m \to \infty$.

Section 5

24. The assumptions of Theorem 6 of Chapter 6 are satisfied for the group (19) applied to the hypothesis $H : \eta = 0$ of Section 5.

25. Let X_{vij} $(i = 1, \ldots, a; \ j = 1, \ldots, b)$, $v = 1, \ldots, n$, be n independent vectors, each having an ab-variate normal distribution with covariance matrix Σ and with means given by

$$E(X_{vij}) = \mu + \alpha_i + \beta_j, \qquad \sum \alpha_i = \sum \beta_j = 0.$$

   (i) For testing the hypothesis $H : \alpha_1 = \cdots = \alpha_a = 0$, give explicit expressions for the matrices Y and Z of (18) and the parameters $\eta = E(Y)$ being tested.

   (ii) Give an example of a situation for which the model of (i) might be appropriate.

26. Generalize both parts of the preceding problem to the two-group case in which $X_{\lambda ij}^{(1)}$ $(\lambda = 1, \ldots, n_1)$ and $X_{vij}^{(2)}$ $(v = 1, \ldots, n_2)$ are $n_1 + n_2$ independent vectors, each having an ab-variate normal distribution with covariance matrix Σ

and with means given by

$$E\left(X_{\lambda i j}^{(1)} \right) = \mu_1 + \alpha_i^{(1)} + \beta_j^{(1)}, \qquad E\left(X_{\nu i j}^{(2)} \right) = \mu_2 + \alpha_i^{(2)} + \beta_j^{(2)},$$

$$\sum \alpha_i^{(1)} = \sum \alpha_i^{(2)} = 0, \qquad \sum \beta_j^{(1)} = \sum \beta_j^{(2)} = 0,$$

and where the hypothesis being tested is

$$H: \alpha_1^{(1)} = \cdots = \alpha_a^{(1)} = \alpha_1^{(2)} = \cdots = \alpha_a^{(2)} = 0.$$

27. As a different generalization, let $(X_{\lambda v1}, \ldots, X_{\lambda v p})$ be independent vectors, each having a p-variate normal distribution with common covariance matrix Σ and with expectation

$$E(X_{\lambda v i}) = \mu^{(i)} + \alpha_\lambda^{(i)} + \beta_v^{(i)}, \quad \sum_\lambda \alpha_\lambda^{(i)} = \sum_v \beta_v^{(i)} = 0 \quad \text{for all } i,$$

and consider the hypothesis that each of $\mu^{(i)}, \alpha_\lambda^{(i)}, \beta_v^{(i)}$ ($\lambda = 1, \ldots, a$; $v = 1, \ldots, b$) is independent of i.

(i) Give explicit expressions for the matrices Y and Z and the parameters $\eta = E(Y)$ being tested.

(ii) Give an example of a situation in which this problem might arise.

28. Let X be an $n \times p$ data matrix satisfying the model assumptions made at the beginning of Sections 1 and 5, and let $X^* = CX$, where C is an orthogonal matrix, the first s rows of which span Π_Ω. If Y^* and Z denote respectively the first s and last $n - s$ rows of X^*, then $E(Y^*) = \eta^*$ say, and $E(Z) = 0$. Consider the hypothesis $H_0 : U'\eta^*V = 0$, where U' and V are constant matrices of dimension $a \times s$ and $p \times b$ and of ranks a and b respectively.

(i) The hypotheses of both Section 1 and Section 5 are special cases of H_0.

(ii) The problem can be put into canonical form Y^{**} ($s \times p$) and Z^{**} ($(n - s) \times p$), where the n rows of Y^{**} and Z^{**} are independent p-variate normal with common covariance matrix and with means $E(Y^{**}) = \eta^{**}$, and where H_0 becomes $H_0 : \eta_{ij}^{**} = 0$ for all $i = 1, \ldots, a$, $j = 1, \ldots, b$.

(iii) Determine groups leaving this problem invariant and for which the first a columns of Y^{**} are maximal invariants, so that the problem reduces to a multivariate linear hypothesis in canonical form.

29. Consider the special case of the preceding problem in which $a = b = 1$, and let $U' = u' = (u_1, \ldots, u_s)$, $V' = v' = (v_1, \ldots, v_p)$. Then for testing $H_0 : u'\eta^*v = 0$ there exists a UMP invariant test which rejects when $u'y^*v/(v'Sv)u'u \geq c$.

Section 6

30. The only simultaneous confidence sets for all $u'\eta v$, $u \in U$, v that are equivariant under the groups G_1–G_3 of the text are those given by (28).

31. Prove that each of the sets of simultaneous confidence intervals (29) and (31) is smallest among all families that are equivariant under a suitable group of transformations.

32. Under the assumptions made at the beginning of Section 6, show that the confidence intervals (33)

   (i)   are uniformly most accurate unbiased,

   (ii) are uniformly most accurate equivariant, and

   (iii) determine the constant k_0.

33. Write the simultaneous confidence sets (23) as explicitly as possible for the following cases:

   (i)   The one-sample problem of Section 3 with $\eta_i = \xi_i$ $(i = 1, \ldots, p)$.

   (ii) The two-sample problem of Section 3 with $\eta_i = \xi_i^{(2)} - \xi_i^{(1)}$.

34. Consider the s-sample situation in which $(X_{v1}^{(k)}, \ldots, X_{vp}^{(k)})$, $v = 1, \ldots, n_k$, $k = 1, \ldots, s$, are independent normal p-vectors with common covariance matrix Σ and with means $(\xi_1^{(k)}, \ldots, \xi_p^{(k)})$. Obtain as explicitly as possible the smallest simultaneous confidence sets for the set of all contrast vectors $(\Sigma u_k \xi_1^{(k)}, \ldots, \Sigma u_k \xi_p^{(k)})$, $\Sigma u_k = 0$.
[Example 10 of Chapter 7 and Problem 16.]

Section 7

35. The problem of testing the hypothesis $H: \eta \in \Pi_\omega$ against $\eta \in \Pi_{\Omega - \omega}$, when the distribution of Y is given by (34), remains invariant under a suitable group of linear transformations, and with respect to this group the test (35) is UMP invariant. The power of this test is given by (37) for all points $(\eta_1, \ldots, \eta_q)$.

36. Let $X_1, \ldots, X_n$ be i.i.d. with cumulative distribution function F, let $a_1 < \cdots < a_{m-1}$ be any given real numbers, and let $a_0 = -\infty$, $a_m = \infty$. If ν_i is the number of X's in (a_{i-1}, a_i), the χ^2-test (43) can be used to test $H: F = F_0$ with $\pi_i = F_0(a_i) - F_0(a_{i-1})$ for $i = 1, \ldots, m$.

   (i)   Unlike the Kolmogorov test, this χ^2-test is not consistent against all $F_1 \neq F_0$ as $n \to \infty$ with the a's remaining fixed.

   (ii) The test is consistent against any F_1 for which

$$F_1(a_i) - F_1(a_{i-1}) \neq F_0(a_i) - F_0(a_{i-1})$$

for at least one i.

Section 8

37. Let the equation of the tangent $\bar{\mathscr{S}}$ at π be $p_i = \pi_i(1 + a_{i1}\xi_1 + \cdots + a_{is}\xi_s)$, and suppose that the vectors $(a_{i1}, \ldots, a_{is})$ are orthogonal in the sense that $\sum a_{ik}a_{il}\pi_i = 0$ for all $k \neq l$.

    (i) If $(\hat{\xi}_1, \ldots, \hat{\xi}_s)$ minimizes $\sum(\nu_i - p_i)^2/\pi_i$ subject to $p \in \bar{\mathscr{S}}$, then $\hat{\xi}_j = \sum_i a_{ij}\nu_i/\sum_i a_{ij}^2\pi_i$.

    (ii) The test statistic (48) for testing $H : p = \pi$ reduces to

$$\frac{n \sum_{j=1}^{s} \left(\sum_{i=1}^{m} a_{ij}\nu_i \right)^2}{\sum_{i=1}^{m} a_{ij}^2\pi_i}.$$

38. In the multinomial model (38), the maximum-likelihood estimators $\hat{p}_i$ of the p's are $\hat{p}_i = x_i/n$.
    [The following are two methods for proving this result: (i) Maximize $\log P(x_1, \ldots, x_m)$ subject to $\sum p_i = 1$ by the method of undetermined multipliers. (ii) Show that $\prod p_i^{x_i} \leq \prod(x_i/n)^{x_i}$ by considering n numbers of which x_i are equal to p_i/x_i for $i = 1, \ldots, m$ and noting that their geometric mean is less than or equal to their arithmetic mean.]

39. In Example 7, show that the maximum-likelihood estimators $\hat{p}_{ij}$, $\hat{\hat{p}}_i$, and $\hat{\hat{p}}'_j$ are as stated.

40. In the situation of Example 7, consider the following model in which the row margins are fixed and which therefore generalizes model (iii) of Chapter 4, Section 7. A sample of n_i subjects is obtained from class A_i ($i = 1, \ldots, a$), the samples from different classes being independent. If n_{ij} is the number of subjects from the ith sample belonging to B_j ($j = 1, \ldots, b$), the joint distribution of $(n_{i1}, \ldots, n_{ib})$ is multinomial, say, $M(n_i; p_{1|i}, \ldots, p_{b|i})$. Determine the likelihood-ratio statistic for testing the hypothesis of *homogeneity* that the vector $(p_{1|i}, \ldots, p_{b|i})$ is independent of i, and specify its asymptotic distribution.

41. The *hypothesis of symmetry* in a square two-way contingency table arises when one of the responses $A_1, \ldots, A_a$ is observed for each of N subjects on two occasions (e.g. before and after some intervention). If n_{ij} is the number of subjects whose responses on the two occasions are (A_i, A_j), the joint distribution of the n_{ij} is given by (59) with $a = b$. The hypothesis H of *symmetry* states that $p_{ij} = p_{ji}$ for all i, j, that is, that the intervention has not changed the probabilities. Determine the likelihood-ratio statistic for testing H, and specify its asymptotic distribution. [Bowker (1948).]

42. In the situation of the preceding problem, consider the *hypothesis of marginal homogeneity* $H' : p_{i+} = p_{+i}$ for all i, where $p_{i+} = \sum_{j=1}^{a} p_{ij}$, $p_{+i} = \sum_{j=1}^{a} p_{ji}$.

    (i) The maximum-likelihood estimates of the p_{ij} under H' are given by $\hat{\hat{p}}_{ij} = n_{ij}/(1 + \lambda_i - \lambda_j)$, where the λ's are the solutions of the equations $\sum_j n_{ij}/(1 + \lambda_i - \lambda_j) = \sum_j n_{ij}/(1 + \lambda_j - \lambda_i)$. (These equations have no explicit solutions.)

    (ii) Determine the number of degrees of freedom of the limiting χ^2-distribution of the likelihood-ratio criterion.

43. Consider the third of the three sampling schemes for a $2 \times 2 \times K$ table discussed in Chapter 4, Section 8, and the two hypotheses

$$H_1 : \Delta_1 = \cdots = \Delta_K = 1 \quad \text{and} \quad H_2 : \Delta_1 = \cdots = \Delta_K.$$

    (i) Obtain the likelihood-ratio test statistic for testing H_1.

    (ii) Obtain equations that determine the maximum-likelihood estimates of the parameters under H_2. (These equations cannot be solved explicitly.)

    (iii) Determine the number of degrees of freedom of the limiting χ^2-distribution of the likelihood-ratio criterion for testing (a) H_1, (b) H_2.

[For a discussion of these and related hypotheses, see for example Shaffer (1973), Plackett (1981), or Bishop, Fienberg, and Holland (1975), and the recent study by Liang and Self (1985).]

Additional Problems

44. In generalization of Problem 8 of Chapter 7, let $(X_{v1}, \ldots, X_{vp})$, $v = 1, \ldots, n$, be independent normal p-vectors with common covariance matrix Σ and with means

$$\xi_{vi} = \sum_{j=1}^{s} a_{vj} \beta_j^{(i)},$$

where $A = (a_{vj})$ is a constant matrix of rank s and where the β's are unknown parameters. If $\theta_i = \sum e_j \beta_j^{(i)}$, give explicit expressions for the elements of V and S for testing the hypothesis $H : \theta_i = \theta_{i0}$ $(i = 1, \ldots, p)$.

45. *Testing for independence.* Let $X = (X_{\alpha i})$, $i = 1, \ldots, p$, $\alpha = 1, \ldots, N$, be a sample from a p-variate normal distribution; let $q < p$, $\max(q, p - q) \leq N$; and consider the hypothesis H that $(X_{11}, \ldots, X_{1q})$ is independent of $(X_{1\,q+1}, \ldots, X_{1p})$, that is, that the covariances $\sigma_{ij} = E(X_{\alpha i} - \xi_i)(X_{\alpha j} - \xi_j)$ are zero for all $i \leq q$, $j > q$. The problem of testing H remains invariant under the transformations $X_{\alpha i}^* = X_{\alpha i} + b_i$ and $X^* = XC$, where C is any nonsingu-

lar $p \times p$ matrix of the structure

$$C = \begin{pmatrix} C_{11} & 0 \\ 0 & C_{22} \end{pmatrix}$$

with C_{11} and C_{22} being $q \times q$ and $(p - q) \times (p - q)$ respectively.

(i) A set of maximal invariants under the induced transformations in the space of the sufficient statistics $X_{\cdot i}$ and the matrix S, partitioned as

$$S = \begin{pmatrix} S_{11} & S_{12} \\ S_{21} & S_{22} \end{pmatrix},$$

are the q roots of the equation

$$|S_{12} S_{22}^{-1} S_{21} - \lambda S_{11}| = 0.$$

(ii) In the case $q = 1$, a maximal invariant is the statistic $R^2 = S_{12} S_{22}^{-1} S_{21} / S_{11}$, which is the square of the *multiple correlation coefficient* between X_{11} and $(X_{12}, \ldots, X_{1p})$. The distribution of R^2 depends only on the square ρ^2 of the population multiple correlation coefficient, which is obtained from R^2 by replacing the elements of S with their expected values σ_{ij}.

(iii) Using the fact that the distribution of R^2 has the density [see for example Anderson (1984)]

$$\frac{(1 - R^2)^{\frac{1}{2}(N-p-2)} (R^2)^{\frac{1}{2}(p-1)-1} (1 - \rho^2)^{\frac{1}{2}(N-1)}}{\Gamma\left[\frac{1}{2}(N - 1)\right] \Gamma\left[\frac{1}{2}(N - p)\right]}$$

$$\times \sum_{h=0}^{\infty} \frac{(\rho^2)^h (R^2)^h \Gamma^2\left[\frac{1}{2}(N - 1) + h\right]}{h! \Gamma\left[\frac{1}{2}(p - 1) + h\right]}$$

and that the hypothesis H for $q = 1$ is equivalent to $\rho = 0$, show that the UMP invariant test rejects this hypothesis when $R^2 > C_0$.

(iv) When $\rho = 0$, the statistic

$$\frac{R^2}{1 - R^2} \cdot \frac{N - p}{p - 1}$$

has the F-distribution with $p - 1$ and $N - p$ degrees of freedom.

[(i): The transformations $X^* = XC$ with $C_{22} = I$ induce on S the transformations

$$(S_{11}, S_{12}, S_{22}) \rightarrow (S_{11}, C_{11} S_{12}, C_{11} S_{22} C_{11}')$$

with the maximal invariants $(S_{11}, S_{12} S_{22}^{-1} S_{21})$. Application to these invariants of the transformations $X^* = XC$ with $C_{11} = I$ completes the proof.]

46. The UMP invariant test of independence in part (ii) of the preceding problem is asymptotically robust against nonnormality.

47. *Bayes character and admissibility of Hotelling's T^2.*

   (i) Let $(X_{\alpha 1}, \ldots, X_{\alpha p})$, $\alpha = 1, \ldots, n$, be a sample from a p-variate normal distribution with unknown mean $\xi = (\xi_1, \ldots, \xi_p)$ and covariance matrix $\Sigma = A^{-1}$, and with $p \leq n - 1$. Then the one-sample T^2-test of $H : \xi = 0$ against $K : \xi \neq 0$ is a Bayes test with respect to prior distributions Λ_0 and Λ_1 which generalize those of Chapter 6, Example 13 (continued).

   (ii) The test of part (i) is admissible for testing H against the alternatives $\psi^2 \leq c$ for any $c > 0$.

[If ω is the subset of points $(0, \Sigma)$ of Ω_H satisfying $\Sigma^{-1} = A + \eta'\eta$ for some fixed positive definite $p \times p$ matrix A and arbitrary $\eta = (\eta_1, \ldots, \eta_p)$, and $\Omega'_{A,b}$ is the subset of points (ξ, Σ) of Ω_K satisfying $\Sigma^{-1} = A + \eta'\eta$, $\xi' = b\Sigma\eta'$ for the same A and some fixed $b > 0$, let Λ_0 and Λ_1 have densities defined over ω and $\Omega_{A,b}$ respectively by

$$\lambda_0(\eta) = C_0 |A + \eta'\eta|^{-n/2}$$

and

$$\lambda_1(\eta) = C_1 |A + \eta'\eta|^{-n/2} \exp\left\{ \frac{nb^2}{2} \left[\eta(A + \eta'\eta)^{-1}\eta' \right] \right\}.$$

(Kiefer and Schwartz, 1965).]

10. REFERENCES

Tests of multivariate linear hypotheses and the associated confidence sets have their origin in the work of Hotelling (1931). The simultaneous confidence intervals of Section 6 were proposed by Roy and Bose (1953), and shown to be smallest equivariant by Wijsman (1979). More details on these procedures and discussion of other multivariate techniques can be found in the comprehensive books by Anderson (1984) and Seber (1984). [A more geometric approach stressing invariance is provided by Eaton (1983).]

Anderson, T. W.
   (1984). *An Introduction to Multivariate Analysis*, 2nd ed., Wiley, New York.
Arnold, S. F.
   (1981). *The Theory of Linear Models and Multivariate Analysis*, Wiley, New York.
Barlow, R. E., Bartholomew, D. J., Bremner, J. M., and Brunk, H. D.
   (1972). *Statistical Inference under Order Restrictions*, Wiley, New York.
Bartlett, M. S.
   (1939). "A note on tests of significance in multivariate analysis." *Proc. Cambridge Philos. Soc.* **35**, 180–185.
   [Proposes the trace test (iii) of Section 4. See also Pillai (1955).]

Bauer, P.
   (1981). "On the robustness of Hotelling's T^2." *Biom. J.* **23**, 405–412.

Bishop, Y. M. M., Fienberg, S. E., and Holland, P. W.
   (1975). *Discrete Multivariate Analysis*, M.I.T. Press, Cambridge, Mass.

Bowker, A. H.
   (1948). "A test for symmetry in contingency tables." *J. Amer. Statist. Assoc.* **43**, 572–574.
   (1960). "A representation of Hotelling's T^2 and Anderson's classification statistic W in terms of simple statistics." In *Contributions to Probability and Statistics* (Olkin et al., eds.), Stanford Univ., Stanford, Calif.

Box, G. E. P.
   (1949). "A general distribution theory for a class of likelihood ratio criteria." *Biometrika* **36**, 317–346.

Box, G. E. P. and Tiao, G. C.
   (1973). *Bayesian Inference in Statistical Analysis*. Addison-Wesley, Reading, Mass.

Chernoff, H.
   (1954). "On the distribution of the likelihood ratio." *Ann. Math. Statist.* **25**, 573–578.

Cochran, W. G.
   (1952). "The χ^2 test of goodness of fit." *Ann. Math. Statist.* **2**, 315–345.
   (1954). "Some methods for strengthening the common χ^2 tests." *Biometrics* **10**, 417–451.

Cohen, A. and Sackrowitz, H. B.
   (1975). "Unbiasedness of the chi-square, likelihood ratio and other goodness of fit tests for the equal cell case." *Ann. Statist.* **3**, 959–964.

Cramér, H.
   (1946). *Mathematical Methods of Statistics*, Princeton U.P.

Dalal, S. R.
   (1978). "Simultaneous confidence procedures for univariate and multivariate Behrens–Fisher type problems." *Biometrika* **65**, 221–225.

Dalal, S. R. and Fortini, P.
   (1982). "An inequality comparing sums and maxima with application to Behrens–Fisher type problems." *Ann. Statist.* **10**, 297–301.

Davis, A. W.
   (1982). "On the effects of moderate multivariate nonnormality on Roy's largest root test." *J. Amer. Statist. Assoc.* **77**, 896–900.

Dempster, A. P.
   (1958). "A high dimensional two-sample significance test." *Ann. Math. Statist.* **29**, 995–1010.

Eaton, M. L.
   (1983). *Multivariate Statistics: A Vector Space Approach*, Wiley, New York.

Everitt, B. S.
   (1979). "A Monte Carlo investigation of the robustness of Hotelling's one and two-sample T^2-statistic." *J. Amer. Statist. Assoc.* **74**, 48–51.

Fisher, R. A.
   (1924a). "The conditions under which chi square measures the discrepancy between observation and hypothesis." *J. Roy. Statist. Soc.* **87**, 442–450.
   [Obtains the limiting distribution (under the hypothesis) of the χ^2 statistic for the case of composite hypotheses and discusses the dependence of this distribution on the method used to estimate the parameters.]
   (1924b). "On a distribution yielding the error functions of several well-known statistics." In

Proc. Int. Math. Congress., Toronto, 805–813.

[Obtains the distribution of the sample multiple correlation coefficient when the population multiple correlation coefficient is zero.]

(1928). "The general sampling distribution of the multiple correlation coefficient." *Proc. Roy. Soc., Ser. A* **121**, 654–673.

[Derives the noncentral χ^2- and noncentral beta-distribution and the distribution of the sample multiple correlation coefficient for arbitrary values of the population multiple correlation coefficient.]

(1935). *The Design of Experiments*, 1st ed., Oliver and Boyd, Edinburgh.

Fix, E., Hodges, J. L., Jr., and Lehmann, E. L.

(1959). "The restricted χ^2 test." In *Studies in Probability and Statistics Dedicated to Harald Cramér*, Almquist and Wiksell, Stockholm.

[Example 6.]

Goodman, L. A.

(1985). "The analysis of cross-classified data having ordered and/or unordered categories: Association models, correlation models and asymmetry models for contingency tables with or without missing entries." *Ann. Statist.* **13**, 10–69.

Haberman, S. J.

(1974). *The Analysis of Frequency Data*. Univ. of Chicago Press, Chicago.

Hoeffding, W.

(1965). "Asymptotically optimal tests for multinomial distributions" (with discussion). *Ann. Math. Statist.* **36**, 369–408.

Hotelling, H.

(1931). "The generalization of Student's ratio." *Ann. Math. Statist.* **2**, 360–378.

[Proposes the statistic (6) as a multivariate extension of Student's t, and obtains the distribution of the statistic under the hypothesis.]

(1951). "A generalized T test and measure of multivariate dispersion." In *Proc. Second Berkeley Symposium on Math. Statistics and Probability*, Univ. of California Press.

Hsu, P. L.

(1938). "Notes on Hotelling's generalized T^2." *Ann. Math. Statist.* **9**, 231–243.

[Obtains the distribution of T^2 in the noncentral case and applies the statistic to the class of problems described in Section 5. The derivation of the T^2-distribution indicated in Problems 6 and 7 is that of Wijsman (1957), which was noted also by Stein (cf. Wijsman, p. 416) and by Bowker (1960).]

(1941). "Canonical reduction of the general regression problem." *Ann. Eugenics* **11**, 42–46.

[Obtains the canonical form of the general linear multivariate hypothesis.]

(1945). "On the former function of the E^2-test and the T^2-test." *Ann. Math. Statist.* **16**, 278–286.

[Obtains a result on best average power for the T^2-test analogous to that of Chapter 7, Problem 5.]

Hunt, G. and Stein, C. M.

(1946). "Most stringent tests of statistical hypotheses." Unpublished.

[Proves the test (9) to be UMP almost invariant, and the roots of (5) to constitute a maximal set of invariants.]

James, G. S.

(1954). "Tests of linear hypotheses in univariate and multivariate analysis when the ratio of the population variances are unknown." *Biometrika* **41**, 19–43.

Johansen, S.

(1980). "The Welch–James approximation to the distribution of the residual sum of squares in weighted linear regression." *Biometrika* **67**, 85–92.

Johnson, N. L. and Kotz, S.

(1970). *Distributions in Statistics: Continuous Distributions*, Vol. 2, Houghton Mifflin, New York. Chapter 30.

Kallenberg, W. C. M., Oosterhoff, J., and Schriever, B. F.

(1985). "The number of classes in chi-squared goodness-of-fit tests." *J. Amer. Statist. Assoc.* **80**, 959–968.

Kariya, T.

(1978). "The general MANOVA problem." *Ann. Statist.* **6**, 220–214.

(1985). *Testing in the Multivariate Linear Model*. Kinokuniya, Tokyo.

Kendall, M. G. and Stuart, A.

(1979). *The Advanced Theory of Statistics*, 4th ed., Vol. 2, Griffin, London.

Kiefer, J. and Schwartz, R.

(1965). "Admissible Bayes character of T^2-, R^2-, and other fully invariant tests for classical multivariate normal problems." *Ann. Math. Statist.* **36**, 747–770.

Lawal, H. B. and Upton, G. J.

(1984). "On the use of χ^2 as test of independence in contingency tables with small expectations." *Austr. J. Statist.* **26**, 75–85.

Lawley, D. N.

(1938). "A generalization of Fisher's z-test." *Biometrika* **30**, 180–187.

[Proposes the trace test (i) of Section 4. See also Bartlett (1939) and Hotelling (1951)].

LeCam, L.

(1956). "On the asymptotic theory of estimation and testing hypotheses." In *Proc. Third Berkeley Symp. on Math. Statist. and Probab.*, Univ. of Calif. Press.

(1960). "Locally asymptotically normal families of distributions." *Univ. of Calif. Publ. in Statist.* **3**, 37–98.

Lewis, T., Saunders, I. W., and Westcott, M.

(1984). "Testing independence in a two-way contingency table; the moments of the chi-squared statistic and the minimum expected value." *Biometrika* **71**, 515–522.

Liang, K.-Y. and Self, S. G.

(1985). "Tests for homogeneity of odds ratio when the data are sparse." *Biometrika* **72**, 353–358.

Marden, J. I.

(1982). "Minimal complete classes of tests of hypotheses with multivariate one-sided alternatives." *Ann. Statist.* **10**, 962–970.

(1983). "Admissibility of invariant tests in the general multivariate analysis of variance problem." *Ann. Statist.* **11**, 1086–1099.

Marden, J. I. and Perlman, M. E.

(1980). "Invariant tests for means with covariates." *Ann. Statist.* **8**, 25–63.

McCullagh, P.

(1985). "On the asymptotic distribution of Pearson's statistic in linear exponential-family models." *Int. Statist. Rev.* **53**, 61–67.

Neyman, J.

(1949). "Contribution to the theory of the χ^2 test." In *Proc. Berkeley Symposium on Mathematical Statistics and Probability*, Univ. of California Press, Berkeley, 239–273.

[Gives a theory of χ^2 tests with restricted alternatives.]

Olson, C. L.

(1974). "Comparative robustness of six tests in multivariate analysis of variance." *J. Amer. Statist. Assoc.* **69**, 894–908.

(1976). "On choosing a test statistic in multivariate analysis of variance." *Psych. Bull.* **83**, 579–586.

Pearson, K.
(1900). "On the criterion that a given system of deviations from the probable in the case of a correlated system of variables is such that it can be reasonably supposed to have arisen from a random sampling." *Phil. Mag. Ser. 5* **50**, 157–172.
[The χ^2-test (43) is proposed for testing a simple multinomial hypothesis, and the limiting distribution of the test criterion is obtained under the hypothesis. The test is extended to composite hypotheses but contains an error in the degrees of freedom of the limiting distribution; a correct solution for the general case was found by Fisher (1924a). Applications.]

Perlman, M. D.
(1969). "One-sided testing problems in multivariate analysis." *Ann. Math. Statist.* **40**, 549–567. [Correction: *Ann. Math. Statist.* **42** (1971), 1777.]

Pillai, K. C. S.
(1955). "Some new test criteria in multivariate analysis." *Ann. Math. Statist.* **26**, 117–121.

Pillai, K. C. S. and Jayachandran, K.
(1967). "Power comparisons of tests of two multivariate hypotheses based on four criteria." *Biometrika* **54**, 195–210.

Plackett, R. L.
(1981). *The Analysis of Categorical Data*, 2nd ed., Macmillan, New York.

Press, S. J. and Shigemasu, K.
(1985). "Bayesian MANOVA and MANOCOVA under exchangeablility." *Comm. Statist.—Theor. Meth.* **14**, 1053–1078.

Quine, M. P. and Robinson, J.
(1985). "Efficiencies of chi-square and likelihood ratio goodness-of-fit tests." *Ann. Statist.* **13**, 727–742.

Radlow, R. and Alf, E. F., Jr.
(1975). "An alternative multinomial assessment of the accuracy of the chi-squared test of goodness of fit." *J. Amer. Statist. Assoc.* **70**, 811–813.

Roy, S. N.
(1953). "On a heuristic method of test construction and its use in multivariate analysis." *Ann. Math. Statist.* **24**, 220–238.
[Proposes the maximum root test (iv) of Section 4.]

Roy, S. N. and Bose, R. C.
(1953). "Simultaneous confidence interval estimation." *Ann. Math. Statist.* **24**, 513–536.
[Proposes the simultaneous confidence interval of Section 6.]

Scheffé, H.
(1956). "A 'mixed model' for the analysis of variance." *Ann. Math. Statist.* **27**, 23–36.
[Example 2.]

Schwartz, R.
(1967). "Admissible tests in multivariate analysis of variance." *Ann. Math. Statist.* **38**, 698–710.
(1969). "Invariant proper Bayes tests for exponential families." *Amer. Math. Statist.* **40**, 270–283.

Seber, G. A. F.
(1984). *Multivariate Observations*, Wiley, New York.

Serfling, R. J.
(1980). *Approximation Theorems of Mathematical Statistics*, Wiley, New York.

Shaffer, J. P.
(1973). "Defining and testing hypotheses in multi-dimensional contingency tables." *Psych. Bull.* **79**, 127–141.

Simaika, J. B.
(1941). "An optimum property of two statistical tests." *Biometrika* **32**, 70–80.
[Shows that the test (9) is UMP among all tests whose power function depends only on the noncentrality parameter (7), and establishes the corresponding property for the test of multiple correlation given in Problem 45(iii).]

Srivastava, M. S. and Awan, H. M.
(1982). "On the robustness of Hotelling's T^2-test and distribution of linear and quadratic forms in sampling from a mixture of two multivariate distributions." *Comm. Statist. Theor. Meth.* **11**, 81–107.

Stein, C.
(1956) "The admissibility of Hotelling's T^2-test." *Ann. Math. Statist.* **27**, 616–623.

Stevens, J. P.
(1980). "Power of the multivariate analysis of variance tests." *Psych. Bull.* **88**, 728–737.

Subrahmaniam, K. and Subrahmaniam, K.
(1973). "On the multivariate Behrens–Fisher problem." *Biometrika* **60**, 107–111.
(1975). "On the confidence region comparison of some solutions for the multivariate Behrens–Fisher problem." *Comm. Statist.* **4**, 57–67.

Wald, A.
(1942). "On the power function of analysis of variance test." *Ann. Math. Statist.* **13**, 434–439.
[Problem 5. This problem is also treated by Hsu (1945).]
(1943). "Tests of statistical hypotheses concerning several parameters when the number of observations is large." *Trans. Amer. Math. Soc.* **54**, 426–482.
[General asymptotic distribution and optimum theory of likelihood ratio (and asymptotically equivalent) tests.]

Wijsman, R.
(1957). "Random orthogonal transformations and their use in some classical distribution problems in multivariate analysis." *Ann. Math. Statist.* **28**, 415–423.
(1979). "Constructing all smallest simultaneous confidence sets in a given class, with applications to MANOVA." *Ann. Statist.* **7**, 1003–1018.
[Optimality results for simultaneous confidence sets including those of Section 6.]
(1980). "Smallest simultaneous confidence sets with applications in multivariate analysis." *Multivariate Anal.* **V**, 483–498.

Wilks, S. S.
(1932). "Certain generalizations in the analysis of variance." *Ann. Math. Statist.* **24**, 471–494.
[Obtains the likelihood-ratio test (Wilks' Λ) for the s-sample problem of Section 4.]
(1938). "The large-sample distribution of the likelihood ratio for testing composite hypotheses." *Ann. Math. Statist.* **9**, 60–62.
[Derives the asymptotic distribution of the likelihood ratio when the hypothesis is true.]

Yao, Y.
(1965). "An approximate degrees of freedom solution to the multivariate Behrens–Fisher problem," *Biometrika* **52**, 139–147.

CHAPTER 9

The Minimax Principle

1. TESTS WITH GUARANTEED POWER

The criteria discussed so far, unbiasedness and invariance, suffer from the disadvantage of being applicable, or leading to optimum solutions, only in rather restricted classes of problems. We shall therefore turn now to an alternative approach, which potentially is of much wider applicability. Unfortunately, its application to specific problems is in general not easy, and has so far been carried out successfully mainly in cases in which there exists a UMP invariant test.

One of the important considerations in planning an experiment is the number of observations required to insure that the resulting statistical procedure will have the desired precision or sensitivity. For problems of hypothesis testing this means that the probabilities of the two kinds of errors should not exceed certain preassigned bounds, say α and $1 - \beta$, so that the tests must satisfy the conditions

(1)
$$E_\theta \varphi(X) \leq \alpha \quad \text{for} \quad \theta \in \Omega_H,$$
$$E_\theta \varphi(X) \geq \beta \quad \text{for} \quad \theta \in \Omega_K.$$

If the power function $E_\theta \varphi(X)$ is continuous and if $\alpha < \beta$, (1) cannot hold when the sets Ω_H and Ω_K are contiguous. This mathematical difficulty corresponds in part to the fact that the division of the parameter values θ into the classes Ω_H and Ω_K for which the two different decisions are appropriate is frequently not sharp. Between the values for which one or the other of the decisions is clearly correct there may lie others for which the relative advantages and disadvantages of acceptance and rejection are approximately in balance. Accordingly we shall assume that Ω is partitioned into three sets

$$\Omega = \Omega_H + \Omega_I + \Omega_K,$$

of which Ω_I designates the *indifference zone*, and Ω_K the class of parameter values differing so widely from those postulated by the hypothesis that false acceptance of H is a serious error, which should occur with probability at most $1 - \beta$.

To see how the sample size is determined in this situation, suppose that $X_1, X_2, \ldots$ constitute the sequence of available random variables, and for a moment let n be fixed and let $X = (X_1, \ldots, X_n)$. In the usual applicational situations (for a more precise statement, see Problem 1) there exists a test φ_n which maximizes

$$(2) \qquad \inf_{\Omega_K} E_\theta \varphi(X)$$

among all level-α tests based on X. Let $\beta_n = \inf_{\Omega_K} E_\theta \varphi_n(X)$, and suppose that for sufficiently large n there exists a test satisfying (1). [Conditions under which this is the case are given by Berger (1951) and Kraft (1955).] The desired sample size, which is the smallest value of n for which $\beta_n \geq \beta$, is then obtained by trail and error. This requires the ability of determining for each fixed n the test that maximizes (2) subject to

$$(3) \qquad E_\theta \varphi(X) \leq \alpha \qquad \text{for} \quad \theta \in \Omega_H.$$

A method for determining a test with this *maximin* property (of maximizing the minimum power over Ω_K) is obtained by generalizing Theorem 7 of Chapter 3. It will be convenient in this discussion to make a change of notation, and to denote by ω and ω' the subsets of Ω previously denoted by Ω_H and Ω_K. Let $\mathscr{P} = \{P_\theta, \theta \in \omega \cup \omega'\}$ be a family of probability distributions over a sample space $(\mathscr{X}, \mathscr{A})$ with densities $p_\theta = dP_\theta/d\mu$ with respect to a σ-finite measure μ, and suppose that the densities $p_\theta(x)$ considered as functions of the two variables (x, θ) are measurable $(\mathscr{A} \times \mathscr{B})$ and $(\mathscr{A} \times \mathscr{B}')$, where $\mathscr{B}$ and $\mathscr{B}'$ are given σ-fields over ω and ω'. Under these assumptions, the following theorem gives conditions under which a solution of a suitable Bayes problem provides a test with the required properties.

Theorem 1. *For any distributions Λ and Λ' over $\mathscr{B}$ and $\mathscr{B}'$, let $\varphi_{\Lambda, \Lambda'}$ be the most powerful test for testing*

$$h(x) = \int_\omega p_\theta(x) \, d\Lambda(\theta)$$

at level α against

$$h'(x) = \int_{\omega'} p_\theta(x) \, d\Lambda'(\theta)$$

and let $\beta_{\Lambda, \Lambda'}$ be its power against the alternative h'. If there exist Λ and Λ' such that

(4)
$$\sup_\omega E_\theta \varphi_{\Lambda, \Lambda'}(X) \leq \alpha,$$

$$\inf_{\omega'} E_\theta \varphi_{\Lambda, \Lambda'}(X) = \beta_{\Lambda, \Lambda'},$$

then:

(i) $\varphi_{\Lambda, \Lambda'}$ *maximizes* $\inf_{\omega'} E_\theta \varphi(X)$ *among all level-α tests of the hypothesis $H : \theta \in \omega$ and is the unique test with this property if it is the unique most powerful level-α test for testing h against h'.*

(ii) *The pair of distributions Λ, Λ' is least favorable in the sense that for any other pair ν, ν' we have*

$$\beta_{\Lambda, \Lambda'} \leq \beta_{\nu, \nu'}.$$

Proof. (i): If φ^* is any other level-α test of H, it is also of level α for testing the simply hypothesis that the density of X is h, and the power of φ^* against h' therefore cannot exceed $\beta_{\Lambda, \Lambda'}$. It follows that

$$\inf_{\omega'} E_\theta \varphi^*(X) \leq \int_{\omega'} E_\theta \varphi^*(X)\, d\Lambda'(\theta) \leq \beta_{\Lambda, \Lambda'} = \inf_{\omega'} E_\theta \varphi_{\Lambda, \Lambda'}(X),$$

and the second inequality is strict if $\varphi_{\Lambda, \Lambda'}$ is unique.

(ii): Let ν, ν' be any other distributions over $(\omega, \mathcal{B})$ and $(\omega', \mathcal{B}')$, and let

$$g(x) = \int_\omega p_\theta(x)\, d\nu(\theta), \qquad g'(x) = \int_{\omega'} p_\theta(x)\, d\nu'(\theta).$$

Since both $\varphi_{\Lambda, \Lambda'}$ and $\varphi_{\nu, \nu'}$ are level-α tests of the hypothesis that $g(x)$ is the density of X, it follows that

$$\beta_{\nu, \nu'} \geq \int \varphi_{\Lambda, \Lambda'}(x) g'(x)\, d\mu(x) \geq \inf_{\omega'} E_\theta \varphi_{\Lambda, \Lambda'}(X) = \beta_{\Lambda, \Lambda'}.$$

Corollary 1. *Let Λ, Λ' be two probability distributions and C a constant such that*

(5) $$\varphi_{\Lambda, \Lambda'}(x) = \begin{cases} 1 & \text{if } \int_{\omega'} p_\theta(x)\, d\Lambda'(\theta) > C \int_\omega p_\theta(x)\, d\Lambda(\theta) \\[2mm] \gamma & \text{if } \int_{\omega'} p_\theta(x)\, d\Lambda'(\theta) = C \int_\omega p_\theta(x)\, d\Lambda(\theta) \\[2mm] 0 & \text{if } \int_{\omega'} p_\theta(x)\, d\Lambda'(\theta) < C \int_\omega p_\theta(x)\, d\Lambda(\theta) \end{cases}$$

is a size-α test for testing that the density of X is $\int_\omega p_\theta(x)\, d\Lambda(\theta)$ and such that

$$(6) \qquad\qquad \Lambda(\omega_0) = \Lambda'(\omega_0') = 1,$$

where

$$\omega_0 = \left\{ \theta : \theta \in \omega \text{ and } E_\theta \varphi_{\Lambda,\,\Lambda'}(X) = \sup_{\theta' \in \omega} E_{\theta'} \varphi_{\Lambda,\,\Lambda'}(X) \right\}$$

$$\omega_0' = \left\{ \theta : \theta \in \omega' \text{ and } E_\theta \varphi_{\Lambda,\,\Lambda'}(X) = \inf_{\theta' \in \omega'} E_{\theta'} \varphi_{\Lambda,\,\Lambda'}(X) \right\}.$$

Then the conclusions of Theorem 1 hold.

Proof. If h, h', and $\beta_{\Lambda,\,\Lambda'}$ are defined as in Theorem 1, the assumptions imply that $\varphi_{\Lambda,\,\Lambda'}$ is a most powerful level-α test for testing h against h', that

$$\sup_\omega E_\theta \varphi_{\Lambda,\,\Lambda'}(X) = \int_\omega E_\theta \varphi_{\Lambda,\,\Lambda'}(X)\, d\Lambda(\theta) = \alpha,$$

and that

$$\inf_{\omega'} E_\theta \varphi_{\Lambda,\,\Lambda'}(X) = \int_{\omega'} E_\theta \varphi_{\Lambda,\,\Lambda'}(X)\, d\Lambda'(\theta) = \beta_{\Lambda,\,\Lambda'}.$$

The condition (4) is thus satisfied and Theorem 1 applies.

Suppose that the sets Ω_H, Ω_I, and Ω_K are defined in terms of a nonnegative function d, which is a measure of the distance of θ from H, by

$$\Omega_H = \{\theta : d(\theta) = 0\}, \qquad \Omega_I = \{\theta : 0 < d(\theta) < \Delta\},$$

$$\Omega_K = \{\theta : d(\theta) \geq \Delta\}.$$

Suppose also that the power function of any test is continuous in θ. In the limit as $\Delta = 0$, there is no indifference zone. Then Ω_K becomes the set $\{\theta : d(\theta) > 0\}$, and the infimum of $\beta(\theta)$ over Ω_K is $\leq \alpha$ for any level-α test. This infimum is therefore maximized by any test satisfying $\beta(\theta) \geq \alpha$ for all $\theta \in \Omega_K$, that is, by any unbiased test, so that unbiasedness is seen to be a limiting form of the maximin criterion. A more useful limiting form, since it will typically lead to a unique test, is given by the following definition. A test φ_0 is said to *maximize the minimum power locally** if, given

---

*A different definition of local minimaxity is given by Giri and Kiefer (1964).

any other test φ, there exists Δ_0 such that

$$(7) \qquad \inf_{\omega_\Delta} \beta_{\varphi_0}(\theta) \geq \inf_{\omega_\Delta} \beta_\varphi(\theta) \qquad \text{for all} \quad 0 < \Delta < \Delta_0,$$

where ω_Δ is the set of θ's for which $d(\theta) \geq \Delta$.

2. EXAMPLES

In Chapter 3 it was shown for a family of probability densities depending on a real parameter θ that a UMP test exists for testing $H: \theta \leq \theta_0$ against $\theta > \theta_0$ provided for all $\theta < \theta'$ the ratio $p_{\theta'}(x)/p_\theta(x)$ is a monotone function of some real-valued statistic. This assumption, although satisfied for a one-parameter exponential family, is quite restrictive, and a UMP test of H will in fact exist only rarely. A more general approach is furnished by the formulation of the preceding section. If the indifference zone is the set of θ's with $\theta_0 < \theta < \theta_1$, the problem becomes that of maximizing the minimum power over the class of alternatives $\omega': \theta \geq \theta_1$. Under appropriate assumptions, one would expect the least favorable distributions Λ and Λ' of Theorem 1 to assign probability 1 to the points θ_0 and θ_1, and hence the maximin test to be given by the rejection region $p_{\theta_1}(x)/p_{\theta_0}(x) > C$. The following lemma gives sufficient conditions for this to be the case.

Lemma 1. *Let $X_1, \ldots, X_n$ be identically and independently distributed with probability density $f_\theta(x)$, where θ and x are real-valued, and suppose that for any $\theta < \theta'$ the ratio $f_{\theta'}(x)/f_\theta(x)$ is a nondecreasing function of x. Then the level-α test φ of H which maximizes the minimum power over ω' is given by*

$$(8) \qquad \varphi(x_1, \ldots, x_n) = \begin{cases} 1 & \text{if} \quad r(x_1, \ldots, x_n) > C, \\ \gamma & \text{if} \quad r(x_1, \ldots, x_n) = C, \\ 0 & \text{if} \quad r(x_1, \ldots, x_n) < C, \end{cases}$$

where $r(x_1, \ldots, x_n) = f_{\theta_1}(x_1) \ldots f_{\theta_1}(x_n)/f_{\theta_0}(x_1) \ldots f_{\theta_0}(x_n)$ and where C and γ are determined by

$$(9) \qquad E_{\theta_0}\varphi(X_1, \ldots, X_n) = \alpha.$$

Proof. The function $\varphi(x_1, \ldots, x_n)$ is nondecreasing in each of its arguments, so that by Lemma 2 of Chapter 3

$$E_\theta \varphi(X_1, \ldots, X_n) \leq E_{\theta'} \varphi(X_1, \ldots, X_n)$$

when $\theta < \theta'$. Hence the power function of φ is monotone and φ is a level-α test. Since $\varphi = \varphi_{\Lambda, \Lambda'}$, where Λ and Λ' are the distributions assigning probability 1 to the points θ_0 and θ_1, the condition (4) is satisfied, which proves the desired result as well as the fact that the pair of distributions (Λ, Λ') is least favorable.

Example 1. Let θ be a location parameter, so that $f_\theta(x) = g(x - \theta)$, and suppose for simplicity that $g(x) > 0$ for all x. We will show that a necessary and sufficient condition for $f_\theta(x)$ to have monotone likelihood ratio in x is that $-\log g$ is convex. The condition of monotone likelihood ratio in x,

$$\frac{g(x - \theta')}{g(x - \theta)} \le \frac{g(x' - \theta')}{g(x' - \theta)} \quad \text{for all} \quad x < x', \quad \theta < \theta',$$

is equivalent to

$$\log g(x' - \theta) + \log g(x - \theta') \le \log g(x - \theta) + \log g(x' - \theta').$$

Since $x - \theta = t(x - \theta') + (1 - t)(x' - \theta)$ and $x' - \theta' = (1 - t)(x - \theta') + t(x' - \theta)$, where $t = (x' - x)/(x' - x + \theta' - \theta)$, a sufficient condition for this to hold is that the function $-\log g$ is convex. To see that this condition is also necessary, let $a < b$ be any real numbers, and let $x - \theta' = a$, $x' - \theta = b$, and $x' - \theta' = x - \theta$. Then $x - \theta = \frac{1}{2}(x' - \theta + x - \theta') = \frac{1}{2}(a + b)$, and the condition of monotone likelihood ratio implies

$$\tfrac{1}{2}[\log g(a) + \log g(b)] \le \log g[\tfrac{1}{2}(a + b)].$$

Since $\log g$ is measurable, this in turn implies that $-\log g$ is convex.*

A density g for which $-\log g$ is convex is called *strongly unimodal*. Basic properties of such densities were obtained by Ibragimov (1956). Strong unimodality is a special case of total positivity. A density of the form $g(x - \theta)$ which is totally positive of order r is said to be a *Polya frequency* function of order r. It follows from Example 1 that $g(x - \theta)$ is a Polya frequency function of order 2 if and only if it is strongly unimodal. [For further results concerning Polya frequency functions and strongly unimodal densities, see Karlin (1968), Marshall and Olkin (1979), Huang and Ghosh (1982), and Loh (1984a, b).]

Two distributions which satisfy the above condition [besides the normal distribution, for which the resulting densities $p_\theta(x_1, \ldots, x_n)$ form an exponential family] are the *double exponential distribution* with

$$g(x) = \tfrac{1}{2}e^{-|x|}$$

*See Sierpinski (1920).

and the *logistic distribution*, whose cumulative distribution function is

$$G(x) = \frac{1}{1 + e^{-x}},$$

so that the density is $g(x) = e^{-x}/(1 + e^{-x})^2$.

Example 2. To consider the corresponding problem for a scale parameter, let $f_\theta(x) = \theta^{-1}h(x/\theta)$ where h is an even function. Without loss of generality one may then restrict x to be nonnegative, since the absolute values $|X_1|, \ldots, |X_n|$ form a set of sufficient statistics for θ. If $Y_i = \log X_i$ and $\eta = \log \theta$, the density of Y_i is

$$h(e^{y-\eta})e^{y-\eta}.$$

By Example 1, if $h(x) > 0$ for all $x \geq 0$, a necessary and sufficient condition for $f_{\theta'}(x)/f_\theta(x)$ to be a nondecreasing function of x for all $\theta < \theta'$ is that $-\log[e^y h(e^y)]$ or equivalently $-\log h(e^y)$ is a convex function of y. An example in which this holds—in addition to the normal and double-exponential distributions, where the resulting densities form an exponential family—is the *Cauchy distribution* with

$$h(x) = \frac{1}{\pi} \frac{1}{1 + x^2}.$$

Since the convexity of $-\log h(y)$ implies that of $-\log h(e^y)$, it follows that if h is an even function and $h(x - \theta)$ has monotone likelihood ratio, so does $h(x/\theta)$. When h is the normal or double-exponential distribution, this property of $h(x/\theta)$ follows therefore also from Example 1. That monotone likelihood ratio for the scale-parameter family does not conversely imply the same property for the associated location parameter family is illustrated by the Cauchy distribution. The condition is therefore more restrictive for a location than for a scale parameter.

The chief difficulty in the application of Theorem 1 to specific problems is the necessity of knowing, or at least being able to guess correctly, a pair of least favorable distributions (Λ, Λ'). Guidance for obtaining these distributions is sometimes provided by invariance considerations. If there exists a group G of transformations of X such that the induced group $\bar{G}$ leaves both ω and ω' invariant, the problem is symmetric in the various θ's that can be transformed into each other under $\bar{G}$. It then seems plausible that unless Λ and Λ' exhibit the same symmetries, they will make the statistician's task easier, and hence will not be least favorable.

Example 3. In the problem of paired comparisons considered in Example 7 of Chapter 6, the observations X_i ($i = 1, \ldots, n$) are independent variables taking on the values 1 and 0 with probabilities p_i and $q_i = 1 - p_i$. The hypothesis H to be tested specifies the set $\omega : \max p_i \leq \frac{1}{2}$. Only alternatives with $p_i \geq \frac{1}{2}$ for all i are considered, and as ω' we take the subset of those alternatives for which $\max p_i \geq \frac{1}{2} + \delta$. One would expect Λ to assign probability 1 to the point $p_1 = \cdots p_n = \frac{1}{2}$, and Λ' to assign positive probability only to the n points $(p_1, \ldots, p_n)$ which have $n - 1$ coordinates equal to $\frac{1}{2}$ and the remaining coordinate equal to $\frac{1}{2} + \delta$. Because of the symmetry with regard to the n variables, it seems plausible that Λ' should assign equal probability $1/n$ to each of these n points. With these choices, the test $\varphi_{\Lambda, \Lambda'}$ rejects when

$$\sum_{i=1}^{n} \left(\frac{\frac{1}{2} + \delta}{\frac{1}{2}} \right)^{x_i} > C.$$

This is equivalent to

$$\sum_{i=1}^{n} x_i > C,$$

which had previously been seen to be UMP invariant for this problem. Since the critical function $\varphi_{\Lambda, \Lambda'}(x_1, \ldots, x_n)$ is nondecreasing in each of its arguments, it follows from Lemma 2 of Chapter 3 that $p_i \le p'_i$ for $i = 1, \ldots, n$ implies

$$E_{p_1, \ldots, p_n} \varphi_{\Lambda, \Lambda'}(X_1, \ldots, X_n) \le E_{p'_1, \ldots, p'_n} \varphi_{\Lambda, \Lambda'}(X_1, \ldots, X_n)$$

and hence the conditions of Theorem 1 are satisfied.

Example 4. Let $X = (X_1, \ldots, X_n)$ be a sample from $N(\xi, \sigma^2)$, and consider the problem of testing $H : \sigma = \sigma_0$ against the set of alternatives $\omega' : \sigma \le \sigma_1$ or $\sigma \ge \sigma_2$ ($\sigma_1 < \sigma_0 < \sigma_2$). This problem remains invariant under the transformations $X'_i = X_i + c$ which in the parameter space induce the group $\bar{G}$ of transformations $\xi' = \xi + c$, $\sigma' = \sigma$. One would therefore expect the least favorable distribution Λ over the line $\omega : -\infty < \xi < \infty$, $\sigma = \sigma_0$, to be invariant under $\bar{G}$. Such invariance implies that Λ assigns to any interval a measure proportional to the length of the interval. Hence Λ cannot be a probability measure and Theorem 1 is not directly applicable. The difficulty can be avoided by approximating Λ by a sequence of probability distributions, in the present case for example by the sequence of normal distributions $N(0, k)$, $k = 1, 2, \ldots$.

In the particular problem under consideration, it happens that there also exist least favorable distributions Λ and Λ', which are true probability distributions and therefore not invariant. These distributions can be obtained by an examination of the corresponding one-sided problem in Chapter 3, Section 9, as follows. On ω, where the only variable is ξ, the distribution Λ of ξ is taken as the normal distribution with an arbitrary mean ξ_1 and with variance $(\sigma_2^2 - \sigma_0^2)/n$. Under Λ' all probability should be concentrated on the two lines $\sigma = \sigma_1$ and $\sigma = \sigma_2$ in the (ξ, σ) plane, and we put $\Lambda' = p\Lambda'_1 + q\Lambda'_2$, where Λ'_1 is the normal distribution with mean ξ_1 and variance $(\sigma_2^2 - \sigma_1^2)/n$, while Λ'_2 assigns probability 1 to the point (ξ_1, σ_2). A computation analogous to that carried out in Chapter 3, Section 9, then shows the acceptance region to be given by

$$\frac{\dfrac{p}{\sigma_1^{n-1}\sigma_2} \exp\left[\dfrac{-1}{2\sigma_1^2} \Sigma(x_i - \bar{x})^2 - \dfrac{n}{2\sigma_2^2}(\bar{x} - \xi_1)^2 \right] + \dfrac{q}{\sigma_2^n} \exp\left[\dfrac{-1}{2\sigma_2^2}\left\{ \Sigma(x_i - \bar{x})^2 + n(\bar{x} - \xi_1)^2 \right\} \right]}{\dfrac{1}{\sigma_0^{n-1}\sigma_2} \exp\left[\dfrac{-1}{2\sigma_0^2}\Sigma(x_i - \bar{x})^2 - \dfrac{n}{2\sigma_2^2}(\bar{x} - \xi_1)^2 \right]} < C,$$

which is equivalent to

$$C_1 \le \sum (x_i - \bar{x})^2 \le C_2 .$$

The probability of this inequality is independent of ξ, and hence C_1 and C_2 can be determined so that the probability of acceptance is $1 - \alpha$ when $\sigma = \sigma_0$, and is equal for the two values $\sigma = \sigma_1$ and $\sigma = \sigma_2$.

It follows from Section 7 of Chapter 3 that there exist p and C which lead to these values of C_1 and C_2 and that the above test satisfies the conditions of Corollary 1 with $\omega_0 = \omega$, and with ω_0' consisting of the two lines $\sigma = \sigma_1$ and $\sigma = \sigma_2$.

3. COMPARING TWO APPROXIMATE HYPOTHESES

As in Chapter 3, Section 2, let $P_0 \ne P_1$ be two distributions possessing densities p_0 and p_1 with respect to a measure μ. Since distributions even at best are known only approximately, let us assume that the true distributions are approximately P_0 or P_1 in the sense that they lie in one of the families

$$(10) \qquad \mathscr{P}_i = \{ Q : Q = (1 - \epsilon_i) P_i + \epsilon_i G_i \}, \qquad i = 0, 1,$$

with ϵ_0, ϵ_1 given and the G_i arbitrary unknown distributions. We wish to find the level-α test of the hypothesis H that the true distribution lies in $\mathscr{P}_0$, which maximizes the minimum power over $\mathscr{P}_1$. This is the problem considered in Section 1 with θ indicating the true distribution, $\Omega_H = \mathscr{P}_0$, and $\Omega_K = \mathscr{P}_1$.

The following theorem shows the existence of a pair of least favorable distributions Λ and Λ' satisfying the conditions of Theorem 1, each assigning probability 1 to a single distribution, Λ to $Q_0 \in \mathscr{P}_0$ and Λ' to $Q_1 \in \mathscr{P}_1$, and exhibits the Q_i explicitly.

Theorem 2. *Let*

$$q_0(x) = \begin{cases} (1 - \epsilon_0) p_0(x) & \text{if } \dfrac{p_1(x)}{p_0(x)} < b, \\[2ex] \dfrac{(1 - \epsilon_0) p_1(x)}{b} & \text{if } \dfrac{p_1(x)}{p_0(x)} \ge b, \end{cases}$$

$$(11)$$

$$q_1(x) = \begin{cases} (1 - \epsilon_1) p_1(x) & \text{if } \dfrac{p_1(x)}{p_0(x)} > a, \\[2ex] a(1 - \epsilon_1) p_0(x) & \text{if } \dfrac{p_1(x)}{p_0(x)} \le a. \end{cases}$$

(i)   *For all $0 < \epsilon_i < 1$, there exist unique constants a and b such that q_0 and q_1 are probability densities with respect to μ; the resulting q_i are members of $\mathscr{P}_i$ $(i = 0, 1)$.*

(ii) *There exist δ_0, δ_1 such that for all $\epsilon_i \leq \delta_i$ the constants a and b satisfy $a < b$ and that the resulting q_0 and q_1 are distinct.*

(iii) *If $\epsilon_i \leq \delta_i$ for $i = 0, 1$, the families $\mathscr{P}_0$ and $\mathscr{P}_1$ are nonoverlapping and the pair (q_0, q_1) is least favorable, so that the maximin test of $\mathscr{P}_0$ against $\mathscr{P}_1$ rejects when $q_1(x)/q_0(x)$ is sufficiently large.*

Note. *Suppose $a < b$, and let*

$$r(x) = \frac{p_1(x)}{p_0(x)}, \quad r^*(x) = \frac{q_1(x)}{q_0(x)}, \quad \text{and} \quad k = \frac{1 - \epsilon_1}{1 - \epsilon_0}.$$

Then

(12)
$$r^*(x) = \begin{cases} ka & \text{when } r(x) \leq a, \\ kr(x) & \text{when } a < r(x) < b, \\ kb & \text{when } b \leq r(x). \end{cases}$$

The maximin test thus replaces the original probability ratio with a censored version.

Proof. The proof will be given under the simplifying assumption that $p_0(x)$ and $p_1(x)$ are positive for all x in the sample space.

(i): For q_1 to be a probability density, a must satisfy the equation

(13)
$$P_1[r(X) > a] + aP_0[r(X) \leq a] = \frac{1}{1 - \epsilon_1}.$$

If (13) holds, it is easily checked that $q_1 \in \mathscr{P}_1$ (Problem 10). To prove existence and uniqueness of a solution a of (13), let

$$\gamma(c) = P_1[r(X) > c] + cP_0[r(X) \leq c].$$

Then

(14)
$$\gamma(0) = 1 \quad \text{and} \quad \gamma(c) \to \infty \quad \text{as } c \to \infty.$$

Furthermore (Problem 12)

(15) $\quad \gamma(c + \Delta) - \gamma(c) = \Delta \displaystyle\int_{r(x) \leq c} p_0(x) \, d\mu(x)$

$$+ \int_{c < r(x) \leq c + \Delta} [c + \Delta - r(x)] p_0(x) \, d\mu(x).$$

It follows from (15) that $0 \le \gamma(c + \Delta) - \gamma(c) \le \Delta$, so that γ is continuous and nondecreasing. Together with (14) this establishes the existence of a solution. To prove uniqueness, note that

$$(16) \qquad \gamma(c + \Delta) - \gamma(c) \ge \Delta \int_{r(x) < c} p_0(x)\, d\mu(x)$$

and that $\gamma(c) = 1$ for all c for which

$$(17) \qquad P_i[r(x) \le c] = 0 \qquad (i = 0, 1).$$

If c_0 is the supremum of the values for which (17) holds, (16) shows that γ is strictly increasing for $c > c_0$ and this proves uniqueness. The proof for b is exactly analogous (Problem 11).

(ii): As $\epsilon_1 \to 0$, the solution a of (13) tends to c_0. Analogously, as $\epsilon_1 \to 0$, $b \to \infty$ (Problem 11).

(iii): This will follow from the following facts:

(a) When X is distributed according to a distribution in $\mathscr{P}_0$, the statistic $r^*(X)$ is stochastically largest when the distribution of X is Q_0.

(b) When X is distributed according to a distribution in $\mathscr{P}_1$, $r^*(X)$ is stochastically smallest for Q_1.

(c) $r^*(X)$ is stochastically larger when the distribution of X is Q_1 than when it is Q_0.

These statements are summarized in the inequalities

$$(18)$$

$$Q_0'[r^*(X) < t] \ge Q_0[r^*(X) < t] \ge Q_1[r^*(X) < t] \ge Q_1'[r^*(X) < t]$$

for all t and all $Q_i' \in \mathscr{P}_i$.

From (12), it is seen that (18) is obvious when $t \le ka$ or $t > kb$. Suppose therefore that $ak < t \le bk$, and denote the event $r^*(X) < t$ by E. Then $Q_0'(E) \ge (1 - \epsilon_0)P_0(E)$ by (10). But $r^*(x) < t \le kb$ implies $r(X) < b$ and hence $Q_0(E) = (1 - \epsilon_0)P_0(E)$. Thus $Q_0'(E) \ge Q_0(E)$, and analogously $Q_1'(E) \le Q_1(E)$. Finally, the middle inequality of (18) follows from Corollary 1 of Chapter 3.

If the ϵ's are sufficiently small so that $Q_0 \ne Q_1$, it follows from (a)–(c) that $\mathscr{P}_0$ and $\mathscr{P}_1$ are nonoverlapping.

That (Q_0, Q_1) is least favorable and the associated test φ is maximin now follows from Theorem 1, since the most powerful test φ for testing Q_0

against Q_1 is a nondecreasing function of $q_1(X)/q_0(X)$. This shows that $E\varphi(X)$ takes on its sup over $\mathcal{P}_0$ at Q_0 and its inf over $\mathcal{P}_1$ at Q_1, and this completes the proof.

Generalizations of this theorem are given by Huber and Strassen (1973, 1974). See also Rieder (1977) and Bednarski (1984). An optimum permutation test, with generalizations to the case of unknown location and scale parameters, is discussed by Lambert (1985).

When the data consist of n identically, independently distributed random variables $X_1, \ldots, X_n$, the neighborhoods (10) may not be appropriate, since they do not preserve the assumption of independence. If P_i has density

$$(19) \qquad p_i(x_1, \ldots, x_n) = f_i(x_1) \ldots f_i(x_n) \qquad (i = 0, 1),$$

a more appropriate model approximating (19) may then assign to $X = (X_1, \ldots, X_n)$ the family $\mathcal{P}_i^*$ of distributions according to which the X_j are independently distributed, each with distribution

$$(20) \qquad (1 - \epsilon_i)F_i(x_j) + \epsilon_i G_i(x_j),$$

where F_i has density f_i and where as before the G_i are arbitrary.

Corollary 2. *Suppose q_0 and q_1 defined by (11) with $x = x_j$ satisfy (18) and hence are a least favorable pair for testing $\mathcal{P}_0$ against $\mathcal{P}_1$ on the basis of the single observation X_j. Then the pair of distributions with densities $q_i(x_1) \ldots q_i(x_n)$ $(i = 0, 1)$ is least favorable for testing $\mathcal{P}_0^*$ against $\mathcal{P}_1^*$, so that the maximin test is given by*

$$(21) \qquad \varphi(x_1, \ldots, x_n) = \begin{cases} 1 \\ \gamma \\ 0 \end{cases} \quad \text{if} \quad \prod_{j=1}^{n} \left[\frac{q_1(x_j)}{q_0(x_j)} \right] \gtreqless c.$$

Proof. By assumption, the random variables $Y_j = q_1(X_j)/q_0(X_j)$ are stochastically increasing as one moves successively from $Q_0' \in \mathcal{P}_0$ to Q_0 to Q_1 to $Q_1' \in \mathcal{P}_1$. The same is then true of any function $\psi(Y_1, \ldots, Y_n)$ which is nondecreasing in each of its arguments by Lemma 1 of Chapter 3, and hence of φ defined by (21). The proof now follows from Theorem 2.

Instead of the problem of testing P_0 against P_1, consider now the situation of Lemma 1 where $H: \theta \leq \theta_0$ is to be tested against $\theta \geq \theta_1$ $(\theta_0 < \theta_1)$ on the basis of n independent observations X_j, each distributed according to a distribution $F_\theta(x_j)$ whose density $f_\theta(x_j)$ is assumed to have monotone likelihood ratio in x_j.

A robust version of this problem is obtained by replacing F_θ with

$$(22) \qquad (1 - \epsilon)F_\theta(x_j) + \epsilon G(x_j), \qquad j = 1, \ldots, n,$$

where ϵ is given and for each θ the distribution G is arbitrary. Let $\mathscr{P}_0^{**}$ and $\mathscr{P}_1^{**}$ be the classes of distributions (22) with $\theta \leq \theta_0$ and $\theta \geq \theta_1$ respectively; and let $\mathscr{P}_0^*$ and $\mathscr{P}_1^*$ be defined as in Corollary 2 with f_{θ_i} in place of f_i. Then the maximin test (21) of $\mathscr{P}_0^*$ against $\mathscr{P}_1^*$ retains this property for testing $\mathscr{P}_0^{**}$ against $\mathscr{P}_1^{**}$.

This is proved in the same way as Corollary 2, using the additional fact that if $F_{\theta'}$ is stochastically larger than F_θ, then $(1 - \epsilon)F_{\theta'} + \epsilon G$ is stochastically larger than $(1 - \epsilon)F_\theta + \epsilon G$.

4. MAXIMIN TESTS AND INVARIANCE

When the problem of testing Ω_H against Ω_K remains invariant under a certain group of transformations, it seems reasonable to expect the existence of an invariant pair of least favorable distributions (or at least of sequences of distributions which in some sense are least favorable and invariant in the limit), and hence also of a maximin test which is invariant. This suggests the possibility of bypassing the somewhat cumbersome approach of the preceding sections. If it could be proved that for an invariant problem there always exists an invariant test that maximizes the minimum power over Ω_K, attention could be restricted to invariant tests; in particular, a UMP invariant test would then automatically have the desired maximin property (although it would not necessarily be admissible). These speculations turn out to be correct for an important class of problems, although unfortunately not in general. To find out under what conditions they hold, it is convenient first to separate out the statistical aspects of the problem from the group-theoretic ones by means of the following lemma.

Lemma 2. Let $\mathscr{P} = \{ P_\theta, \theta \in \Omega \}$ be a dominated family of distributions on $(\mathscr{X}, \mathscr{A})$, and let G be a group of transformations of $(\mathscr{X}, \mathscr{A})$, such that the induced group $\bar{G}$ leaves the two subsets Ω_H and Ω_K of Ω invariant. Suppose that for any critical function φ there exists an (almost) invariant critical function ψ satisfying

$$(23) \qquad \inf_{\bar{G}} E_{\bar{g}\theta}\varphi(X) \leq E_\theta\psi(X) \leq \sup_{\bar{G}} E_{\bar{g}\theta}\varphi(X)$$

for all $\theta \in \Omega$. Then if there exists a level-α test φ_0 maximizing $\inf_{\Omega_K} E_\theta\varphi(X)$, there also exists an (almost) invariant test with this property.

Proof. Let $\inf_{\Omega_K} E_\theta \varphi_0(X) = \beta$, and let ψ_0 be an (almost) invariant test such that (23) holds with $\varphi = \varphi_0$, $\psi = \psi_0$. Then

$$E_\theta \psi_0(X) \leq \sup_{\bar{G}} E_{\bar{g}\theta} \varphi_0(X) \leq \alpha \qquad \text{for all} \quad \theta \in \Omega_H$$

and

$$E_\theta \psi_0(X) \geq \inf_{\bar{G}} E_{\bar{g}\theta} \varphi_0(X) \geq \beta \qquad \text{for all} \quad \theta \in \Omega_K,$$

as was to be proved.

To determine conditions under which there exists an invariant or almost invariant test ψ satisfying (23), consider first the simplest case that G is a finite group, $G = \{g_1, \ldots, g_N\}$ say. If ψ is then defined by

$$(24) \qquad \qquad \psi(x) = \frac{1}{N} \sum_{i=1}^{N} \varphi(g_i x),$$

it is clear that ψ is again a critical function, and that it is invariant under G. It also satisfies (23), since $E_\theta \varphi(gX) = E_{\bar{g}\theta} \varphi(X)$ so that $E_\theta \psi(X)$ is the average of a number of terms of which the first and last member of (23) are the minimum and maximum respectively.

An illustration of the finite case is furnished by Example 3. Here the problem remains invariant under the $n!$ permutations of the variables $(X_1, \ldots, X_n)$. Lemma 2 is applicable and shows that there exists an invariant test maximizing $\inf_{\Omega_K} E_\theta \varphi(X)$. Thus in particular the UMP invariant test obtained in Example 7 of Chapter 6 has this maximin property and therefore constitutes a solution of the problem.

The definition (24) suggests the possibility of obtaining $\psi(x)$ also in other cases by averaging the values of $\varphi(gx)$ with respect to a suitable probability distribution over the group G. To see what conditions would be required of this distribution, let $\mathscr{B}$ be a σ-field of subsets of G and ν a probability distribution over $(G, \mathscr{B})$. Disregarding measurability problems for the moment, let ψ be defined by

$$(25) \qquad \qquad \psi(x) = \int \varphi(gx) \, d\nu(g).$$

Then $0 \leq \psi \leq 1$, and (23) is seen to hold by applying Fubini's theorem (Theorem 3 of Chapter 2) to the integral of ψ with respect to the distribution P_θ. For any $g_0 \in G$,

$$\psi(g_0 x) = \int \varphi(gg_0 x) \, d\nu(g) = \int \varphi(hx) \, d\nu^*(h)$$

where $h = gg_0$ and where ν^* is the measure defined by

$$\nu^*(B) = \nu\left(Bg_0^{-1}\right) \qquad \text{for all} \quad B \in \mathscr{B},$$

into which ν is transformed by the transformation $h = gg_0$. Thus ψ will have the desired invariance property, $\psi(g_0 x) = \psi(x)$ for all $g_0 \in G$, if ν is *right invariant*, that is, if it satisfies

(26) $\nu(Bg) = \nu(B) \qquad \text{for all} \quad B \in \mathscr{B}, \quad g \in G.$

The measurability assumptions required for the above argument are: (i) For any $A \in \mathscr{A}$, the set of pairs (x, g) with $gx \in A$ is measurable ($\mathscr{A} \times \mathscr{B}$). This insures that the function ψ defined by (25) is again measurable. (ii) For any $B \in \mathscr{B}$, $g \in G$, the set Bg belongs to $\mathscr{B}$.

Example 5. If G is a finite group with elements $g_1, \ldots, g_N$, let $\mathscr{B}$ be the class of all subsets of G and ν the probability measure assigning probability $1/N$ to each of the N elements. The condition (26) is then satisfied, and the definition (25) of ψ in this case reduces to (24).

Example 6. Consider the group G of orthogonal $n \times n$ matrices Γ, with the group product $\Gamma_1 \Gamma_2$ defined as the corresponding matrix product. Each matrix can be interpreted as the point in n^2-dimensional Euclidean space whose coordinates are the n^2 elements of the matrix. The group then defines a subset of this space; the Borel subsets of G will be taken as the σ-field $\mathscr{B}$. To prove the existence of a right invariant probability measure over $(G, \mathscr{B})$,[*] we shall define a random orthogonal matrix whose probability distribution satisfies (26) and is therefore the required measure. With any nonsingular matrix $x = (x_{ij})$, associate the orthogonal matrix $y = f(x)$ obtained by applying the following Gram–Schmidt orthogonalization process to the n row vectors $x_i = (x_{i1}, \ldots, x_{in})$ of x: y_1 is the unit vector in the direction of x_1; y_2 the unit vector in the plane spanned by x_1 and x_2, which is orthogonal to y_1 and forms an acute angle with x_2; and so on. Let $y = (y_{ij})$ be the matrix whose ith row is y_i.

Suppose now that the variables X_{ij} $(i, j = 1, \ldots, n)$ are independently distributed as $N(0, 1)$, let X denote the random matrix (X_{ij}), and let $Y = f(X)$. To show that the distribution of the random orthogonal matrix Y satisfies (26), consider any fixed orthogonal matrix Γ and any fixed set $B \in \mathscr{B}$. Then $P\{Y \in B\Gamma\} = P\{Y\Gamma' \in B\}$ and from the definition of f it is seen that $Y\Gamma' = f(X\Gamma')$. Since the n^2 elements of the matrix $X\Gamma'$ have the same joint distribution as those of the matrix X, the matrices $f(X\Gamma')$ and $f(X)$ also have the same distribution, as was to be proved.

Examples 5 and 6 are sufficient for the applications to be made here. General conditions for the existence of an invariant probability measure, of which these examples are simple special cases, are given in the theory of Haar measure. [This is treated, for example, in the books by Halmos (1974),

---

[*]A more detailed discussion of this invariant measure is given by James (1954).

Loomis (1953), and Nachbin (1965). For a discussion in a statistical setting, see Eaton (1983), Farrell (1985), and for a more elementary treatment Berger (1985).]

5. THE HUNT–STEIN THEOREM

Invariant measures exist (and are essentially unique) for a large class of groups, but unfortunately they are frequently not finite and hence cannot be taken to be probability measures. The situation is similar and related to that of the nonexistence of a least favorable pair of distributions in Theorem 1. There it is usually possible to overcome the difficulty by considering instead a sequence of distributions, which has the desired property in the limit. Analogously we shall now generalize the construction of ψ as an average with respect to a right-invariant probability distribution, by considering a sequence of distributions over G which are approximately right-invariant for n sufficiently large.

Let $\mathscr{P} = \{ P_\theta, \ \theta \in \Omega \}$ be a family of distributions over a Euclidean space $(\mathscr{X}, \mathscr{A})$ dominated by a σ-finite measure μ, and let G be a group of transformations of $(\mathscr{X}, \mathscr{A})$ such that the induced group $\bar{G}$ leaves Ω invariant.

Theorem 3. (*Hunt–Stein.*) *Let $\mathscr{B}$ be a σ-field of subsets of G such that for any $A \in \mathscr{A}$ the set of pairs (x, g) with $gx \in A$ is in $\mathscr{A} \times \mathscr{B}$ and for any $B \in \mathscr{B}$ and $g \in G$ the set Bg is in $\mathscr{B}$. Suppose that there exists a sequence of probability distributions ν_n over $(G, \mathscr{B})$ which is asymptotically right-invariant in the sense that for any $g \in G$, $B \in \mathscr{B}$*

$$(27) \qquad \lim_{n \to \infty} |\nu_n(Bg) - \nu_n(B)| = 0.$$

Then given any critical function φ, there exists a critical function ψ which is almost invariant and satisfies (23).

Proof. Let

$$\psi_n(x) = \int \varphi(gx) \, d\nu_n(g),$$

which as before is measurable and between 0 and 1. By the weak compactness theorem (Theorem 3 of the Appendix) there exists a subsequence $\{ \psi_{n_i} \}$ and a measurable function ψ between 0 and 1 satisfying

$$\lim_{i \to \infty} \int \psi_{n_i} p \, d\mu = \int \psi p \, d\mu$$

for all μ-integrable functions p, so that in particular

$$\lim_{i \to \infty} E_\theta \psi_{n_i}(X) = E_\theta \psi(X)$$

for all $\theta \in \Omega$. By Fubini's theorem

$$E_\theta \psi_{n_i}(X) = \int [E_\theta \varphi(gX)] \, d\nu_{n_i}(g) = \int E_{\bar{g}\theta} \varphi(X) \, d\nu_{n_i}(g)$$

so that

$$\inf_{\overline{G}} E_{\bar{g}\theta} \varphi(X) \le E_\theta \psi_{n_i}(X) \le \sup_{\overline{G}} E_{\bar{g}\theta} \varphi(X),$$

and ψ satisfies (23).

In order to prove that ψ is almost invariant we shall show below that for all x and g,

(28) $$\psi_{n_i}(gx) - \psi_{n_i}(x) \to 0.$$

Let $I_A(x)$ denote the indicator function of a set $A \in \mathscr{A}$. Using the fact that $I_{gA}(gx) = I_A(x)$, we see that (28) implies

$$\int_A \psi(x) \, dP_\theta(x) = \lim_{i \to \infty} \int \psi_{n_i}(x) I_A(x) \, dP_\theta(x)$$

$$= \lim_{i \to \infty} \int \psi_{n_i}(gx) I_{gA}(gx) \, dP_\theta(x)$$

$$= \int \psi(x) I_{gA}(x) \, dP_{\bar{g}\theta}(x) = \int_A \psi(gx) \, dP_\theta(x)$$

and hence $\psi(gx) = \psi(x)$ (a.e. $\mathscr{P}$), as was to be proved.

To prove (28), consider any fixed x and any integer m, and let G be partitioned into the mutually exclusive sets

$$B_k = \left\{ h \in G : a_k < \varphi(hx) \le a_k + \frac{1}{m} \right\}, \qquad k = 0, \dots, m,$$

where $a_k = (k - 1)/m$. In particular, B_0 is the set $\{ h \in G : \varphi(hx) = 0 \}$. It is seen from the definition of the sets B_k that

$$\sum_{k=0}^{m} a_k \nu_{n_i}(B_k) \le \sum_{k=0}^{m} \int_{B_k} \varphi(hx) \, d\nu_{n_i}(h) \le \sum_{k=0}^{m} \left(a_k + \frac{1}{m} \right) \nu_{n_i}(B_k)$$

$$\le \sum_{k=0}^{m} a_k \nu_{n_i}(B_k) + \frac{1}{m}$$

and analogously that

$$\left| \sum_{k=0}^{m} \int_{B_k g^{-1}} \varphi(hgx)\, d\nu_{n_i}(h) - \sum_{k=0}^{m} a_k \nu_{n_i}\left(B_k g^{-1}\right) \right| \le \frac{1}{m},$$

from which it follows that

$$\left| \psi_{n_i}(gx) - \psi_{n_i}(x) \right| \le \sum |a_k| \cdot \left| \nu_{n_i}\left(B_k g^{-1}\right) - \nu_{n_i}(B_k) \right| + \frac{2}{m}.$$

By (27) the first term of the right-hand side tends to zero as i tends to infinity, and this completes the proof.

When there exist a right-invariant measure ν over G and a sequence of subsets G_n of G with $G_n \subseteq G_{n+1}$, $\bigcup G_n = G$, and $\nu(G_n) = c_n < \infty$, it is suggestive to take for the probability measures ν_n of Theorem 3 the measures ν/c_n truncated on G_n. This leads to the desired result in the example below. On the other hand, there are cases in which there exists such a sequence of subsets of G_n but no invariant test satisfying (23) and hence no sequence ν_n satisfying (27).

Example 7. Let $x = (x_1, \ldots, x_n)$, $\mathscr{A}$ be the class of Borel sets in n-space, and G the group of translations $(x_1 + g, \ldots, x_n + g)$, $-\infty < g < \infty$. The elements of G can be represented by the real numbers, and the group product gg' is then the sum $g + g'$. If $\mathscr{B}$ is the class of Borel sets on the real line, the measurability assumptions of Theorem 3 are satisfied. Let ν be Lebesgue measure, which is clearly invariant under G, and define ν_n to be the uniform distribution on the interval $I(-n, n) = \{ g : -n \le g \le n \}$. Then for all $B \in \mathscr{B}$, $g \in G$,

$$|\nu_n(B) - \nu_n(Bg)| = \frac{1}{2n} |\nu[B \cap I(-n, n)] - \nu[B \cap I(-n - g, n - g)]| \le \frac{|g|}{2n},$$

so that (27) is satisfied.

This argument also covers the group of scale transformations $(ax_1, \ldots, ax_n)$, $0 < a < \infty$, which can be transformed into the translation group by taking logarithms.

When applying the Hunt–Stein theorem to obtain invariant minimax tests, it is frequently convenient to carry out the calculation in steps, as was done in Theorem 7 of Chapter 6. Suppose that the problem remains invariant under two groups D and E, and denote by $y = s(x)$ a maximal invariant with respect to D and by E^* the group defined in Theorem 2, Chapter 6, which E induces in y-space. If D and E^* satisfy the conditions of the Hunt–Stein theorem, it follows first that there exists a maximin test depending only on $y = s(x)$, and then that there exists a maximin test depending only on a maximal invariant $z = t(y)$ under E^*.

Example 8. Consider a univariate linear hypothesis in the canonical form in which $Y_1, \ldots, Y_n$ are independently distributed as $N(\eta_i, \sigma^2)$, where it is given that $\eta_{s+1} = \cdots = \eta_n = 0$, and where the hypothesis to be tested is $\eta_1 = \cdots = \eta_r = 0$. It was shown in Section 1 of Chapter 7 that this problem remains invariant under certain groups of transformations and that with respect to these groups there exists a UMP invariant test. The groups involved are the group of orthogonal transformations, translation groups of the kind considered in Example 7, and a group of scale changes. Since each of these satisfies the assumptions of the Hunt–Stein theorem, and since they leave invariant the problem of maximizing the minimum power over the set of alternatives

$$(29) \qquad \sum_{i=1}^{r} \frac{\eta_i^2}{\sigma^2} \geq \psi_1^2 \qquad (\psi_1 > 0),$$

it follows that the UMP invariant test of Chapter 7 is also the solution of this maximin problem. It is also seen slightly more generally that the test which is UMP invariant under the same groups for testing

$$\sum_{i=1}^{r} \frac{\eta_i^2}{\sigma^2} \leq \psi_0^2$$

(Problem 4 of Chapter 7) maximizes the minimum power over the alternatives (29) for $\psi_0 < \psi_1$.

Example 9. (Stein.) Let G be the group of all nonsingular linear transformations of p-space. That for $p > 1$ this does not satisfy the conditions of Theorem 3 is shown by the following problem, which is invariant under G but for which the UMP invariant test does not maximize the minimum power. Generalizing Example 1 of Chapter 6, let $X = (X_1, \ldots, X_p)$, $Y = (Y_1, \ldots, Y_p)$ be independently distributed according to p-variate normal distributions with zero means and nonsingular covariance matrices $E(X_i X_j) = \sigma_{ij}$ and $E(Y_i Y_j) = \Delta \sigma_{ij}$, and let $H: \Delta \leq \Delta_0$ be tested against $\Delta \geq \Delta_1$ ($\Delta_0 < \Delta_1$), the σ_{ij} being unknown.

This problem remains invariant if the two vectors are subjected to any common nonsingular transformation, and since with probability 1 this group is transitive over the sample space, the UMP invariant test is trivially $\varphi(x, y) \equiv \alpha$. The maximin power against the alternatives $\Delta \geq \Delta_1$ that can be achieved by invariant tests is therefore α. On the other hand, the test with rejection region $Y_1^2/X_1^2 > C$ has a strictly increasing power function $\beta(\Delta)$, whose minimum over the set of alternatives $\Delta \geq \Delta_1$ is $\beta(\Delta_1) > \beta(\Delta_0) = \alpha$.

It is a remarkable feature of Theorem 3 that its assumptions concern only the group G and not the distributions P_θ.* When these assumptions hold for a certain G it follows from (23) as in the proof of Lemma 2 that for any

---

*These assumptions are essentially equivalent to the condition that the group G is *amenable.* Amenability and its relationship to the Hunt–Stein theorem are discussed by Bondar and Milnes (1982) and (with a different terminology) by Stone and von Randow (1968).

testing problem which remains invariant under G and possesses a UMP invariant test, this test maximizes the minimum power over any invariant class of alternatives. Suppose conversely that a UMP invariant test under G has been shown in a particular problem not to maximize the minimum power, as was the case for the group of linear transformations in Example 9. Then the assumptions of Theorem 3 cannot be satisfied. However, this does not rule out the possibility that for another problem remaining invariant under G, the UMP invariant test may maximize the minimum power. Whether or not it does is no longer a property of the group alone but will in general depend also on the particular distributions.

Consider in particular the problem of testing $H: \xi_1 = \cdots = \xi_p = 0$ on the basis of a sample $(X_{\alpha 1}, \ldots, X_{\alpha p})$, $\alpha = 1, \ldots, n$, from a p-variate normal distribution with mean $E(X_{\alpha i}) = \xi_i$ and common covariance matrix $(\sigma_{ij}) = (a_{ij})^{-1}$. This was seen in Section 3 of Chapter 8 to be invariant under a number of groups, including that of all nonsingular linear transformations of p-space, and a UMP invariant test was found to exist. An invariant class of alternatives under these groups is

$$(30) \qquad \sum\sum \frac{a_{ij}\xi_i\xi_j}{\sigma^2} \geq \psi_1^2.$$

Here Theorem 3 is not applicable, and the question whether the T^2-test of $H: \psi = 0$ maximizes the minimum power over the alternatives

$$(31) \qquad \sum\sum a_{ij}\xi_i\xi_j = \psi_1^2$$

[and hence a fortiori over the alternatives (30)] presents formidable difficulties. The minimax property was proved for the case $p = 2$, $n = 3$ by Giri, Kiefer, and Stein (1963), for the case $p = 2$, $n = 4$ by Linnik, Pliss, and Salaevskii (1968), and for $p = 2$ and all $n \geq 3$ by Salaevskii (1971). The proof is effected by first reducing the problem through invariance under the group G_1 of Example 11 of Chapter 6, to which Theorem 3 is applicable, and then applying Theorem 1 to the reduced problem. It is a consequence of this approach that it also establishes the admissibility of T^2 as a test of H against the alternatives (31). In view of the inadmissibility results for point estimation when $p \geq 3$ (see *TPE*, Sections 4.5 and 4.6), it seems unlikely that T^2 is admissible for $p \geq 3$, and hence that the same method can be used to prove the minimax property in this situation.

The problem becomes much easier when the minimax property is considered against local or distant alternatives rather than against (31). Precise definitions and proofs of the fact that T^2 possesses these properties for all p and n are provided by Giri and Kiefer (1964) and in the references given in Chapter 8, Section 3.

The theory of this and the preceding section can be extended to confidence sets if the accuracy of a confidence set at level $1 - \alpha$ is assessed by its volume or some other appropriate measure of its size. Suppose that the distribution of X depends on the parameters θ to be estimated and on nuisance parameters ϑ, and that μ is a σ-finite measure over the parameter set $\omega = \{\theta : (\theta, \vartheta) \in \Omega\}$, with ω assumed to be independent of ϑ. Then the confidence sets $S(X)$ for θ are minimax with respect to μ at level $1 - \alpha$ if they minimize

$$\sup E_{\theta, \vartheta} \mu[S(X)]$$

among all confidence sets at the given level.

The problem of minimizing $E\mu[S(X)]$ is related to that of minimizing the probability of covering false values (the criterion for accuracy used so far) by the relation (Problem 26)

$$(32) \qquad E_{\theta_0, \vartheta} \mu[S(X)] = \int_{\theta \neq \theta_0} P_{\theta_0, \vartheta}[\theta \in S(X)] \, d\mu(\theta),$$

which holds provided μ assigns measure zero to the set $\{\theta = \theta_0\}$. (For the special case that θ is real-valued and μ Lebesgue measure, see Problem 29 of Chapter 5.)

Suppose now that the problem of estimating θ is invariant under a group G in the sense of Chapter 6, Section 11 and that μ satisfies the invariance condition

$$(33) \qquad \mu[S(gx)] = \mu[S(x)].$$

If uniformly most accurate equivariant confidence sets exist, they minimize (32) among all equivariant confidence sets at the given level, and one may hope that under the assumptions of the Hunt–Stein theorem, they will also be minimax with respect to μ among the class of all (not necessarily equivariant) confidence sets at the given level. Such a result does hold and can be used to show for example that the most accurate equivariant confidence sets of Examples 17 and 18 of Chapter 6 minimize their maximum expected Lebesgue measure. A more general class of examples is provided by the confidence intervals derived from the UMP invariant tests of univariate linear hypotheses such as the confidence spheres for $\theta_i = \mu + \alpha_i$ or for α_i given in Section 5 of Chapter 7.

Minimax confidence sets $S(x)$ are not necessarily admissible; that is, there may exist sets $S'(x)$ having the same confidence level but such that

$$E_{\theta, \vartheta} \mu[S'(X)] \leq E_{\theta, \vartheta} \mu[S(X)] \qquad \text{for all } \theta, \vartheta$$

with strict inequality holding for at least some (θ, ϑ).

Example 10. Let X_i $(i = 1, \ldots, s)$ be independently normally distributed with mean $E(X_i) = \theta_i$ and variance 1, and let G be the group generated by translations $X_i + c_i$ $(i = 1, \ldots, s)$ and orthogonal transformations of $(X_1, \ldots, X_s)$. (G is the Euclidean group of rigid motions in s-space.) A slight generalization of Example 17 of Chapter 6 shows the confidence sets

$$\text{(34)} \qquad\qquad \sum (\theta_i - X_i)^2 \le c$$

to be uniformly most accurate equivariant. The volume $\mu[S(X)]$ of any confidence set $S(X)$ remains invariant under the transformations $g \in G$, and it follows from the results of Problems 30 and 31 and Examples 7 and 8 that the confidence sets (34) minimize the maximum expected volume. However, very surprisingly, they are not admissible unless $s = 1$ or 2. This result, which will not be proved here, is closely related to the inadmissibility of $X_1, \ldots, X_s$ as a point estimator of $(\theta_1, \ldots, \theta_s)$ for a wide variety of loss functions. The work on point estimation, which is discussed in *TPE*, Sections 4.5 and 4.6, for squared error loss, provides an easier access to these ideas than the present setting. A convenient entry into the literature on admissibility of confidence sets is Hwang and Casella (1982).

The inadmissibility of the confidence sets (34) is particularly surprising in that the associated UMP invariant tests of the hypotheses $H : \theta_i = \theta_{i_0}$ $(i = 1, \ldots, s)$ are admissible (Problems 28, 29).

6. MOST STRINGENT TESTS

One of the practical difficulties in the consideration of tests that maximize the minimum power over a class Ω_K of alternatives is the determination of an appropriate Ω_K. If no information is available on which to base the choice of this set and if a natural definition is not imposed by invariance arguments, a frequently reasonable definition can be given in terms of the power that can be achieved against the various alternatives. The *envelope power function* β_α^* was defined in Chapter 6, Problem 15, by

$$\beta_\alpha^*(\theta) = \sup \beta_\varphi(\theta),$$

where β_φ denotes the power of a test φ and where the supremum is taken over all level-α tests of H. Thus $\beta_\alpha^*(\theta)$ is the maximum power that can be attained at level α against the alternative θ. (That it can be attained follows under mild restrictions from Theorem 3 of the Appendix.) If

$$S_\Delta^* = \{ \theta : \beta_\alpha^*(\theta) = \Delta \},$$

then of two alternatives $\theta_1 \in S_{\Delta_1}^*$, $\theta_2 \in S_{\Delta_2}^*$, θ_1 can be considered closer to H, equidistant, or further away than θ_2 as Δ_1 is $<$, $=$, or $> \Delta_2$.

The idea of measuring the distance of an alternative from H in terms of the available information has been encountered before. If for example

$X_1, \ldots, X_n$ is a sample from $N(\xi, \sigma^2)$, the problem of testing $H: \xi \leq 0$ was discussed (Chapter 5, Section 2) both when the alternatives ξ are measured in absolute units and when they are measured in σ-units. The latter possibility corresponds to the present proposal, since it follows from invariance considerations (Problem 15 of Chapter 6) that $\beta_\alpha^*(\xi, \sigma)$ is constant on the lines $\xi/\sigma = $ constant.

Fixing a value of Δ and taking as Ω_K the class of alternatives θ for which $\beta_\alpha^*(\theta) \geq \Delta$, one can determine the test that maximizes the minimum power over Ω_K. Another possibility, which eliminates the need of selecting a value of Δ, is to consider for any test φ the difference $\beta_\alpha^*(\theta) - \beta_\varphi(\theta)$. This difference measures the amount by which the actual power $\beta_\varphi(\theta)$ falls short of the maximum power attainable. A test that minimizes

$$(35) \qquad \sup_{\Omega - \omega} \left[\beta_\alpha^*(\theta) - \beta_\varphi(\theta) \right]$$

is said to be *most stringent*. Thus a test is most stringent if it minimizes its maximum shortcoming.

Let φ_Δ be a test that maximizes the minimum power over S_Δ^*, and hence minimizes the maximum difference between $\beta_\alpha^*(\theta)$ and $\beta_\varphi(\theta)$ over S_Δ^*. If φ_Δ happens to be independent of Δ, it is most stringent. This remark makes it possible to apply the results of the preceding sections to the determination of most stringent tests. Suppose that the problem of testing $H: \theta \in \omega$ against the alternatives $\theta \in \Omega - \omega$ remains invariant under a group G, that there exists a UMP almost invariant test φ_0 with respect to G, and that the assumptions of Theorem 3 hold. Since $\beta_\alpha^*(\theta)$ and hence the set S_Δ^* is invariant under $\bar{G}$ (Problem 15 of Chapter 6), it follows that φ_0 maximizes the minimum power over S_Δ^* for each Δ, and φ_0 is therefore most stringent.

As an example of this method consider the problem of testing $H: p_1, \ldots, p_n \leq \frac{1}{2}$ against the alternative $K: p_i > \frac{1}{2}$ for all i, where p_i is the probability of success in the ith trial of a sequence of n independent trials. If X_i is 1 or 0 as the ith trial is a success or failure, then the problem remains invariant under permutations of the X's, and the UMP invariant test rejects (Example 7 of Chapter 6) when $\Sigma X_i > C$. It now follows from the remarks above that this test is also most stringent.

Another illustration is furnished by the general univariate linear hypothesis. Here it follows from the discussion in Example 8 that the standard test for testing $H: \eta_1 = \cdots = \eta_r = 0$ or $H': \Sigma_{i=1}^r \eta_i^2/\sigma^2 \leq \psi_0^2$ is most stringent.

When the invariance approach is not applicable, the explicit determination of most stringent tests typically is difficult. The following is a class of problems for which they are easily obtained by a direct approach. Let the distributions of X constitute a one-parameter exponential family, the density of which is given by (12) of Chapter 3, and consider the hypothesis $H : \theta = \theta_0$. Then according as $\theta > \theta_0$ or $\theta < \theta_0$, the envelope power $\beta_\alpha^*(\theta)$ is the power of the UMP one-sided test for testing H against $\theta > \theta_0$ or $\theta < \theta_0$. Suppose that there exists a two-sided test φ_0 given by (3) of Chapter 4, such that

$$(36) \qquad \sup_{\theta < \theta_0} \left[\beta_\alpha^*(\theta) - \beta_{\varphi_0}(\theta) \right] = \sup_{\theta > \theta_0} \left[\beta_\alpha^*(\theta) - \beta_{\varphi_0}(\theta) \right],$$

and that the supremum is attained on both sides, say at points $\theta_1 < \theta_0 < \theta_2$. If $\beta_{\varphi_0}(\theta_i) = \beta_i$, $i = 1, 2$, an application of the fundamental lemma [Theorem 5(iii) of Chapter 3] to the three points $\theta_1, \theta_2, \theta_0$ shows that among all tests φ with $\beta_\varphi(\theta_1) \geq \beta_1$ and $\beta_\varphi(\theta_2) \geq \beta_2$, only φ_0 satisfies $\beta_\varphi(\theta_0) \leq \alpha$. For any other level-$\alpha$ test, therefore, either $\beta_\varphi(\theta_1) < \beta_1$ or $\beta_\varphi(\theta_2) < \beta_2$, and it follows that φ_0 is the unique most stringent test. The existence of a test satisfying (36) can be proved by a continuity consideration [with respect to variation of the constants C_i and γ_i which define the boundary of the test (3) of Chapter 4] from the fact that for the UMP one-sided test against the alternatives $\theta > \theta_0$ the right-hand side of (36) is zero and the left-hand side positive, while the situation is reversed for the other one-sided test.

7. PROBLEMS

Section 1

1. *Existence of maximin tests.* Let $(\mathscr{X}, \mathscr{A})$ be a Euclidean sample space, and let the distributions P_θ, $\theta \in \Omega$, be dominated by a σ-finite measure over $(\mathscr{X}, \mathscr{A})$. For any mutually exclusive subsets Ω_H, Ω_K of Ω there exists a level-α test maximizing (2).
   [Let $\beta = \sup[\inf_{\Omega_K} E_\theta \varphi(X)]$, where the supremum is taken over all level-α tests of $H : \theta \in \Omega_H$. Let φ_n be a sequence of level-α tests such that $\inf_{\Omega_K} E_\theta \varphi_n(X)$ tends to β. If φ_{n_i} is a subsequence and φ a test (guaranteed by Theorem 3 of the Appendix) such that $E_\theta \varphi_{n_i}(X)$ tends to $E_\theta \varphi(X)$ for all $\theta \in \Omega$, then φ is a level-α test and $\inf_{\Omega_K} E_\theta \varphi(X) = \beta$.]

2. *Locally most powerful tests.* Let d be a measure of the distance of an alternative θ from a given hypothesis H. A level-α test φ_0 is said to be *locally most powerful* (LMP) if, given any other level-α test φ, there exists Δ such that

   $$(37) \qquad \beta_{\varphi_0}(\theta) \geq \beta_\varphi(\theta) \qquad \text{for all } \theta \text{ with } 0 < d(\theta) < \Delta.$$

   Suppose that θ is real-valued and that the power function of every test is continuously differentiable at θ_0.

(i) If there exists a unique level-α test φ_0 of $H: \theta = \theta_0$ maximizing $\beta'_\varphi(\theta_0)$, then φ_0 is the unique LMP level-α test of H against $\theta > \theta_0$ for $d(\theta) = \theta - \theta_0$.

(ii) To see that (i) is not correct without the uniqueness assumption, let X take on the values 0 and 1 with probabilities $P_\theta(0) = \frac{1}{2} - \theta^3$, $P_\theta(1) = \frac{1}{2} + \theta^3$, $-\frac{1}{2} < \theta^3 < \frac{1}{2}$, and consider testing $H: \theta = 0$ against $K: \theta > 0$. Then every test φ of size α maximizes $\beta'_\varphi(0)$, but not every such test is LMP. [Kallenberg et al. (1984).]

(iii) The following* is another counterexample to (i) without uniqueness, in which in fact no LMP test exists. Let X take on the values $0, 1, 2$ with probabilities

$$P_\theta(x) = \alpha + \epsilon\left[\theta + \theta^2 \sin\left(\frac{x}{\theta}\right)\right] \quad \text{for} \quad x = 1, 2,$$

$$P_\theta(0) = 1 - p_\theta(1) - p_\theta(2),$$

where $-1 \leq \theta \leq 1$ and ϵ is a sufficiently small number. Then a test φ at level α maximizes $\beta'(0)$ provided

$$\varphi(1) + \varphi(2) = 1;$$

but no LMP test exists.

(iv) A unique LMP test maximizes the minimum power locally provided its power function is bounded away from α for every set of alternatives which is bounded away from H.

(v) Let $X_1, \ldots, X_n$ be a sample from a Cauchy distribution with unknown location parameter θ, so that the joint density of the X's is $\pi^{-n}\prod_{i=1}^{n}[1 + (x_i - \theta)^2]^{-1}$. The LMP test for testing $\theta = 0$ against $\theta > 0$ at level $\alpha < \frac{1}{2}$ is not unbiased and hence does not maximize the minimum power locally.

[(iii): The unique most powerful test against θ is

$$\begin{cases} \varphi(1) \\ \varphi(2) \end{cases} = 1 \quad \text{if} \quad \sin\left(\frac{1}{\theta}\right) \gtrless \sin\left(\frac{2}{\theta}\right),$$

and each of these inequalities holds at values of θ arbitrarily close to 0.
(v): There exists M so large that any point with $x_i \geq M$ for all $i = 1, \ldots, n$ lies in the acceptance region of the LMP test. Hence the power of the test tends to zero as θ tends to infinity.]

3. A level-α test φ_0 is locally unbiased (loc. unb.) if there exists $\Delta_0 > 0$ such that $\beta_{\varphi_0}(\theta) \geq \alpha$ for all θ with $0 < d(\theta) < \Delta_0$; it is LMP loc. unb. if it is loc. unb.

*Due to John Pratt.

and if, given any other loc. unb. level-α test φ, there exists Δ such that (37) holds. Suppose that θ is real-valued and that $d(\theta) = |\theta - \theta_0|$, and that the power function of every test is twice continuously differentiable at $\theta = \theta_0$.

(i) If there exists a unique test φ_0 of $H : \theta = \theta_0$ against $K : \theta \neq \theta_0$ which among all loc. unb. tests maximizes $\beta''(\theta_0)$, then φ_0 is the unique LMP loc. unb. level-α test of H against K.

(ii) The test of part (i) maximizes the minimum power locally provided its power function is bounded away from α for every set of alternatives that is bounded away from H.

[(ii): A necessary condition for a test to be locally minimax is that it is loc. unb.]

Section 2

4. Let the distribution of X depend on the parameters $(\theta, \vartheta) = (\theta_1, \ldots, \theta_r, \vartheta_1, \ldots, \vartheta_s)$. A test of $H : \theta = \theta^0$ is *locally strictly unbiased* if for each ϑ, (a) $\beta_\varphi(\theta^0, \vartheta) = \alpha$, (b) there exists a θ-neighborhood of θ^0 in which $\beta_\varphi(\theta, \vartheta) > \alpha$ for $\theta \neq \theta^0$.

(i) Suppose that the first and second derivatives

$$\beta_\varphi^i(\vartheta) = \frac{\partial}{\partial \theta_i} \beta_\varphi(\theta, \vartheta)\Big|_{\theta^0} \quad \text{and} \quad \beta_\varphi^{ij}(\vartheta) = \frac{\partial^2}{\partial \theta_i\, \partial \theta_j} \beta_\varphi(\theta, \vartheta)\Big|_{\theta^0}$$

exist for all critical functions φ and all ϑ. Then a necessary and sufficient condition for φ to be locally strictly unbiased is that $\beta_\varphi^i(\vartheta) = 0$ for all i and ϑ, and that the matrix $(\beta_\varphi^{ij}(\vartheta))$ is positive definite for all ϑ.

(ii) A test of H is said to be of *type* E (*type* D is $s = 0$ so that there are no nuisance parameters) if it is locally strictly unbiased and among all tests with this property maximizes the determinant $|(\beta_\varphi^{ij})|$.* (This determinant under the stated conditions turns out to be equal to the Gaussian curvature of the power surface at θ^0.) Then the test φ_0 given by (7) of Chapter 7 testing the general linear univariate hypothesis (3) of Chapter 7 is of type E.

[(ii): With $\theta = (\eta_1, \ldots, \eta_r)$ and $\vartheta = (\eta_{r+1}, \ldots, \eta_s, \sigma)$, the test φ_0, by Problem 5 of Chapter 7, has the property of maximizing the surface integral

$$\int_S \left[\beta_\varphi(\eta, \sigma^2) - \alpha \right] dA$$

---

*An interesting example of a type-D test is provided by Cohen and Sackrowitz (1975), who show that the χ^2-test of Chapter 8, Example 5 has this property.

among all similar (and hence all locally unbiased) tests where $S = \{(\eta_1, \ldots, \eta_r) : \sum_{i=1}^r \eta_i^2 = \rho^2 \sigma^2\}$. Letting ρ tend to zero and utilizing the conditions

$$\beta_\varphi^i(\vartheta) = 0, \qquad \int_S \eta_i \eta_j \, dA = 0 \quad \text{for } i \neq j, \qquad \int_S \eta_i^2 \, dA = k(\rho \sigma),$$

one finds that φ_0 maximizes $\sum_{i=1}^r \beta_\varphi^{ii}(\eta, \sigma^2)$ among all locally unbiased tests. Since for any positive definite matrix, $|(\beta_\varphi^{ij})| \leq \prod \beta_\varphi^{ii}$, it follows that for any locally strictly unbiased test φ,

$$\left|\left(\beta_\varphi^{ij}\right)\right| \leq \prod \beta_\varphi^{ii} \leq \left[\frac{\sum \beta_\varphi^{ii}}{r}\right]^r \leq \left[\frac{\sum \beta_{\varphi_0}^{ii}}{r}\right]^r = \left[\beta_{\varphi_0}^{11}\right]^r = \left|\left(\beta_{\varphi_0}^{ij}\right)\right|.]$$

5. Let $Z_1, \ldots, Z_n$ be identically independently distributed according to a continuous distribution D, of which it is assumed only that it is symmetric about some (unknown) point. For testing the hypothesis $H : D(0) = \frac{1}{2}$, the sign test maximizes the minimum power against the alternatives $K : D(0) \leq q$ $(q < \frac{1}{2})$. [A pair of least favorable distributions assign probability 1 respectively to the distributions $F \in H$, $G \in K$ with densities

$$f(x) = \frac{1 - 2q}{2(1 - q)} \left(\frac{q}{1 - q}\right)^{[|x|]}, \qquad g(x) = (1 - 2q) \left(\frac{q}{1 - q}\right)^{[|x|]}$$

where for all x (positive, negative, or zero) $[x]$ denotes the largest integer $\leq x$.]

6. Let $f_\theta(x) = \theta g(x) + (1 - \theta) h(x)$ with $0 \leq \theta \leq 1$. Then $f_\theta(x)$ satisfies the assumptions of Lemma 1 provided $g(x)/h(x)$ is a nondecreasing function of x.

7. Let $x = (x_1, \ldots, x_n)$, and let $g_\theta(x, \xi)$ be a family of probability densities depending on $\theta = (\theta_1, \ldots, \theta_r)$ and the real parameter ξ, and jointly measurable in x and ξ. For each θ, let $h_\theta(\xi)$ be a probability density with respect to a σ-finite measure ν such that $p_\theta(x) = \int g_\theta(x, \xi) h_\theta(\xi) \, d\nu(\xi)$ exists. We shall say that a function f of two arguments $u = (u_1, \ldots, u_r)$, $v = (v_1, \ldots, v_s)$ is nondecreasing in (u, v) if $f(u', v)/f(u, v) \leq f(u', v')/f(u, v')$ for all (u, v) satisfying $u_i \leq u_i'$, $v_j \leq v_j'$ $(i = 1, \ldots, r; \ j = 1, \ldots, s)$. Then $p_\theta(x)$ is nondecreasing in (x, θ) provided the product $g_\theta(x, \xi) h_\theta(\xi)$ is (a) nondecreasing in (x, θ) for each fixed ξ; (b) nondecreasing in (θ, ξ) for each fixed x; (c) nondecreasing in (x, ξ) for each fixed θ.

[Interpreting $g_\theta(x, \xi)$ as the conditional density of x given ξ, and $h_\theta(\xi)$ as the a priori density of ξ, let $\rho(\xi)$ denote the a posteriori density of ξ given x, and let $\rho'(\xi)$ be defined analogously with θ' in place of θ. That $p_\theta(x)$ is nondecreasing in its two arguments is equivalent to

$$\int \frac{g_\theta(x',\xi)}{g_\theta(x,\xi)} \rho(\xi)\, d\nu(\xi) \le \int \frac{g_{\theta'}(x',\xi)}{g_{\theta'}(x,\xi)} \rho'(\xi)\, d\nu(\xi).$$

By (a) it is enough to prove that

$$D = \int \frac{g_\theta(x',\xi)}{g_\theta(x,\xi)} [\rho'(\xi) - \rho(\xi)]\, d\nu(\xi) \ge 0.$$

Let $S_- = \{\xi : \rho'(\xi)/\rho(\xi) < 1\}$ and $S_+ = \{\xi : \rho'(\xi)/\rho(\xi) \ge 1\}$. By (b) the set S_- lies entirely to the left of S_+. It follows from (c) that there exists $a \le b$ such that

$$D = a\int_{S_-} [\rho'(\xi) - \rho(\xi)]\, d\nu(\xi) + b\int_{S_+} [\rho'(\xi) - \rho(\xi)]\, d\nu(\xi),$$

and hence that $D = (b - a)\int_{S_+} [\rho'(\xi) - \rho(\xi)]\, d\nu(\xi) \ge 0.]$

8. (i) Let X have binomial distribution $b(p, n)$, and consider testing $H: p = p_0$ at level α against the alternatives $\Omega_K : p/q \le \frac{1}{2}p_0/q_0$ or $\ge 2p_0/q_0$. For $\alpha = .05$ determine the smallest sample size for which there exists a test with power $\ge .8$ against Ω_K if $p_0 = .1, .2, .3, .4, .5$.

   (ii) Let $X_1,\ldots,X_n$ be independently distributed as $N(\xi, \sigma^2)$. For testing $\sigma = 1$ at level $\alpha = .05$, determine the smallest sample size for which there exists a test with power $\ge .9$ against the alternatives $\sigma^2 \le \frac{1}{2}$ and $\sigma^2 \ge 2$.

[See Problem 5 of Chapter 4.]

9. *Double-exponential distribution.* Let $X_1,\ldots,X_n$ be a sample from the double-exponential distribution with density $\frac{1}{2}e^{-|x-\theta|}$. The LMP test for testing $\theta \le 0$ against $\theta > 0$ is the sign test, provided the level is of the form

$$\alpha = \frac{1}{2^n} \sum_{k=0}^m \binom{n}{k},$$

so that the level-α sign test is nonrandomized.

[Let R_k $(k = 0,\ldots,n)$ be the subset of the sample space in which k of the X's are positive and $n - k$ are negative. Let $0 \le k < l < n$, and let S_k, S_l be subsets of R_k, R_l such that $P_0(S_k) = P_0(S_l) \ne 0$. Then it follows from a consideration of $P_\theta(S_k)$ and $P_\theta(S_l)$ for small θ that there exists Δ such that $P_\theta(S_k) < P_\theta(S_l)$ for $0 < \theta < \Delta$. Suppose now that the rejection region of a nonrandomized test of $\theta = 0$ against $\theta > 0$ does not consist of the upper tail of a sign test. Then it can be converted into a sign test of the same size by a finite number of steps, each of which consists in replacing an S_k by an S_l with $k < l$, and each of which therefore increases the power for θ sufficiently small.]

Section 3

10. If (13) holds, show that q_1 defined by (11) belongs to $\mathcal{P}_1$.

11. Show that there exists a unique constant b for which q_0 defined by (11) is a probability density with respect to μ, that the resulting q_0 belongs to $\mathcal{P}_0$, and that $b \to \infty$ as $\epsilon_0 \to 0$.

12. Prove the formula (15).

13. Show that if $\mathcal{P}_0 \neq \mathcal{P}_1$ and ϵ_0, ϵ_1 are sufficiently small, then $Q_0 \neq Q_1$.

14. Evaluate the test (21) explicitly for the case that P_i is the normal distribution with mean ξ_i and known variance σ^2, and when $\epsilon_0 = \epsilon_1$.

15. Determine whether (21) remains the maximin test if in the model (20) G_i is replaced by G_{ij}.

16. Write out a formal proof of the maximin property outlined in the last paragraph of Section 3.

Section 4

17. Let $X_1, \ldots, X_n$ be independently normally distributed with means $E(X_i) = \mu_i$ and variance 1. The test of $H: \mu_1 = \cdots = \mu_n = 0$ that maximizes the minimum power over $\omega': \Sigma \mu_i \geq d$ rejects when $\Sigma X_i \geq C$.
[If the least favorable distribution assigns probability 1 to a single point, invariance under permutations suggests that this point will be $\mu_1 = \cdots = \mu_n = d/n$].

18.* (i) In the preceding problem determine the maximin test if ω' is replaced by $\Sigma a_i \mu_i \geq d$, where the a's are given positive constants.

  (ii) Solve part (i) with Var$(X_i) = 1$ replaced by Var$(X_i) = \sigma_i^2$ (known).

[(i): Determine the point $(\mu_1^*, \ldots, \mu_n^*)$ in ω' for which the MP test of H against $K: (\mu_1^*, \ldots, \mu_n^*)$ has the smallest power, and show that the MP test of H against K is a maximin solution.]

Section 5

19. Let $X = (X_1, \ldots, X_p)$ and $Y = (Y_1, \ldots, Y_p)$ be independently distributed according to p-variate normal distributions with zero means and covariance matrices $E(X_i X_j) = \sigma_{ij}$ and $E(Y_i Y_j) = \Delta \sigma_{ij}$.

  (i) The problem of testing $H: \Delta \leq \Delta_0$ remains invariant under the group G of transformations $X^* = XA$, $Y^* = YA$, where $A = (a_{ij})$ is any nonsingular $p \times p$ matrix with $a_{ij} = 0$ for $i > j$, and there exists a UMP invariant test under G with rejection region $Y_1^2 / X_1^2 > C$.

*Due to Fritz Scholz.

(ii)   The test with rejection region $Y_1^2/X_1^2 > C$ maximizes the minimum power for testing $\Delta \leq \Delta_0$ against $\Delta \geq \Delta_1$ ($\Delta_0 < \Delta_1$).

[(ii): That the Hunt–Stein theorem is applicable to G can be proved in steps by considering the group G_q of transformations $X_q' = \alpha_1 X_1 + \cdots + \alpha_q X_q$, $X_i' = X_i$ for $i = 1,\ldots, q - 1, q + 1,\ldots, p$, successively for $q = 1,\ldots, p - 1$. Here $\alpha_q \neq 0$, since the matrix A is nonsingular if and only if $a_{ii} \neq 0$ for all i. The group product $(\gamma_1,\ldots, \gamma_q)$ of two such transformations $(\alpha_1,\ldots, \alpha_q)$ and $(\beta_1,\ldots, \beta_q)$ is given by $\gamma_1 = \alpha_1\beta_q + \beta_1$, $\gamma_2 = \alpha_2\beta_q + \beta_2,\ldots, \gamma_{q-1} = \alpha_{q-1}\beta_q + \beta_{q-1}$, $\gamma_q = \alpha_q\beta_q$, which shows G_q to be isomorphic to a group of scale changes (multiplication of all components by β_q) and translations [addition of $(\beta_1,\ldots, \beta_{q-1}, 0)$]. The result now follows from the Hunt–Stein theorem and Example 7, since the assumptions of the Hunt–Stein theorem, except for the easily verifiable measurability conditions, concern only the abstract structure $(G, \mathscr{B})$, and not the specific realization of the elements of G as transformations of some space.]

20.   Suppose that the problem of testing $\theta \in \Omega_H$ against $\theta \in \Omega_K$ remains invariant under G, that there exists a UMP almost invariant test φ_0 with respect to G, and that the assumptions of Theorem 3 hold. Then φ_0 maximizes $\inf_{\Omega_K}[w(\theta)E_\theta\varphi(X) + u(\theta)]$ for any weight functions $w(\theta) \geq 0$, $u(\theta)$ that are invariant under $\bar{G}$.

Section 6

21.   *Existence of most stringent tests.*   Under the assumptions of Problem 1 there exists a most stringent test for testing $\theta \in \Omega_H$ against $\theta \in \Omega - \Omega_H$.

22.   Let $\{\Omega_\Delta\}$ be a class of mutually exclusive sets of alternatives such that the envelope power function is constant over each Ω_Δ and that $\cup\Omega_\Delta = \Omega - \Omega_H$, and let φ_Δ maximize the minimum power over Ω_Δ. If $\varphi_\Delta = \varphi$ is independent of Δ, then φ is most stringent for testing $\theta \in \Omega_H$.

23.   Let $(Z_1,\ldots, Z_N) = (X_1,\ldots, X_m, Y_1,\ldots, Y_n)$ be distributed according to the joint density (56) of Chapter 5, and consider the problem of testing $H: \eta = \xi$ against the alternatives that the X's and Y's are independently normally distributed with common variance σ^2 and means $\eta \neq \xi$. Then the permutation test with rejection region $|\bar{Y} - \bar{X}| > C[T(Z)]$, the two-sided version of the test (55) of Chapter 5, is most stringent.

[Apply Problem 22 with each of the sets Ω_Δ consisting of two points (ξ_1, η_1, σ), (ξ_2, η_2, σ) such that

$$\xi_1 = \zeta - \frac{n}{m + n}\delta, \qquad \eta_1 = \zeta + \frac{m}{m + n}\delta;$$

$$\xi_2 = \zeta + \frac{n}{m + n}\delta, \qquad \eta_2 = \zeta - \frac{m}{m + n}\delta$$

for some ζ and δ.]

Additional Problems

24. Let $X_1, \ldots, X_n$ be independent normal variables with variance 1 and means $\xi_1, \ldots, \xi_n$, and consider the problem of testing $H: \xi_1 = \cdots = \xi_n = 0$ against the alternatives $K = \{K_1, \ldots, K_n\}$, where $K_i: \xi_j = 0$ for $j \neq i$, $\xi_i = \xi$ (known and positive). Show that the problem remains invariant under permutation of the X's and that there exists a UMP invariant test ϕ_0 which rejects when $\sum e^{-\xi x_j} > C$, by the following two methods.

   (i) The order statistics $X_{(1)} < \cdots < X_{(n)}$ constitute a maximal invariant.

   (ii) Let f_0 and f_i denote the densities under H and K_i respectively. Then the level-α test ϕ_0 of H vs. $K': f = (1/n)\sum f_i$ is UMP invariant for testing H vs. K.

[(ii): If ϕ_0 is not UMP invariant for H vs. K, there exists an invariant test ϕ_1 whose (constant) power against K exceeds that of ϕ_0. Then ϕ_1 is also more powerful against K'.]

25. The UMP invariant test ϕ_0 of Problem 24

   (i)   maximizes the minimum power over K;

   (ii) is admissible.

   (iii) For testing the hypothesis H of Problem 24 against the alternatives $K' = \{K_1, \ldots, K_n, K_1', \ldots, K_n'\}$, where under $K_i': \xi_j = 0$ for all $j \neq i$, $\xi_i = -\xi$, determine the UMP test under a suitable group G', and show that it is both maximin and invariant.

[(ii): Suppose ϕ' is uniformly at least as powerful as ϕ_0, and more powerful for at least one K_i, and let

$$\phi^*(x_1, \ldots, x_n) = \frac{\sum \phi'(x_{i_1}, \ldots, x_{i_n})}{n!},$$

where the summation extends over all permutations. Then ϕ^* is invariant, and its power is independent of i and exceeds that of ϕ_0.]

26. Show that the UMP invariant test of Problem 24 is most stringent.

27. For testing $H: f_0$ against $K: \{f_1, \ldots, f_s\}$, suppose there exists a finite group $G = \{g_1, \ldots, g_N\}$ which leaves H and K invariant and which is transitive in the sense that given $f_j, f_{j'}$ $(1 \leq j, j')$ there exists $g \in G$ such that $\bar{g}f_j = f_{j'}$. In generalization of Problems 24, 25, determine a UMP invariant test, and show that it is both maximin against K and admissible.

28. To generalize the results of the preceding problem to the testing of $H: f$ vs. $K: \{f_\theta, \theta \in \omega\}$, assume:

   (i)   There exists a group G that leaves H and K invariant.

   (ii) $\bar{G}$ is transitive over ω.

   (iii) There exists a probability distribution Q over G which is right-invariant in the sense of Section 4.

Determine a UMP invariant test, and show that it is both maximin against K and admissible.

29. Let $X_1, \ldots, X_n$ be independent normal with means $\theta_1, \ldots, \theta_n$ and variance 1.

    (i)   Apply the results of the preceding problem to the testing of $H: \theta_1 = \cdots = \theta_n = 0$ against $K: \Sigma \theta_i^2 = r^2$, for any fixed $r > 0$.

    (ii) Show that the results of (i) remain valid if H and K are replaced by $H': \Sigma \theta_i^2 \leq r_0^2$, $K': \Sigma \theta_i^2 \geq r_1^2$ $(r_0 < r_1)$.

30. Suppose in Problem 29(i) the variance σ^2 is unknown and that the data consist of $X_1, \ldots, X_n$ together with an independent random variable S^2 for which S^2/σ^2 has a χ^2-distribution. If K is replaced by $\Sigma \theta_i^2/\sigma^2 = r^2$, then

    (i)   the confidence sets $\Sigma(\theta_i - X_i)^2/S^2 \leq C$ are uniformly most accurate equivariant under the group generated by the n-dimensional generalization of the group G_0 of Example 17 of Chapter 6, and the scale changes $X_i' = cX_i$, $S'^2 = c^2 S^2$.

    (ii) The confidence sets of (i) are minimax with respect to the measure μ given by

$$\mu\left[C(X, S^2)\right] = \frac{1}{\sigma^2}\left[\text{volume of } C(X, S^2)\right].$$

[Use polar coordinates with $\theta^2 = \Sigma \theta_i^2$.]

31. *Locally uniformly most powerful tests.* If the sample space is finite and independent of θ, the test φ_0 of Problem 2(i) is not only LMP but also locally uniformly most powerful (LUMP) in the sense that there exists a value $\Delta > 0$ such that φ_0 maximizes $\beta_\varphi(\theta)$ for all θ with $0 < \theta - \theta_0 < \Delta$.
[See the argument following (19) of Chapter 6, Section 9.]

32. The following two examples show that the assumption of a finite sample space is needed in Problem 31.

    (i)   Let $X_1, \ldots, X_n$ be i.i.d. according to a normal distribution $N(\sigma, \sigma^2)$ and test $H: \sigma = \sigma_0$ against $K: \sigma > \sigma_0$.

    (ii) Let X and Y be independent Poisson variables with $E(X) = \lambda$ and $E(Y) = \lambda + 1$, and test $H: \lambda = \lambda_0$ against $K: \lambda > \lambda_0$. In each case, determine the LMP test and show that it is not LUMP.

    [Compare the LMP test with the most powerful test against a simple alternative.]

8. REFERENCES

The concepts and results of Section 1 are essentially contained in the minimax theory developed by Wald for general decision problems. An exposition of this theory and some of its applications is given in Wald's book (1950). The ideas of Section 3, and in particular Theorem 2, are due to Huber (1965) and form the core of his theory of robust tests [Huber (1981, Chapter 10)]. The material of sections 4 and 5, including Lemma 2, Theorem 3, and Example 8, constitutes the main part of an unpublished paper of Hunt and Stein (1946).

Bednarski, T.
  (1982). "Binary experiments, minimax tests and 2-alternating capacities." *Ann. Statist.* **10**, 226–232.
  (1984). "Minimax testing between Prohorov neighbourhoods." *Statist. and Decisions* **2**, 281–292.

Berger, A.
  (1951). "On uniformly consistent tests." *Ann. Math. Statist.* **22**, 289–293.

Berger, J. O.
  (1985). *Statistical Decision Theory and Bayesian Analysis.* 2nd Ed., Springer, New York.

Bondar, J. V. and Milnes, P.
  (1981). "Amenability: A survey for statistical applications of Hunt–Stein and related conditions on groups." *Z. Wahrsch.* **57**, 103–128.
  (1982). "A converse to the Hunt–Stein theorem." Unpublished.

Cohen, A. and Sackrowitz, H. B.
  (1975). "Unbiasedness of the chi square, likelihood ratio, and other goodness of fit tests for the equal cell case." *Ann. Statist.* **3**, 959–964.

Eaton, M. L.
  (1985). *Multivariate Statistics.* Wiley, New York.

Farrell, R. H.
  (1985). *Techniques of Multivariate Calculation*, Springer, Berlin.

Giri, N. and Kiefer, J.
  (1964). "Local and asymptotic minimax properties of multivariate tests." *Ann. Math. Statist.* **35**, 21–35.

Giri, N., Kiefer, J., and Stein, C.
  (1963). "Minimax character of Hotelling's T^2 test in the simplest case." *Ann. Math. Statist.* **34**, 1524–1535.

Halmos, P.
  (1974). *Measure Theory*, Springer, New York.

Huang J. S. and Ghosh, M.
  (1982). "A note on strong unimodality of order statistics." *J. Amer. Statist. Assoc.* **77**, 929–930.

Huber, P. J.
  (1965). "A robust version of the probability ratio test." *Ann. Math. Statist.* **36**, 1753–1758.
  (1981). *Robust Statistics*, Wiley, New York.

Huber, P. J. and Strassen, V.
  (1973, 1974). "Minimax tests and the Neyman–Pearson lemma for capacities." *Ann. Statist.* **1**, 251–263; **2**, 223–224.

Hunt, G. and Stein, C.
  (1946). "Most stringent tests of statistical hypotheses." Unpublished.

Hwang, J. T. and Casella, G.
(1982). "Minimax confidence sets for the mean of a multivariate normal distribution." *Ann. Statist.* **10**, 868–881.

Ibragimov, J. A.
(1956). "On the composition of unimodal distributions" (Russian). *Teoriya Veroyatnostey* **1**, 283–288; Engl. transl., *Theor. Probab. Appl.* **1** (1956), 255–260.

Isaacson, S. L.
(1951). "On the theory of unbiased tests of simple statistical hypotheses specifying the values of two or more parameters." *Ann. Math. Statist.* **22**, 217–234.
[Introduces type *D* and *E* tests.]

James, A. T.
(1954). "Normal multivariate analysis and the orthogonal group." *Ann. Math. Statist.* **25**, 40–75.

Kallenberg, W. C. M. et al.
(1984). *Testing Statistical Hypotheses: Worked Solutions*, CWI Syllabus No. 3, Centrum voor Wiskunde en Informatien, Amsterdam.

Karlin, S.
(1968). *Total Positivity*, Stanford Univ. Press, Stanford, Calif.

Kiefer, J.
(1958). "On the nonrandomized optimality and randomized nonoptimality of symmetrical designs." *Ann. Math. Statist.* **29**, 675–699.
[Problem 4(ii).]

Kraft, C.
(1955). "Some conditions for consistency and uniform consistency of statistical procedures." *Univ. of Calif. Publ. in Statist.* **2**, 125–142.

Lambert, D.
(1985). Robust two-sample permutation tests. *Ann. Statist.* **13**, 606–625.

Lehmann, E. L.
(1947). "On families of admissible tests." *Ann. Math. Statist.* **18**, 97–104.
[Last example of Section 6.]
(1950). "Some principles of the theory of testing hypotheses." *Ann. Math. Statist.* **21**, 1–26.
[Theorem 1; Problem 19.]
(1955). "Ordered families of distributions." *Ann. Math. Statist.* **26**, 399–419.
[Lemma 1; Problems 2, 7,* and 8.]

Lehmann, E. L. and Stein, C.
(1949). "On the theory of some nonparametric hypotheses." *Ann. Math. Statist.* **20**, 28–45.
[Problem 23.]

Linnik, Yu. V., Pliss, V. A., and Salaevskii, O. V.
(1968). "On the theory of Hotelling's test" (Russian). *Dok. AN SSSR* **168**, 743–746.

Loh, W. Y.
(1984a). "Strong unimodality and scale mixtures." *Ann. Inst. Statist. Math.* **36**, 441–450.
(1984b). "Bounds on ARE's for restricted classes of distributions defined via tail-orderings." *Ann. Statist.* **12**, 685–701.

*This problem is a corrected version of Theorem 3 of the paper in question. I am grateful to R. Blumenthal for pointing out an error in the statement of this theorem in the paper.

Loomis, L. H.

(1953). *An Introduction to Abstract Harmonic Analysis*, Van Nostrand, New York.

Marshall, A. W. and Olkin, I.

(1979). *Inequalities: Theory of Majorization and its Applications*, Academic, New York.

Nachbin, L.

(1965). *The Haar Integral*, Van Nostrand, New York.

Neyman, J.

(1935). "Sur la vérification des hypothèses statistiques composées." *Bull. Soc. Math. France* **63**, 246–266.

[Defines, and shows how to derive, tests of type B, that is, tests which are LMP among locally unbiased tests in the presence of nuisance parameters.]

Neyman, J. and Pearson, E. S.

(1936, 1938). "Contributions to the theory of testing statistical hypotheses." *Statist. Res. Mem.* **1**, 1–37; **2**, 25–57.

[Discusses tests of types A, that is, tests which are LMP among locally unbiased tests when no nuisance parameters are present.]

Rieder, H.

(1977). "Least favorable pairs for special capacities." *Ann. Statist.* **5**, 909–921.

Ruist, E.

(1954). "Comparison of tests for non-parametric hypotheses." *Arkiv Mat.* **3**, 133–136. [Problem 5.]

Salaevskii, Y.

(1971). *Essay in Investigations in Classical Problems of Probability Theory and Mathematical Statistics* (V. M. Kalinin and O. V. Salaevskii, eds.) (Russian), Leningrad Seminars in Math., Vol. 13, Steklov Math. Inst.; Engl. transl., Consultants Bureau, New York.

Schoenberg, I. J.

(1951). "On Pólya frequency functions. I." *J. Analyse Math.* **1**, 331–374. [Example 1.]

Schwartz, R. E.

(1967). "Locally minimax tests." *Ann. Math. Statist.* **38**, 340–360.

Serfling, R. J.

(1980). *Approximation Theorems of Mathematical Statistics*, Wiley, New York.

Sierpinski, W.

(1920). "Sur les fonctions convexes measurables." *Fundamenta Math.* **1**, 125–129.

Stone, M. and von Randow, R.

(1968). "Statistically inspired conditions on the group structure of invariant experiments and their relationships with other conditions on locally compact topological groups." *Z. Wahrsch.* **10**, 70–78.

Wald, A.

(1942). *On the Principles of Statistical Inference*, Notre Dame Math. Lectures No. 1, Notre Dame, Ind.

[Definition of most stringent tests.]

(1950). *Statistical Decision Functions*. Wiley, New York.

Wolfowitz, J.

(1949). "The power of the classical tests associated with the normal distribution." *Ann. Math. Statist.* **20**, 540–551.

[Proves that the standard tests of the univariate linear hypothesis and for testing the absence of multiple correlation are most stringent among all similar tests and possess certain related optimum properties.]

CHAPTER 10

Conditional Inference

1. MIXTURES OF EXPERIMENTS

The present chapter has a somewhat different character from the preceding ones. It is concerned with problems regarding the proper choice and interpretation of tests and confidence procedures, problems which—despite a large literature—have not found a definitive solution. The discussion will thus be more tentative than in earlier chapters, and will focus on conceptual aspects more than on technical ones.

Consider the situation in which either the experiment $\mathscr{E}$ of observing a random quantity X with density p_θ (with respect to μ) or the experiment $\mathscr{F}$ of observing an X with density q_θ (with respect to ν) is performed with probability p and $q = 1 - p$ respectively. On the basis of X, and knowledge of which of the two experiments was performed, it is desired to test $H_0: \theta = \theta_0$ against $H_1: \theta = \theta_1$. For the sake of convenience it will be assumed that the two experiments have the same sample space and the same σ-field of measurable sets. The sample space of the overall experiment consists of the union of the sets

$$\mathscr{X}_0 = \{(I, x): I = 0, x \in \mathscr{X}\} \quad \text{and} \quad \mathscr{X}_1 = \{(I, x): I = 1, x \in \mathscr{X}\}$$

where I is 0 or 1 as $\mathscr{E}$ or $\mathscr{F}$ is performed.

A level-α test of H_0 is defined by its critical function

$$\phi_i(x) = \phi(i, x)$$

and must satisfy

$$(1) \quad pE_0[\phi_0(X)|\mathscr{E}] + qE_0[\phi_1(X)|\mathscr{F}] = p\int \phi_0 p_{\theta_0}\, d\mu + q\int \phi_1 q_{\theta_0}\, d\nu \le \alpha.$$

Suppose that p is unknown, so that H_0 is composite. Then a level-α test of

539

H_0 satisfies (1) for all $0 < p < 1$, and must therefore satisfy

$$(2) \qquad \alpha_0 = \int \phi_0 p_{\theta_0} \, d\mu \le \alpha \quad \text{and} \quad \alpha_1 = \int \phi_1 q_{\theta_0} \, d\nu \le \alpha.$$

As a result, a UMP test against H_1 exists and is given by

$$(3) \quad \phi_0(x) = \begin{cases} 1 \\ \gamma_0 \\ 0 \end{cases} \text{if} \quad \frac{p_{\theta_1}(x)}{p_{\theta_0}(x)} \gtreqless c_0, \qquad \phi_1(x) = \begin{cases} 1 \\ \gamma_1 \\ 0 \end{cases} \text{if} \quad \frac{q_{\theta_1}(x)}{q_{\theta_0}(x)} \gtreqless c_1,$$

where the c_i and γ_i are determined by

$$(4) \qquad E_{\theta_0}[\phi_0(X)|\mathscr{E}] = E_{\theta_0}[\phi_1(X)|\mathscr{F}] = \alpha.$$

The power of this test against H_1 is

$$(5) \qquad \beta(p) = p\beta_0 + q\beta_1$$

with

$$(6) \qquad \beta_0 = E_{\theta_1}[\phi_0(X)|\mathscr{E}], \qquad \beta_1 = E_{\theta_1}[\phi_1(X)|\mathscr{F}].$$

The situation is analogous to that of Chapter 4, Section 4, and, as was discussed there, it may be more appropriate to consider the conditional power β_i when $I = i$, since this is the power pertaining to the experiment that has been performed. As in the earlier case, the conditional power β_I can also be interpreted as an estimate of the unknown $\beta(p)$, which is unbiased, since

$$E(\beta_I) = p\beta_0 + q\beta_1 = \beta(p).$$

So far, the probability p of performing experiment $\mathscr{E}$ has been assumed to be unknown. Suppose instead that the value of p is known, say $p = \frac{1}{2}$. The hypothesis H can be tested at level α by means of (3) as before, but the power of the test is now known to be $\frac{1}{2}(\beta_0 + \beta_1)$. Suppose that $\beta_0 = .3$, $\beta_1 = .9$, so that at the start of the experiment the power is $\frac{1}{2}(.3 + .9) = .6$. Now a fair coin is tossed to decide whether to perform $\mathscr{E}$ (in case of heads) or $\mathscr{F}$ (in case of tails). If the coin shows heads, should the power be reassessed and scaled down to .3?

Let us postpone the answer and first consider another change resulting from the knowledge of p. A level-α test of H now no longer needs to satisfy

(2) but only the weaker condition

$$(7) \qquad \frac{1}{2}\left[\int \phi_0 p_{\theta_0}\, d\mu + \int \phi_1 q_{\theta_0}\, d\nu\right] \leq \alpha.$$

The most powerful test against K is then again given by (3), but now with $c_0 = c_1 = c$ and $\gamma_0 = \gamma_1 = \gamma$ determined by (Problem 3)

$$(8) \qquad \tfrac{1}{2}(\alpha_0 + \alpha_1) = \alpha,$$

where

$$(9) \qquad \alpha_0 = E_{\theta_0}[\phi_0(X)|\mathscr{E}], \qquad \alpha_1 = E_{\theta_0}[\phi_1(X)|\mathscr{F}].$$

As an illustration of the change, suppose that experiment $\mathscr{F}$ is reasonably informative, say that the power β_1 given by (6), is .8, but that $\mathscr{E}$ has little ability to distinguish between p_{θ_0} and p_{θ_1}. Then it will typically not pay to put much of the rejection probability into α_0; if β_0 [given by (6)] is sufficiently small, the best choice of α_0 and α_1 satisfying (8) is approximately $\alpha_0 \approx 0$, $\alpha_1 \approx 2\alpha$. The situation will be reversed if $\mathscr{F}$ is so informative that $\mathscr{F}$ can attain power close to 1 with an α_1 much smaller than $\alpha/2$.

When p is known, there are therefore two issues. Should the procedure be chosen which is best on the average over both experiments, or should the best conditional procedure be preferred; and, for a given test or confidence procedure, should probabilities such as level, power, and confidence coefficient be calculated conditionally, given the experiment that has been selected, or unconditionally? The underlying question is of course the same: Is a conditional or unconditional point of view more appropriate?

The answer cannot be found within the model but depends on the context. If the overall experiment will be performed many times, for example in an industrial or agricultural setting, the average performance may be the principal feature of interest, and an unconditional approach suitable. However, if repetitions refer to different clients, or are potential rather than actual, interest will focus on the particular event at hand, and conditioning seems more appropriate. Unfortunately, as will be seen in later sections, it is then often not clear how the conditioning events should be chosen.

The difference between the conditional and the unconditional approach tends to be most striking, and a choice between them therefore most pressing, when the two experiments $\mathscr{E}$ and $\mathscr{F}$ differ sharply in the amount of information they contain, if for example the difference $|\beta_1 - \beta_0|$ in (6) is large. To illustrate an extreme situation in which this is not the case, suppose that $\mathscr{E}$ and $\mathscr{F}$ consist in observing X with distribution $N(\theta, 1)$ and $N(-\theta, 1)$ respectively, that one of them is selected with known probabilities p and q respectively, and that it is desired to test $H : \theta = 0$ against $K : \theta > 0$. Here $\mathscr{E}$ and $\mathscr{F}$ contain exactly the same amount of information about θ. The unconditional most powerful level-α test of H against $\theta_1 > 0$ is seen to reject (Problem 5) when $X > c$ if $\mathscr{E}$ is performed, and when $X < -c$ if $\mathscr{F}$ is performed, where $P_0(X > c) = \alpha$. The test is UMP against $\theta > 0$, and happens to coincide with the UMP conditional test.

The issues raised here extend in an obvious way to mixtures of more than two experiments. As an illustration of a mixture over a continuum, consider a regression situation. Suppose that $X_1, \ldots, X_n$ are independent, and that the conditional density of X_i given t_i is

$$\frac{1}{\sigma} f\left(\frac{x_i - \alpha - \beta t_i}{\sigma} \right).$$

The t_i themselves are obtained with error. They may for example be independently normally distributed with mean c_i and known variance τ^2, where the c_i are the intended values of the t_i. Then it will again often be the case that the most appropriate inference concerning α, β, and σ is conditional on the observed values of the t's (which represent the experiment actually being performed). Whether this is the case will, as before, depend on the context.

The argument for conditioning also applies when the probabilities of performing the various experiments are unknown, say depend on a parameter ϑ, provided ϑ is unrelated to θ, so that which experiment is chosen provides no information concerning θ. A more precise statement of this generalization is given at the end of the next section.

2. ANCILLARY STATISTICS

Mixture models can be described in the following general terms. Let $\{\mathscr{E}_z, z \in \mathscr{Z}\}$ denote a collection of experiments of which one is selected according to a known probability distribution over $\mathscr{Z}$. For any given z, the experiment $\mathscr{E}_z$ consists in observing a random quantity X, which has a distribution $P_\theta(\cdot | z)$. Although this structure seems rather special, it is common to many statistical models.

Consider a general statistical model in which the observations X are distributed according to P_θ, $\theta \in \Omega$, and suppose there exists an *ancillary statistic*, that is, a statistic Z whose distribution F does not depend on θ. Then one can think of X as being obtained by a two-stage experiment: Observe first a random quantity Z with distribution F; given $Z = z$, observe a quantity X with distribution $P_\theta(\cdot|z)$. The resulting X is distributed according to the original distribution P_θ. Under these circumstances, the argument of the preceding section suggests that it will frequently be appropriate to take the conditional point of view.* (Unless Z is discrete, these definitions involve technical difficulties concerning sets of measure zero and the existence of conditional distributions, which we shall disregard.)

An important class of models in which ancillary statistics exist is obtained by invariance considerations. Suppose the model $\mathscr{P} = \{ P_\theta, \ \theta \in \Omega \}$ remains invariant under the transformations

$$X \to gX, \quad \theta \to \bar{g}\theta; \quad g \in G, \quad \bar{g} \in \bar{G},$$

and that $\bar{G}$ is transitive over Ω.[†]

Theorem 1. *If $\mathscr{P}$ remains invariant under G and if $\bar{G}$ is transitive over Ω, then a maximal invariant T (and hence any invariant) is ancillary.*

Proof. It follows from Theorem 3 of Chapter 6 that the distribution of a maximal invariant under G is invariant under $\bar{G}$. Since $\bar{G}$ is transitive, only constants are invariant under $\bar{G}$. The probability $P_\theta(T \in B)$ is therefore constant, independent of θ, for all B, as was to be proved.

As an example, suppose that $X = (X_1, \ldots, X_n)$ is distributed according to a location family with joint density $f(x_1 - \theta, \ldots, x_n - \theta)$. The most powerful test of $H : \theta = \theta_0$ against $K : \theta = \theta_1 > \theta_0$ rejects when

$$(10) \qquad \frac{f(x_1 - \theta_1, \ldots, x_n - \theta_1)}{f(x_1 - \theta_0, \ldots, x_n - \theta_0)} \geq c.$$

Here the set of differences $Y_i = X_i - X_n$ $(i = 1, \ldots, n - 1)$ is ancillary. This is obvious by inspection and follows from Theorem 1 in conjunction with Example 1(i) of Chapter 6. It may therefore be more appropriate to consider the testing problem conditionally given $Y_1 = y_1, \ldots, Y_{n-1} = y_{n-1}$. To determine the most powerful conditional test, transform to $Y_1, \ldots, Y_n$, where $Y_n = X_n$. The conditional density of Y_n given $y_1, \ldots, y_{n-1}$ is

$$(11) \quad p_\theta(y_n|y_1, \ldots, y_{n-1}) = \frac{f(y_1 + y_n - \theta, \ldots, y_{n-1} + y_n - \theta, y_n - \theta)}{\int f(y_1 + u, \ldots, y_{n-1} + u, u) \, du},$$

---

*A distinction between experimental mixtures and the present situation, relying on aspects outside the model, is discussed by Basu (1964) and Kalbfleisch (1975).

[†] The family $\mathscr{P}$ is then a group family; see *TPE*, Chapter 1, Section 3.

and the most powerful conditional test rejects when

(12)
$$\frac{p_{\theta_1}(y_n | y_1, \ldots, y_{n-1})}{p_{\theta_0}(y_n | y_1, \ldots, y_{n-1})} > c(y_1, \ldots, y_{n-1}).$$

In terms of the original variables this becomes

(13)
$$\frac{f(x_1 - \theta_1, \ldots, x_n - \theta_1)}{f(x_1 - \theta_0, \ldots, x_n - \theta_0)} > c(x_1 - x_n, \ldots, x_{n-1} - x_n).$$

The constant $c(x_1 - x_n, \ldots, x_{n-1} - x_n)$ is determined by the fact that the conditional probability of (13), given the differences of the x's, is equal to α when $\theta = \theta_0$.

For describing the conditional test (12) and calculating the critical value $c(y_1, \ldots, y_{n-1})$, it is useful to note that the statistic $Y_n = X_n$ could be replaced by any other Y_n satisfying the equivariance condition*

(14) $Y_n(x_1 + a, \ldots, x_n + a) = Y_n(x_1, \ldots, x_n) + a$ for all a.

This condition is satisfied for example by the mean of the X's, the median, or any of the order statistics. As will be shown in the following Lemma 1, any two statistics Y_n and Y_n' satisfying (14) differ only by a function of the differences $Y_i = X_i - X_n$ $(i = 1, \ldots, n - 1)$. Thus conditionally, given the values $y_1, \ldots, y_{n-1}$, Y_n and Y_n' differ only by a constant, and their conditional distributions (and the critical values $c(y_1, \ldots, y_{n-1})$ differ by the same constant. One can therefore choose Y_n, subject to (14), to make the conditional calculations as convenient as possible.

Lemma 1. *If Y_n and Y_n' both satisfy* (14), *then their difference $\Delta = Y_n' - Y_n$ depends on $(x_1, \ldots, x_n)$ only through the differences $(x_1 - x_n, \ldots, x_{n-1} - x_n)$.*

Proof. Since Y_n and Y_n' satisfy (14),

$$\Delta(x_1 + a, \ldots, x_n + a) = \Delta(x_1, \ldots, x_n) \quad \text{for all } a.$$

Putting $a = -x_n$, one finds

$$\Delta(x_1, \ldots, x_n) = \Delta(x_1 - x_n, \ldots, x_{n-1} - x_n, 0),$$

which is a function of the differences.

---

*For a more detailed discussion of equivariance, see *TPE*, Chapter 3.

The existence of ancillary statistics is not confined to models that remain invariant under a transitive group $\overline{G}$. The mixture and regression examples of Section 1 provide illustrations of ancillaries without the benefit of invariance. Further examples are given in Problems 8–13.

If conditioning on an ancillary statistic is considered appropriate because it makes the inference more relevant to the situation at hand, it is desirable to carry the process as far as possible and hence to condition on a *maximal* ancillary. An ancillary Z is said to be maximal if there does not exist an ancillary U such that $Z = f(U)$ without Z and U being equivalent. [For a more detailed treatment, which takes account of the possibility of modifying statistics on sets of measure zero without changing their probabilistic properties, see Basu (1959).]

Conditioning, like sufficiency and invariance, leads to a reduction of the data. In the conditional model, the ancillary is no longer part of the random data but has become a constant. As a result, conditioning often leads to a great simplification of the inference. Choosing a maximal ancillary for conditioning thus has the additional advantage of providing the greatest reduction of the data.

Unfortunately, maximal ancillaries are not always unique, and one must then decide which maximal ancillary to choose for conditioning. [This problem is discussed by Cox (1971) and Becker and Gordon (1983).] If attention is restricted to ancillary statistics that are invariant under a given group G, the maximal ancillary of course coincides with the maximal invariant.

Another issue concerns the order in which to apply reduction by sufficiency and ancillarity.

Example 1. Let (X_i, Y_i), $i = 1, \ldots, n$, be independently distributed according to a bivariate normal distribution with $E(X_i) = E(Y_i) = 0$, $\text{Var}(X_i) = \text{Var}(Y_i) = 1$, and unknown correlation coefficient ρ. Then $X_1, \ldots, X_n$ are independently distributed as $N(0, 1)$ and are therefore ancillary. The conditional density of the Y's given $X_1 = x_1, \ldots, X_n = x_n$ is

$$C \exp\left(-\frac{1}{2(1 - \rho^2)} \sum (y_i - \rho x_i)^2 \right),$$

with the sufficient statistics $(\sum Y_i^2, \sum x_i Y_i)$.

Alternatively, one could begin by noticing that $(Y_1, \ldots, Y_n)$ is ancillary. The conditional distribution of the X's given $Y_1 = y_1, \ldots, Y_n = y_n$ then admits the sufficient statistics $(\sum X_i^2, \sum X_i y_i)$. A unique maximal ancillary V does not exist in this case, since both the X's and Y's would have to be functions of V. Thus V would have to be equivalent to the full sample $(X_1, Y_1), \ldots, (X_n, Y_n)$, which is not ancillary.

Suppose instead that the data are first reduced to the sufficient statistics $T = (\Sigma X_i^2 + \Sigma Y_i^2, \Sigma X_i Y_i)$. Based on T, no nonconstant ancillaries appear to exist.* This example and others like it suggest that it is desirable to reduce the data as far as possible through sufficiency, before attempting further reduction by means of ancillary statistics.

Note that contrary to this suggestion, in the location example at the beginning of the section, the problem was not first reduced to the sufficient statistics $X_{(1)} < \cdots < X_{(n)}$. The omission can be justified in hindsight by the fact that the optimal conditional tests are the same whether or not the observations are first reduced to the order statistics.

In the structure described at the beginning of the section, the variable Z that labels the experiment was assumed to have a known distribution. The argument for conditioning on the observed value of Z does not depend on this assumption. It applies also when the distribution of Z depends on an unknown parameter ϑ, which is independent of θ and hence by itself contains no information about θ, that is, when the distribution of Z depends only on ϑ, the conditional distribution of X given $Z = z$ depends only on θ, and the parameter space Ω for (θ, ϑ) is a Cartesian product $\Omega = \Omega_\theta \times \Omega_\vartheta$, with

$$(15) \qquad (\theta, \vartheta) \in \Omega \quad \Leftrightarrow \quad \theta \in \Omega_\theta \text{ and } \vartheta \in \Omega_\vartheta.$$

(the parameters θ and ϑ are then said to be *variation-independent*, or unrelated.)

Statistics Z satisfying this more general definition are called *partial ancillary* or *S-ancillary*. (The term ancillary without modification will be reserved here for a statistic that has a known distribution.) Note that if $X = (T, Z)$ and Z is a partial ancillary, then T is a partial sufficient statistic in the sense of Chapter 3, Problem 36. For a more detailed discussion of this and related concepts of partial ancillarity, see for example Basu (1978) and Barndorff-Nielsen (1978).

Example 2. Let X and Y be independent with Poisson distributions $P(\lambda)$ and $P(\mu)$, and let the parameter of interest be $\theta = \mu/\lambda$. It was seen in Chapter 4, Section 4 that the conditional distribution of Y given $Z = X + Y = z$ is binomial $b(p, z)$ with $p = \mu/(\lambda + \mu) = \theta/(\theta + 1)$ and therefore depends only on θ, while the distribution of Z is Poisson with mean $\vartheta = \lambda + \mu$. Since the parameter space $0 < \lambda, \mu < \infty$ is equivalent to the Cartesian product of $0 < \theta < \infty, 0 < \vartheta < \infty$, it follows that Z is S-ancillary for θ.

The UMP unbiased level-α test of $H: \mu \leq \lambda$ against $\mu > \lambda$ is UMP also among all tests whose conditional level given z is α for all z. (The class of conditional tests coincides exactly with the class of all tests that are similar on the boundary $\mu = \lambda$.)

---

*So far, nonexistence has not been proved. It seems likely that a proof can be obtained by the methods of Unni (1978).

When Z is S-ancillary for θ in the presence of a nuisance parameter ϑ, the unconditional power $\beta(\theta, \vartheta)$ of a test φ of $H: \theta = \theta_0$ may depend on ϑ as well as on θ. The conditional power $\beta(\vartheta|z) = E_\theta[\varphi(X)|z]$ can then be viewed as an unbiased estimator of the (unknown) $\beta(\theta, \vartheta)$, as was discussed at the end of Chapter 4, Section 4. On the other hand, if no nuisance parameters ϑ are present and Z is ancillary for θ, the unconditional power $\beta(\theta) = E_\theta \varphi(X)$ and the conditional power $\beta(\theta|z)$ provide two alternative evaluations of the power of φ against θ, which refer to different sampling frameworks, and of which the latter of course becomes available only after the data have been obtained.

Surprisingly, the S-ancillarity of $X + Y$ in Example 2 does not extend to the corresponding binomial problem.

Example 3. Let X and Y have independent binomial distributions $b(p_1, m)$ and $b(p_2, n)$ respectively. Then it was seen in Chapter 4, Section 5 that the conditional distribution of Y given $Z = X + Y = z$ depends only on the cross-product ratio $\Delta = p_2 q_1 / p_1 q_2$ ($q_i = 1 - p_i$). However, Z is not S-ancillary for Δ. To see this, note that S-ancillarity of Z implies the existence of a parameter ϑ unrelated to Δ and such that the distribution of Z depends only on ϑ. As Δ changes, the family of distributions $\{P_\vartheta, \vartheta \in \Omega_\vartheta\}$ of Z would remain unchanged. This is not the case, since Z is binomial when $\Delta = 1$ and not otherwise (Problem 15). Thus Z is not S-ancillary.

In this example, all unbiased tests of $H: \Delta = \Delta_0$ have a conditional level given z that is independent of z, but conditioning on z cannot be justified by S-ancillarity.

Closely related to this example is the situation of the multinomial 2×2 table discussed from the point of view of unbiasedness in Chapter 4, Section 6.

Example 4. In the notation of Chapter 4, Section 6, let the four cell entries of a 2×2 table be X, X', Y, Y' with row totals $X + X' = M$, $Y + Y' = N$, and column totals $X + Y = T$, $X' + Y' = T'$, and with total sample size $M + N = T + T' = s$. Here it is easy to check that (M, N) is S-ancillary for $\theta = (\theta_1, \theta_2) = (p_{AB}/p_B, p_{A\bar{B}}/p_{\bar{B}})$ with $\vartheta = p_B$. Since the cross-product ratio Δ can be expressed as a function of (θ_1, θ_2), it may be appropriate to condition a test of $H: \Delta = \Delta_0$ on (M, N). Exactly analogously one finds that (T, T') is S-ancillary for $\theta' = (\theta'_1, \theta'_2) = (p_{AB}/p_A, p_{\bar{A}B}/p_{\bar{A}})$, and since Δ is also a function of (θ'_1, θ'_2), it may be equally appropriate to condition a test of H on (T, T'). One might hope that the set of all four marginals $(M, N, T, T') = Z$ would be S-ancillary for Δ. However, it is seen from the preceding example that this is not the case.

Here, all unbiased tests have a constant conditional level given z. However, S-ancillarity permits conditioning on only one set of margins (without giving any guidance as to which of the two to choose), not on both.

Despite such difficulties, the principle of carrying out tests and confidence estimation conditionally on ancillaries or S-ancillaries frequently provides an attractive alternative to the corresponding unconditional procedures, primarily because it is more appropriate for the situation at hand. However, insistence on such conditioning leads to another difficulty, which is illustrated by the following example.

Example 5. Consider N populations Π_i, and suppose that an observation X_i from Π_i has a normal distribution $N(\xi_i, 1)$. The hypothesis to be tested is $H: \xi_1 = \cdots = \xi_N$. Unfortunately, N is so large that it is not practicable to take an observation from each of the populations; the total sample size is restricted to be $n < N$. A sample $\Pi_{J_1}, \ldots, \Pi_{J_n}$ of n of the N populations is therefore selected at random, with probability $1 / \binom{N}{n}$ for each set of n, and an observation X_{J_i} is obtained from each of the populations Π_{J_i} in the sample.

Here the variables $J_1, \ldots, J_n$ are ancillary, and the requirement of conditioning on ancillaries would restrict any inference to the n populations from which observations are taken. Systematic adherence to this requirement would therefore make it impossible to test the original hypothesis H.* Of course, rejection of the partial hypothesis $H_{j_1, \ldots, j_n}: \xi_{j_1} = \cdots = \xi_{j_n}$ would imply rejection of the original H. However, acceptance of $H_{j_1, \ldots, j_n}$ would permit no inference concerning H.

The requirement to condition in this case runs counter to the belief that a sample may permit inferences concerning the whole set of populations, which underlies much of statistical practice.

With an unconditional approach such an inference is provided by the test with rejection region

$$\sum \left[X_{J_i} - \left(\frac{1}{n} \sum_{k=1}^{n} X_{J_k} \right) \right]^2 \geq c,$$

where c is the upper α-percentage point of χ^2 with $n - 1$ degrees of freedom. Not only does this test actually have unconditional level α, but its conditional level given $J_1 = j_1, \ldots, J_n = j_n$ also equals α for all $(j_1, \ldots, j_n)$. There is in fact no difference in the present case between the conditional and the unconditional test: they will accept or reject for the same sample points. However, as has been pointed out, there is a crucial difference between the conditional and unconditional interpretations of the results.

If $\beta_{j_1, \ldots, j_n}(\xi_{j_1}, \ldots, \xi_{j_n})$ denotes the conditional power of this test given $J_1 = j_1, \ldots, J_n = j_n$, its unconditional power is

$$\frac{\sum \beta_{j_1, \ldots, j_n}(\xi_{j_1}, \ldots, \xi_{j_n})}{\binom{N}{n}}$$

summed over all $\binom{N}{n}$ n-tuples $j_1 < \cdots < j_n$. As in the case with any test, the conditional power given an ancillary (in the present case $J_1, \ldots, J_n$) can be viewed as an unbiased estimate of the unconditional power.

---

*For other implications of this requirement, called the weak conditionality principle, see Birnbaum (1962) and Berger and Wolpert (1984).

3. OPTIMAL CONDITIONAL TESTS

Although conditional tests are often sensible and are beginning to be employed in practice [see for example Lawless (1972, 1973, 1978) and Kappenman (1975)], not much theory has been developed for the resulting conditional models. Since the conditional model tends to be simpler than the original unconditional one, the conditional point of view will frequently bring about a simplification of the theory. This possibility will be illustrated in the present section on some simple examples.

Example 6. Specializing the example discussed at the beginning of Section 1, suppose that a random variable is distributed according to $N(\theta, \sigma_1^2)$ or $N(\theta, \sigma_0^2)$ as $I = 1$ or 0, and that $P(I = 1) = P(I = 0) = \frac{1}{2}$. Then the most powerful test of $H : \theta = \theta_0$ against $\theta = \theta_1$ $(> \theta_0)$ based on (I, X) rejects when

$$\frac{x - \frac{1}{2}(\theta_0 + \theta_1)}{2\sigma_i^2} \geq k.$$

A UMP test against the alternatives $\theta > \theta_0$ therefore does not exist. On the other hand, if H is tested conditionally given $I = i$, a UMP conditional test exists and rejects when $X > c_i$ where $P(X > c_i \mid I = i) = \alpha$ for $i = 0, 1$.

The nonexistence of UMP unconditional tests found in this example is typical for mixtures with known probabilities of two or more families with monotone likelihood ratio, despite the existence of UMP conditional tests in these cases.

Example 7. Let $X_1, \ldots, X_n$ be a sample from a normal distribution $N(\xi, a^2\xi^2)$, $\xi > 0$, with known coefficient of variation $a > 0$, and consider the problem of testing $H : \xi = \xi_0$ against $K : \xi > \xi_0$. Here $T = (T_1, T_2)$ with $T_1 = \bar{X}$, $T_2 = \sqrt{(1/n)\Sigma X_i^2}$ is sufficient, and $Z = T_1/T_2$ is ancillary. If we let $V = \sqrt{n}\, T_2/a$, the conditional density of V given $Z = z$ is equal to (Problem 18)

$$(16) \qquad p_\xi(v|z) = \frac{k}{\xi^n} v^{n-1} \exp\left\{ -\frac{1}{2}\left[\frac{v}{\xi} - \frac{z\sqrt{n}}{a} \right]^2 \right\}.$$

The density has monotone likelihood ratio, so that the rejection region $V > C(z)$ constitutes a UMP conditional test.

Unconditionally, $Y = \bar{X}$ and $S^2 = \Sigma(X_i - \bar{X})^2$ are independent with joint density

$$(17) \qquad cs^{(n-3)/2} \exp\left(-\frac{n}{2a^2\xi^2}(y - \xi)^2 - \frac{1}{2a^2\xi^2}s^2 \right),$$

and a UMP test does not exist. [For further discussion of this example, see Hinkley (1977).]

An important class of examples is obtained from situations in which the model remains invariant under a group of transformations that is transitive over the parameter space, that is, when the given class of distributions constitutes a group family. The maximal invariant V then provides a natural ancillary on which to condition, and an optimal conditional test may exist even when such a test does not exist unconditionally. Perhaps the simplest class of examples of this kind are provided by location families under the conditions of the following lemma.

Lemma 2. *Let $X_1, \ldots, X_n$ be independently distributed according to $f(x_i - \theta)$, with f strongly unimodal. Then the family of conditional densities of $Y_n = X_n$ given $Y_i = X_i - X_n$ $(i = 1, \ldots, n - 1)$ has monotone likelihood ratio.*

Proof. The conditional density (11) is proportional to

$$(18) \qquad f(y_n + y_1 - \theta) \cdots f(y_n + y_{n-1} - \theta)f(y_n - \theta).$$

By taking logarithms and using the fact that each factor is strongly unimodal, it is seen that the product is also strongly unimodal, and the result follows from Example 1 of Chapter 9.

Lemma 2 shows that for strongly unimodal f there exists a UMP conditional test of $H : \theta \le \theta_0$ against $K : \theta > \theta_0$, which rejects when

$$(19) \qquad X_n > c(X_1 - X_n, \ldots, X_{n-1} - X_n).$$

Conditioning has reduced the model to a location family with sample size one. The double-exponential and logistic distributions are both strongly unimodal (Section 9.2), and thus provide examples of UMP conditional tests. In neither case does there exist a UMP unconditional test unless $n = 1$.

As a last class of examples, we shall consider a situation with a nuisance parameter. Let $X_1, \ldots, X_m$ and $Y_1, \ldots, Y_n$ be independent samples from location families with densities $f(x_1 - \xi, \ldots, x_m - \xi)$ and $g(y_1 - \eta, \ldots, y_n - \eta)$ respectively, and consider the problem of testing $H : \eta \le \xi$ against $K : \eta > \xi$. Here the differences $U_i = X_i - X_m$ and $V_j = Y_j - Y_n$ are ancillary. The conditional density of $X = X_m$ and $Y = Y_n$ given the u's and v's is seen from (18) to be of the form

$$(20) \qquad f_u^*(x - \xi)g_v^*(y - \eta),$$

where the subscripts u and v indicate that f^* and g^* depend on the u's and v's respectively. The problem of testing H in the conditional model remains invariant under the transformations: $x' = x + c$, $y' = y + c$, for which $Y - X$ is maximal invariant. A UMP invariant conditional test will then exist provided the distribution of $Z = Y - X$, which depends only on $\Delta = \eta - \xi$, has monotone likelihood ratio. The following lemma shows that a sufficient condition for this to be the case is that f_u^* and g_v^* have monotone likelihood ratio in x and y respectively.

Lemma 3. *Let X, Y be independently distributed with densities $f^*(x - \xi)$, $g^*(y - \eta)$ respectively. If f^* and g^* have monotone likelihood with respect to ξ and η, then the family of densities of $Z = Y - X$ has monotone likelihood ratio with respect to $\Delta = \eta - \xi$.*

Proof. The density of Z is

$$(21) \qquad h_\Delta(z) = \int g^*(y - \Delta) f^*(y - z)\, dy.$$

To see that $h_\Delta(z)$ has monotone likelihood ratio, one must show that for any $\Delta < \Delta'$, $h_{\Delta'}(z)/h_\Delta(z)$ is an increasing function of z. For this purpose, write

$$\frac{h_{\Delta'}(z)}{h_\Delta(z)} = \int \frac{g^*(y - \Delta')}{g^*(y - \Delta)} \cdot \frac{g^*(y - \Delta) f^*(y - z)}{\int g^*(u - \Delta) f(u - z)\, du}\, dy.$$

The second factor is a probability density for Y,

$$(22) \qquad p_z(y) = C_z g^*(y - \Delta) f^*(y - z),$$

which has monotone likelihood ratio in the parameter z by the assumption made about f^*. The ratio

$$(23) \qquad \frac{h_{\Delta'}(z)}{h_\Delta(z)} = \int \frac{g^*(y - \Delta')}{g^*(y - \Delta)} p_z(y)\, dy$$

is the expectation of $g^*(Y - \Delta')/g^*(Y - \Delta)$ under the distribution $p_z(y)$. By the assumption about g^*, $g^*(y - \Delta')/g^*(y - \Delta)$ is an increasing function of y, and it follows from Lemma 2 of Chapter 3 that its expectation is an increasing function of z.

It follows from (18) that $f_u^*(x - \xi)$ and $g_v^*(y - \eta)$ have monotone likelihood ratio provided this condition holds for $f(x - \xi)$ and $g(y - \eta)$, i.e. provided f and g are strongly unimodal. Under this assumption, the conditional distribution $h_\Delta(z)$ then has monotone likelihood ratio by Lemma

3, and a UMP conditional test exists and rejects for large values of Z. (This result also follows from Problem 7 of Chapter 9).

The difference between conditional tests of the kind considered in this section and the corresponding (e.g., locally most powerful) unconditional tests typically disappears as the sample size(s) tend(s) to infinity. Some results in this direction are given by Liang (1984); see also Barndorff-Nielsen (1983).

The following multivariate example provides one more illustration of a UMP conditional test when unconditionally no UMP test exists. The results will only be sketched. The details of this and related problems can be found in the original literature reviewed by Marden and Perlman (1980) and Marden (1983).

Example 8. The normal multivariate two-sample problem with covariates was seen in Chapter 8, Example 3, to reduce to the canonical form (the notation has been changed) of $m + 1$ independent normal vectors of dimension $p = p_1 + p_2$,

$$Y = \begin{pmatrix} Y_1 & Y_2 \end{pmatrix} \quad \text{and} \quad Z_1, \ldots, Z_m,$$

with common covariance matrix Σ and expectations

$$E(Y_1) = \eta_1, \qquad E(Y_2) = E(Z_1) = \cdots = E(Z_m) = 0.$$

The hypothesis being tested is $H : \eta_1 = 0$. Without the restriction $E(Y_2) = 0$, the model would remain invariant under the group G_3 of transformations (Chapter 8, Section 2): $Y^* = YB$, $Z^* = ZB$, where B is any nonsingular $p \times p$ matrix. However, the stated problem remains invariant only under the subgroup G' in which B is of the form [Problem 22(i)]

$$B = \begin{pmatrix} B_{11} & 0 \\ B_{21} & B_{22} \end{pmatrix} \begin{matrix} p_1 \\ p_2 \end{matrix}.$$

$$\qquad p_1 \qquad p_2$$

If

$$Z'Z = S = \begin{pmatrix} S_{11} & S_{12} \\ S_{21} & S_{22} \end{pmatrix} \quad \text{and} \quad \Sigma = \begin{pmatrix} \Sigma_{11} & \Sigma_{12} \\ \Sigma_{21} & \Sigma_{22} \end{pmatrix},$$

the maximal invariants under G' are the two statistics $D = Y_2 S_{22}^{-1} Y_2'$ and

$$N = \frac{\left(Y_1 - S_{12}S_{22}^{-1}Y_2\right)\left(S_{11} - S_{12}S_{22}^{-1}S_{21}\right)^{-1}\left(Y_1 - S_{12}S_{22}^{-1}Y_2\right)'}{1 + D},$$

and the joint distribution of (N, D) depends only on the maximal invariant under G',

$$\Delta = \eta_1 \big(\Sigma_{11} - \Sigma_{12}\Sigma_{22}^{-1}\Sigma_{21} \big)^{-1} \eta_1'.$$

The statistic D is ancillary [Problem 22(ii)], and the conditional distribution of N given $D = d$ is that of the ratio of two independent χ^2-variables: the numerator noncentral χ^2 with p degrees of freedom and noncentrality parameter $\Delta/(1 + d)$, and the denominator central χ^2 with $m + 1 - p$ degrees of freedom. It follows from Chapter 7, Section 1, that the conditional density has monotone likelihood ratio. A conditionally UMP invariant test therefore exists, and rejects H when $(m + 1 - p)N/p > C$, where C is the critical value of the F-distribution with p and $m + 1 - p$ degrees of freedom. On the other hand, a UMP invariant (unconditional) test does not exist; comparisons of the optimal conditional test with various competitors are provided by Marden and Perlman (1980).

4. RELEVANT SUBSETS

The conditioning variables considered so far have been ancillary statistics, i.e. random variables whose distribution is fixed, independent of the parameters governing the distribution of X, or at least of the parameter of interest. We shall now examine briefly some implications of conditioning without this constraint. Throughout most of the section we shall be concerned with the simple case in which the conditioning variable is the indicator of some subset C of the sample space, so that there are only two conditioning events $I = 1$ (i.e. $X \in C$) and $I = 0$ (i.e. $X \in \bar{C}$, the complement of C). The mixture problem at the beginning of Section 1, with $\mathscr{X}_1 = C$ and $\mathscr{X}_0 = \bar{C}$, is of this type.

Suppose X is distributed with density p_θ, and R is a level-α rejection region for testing the simple hypothesis $H: \theta = \theta_0$ against some class of alternatives. For any subset C of the sample space, consider the conditional rejection probabilities

$$(24) \qquad \alpha_C = P_{\theta_0}(X \in R | C) \quad \text{and} \quad \alpha_{\bar{C}} = P_{\theta_0}(X \in R | \bar{C}),$$

and suppose that $\alpha_C > \alpha$ and $\alpha_{\bar{C}} < \alpha$. Then we are in the difficulty described in Section 1. Before X was observed, the probability of falsely rejecting H was stated to be α. Now that X is known to have fallen into C (or $\bar{C}$), should the original statement be adjusted and the higher value α_C (or lower value $\alpha_{\bar{C}}$) be quoted? An extreme case of this possibility occurs when C is a subset of R or $\bar{R}$, since then $P(X \in R \mid X \in C) = 1$ or 0.

It is clearly always possible to chose C so that the conditional level α_C exceeds the stated α. It is not so clear whether the corresponding possibility always exists for the levels of a family of confidence sets for θ, since the inequality must now hold for all θ.

Definition. A subset C of the sample space is said to be a *negatively biased relevant subset* for a family of confidence sets $S(X)$ with unconditional confidence level $\gamma = 1 - \alpha$ if for some $\epsilon > 0$

$$(25) \qquad \gamma_C(\theta) = P_\theta[\theta \in S(X)|X \in C] \leq \gamma - \epsilon \qquad \text{for all } \theta,$$

and a *positively biased relevant subset* if

$$(26) \qquad P_\theta[\theta \in S(X)|X \in C] \geq \gamma + \epsilon \qquad \text{for all } \theta.$$

The set C is *semirelevant, negatively or positively biased*, if respectively

$$(27) \qquad P_\theta[\theta \in S(X)|X \in C] \leq \gamma \qquad \text{for all } \theta$$

or

$$(28) \qquad P_\theta[\theta \in S(X)|X \in C] \geq \gamma \qquad \text{for all } \theta,$$

with strict inequality holding for at least some θ.

Obvious examples of relevant subsets are provided by the subsets $\mathcal{X}_0$ and $\mathcal{X}_1$ of the two-experiment example of Section 1.

Relevant subsets do not always exist. The following four examples illustrate the various possibilities.

Example 9. Let X be distributed as $N(\theta,1)$, and consider the standard confidence intervals for θ:

$$S(X) = \{\theta : X - c < \theta < X + c\},$$

where $\Phi(c) - \Phi(-c) = \gamma$. In this case, there exists not even a semirelevant subset.

To see this, suppose first that a positively biased semirelevant subset C exists, so that

$$A(\theta) = P_\theta[X - c < \theta < X + c \text{ and } X \in C] - \gamma P_\theta[X \in C] \geq 0$$

for all θ, with strict inequality for some θ_0. Consider a prior normal density $\lambda(\theta)$ for θ with mean 0 and variance τ^2, and let

$$\beta(x) = P[x - c < \Theta < x + c|x],$$

where Θ has density $\lambda(\theta)$. The posterior distribution of Θ given x is then normal with mean $\tau^2 x/(1 + \tau^2)$ and variance $\tau^2/(1 + \tau^2)$ [Problem 24(i)], and it follows that

$$\beta(x) = \Phi\left[\frac{x}{\tau\sqrt{1+\tau^2}} + \frac{c\sqrt{1+\tau^2}}{\tau}\right] - \Phi\left[\frac{x}{\tau\sqrt{1+\tau^2}} - \frac{c\sqrt{1+\tau^2}}{\tau}\right]$$

$$\leq \Phi\left[\frac{c\sqrt{1+\tau^2}}{\tau}\right] - \Phi\left[\frac{-c\sqrt{1+\tau^2}}{\tau}\right] \leq \gamma + \frac{c}{\sqrt{2\pi}\,\tau^2}.$$

Next let $h(\theta) = \sqrt{2\pi}\,\tau\lambda(\theta) = e^{-\theta^2/2\tau^2}$ and

$$D = \int h(\theta) A(\theta) \, d\theta \leq \sqrt{2\pi}\,\tau \int \lambda(\theta) \{ P_\theta[X - c < \theta < X + c \text{ and } X \in C]$$

$$- E_\theta[\beta(X) I_C(X)] \} \, d\theta + \frac{c}{\tau}.$$

The integral on the right side is the difference of two integrals each of which equals $P[X - c < \Theta < X + c$ and $X \in C]$, and is therefore 0, so that $D \leq c/\tau$.

Consider now a sequence of normal priors $\lambda_m(\theta)$ with variances $\tau_m^2 \to \infty$, and the corresponding sequences $h_m(\theta)$ and D_m. Then $0 \leq D_m \leq c/\tau_m$ and hence $D_m \to 0$. On the other hand, D_m is of the form $D_m = \int_{-\infty}^{\infty} A(\theta) h_m(\theta) \, d\theta$, where $A(\theta)$ is continuous, nonnegative, and > 0 for some θ_0. There exists $\delta > 0$ such that $A(\theta) \geq \frac{1}{2} A(\theta_0)$ for $|\theta - \theta_0| < \delta$ and hence

$$D_m \geq \int_{\theta_0 - \delta}^{\theta_0 + \delta} \frac{1}{2} A(\theta_0) h_m(\theta) \, d\theta \to \delta A(\theta_0) > 0 \quad \text{as} \quad m \to \infty.$$

This provides the desired contradiction.

That also no negatively semirelevant subsets exist is a consequence of the following result.

Theorem 2. *Let $S(x)$ be a family of confidence sets for θ such that $P_\theta[\theta \in S(X)] = \gamma$ for all θ, and suppose that $0 < P_\theta(C) < 1$ for all θ.*

(i) If C is semirelevant, then its complement $\overline{C}$ is semirelevant with opposite bias.

(ii) If there exists a constant a such that

$$1 > P_\theta(C) > a > 0 \quad \text{for all } \theta$$

and C is relevant, then $\overline{C}$ is relevant with opposite bias.

Proof. The result is an immediate consequence of the identity

$$P_\theta(C)[\gamma_C(\theta) - \gamma] = [1 - P_\theta(C)][\gamma - \gamma_{\overline{C}}(\theta)].$$

The next example illustrates the situation in which a semirelevant subset exists but no relevant one.

Example 10. Let X be $N(\theta, 1)$, and consider the uniformly most accurate lower confidence bounds $\underline{\theta} = X - c$ for θ, where $\Phi(c) = \gamma$. Here $S(X)$ is the interval $[X - c, \infty)$ and it seems plausible that the conditional probability of $\theta \in S(X)$ will be lowered for a set C of the form $X \geq k$. In fact

$$(29) \quad P_\theta(X - c \leq \theta | X \geq k) = \begin{cases} \dfrac{\Phi(c) - \Phi(k - \theta)}{1 - \Phi(k - \theta)} & \text{when} \quad \theta > k - c, \\ 0 & \text{when} \quad \theta < k - c. \end{cases}$$

The probability (29) is always $< \gamma$, and tends to γ as $\theta \to \infty$. The set $X \geq k$ is therefore semirelevant negatively biased for the confidence sets $S(X)$.

We shall now show that no relevant subset C with $P_\theta(C) > 0$ exists in this case. It is enough to prove the result for negatively biased sets; the proof for positive bias is exactly analogous. Let A be the set of x-values $-\infty < x < c + \theta$, and suppose that C is negatively biased and relevant, so that

$$P_\theta[X \in A | C] \leq \gamma - \epsilon \qquad \text{for all } \theta.$$

If

$$a(\theta) = P_\theta(X \in C), \qquad b(\theta) = P_\theta(X \in A \cap C),$$

then

$$(30) \qquad\qquad b(\theta) \leq (\gamma - \epsilon) a(\theta) \qquad \text{for all } \theta.$$

The result is proved by comparing the integrated coverage probabilities

$$A(R) = \int_{-R}^{R} a(\theta) \, d\theta, \qquad B(R) = \int_{-R}^{R} b(\theta) \, d\theta$$

with the Lebesgue measure of the intersection $C \cap (-R, R)$,

$$\mu(R) = \int_{-R}^{R} I_C(x) \, dx,$$

where $I_C(x)$ is the indicator of C, and showing that

$$(31) \qquad\qquad \frac{A(R)}{\mu(R)} \to 1, \qquad \frac{B(R)}{\mu(R)} \to \gamma \qquad \text{as} \quad R \to \infty.$$

This contradicts the fact that by (30),

$$B(R) \leq (\gamma - \epsilon) A(R) \qquad \text{for all } R,$$

and so proves the desired result.

To prove (31), suppose first that $\mu(\infty) < \infty$. Then if ϕ is the standard normal density

$$A(\infty) = \int_{-\infty}^{\infty} d\theta \int_C \phi(x - \theta)\, dx = \int_C dx = \mu(\infty),$$

and analogously $B(\infty) = \gamma\mu(\infty)$, which establishes (31).

When $\mu(\infty) = \infty$, (31) will be proved by showing that

$$(32) \qquad A(R) = \mu(R) + K_1(R), \qquad B(R) = \gamma\mu(R) + K_2(R),$$

where $K_1(R)$ and $K_2(R)$ are bounded. To see (32), note that

$$\mu(R) = \int_{-R}^{R} I_C(x)\, dx = \int_{-R}^{R} I_C(x)\left[\int_{-\infty}^{\infty} \phi(x - \theta)\, d\theta\right] dx$$

$$= \int_{-\infty}^{\infty}\left[\int_{-R}^{R} I_C(x)\phi(x - \theta)\, dx\right] d\theta,$$

while

$$(33) \qquad A(R) = \int_{-R}^{R}\left[\int_{-\infty}^{\infty} I_C(x)\phi(x - \theta)\, dx\right] d\theta.$$

A comparison of each of these double integrals with that over the region $-R < x < R$, $-R < \theta < R$, shows that the difference $A(R) - \mu(R)$ is made up of four integrals, each of which can be seen to be bounded by using the fact that $\int |t|\phi(t)\, dt < \infty$ [Problem 24(ii)]. This completes the proof.

Example 11. Let $X_1, \ldots, X_n$ be independently normally distributed as $N(\xi, \sigma^2)$, and consider the uniformly most accurate equivariant (and unbiased) confidence intervals for ξ given by (28) of Chapter 6.

It was shown by Buehler and Feddersen (1963) and Brown (1967) that in this case there exist positively biased relevant subsets of the form

$$(34) \qquad C: \frac{|\bar{X}|}{S} \leq k.$$

In particular, for confidence level $\gamma = .5$ and $n = 2$, Brown shows that with $C: |\bar{X}|/|X_2 - X_1| \leq \frac{1}{2}(1 + \sqrt{2})$, the conditional level is $> \frac{2}{3}$ for all values of ξ and σ. It follows from Theorem 2 that $\bar{C}$ is negatively biased semirelevant, and Buehler (1959) shows that any set $C^*: S \leq k$ has the same property. These results are intuitively plausible, since the length of the confidence intervals is proportional to S, and one would expect short intervals to cover the true value less often than long ones.

Theorem 2 does not show that $\bar{C}$ is negatively biased relevant, since the probability of the set (34) tends to zero as $\xi/\sigma \to \infty$. It was in fact proved by Robinson (1976) that no negatively biased relevant subset exists in this case.

The calculations for $\bar{C}$ throw some light on the common practice of stating confidence intervals for ξ only when a preliminary test of $H : \xi = 0$ rejects the hypothesis. For a discussion of this practice see Olshen (1973), and Meeks and D'Agostino (1983).

The only type of example still missing is that of a positively biased relevant subset. It was pointed out by Fisher (1956a, b) that the Welch–Aspin solution of the Behrens–Fisher problem (discussed in Chapter 6, Section 6) provides an illustration of this possibility. The following are much simpler examples of both negatively and positively biased relevant subsets.

Example 12. An extreme form of both positively and negatively biased subsets was encountered in Chapter 7, Section 11, where lower and upper confidence bounds $\underline{\Delta} \leq \Delta$ and $\Delta \leq \bar{\Delta}$ were obtained in (98) and (99) for the ratio $\Delta = \sigma_A^2/\sigma^2$ in a model II one-way classification. Since

$$P(\underline{\Delta} \leq \Delta | \underline{\Delta} < 0) = 1 \quad \text{and} \quad P(\Delta \leq \bar{\Delta} | \bar{\Delta} < 0) = 0,$$

the sets $C_1 : \underline{\Delta} < 0$ and $C_2 : \bar{\Delta} < 0$ are relevant subsets with positive and negative bias respectively.

The existence of conditioning sets C for which the conditional coverage probability of level-γ confidence sets is 0 or 1, such as in Example 12 or Problems 27, 28 are an embarrassment to confidence theory, but fortunately they are rare. The significance of more general relevant subsets is less clear,[*] particularly when a number of such subsets are available. Especially awkward in this connection is the possibility [discussed by Buehler (1959)] of the existence of two relevant subsets C and C' with nonempty intersection and opposite bias.

If a conditional confidence level is to be cited for some relevant subset C, it seems appropriate to take account also of the possibility that X may fall into $\bar{C}$ and to state in advance the three confidence coefficients γ, γ_C, and $\gamma_{\bar{C}}$. The (unknown) probabilities $P_\theta(C)$ and $P_\theta(\bar{C})$ should also be considered. These points have been stressed by Kiefer, who has also suggested the extension to a partition of the sample space into more than two sets. For an account of these ideas see Kiefer (1977a, b), Brownie and Kiefer (1977), and Brown (1978).

Kiefer's theory does not consider the choice of conditioning set or statistic. The same question arose in Section 2 with respect to conditioning on ancillaries. The problem is similar to that of the choice of model. The answer depends on the context and purpose of the analysis, and must be determined from case to case.

---

[*]For a discussion of this issue, see Buehler (1959), Robinson (1976, 1979a), and Bondar (1977).

5. PROBLEMS

Section 1

1. Let the experiments $\mathscr{E}$ and $\mathscr{F}$ consist in observing $X: N(\xi, \sigma_0^2)$ and $X: N(\xi, \sigma_1^2)$ respectively ($\sigma_0 < \sigma_1$), and let one of the two experiments be performed, with $P(\mathscr{E}) = P(\mathscr{F}) = \frac{1}{2}$. For testing $H: \xi = 0$ against $\xi = \xi_1$, determine values σ_0, σ_1, ξ_1, and α such that

$$\text{(i)} \quad \alpha_0 < \alpha_1; \qquad \text{(ii)} \quad \alpha_0 > \alpha_1,$$

where the α_i are defined by (9).

2. Under the assumptions of Problem 1, determine the most accurate invariant (under the transformation $X' = -X$) confidence sets $S(X)$ with

$$P\big(\xi \in S(X)|\mathscr{E}\big) + P\big(\xi \in S(X)|\mathscr{F}\big) = 2\gamma.$$

Find examples in which the conditional confidence coefficients γ_0 given $\mathscr{E}$ and γ_1 given $\mathscr{F}$ satisfy

$$\text{(i)} \quad \gamma_0 < \gamma_1; \qquad \text{(ii)} \quad \gamma_0 > \gamma_1.$$

3. The test given by (3), (8), and (9) is most powerful under the stated assumptions.

4. Let $X_1, \ldots, X_n$ be independently distributed, each with probability p or q as $N(\xi, \sigma_0^2)$ or $N(\xi, \sigma_1^2)$.

   (i) If p is unknown, determine the UMP unbiased test of $H: \xi = 0$ against $K: \xi > 0$.

   (ii) Determine the most powerful test of H against the alternative ξ_1 when it is known that $p = \frac{1}{2}$, and show that a UMP unbiased test does not exist in this case.

   (iii) Let α_k ($k = 0, \ldots, n$) be the conditional level of the unconditional most powerful test of part (ii) given that k of the X's came from $N(\xi, \sigma_0^2)$ and $n - k$ from $N(\xi, \sigma_1^2)$. Investigate the possible values $\alpha_0, \alpha_1, \ldots, \alpha_n$.

5. With known probabilities p and q perform either $\mathscr{E}$ or $\mathscr{F}$, with X distributed as $N(\theta, 1)$ under $\mathscr{E}$ or $N(-\theta, 1)$ under $\mathscr{F}$. For testing $H: \theta = 0$ against $\theta > 0$ there exist a UMP unconditional and a UMP conditional level-α test. These coincide and do not depend on the value of p.

6. In the preceding problem, suppose that the densities of X under $\mathscr{E}$ and $\mathscr{F}$ are $\theta e^{-\theta x}$ and $(1/\theta)e^{-x/\theta}$ respectively. Compare the UMP conditional and unconditional tests of $H: \theta = 1$ against $K: \theta > 1$.

Section 2

7. Let X, Y be independently normally distributed as $N(\theta, 1)$, and let

$$V = Y - X$$

and

$$W = \begin{cases} Y - X & \text{if} \quad X + Y > 0, \\ X - Y & \text{if} \quad X + Y \le 0. \end{cases}$$

   (i) Both V and W are ancillary, but neither is a function of the other.

   (ii) (V, W) is not ancillary.

   [Basu (1959).]

8. An experiment with n observations $X_1, \ldots, X_n$ is planned, with each X_i distributed as $N(\theta, 1)$. However, some of the observations do not materialize (for example, some of the subjects die, move away, or turn out to be unsuitable). Let $I_j = 1$ or 0 as X_j is observed or not, and suppose the I_j are independent of the X's and of each other and that $P(I_j = 1) = p$ for all j.

   (i) If p is known, the effective sample size $M = \Sigma I_j$ is ancillary.

   (ii) If p is unknown, there exists a UMP unbiased level-α test of $H: \theta \le 0$ vs. $K: \theta > 0$. Its conditional level (given $M = m$) is $\alpha_m = \alpha$ for all $m = 0, \ldots, n$.

9. Consider n tosses with a biased die, for which the probabilities of $1, \ldots, 6$ points are given by

1	2	3	4	5	6
$\dfrac{1 - \theta}{12}$	$\dfrac{2 - \theta}{12}$	$\dfrac{3 - \theta}{12}$	$\dfrac{1 + \theta}{12}$	$\dfrac{2 + \theta}{12}$	$\dfrac{3 + \theta}{12}$

   and let X_i be the number of tosses showing i points.

   (i) Show that the triple $Z_1 = X_1 + X_5$, $Z_2 = X_2 + X_4$, $Z_3 = X_3 + X_6$ is a maximal ancillary; determine its distribution and the distribution of $X_1, \ldots, X_6$ given $Z_1 = z_1$, $Z_2 = z_2$, $Z_3 = z_3$.

   (ii) Exhibit five other maximal ancillaries.

   [Basu (1964).]

10. In the preceding problem, suppose the probabilities are given by

1	2	3	4	5	6
$\dfrac{1 - \theta}{6}$	$\dfrac{1 - 2\theta}{6}$	$\dfrac{1 - 3\theta}{6}$	$\dfrac{1 + \theta}{6}$	$\dfrac{1 + 2\theta}{6}$	$\dfrac{1 + 3\theta}{6}$

   Exhibit two different maximal ancillaries.

11. Let X be uniformly distributed on $(\theta, \theta + 1)$, $0 < \theta < \infty$, let $[X]$ denote the largest integer $\leq X$, and let $V = X - [X]$.

   (i) The statistic $V(X)$ is uniformly distributed on $(0, 1)$ and is therefore ancillary.

   (ii) The marginal distribution of $[X]$ is given by

   $$[X] = \begin{cases} [\theta] & \text{with probability } 1 - V(\theta), \\ [\theta] + 1 & \text{with probability } V(\theta). \end{cases}$$

   (iii) Conditionally, given that $V = v$, $[X]$ assigns probability 1 to the value $[\theta]$ if $V(\theta) \leq v$ and to the value $[\theta] + 1$ if $V(\theta) > v$.

   [Basu (1964).]

12. Let X, Y have joint density

   $$p(x, y) = 2f(x)f(y)F(\theta xy),$$

   where f is a known probability density symmetric about 0, and F its cumulative distribution function. Then

   (i) $p(x, y)$ is a probability density.

   (ii) X and Y each have marginal density f and are therefore ancillary, but (X, Y) is not.

   (iii) $X \cdot Y$ is a sufficient statistic for θ.

   [Dawid (1977).]

13. A sample of size n is drawn with replacement from a population consisting of N distinct unknown values $\{a_1, \ldots, a_N\}$. The number of distinct values in the sample is ancillary.

14. Assuming the distribution (22) of Chapter 4, Section 9, show that Z is S-ancillary for $p = p_+/(p_+ + p_-)$.

15. In the situation of Example 3, $X + Y$ is binomial if and only if $\Delta = 1$.

16. In the situation of Example 2, the statistic Z remains S-ancillary when the parameter space is $\Omega = \{(\lambda, \mu) : \mu \leq \lambda\}$.

17. Suppose $X = (U, Z)$, the density of X factors into

   $$p_{\theta, \vartheta}(x) = c(\theta, \vartheta) g_\theta(u; z) h_\vartheta(z) k(u, z),$$

   and the parameters θ, ϑ are unrelated. To see that these assumptions are not enough to insure that Z is S-ancillary for θ, consider the joint density

   $$C(\theta, \vartheta) e^{-\frac{1}{2}(u-\theta)^2 - \frac{1}{2}(z-\vartheta)^2} I(u, z),$$

where $I(u, z)$ is the indicator of the set $\{(u, z): u \le z\}$.
[Basu (1978).]

Section 3

18. Verify the density (16) of Example 7.

19. Let the real-valued function f be defined on an open interval.

    (i) If f is logconvex, it is convex.

    (ii) If f is strongly unimodal, it is unimodal.

20. Let $X_1, \ldots, X_m$ and $Y_1, \ldots, Y_n$ be positive, independent random variables distributed with densities $f(x/\sigma)$ and $g(y/\tau)$ respectively. If f and g have monotone likelihood ratios in (x, σ) and (y, τ) respectively, there exists a UMP conditional test of $H: \tau/\sigma \le \Delta_0$ against $\tau/\sigma > \Delta_0$ given the ancillary statistics $U_i = X_i/X_m$ and $V_j = Y_j/Y_n$ ($i = 1, \ldots, m - 1$; $j = 1, \ldots, n - 1$).

21. Let $V_1, \ldots, V_n$ be independently distributed as $N(0, 1)$, and given $V_1 = v_1, \ldots, V_n = v_n$, let X_i ($i = 1, \ldots, n$) be independently distributed as $N(\theta v_i, 1)$.

    (i) There does not exist a UMP test of $H: \theta = 0$ against $K: \theta > 0$.

    (ii) There does exist a UMP conditional test of H against K given the ancillary $(V_1, \ldots, V_n)$.

    [Buehler (1982).]

22. In Example 8,

    (i) the problem remains invariant under G' but not under G_3;

    (ii) the statistic D is ancillary.

Section 4

23. In Example 9, check directly that the set $C = \{x : x \le -k \text{ or } x \ge k\}$ is not a negatively biased semirelevant subset for the confidence intervals $(X - c, X + c)$.

24. (i) Verify the posterior distribution of Θ given x claimed in Example 9.

    (ii) Complete the proof of (32).

25. Let X be a random variable with cumulative distribution function F. If $E|X| < \infty$, then $\int_{-\infty}^{0} F(x)\, dx$ and $\int_{0}^{\infty}[1 - F(x)]\, dx$ are both finite.
    [Apply integration by parts to the two integrals.]

26. Let X have probability density $f(x - \theta)$, and suppose that $E|X| < \infty$. For the confidence intervals $X - c < \theta$ there exist semirelevant but no relevant subsets.
    [Buehler (1959).]

27. Let $X_1, \ldots, X_n$ be independently distributed according to the uniform distribution $U(\theta, \theta + 1)$.

   (i) Uniformly most accurate lower confidence bounds $\underline{\theta}$ for θ at confidence level $1 - \alpha$ exist and are given by

$$\underline{\theta} = \max(X_{(1)} - k, X_{(n)} - 1),$$

where $X_{(1)} = \min(X_1, \ldots, X_n)$, $X_{(n)} = \max(X_1, \ldots, X_n)$, and $(1 - k)^n = \alpha$.

   (ii) The set $C: x_{(n)} - x_{(1)} \geq 1 - k$ is a relevant subset with $P_\theta(\underline{\theta} \leq \theta | C) = 1$ for all θ.

   (iii) Determine the uniformly most accurate conditional lower confidence bounds $\underline{\theta}(v)$ given the ancillary statistic $V = X_{(n)} - X_{(1)} = v$, and compare them with $\underline{\theta}$.

[The conditional distribution of $Y = X_{(1)}$ given $V = v$ is $U(\theta, \theta + 1 - v)$.]
[Pratt (1961), Barnard (1976).]

28.   (i) Under the assumptions of the preceding problem, the uniformly most accurate unbiased (or invariant) confidence intervals for θ at confidence level $1 - \alpha$ are

$$\underline{\theta} = \max(X_{(1)} + d, X_{(n)}) - 1 < \theta < \min(X_{(1)}, X_{(n)} - d) = \bar{\theta},$$

where d is the solution of the equation

$$2d^n = \alpha \qquad \text{if} \quad \alpha < 1/2^{n-1},$$

$$2d^n - (2d - 1)^n = \alpha \qquad \text{if} \quad \alpha > 1/2^{n-1}.$$

   (ii) The sets $C_1: X_{(n)} - X_{(1)} > d$ and $C_2: X_{(n)} - X_{(1)} < 2d - 1$ are relevant subsets with coverage probability

$$P_\theta\left[\underline{\theta} < \theta < \bar{\theta} | C_1\right] = 1 \quad \text{and} \quad P_\theta\left[\underline{\theta} < \theta < \bar{\theta} | C_2\right] = 0.$$

   (iii) Determine the uniformly most accurate unbiased (or invariant) conditional confidence intervals $\underline{\theta}(v) < \theta < \bar{\theta}(v)$ given $V = v$ at confidence level $1 - \alpha$, and compare $\underline{\theta}(v)$, $\bar{\theta}(v)$, and $\bar{\theta}(v) - \underline{\theta}(v)$ with the corresponding unconditional quantities.

[Welch (1939), Pratt (1961), Kiefer (1977a).]

29. Instead of conditioning the confidence sets $\theta \in S(X)$ on a set C, consider a randomized procedure which assigns to each point x a probability $\psi(x)$ and makes the confidence statement $\theta \in S(x)$ with probability $\psi(x)$ when x is observed.*

---

*Randomized and nonrandomized conditioning is interpreted in terms of betting strategies by Buehler (1959) and Pierce (1973).

(i) The randomized procedure can be represented by a nonrandomized conditioning set for the observations (X, U), where U is uniformly distributed on $(0, 1)$ and independent of X, by letting $C = \{(x, u) : u < \psi(x)\}$.

(ii) Extend the definition of relevant and semirelevant subsets to randomized conditioning (without the use of U).

(iii) Let $\theta \in S(X)$ be equivalent to the statement $X \in A(\theta)$. Show that ψ is positively biased semirelevant if and only if the random variables $\psi(X)$ and $I_{A(\theta)}(X)$ are positively correlated, where I_A denotes the indicator of the set A.

30. The nonexistence of (i) semirelevant subsets in Example 9 and (ii) relevant subsets in Example 10 extends to randomized conditioning procedures.

6. REFERENCES

Conditioning on ancillary statistics was introduced by Fisher (1934, 1935, 1936).* The idea was emphasized in Fisher (1956b) and by Cox (1958), who motivated it in terms of mixtures of experiments providing different amounts of information. The consequences of adopting a general principle of conditioning in mixture situations were explored by Birnbaum (1962) and Durbin (1970). Following Fisher's suggestion (1934), Pitman (1938) developed a theory of conditional tests and confidence intervals for location and scale parameters.

The possibility of relevant subsets was pointed out by Fisher (1956a, b). Its implications (in terms of betting procedures) were developed by Buehler (1959), who in particular introduced the distinction between relevant and semirelevant, positively and negatively biased subsets, and proved the nonexistence of relevant subsets in location models. The role of relevant subsets in statistical inference, and their relationship to Bayes and admissibility properties, was discussed by Pierce (1973), Robinson (1976, 1979a, b), and Bondar (1977) among others.

Fisher (1956a, b) introduced the idea of relevant subsets in the context of the Behrens–Fisher problem. As a criticism of the Welch–Aspin solution, he established the existence of negatively biased relevant subsets for that procedure. It was later shown by Robinson (1976) that no such subsets exist for Fisher's preferred solution, the so-called Behrens–Fisher intervals. This fact may be related to the conjecture [supported by substantial numerical evidence in Robinson (1976) but so far unproved] that the unconditional coverage probability of the Behrens–Fisher intervals always exceeds the

---

*Fisher's contributions to this topic are discussed in Savage (1976, pp. 467–469).

nominal level. For a review of these issues, see Wallace (1980) and Robinson (1982).

Barnard, G. A.
(1976). "Conditional inference is not inefficient." *Scand. J. Statist.* 3, 132–134. [Problem 27.]

Barnard, G. A. and Sprott, D. A.
(1971). "A note on Basu's examples of anomalous ancillary statistics." In *Foundations of Statistical Inference* (Godambe and Sprott, eds.), Holt, Rinehart, and Winston, Toronto, 163–176.

Barndorff-Nielsen, O.
(1978). *Information and Exponential Families in Statistical Theory*, Wiley, New York. [Provides a systematic discussion of various concepts of ancillarity with many examples]. (1980). "Conditionality resolutions." *Biometrika* 67, 293–310. (1983). "On a formula for the distribution of the maximum likelihood estimator." *Biometrika* 70, 343–365.

Bartholomew, D. J.
(1967). "Hypothesis testing when the sample size is treated as a random variable" (with discussion). *J. Roy. Statist. Soc.* 29, 53–82.

Bartlett, M. S.
(1940). "A note on the interpretation of quasi-sufficiency." *Biometrika* 31, 391–392. (1956). "Comment on Sir Ronald Fisher's paper: On a test of significance in Pearson's Biometrika Tables No. 11." *J. Roy. Statist. Soc.* (*B*) 18, 295–296.

Basu, D.
(1959). "The family of ancillary statistics." *Sankhyā* (*A*) 21, 247–256. [Problem 7.] (1964). "Recovery of ancillary information." *Sankhyā* (*A*) 26, 3–16. [Problems 9, 11.] (1977). "On the elimination of nuisance parameters." *J. Amer. Statist. Assoc.* 72, 355–366. [A systematic review of various strategies (including the use of ancillaries) for eliminating nuisance parameters.] (1978). "On partial sufficiency: A review." *J. Statist. Planning and Inference* 2, 1–13.

Becker, N. and Gordon, I.
(1983). "On Cox's criterion for discriminating between alternative ancillary statistics." *Int. Statist. Rev.* 51, 89–92.

Berger, J.
(1984). "A review of J. Kiefer's work on conditional frequentist statistics." In *The Collected Works of Jack Kiefer* (Brown, Olkin, and Sacks, eds.), Springer. (1985). "The frequentist viewpoint of conditioning." In *Proc. Berkeley Conf. in Honor of J. Neyman and J. Kiefer* (Le Cam and Olshen, eds.), Wadsworth, Belmont, Calif.

Berger, J. and Wolpert, R.
(1984). *The Likelihood Principle*, IMS Lecture Notes–Monograph Series.

Birnbaum, A.
(1962). "On the foundations of statistical inference" (with discussion). *J. Amer. Statist. Assoc.* 57, 269–326.

Bondar, J. V.
(1977). "A conditional confidence principle." *Ann. Statist.* 5, 881–891.

Brown, L. D.

(1967). "The conditional level of Student's t-test." *Ann. Math. Statist.* **38**, 1068–1071.

(1978). "An extension of Kiefer's theory of conditional confidence procedures." *Ann. Statist.* **6**, 59–71.

Brownie, C. and Kiefer, J.

(1977). "The ideas of conditional confidence in the simplest setting." *Comm. Statist.* **A6**(8), 691–751.

Buehler, R. J.

(1959). "Some validity criteria for statistical inferences." *Ann. Math. Statist.* **30**, 845–863.
[The first systematic treatment of relevant subsets, including Example 9.]

(1982). "Some ancillary statistics and their properties." *J. Amer. Statist. Assoc.* **77**, 581–589.
[A review of the principal examples of ancillaries.]

Buehler, R. J. and Feddersen, A. P.

(1963). "Note on a conditional property of Student's t." *Ann. Math. Statist.* **34**, 1098–1100.

Cox, D. R.

(1958). "Some problems connected with statistical inference." *Ann. Math. Statist.* **29**, 357–372.

(1971). "The choice between ancillary statistics." *J. Roy. Statist. Soc.* (*B*) **33**, 251–255.

Cox, D. R. and Hinkley, D. V.

(1974). *Theoretical Statistics*, Chapman and Hall, London.
[Discusses many of the ancillary examples given here.]

Dawid, A. P.

(1975). "On the concepts of sufficiency and ancillarity in the presence of nuisance parameters." *J. Roy. Statist. Soc.* (*B*) **37**, 248–258.

(1977). "Discussion of Wilkinson: On resolving the controversy in statistical inference." *J. Roy. Statist. Soc.* **39**, 151–152.
[Problem 12.]

Durbin, J.

(1970). "On Birnbaum's theorem on the relation between sufficiency, conditionality, and likelihood." *J. Amer. Statist. Assoc.* **65**, 395–398.

Fisher, R. A.

(1925). "Theory of statistical estimation." *Proc. Cambridge Phil. Soc.* **22**, 700–725.
[First use of the term "ancillary".]

(1934). "Two new properties of mathematical likelihood." *Proc. Roy. Soc.* (*A*) **144**, 285–307.
[Introduces the idea of conditioning on ancillary statistics and applies it to the estimation of location parameters.]

(1935). "The logic of inductive inference" (with discussion). *J. Roy. Statist. Soc.* **98**, 39–82.

(1936). "Uncertain inference." *Proc. Amer. Acad. Arts and Sci.* **71**, 245–258.

(1956a). "On a test of significance in Pearson's Biometrika tables (No. 11)." *J. Roy. Statist. Soc.* (*B*) **18**, 56–60. (See also the discussion of this paper by Neyman, Bartlett, and Welch in the same volume, pp. 288–302).
[Exhibits a negatively biased relevant subset for the Welch–Aspin solution of the Behrens–Fisher problem.]

(1956b). *Statistical Methods and Scientific Inference* (3rd ed., 1973), Oliver and Boyd, Edinburgh.
[Contains Fisher's last comprehensive statement of his views on many topics, including ancillarity and the Behrens–Fisher problem.]

Frisén, M.

(1980). "Consequences of the use of conditional inference in the analysis of a correlated contingency table." *Biometrika* **67**, 23–30.

Hájek, J.

(1967). "On basic concepts of statistics." In *Proc. Fifth Berkeley Symp. Math. Statist. and Probab.*, Univ. of Calif. Press, Berkeley.

Hinkley, D. V.

(1977). "Conditional inference about a normal mean with known coefficient of variation." *Biometrika*, **64**, 105–108.

Huang, J. S. and Ghosh, M.

(1982). "A note on strong unimodality of order statistics." *J. Amer. Statist. Assoc.* **77**, 929–930.

Kalbfleisch, J. D.

(1975). "Sufficiency and conditionality" (with discussion). *Biometrika* **62**, 251–259.

Kappenman, R. F.

(1975). "Conditional confidence intervals for the double exponential distribution parameters." *Technometrics* **17**, 233–235.

Kiefer, J.

(1977a). "Conditional confidence statements and confidence estimators" (with discussion). *J. Amer. Statist. Assoc.* **72**, 789–827.
[The key paper in Kiefer's proposed conditional confidence approach.]
(1977b). "Conditional confidence and estimated confidence in multi-decision problems (with applications to selections and ranking)." *Multiv. Anal.* **IV**, 143–158.

Lawless, J. F.

(1972). "Conditional confidence interval procedures for the location and scale parameters of the Cauchy and logistic distributions." *Biometrika* **59**, 377–386.
(1973). "Conditional versus unconditional confidence intervals for the parameters of the Weibull distribution." *J. Amer. Statist. Assoc.* **68**, 655–669.
(1978). "Confidence interval estimation for the Weibull and extreme value distributions." *Technometrics* **20**, 355–368.

Le Cam, L.

(1958). "Les propiétés asymptotiques des solutions de Bayes." *Publ. Inst. Statist. Univ. Paris.* **VII** (3–4), 17–35.

Liang, K. Y.

(1984). "The asymptotic efficiency of conditional likelihood methods." *Biometrika* **71**, 305–313.

Marden, J.

(1983). "Admissibility of invariant tests in the general multivariate analysis of variance problem." *Ann. Statist.* **11**, 1086–1099.

Marden, J. and Perlman, M. D.

(1980). "Invariant tests for means with covariates." *Ann. Statist.* **8**, 25–63.

Meeks, S. L. and D'Agostino, R. B.

(1983). "A note on the use of confidence limits following rejection of a null hypothesis." *Amer. Statist.* **37**, 134–136.

Olshen, R. A.

(1973). "The conditional level of the *F*-test." *J. Amer. Statist. Assoc.* **68**, 692–698.

Pierce, D. A.

(1973). "On some difficulties in a frequency theory of inference." *Ann. Statist.* **1**, 241–250.

Pitman, E. J. G.

(1938). "The estimation of the location and scale parameters of a continuous population of any given form." *Biometrika* **30**, 391–421.

Plackett, R. L.

(1977). "The marginal totals of a 2 × 2 table." *Biometrika* **64**, 37–42.

[Discusses the fact that the marginals of a 2 × 2 table supply some, but only little, information concerning the odds ratio. See also Barndorff-Nielsen (1978), Example 10.8.]

Pratt, J. W.

(1961). "Review of Testing Statistical Hypotheses by E. L. Lehmann." *J. Amer. Statist. Assoc.* **56**, 163–167.

[Problems 27, 28.]

(1981). "Concavity of the log likelihood." *J. Amer. Statist. Assoc.* **76**, 103–106.

Robinson, G. K.

(1975). "Some counterexamples to the theory of confidence intervals." *Biometrika* **62**, 155–161. [Correction (1977), *Biometrika* **64**, 655.]

(1976). "Properties of Student's *t* and of the Behrens–Fisher solution to the two means problem." *Ann. Statist.* **4**, 963–971. [Correction (1982), *Ann. Statist.* **10**, 321.]

(1979a). "Conditional properties of statistical procedures." *Ann. Statist.* **7**, 742–755.

(1979b). "Conditional properties of statistical procedures for location and scale parameters." *Ann. Statist.* **7**, 756–771.

[Basic results concerning the existence of relevant and semirelevant subsets for location and scale parameters, including Example 9.]

(1982). "Behrens–Fisher problem." In *Encyclopedia of the Statistical Sciences*, Vol. 1, Wiley, New York.

Sandved, E.

(1967). "A principle for conditioning on an ancillary statistic." *Skand. Aktuar. Tidskr.* **50**, 39–47.

(1972). "Ancillary statistics in models without and with nuisance parameters." *Skand. Aktuar. Tidskr.* **55**, 81–91.

Savage, L. J.

(1976). "On rereading R. A. Fisher" (with discussion). *Ann. Statist.* **4**, 441–500.

Sprott, D. A.

(1975). "Marginal and conditional sufficiency." *Biometrika* **62**, 599–605.

Sverdrup, E.

(1966). "The present state of the decision theory and the Neyman–Pearson theory." *Rev. Int. Statist. Inst.* **34**. 309–333.

Unni, K.

(1978). *The Theory of Estimation in Algebraic and Analytic Exponential Families with Applications to Variance Components Models*, unpublished Ph.D. Thesis, Indian Statistical Institute.

Wallace, D.

(1959). "Conditional confidence level properties." *Ann. Math. Statist.* **30**, 864–876.

Wallace, D. L.

(1980). "The Behrens–Fisher and Fieller–Creasy problems." In *R. A. Fisher: An Appreciation* (Fienberg and Hinkley, eds.) Springer, New York, pp. 119–147.

Welch, B. L.

(1939). "On confidence limits and sufficiency with particular reference to parameters of location." *Ann. Math. Statist.* **10**, 58–69.

(1956). "Note on some criticisms made by Sir Ronald Fisher." *J. Roy. Statist. Soc.* **18**, 297–302.

Appendix

1. EQUIVALENCE RELATIONS; GROUPS

A relation: $x \sim y$ among the points of a space $\mathscr{X}$ is an equivalence relation if it is reflexive, symmetric, and transitive, that is, if

(i) $x \sim x$ for all $x \in \mathscr{X}$;

(ii) $x \sim y$ implies $y \sim x$;

(iii) $x \sim y$, $y \sim z$ implies $x \sim z$.

Example 1. Consider a class of statistical decision procedures as a space, of which the individual procedures are the points. Then the relation defined by $\delta \sim \delta'$ if the procedures δ and δ' have the same risk function is an equivalence relation. As another example consider all real-valued functions defined over the real line as points of a space. Then $f \sim g$ if $f(x) = g(x)$ a.e. is an equivalence relation.

Given an equivalence relation, let D_x denote the set of points of the space that are equivalent to x. Then $D_x = D_y$ if $x \sim y$, and $D_x \cap D_y = 0$ otherwise. Since by (i) each point of the space lies in at least one of the sets D_x, it follows that these sets, the *equivalence classes* defined by the relation $\sim$, constitute a partition of the space.

A set G of elements is called a *group* if it satisfies the following conditions.

(i) There is defined an operation, group multiplication, which with any two elements $a, b \in G$ associates an element c of G. The element c is called the product of a and b and is denoted by ab.

(ii) Group multiplication obeys the associative law

$$(ab)c = a(bc).$$

(iii) There exists an element $e \in G$, called the *identity*, such that

$$ae = ea = a \qquad \text{for all} \quad a \in G.$$

(iv)   For each element $a \in G$, there exists an element $a^{-1} \in G$, its
inverse, such that

$$aa^{-1} = a^{-1}a = e.$$

Both the identity element and the inverse a^{-1} of any element a can
be shown to be unique.

Example 2.   The set of all $n \times n$ orthogonal matrices constitutes a group if
matrix multiplication and inverse are taken as group multiplication and inverse
respectively, and if the identity matrix is taken as the identity element of the group.
With the same specification of the group operations, the class of all nonsingular
$n \times n$ matrices also forms a group. On the other hand, the class of all $n \times n$
matrices fails to satisfy condition (iv).

If the elements of G are transformations of some space onto itself, with
the group product ba defined as the result of applying first transformation a
and following it by b, then G is called a *transformation group*. Assumption
(ii) is then satisfied automatically. For any transformation group defined
over a space $\mathscr{X}$ the relation between points of X given by

$$x \sim y \quad \text{if} \quad \text{there exists } a \in G \text{ such that } y = ax$$

is an equivalence relation. That it satisfies conditions (i), (ii), and (iii)
required of an equivalence follows respectively from the defining properties
(iii), (iv), and (i) of a group.

Let $\mathscr{C}$ be any class of $1:1$ transformations of a space, and let G be the
class of all finite products $a_1^{\pm 1} a_2^{\pm 1} \ldots a_m^{\pm 1}$, with $a_1, \ldots, a_m \in \mathscr{C}$, $m =$
$1, 2, \ldots$, where each of the exponents can be $+1$ or -1 and where the
elements $a_1, a_2, \ldots$ need not be distinct. Then it is easily checked that G is
a group, and is in fact the smallest group containing $\mathscr{C}$.

2. CONVERGENCE OF DISTRIBUTIONS

When studying convergence properties of functions it is frequently convenient to consider a class of functions as a realization of an abstract space $\mathscr{F}$
of points f in which convergence of a sequence f_n to a limit f, denoted by
$f_n \to f$, has been defined.

Example 3.   Let μ be a measure over a measurable space $(\mathscr{X}, \mathscr{A})$.

(i) Let $\mathscr{F}$ be the class of integrable functions. Then f_n converges to f *in the
mean* if*

$$(1) \qquad\qquad\qquad \int |f_n - f| \, d\mu \to 0.$$

---

*Here and in the examples that follow, the limit f is not unique. More specifically, if
$f_n \to f$, then $f_n \to g$ if and only if $f = g$ (a.e. μ). Putting $f \sim g$ when $f = g$ (a.e. μ),
uniqueness can be obtained by working with the resulting equivalence classes of functions
rather than with the functions themselves.

(ii) Let $\mathscr{F}$ be a uniformly bounded class of measurable functions. The sequence f_n is said to converge to f *weakly* if

$$\text{(2)} \qquad \int f_n p \, d\mu \to \int fp \, d\mu.$$

for all functions p that are integrable μ.

(iii) Let $\mathscr{F}$ be the class of measurable functions. Then f_n converges to f *pointwise* if

$$\text{(3)} \qquad f_n(x) \to f(x) \qquad \text{a.e. } \mu.$$

A subset $\mathscr{F}_0$ of $\mathscr{F}$ is *dense* in $\mathscr{F}$ if, given any $f \in \mathscr{F}$, there exists a sequence in $\mathscr{F}_0$ having f as its limit point. A space $\mathscr{F}$ is *separable* if there exists a countable dense subset of $\mathscr{F}$. A space $\mathscr{F}$ such that every sequence has a convergent subsequence whose limit point is in $\mathscr{F}$ is *compact*.* A space $\mathscr{F}$ is a *metric space* if for every pair of points f, g in $\mathscr{F}$ there is defined a distance $d(f, g) \geq 0$ such that

(i) $d(f, g) = 0$ if and only if $f = g$;

(ii) $d(f, g) = d(g, f)$;

(iii) $d(f, g) + d(g, h) \geq d(f, h)$ for all f, g, h.

The space is *pseudometric* if (i) is replaced by

(i′) $d(f, f) = 0$ for all $f \in \mathscr{F}$.

A pseudometric space can be converted into a metric space by introducing the equivalence relation $f \sim g$ if $d(f, g) = 0$. The equivalence classes $F, G, \ldots$ then constitute a metric space with respect to the distance $D(F, G) = d(f, g)$ where $f \in F$, $g \in G$.

In any pseudometric space a natural convergence definition is obtained by putting $f_n \to f$ if $d(f_n, f) \to 0$.

Example 4. The space of integrable functions of Example 3(i) becomes a pseudometric space if we put

$$d(f, g) = \int |f - g| \, d\mu$$

and the induced convergence definition is that given by (1).

Example 5. Let $\mathscr{P}$ be a family of probability distributions over $(\mathscr{X}, \mathscr{A})$. Then $\mathscr{P}$ is a metric space with respect to the metric

$$\text{(4)} \qquad d(P, Q) = \sup_{A \in \mathscr{A}} |P(A) - Q(A)|.$$

---

*The term *compactness* is more commonly used for an alternative concept, which coincides with the one given here in metric spaces. The distinguishing term *sequential compactness* is then sometimes given to the notion defined here.

Lemma 1. *If $\mathscr{F}$ is a separable pseudometric space, then every subset of $\mathscr{F}$ is also separable.*

Proof.   By assumption there exists a dense countable subset $\{f_n\}$ of $\mathscr{F}$. Let

$$S_{m,n} = \left\{ f : d(f,f_n) < \frac{1}{m} \right\},$$

and let A be any subset of $\mathscr{F}$. Select one element from each of the intersections $A \cap S_{m,n}$ that is nonempty, and denote this countable collection of elements by A_0. If a is any element of A and m any positive integer, there exists an element f_{n_m} such that $d(a,f_{n_m}) < 1/m$. Therefore a belongs to S_{m,n_m}, the intersection $A \cap S_{m,n_m}$ is nonempty, and there exists therefore an element of A_0 whose distance to a is $< 2/m$. This shows that A_0 is dense in A, and hence that A is separable.

Lemma 2. *A sequence f_n of integrable functions converges to f in the mean if and only if*

$$(5) \qquad \int_A f_n \, d\mu \to \int_A f \, d\mu \qquad \text{uniformly for} \quad A \in \mathscr{A}.$$

Proof.   That (1) implies (5) is obvious, since for all $A \in \mathscr{A}$

$$\left| \int_A f_n \, d\mu - \int_A f \, d\mu \right| \le \int |f_n - f| \, d\mu.$$

Conversely, suppose that (5) holds, and denote by A_n and A_n' the set of points x for which $f_n(x) > f(x)$ and $f_n(x) < f(x)$ respectively. Then

$$\int |f_n - f| \, d\mu = \int_{A_n} (f_n - f) \, d\mu - \int_{A_n'} (f_n - f) \, d\mu \to 0.$$

Lemma 3. *A sequence f_n of uniformly bounded functions converges to a bounded function f weakly if and only if*

$$(6) \qquad \int_A f_n \, d\mu \to \int_A f \, d\mu \qquad \text{for all } A \text{ with } \mu(A) < \infty.$$

Proof.   That weak convergence implies (6) is seen by taking for p in (2) the indicator function of a set A, which is integrable if $\mu(A) < \infty$. Conversely (6) implies that (2) holds if p is any simple function $s = \sum a_i I_{A_i}$ with all the $\mu(A_i) < \infty$. Given any integrable function p, there exists, by the definition of the integral, such a simple function s for which $\int |p - s| \, d\mu < \epsilon/3M$, where M is a bound on the $|f|$'s. We then have

$$\left| \int (f_n - f) p \, d\mu \right| \le \left| \int f_n (p - s) \, d\mu \right| + \left| \int f(s - p) \, d\mu \right| + \left| \int (f_n - f) s \, d\mu \right|.$$

The first two terms on the right-hand side are $< \epsilon/3$, and the third term tends to zero as n tends to infinity. Thus the left-hand side is $< \epsilon$ for n sufficiently large, as was to be proved.

Lemma 4.*   *Let f and f_n, $n = 1, 2, \ldots$, be nonnegative integrable functions with*

$$\int f \, d\mu = \int f_n \, d\mu = 1.$$

Then pointwise convergence of f_n to f implies that $f_n \to f$ in the mean.

Proof.   If $g_n = f_n - f$, then $g_n \ge -f$, and the negative part $g_n^- = \max(-g_n, 0)$ satisfies $|g_n^-| \le f$. Since $g_n(x) \to 0$ (a.e. μ), it follows from Theorem 1(ii) of Chapter 2 that $\int g_n^- \, d\mu \to 0$, and $\int g_n^+ \, d\mu$ then also tends to zero, since $\int g_n \, d\mu = 0$. Therefore $\int |g_n| \, d\mu = \int (g_n^+ + g_n^-) \, d\mu \to 0$, as was to be proved.

Let P and P_n, $n = 1, 2, \ldots$, be probability distributions over $(\mathscr{X}, \mathscr{A})$ with densities p_n and p with respect to μ. Consider the convergence definitions

(a) $p_n \to p$ (a.e. μ);
(b) $\int |p_n - p| \, d\mu \to 0$;
(c) $\int g p_n \, d\mu \to \int g p \, d\mu$ for all bounded measurable g;

and

(b') $P_n(A) \to P(A)$ uniformly for all $A \in \mathscr{A}$;
(c') $P_n(A) \to P(A)$ for all $A \in \mathscr{A}$.

Then Lemmas 2 and 4 together with a slight modification of Lemma 3 show that (a) implies (b) and (b) implies (c), and that (b) is equivalent to (b') and (c) to (c'). It can further be shown that neither (a) and (b) nor (b) and (c) are equivalent.[†]

*Scheffé (1947).
[†] Robbins, (1948).

3. DOMINATED FAMILIES OF DISTRIBUTIONS

Let $\mathcal{M}$ be a family of measures defined over a measurable space $(\mathcal{X}, \mathcal{A})$. Then $\mathcal{M}$ is said to be *dominated* by a σ-finite measure μ defined over $(\mathcal{X}, \mathcal{A})$ if each member of $\mathcal{M}$ is absolutely continuous with respect to μ. The family $\mathcal{M}$ is said to be *dominated* if there exists a σ-finite measure dominating it. Actually, if $\mathcal{M}$ is dominated there always exists a finite dominating measure. For suppose that $\mathcal{M}$ is dominated by μ and that $\mathcal{X} = \bigcup A_i$ with $\mu(A_i)$ finite for all i. If the sets A_i are taken to be mutually exclusive, the measure $\nu(A) = \Sigma \mu(A \cap A_i)/2^i \mu(A_i)$ also dominates $\mathcal{M}$ and is finite.

Theorem 1.* *A family $\mathcal{P}$ of probability measures over a Euclidean space $(\mathcal{X}, \mathcal{A})$ is dominated if and only if it is separable with respect to the metric (4) or equivalently with respect to the convergence definition*

$$P_n \to P \qquad \text{if} \quad P_n(A) \to P(A) \quad \text{uniformly for } A \in \mathcal{A}.$$

Proof. Suppose first that $\mathcal{P}$ is separable and that the sequence $\{P_n\}$ is dense in $\mathcal{P}$, and let $\mu = \Sigma P_n/2^n$. Then $\mu(A) = 0$ implies $P_n(A) = 0$ for all n, and hence $P(A) = 0$ for all $P \in \mathcal{P}$. Conversely suppose that $\mathcal{P}$ is dominated by a measure μ, which without loss of generality can be assumed to be finite. Then we must show that the set of integrable functions $dP/d\mu$ is separable with respect to the convergence definition (5) or, because of Lemma 2, with respect to convergence in the mean. It follows from Lemma 1 that it suffices to prove this separability for the class $\mathcal{F}$ of all functions f that are integrable μ. Since by the definition of the integral every integrable function can be approximated in the mean by simple functions, it is enough to prove this for the case that $\mathcal{F}$ is the class of all simple integrable functions. Any simple function can be approximated in the mean by simple functions taking on only rational values, so that it is sufficient to prove separability of the class of functions $\Sigma r_i I_{A_i}$ where the r's are rational and the A's are Borel sets, with finite μ-measure since the f's are integrable. It is therefore finally enough to take for $\mathcal{F}$ the class of functions I_A, which are indicator functions of Borel sets with finite measure. However, any such set can be approximated by finite unions of disjoint rectangles with rational end points. The class of all such unions is denumerable, and the associated indicator functions will therefore serve as the required countable dense subset of $\mathcal{F}$.

*Berger, (1951).

An examination of the proof shows that the Euclidean nature of the space $(\mathscr{X}, \mathscr{A})$ was used only to establish the existence of a countable number of sets $A_i \in \mathscr{A}$ such that for any $A \in \mathscr{A}$ with finite measure there exists a subsequence A_{i_j} with $\mu(A_{i_j}) \to \mu(A)$. This property holds quite generally for any σ-field $\mathscr{A}$ which has a *countable number of generators*, that is, for which there exists a countable number of sets B_i such that $\mathscr{A}$ is the smallest σ-field containing the B_i.[†] It follows that Theorem 1 holds for any σ-field with this property. Statistical applications of such σ-fields occur in sequential analysis, where the sample space $\mathscr{X}$ is the union $\mathscr{X} = \bigcup_i \mathscr{X}_i$ of Borel subsets $\mathscr{X}_i$ of i-dimensional Euclidean space. In these problems, $\mathscr{X}_i$ is the set of points $(x_1, \ldots, x_i)$ for which exactly i observations are taken. If $\mathscr{A}_i$ is the σ-field of Borel subsets of $\mathscr{X}_i$, one can take for $\mathscr{A}$ the σ-field generated by the $\mathscr{A}_i$, and since each $\mathscr{A}_i$ possesses a countable number of generators, so does $\mathscr{A}$.

If $\mathscr{A}$ does not possess a countable number of generators, a somewhat weaker conclusion can be asserted. Two families of measures $\mathscr{M}$ and $\mathscr{N}$ are *equivalent* if $\mu(A) = 0$ for all $\mu \in \mathscr{M}$ implies $\nu(A) = 0$ for all $\nu \in \mathscr{N}$ and vice versa.

Theorem 2.[‡] *A family $\mathscr{P}$ of probability measures is dominated by a σ-finite measure if and only if $\mathscr{P}$ has a countable equivalent subset.*

Proof. Suppose first that $\mathscr{P}$ has a countable equivalent subset $\{P_1, P_2, \ldots\}$. Then $\mathscr{P}$ is dominated by $\mu = \Sigma P_n/2^n$. Conversely, let $\mathscr{P}$ be dominated by a σ-finite measure μ, which without loss of generality can be assumed to be finite. Let $\mathscr{Q}$ be the class of all probability measures Q of the form $\Sigma c_i P_i$, where $P_i \in \mathscr{P}$, the c's are positive, and $\Sigma c_i = 1$. The class $\mathscr{Q}$ is also dominated by μ, and we denote by q a fixed version of the density $dQ/d\mu$. We shall prove the fact, equivalent to the theorem, that there exists Q_0 in $\mathscr{Q}$ such that $Q_0(A) = 0$ implies $Q(\mathscr{A}) = 0$ for all $Q \in \mathscr{Q}$.

Consider the class $\mathscr{C}$ of sets C in $\mathscr{A}$ for which there exists $Q \in \mathscr{Q}$ such that $q(x) > 0$ a.e. μ on C and $Q(C) > 0$. Let $\mu(C_i)$ tend to $\sup_{\mathscr{C}} \mu(C)$, let $q_i(x) > 0$ a.e. on C_i, and denote the union of the C_i by C_0. Then $q_0^*(x) = \Sigma c_i q_i(x)$ agrees a.e. with the density of $Q_0 = \Sigma c_i Q_i$ and is positive a.e. on C_0, so that $C_0 \in \mathscr{C}$. Suppose now that $Q_0(A) = 0$, let Q be any other member of $\mathscr{Q}$, and let $C = \{x: q(x) > 0\}$. Then $Q_0(A \cap C_0) = 0$, and therefore $\mu(A \cap C_0) = 0$ and $Q(A \cap C_0) = 0$. Also $Q(A \cap \tilde{C}_0 \cap \tilde{C}) = 0$. Finally, $Q(A \cap \tilde{C}_0 \cap C) > 0$ would lead to $\mu(C_0 \cup [A \cap \tilde{C}_0 \cap C]) > \mu(C_0)$ and hence to a contradiction of the relation $\mu(C_0) = \sup_{\mathscr{C}} \mu(C)$, since $A \cap \tilde{C}_0 \cap C$ and therefore $C_0 \cup [A \cap \tilde{C}_0 \cap C]$ belongs to $\mathscr{C}$.

[†] A proof of this is given for example by Halmos (1974, Theorem B of Section 40).
[‡] Halmos and Savage (1948).

4. THE WEAK COMPACTNESS THEOREM

The following theorem forms the basis for proving the existence of most powerful tests, most stringent tests, and so on.

Theorem 3.[†]   (*Weak compactness theorem.*)   *Let μ be a σ-finite measure over a Euclidean space, or more generally over any measurable space $(\mathfrak{X}, \mathscr{A})$ for which $\mathscr{A}$ has a countable number of generators. Then the set of measurable functions ϕ with $0 \le \phi \le 1$ is compact with respect to the weak convergence* (2).

Proof.   Given any sequence $\{\phi_n\}$, we must prove the existence of a subsequence $\{\phi_{n_i}\}$ and a function ϕ such that

$$\lim \int \phi_{n_i} p \, d\mu = \int \phi p \, d\mu$$

for all integrable p. If μ^* is a finite measure equivalent to μ, then p^* is integrable μ^* if and only if $p = (d\mu^*/d\mu)p^*$ is integrable μ, and $\int \phi p \, d\mu = \int \phi p^* \, d\mu^*$ for all ϕ. We may therefore assume without loss of generality that μ is finite. Let $\{p_n\}$ be a sequence of p's which is dense in the p's with respect to convergence in the mean. The existence of such a sequence is guaranteed by Theorem 1 and the remark following it. If

$$\Phi_n(p) = \int \phi_n p \, d\mu,$$

the sequence $\Phi_n(p)$ is bounded for each p. A subsequence Φ_{n_k} can be extracted such that $\Phi_{n_k}(p_m)$ converges for each p_m by the following diagonal process. Consider first the sequence of numbers $\{\Phi_n(p_1)\}$ which possesses a convergent subsequence $\Phi_{n_1'}(p_1), \Phi_{n_2'}(p_1), \ldots$. Next the sequence $\Phi_{n_1'}(p_2), \Phi_{n_2'}(p_2), \ldots$ has a convergent subsequence $\Phi_{n_1''}(p_2), \Phi_{n_2''}(p_2), \ldots$. Continuing in this way, let $n_1 = n_1'$, $n_2 = n_2''$, $n_3 = n_3'''$, ... Then $n_1 < n_2 < \ldots$, and the sequence $\{\Phi_{n_i}\}$ converges for each p_m. It follows from the inequality

$$\left| \int (\phi_{n_j} - \phi_{n_i}) p \, d\mu \right| \le \left| \int (\phi_{n_j} - \phi_{n_i}) p_m \, d\mu \right| + 2 \int |p - p_m| \, d\mu$$

that $\Phi_{n_i}(p)$ converges for all p. Denote its limit by $\Phi(p)$, and define a set

[†] Banach (1932). The theorem is valid even without the assumption of a countable number of generators; see Nölle and Plachky (1967), and Aloaglu's theorem, given for example in Royden (1968, Chapter 10, Theorem 17).

function Φ^* over $\mathscr{A}$ by putting

$$\Phi^*(A) = \Phi(I_A).$$

Then Φ^* is nonnegative and bounded, since for all A, $\Phi^*(A) \leq \mu(A)$. To see that it is also countably additive let $A = \bigcup A_k$ where the A_k are disjoint. Then $\Phi^*(A) = \lim \Phi^*_{n_i}(\bigcup A_k)$ and

$$\left| \int_{\bigcup A_k} \phi_{n_i} \, d\mu - \sum \Phi^*(A_k) \right| \leq \left| \int_{\bigcup_{k=1}^m A_k} \phi_{n_i} \, d\mu - \sum_{k=1}^m \Phi^*(A_k) \right|$$

$$+ \left| \int_{\bigcup_{k=m+1}^\infty A_k} \phi_{n_i} \, d\mu - \sum_{k=m+1}^\infty \Phi^*(A_k) \right|.$$

Here the second term is to be taken as zero in the case of a finite sum $A = \bigcup_{k=1}^m A_k$, and otherwise does not exceed $2\mu(\bigcup_{k=m+1}^\infty A_k)$, which can be made arbitrarily small by taking m sufficiently large. For any fixed m the first term tends to zero as i tends to infinity. Thus Φ^* is a finite measure over $(\mathscr{X}, \mathscr{A})$. It is furthermore absolutely continuous with respect to μ, since $\mu(A) = 0$ implies $\Phi_{n_i}(I_A) = 0$ for all i, and therefore $\Phi(I_A) = \Phi^*(A) = 0$. We can now apply the Radon–Nikodym theorem to get

$$\Phi^*(A) = \int_A \phi \, d\mu \qquad \text{for all } A,$$

with $0 \leq \phi \leq 1$. We then have

$$\int_A \phi_{n_i} \, d\mu \to \int_A \phi \, d\mu \qquad \text{for all } A,$$

and weak convergence of the ϕ_{n_i} to ϕ follows from Lemma 3.

5. REFERENCES

Banach, S.
(1932). *Théorie des Operations Linéaires*, Funduszu Kultury Narodowej, Warszawa.

Berger, A.
(1951). "Remark on separable spaces of probability measures." *Ann. Math. Statist.* **22**, 119–120.

Halmos, P.
(1974). *Measure Theory*, Springer, New York.

Halmos, P. and Savage, L. J.
(1948). "Application of the Radon–Nikodym theorem to the theory of sufficient statistics."
Ann. Math. Statist. **20**, 225–241.

Nölle, G. and Plachky, D.
(1967). "Zur schwachen Folgenkompaktheit von Testfunktionen." *Z. Wahrsch. und verw. Geb.* **8**, 182–184.

Robbins, H.
(1948). "Convergence of distributions." *Ann. Math. Statist.* **19**, 72–76.

Royden, H. L.
(1968). *Real Analysis*, 2nd ed., Macmillan, New York.

Scheffé, H.
(1947). "A useful convergence theorem for probability distribution functions." *Ann. Math. Statist.* **18**, 434–438.

Author Index

Aaberge, R., 360
Adyanthaya, N.K., 279
Agresti, A., 165, 181
Aiyar, R.J., 351, 358, 403, 444
Albers, W., 213, 274, 351, 358, 403, 444
Albert, A., 375, 444
Alf, E.F., Jr., 479, 502
Andersen, S.L., 273, 275
Anderson, T.W., 176, 181, 213, 274, 358,
  462, 465, 471, 497, 498
Andersson, S., 289, 358
Anscombe, F., 433, 444
Antille, A., 326, 358
Arbuthnot, J., 126, 127
Armsen, P., 155, 181
Arnold, S.F., 313, 357, 358, 388, 396, 444,
  462, 465, 491, 498
Arrow, K., 70, 127
Arthur, K.H., 427, 445
Arvesen, J.N., 422, 444
Awan, H.M., 460, 503

Bahadur, R.R., 64, 66, 152, 182, 191, 273,
  274
Bain, L.J., 272, 274, 276
Balakrishnan, N., 207, 281
Banach, S., 576, 577
Barankin, E.W., 57, 66
Bar-Lev, S.K., 272, 274
Barlow, R.E., 380, 444, 460, 498
Barnard, G.A., 156, 182, 358, 563, 565
Barndorff–Nielsen, O., 57, 66, 67, 122,
  127, 546, 552, 565
Barnett, V., 5, 29, 127
Bartholomew, D.J., 444, 498

Bartlett, M.S., 125, 127, 182, 445, 498, 501,
  565, 566
Basu, D., 122, 127, 191, 274, 543, 545, 546,
  560, 561, 562, 565
Bauer, P., 462, 499
Becker, N., 545, 565
Bednarski, T., 515, 536
Bell, C.B., 144, 182, 274, 316, 358
Benjamini, Y., 206, 274
Bennett, B., 155, 183
Bennett, B.M., 225, 274
Beran, R., 251, 274, 322, 358
Berger, A., 505, 536, 574, 577
Berger, J.O., x, xi, 15, 18, 29, 125, 127,
  227, 274, 519, 536, 548, 565
Berk, R.H., 290, 298, 316, 358
Berkson, J., 159
Bernoulli, D., 126, 127
Bhapkar, V.P., 166, 187
Bhattacharya, P.K., 326, 358
Bickel, P., x, 10, 29, 162, 182, 236, 274,
  298, 316, 322, 358, 433, 445
Billingsley, P., 50, 66, 67, 143, 176, 178,
  182, 213, 274, 336, 358, 405, 445
Birch, M.W., 166, 182
Birnbaum, A., 115, 127, 154, 182, 358,
  359, 548, 565
Birnbaum, Z.W., 78, 127, 336, 359
Bishop, Y.M.M., 165, 182, 487, 496, 499
Blackwell, D., 15, 22, 29, 48, 67, 87, 113,
  127, 144, 182
Blair, R.C., 321, 359
Bloomfield, P., 409, 417, 452
Blyth, C.R., x, xiii, 5, 29, 94, 128, 221, 275
Bohrer, R., 130, 417, 445

579

Bondar, J.V., 359, 522, 536, 558, 564, 565
Bondessen, L., 12, 29
Boos, D.D., 326, 359
Boschloo, R.D., 155, 182
Bose, R.C., 498, 502
Bowker, A., 495, 499, 500
Box, G., 208, 213, 227, 273, 275, 396, 427, 433, 445, 465, 499
Box, J.F., 29
Brain, C.W., 355, 359
Breiman, L., 144, 182
Bremner, J.M., 444, 498
Broemeling, L.D., 427, 445
Bross, I.D.J., 155, 182
Brown, K.G., 427, 445
Brown, L.D., 57, 67, 82, 86, 128, 130, 140, 182, 311, 359, 429, 445, 557, 558, 566
Brown, M.B., 379, 434, 445
Brownie, C., 558, 566
Brunk, H.D., 444, 498
Buehler, R., 128, 230, 272, 275, 278, 557, 558, 562, 563, 564, 566
Burkholder, D.L., 64, 67

Casella, G., 525, 537
Chalmers, T.C., 129
Chambers, E.A., 165, 182
Chapman, D.G., 78, 127
Chapman, J.W., 182
Chen, H.J., 182, 355, 363
Chernoff, H., 8, 30, 128, 304, 359, 486, 499
Chhikara, R.S., 124, 128, 272, 275, 276
Chmielewski, M.A., 440, 445
Chou, Y.-M., 427, 445
Cima, J.A., 417, 445
Clinch, J.C., 379, 445
Cochran, W.G., 210, 275, 445, 499
Cohen, A., 82, 128, 166, 170, 182, 313, 359, 441, 446, 480, 499, 529, 536
Cohen, J., 70, 128, 130, 369, 446
Cohen, L., 113, 128
Conover, W.J., 156, 183, 207, 275, 378, 446
Cox, D.R., x, xi, 5, 17, 29, 30, 70, 128, 164, 165, 183, 290, 359, 396, 433, 445, 446, 545, 564, 566
Cox, G.H., 445
Cox, M.A., 156, 183
Cramér, H., 29, 30, 204, 275, 479, 499
Cressie, N., 155, 183, 206, 275
Cyr, J.L., 378, 446

D'Agostino, R.B., 355, 359, 558, 567
Dalal, S.R., 462, 499
Dantzig, G.B., 97, 128
Darroch, J.N., 165, 183, 446
Das Gupta, S., 446
Davenport, J.M., 304, 359
David, F.N., 183
David, H.A., 318, 359
Davis, A.W., 460, 499
Davis, B.M., 155, 186
Davis, C.E., 359
Dawid, A.P., 122, 128, 230, 275, 561, 566
Dempster, A.P., 70, 128, 230, 461, 499
Denny, J.L., 176, 183
Deshpande, J.V., 355, 359
Deuchler, G., 359, 360
Doksum, K., 29, 355, 360, 433, 445
Draper, D., 380, 446
Dubins, L.E., 48, 67
Duncan, D.B., 446
Durbin, J., 323, 337, 360, 564, 566
Dvoretzky, A., 78, 113, 128

Eaton, M.L., 289, 360, 498, 499, 519, 536
Eberhardt, K.R., 183
Edgell, S.E., 251, 275
Edgeworth, F.Y., 30, 126, 129
Edgington, E.S., 378, 446
Edwards, A.W.F., 129, 157, 183, 230, 275
Efron, B., 248, 258, 275
Ehrenfeld, 396
Eicker, F., 446, 448
Einot, I., 383, 446
Eisenberg, H., x
Eisenhart, C., 177, 187, 446
Elfving, G., 64, 67
Englehardt, M.E., 272, 274, 276
Epps, T.W., 360
Epstein, B., 67, 129, 360
Everitt, B.S., 460, 499

Fabius, J., x
Falk, M., 213, 276
Farrell, R., x, 289, 360, 519, 536
Feddersen, A.P., 557, 566
Feller, W., 3, 30, 183, 336, 360
Fenstad, G.U., 209, 276, 360
Ferguson, T.S., x, xi, 4, 15, 18, 29, 30
Fienberg, S.E., 165, 182, 183, 487, 496, 499
Finch, P.D., 162, 183

Finney, D.J., 155, 183
Fisher, R.A., 28, 30, 127, 129, 155, 183, 229, 273, 276, 280, 360, 444, 446, 499, 500, 502, 558, 564, 566, 568
Fix, E., 225, 485, 500
Fligner, M.A., 183, 363
Folks, J.L., 124, 128, 272, 276
Forsythe, A.B., 248, 267, 276, 379, 434, 445
Fortini, P., 462, 499
Fourier, J.B.J., 126, 129
Franck, W.E., 355, 360
Fraser, D.A.S., 129, 184, 230, 276, 360
Freedman, D.A., 161, 184
Freeman, M.F., 433, 447
Freiman, J., 70, 129
Frisén, M., 170, 184, 566

Gabriel, K.R., 246, 248, 276, 382, 383, 388, 446, 447, 449
Galambos, J., 355, 360
Garside, G.R., 155, 184
Gart, J.J., 164, 184
Gastwirth, J.L., 212, 276, 326, 358, 360
Gatsonis, C., 166, 182
Gauss, C.F., 28, 126, 129, 444
Gautschi, W., xiii
Gavarret, J., 126, 129
George, E.O., 170, 186
Ghosh, B.K., 8, 30, 78, 129, 221, 276
Ghosh, J.K., 56, 67, 290, 361
Ghosh, M., 184, 509, 536, 567
Gibbons, J.D., 70, 129, 132
Giri, N., 507, 523, 536
Girshick, M.A., 15, 29, 184
Glaser, R.E., 272, 277
Glick, I., x
Godambe, V.P., 129
Gokhale, D.V., 161, 184
Goodman, L., 151, 176, 181, 184, 487, 500
Gordon, I., 545, 565
Gosset, W.S. (Student), 210, 280, 380
Govindarajulu, Z., 8, 30
Graybill, F.A., 369, 447
Green, B.F., 236, 277
Green, J.R., 355, 361
Greenwald, A.G., 71, 129
Grenander, U., 78, 129
Guenther, W.C., 177, 184
Guillier, C.L., 351, 358, 403, 444

Haber, M., 156, 184
Haberman, S.J., 157, 165, 184, 487, 500
Hackney, O.P., 396, 451
Hahn, G.J., 447
Hajek, J., 323, 336, 361, 380, 447
Hakstian, A.R., 251, 280
Haldane, J.B.S., 184
Hall, I.J., 78, 129
Hall, P., 94, 129, 355, 361
Hall, W.J., 70, 130, 248, 276, 290, 361
Halmos, P.R., 67, 113, 129, 293, 361, 518, 536, 575
Hammel, E.A., 182
Hansen, O.H., 126, 130, 432, 447
Hartigan, J., 248, 267, 274, 276, 277
Hartley, H.O., 369, 449
Harville, D.A., 427, 447
Hegazy, Y.A.S., 355, 361
Hegemann, V., 395, 447
Helmert, F.R., 277
Hemelrijk, J., x, 361
Herbach, L.H., 447
Hettsmansperger, T.P., 380, 392, 447
Heyman, E.R., 165, 183
Higgins, J.J., 321, 359
Hill, D.L., 361
Hillier, G.H., 379, 448
Hinkley, D., x, xi, 17, 29, 30, 433, 447, 549, 566, 567
Hipp, C., 57, 67
Hobson, E.W., 242, 277
Hochberg, Y., 417, 445
Hocking, R.R., 396, 441, 447, 451
Hodges, J.L., Jr., 30, 89, 130, 197, 277, 485, 500
Hoeffding, W., x, 236, 273, 277, 361, 478, 500
Hoel, P.G., 96, 130, 185
Hogg, R.V., 355, 361
Holland, P.W., 165, 182, 487, 496, 499
Holm, S., 388, 447
Hooper, P.M., 290, 313, 361
Horst, C., 155, 183
Hotelling, H., 126, 130, 133, 210, 273, 277, 357, 433, 447, 452, 498, 500, 501
Houtman, A.M., 447
Høyland, A., xiii
Hoyle, M.H., 361
Hsu, C.F., 246, 276
Hsu, C.T., 277

Hsu, P., 155, 183
Hsu, P.L., 251, 277, 361, 448, 451, 500, 503
Huang, J.S., 509, 536, 567
Huber, P., 78, 448, 515, 536
Hunt, G., 31, 358, 361, 448, 500, 536
Hunter, J.S., 396, 445
Hunter, W.G., 396, 445
Hutchinson, D.W., 221, 275
Hwang, J.T., 525, 537

Ibragimov, J.A., 509, 537
Isaacson, S.L., 537

Jagers, P., 367, 448
James, A.T., 518, 537
James, G.S., 379, 448, 462, 500
Jayachandran, K., 463, 502
Jenkins, G., 213, 275
Jogdeo, (Joag-Dev), K., x, 130
Johansen, S., 67, 145, 185, 379, 448, 462, 500
John, A., 396, 448
John, P.W., 396, 448
John, R.D., 236, 246, 277
Johnson, D.E., 392, 394, 447, 449
Johnson, M.E., 207, 275, 378, 446
Johnson, M.M., 207, 275, 378, 446
Johnson, N.L., x, xi, 114, 123, 130, 155, 185, 196, 200, 221, 254, 270, 277, 369, 448, 501
Johnson, N.S., 161, 184
Johnstone, I.M., 86, 128, 140, 182, 429, 445

Kabe, D.G., 112, 130
Kalbfleisch, J.D., 543, 567
Kallenberg, W.C.M., x, xi, 171, 185, 528, 537
Kanoh, S., 444, 448
Kappenman, R.F., 549, 567
Kariya, T., 439, 448, 471, 501
Karlin, S., 3, 31, 78, 82, 86, 130, 303, 304, 361, 509, 537
Kasten, E.L., 155, 182
Kempthorne, O., 156, 185, 396, 448, 452
Kendall, M.G., x, xi, 29, 31, 337, 351, 361, 441, 448, 480, 501
Kent, J., 355, 362
Kersting, G., 326, 358
Kesselman, H.J., 379, 445

Kiefer, J., x, 31, 32, 78, 128, 358, 362, 396, 448, 460, 498, 501, 507, 523, 536, 537, 558, 563, 566, 567
King, M.L., 379, 448
Klett, G.W., 218, 281, 331, 364
Koch, G.G., 165, 183
Koehn, Y., 191, 277
Kohne, W., 213, 276
Kolmogorov, A., 21, 31, 336
Kolodziejcyk, S., 449
Kotz, S., x, xi, 114, 123, 130, 155, 185, 196, 200, 221, 254, 270, 277, 369, 448, 501
Kowalski, C.J., 251, 277
Koziol, J.A., 170, 185, 326, 362
Krafft, O., x, 107, 130
Kraft, C., 505, 536
Kruskal, W., x, 71, 113, 130, 157, 184, 185, 360, 362
Kudo, A., 78, 129
Kudo, H., 31, 32
Kuebler, R.R., 129
Kusunoki, U., 444, 448

Lambert, D., 70, 130, 208, 236, 278, 515, 537
Landers, D., 290, 362
Landis, J.R., 165, 183
Lane, D., 161, 184
Laplace, P..S., 28, 31, 126, 130, 380
Latscha, R., 155, 183
Laurent, A.G., 112, 130
Lawal, H.B., 487, 501
Lawless, J.F., 549, 567
Lawley, D.N., 501
Layard, M.W.J., 422, 444
Le Cam, L., 17, 22, 31, 64, 67, 89, 131, 485, 501, 567
Legendre, A.M., 444
Lentner, M.M., 272, 278
Levy, K.J., 334, 362
Lewis, T., 487, 501
Lexis, W., 126, 131
Liang, K.Y., 496, 501, 552, 567
Lieberman, G.J., 155, 185
Lindley, D., 125, 131, 162, 185, 227, 278
Ling, R.F., 155, 185
Linnik, Y.V., 304, 523, 537
Littell, R.C., 170, 185

Loève, M., 66, 67
Loh, W.-Y., x, 273, 278, 290, 321, 362, 509, 537
Loomis, L.H., 519, 538
Loranger, M., x
Lorenzen, T.J., 396, 449
Louv, W.C., 170, 185
Lyapounov, A.M., 113, 131

MacGibbon, K.B., 86, 128, 140, 182, 429, 445
Mack, C., 155, 184
Mack, G.A., 392, 449
Madansky, A., 261, 278
Maitra, A.P., 57, 66
Manoukian, E.B., 378, 446
Marasinghe, M.C., 392, 449
Marcus, R., 388, 449
Marden, J.I., x, 166, 170, 182, 186, 362, 461, 471, 501, 552, 553, 567
Mardia, K.V., 200, 278, 449
Maritz, J.S., 278
Marshall, A.W., 429, 449, 509, 538
Massey, F.J., 356, 362
McCullagh, P., 479, 501
McDonald, L.L., 155, 186
McKean, J.W., 380, 449
McLaughlin, D.H., 208, 281
Meeks, S.L., 558, 567
Mehta, R.C., 156, 186
Michel, R., 147, 186
Miller, J.J., 427, 441, 446, 449
Miller, R.G., 396, 444, 449
Milliken, G.A., 155, 186
Milnes, P., 522, 536
Morgan, W.A., 278
Morgenstern, D., 278
Morimoto, H., 56, 67
Moses, L.E., 362
Mosteller, F., 184
Mudholkar, G.S., 170, 186
Murray, L.W., 421, 451

Nachbin, L., 519, 538
Naiman, D.Q., 417, 449
Nandi, H.K., 186
Narula, S.C., 334, 362
Neuhaus, G., 336, 362
Neyman, J., xii, 28, 29, 31, 32, 126, 127, 131, 159, 186, 225, 273, 278, 501, 538, 566
Nölle, G., 576, 578
Noon, S.M., 251, 275
Novick, M., 162, 185

O'Brien, K.F., 165, 187
O'Connell, W., 182
Odén, A., 234, 279
Olkin, I., 429, 449, 509, 538
Olshen, R.A., 449, 450, 558, 567
Olson, C.L., 463, 465, 501, 502
Oosterhoff, J., 480, 501
Ord, J.K., xi, 441, 448
Overall, J.E., 156, 186
Owen, D.B., 155, 185, 209, 254, 279, 445

Pachares, J., 140, 186
Patel, J.K., 270, 279
Patel, N.R., 156, 186
Paulson, E., 131, 279
Pearson, E.S., xii, 29, 32, 126, 131, 186, 273, 278, 369, 449, 538
Pearson, K., 28, 126, 132, 502
Pedersen, J.G., 230, 279
Pedersen, K., 57, 67, 132
Peisakoff, M., 32
Pereira, B. de B., 290, 363
Peritz, E., 388, 411, 449, 450
Perlman, M., 170, 185, 446, 460, 471, 501, 502, 552, 553, 567
Peterson, R.P., 96, 130
Pfanzagl, J., 78, 80, 132, 214, 279, 304, 363
Piegorsch, W.W., 417, 450
Pierce, D.A., 563, 564, 567
Pillai, K.C.S., 463, 498, 502
Pitman, E.J.G., 32, 273, 279, 357, 363, 564, 567
Plachky, D., 576, 578
Plackett, R.L., 156, 165, 183, 186, 496, 502, 568
Please, N.W., 279
Pliss, V.A., 523, 537
Policello, G.E., 363
Posten, H.O., 209, 279
Pratt, J., x, 30, 32, 70, 115, 129, 132, 155, 185, 189, 209, 279, 323, 363, 528, 563, 568
Prescott, P., 380, 450

Press, S.J., 465, 502
Przyborowski, J., 186
Pulley, L.B., 360
Putter, J., 186

Quade, D., 359
Quenouille, M.H., 396, 448
Quesenberry, C.P., 355, 356, 362, 363
Quine, M.P., 480, 502

Radlow, R., 479, 502
Ramachandran, K.V., 219, 279
Ramamoorthi, R.V., 22, 29, 32, 64, 67
Ramsey, P.H., 209, 279
Randles, R.H., 321, 329, 363, 380, 450
Rao, P.V., 361
Ratcliffe, J.F., 279
Rayner, J., x
Read, C.B., 270, 279
Reid, C., 29, 32, 126, 132
Reinhardt, H.E., 107, 109
Reiser, B., 272, 274
Richmond, J., 417, 444, 450
Rieder, H., 515, 538
Robbins, H., 573, 578
Robinson, G.K., 262, 280, 557, 558,
    564, 568
Robinson, J., 236, 246, 277, 280, 378, 450,
    480, 502
Rogge, L., 290, 362
Rojo, J., 24, 32
Ronchetti, E., 380, 450
Rosenstein, R.B., 427, 445
Rosenthal, R., 70, 132
Ross, S., 3, 32
Rothenberg, T., 197, 280, 379, 450
Roy, K.K., 22, 32, 64, 67
Roy, S.N., 498, 502
Royden, H.L., 576, 578
Rubin, D.B., 70, 132, 227, 280
Rubin, H., 78, 82, 130, 212, 280, 326, 360
Ruist, E., 538
Runger, G., 433, 447
Ryll–Nardzewski, C., 67

Sackrowitz, H., 311, 359, 480, 499, 529, 536
Salaevskii, O.V., 523, 537, 538
Sanathanan, L., 70, 132
Sandved, E., 568
Saunders, I.W., 487, 501

Savage, I.R., x
Savage, L.J., xiii, 15, 29, 32, 67, 70, 132,
    184, 273, 274, 564, 568, 575, 578
Schaper, C., x
Schatzoff, M., 70, 128
Scheffé, H., 128, 185, 186, 209, 219, 273,
    280, 304, 363, 379, 388, 396, 441, 444,
    450, 502, 573, 578
Schmetterer, L., 29, 33
Schoenberg, I.J., 538
Scholz, F.W., x, 170, 187, 532
Schrader, R.M., 380, 449
Schriever, B.F., 480, 501
Schwartz, R., 358, 362, 460, 498, 501, 502,
    538
Schweder, T., 70, 132
Seal, H.L., 444, 450
Seber, G.A.F., 450, 460, 462, 465, 490, 498,
    502
Self, S.G., 496, 501
Sellke, T., 125, 127
Sen, P.K., 274
Serfling, R.J., 204, 280, 323, 356, 363, 450,
    485, 486, 487, 491, 503
Shaffer, J., x, 387, 388, 434, 444, 449, 451,
    496, 503
Shapiro, S.S., 355, 359, 363
Shewart, W.A., 273, 280
Shigemasu, K., 465, 502
Shorack, G., 272, 280
Shuster, J., 124, 132
Sidak, Z., 323, 336, 361, 380, 447
Sierpinski, W., 509, 538
Silvey, S.D., 33, 396, 451
Simaika, J.B., 503
Singh, M., 209, 281
Singleton, K.J., 360
Skillings, J.H., 392, 449
Smirnov, N.V., 336, 363
Smith, C.A.B., 184
Smith, D.W., 421, 451
Smith, H., 129
Sobel, M., 67, 129
Somes, G.W., 165, 187
Sophister (G.E.F. Story), 273, 280
Speed, F.M., 396, 447, 451
Speed, T.P., 447
Spiegelhalter, D.J., 355, 363
Spjotvøll, E., 70, 78, 132, 383, 385, 388,
    396, 422, 451

Sprott, D.A., 122, 132, 565, 568
Spurrier, J.D., 355, 363
Srivastava, M.S., 251, 274, 460
Starbuck, R.R., 156, 186, 355, 356, 363
Steiger, J.H., 251, 280
Stein, C., xiii, 31, 111, 131, 132, 273, 278, 280, 305, 311, 322, 363, 370, 448, 449, 500, 503, 523, 536, 537
Sterling, T.D., 71, 133
Stevens, J.P., 463, 503
Stigler, S.M., 30, 33, 126, 133
Still, H.A., 221, 275
Stone, C.J., 10, 33, 322, 364, 522, 538
Stone, M., 70, 128, 133, 230, 275, 280, 522, 538
Story, G.E.F. (Sophister), 273, 280
Strassen, V., 515, 536
Strawderman, W.E., 82, 128, 130, 313, 359
Striebel, C., xiii
Stuart, A., x, xi, 29, 31, 337, 361, 441, 448, 480, 501
Student (W.S. Gosset), 210, 280, 380
Subrahmanian, K. and K., 462, 503
Sugiura, N., 322, 364
Sukhatme, P.V., 364
Sverdrup, E., 187, 568
Swed, F.S., 177, 187

Takeuchi, K., 112, 133
Tallis, G.M., 336, 364
Tan, W.Y., 206, 281
Tate, R.F., 218, 281, 331, 364
Taylor, H.M., 3, 31
Terry, M.E., 361
Thomas, D.L., 191, 277
Thompson, W.A., Jr., 70, 133
Thompson, W.R., 133
Tiao, G.C., 208, 227, 275, 427, 445, 465, 499
Tiku, M.L., 207, 209, 281, 369, 427, 451
Tocher, K.D., 187
Tong, Y.L., 176, 187
Torgersen, E.N., 89, 126, 130, 133, 432, 447
Tritchler, D., 248, 281
Truax, D.R., 78, 130
Tsao, C.K., 360
Tukey, J.W., 20, 33, 96, 133, 208, 273, 281, 364, 433, 447, 451
Tweedie, M.C.K., 273, 281

Unni, K., 546, 568
Upton, G.J., 487, 501
Uthoff, V.A., 355, 364

Vadiveloo, J., 248, 281
van Zwet, W.R., 236, 274
Venable, T.C., 166, 187
Von Neumann, J., 29
von Randow, R., 522, 538

Wacholder, S., 180, 187
Wald, A., 17, 18, 29, 33, 78, 97, 113, 128, 133, 396, 451, 485, 503, 535, 538
Wallace, D.L., 304, 364, 565, 568
Walsh, J.E., 187, 364
Wang, Y.Y., 209, 281, 304, 364
Webster, J.T., 304, 359
Wedel, H., 234, 279
Weinberg, C.R., 180, 187
Welch, B.L., 379, 451, 563, 566, 568
Welsh, A.H., 355, 361
Wescott, M., 487, 501
Wijsman, R.A., x, 289, 290, 361, 364, 409, 444, 451, 452, 471, 498, 500, 503
Wilcoxon, F., 364
Wilenski, H., 186
Wilk, M.B., 355, 363, 396, 452
Wilkinson, G.N., 133
Wilks, S.S., 503
Wilson, E.B., 126, 133
Winters, F.W., 273, 280
Witting, H., 107, 130
Wolfe, D.A., 321, 329, 363, 380, 450
Wolfowitz, J., 78, 95, 113, 128, 133, 177, 187, 364, 538
Wolpert, R., 548, 565
Working, H., 126, 133, 452
Wright, A.L., 176, 183, 326, 358
Wynn, H.P., 409, 417, 452

Yakowitz, S.J., 176, 183
Yamada, S., 56, 67
Yandell, B.S., 355, 360
Yao, Y., 462, 503
Yates, F., 156, 187, 360
Yeh, H.C., 209, 279
Yuen, K.K., 230, 281

Zemroch, P.J., 200, 278, 449
Zucchini, W., 326, 358

Subject Index

Absolute continuity (of one measure with respect to another), 40. *See also* Equivalence, of two measures; Radon-Nikodym derivative
Action problem, 4
Adaptive test, 322
Additivity of effects, 388; in model II, 418; test for, 392
Admissibility, 17; Bayes method for proving, 309; of confidence sets, 313; in exponential families, 307; of invariant procedures, 28, 311; of multiple comparison procedures, 384; of UMP invariant tests, 305; of UMP unbiased tests, 170; of unbiased procedures, 27, 305. *See also* Alpha-admissibility; d-admissibility; Inadmissibility
a. e., *see* Almost everywhere
Aggregation (of several contingency tables), 162
Almost everywhere (a. e.), 40, 140
Almost invariance: of decision procedures, 24; of likelihood ratio, 341; relation to invariance, 297, 298, 316, 340; relation to invariance of power function, 300; relation to maximin tests, 516; relation to unbiasedness, 302; of sets, 342; of tests, 297, 298. *See also* Invariance
Aloaglu's theorem, 576
Alpha-admissibility, 306, 342, 384
Alternatives (to a hypothesis), 68
Amenable group, 522, 536
Analysis of covariance, 401
Analysis of variance, 375, 395, 444, 446; different models for, 418; for one-way classification, 375; in random effects model, 425; robustness of F-tests, 401;

for two-way classification, 390, 395. *See also* Linear hypothesis; Linear model
Ancillary statistic, 542, 560, 564, 565, 566; and invariance, 543; maximal, 545, 560; and sufficiency, 545. *See also* Partial ancillarity
Approximate hypotheses: extended Neyman-Pearson lemma for, 512, 515
Arcsine transformation for binomial variables, 432, 445
Association, 162; spurious, 162; Yule's measure of, 157. *See also* Dependence, positive
Asymptotic (relative) efficiency, 321
Asymptotic normality: of functions of asymptotically normal variables, 205; of mean, 204. *See also* Central limit theorem
Asymptotic optimality, vii, 477, 485
Attributes: paired comparisons by, 169, 291, 510, 526; sample inspection by, 80, 293
Autoregressive process (first order), 212
Average power, maximum, 429

Bartlett's test for variances, 378
Basu's theorem, 191
Bayesian confidence sets, *see* Credible region
Bayesian inference, 15, 70, 227, 427, 465, 511, 564
Bayes risk, 14
Bayes solution, 14, 18, 25, 33; to maximize minimum power, 505; to prove admissibility, 309; restricted, 15. *See also* Credible region; Prior distribution
Bayes sufficiency, 21, 22, 31
Bayes test, 125, 343, 430, 465, 498
Behrens-Fisher distribution, 262

Behrens-Fisher problem, 209, 262, 304, 360, 361, 558, 564, 566; for many samples, 379; multivariate, 462; nonparametric, 323. *See also* Welch-Aspin test

Beta distribution, 200, 272; as distribution of order statistics, 345; noncentral, 369, 428; relation to F-distribution, 200; relation to gamma distribution, 272; in testing linear hypotheses, 369; in testing ratio of variances, 200, 255

Bimeasurable transformation, 284

Binomial distribution b(p,n), 2; in comparing two Poisson distributions, 153; completeness of, 141; as exponential family, 56, 81; as log-linear model in bio-assay, 178; variance stabilizing transformation for, 432, 445. *See also* Contingency tables; Multinomial distribution; Negative binomial distribution Nb; Two by two table

Binomial probabilities: comparison of two, 121, 154, 159, 161, 175, 180, 183, 261; confidence bounds for, 93, 117; confidence intervals for, 219, 221; credible region for, 227; one-sided test for, 93, 113, 167; two-sided test for, 118, 138, 167, 171. *See also* Contingency tables; Independence, test for; Median; Paired comparisons; Sample inspection; Sign test

Binomial trials, 7; obtained by dichotomizing continuous variables, 164; sufficient statistics for, 19, 28. *See also* Inverse sampling

Bioassay, 178

Bivariate distribution(general): class of one-parametric families of, 251; testing for independence or lack of correlation in, 250, 350. *See also* Dependence, positive

Bivariate normal correlation coefficient: confidence bounds for, 353; distribution of, 267, 270; test for, 249, 304, 340

Bivariate normal distribution, 249, 267, 271; ancillary statistics in, 545; joint distribution of second moments in, 268; test for independence in, 249, 253, 271; testing parameters in, 268, 305

Borel set, 35

Bounded completeness, 144, 172, 191, 300; example of, without completeness, 173. *See also* Completeness of family of distributions

Canonical form: for model II two-way layout, 438, 441; for multivariate linear hypothesis,

454; for multivariate linear hypothesis with covariates, 471; for nested classification in model II, 423, 438; for repeated measurement model, 467; for univariate linear hypothesis, 366, 370

Cartesian product, 40

Cauchy distribution, 86, 115, 510, 567

Causal influence, 162

CDF, *see* Cumulative distribution function

Center of symmetry: confidence intervals for, 263. *See also* Symmetry

Central limit theorem, 204; for dependent variables, 213; Lindeberg form of, 402

Chebyshev inequality, 257

Chi-squared distribution, 56, 139; in estimating normal variance, 218, 229; as exponential family, 56; as limit for likelihood ratio, 487; in multivariate distribution theory, 490; non-central, 427, 428, 434, 447, 500; relation to beta-distribution, 200; relation to exponential distribution, 64, 82, 114; relation to F-distribution, 199; relation to t-distribution, 196; for testing linear hypotheses with known variance or covariance matrix, 431, 477; in testing normal variance, 110, 139, 194, 290; for total waiting time in Poisson process, 92. *See also* Gamma distribution; Normal one-sample problem, the variance; Wishart distribution

Chi-squared test, 477, 480, 500, 502; restricted, 481, 500, 501; in r × c contingency tables, 487; for testing goodness of fit, 480, 494; for testing uniform distribution, 480, 482

Cluster sampling, 211

Cochran-Mantel-Haenszel test, 165

Coefficient of variation, 549; confidence bounds for, 352, 356; tests for, 294, 303

Comparison of experiments, 86, 114, 116, 159, 167, 223, 264, 339

Completeness of a class of decision procedures, 17, 18; of classes of one-sided tests, 82, 83, 461; of class of two-sided tests, 172; relation to sufficiency, 64. *See also* Admissibility

Completeness of family of distributions, 141, 172, 173, 180; of binomial distributions, 141; for exponential distributions, 256; of exponential families, 142; of normal distributions, 142, 172; of order statistics, 163, 173, 183, 187; relations to bounded

completeness, 144, 173; of uniform distributions, 141, 172
Completion of measure, 35
Complexity: of multiple comparison procedure, 387
Components of variance, 425, 558. *See also* Random effects model
Composite hypothesis, 72; large-sample tests for, 483; *vs.* simple alternative, 104
Conditional distribution, 48; in bivariate normal distributions, 267; example of nonexistence, 48, 67; in exponential families, 58, 146; in Poisson distribution, 65
Conditional expectation, 44, 47, 50
Conditional independence, 162; test of, 163
Conditional inference, ix, 541, 558, 564, 566
Conditionality principle, weak, 548
Conditional power, 151, 170, 246, 541, 547
Conditional probability, 43, 47, 48, 66
Conditional test, 182, 549; most powerful, 540, 543
Confidence bands: for cumulative distribution function, 334, 354; in linear models, 406; for regression line, 417, 444; for regression surface, 444. *See also* Simultaneous confidence intervals
Confidence bounds, 89; impossible, 421, 558; with minimum risk, 117; in monotone likelihood ratio families, 91; in presence of nuisance parameters, 213; randomized, 93; relation to median unbiased estimates, 95, 214; relation to one-sided tests, 214; standard, 96, 229; uniformly most accurate, 90
Confidence coefficient, 90, 213; conditional, 558
Confidence ellipsoids, 461, 490
Confidence intervals, ix, 68, 94; of bounded length, 258, 259; for center of symmetry, 263; distribution-free, 247, 263, 329; empty, 421, 558; history of, 126; interpretation of, 214, 225; logarithmically shortest, 331; loss functions for, 6, 24, 94, 95; minimax, 524; for parameters suggested by data, 410; in randomization models, 247; randomized, 219; unbiased, 13, 24, 217. *See also* Simultaneous confidence intervals
Confidence level, 89
Confidence sets, 90; admissibility of, 313; average smallest, 330; conditional, 541; derived from a pivotal quantity, 333, 357;

equivariant, 327, 333, 524; example of inadmissible, 525; minimax, 524; relation with tests, 90, 214, 216; of smallest Lebesgue measure, 261, 330, 524; unbiased, 217; which are not intervals, 225. *See also* Credible region; Equivariant confidence sets; Relevant and semirelevant subsets; Simultaneous confidence sets
Conservative test, 155
Consistency of sequence of tests, 356, 478, 494
Consumer preferences, 166, 167
Contingency tables: general, 165; loglinear models for, 165; models for, 161, 495; $r \times c$ tables, 156, 487, 495; three factor, 162; $2 \times 2 \times K$, 162, 165, 179; $2 \times 2 \times 2 \times L$, 179. *See also* Two by two tables
Continuity correction, 155
Contrasts, 388, 415; in multivariate case, 472, 494
Convergence: in law, 204; in mean, 570; pointwise, 571; in probability, 257; weak, 571
Convergence theorem: for densities, 573; dominated, 39; for functions of random variables, 205; monotone, 39. *See also* Cramér-Wold theorem
Correlation coefficient: in bivariate normal distribution, 249; confidence bounds for, 353; intraclass, 438; testing value of, 249, 304, 340. *See also* Bivariate distribution; Dependence, positive; Multiple correlation coefficient; Rank correlation coefficient; Sample correlation coefficient R
Countable additivity, 34
Countable generators of σ-field, 575
Counting measure, 35
Covariance matrix, 453; estimation of, 488; special structure, 440, 441; tests for, 379, 462
Covariates, 470, 552
Cramér-Wold theorem, 491
Credible region, 226; equal tails, 229; highest probability density, 227, 262
Critical function, 71
Critical region, 68
Cross product ratio, *see* Odds ratio
Cumulative distribution function (cdf), 36, 62; confidence bands for, 334, 354; empirical, 323, 335; inverse of, 344. *See also* Kolmogorov test for goodness of fit

d-admissibility, 306, 342. *See also* Admissibility

Data Snooping, 410, 476

Decision problem: specification of, 2

Decision space, 2, 3

Decision theory, 29, 33; and inference, 4, 5, 71

Deficiency, 197

Dependence, positive, 157, 176, 210, 251, 271, 315, 350; measures of, 157. *See also* Correlation coefficient; Independence

Design of experiments, 7, 8, 159, 396, 447. *See also* Random assignment; Sample size

Directional error, 387

Direct product, 40

Dirichlet distribution, 262

Distribution, *see the following families of distributions:* Beta, Binomial, Bivariate normal, Cauchy, Chi-squared, Dirichlet, Double exponential, Exponential, F, Gamma, Hypergeometric, Inverse Gaussian, Logistic, Multinomial, Multivariate normal, Negative binomial, Noncentral, Normal, Pareto, Poisson, Polya, t, Hotelling's T^2, Triangular, Uniform, Weibull, Wishart. *See also* Exponential family; Monotone likelihood ratio; Total positivity; Variation diminishing

Dominated convergence theorem, 39

Dominated family of distributions, 53, 574, 575

Domination: of one procedure over another, 17. *See also* Admissibility; Inadmissibility

Double exponential distribution, 355, 509, 567; locally most powerful test in, 531; UMP conditional test in, 550

Duncan multiple comparison procedure, 383, 385

Dunnett's multiple comparison method, 443

EDF, *see* Empirical distribution function

Efficiency, relative asymptotic, 321

Efficiency robustness, 208, 322. *See also* Robustness

Empirical distribution function(EDF), 323, 335

Envelope power function, 341, 525. *See also* Most stringent test

Equivalence: of family of distributions or measures, 54, 575; of statistics, 43; of two measures, 61

Equivalence classes, 569

Equivalence relation, 569

Equivariance, 12, 544. *See also* Invariance

Equivariant confidence bands, 335, 406, 417, 472

Equivariant confidence sets, 327, 330; and pivotal quantities, 333, 357. *See also* Uniformly most accurate confidence sets

Error of first and second kind, 69, 70

Error rate per experiment, 388

Essentially complete class, 18, 64, 82, 113. *See also* Completeness of a class of decision procedures

Estimation, *see* Confidence bands; Confidence bounds; Confidence intervals; Confidence sets; Equivariance; Maximum likelihood; Median: Point estimation; Unbiasedness

Euclidean sample space, 49

Expectation (of a random variable), 38; conditional, 44, 47, 50

Expected normal order statistics, 318

Experimental design, *see* Design of experiments

Exponential distribution, 23, 360; completeness in, 256; confidence bounds and intervals in, 92, 261, 354; order statistics from, 65; other tests for, 355; relation to Pareto distribution, 123; relation to Poisson process, 23, 65, 82, 154; r-sample problem for, 354, 364; sufficient statistics in, 28; testing against gamma distribution, 272; testing against normal or uniform distribution, 355; tests in, 93, 112, 255; two-sample problem for, 338. *See also* Chi-squared distribution; Gamma distribution; Life testing

Exponential family, 56, 59, 66; admissibility of tests in, 307; completeness of, 142; equivalent forms for, 150; median unbiased estimators in, 214; moments of sufficient statistics, 66; monotone likelihood ratio of, 80, 119; natural parameter space of, 57, 66; testing in multiparameter, 145, 171, 181, 188; testing in one-parameter, 80, 120, 135, 172; total positivity of, 119. *See also* One-parameter exponential family

Exponential waiting times, 23, 65, 82, 92. *See also* Exponential distribution

Factorization criterion for sufficient statistics, 19, 30, 31, 55, 66, 67

F-distribution, 199, 446, 449; in confidence intervals for ratio of variances, 219, 421; in Hotelling's T^2-test, 459; noncentral, 428;

relation to beta distribution, 200; relation to distribution of multiple correlation coefficient, 497; for simultaneous confidence sets, 475. See also F-test for linear hypothesis; F-test for ratio of variances

Fiducial probability, 127, 131, 133, 229; distribution, 129, 229, 230

Field, 60

Finite decision problem, 64

Fisher's exact test, 155, 158, 180, 187. See also Two by two tables

Fisher's least significant difference, 382, 386

Fixed effects model, 418. See also Linear model; Model I and II

Free Group, 26

Friedman's rank test, 392

F-test for linear hypothesis, 369; admissibility of, 370; as Bayes test, 430; has best average power, 429; in Fisher's least significant difference method, 382; in Gabriel's simultaneous test procedure, 382, 416; in mixed models, 426; permutation version of, 450; power of, 369; robustness of, 378, 379, 401. See also F-distribution

F-test for ratio of variances, 122, 199; admissibility of, 313; in mixed models, 426; in model II analysis of variance, 420, 424; nonrobustness of, 207, 378; power of, 200. See also F-distribution; Normal two-sample problem, ratio of variances

Fubini's theorem, 40

Fully informative statistics, 113

Fundamental lemma, see Neyman-Pearson fundamental lemma

Gabriel's simultaneous test procedure, 382, 416

Gamma distribution $\Gamma(g, b)$, 123, 271, 272, 356. See also Beta distribution; Chi-squared distribution; Exponential distribution

Goodness of fit, 336, 355, 480, 482, 494. See also separate families

Group, 569; amenable, 522; finite, 518; free, 26; generated by subgroups, 288; linear, 286, 299, 522; orthogonal, 286, 522, 525; permutation, 286, 298, 356; of rigid motions, 525; scale, 285, 337; transitive, 285, 543, 550; transformation, 282, 570; translation, 285, 521; triangular, 305. See also Equivariance; Invariance

Group family, 543, 550

Guaranteed power: achieved through sequential procedure, 151, 153, 260; with minimal sample size, 505

Haar measure, 299

Homogeneity, tests of: against ordered alternatives, 380; for exponential distributions, 364; for K two-by-two tables, 165; for multinomial distributions, 495, 496; for multivariate normal means, 463; nonparametric, 380, 392; for normal means, 374, 378, 379, 381, 389, 394; for normal variances, 376; for subsets of means, 381. See also Multiple comparisons; Normal many-sample problem

Hotelling's T^2-distribution, 459, 500; derivation of, 489; noncentral, 460, 500; χ^2-limit of, 490

Hotelling's T^2-test, 459, 460, 500; admissibility of, 460, 498, 523; application to one- and two-sample problems, 459, 461, 462, 471; application to two-factor mixed model, 466; as Bayes solution, 498; best average power of, 500; minimaxity of, 523; in multivariate regression, 462, 490; in repeated measurements, 466, 469; robustness of, 460, 462

HPD (Highest probability density) credible region, 227, 262

Huber condition(for robustness), 404, 436, 448

Hunt-Stein theorem, 519

Hypergeometric distribution, 80; monotone likelihood ratio of, 80; relation to distribution of runs, 177; in testing equality of two binomials, 155; in testing for independence in a two by two table, 158, 161; UMP one-sided test for testing mean of, 80. See also Fisher's exact test; Two by two tables

Hypothesis testing, 3, 68; conditional, 539; history of, 126, 131; large-sample approach, ix, 477; loss functions for, 72, 82, 172, 292; without stochastic basis, 162

Improper prior distribution, 226

Inadmissibility, 17; of confidence sets for vector means, 525; of likelihood ratio test, 341; of UMP invariant test, 305. See also Admissibility

Independence: conditional, 162; of normal correlation coefficient from sample means

Independence (*Continued*)
and variances, 192; relation to absence of correlation, 250; of sample mean from function of differences in normal samples, 191; of statistic from a complete sufficient statistic, 191; of sum and ratio of independent χ^2 variables, 192; of two random variables, 40

Independence, test for: in bivariate normal distribution, 248; in multivariate normal distribution, 462, 496; in nonparametric models, 251, 314, 350; in r $\times$ c contingency tables, 487; *vs.* tests for absence of correlation, 250; in two by two tables, 156, 161

Indicator function of a set, 39

Indifference zone, 505

Inference, statistical, 1, 4, 71. *See also* Decision theory

Integrable function, 38

Integration, 37

Interaction, 393, 396, 444; in random effects and mixed models, 440, 441; test for absence of, 392, 394, 434

Interval estimation, *see* Confidence intervals

Into, *see* Transformation

Intraclass correlation coefficient, 438

Invariance: of decision procedure, 11, 12, 31, 32; of likelihood ratio, 341; of measure, 299, 518, 519; of power functions, 299, 300; relation to equivariance, 12; relation to minimax principle, 26, 516, 519; relation to sufficiency, 290, 301; relation to unbiasedness, 24, 302; of test, 284, 357; warning against inappropriate use of, 377. *See also* Almost invariance; Equivariance

Invariant measure, 299, 518, 519; over orthogonal group, 518; over translation group, 521

Inverse Gaussian distribution, 124, 272

Inverse sampling: for binomial trials, 81; for Poisson variables, 82. *See also* Negative binomial distribution Nb; Poisson process

Kendall's t-statistic, 351

Kolmogorov test for goodness of fit, 336, 356, 480, 494. *See also* Goodness of fit

Kruskal-Wallis test, 380

Large-sample tests, ix, 204, 380, 477, 480, 503; for composite hypotheses, 483

Latin square design, 396, 434

Lawley-Hotelling trace test, 463; robustness of, 465; simultaneous confidence intervals based on, 471

Least favorable distribution, 18, 104, 107, 506, 510, 512, 516, 519

Least squares estimates, 370, 374

Lebesgue convergence theorems, 39

Lebesgue integral, 38

Lebesgue measure, 35

Level of significance, *see* Significance level

Life testing, 65, 114. *See also* Exponential distribution; Poisson process

Likelihood, 16. *See also* Maximum likelihood

Likelihood principle, 565

Likelihood ratio: censored, 513; invariance of, 341; large-sample theory of, 486, 503; preference order based on, 73, 79; procedure, 16; sufficiency of, 63

Likelihood ratio test, 16, 126; example of inadmissible, 341; large-sample theory of, 486, 503

Lindley's Paradox, 125

Linear hypothesis, multivariate, 453, 465, 498; Bayesian treatment of, 465; canonical form of, 454, 500; concerning row vectors of a matrix of means, 467, 470; with covariates, 470; invariant test for when r = 1, 459; with known covariance matrix, 477; reduction through invariance of, 456, 488; robustness of tests for, 491; suggested by the data, 476; tests for when r > 1, 463. *See also* Hotelling's T^2-test; Multivariate analysis of variance (MANOVA); Multivariate normal distribution; Multivariate one-sample problem; Multivariate two-sample problem; Regression, multivariate; Repeated measurements

Linear hypothesis, univariate, 365, 449; admissibility of test for, 370; canonical form for, 366; inhomogeneous form of, 372; with known variance, 431; more efficient tests for, 380; parametric form of, 373; power of test for, 369; properties of test for, 369, 429, 522, 529, 538; reduction of, through invariance, 367; robustness of test for, 378, 379, 401; suggested by the data, 411. *See also* Analysis of variance; Homogeneity, tests of; Mixed model; Model I and II; One-way classification; Regression; Two-way classification

Linear model, 365, 444; Bayesian inference for, 427; confidence intervals in, 391, 430; simultaneous confidence intervals in, 406, 411, 417; testing set of linear functions in, 483. *See also* Simultaneous confidence intervals and sets

Locally optimal tests, 186, 507, 527, 528, 529, 535, 538

Location families, 84, 543; comparing two, 289; conditional inference in, 543, 550, 564, 566; condition for monotone likelihood ratio, 509; dichotomization of, 164; example lacking monotone likelihood ratio, 86; existence of semi-relevant but not of relevant subsets for, 562, 567; are stochastically increasing, 84

Location-scale families, 11, 32; comparing two, 338, 355. *See also* Normality, testing for

Logistic distribution, 164, 165, 318, 320, 510, 550, 567

Logistic response model, 165

Loglinear model, 165, 178

Loss function, 1, 28; in confidence estimation, 6, 24, 90, 94, 95; in hypothesis testing, 72, 82, 172, 292; monotone, 95; specification of, 5

L-unbiased, 13. *See also* Unbiasedness

McNemar's test, 169, 180

Main effects, 389, 396, 433; confidence sets for, 391; tests for, 390, 394, 395. *See also* Two-way classification

Mantel-Haenszel test, 165

Markov chain, 176

Markov property, 176

Matched pairs: by attributes, 169, 179, 291, 510, 526; comparison with complete randomization, 180, 264; confidence intervals for, 246, 264; generalization of, 241; normal theory and permutation tests for, 239, 264; rank tests for, 314, 323

Maximal invariant, 285; ancillarity of, 543; distribution of, 289; method for determining, 287; obtained in steps, 287, 288

Maximin test, 505, 512, 515; existence of, 527; local, 507; relation to invariance, 516, 519, 533. *See also* Least favorable distribution; Minimax principle; Most stringent test

Maximum likelihood, 16, 17, 30, 31, 485, 495. *See also* Likelihood ratio test

Maximum modulus confidence intervals, 411

Measurable: function, 36, 42; set, 35; space, 35; transformation, 36

Measure theory, xiii, 34, 66

Median, 23; confidence bounds for, 120, 133; test for, 187, 530

Median unbiasedness, 23, 29; examples of, 216, 219; relation to confidence bounds, 95, 214

Metric space, 571

Minimal complete class of decision procedures, 17. *See also* Completeness of family of distributions; Essentially complete class

Minimal sufficient statistic, 22, 28, 66

Minimax principle, 14, 18, 32, 33, 535; in confidence estimation, 524; in hypothesis testing, 505; relation to invariance, 26, 516, 519; relation to unbiasedness, 26, 507. *See also* Maximin test; Restricted Bayes solution

Mixed model, 418, 427; for nested classification, 425; for two-way layout, 427, 439, 440, 441. *See also* Model I and II

Mixtures of experiments, 539, 542, 559, 564

MLR, *see* Monotone likelihood ratio

Model I and II, 418, 446, 452. *See also* Fixed effects model; Mixed model; Random effects model

Model selection, 10

Monotone class of sets, 60

Monotone convergence theorem, 39

Monotone likelihood ratio, 78, 130; approximate, 516; conditional tests based on samples from a distribution with, 549, 550, 551, 562; conditions for, 114; of distribution of correlation coefficient, 340; of exponential family, 80, 120; of hypergeometric distribution, 80; implications of, 85, 103, 115; of location parameter families, 104, 115, 509; mixtures of families with, 530, 549, 551; of noncentral t, 295; of noncentral χ^2 and F, 428; relation to total positivity, 119; tests and confidence procedures in the presence of, 78, 82, 91

Most stringent test, 358, 525, 538; existence of, 533

Moving average process, 211

Multinomial distribution, 56; as conditional distribution, 65; Dirichlet prior for, 262; for

Multinomial distribution (*Continued*)
entries of 2 × 2 table, 157, 169; limit
distribution of, 479; in testing consumer
preferences, 166; for 2 × 2 × K table, 162
Multinomial model: maximum likelihood
estimation in, 495; for r × c table, 487;
testing a composite hypothesis in, 483;
testing a simple hypothesis in, 478, 481; for
three-factor contingency table, 162, 163; for
2 × 2 table, 157, 159, 161, 169. *See also*
Chi-squared test; Contingency tables
Multiple comparisons, 4, 380, 396, 446, 451;
complexity of, 387; significance levels for,
382. *See also* Duncan and Dunnett multiple
comparison methods; Newman-Keuls
multiple comparison procedure;
Simultaneous confidence intervals; Tukey
levels; Tukey's T-method
Multiple correlation coefficient, 497;
distribution of, 446, 497, 500; optimum
test for, 497, 503, 538
Multiple decision procedures, 4, 27. *See also*
Multiple comparisons; Three-decision
problems
Multivariate analysis of variance
(MANOVA), 462. *See also* Linear
hypothesis, multivariate
Multivariate linear hypothesis, *see* Linear
hypothesis, multivariate
Multivariate normal distribution, 440, 441,
453; as limit of multinomial distributions,
479. *See also* Bivariate normal distribution
Multivariate (normal) one-sample problem:
simultaneous confidence sets in, 494;
testing the covariance matrix, 462; testing
independence of two sets of variates in, 496;
testing the mean vector, 459, 466, 523. *See
also* Hotelling's T²-test; Simultaneous
confidence ellipsoids; Simultaneous
confidence sets
Multivariate (normal) two-sample problem,
461, 532; Behrens-Fisher problem, 462;
with covariates, 470, 552; robustness of
tests for, 490; simultaneous confidence
sets in, 494
Multivariate regression, 462, 490, 496
Multivariate t-distribution, 353

Natural parameterspace of an exponential
family, 57, 66

Negative binomial distribution Nb(p,m), 22,
81, 181
Neighborhood model, 512, 515, 516
Nested classification, 422, 438
Newman-Keuls multiple comparison
procedure, 382, 386
Newton's identities, 47
Neyman-Pearson fundamental lemma, 74,
131; approximate version of, 512; censored
version of, 513; generalized, 77, 96, 118,
128
Neyman structure, 141, 144
Noncentral: beta distribution, 369, 428, 447,
500; F-distribution, 426, 428, 429, 446;
t-distribution, 196, 253, 276, 295, 303; χ²-
distribution, 427, 428, 429, 434, 447, 500
Noninformative prior, 226
Nonparametric: alternative approach to, 380;
independence problem, 252, 317; many-
sample problem, 380; one-sample problem,
143, 263; test, 107; test in two-way layout,
392. *See also* Permutation test; Rank tests;
Sign test
Nonparametric two-sample problem, 232,
317; confidence intervals in, 246, 263, 347,
362; omnibus alternatives, 322; universally
unbiased test in, 348. *See also* Normal
scores test; Wilcoxon two-sample test
Normal distribution N(ξ,σ²), 3, 56;
tests of, 355; testing against Cauchy,
double exponential, exponential, or uniform
distribution, 355. *See also* Bivariate normal
distribution; Multivariate normal
distribution
Normality, testing for, 355. *See also* Normal
distribution
Normal many-sample problem: confidence
sets for vector means, 331, 332, 406, 409,
525, 535; tests for means in, 374, 377, 378,
532, 548; tests for variances in, 376, 378.
See also Homogeneity, tests of
Normal one-sample problem, the coefficient
of variation: confidence intervals for, 352,
356; test for, 294, 303
Normal one-sample problem, the mean:
admissibility of test for, 309, 310;
confidence intervals for, 215, 329, 554,
557; credible region for, 226, 228;
likelihood ratio test for, 108; median
unbiased estimate of, 216; nonexistence of

test with controlled power, 253; nonexistence
of UMP test for, 111; optimum test for, 111,
195, 254, 255, 294, 303, 339, 372, 549; test
for, based on random sample size, 112;
two-stage confidence intervals for, of
fixed length, 259; two-stage test for, with
controlled power, 260. *See also* Matched
pairs; t-test
Normal one-sample problem, the variance:
admissibility of test for, 312; confidence
intervals for, 217, 352; credible region for,
229; likelihood ratio test for, 108; nonrobustness of test for, 206; optimum test for,
108, 139, 193, 290, 511
Normal response model, 165
Normal scores test, 318, 322, 323, 324, 357,
360; comparison with t-test, 321; optimality
of, 320
Normal subgroup, 337
Normal two-sample problem, difference of
means: comparison with matched pairs,
264; confidence intervals for, 218, 353;
credible region for, 262; test for (variances
equal), 122, 201, 204, 208, 255, 296, 373.
See also Behrens-Fisher problem;
Homogeneity, tests of; t-distribution; t-test;
Two-sample problem
Normal two-sample problem, ratio of
variances: confidence intervals for, 218,
333, 351; credible region for, 262;
nonrobustness of test for, 207; test for, 122,
198, 290. *See also* F-test for ratio of
variances; Ratio of variances
Null set, 48, 61, 140

Odds ratio, 154, 163, 164, 547; most
accurate unbiased confidence intervals for,
261. *See also* Binomial probabilities;
Contingency table; Two by two tables
One-parameter exponential family, 80, 101;
most stringent test in, 527. *See also*
Exponential family
One-sided hypotheses, 78, 151, 167;
multivariate, 460. *See also* Confidence
bounds
One-way classification, 374; Bayesian
inference for, 427; model II for, 418;
multivariate, 463; nonparametric, 380.
See also Homogeneity, tests of; Normal
many-sample problem

Onto, *see* Transformation
Optimality, ix, xii, 8, 9
Orbit of transformation group, 285
Ordered alternatives, 380
Order statistics, 46; completeness of, 143,
173, 183, 187; distribution of, 345;
equivalent to sums of powers, 46;
expected values of, 318; as maximal
invariants, 286; in permutation tests, 231;
as sufficient statistics, 63, 231
Orthogonal group, 286, 366, 518

Paired comparisons, *see* Matched pairs
Pairwise sufficiency, 64
Parameters, unrelated, *see* Variation
independent parameters
Parameter space, 1
Pareto distribution, 123, 272
Partial ancillarity, 546, 547, 561
Partial sufficiency, 122, 565
Performance robustness, 208, 321. *See also*
Robustness
Permutation test, 208, 232, 265, 273, 276,
278, 279, 450; approximated by standard
t-test, 236, 253; complete class, 243;
confidence intervals based on, 246, 263,
266, 267; most powerful for nonparametric
hypotheses, 232, 252; as randomization
test, 238; robustness of, 321; most
stringent, 533; for testing independence,
252; for variances, 378. *See also*
Nonparametric; Randomization model
Pillai-Bartlett trace test, 463; robustness of,
465
Pivotal quantity, 333, 357
Point estimation, 4, 30; equivariant, 12;
unbiased, 13, 14, 23. *See also* Median
unbiasedness
Poisson distribution $P(\tau)$, 2, 56, 65, 171; as
distribution of sum of Poisson variables,
65; relation to exponential distribution,
23, 82, 88, 114; square root transformation
for, 432, 445; sufficient statistics for, 20.
See also Exponential distribution; Poisson
parameters; Poisson process
Poisson model: for 2×2 table, 159, 161;
for $2 \times 2 \times K$ table, 163, 181
Poisson parameters: comparing k, 364;
comparing two, 151, 152, 186, 221, 546;
confidence intervals for the ratio of two,

Poisson parameters (*Continued*)
221; one-sided test for, 81, 114;
one-sided test for sum of, 120
Poisson process, 3, 65, 88; comparison of
experiments for, 88; confidence bounds
for scale parameter, 92; distribution of
waiting times in, 23; test for scale
parameter in, 81, 114; and 2 × 2 tables,
159. *See also* Exponential distribution
Polya frequency function, 509, 538. *See
also* Total positivity
Positive dependence, *see* Dependence,
positive
Positive part of a function, 38
Posterior distribution, 225; percentiles of,
229. *See also* Bayesian inference
Posterior probability, 125
Power function, 69; of invariant test, 300; of
one-sided test, 79, 117; of two-sided test,
102
Power series distribution, 181
Power of a test, 69, 70, 446; conditional,
150, 547; robustness of, 207; unbiased
estimation of, 151, 547
Preference ordering of decision procedures, 9,
14, 15
Prior distribution, 14, 225; improper, 226,
311; noninformative, 226. *See also* Bayesian
inference; Least favorable distribution;
Posterior distribution
Probability density (with respect to μ), 40;
convergence theorem for, 573
Probability distribution of a random variable,
36. *See also* Cumulative distribution
function (cdf)
Probability integral transformation, 320
Probability measure, 35
Probability ratio, *see* Likelihood ratio
Probability theory, 34, 66
Product measure, 40
Projection: as maximal invariant, 287,
374
Pseudometric space, 571
P-value, 70, 114, 170; combination of, from
independent experiments, 170

Quadrant dependence, 176, 251, 271. *See
also* Dependence, positive
Quadrinomial distribution, 163
Quality control, 106, 293

Radon-Nikodym derivative, 40; properties of,
61
Radon-Nikodym theorem, 40
Random assignment, 160, 161, 238, 396
Random breaking of ties, 167
Random effects model, 418, 426, 447; for
nested classifications, 422; for one-way
layout, 418; for two-way layout, 438, 440.
See also Ratio of variances
Randomization, 6, 396; as basis for inference,
238; to lower the maximum risk, 25;
possibility of dispensing with, 113; relation
to permutation test, 240. *See also* Random
assignment; Randomized procedure
Randomization model, 162, 245; confidence
intervals in, 246
Randomized procedure, 6, 25, 113; confidence
intervals, 219; test, 71, 74, 155
Randomness, hypothesis of, 349, 350
Random sample size, 112, 181, 561
Random variable, 36
Rank correlation coefficient, 351
Ranks, 286; distribution under alternative,
344, 345, 361; as maximal invariants, 286,
315; null distribution of, 317. *See also*
Signed ranks
Rank-sum test, 178, 184. *See also* Wilcoxon
test
Rank tests, 316; surveys of, 380. *See also*
Independence, test for; Nonparametric;
Nonparametric two-sample problem;
Symmetry; Trend
Ratio of quadratic forms, maximum of, 474
Ratio of variances: confidence intervals for,
219, 262, 333, 351; in model II, 419,
421, 558; tests for, in two-sample
problems, 122, 198, 207, 290, 339, 562.
See also F-test for ratio of variances;
Homogeneity, tests of; Random effects
model
Rectangular distribution, *see* Uniform
distribution
Reference set, ix. *See also* Conditional
inference
Regression, 222, 446, 450, 542; with both
variables subject to error, 435; comparing
several lines, 399, 435; confidence band for,
417, 444; confidence intervals for
coefficients, 223, 398; confidence sets for
abscissa of line, 224; general linear model

for, 374, 430; as linear model, 365; multivariate, 462, 490, 496; nonparametric, 350; polynomial, 435; robustness of tests for, 401, 436; tests for coefficients, 223, 397, 398, 400. *See also* Trend

Regression dependence, 251, 271, 315. *See also* Dependence, positive

Relevant and semirelevant subsets, 230, 554, 564, 568; randomized version of, 563

Repeated measurements, 462, 466

Restricted Bayes solution, 15, 30

Restricted χ^2-test, 481, 500

Risk function, 2, 28

Robustness, ix, 10, 203, 208, 213, 273, 444, 536; of analysis of variance tests, 401; against dependence, 209; for F-test of means, 378, 379; of general linear models tests, 379, 405; lack of, for F-test of variances, 207, 422; lack of, for χ^2-test of variance, 206; lack of, for Wilcoxon test, 323; of multivariate tests, 465, 491; of regression tests, 401, 405; of test of independence or lack of correlation, 250; for tests in two-way layout, 434, 436; of t-test, 205, 209, 273, 321. *See also* Adaptive test; Behrens-Fisher problem; Efficiency robustness; Huber condition; Performance robustness; Permutation test; Rank tests

Roy's maximum root test, 463, 465; robustness of, 465; simultaneous confidence sets based on, 475

Runs test: power of, 183; for testing independence in a Markov chain, 176, 177

Sample, 3; haphazard, 237; stratified, 231

Sample correlation coefficient R, 249; distribution of, 267, 270, 271, 276; monotone likelihood ratio of distribution, 340; variance stabilizing transformation for, 432. *See also* Bivariate normal distribution; Multiple correlation coefficient; Rank correlation coefficient

Sample distribution function, *see* Empirical distribution function (EDF)

Sample inspection: by attributes, 80, 293, 339; choice of inspection stringency for, 89; for comparing two products, 167, 296;

comparison of two methods, 339; by variables, 106, 293, 339

Sample size: required to achieve specified power, 70, 153, 260, 504

Sample space, 37

S-ancillary, *see* Partial ancillary

Scale families: condition for monotone likelihood ratio, 510

Scheffé's S-method, 382, 388, 405, 411, 444; alternatives to, 417, 437; multivariate extensions, 471

Selection procedures, 117, 127

Separable: family of distributions, 574; space, 571

Separate families of hypotheses, 290, 338, 355, 360, 363

Sequential analysis, ix, 8, 78, 175, 196, 215

Sequential experimentation, 8, 66

Shift, confidence intervals for: based on permutation tests, 246, 263; based on rank tests, 347, 362. *See also* Behrens-Fisher problem; Exponential distribution; Nonparametric two-sample problem; Normal two-sample problem, difference of means

Shift model, 164, 329

σ-field, 35; with countable generators, 575

σ-finite, 35

Signed ranks, 317; distribution under alternatives, 348; null distribution of, 324

Significance level, 69, 71; for multiple comparisons, 382, 385; nominal, 387. *See also* P-value

Significance probability, *See* P-value

Sign test, 106; in double exponential distribution, 531; for matched pairs, 170; for testing consumer preferences, 166; for testing symmetry with respect to a given point, 168, 325, 530; treatment of ties in, 167, 186. *See also* Binomial probabilities; Median; Sample inspection

Similar test, 135, 140, 182, 183, 186; characterization of, 144; relation to unbiased test, 135

Simple: class of distributions, 72; hypothesis, 73, 483

Simple function, 37

Simple hypothesis *vs.* simple alternative, 73; with large samples, 125. *See also* Neyman-Pearson fundamental lemma

Simultaneous confidence ellipsoids, 576
Simultaneous confidence intervals, 388, 406,
411, 444, 452; for the components of a
vector mean, 411; for all contrasts, 388,
415; in multivariate case, 471, 503.
See also Confidence bands; Dunnett's
multiple comparison method; Scheffé's
S-method; Tukey's T-method
Simultaneous confidence sets: for a family of
linear functions, 408; multivariate, 475, 498;
smallest, 409; taut, 409
Simultaneous inference, ix
Simultaneous tests, 70, 415. See also Multiple
comparisons
Smirnov test, 322, 323
Spherically symmetric distributions, 257, 439
Square root transformation, 432, 445
Stagewise tests, 381, 388
Standard confidence bounds, 96, 229
Stationarity, 176
Statistic, 37; equivalent representations of, 41;
fully informative, 113; subfield induced by,
41
Statistical inference, 1; and decision theory,
4, 71
Stein's two-stage procedure, 258
Stochastically increasing, 84; relation to
monotone likelihood ratio, 85
Stochastically larger, 84, 116, 314
Stochastic process, 129. See also Poisson
process
Stratified sampling, 231
Strictly unbiased, 137
Strongly unimodal, 509, 562
Studentization, 209, 213, 380
Studentized range, 381, 443
Student's t-test, see t-test
Subfield, 41
Sufficient statistic, 19, 30, 53, 66, 67, 124;
asymptotically, 485; Bayes definition of, 21,
22; factorization criterion, 19, 30, 31, 53,
54; likelihood ratio as, 63; minimal, 22, 28;
pairwise, 64; in presence of nuisance
parameters, 122; relation to ancillarity, 545;
relation to comparison of experiments, 87;
relation to fully informative statistic, 113;
relation to invariance, 290, 301; statistics
independent of, 191. See also Partial
sufficiency
Symmetric distribution, 63

Symmetry, 10; relation to invariance, 11, 377;
in a square two-way contingency table, 495;
sufficient statistics for distributions with, 63;
testing for, 326, 360, 361; testing, with
respect to given point, 168, 316, 323, 325,
326, 349

Tautness, 409
t-distribution, 196, 257, 258, 280; as
approximation to permutation distribution,
236; as distribution of function of sample
correlation coefficient, 250; monotone
likelihood ratio of, 295; multivariate, 353;
noncentral, 196, 253, 276; normal limit of,
205; as posterior distribution, 228; in two-
stage-sampling, 259
Test, 3, 68; almost invariant, 297; conditional,
541, 549, 552; invariant, 284; locally
maximin, 507; locally most powerful
(LMP), 202, 527, 528, 538; maximin, 505;
most stringent, 526; randomized, 71, 155;
similar, 135; strictly unbiased, 137; of type
A, 131, 538; of type A_1, 131; of type B, 202,
538; of Type B_1, 202; type D,E, 529;
unbiased, 13, 134; uniformly most powerful
(UMP), 32
Three-decision problems, 101, 152
Three factor contingency table, 162
Ties, 167, 186
Time series, 213
Total positivity, 86, 118, 119, 140, 509; of
order three, 119, 120, 303. See also Polya
frequency function
TPE, ix, x
Transformation: of integrals, 43; into, 36;
onto, 36; probability integral, 320; variance
stabilizing, 376, 432, 433
Transformation group, 570. See also
Invariance
Transitive: binary relation, 569; transformation
group, 285
Trend: test for absence of, 349, 403
Triangular distribution, 355
t-test: admissibility of, 309, 310, 343; as
Bayes solution, 311, 343; comparison with
Wilcoxon and Normal scores tests, 321,
324; not efficiency robust, 322; as likelihood
ratio test, 27, 108; in linear hypothesis with
one constraint, 370; for matched pairs, 240,
264; permutation version of, 208, 236;

power of, 196, 203, 207, 253, 256; one-sample, 111, 195, 209, 213, 257, 273, 339, 380; for regression coefficient, 223, 397, 398; relevant subsets for, 557; robustness of, 205, 207, 208, 209, 273; two-sample, 202, 207, 230, 361; two-stage, 258. *See also* Normal one- and two-sample problem; Regression; Welch approximate t-test

Tukey levels for multiple comparisons, 383, 387, 433

Tukey's T-method, 382, 388, 433, 442, 443, 451

Two-sample problem, *see* Behrens-Fisher problem; Binomial probabilities; Exponential distribution; Matched pairs; Nonparametric two-sample problem; Normal two-sample problem; Permutation test; Poisson parameters; Shift, confidence intervals for; Two-by-two tables

Two-sided alternatives, 101, 135, 152, 167

Two-stage procedures, 258, 259

Two by two tables: alternative models for, 159, 161; comparison of experiments for, 87, 159; Fisher's exact test for, 155, 180, 187; for matched pairs, 169, 179, 180; multinomial model for, 157; S-ancillaries for, 547, 568. *See also* Contingency tables

Two by two by two table, 165

Two-way classification: Bayesian inference for, 427; mixed model for, 439, 440, 441; with m observations per cell, 393; multiple comparison procedures for, 396; multivariate, 492, 493; with one observation per cell, 388; random effects model for, 438, 440; rank tests for, 392; reorganization of variables in, 433; robustness of tests in, 434, 436; simultaneous inference in, 416. *See also* Contingency tables; Interaction; Nested classification; Two-by-two tables

Two-way contingency tables, *see* Contingency tables; Two-by-two tables

Two-way layout, *see* Two-way classification

Type A, A_1, B, B_1, D, E test, *see* Test of type A, A_1, B, B_1, D, E

UMP invariant test, 188, 289, 292; admissibility, 305; conditional, 551, 553; conditions to be UMP almost invariant, 297; examples of nonuniqueness, 304, 305; relation with UMP unbiased test, 302. *See also* Invariance; Linear hypothesis, multivariate; Linear hypothesis, univariate

UMP test, 72, 126; conditional, 542, 549, 550, 552; examples involving two parameters, 112; for exponential distributions, 112; for inverse Gaussian distributions, 124; in monotone likelihood ratio families, 78; a nonparametric example, 107; in normal one-sample problem, 108, 111; in one-parameter exponential families, 80; for uniform distributions, 111, 115; in Weibull distributions, 124

UMP unbiased test, 134, 135, 186; admissibility of, 170; example of nonexistence of, 171; via invariance, 188, 302; for multiparameter exponential families, 147, 188; for one-parameter exponential families, 135; for strictly totally positive families, 140. *See also* Unbiasedness

Unbiasedness, 12, 23, 28, 186; for confidence sets, 13, 24, 216; and invariance, 24, 302; and minimax, 26; for point estimation, 13, 23, 28; and similarity, 135; strict, 137; of tests, 134; for two-decision procedures, 13. *See also* UMP unbiased test; Uniformly most accurate confidence sets

Undetermined multipliers, 100, 104, 118

Uniform distribution U(a,b), 7, 21, 23; completeness of, 141, 172; discrete, 123, 180; as distribution of integral transform, 320; distribution of order statistics from, 345; as null distributions of p-value, 170; one-sample problems in, 111, 115, 354, 563; relation to exponential distribution, 112; sufficient statistics for, 21, 28, 172; testing against exponential or triangular distribution, 355; other tests for, 480, 482

Uniformly most accurate confidence sets, 90, 217; equivariant, 327, 524; relation to UMP tests, 91; unbiased, 217; uniformly minimize expected Lebesgue measure, 330. *See also* Confidence bands; Confidence bounds; Confidence intervals; Confidence sets; Simultaneous confidence intervals; Simultaneous confidence sets

Uniformly most powerful, *see* UMP invariant test; UMP test; UMP unbiased test

Unimodal, 562. *See also* Strongly unimodel

Unrelated parameters, *see* Variation
independent parameters

Variance components, *see* Components of
variance
Variance stabilizing transformation, 376, 432
Variation diminishing, 86. *See also* Total
positivity
Variation independent parameters, 546, 561

Waiting times (in a Poisson process), 23, 114.
See also Exponential distribution; Life
testing; Poisson process
Weak compactness theorem, 576
Weak convergence, 571, 572
Weibull distribution W(b,c), 124, 567

Welch approximate t-test, 209, 304
Welch-Aspin test, 304; relevant subsets for,
558, 566
Wilcoxon one-sample test, 324, 326, 348,
349, 364
Wilcoxon signed-rank test, *see* Wilcoxon one-
sample test
Wilcoxon two-sample test, 318, 322, 323,
343, 357; comparison with t-test, 321;
confidence intervals based on, 329; history
of, 360, 364; optimality of, 320, 346
Wilks' Λ, 463; robustness of, 465
Wishart distribution, 490
Working-Hotelling confidence band, 417, 444

Yule's measure of association, 157

Theory of Point Estimation

Contents

CHAPTER		PAGE
1	PREPARATIONS	1
	1 The problem	1
	2 Measure theory and integration	8
	3 Group families	19
	4 Exponential families	26
	5 Sufficient statistics	36
	6 Convex loss functions	48
	7 Problems	57
	8 References	70
2	UNBIASEDNESS	75
	1 UMVU estimators	75
	2 The normal and exponential one- and two-sample problem	83
	3 Discrete distributions	91
	4 Nonparametric families	101
	5 Performance of the estimators	105
	6 The information inequality	115
	7 The multiparameter case and other extensions	123
	8 Problems	130
	9 References	145
3	EQUIVARIANCE	154
	1 Location parameters	154
	2 The principle of equivariance	165
	3 Location-scale families	173
	4 Linear models (Normal)	183
	5 Exponential linear models	196
	6 Sampling from a finite population	207
	7 Problems	218
	8 References	231

CONTENTS

CHAPTER   PAGE

4 GLOBAL PROPERTIES 236
    1 Bayes estimation 236
    2 Minimax estimation 249
    3 Minimaxity and admissibility in exponential families 262
    4 Equivariance, admissibility, and the minimax property 279
    5 Simultaneous estimation 290
    6 Shrinkage estimators.............................. 299
    7 Problems 310
    8 References 320

5 LARGE-SAMPLE THEORY 331
    1 Convergence in probability and in law 331
    2 Large-sample comparisons of estimators 344
    3 The median as an estimator of location 352
    4 Trimmed means 360
    5 Linear combinations of order statistics (L-estimators) 368
    6 M- and R-estimators............................. 376
    7 Problems 388
    8 References 398

6 ASYMPTOTIC OPTIMALITY 403
    1 Asymptotic efficiency 403
    2 Efficient likelihood estimations 409
    3 Likelihood estimation: Multiple roots 420
    4 The multiparameter case 427
    5 Applications 436
    6 Extensions 443
    7 Asymptotic efficiency of Bayes estimators 454
    8 Local asymptotic optimality 465
    9 Problems 472
   10 References 482

AUTHOR INDEX 491

SUBJECT INDEX 497

Applied Probability and Statistics (Continued)

HOGG and KLUGMAN • Loss Distributions

HOLLANDER and WOLFE • Nonparametric Statistical Methods

IMAN and CONOVER • Modern Business Statistics

JAGERS • Branching Processes with Biological Applications

JESSEN • Statistical Survey Techniques

JOHNSON and KOTZ • Distributions in Statistics
Discrete Distributions
Continuous Univariate Distributions—1
Continuous Univariate Distributions—2
Continuous Multivariate Distributions

JOHNSON and KOTZ • Urn Models and Their Application: An Approach to Modern Discrete Probability Theory

JOHNSON and LEONE • Statistics and Experimental Design in Engineering and the Physical Sciences, Volumes I and II, *Second Edition*

JUDGE, HILL, GRIFFITHS, LÜTKEPOHL and LEE • Introduction to the Theory and Practice of Econometrics

JUDGE, GRIFFITHS, HILL, LÜTKEPOHL and LEE • The Theory and Practice of Econometrics, *Second Edition*

KALBFLEISCH and PRENTICE • The Statistical Analysis of Failure Time Data

KISH • Survey Sampling

KUH, NEESE, and HOLLINGER • Structural Sensitivity in Econometric Models

KEENEY and RAIFFA • Decisions with Multiple Objectives

LAWLESS • Statistical Models and Methods for Lifetime Data

LEAMER • Specification Searches: Ad Hoc Inference with Nonexperimental Data

LEBART, MORINEAU, and WARWICK • Multivariate Descriptive Statistical Analysis: Correspondence Analysis and Related Techniques for Large Matrices

LINHART and ZUCCHINI • Model Selection

McNEIL • Interactive Data Analysis

MAINDONALD • Statistical Computation

MANN, SCHAFER and SINGPURWALLA • Methods for Statistical Analysis of Reliability and Life Data

MARTZ and WALLER • Bayesian Reliability Analysis

MIKÉ and STANLEY • Statistics in Medical Research: Methods and Issues with Applications in Cancer Research

MILLER • Beyond ANOVA, Basics of Applied Statistics

MILLER • Survival Analysis

MILLER, EFRON, BROWN, and MOSES • Biostatistics Casebook

MONTGOMERY and PECK • Introduction to Linear Regression Analysis

NELSON • Applied Life Data Analysis

OSBORNE • Finite Algorithms in Optimization and Data Analysis

OTNES and ENOCHSON • Applied Time Series Analysis: Volume I, Basic Techniques

OTNES and ENOCHSON • Digital Time Series Analysis

PANKRATZ • Forecasting with Univariate Box-Jenkins Models: Concepts and Cases

PIELOU • Interpretation of Ecological Data: A Primer on Classification and Ordination

POLLOCK • The Algebra of Econometrics

PRENTER • Splines and Variational Methods

RAO and MITRA • Generalized Inverse of Matrices and Its Applications

RIPLEY • Spatial Statistics

RUBINSTEIN • Monte Carlo Optimization, Simulation, and Sensitivity of Queueing Networks

SCHUSS • Theory and Applications of Stochastic Differential Equations

SEAL • Survival Probabilities: The Goal of Risk Theory

SEARLE • Linear Models

(*continued from front*)